Pathogenic Fungi
in Humans
and Animals

MYCOLOGY SERIES

Editor

J. W. Bennett

Professor
Department of Cell and Molecular Biology
Tulane University
New Orleans, Louisiana

Founding Editor

Paul A. Lemke

Additional Volumes in Preparation

Pathogenic Fungi in Humans and Animals

Second Edition

edited by

Dexter H. Howard

UCLA School of Medicine
Los Angeles, California

CRC Press
Taylor & Francis Group
Boca Raton London New York

CRC Press is an imprint of the
Taylor & Francis Group, an **informa** business

First published 2003 by Marcel Dekker, Inc.
The first edition was published as *Fungi Pathogenic for Humans and Animals (in three parts),* edited by Dexter H. Howard

Published 2019 by CRC Press
Taylor & Francis Group
6000 Broken Sound Parkway NW, Suite 300
Boca Raton, FL 33487-2742

First issued in paperback 2019

No claim to original U.S. Government works

ISBN 13: 978-0-367-45471-5 (pbk)
ISBN 13: 978-0-8247-0683-8 (hbk)

Visit the Taylor & Francis Web site at
http://www.taylorandfrancis.com

and the CRC Press Web site at
http://www.crcpress.com

Library of Congress Cataloging-in-Publication Data
A catalog record for this book is available from the Library of Congress.

Preface to the Second Edition

The major object of the first edition of *Fungi Pathogenic for Humans and Animals* was a thorough review of the basic biology, host-parasite interactions, and current method of detection and characterization of the zoopathogenic fungi. This is a revision of the original book with a minor change in the title. Events since the publication of the First Edition have made necessary the preparation of this second edition.

Some chapters in the first edition remain adequate summaries of the topic (e.g., ultrastructure, conidium formation, and subcellular particles) and will not be repeated here. In other cases the original chapters have been expanded into monographic treatments (e.g., dimorphism, antigens, and cell wall composition) and a consideration of those topics would be inappropriate in a smaller context. In many cases, knowledge in fields covered in the first edition has grown so much that a single chapter is no longer adequate (e.g., nutrition, antifungal drugs, and mycotoxins). Finally, the topics of immunology and pathogenesis have received very recent coverage that makes inclusion of them in the second edition unnecessary. (*The Mycota, VI, Human and Animal Relationship*, D. H. Howard and J. D. Miller, eds. New York: Springer Verlag, 1996; Fungal Pathogenesis, R. A. Calderone and R. L. Cihlar, eds. New York: Marcel Dekker, Inc., 2002.)

The area of coverage from the first edition that has changed remarkably is the topic of classification and nomenclature, for which there is not a current treatment. In the years since the first edition many molecular techniques, upon which taxonomic decisions are now based, have been introduced. Accordingly, Part One of the second edition has been expanded to include peronosporomycetes, a rearrangement of sections on the filamentous Ascomycetes, and a division of the chapter on yeasts into two separate chapters—one on heterobasidiomycetes, and a second covering both the endomycetes and the blastomyces.

The second edition has been further expanded to introduce a topic not considered in the first edition: fungal populations. Four topics are covered: Molecular methods used in taxonomic decisions; Population genetics of Medically Important fungi—phylogeny; population studies—DNA- and PCR-Fingerprinting of Medically Important fungi; and population instabilities—the phenotyic variability of *Candida albicans*.

Dexter H. Howard

Preface to the First Edition

PART A

In 1947, Walter J. Nickerson edited a volume on the *Biology of Pathogenic Fungi*, which covered some aspects of the biology, physiology, and biochemistry of the fungi pathogenic for man. Since then several fine textbooks have appeared, in which are described the essential features of the clinical manifestation, epidemiology, and pathology of fungous infections together with the identifying characteristics of their etiologic agents. In addition, individual mycoses or single aspects of all of the major mycoses have received abundant monographic treatment. However, a comprehensive consideration of the biology of the fungi pathogenic to man and animals has not recently been offered. These volumes are an effort to do so. It was not my object to present a series of "selected topics" or "recent advances" but rather a thorough review of basic biology, host-parasite interactions, and current methods of detection and characterization of the zoopathogenic fungi.

Part A of this series deals with the basic biology of medically important fungi. The first section comprises a discussion of the taxonomy and classification of the zoopathogens. This section is followed by individual chapters on ultrastructural cytology, dimorphism, sporulation, blastoconidium germination, and the transport of metabolites. Some additional topics which suggested themselves for coverage could not be included in Part A because of size limitations. Other basic biology topics will be included in subsequent parts of the series. (Members of the bacterial order Actinomycetales, often included in various types of works on medical mycology, will not be considered in this series.)

So much has been accomplished in each of these fields that it would be impossible for any single author to give an in-depth presentation. Accordingly, a compilation by many individuals was attempted despite the recognized difficulties of variation and potential redundancy. The scope of these volumes is such as to attract the attention of all engaged in research on fundamental aspects of the fungi pathogenic to man and animals. However, the coverage will be of interest to students who wish an up-to-date consideration of the frontiers of investigation on these pathogens.

A special acknowledgment is required. All responsibility for necessary editing with regard to uniformity of format, proper references, and retyping of heavily altered copy was assumed by my wife, Mrs. Lois F. Howard. She was, indeed, the primary copyeditor of this work. She merits recognition for her unswerving pursuit of excellence of presentation and the indefatigable insistence on perfection of textural detail which I had hoped from the outset would characterize these volumes. Her efforts, in no small measure, and those of several of the authors, have been supported by research grant AI 16252 from the National Institute of Allergy and Infectious Diseases, National Institutes of Health, which is used to fund the Collaborative California Universities–Mycology Research Unit (CCU-MRU).

I am, of course, grateful to the splendid group of authors collaborating with me to produce these volumes.

PART B

The original plan for a comprehensive coverage of the zoopathogenic fungi called for a division of the work into three parts. The first part was to include chapters on the basic biology of the fungi with special consideration classification, morphology, and physiology. The second part was to cover aspects of pathogenicity such as mechanisms of pathogenesis, host responses (cellular and humoral), toxins, and antigens. The last part was to have dealt with practical matters of detecting the fungi in nature and in clinical materials and with certain applications such as vaccines and antifungal drugs.

In gathering the material, it became clear that the topical divisions would have produced volumes of exaggerated disproportion. Far too much on the basic biology was at hand and some of the topics in pathogenicity required preparations that exceeded practical deadlines. Therefore, rearrangements were made. The third volume of the series, labeled Part B:II, now contains two sections and an Addendum. The first of these sections contains chapters on pathogenesis, including aspects of the basic biology that have a direct relation to mechanisms of tissue invasion, namely, cell wall composition, subcellular particles, and enzymes involved in in vivo survival or tissue destruction. In addition, the rearrangement has allowed for an updated coverage of cellular defense mechanisms involving phagocytic cells and, in an Addendum, a reconsideration of the genus *Exophiala*, a particularly troublesome member of the dematiaceous group.

The second section of Part B:II is more or less as originally planned and covers such practical matters as epidemiology, detection of fungi in tissue, growth of fungi in culture, and the development of fungal vaccines.

And thus the work is completed. I gratefully acknowledge the indispensable assistance of Mrs. Lois F. Howard, who checked all bibliographical material, read all of the manuscripts, and tried to maintain consistency throughout the three parts of the work. I also thank Ms. Bette Y. Tang for the numerous typings and retypings of edited manuscripts. Some support for this venture has been supplied by research grant AI 16252 from the National Institute of Allergy and Infectious Diseases, National Institutes of Health, which is used to fund the Collaborative California Universities–Mycology Research Unit (CCU-MRU).

I am especially grateful to the group of splendid authors who collaborated with me to produce this volume.

Dexter H. Howard

Contents

Contributors

Teun Boekhout, Ph.D. Centraalbureau voor Schimmelcultures, Institute of the Royal Netherlands Academy of Arts and Sciences, Utrecht, The Netherlands

Michael W. Dick, D.Sc. Department of Botany at Earley Gate, School of Plant Sciences, University of Reading, Reading, England

Frank M. Dugan, Ph.D. USDA-ARS Western Regional Plant Introduction Station, USDA, Pullman, Washington, U.S.A.

Eveline Guého, Ph.D. Mauves sur Huisne, France

Kevin C. Hazen, Ph.D. Department of Pathology, University of Virginia Health System, Charlottesville, Virginia, U.S.A.

Dexter H. Howard, Ph.D. Departments of Microbiology, Immunology, and Molecular Genetics, Center for the Health Sciences, UCLA School of Medicine, Los Angeles, California, U.S.A.

Susan A. Howell, B.Sc., Ph.D. Department of Medical Mycology, St. John's Institute of Dermatology, King's College London, London, England

Shung-Chang Jong, Ph.D. Department of Microbiology, American Type Culture Collection, Manassas, Virginia, U.S.A.

Thomas G. Mitchell, Ph.D. Department of Genetics and Microbiology, Duke University Medical Center, Durham, North Carolina, U.S.A.

Arvind A. Padhye, Ph.D. Mycotic Disease Branch, Division of Bacterial and Mycotic Diseases, National Center for Infectious Disease, Centers for Disease Control and Prevention, Atlanta, Georgia, U.S.A.

Elena Rustchenko, Ph.D. Department of Biochemistry and Biophysics, University of Rochester Medical School, Rochester, New York, U.S.A.

Wiley A. Schell, M.S. Department of Medicine, Duke University Medical Center, Durham, North Carolina, U.S.A.

M. A. A. Schipper, Ph.D. Centraalbureau voor Schimmelcultures, Institute of the Royal Netherlands Academy of Arts and Sciences, Utrecht, The Netherlands

Fred Sherman, Ph.D. Department of Biochemistry and Biophysics, University of Rochester Medical School, Rochester, New York, U.S.A.

Lynne Sigler, M.S. Microfungus Collection and Herbarium, Devonian Botanic Garden and Medical Microbiology and Immunology, University of Alberta, Edmonton, Alberta, Canada

J. A. Stalpers, Ph.D. Centraalbureau voor Schimmelcultures, Institute of the Royal Netherlands Academy of Arts and Sciences, Utrecht, The Netherlands

Richard Summerbell, Ph.D. Centraalbureau voor Schimmelcultures, Institute of the Royal Netherlands Academy of Arts and Sciences, Utrecht, The Netherlands

Irene Weitzman, Ph.D. Department of Pathology, Columbia University College of Physicians and Surgeons, New York, New York, and Department of Microbiology, Arizona State University, Tempe, Arizona, U.S.A.

Jianping Xu, Ph.D. Department of Biology, McMaster University, Hamilton, Ontario, Canada

Pathogenic Fungi
in Humans
and Animals

1

An Introduction to the Taxonomy of Zoopathogenic Fungi

Dexter H. Howard
UCLA School of Medicine, Los Angeles, California, U.S.A.

I. THE CLASSIFICATION OF FUNGI

The subject of this book is the classification of zoopathogenic fungi. The basic units in classification are the species, and these are arranged into hierarchal groups of genera, families, orders, classes, phyla, and kingdoms. The categories may be subdivided (e.g., subphylum, subclass, suborder) to indicate degrees of relationships. Populations within a given species that have some characteristics in common may be set apart as tribes or varieties or some other subset designation. The delineation of the zoopathogen *Ajellomyces capsulatus*, anamorph: *Histoplasma capsulatum*, goes as follows (1):

Kingdom: Fungi
   Phylum: Ascomycota
      Class: Ascomycetes
         Order: Onygenales
            Family: Onygenaceae
               Genus: *Ajellomyces*
                  Species: *Ajellomyces capsulatus*
                     Variety: the varietal state applies to the anamorph (2): *Histoplasma capsulatum* var. *capsulatum*, *H. capsulatum* var. *duboisii*, and *H. capsulatum* var. *farciminosum*

Table 1 The Kingdoms and Phyla
of Microorganisms Covered
in This Book

Kingdom: Fungi
  Phylum: Chytridiomycota
  Phylum: Zygomycota
  Phylum: Ascomycota
  Phylum: Basidiomycota
  Mitosporic Fungi[a]
Kingdom: Straminipila
  Phylum: Heterokonta
Kingdom: Protoctista
  Phylum: Protista

[a] This group of anamorphic fungi—which
lack any known teleomorphic stage—was at
one time collected into a phylum Deuteromycota, but this term, still used for convenience
in some arrangements (see, e.g., Chap. 8), has
been largely discontinued.

II. NOMENCLATURE OF HIERARCHAL GROUPS

The pathogenic organisms traditionally considered as fungi* are now distributed
among three biologic kingdoms: *Fungi*, *Straminipila*, and *Protoctista* (3). The
bulk of this book is devoted to the members of the kingdom Fungi. Also included,
however, is a consideration of the zoopathogenic members of the kingdoms
Straminipila and Protoctista. (See Chap. 2.)

The phyla to be considered in this book are listed in Table 1. Opinions on
the nomenclature of hierarchal groups vary somewhat among taxonomists. For
example, the terms Mycota and Fungi have both been used to designate the kingdom to which most of the human pathogens belong (4). Likewise, at one time
mycologists used the term division instead of phylum (2) but more recently the
word phylum has been widely adopted in fungal taxonomy. Authors in this book
have used one term or the other as alternative words for the two highest hierarchal
categories. I have let stand whichever usage the author chooses for these categories. In this summary chapter, the various taxonomic groups are the ones that

* Fundamental descriptions of fungi may be found in the textbooks on medical mycology by K. J.
  Kwon-Chung and J. E. Bennett (2) and on fundamental mycology by O. J. Alexopoulos, C. W.
  Mims, and M. Blackwell (3). A glossary of terms employed in this introduction will be found at
  the end of the chapter.

contain animal pathogens (zoopathogens) and are covered in this book. The decisions on coverage within these groups have been left up to the authors selected.

III. THE KINGDOM FUNGI

A. Chytridiomycota

This phylum comprises those members of the kingdom Fungi that produce motile cells in their life cycles (3). There are five orders of such pathogens or commensals within the single class of Chytridiomycetes (Table 2). The zoospores, produced by asexual reproduction of members of four of the five orders, have a single, smooth, posteriorly directed whiplash flagellum. A single order, Neocallimasticales, contains forms that are polyflagellate. Only some examples of pathogens are included in this brief sketch of the phylum. For complete coverage of the biology and host–parasite relations among the zoopathogenic chytrids, see Chap. 2.

1. Neocallimasticales. The members of the order differ from other members of the class by being composed of obligate anaerobes that have multiple posterior flagella. These chytrids are commensals of the rumen and cecum of herbivores (3).
2. Blastocladiales. Important pathogens of mosquitoes and other insects (*Coelomomyces* spp.) and of rotifers and nematodes (*Caternaria* spp.) are contained in this order. The pathogenic species of the two genera indicated are listed in Chap. 2.
3. Chytridiales. The genera *Olpidium*, *Endochytrium*, *Rhizophydium*, and *Phlyctochytrium* variously contain important pathogens of nematodes, rotifers, and amphibians.

Table 2 The Phylum
Chytridiomycota

Phylum: Chytridiomycota
  Class: Chytridiomycetes
    Order: Neocallimasticales
    Order: Blastocladiales
    Order: Chytridiales
    Order: Spizellomycetales
    Order: Monoblepharidales

Note: For a more complete treatment of this phylum, see Table 1 of Chap. 2.

4. Spizellomycetales. The members of this order are diverse ecologically, being found in soil and water, and include plant and fungal parasites.
5. Monoblepharidales. This order consists of but a few mostly sapro-phytic species.

B. Zygomycota

The phylum Zygomycota (Table 3) contains those fungi that produce zygospores as a result of sexual reproduction and consists of two classes: the class Trichomy-cetes, which are obligate parasites of arthropods and will not be considered in this book (3), and the class Zygomycetes, which contains several important patho-gens. There are two orders in the class Zygomycetes—Mucorales and Ento-mophthorales, both of which contain human and animal pathogens. The Mucor-ales generally produce nonseptate hyphae, while the Entomophthorales are usually septated.

1. Mucorales. (See Chap. 3.) The members of this order are grouped into six families of agents of disease.
   a. Mucoraceae. This family contains the genera *Rhizopus*, *Absidia*, Apophysomyces, *Mucor*, and *Rhizomucor*, all of which are impor-tant pathogens. These fungi reproduce asexually by means of spo-rangia containing sporangiospores.

Table 3 The Phylum Zygomycota

Phylum: Zygomycota
  Class: Trichomycetes
  Class: Zygomycetes
    Order: Mucorales
      Family: Mucoraceae
      Family: Syncephalastraceae
      Family: Mortierellaceae
      Family: Saksenaeaceae
      Family: Thamnidiaceae
      Family: Cunninghamellaceae
    Order: Entomophthorales
      Family: Basidiobolaceae
      Family: Entomophthoraceae
      Family: Completoriaceae
      Family: Ancylistaceae
      Family: Meristacraceae
      Family: Neozygitaceae

b. Syncephalastraceae. One species, *Syncephalastrum racemosum*, has been reported in clinical materials. Asexual reproduction in this family is different from that of other members of the Mucorales. Sporangiospore-containing merosporangia are produced on vesicles formed at the tips of sporangiophores. The sporangiospores are arranged in a row within the cylindrical merosporangium (4).

c. Mortierellaceae. *Mortierella* spp. produce very small sporangiospores that are contained in a sporangium that lacks a columella. Fungi thought to be species of *Mortierella* have been isolated from human infections. *M. wolfii* has been found in bovine mycotic abortion.

d. Saksenaeaceae. Members of this family produce sporangiospores in flask-shaped sporangia that have long beaks. One species, *Saksenaea vasiformis*, is an important pathogen.

e. Thamnidiaceae. Single or multispored sporangiola formed at the tips of branched sporangiophores characterize members of this family. *Cokeromyces recurvatus* is an uncommon pathogenic species within the family.

f. Cunninghamellaceae. In the genus *Cunninghamella* single-spore sporangiola are formed on swollen, round vesicles that occur on the tips of branched or unbranched sporangiophores. The family contains at least one important pathogen, *C. bertholletiae*.

2. Entomophthorales. The zygospores of members of this order are similar to those of the Mucorales, but differ in morphological detail. The asexual spores are called conidia, though with some species they appear to be monospored sporangia (2). The families within the order Entomophthorales are shown in Table 3. Only the host–parasite relations of the families are given here. The additional biologic bases for family separations are given in Chap. 4.

a. Basidiobolaceae. The family Basidiobolaceae comprises species that are commensals of amphibians and reptiles, and parasites in vertebrates. The human pathogen *Basidiobolus ranorum* is found in this family.

b. Entomophthoraceae. The family Entomophthoraceae contains pathogens of insects and other arthropods.

c. Completoriaceae. The family Completoriaceae has no animal parasites, but certain species are obligate intracellular parasites of fern gametophytes.

d. Ancylistaceae. *Conidiobolus* spp. are contained in the family Ancylistaceae. The genus contains at least two human pathogens, *C. coronatus* and *C. incongruus*.

e. Meristacraceae. The family Meristacraceae contains obligate parasites of nematodes and tardigrades.
f. Neozygitaceae. Obligate pathogens of mites and other insects are
   found in this family.

C. Ascomycota

This phylum is made up of fungi that reproduce sexually by means of ascospores
contained in an ascus. The morphology and arrangement of ascospores within
the ascus and the morphology of an ascus-bearing structure (ascoma), when present, is one approach to the hierarchal arrangement of ascomycetes. The group is
divided into two classes: the Endomycetes and the Ascomycetes. Asexual reproduction in the class Endomycetes is by budding or fission of somatic cells and in
the Ascomycetes by formation of blastic or thallic conidia. Molecular phylogeny
studies have allowed associations to be realized even when a known teleomorph
for a given anamorph has not been revealed. Some of these associations are suggested throughout the coverage in this section.

1. Endomycetes
   a. Saccharomycetales. Ascomata are not formed. Ascospores are of
      various shapes. Asci are formed singly or in chains.
      (1) Saccharomycetaceae. This family contains those yeasts that
          reproduce by budding (blastoconidia). The colonies are ac-
          cordingly mainly unicellular, though some species produce
          pseudohyphae.
      (2) Dipodascaceae. This family includes the genus *Dipodascus*,
          which produces a *Geotrichum* anamorph with arthroconidia.
          Other pathogenic yeasts that reproduce by fission (arthroconi-
          dia) are included in this family. (See Chap. 8.).
2. Ascomycetes.
   a. Onygenales. The order consists of four families. The members of
      these families produce ascomata called cleistothecia, the peridium
      of which is composed of a loose network of hyphae (2). The term
      gymnothecium is sometimes applied to this type of cleistothecium.
      (1) Arthrodermataceae. (See Chap. 5.) This family comprises
          parasites known as the dermatophytes (ringworm fungi) and
          saprophytes that are morphologically similar. The family is
          represented by the single genus *Arthroderma*, whose asexual
          states are in the genera *Microsporum* and *Trichophyton*. An
          additional member of the group known only in its conidial
          state is *Epidermophyton*.
      (2) Onygenaceae. (See Chap. 6.) Two important systemic patho-

gens of humans are found in this family: *Ajellomyces capsulatus* (anamorph, *Histoplasma capsulatum*) and *A. dermatitidis* (anamorph, *Blastomyces dermatitidis*). In addition, one rare pathogen, *Ajellomyces crescens* (anamorph, *Emmonsia crescens*), belongs in this family. A relationship to the Onygenaceae is inferred for other important human pathogens, such as *Coccidioides immitis* and *Paracoccidioides brasiliensis*. *Emmonsia parva*, a pathogen of small mammals, is also thought to belong in the family. Several other pathogens and saprophytes (contaminants of clinical materials) are found in the family. The anamorphic forms of the latter are placed in the genera *Chrysosporium* and *Malbranchia*. (See Chap. 6.)

(3) Gymnoascaceae and Myxotrichaceae. A number of uncommon pathogens and saprophytes encountered as contaminants in clinical materials are found in these two families. (See Chap. 6.).

b. Eurotiales. This order contains a number of cleistothecial fungi whose anamorphs are usually phialidically formed conidia. The most important family is the Trichocomaceae (see Chap. 7), which includes the teleomorphs of *Aspergillus* and *Penicillium*. There are a number of species in these asexual genera that are not known to produce a sexual state. Nevertheless, their affinity to the Eurotiales is clear. There are other important opportunistic pathogens of Eurotialean affinity known only in their anamorphic forms. (See Chap. 7.) The other families in the order Eurotiales (Table 4) contain only rare causes of disease.

c. Ophiostomatales. A perithecium is the ascoma of the sexual state of members of this order. One important pathogen in this group is thought to have close affinity with *Ophiostoma* (Family Ophiostomataceae), but is currently known only in its anamorphic form, *Sporothrix schenckii*.

d. Dothideales. The members of this order produce cleistothecial ascomata. The order comprises six families.

(1) Didymosphaeriaceae—*Neostestudina rosatii*, an agent of mycetoma.

(2) Piedraiaceae—*Piedraia hortae*, the cause of black piedra.

(3) Herpotrichiellaceae—no teleomorphs that are known as zoopathogens but a number of anamorphs whose cells are melanized are thought to belong in this ascomycete family (e.g., *Phialophora*, *Cladophialophora*, *Exophiala*, and *Fonsecaea*). (See Chap. 10 on the Dematiaceous Hyphomycetes.).

Table 4   The Phylum Ascomycota

Phylum: Ascomycota
  Class: Endomycetes
    Order: Saccharomycetales
      Family: Saccharomycetaceae
      Family: Dipodascaceae
  Class: Ascomycetes
    Order: Dothideales
      Family: Didymosphaeriaceae
      Family: Herpotrichiellaceae
      Family: Piedraiaceae
      Family: Dothideaceae
      Family: Botryosphaeriaceae
      Family: Mycosphaerellaceae
    Order: Eurotiales
      Family: Trichocomaceae
      Family: Pseudoeurotiaceae
      Family: Eremomycetaceae
      Family: Thermoascaceae
    Order: Microascales
      Family: Microascaseae
    Order: Onygenales
      Family: Arthrodermataceae
      Family: Onygenaceae
      Family: Gymnoascaceae
      Family: Myxotrichaceae
    Order: Ophiostomatales
      Family: Ophiostomataceae
    Order: Hypocreales
      Family: Hypocreaceae
    Order: Pleosporales
      Family: Leptosphaeraceae
      Family: Pleosporaceae
    Order: Sordariales

---

   e. Microascales. This order, which produces ascomata that are cleis-
totothecia or perithecia, contains a single family of pathogens, *Microascaceae*, which houses the important agent of mycetoma,
Pseudallescheria boydii (anamorph, *Scedosporium apiospermum*),
and common clinical isolates in the anamorphic genus *Scopulariopsis*. (See Chap. 11.)

   f. Hypocreales. The ascoma of this order is a perithecium. This order contains many important opportunistic pathogens of humans and common clinical contaminants. One example is *Fusarium* spp. (teleomorphs *Gibberella* spp. and *Nectria* spp., family Hypocreaceae). (See Chap. 7.).

g. Pleosporales. The ascomata of this order are cleistothecia. The order comprises two families: Leptosphaeraceae, which contains *Leptosphaeria senegalensis* and *L. thompkinsii*, agent of mycetoma, and Pleosporaceae, which contains *Cochliobolus* spp. (anamorphs: *Curvularia* and *Bipolaris*) and other dematiaceous pathogens best known by their anamorphic names. (See Chap. 10.)

h. Sordariales. Ascomata are perithecia or cleistothecia. Perhaps the best-known member of this order is *Neurospora*. Rare human pathogens include *Chaetomium* spp. (2).

D. Basidiomycota

The distinctive feature of the members of the phylum Basidiomycota is the production of basidiospores (sexual spores) on the outside of a club-shaped to elongate structure called the basidium. The members of the basidiomycota considered in this book will be the zoopathogenic yeasts found in the phylum and some rare causes of infectious diseases found among the "mushrooms" (Table 5). Yeast forms are found in all three main phylogenetic lines of basidiomycetes, namely the Hymenomycetes (Cystofilobasidiales, Trichosporonales, Tremellales, and Filobasidiales), Urediniomycetes (Sporidiales), and the Ustilaginomycetes (Malasseziales). Medically important basidiomycetous yeasts belong to the genera *Cryptococcus*, *Trichosporon* (Hymenomycetes), and *Malassezia* (Ustilaginomycetes). Other basidiomycetous yeasts reported in clinical material occur in the genera *Rhodotorula* and *Sporobolomyces* (Urediniomycetes). This topic is covered in Chap. 9.

The class Hymenomycetes also contains forms known colloquially as mushrooms (e.g., the orders Agaricales and Aphyllophorales contain such fungi). The former contains *Coprinus cinereus*, and the latter houses *Schizophyllum commune*, both of which have been reported to cause rare infection in humans. It is ordinarily the toxic or hallucinogenic aspects of mushrooms that involve human disease. These topics were covered in the first edition of this work (5), but will not be considered in the second edition. Two excellent references on the topics are Arora (6) and Lincoff and Mitchell (7).

E. Mitosporic Fungi: The Fungi Imperfecti

This group comprises the fungi that have no known teleomorphic stage and reproduce only by means of asexual conidia. Although molecular methods have sug-

Table 5 The Phylum Basidiomycota

Phylum: Basidiomycota
Class: Hymenomycetes
Order: Cystofilobasidiales
Order: Trichosporonales
Order: Tremellales
Order: Filobasidiales
Order: Aphyllophorales
Order: Agaricales
Class: Urediniomycetes
Order: Sporidiales
Class: Ustilaginomycetes
Order: Malasseziales

Note: Taxonomic ordering of the higher taxa
of basidiomycetous fungi has been subject to
much change over the years (3). The scheme
given here is based on morphology and molecular approaches (3). The major focus in
this book is on the groups containing yeasts
(see Chap. 9), but two orders of mushrooms,
Aphyllophorales and Agaricales, which contain rare human pathogens, are also included.
(See Chap. 11.)

gested teleomorphic association for these anamorphs, it is often convenient to discuss them as a group of anamorphs. (See, e.g., Chap. 10.) At one time these anamorphs were collected into a phylum designated the Deuteromycota and arranged into appropriate hierarchal subdivisions. Such arrangements were subsequently recognized as inappropriate, and the terms "form phylum," "form class," etc. were adopted (2). These artificial names were subsequently abandoned, and the term *Mitosporic Fungi* has been used instead (4,8). There are a number of subdivisions that are nevertheless useful in arranging such imperfect fungi (Table 6).

The Blastomycetes are those unicellular fungi that reproduce by budding or fission but for which no teleomorphic stage has been identified. These forms obviously have ascomycete or basidiomycete affinities but are separated into this group of mitosporic fungi in most treatments (see Chaps. 8 and 9), although some taxonomists indicate apparent teleomorphic associations in their treatment of the subject (9). The Hyphomycetes are hyphal fungi for which no teleomorph stage has yet been revealed. Again, teleomorph associations for such imperfect fungi have been suggested by some investigators (9), but usually they are referred to as mitosporic fungi and are not arranged into a hierarchal organization in the

Table 6 Mitosporic Fungi

A. Blastomycetes—anamorphic yeasts with no known teleomorphic state
B. Coelomycetes—fungi producing a pycnidium or acervulus
C. Hyphomycetes—fungi with no pycnidium or acervulus
    1. Moniliaceous—hyphae and conidia hyaline
    2. Dematiaceous—hyphae and conidia dark
D. Agonomycetes (Mycelia Sterilia)—fungi producing hyphae only. No reproductive propagules

---

Note: The fungi imperfecti, which produced only conidia were gathered in the past, into the phylum Deuteromycota (3). This practice has been abandoned in recent taxonomic treatments of zoopathogenic fungi.

literature on medical mycology (8). It is customary to divide the Hyphomycetes into Dematiaceous (i.e., dark-spored; see Chap. 10) and Moniliaceous (i.e., hyaline-spored; see Chap. 11). The pathogenic Dematiaceous Hyphomycetes are involved in the phaeohyphomycoses. Two other categories are considered among the mitosporic fungi: the Coelomycetes (i.e., those fungi that form their conidia on acervuli or within pycnidia) and Agonomycetes (Mycelia Sterilia; i.e., these fungi that produce no reproductive propagules of any kind). (N.B.: they may form chlamydospores.) One member of this latter group, *Rhizoctonia* sp., was reported in a case of human keratitis. (See Ref. 9.)

Some interesting new work on three pathogens, *Rhinosporidium seeberi* (rhinosporidiosis), *Pneumocystis carinii* (pneumocystosis), and *Lacazia (Loboa) loboi*, has revealed previously unknown taxonomic associations. Rhinosporidiosis presents as tumorlike masses, commonly in the nasal mucosa or conjunctiva of humans and animals (2). The taxonomic position of *R. seeberi* has always been uncertain. Recently studies on 18S r RNA genes from tissue infected with the fungus has shown a phylogenetic relationship to a novel group of protists that infect fish and amphibians. The fungus is "the first known human pathogen from the DRIPs clade, a novel clade of aquatic protistan parasites (Icthyasporea)" (10).

Pneumocystis carinii was originally considered a protozoan, and most of the information about this important opportunistic pathogen is found in the literature on animal parasites. Studies on the ribosomal RNA sequences, however, established that *P. carinii* was a member of the kingdom Fungi (11). The systematic position of *P. carinii* within the kingdom Fungi is uncertain, but the best evidence indicates that it is an ascomycete (12).

Lobomycosis is a chronic cutaneous and subcutaneous infection in humans and dolphins whose etiologic agent *Lacazia (Loboa) loboi* has remained in an uncertain taxonomic position since its original description (2). Recently a phylogenic analysis by molecular means has confirmed earlier studies that indicated the fungus belongs in the family Onygenaceae of the order Onygenales (Ascomycota) (13).

IV. THE KINGDOM STRAMINIPILA

A. Phylum Heterokonta

Only one phylum, Heterokonta, in the kingdom Straminipila, will be considered in this book. The phylum is diverse, and only a brief summary indicating the various hosts of these parasites is given here. (See Chap. 2 for a consideration of their biology.) The phylum comprises fungi whose zoospores have two flagella, one directed anteriorly that is adorned with two rows of tripartite hairs, and one directed posteriorly that is smooth. There are two classes (Labyrinthista and Peronosporomycetes) in the phylum that contain animal parasites. Only one genus, *Labyrinthriloides*, in the class Labyrinthista, that contains pathogens of certain shell-less mollusks, will be considered in this book. The class Peronosporomycetes contains three subclasses: the Peronosporomycetidae, which is mainly restricted to phytopathogens but also contains a mammalian pathogen, *Pythium insidiosum*; the Rhipidiomycetidae, whose members are saprophytes; and the Saprolegniomycetidae, which includes parasites of animals. There are four orders of pathogens within the last subclass (Table 7).

1. Subclass Peronosporomycetidae.
    a. Pythiales. This order contains a pathogen of mammals including humans, *Pythium insidiosum*.
2. Subclass Saprolegniomycetidae.
    a. Saprolegniales. A number of important fish, crustacean, and insect pathogens are contained in this order.
    b. Salilagenidiales. This order comprises parasites of crustacea and mollusks.
    c. Myzocytiopsidiales. Pathogens of nematodes, rotifers, and mosquitoes and other insects are contained in this order.
    d. Leptomitales. This group comprises pathogens of nematodes, rotifers, and insects.

V. KINGDOM PROTOCTISTA

Only the phylum Protista of the kingdom Protoctista will be considered. The members of the class Plasmodiophoromycetes are characterized by zoospores with two unornamented flagella that are of unequal length. A single genus in the

Table 7 The Kingdoms Straminipila and
Protoctista

Kingdom: Straminipila
  Phylum: Heterokonta[a]
    Class: Labyrinthista
      Order: Thraustochytriales
    Class: Peronosporomycetes
      Subclass: Peronosporomycetidae
      Order: Pythiales
      Subclass: Saprolegniomycetidae
      Order: Saprolegniales
      Order: Leptomitales
      Order: Salilagenidiales
      Order: Myzocytiopsidales
Kingdom: Protoctista
  Phylum: Protista[b]
    Class: Plasmodiophoromycetes
      Order: Haptoglossales

---

[a] The only phylum in the kingdom Straminipila to be
considered here. The hierarchal considerations have
been simplified. See Tables 1 and 2 of Chap. 2 for a
complete treatment of this phylum.
[b] The only phylum of the Protoctista to be considered.

order Haptoglossales, *Haptoglossa*, a pathogen of rotifers will be considered in this book.

VI. POPULATIONS

Groups of individuals will be considered in Part 2 of this book. Populations of individual species that are separated from other populations by physical or biological barriers will diverge from one another (14). Examples of such divergence among fungi are accumulating. In one recent example, RFLP analysis of 32 isolates of *Paracoccidioides brasiliensis* indicated at least five geographically distinct groups of strains. The groups of strains corresponded to country borders (15). Such divergence is one possible basis for the emergence of new species (evolution). Molecular techniques applicable to such descriptions are covered in Chap. 12, and the emergence of a new field—molecular epidemiology, which is based on molecular data—is considered in Chap. 13.

Divergence in populations may also have a genetic basis (14). One example is *Candida albicans*. Until recently, this fungus was considered an amictic anamorph. It has been shown that the fungus spontaneously produces a high frequency of chromosomal aberrations (16). Such alterations have been suggested to be a means of achieving genetic variability in an organism that lacks a teleomorphic form (16). Such variation could be a basis for speciation in a group of fungi like the unicellular yeasts in which taxonomic characters are limited and depend to a large extent on utilization of various metabolites (17). Very recently it has been shown that *C. albicans* can be induced to mate (18,19), but the extent to which mating occurs in populations of *C. albicans* remains to be evaluated. The genetic variability of *C. albicans* and the recent studies on mating will be covered in Chap. 14.

GLOSSARY

Acervulus (pl. acervuli).   A fruiting body without a covering tissue, usually a discoid or flat mass of conidiophores producing conidia in a moist mass.

Anamorph.   The forms representing asexual reproduction.

Arthroconidium (pl. arthroconidia).   A conidium produced by septation of a hypha and fragmentation at the septa.

Ascocarp.   *See* ascoma.

Ascoma (pl. ascomata).   A fungus structure containing asci, which in turn bear ascospores. Also called an ascocarp.

Ascospores.   A sexual spore borne within an ascus. The definitive structural element of the phylum Ascomycota.

Ascus (pl. asci).   An ascospore-bearing saclike structure.

Asexual reproduction.   Reproduction that does not involve fusion of nuclei and meiosis.

Basidiospores.   Sexual spores borne on a basidium which define the phylum Basidiomycota.

Basidium (pl. basidia).   The fungus structure upon which basidiospores are borne.

Blastoconidium (pl. blastoconidia).   A conidium produced by budding.

Clade.   Literally a branch. A monophyletic group. The word is used in studies on phylogeny (evolution) of a species.

Cleistothecium.   A completely closed ascoma.

Colony (pl. colonies).   In the unicellular yeasts, a collection of yeast cells (blastoconidia, q.v. or arthroconidia, q.v.). In the molds, a collection of hyphae, also called a mycelium (pl. mycelia) or a thallus (pl. thalli).

Columella.   The sterile inflated end of a sporangiophore that extends into the sporangium.

Conidiophores.   A specialized hypha that bears conidia.

Conidium (pl. conidia).   An asexual reproductive propagule.

Dermatophyte. Literally, "skin plant." A group of pathogenic fungi that cause the clinical conditions known as Tinea (ringworm).

Dimorphic. Having two forms. In medical mycology the term has been used to describe a duality of form associated with the parasitic and saprophytic phases of the life cycle of a zoopathogenic fungus.

Flagellum (pl. flagella). A hairlike structure that propels a motile cell.

Gymnothecium (pl. gymnothecia). An ascoma whose wall is composed of loosely woven hyphal elements.

Hypha (pl. hyphae). A tubular filament—usually branched—that makes up the vegetative portion of a mold. Two forms of hyphae occur: nonseptate, in which the cytoplasm of the hypha is uninterrupted by cross walls (septa), and septate, in which the cytoplasm of the hyphae is regularly interrupted by septa.

Merosporangium (pl. merosporangia). A cylindrical sporangium in which the sporangiospores are arranged in a row.

Mitospore. A nucleated spore formed by mitosis.

Ostiole. A hole in an ascoma or a pycnidium.

Peridium (pl. peridia). The outer wall of a fruiting body.

Perithecium. A closed ascoma with a hole (ostiole) at the top and a wall of its own.

Pycnidium (pl. pycnidia). An asexually produced flasklike body with an ostiole and containing conidia.

Sexual reproduction. Reproduction that includes fusion of nuclei and meiosis.

Sporangiolum (pl. sporangiola). A small sporangium containing a few sporangiospores.

Sporangiophores. The modified hyphal element that bears the sporangium.

Sporangiospores. Asexual spores produced within a sporangium.

Sporangium (pl. sporangia). A saclike structure within which asexual spores (sporangiospores) are formed by progressive cleavage.

Teleomorph. The forms representing sexual reproduction.

Zoospore. A motile asexual spore.

Zygospore. A resting spore that develops from a zygote formed by the union of two similar gametes. The spore that defines the phylum zygomycota.

REFERENCES

1. DH Howard, HM Kokkinos. Classification of fungi. In: RD Fegin, JD Cherry, eds. Textbook of Pediatric Infectious Diseases. vol 2. 4th ed. Philadelphia: Saunders, 1998, pp. 2287–2288.
2. KJ Kwon-Chung, JE Bennett. Medical Mycology. Philadelphia: Lea & Febiger, 1992.

3. CJ Alexopoulos, CW Mims, M Blackwell. Introductory Mycology. 4th ed. New York: Wiley, 1996.

4. DL Hawksworth, PM Kirk, BC Sutton, DN Pegler. Ainsworth and Bisby's Dictionary of the Fungi. 8th ed. Cambridge, UK: International Mycological Institute, CAB International., University Press, 1996.

5. PP Vergeer. Poisonous fungi: Mushrooms. In: DH Howard, ed. Fungi Pathogenic for Humans and Animals. part B. New York: Marcel Dekker, 1983, pp. 373–412.

6. D Arora. Mushrooms Demystified. 2nd ed. Berkeley, CA: Ten Speed Press, 1986.

7. G Lincoff, DH Mitchell. Toxic and Hallucinogenic Mushroom Poisoning. New York: Van Nostrand Reinhold, 1977.

8. DA Sutton, AW Fothergill, MG Rinaldi. Guide to Clinically Significant Fungi. Baltimore: Williams and Wilkins, 1998.

9. GS de Hoog, J Guarro, J Gené, MJ Figueras. Atlas of Clinical Fungi. Utrecht, the Netherlands: Centraalbureaus voor Schimmelcultures, 2000.

10. DN Fredricks, JA Jolley, PW Lepp, JC Kosek, DA Relman. *Rhinosporidium seeberi*: A human pathogen from a novel group of aquatic protistan parasites. Emerging Infec Dis 6:273–282, 2000.

11. JC Edman, JA Kovacs, H Mansur, DV Santi, HJ Elwood, ML Sogin. Ribosomal RNA sequence shows *Pneumocystis carinii* to be a member of the Fungi. Nature 334:519–522, 1988.

12. JW Taylor, E Swann, ML Berbee. Molecular evoluation of ascomycete fungi. In: DL Hawksworth, ed. Ascomycetes Systematics: Problems and Perspectives in the Nineties. New York: Plenum, 1994.

13. RJ Tarcha, RA Herr, JW Taylor, L Ajello, L Mendoza. Phylogenetic analysis of the ITS 1, 5.85 and ITS 2 rDNA sequences of *Lacazia loboi* confirms placement of this unique fungal pathogen within the Onygenales. Abstr. 101st Meeting of the American Society for Microbiology, F 131:381, 2001.

14. E Mayr. Systematics and the origin of species from the viewpoint of a zoologist: A reprint with a new introduction by the author. Cambridge, MA: Howard University Press, 1999.

15. GA Niño-Vega, AM Calcagno, G San-Blas, F San-Blas, GW Gooday, NAR Gow. RFLP analysis reveals marked geographical isolation between strains of *Paracoccidioides brasiliensis*. Med Mycol 38:437–441, 2000.

16. EP Rustchenko-Bulgac, F Sherman, JB Hicks. Chromosomal rearrangements associated with morphological mutants provide a means for genetic variation in *Candida albicans*. J Bacteriol 172:1276–1283, 1990.

17. NJW Kreger-van Rij, ed. The Yeasts. Amsterdam: Elsevier, 1984.

18. CM Hull, RM Rainer, AD Johnson. Evidence for mating of the ''asexual'' yeast *Candida albicans* in a mammalian host. Science 289:307–310, 2000.

19. BB Magee, PT Magee. Induction of mating in *Candida albicans* by construction of MTLa and MTLα strains. Science 289:310–313, 2000.

2

The Peronosporomycetes and Other Flagellate Fungi

Michael W. Dick
School of Plant Sciences, University of Reading, Reading, England

I. BASIC BIOLOGY

There are two major phylogenetic lines of flagellate fungi: those with zoospores having a single, posteriorly directed, smooth whiplash flagellum and those with zoospores having two (heterokont and anisokont) flagella, one directed anteriorly and clothed with two rows of tubular tripartite hairs (straminipilous ornamentation) and one directed posteriorly, smooth and with an acronema. The first line (the Chytridiomycetes) constitutes a near-basal clade on the animal/mycote phylogeny; the second line (the Peronosporomycetes) is a major clade with the chromophyte algae (Table 1). The Plasmodiophoromycetes (*Haptoglossa*) is characterized by zoospores with two homokont but anisokont unornamented flagella.

The Peronosporomycetes are probably the largest and are certainly the most diverse monophyletic class of flagellate fungi. Originally separated from other flagellate fungi by their oogamous sexual reproduction, the Peronosporomycetes are now primarily distinguished from other fungi by the distinctive biflagellation of the zoospore. Peronosporomycetes are fungi on both physiological and morphological criteria; that is, they are eukaryotic with uninucleate or coenocytic protoplasts bounded by cell walls in their assimilatory states and are thus obligately osmotrophic heterotrophs (1,7,8).

The peronosporomycete/chromophyte algae/straminipilous-heterotroph monophyletic line includes gut commensals such as *Blastocystis* and *Proteromonas* (Slopalinida), which are not fungi (2), and animal parasites such as *Labyrinthuloides* (Labyrinthista), which are occasionally treated as fungi but which I do not regard as constituting a fungal class (3,8). A few organisms, such as

17

Table 1   Classification of Flagellate Fungal Parasites and Commensals of Animals

Kingdom: Mycota
Phylum: Chytridiomycota
  Class: Chytridiomycetes
    Order: Chytridiales
    Order: Spizellomycetales
    Order: Neocallimasticales: Neocallimaticaceae; *Neocallimastix* IB Heath, *Caeco-*
      myces JJ Gold, *Piromyces* JJ Gold et al., *Anaeromyces* Breton et al., *Orpino-*
      myces Barr *et al.*
    Order: Blastocladiales: Coelomomycetaceae; Catenariaceae; Oedogoniomycetaceae
      Coelomomyces Couch, *Catenaria* Sorokin, *Oedogoniomyces* Kobayasi & M.
      Ôkubo
    Order: Monoblepharidales
Kingdom: Straminipila (subkingdom Chromophyta)
Phylum: Heterokonta (other phyla omitted from consideration in this chapter)
  Class: Labyrinthista
    Order: Labyrinthulales (one doubtful species—*L. thais*)
    Order: Thraustochytriales
      Thraustochytriaceae: *Labyrinthuloides* F. O. Perkins
Subphylum: Peronosporomycotina
    Order: Lagenismatales
  Class: Hyphochytriomycetes
  Class: Peronosporomycetes
Kingdom: Protoctista
Phylum: Protista (other phyla omitted from consideration)
  Class: Plasmodiophoromycetes
    Order: Plasmodiophorales
    Order: Haptoglossales
      Haptoglossaceae: *Haptoglossa* Drechsler

Coelomomyces (Chytridiomycetes), do not fit the above definition of a fungus, lacking a cell wall in the assimilatory phase (4). The Neocallimastigales (Chytridiomycetes) are divergent within their class, being obligate anaerobes with zoospores possessing multiple posterior flagella (5).

Within the Peronosporomycetes three subclasses can be recognized (Table 2). The *Peronosporomycetidae* are species-rich (60% of the total number of species in the class) but mainly restricted to phytopathogens. The *Rhipidiomycetidae* are saprobic on vegetable substrata, but the taxa of the *Saprolegniomycetidae* have diverse habitats and substrata and include the most important of the flagellate fungal parasites in animals. In addition, the peronosporomycete order *incer-*

Table 2 Classification of the Class Peronosporomycetes (orders and families containing parasites of animals in boldface)

Peronosporomycetes
Subclass: **Peronosporomycetidae**
  Order: Peronosporales
  Order: **Pythiales**
    Pythiaceae: *Pythium* Pringsh., *Lagenidium* Zopf
Subclass: Rhipidiomycetidae
  Order: Rhipidiales
Subclass: **Saprolegniomycetidae**
  Order: **Saprolegniales**
    Saprolegniaceae: *Saprolegnia* Nees, *Achlya* Nees, *Sommerstorffia* Arnautov, *Hydatinophagus* Valkanov, *Couchia* W. W. Martin
    Leptolegniaceae: *Leptolegnia* de Bary, *Aphanomyces* de Bary
  Order: Sclerosporales
  Order: **Leptomitales**
    Apodachlyellaceae: *Eurychasmopsis* Canter & M. W. Dick
    Leptolegniellaceae: *Aphanomycopsis* Scherff., *Nematophthora* Kerry & D. H. Crump
  Order: **Salilagenidiales**
    Salilagenidiaceae: *Salilagenidium* M. W. Dick (36)
    Haliphthoraceae: *Haliphthoros* Vishniac, *Atkinsiella* Vishniac, *Halodaphnea* M. W. Dick
  Order: Olpidiopsidales
  Order: **Myzocytiopsidales**
    Myzocytiopsidaceae: *Myzocytiopsis* M. W. Dick, *Gonimochaete* Drechsler, *Chlamydomyzium* M. W. Dick
    Crypticolaceae: *Crypticola* Humber *et al.*
Genera *Incertae Sedis:* *Blastulidium* Pérez, *Ciliomyces* I. Foissner & W. Foissner, *Endosphaerium* D'Eliscu

---

tae sedis, the Myzocytiopsidales, contains important parasites of nematodes. Although most species of Peronosporomycetes are freshwater or terrestrial, a few are oligohaline or marine, but it is yet to be established whether these taxa represent primarily or secondarily marine evolutionary developments.

The evolution of the Peronosporomycetes differs from that of eumycote fungi and the uniflagellate class Chytridiomycetes in that it is based on vegetative diploidy, not haploidy. The nuclear cycle is haplomitotic B (6), in which mitosis is confined to the diploid phase. Haploid mitosis does not occur. Both haplomitotic A and diplomitotic ploidy cycles are absent from the Peronosporomycetes, although they occur elsewhere in the phylum and kingdom (e.g., Fucophyceae). Diplomitotic ploidy cycles occur in the Blastocladiales (Chytridiomycetes).

II. HOST–PARASITE RELATIONSHIPS AND INTERACTIONS

Parasitism of aquatic animals and the saprobic existence of these fungi on dead animals and sloughed animal remains has been recognized for 250 years (7,8). However, the range of animals parasitized is both wide yet restricted, including vertebrates, crustaceans, insects, and aschelminths (nematodes and rotifers). Parasites of crustaceans bridge the freshwater and marine environments, albeit with different and possibly phylogenetically distantly related species.

Fungus/host-animal relationships are sometimes fairly tightly circumscribed, as with the Salilagenidiales parasitic in marine crustaceans and the Myzocytiopsidales parasitic in aschelminths. However, in both groups there are well-documented occurrences of parasitism outside this normal range. *Myzocytiopsis* can infect tardigrades (9). Another species of the same genus has been reported on a gasterotrich protoctist (see Table 3). One species of *Halodaphnea* has been described from a marine rotifer rather than a crustacean (see Key in Section IV,B,6), and other within-habitat/cross-host boundary parasitisms can be found in the literature. On the other hand, animal groups are often parasitized by a range of unrelated fungi: for example, nematodes by Ascomycetes, Peronosporomycetes, and Plasmodiophoromycetes; insects (Diptera) by Peronosporomycetes (Pythiales and Saprolegniales) and Chytridiomycetes (Blastocladiales).

By far the most noteworthy crustacean parasites are *Aphanomyces* (Saprolegniales) on freshwater crayfish (10,11,220) and *Salilagenidium* and *Halodaphnea* (Salilagenidiales) on marine crabs and prawns (12–14). Entire populations of European crayfish have been eliminated from many river systems in Europe following the introduction and spread of *Aphanomyces astaci* (Krebspest disease), and recovery is improbable (11,15,220). Mariculture of crabs, prawns, and shrimps in Asian coastal waters is subject to epidemics caused by various species of the Salilagenidales (12,13).

Although there are several examples of insect parasitism, *Lagenidium* (Pythiales), which is endoparasitic in mosquito larvae, has received a considerable amount of research funding for biological control of mosquito populations (see, e.g., 16–19), but *Crypticola* (Myzocytiopsidales), on mosquito and blackfly larvae (see Ref. 14), and *Leptolegnia chapmanii* (Saprolegniales), also in mosquito larvae, have not been studied as widely (20,21). *Coelomomyces* (Blastocladiales) was the subject of a number of papers in the early quest for biological control of mosquitoes (22–24).

The disease of salmonid fish commonly referred to as UDN (ulcerative dermal

Table 3 *Peronosporomycetes, Chytridiomycetes, and Plasmodiophoromycetes associated with Aschelminthes*

	Nematode hosts	Rotifer hosts
Catenaria allomycis (175)	In *Allomyces anomalus* R. Emers. (Blastocladiales) from soil	
Catenaria anguillulae (176)	In sheep liver fluke eggs (and nematodes—later reports)	
Catenaria auxiliaris (J. G. Kühn) (177)	In *Heterodera schachtii* Schmidt	
Catenaria vermicola (178)	In *Xiphinema chambersi* Thorne (from St. Augustine grass)	
Chlamydomyzium anomalum (G. L. Barron) (171)	In an unidentified nematode (from barnyard soil)	
Chlamydomyzium aplanosporum (171)		In *Distylae* rotifers (from rotting straw)
Chlamydomyzium internum (171)		In *Distylae* rotifers (from organic debris with pine needles)
Chlamydomyzium oviparasiticum (171)		In eggs of *Adineta* spp. [from farmyard soil]
Chlamydomyzium septatum (Karling) (171)		In rotifer eggs (and sporangia of *Catenaria anguillulae* Sorokin on snake skin, from soil sample)
Chlamydomyzium sphaericum (55)	Adult *Rhabditis* nematodes and rarely, tardigrades	
Endochytrium oophilum (De Wild) (179)	In various algae, protists, and rotifer and nematode eggs (according to later reports)	
Endochytrium operculatum (De Wild) (179)	In various algae, protists, and rotifer and nematode eggs (according to later reports)	
Gonimochaete horridula (180)	In *Acrobeloides* sp. [affin. *A. buetschlii* (De Man) Thorne on decaying leaves of *Acer rubrum* L.]	
Gonimochaete latutubus (181)	In *Rhabditis marina* Bastian	
Gonimochaete lignicola (182)	In nematodes (from riverbank soil)	

Table 3 Continued

	Nematode hosts	Rotifer hosts
Gonimochaete pyriformis (183)	In *Diploscapter* spp. (from soil in a topical greenhouse)	
Halodaphnea parasitica (K. Nakam. & Hatai) (14)		In *Brachionus plicatus* Müller (marine)
Haptoglossa elegans (184)		In bdelloid rotifers (from debris under a tree fern)
Haptoglossa heterospora (185)	In many species of nematodes found in agar–water–soil litter cultures	
Haptoglossa humicola (186)		In bdelloid rotifers (from soil in a tropical greenhouse)
Haptoglossa intermedia (187)		In bdelloid rotifers and nematodes (from woodland soil)
Haptoglossa mirabilis (188)		In *Adineta* spp. (from soil in a tropical greenhouse)
Haptoglossa zoospora (189)	In nematodes (from farmyard soil)	
Hydatinophagus americanus (190) *Aphanomyces americanus* (115,190)		In *Monostyla* sp.
Hydatinophagus apsteinii (115,191) *Aphanomyces hydatinae* Valkanov, *Archiv für Protistenkunde* 74: 17 (1931)		In *Epiphanes senta* (Müller as its synonym, *Hydatina senta* Ehrenb.)
Myzocytiopsis bolata (55)	In adult *Rhabditis* nematodes; some eggs parasitized	
Myzocytiopsis distylae (Karling) (171)		In eggs of *Distyla* sp. (from a water sample)

Species		
Myzocytiopsis elegans (Perronc) (171)		In *Philodina roseola* Ehrenb.
Myzocytiopsis fijiensis (Karling) (171)		In eggs of *Distyla* sp. (soil sample)
Myzocytiopsis glutinospora (G. L. Barron) (171)	In *Rhabditis terricola* Dujardin (from barnyard soil)	
Myzocytiopsis humicola (G. L. Barron & Percy) (171)	From soil in a cattle pen	
Myzocytiopsis indica (U. P. Singh) (171)		In rotifers
Myzocytiopsis intermedia (G. L. Barron) (171)	In *Rhabditis* sp. (soil sample from a birch–maple–poplar wood)	
Myzocytiopsis lenticularis (G. L. Barron) (171)	In nematodes (from soil in a cattle pen)	
Myzocytiopsis microspora (Karling) (171)		In bodies of *Distyla* sp. (from a soil sample)
Myzocytiopsis oophila (Sparrow) (171)		In eggs and embryos of rotifers (from aquatic debris)
Myzocytiopsis osiris (36)	Infecting a large proportion of the population of adult *Rhabditis* nematodes and some juveniles	
Myzocytiopsis papillata (G. L. Barron) (171)	In *Rhabditis terricola* Dujardin (from barnyard soil)	
Myzocytiopsis parthenospora (Karling) (171)	Bodies of *Heterodera* sp. and eggs of *Chaetonotus larus* O. Müller (*Gasterotricha* from a soil sample)	In eggs and bodies of *Distyla* sp. and *Philodina* sp.
Myzocytiopsis subuliformis (E. Maupas) (171)	In *Rhabditis teres* Schn. and *Rhabditis giardii* E. Maupas	
Myzocytiopsis vermicola (Zopf) (171)	In *Anguillula* sp.	

Table 3 Continued

	Nematode hosts	Rotifer hosts
Myzocytiopsis zoophthora (Sparrow) (171)		in rotifers and rotifer eggs
Nematophthora gynophila (192)	In females of *Heterodera avenae* Wollenweber	
Olpidium entophytum (A. Braun) Rabench., *Flora Europaea Algarum* 3: 283 (1868) var. *intermedium* J. C. Constantineanu, *Revue Générale de Botanique* 13: 371 (1901) (37:137)		In rotifer eggs
Olpidium granulatum (37:145)		In rotifer eggs
Olpidium gregarium (Nowak.) J. Schröt, *Kryptogamen-Flora von Schlesien* 3(1): 182 (1885) (37: 144)		In rotifer eggs
Olpidium incognitum (193)		"Parasitic in the egg of a rotifer"
Olpidium longum (193)		"Parasitic in the eggs of certain [aquatic] animalcules"
Olpidium macrosporum (Nowak.) J. Schröt, *Kryptogamen-Flora von Schlesien* 3(1): 182 (1885) (37: 155)		In rotifer eggs
Olpidium nematodeae Skvortzov, *Archiv fur Protistenkunde* 57:204 (1927) (37:147)	In nematodes	
Olpidium paradoxum (194)		From pond water
Olpidium poreferum (195) [see also *Olpidium allomycetos* Karling, *American Journal of Botany* 35: 508, 1948 (37: 143)]		Parasitic in rotifers
Olpidium rotiferum Karling, *Lloydia* 9: 6 (1946) (37:146)		Rarely (experimentally) in rotifer eggs In adult rotifers and rotifer eggs

Olpidium sparrowii (196) — In eggs of *Lecane* spp. (from cultivated soil)

Olpidium vermicola (197) — Endoparasitic in nematode eggs (from rotting wood)

Olpidium zootoctum (A. Braun) (37: 155) — In *Anguillula* sp.

Phlyctochytrium nematodae Karling, *Lloydia* 9: 7 (1946) (37: 337) — Parasitic in adults, eggs, and cysts of nematodes

Pseudosporopsis rotatoriorum (198) — An anisokont biflagellate monad parasitic in nematodes [probably a *Haptoglossa*]

Pythium caudatum (G. L. Barron) (36, 55) — In nematodes (from soil in a cattle pen)

Legenidium caudasum G. L. Barron, *Antonie van Leeuwenhock* 42: 134 (1976)

Sommerstorffa spinosa (199) — Rotifers [thallus epiphytic on *Cladophora* (*Chlorophyceae*)]

Synchaetophagus balticus (200) — In *Synchaeta monopus* Plate

Rhizophydium gibbosum (Zopf) A. Fisch., *Rabenhorst's Kryptogamen-Flora, 2 Aufl., Bd 1, Abt. 4: 102* (1892) (37: 237) — Parasitic on rotifer eggs (and in various nematodes and algae according to other authors)

Rhizophydium vermicola (37: 277) — Nematodes infected by *Harposporium, Aphanomyces*, or *Lagenidium* (= *Myzocytiopsis*)

Rhizophydium zoophthorum (P. A. Dang.) A. Fisch., *Rabenhorst's Kryptogamen-Flora, 2 Aufl., Bd 1, Abt. 4: 94* (1892) (37: 285) — In cultures of rotifers (adults and eggs)

Note: listed in alphabetic order with nomenclatural citation (and reference number, page number appended for Sparrow, 1960) with the type host and habitats where noted. Many of the species of *Myzocytiopsidales* and *Haptoglossales* can be transferred between nematodes and rotifers in laboratory culture.

Key to the Species of *Salilagenidiaceae* and *Haliphthoraceae*

1   Thallus more or less mycelial, branched, mean hyphal diam. <24 μm, usually <15 μm, septate or sparingly septate; non-rhizoidal, culturable; eucarpic or holocarpic with time; sexual reproduction present or absent.   7

1'   Thallus more or less inflated, lobed, branches with mean diam. >25 μm; septa rare or absent; rarely rhizoidal (in culture), culturable; holocarpic with time (doubtfully eucarpic); sexual reproduction absent.   2

2(1')   Thallus irregularly and broadly tubular, rhizoids absent; sporangiogenesis not reported to have a centripetal contraction phase; zoospores with lateral flagellar insertion; mean diameter of zoospore cysts <6 μm (volume equivalent <120 μm³), or if greater then parasitic in mollusks [these species may be closer to *Haliphthoros*, but resemble *A. dubia* in their habit and the long tapering exit tubes].   (Halodaphnea) 3

2'(1')   Thallus inflated, septa normally absent; sporangiogenesis intrasporangial, often with a late, marked centripetal contraction, initials amoeboid at first on fine cytoplasmic strands; mean diameter of zoospore cysts >7 μm (volume equivalent >180 μm³) [parasitic in eggs of crabs *Pinnotheres pisum* (L.) and other crustaceans (*Crangon, Gonoplax, Leander, Macropodia, Paguristes, Portunus*, and *Typton*; sometimes regarded as saprobic; rhizoids cut off by septa occasionally present in culture but not observed on natural substrata; thallus lobes stout, 27–50 μm, tips swollen, up to 100 μm diam., zoosporangia 50–400 × 10–30 μm, with one or more exit tubes, distal part (up to 50 μm) hyaline; occasionally proliferous; zoospores pyriform or slipper-shaped (with lateral flagellation?), diplanetic (but not dimorphic?), 10–12 μm long (4.0–6.0 × 6.0–8.2 μm; (4.0)4.8(6.0) × 8.7(10.0) μm), first-formed cysts 7.0–8.0(9.0) μm (7.4 μm) diam., second-formed cysts 6.0–7.0 μm (6.8 μm) diam., germ tube 1.7 μm diam., gemmae present] [this genus may not belong in the Haliphthoraceae].   *Atkinsiella dubia* (D. Atkins) Vishniac (monotypic genus in this text)

3(2)   Zoospore cysts approximately 5.0 μm diam. (volume equivalent <75 μm³); colonies more or less saccate with 1–3 zoosporangial exit tube(s); optimum temperature for growth >24°C [parasitic in various Crustacea].   4

3'(2)   Zoospore cysts 8.0 μm diam. (volume equivalent >250 μm³); colonies filamentous with 1 (rarely 2) exit tube(s) produced from each zoosporangium; optimum temperature for growth 20°C (5–25°C) [parasitic in *Haliotis sieboldii* Reeve (abalone); hyphae stout, irregular 16–41(140) μm diam., branched, nonseptate, becoming septate to delimit zoosporangia; zoospores pyriform 4.0–8.0 × 7.0–12.0 μm, diplanetic, isokont; zoospore cysts germinating by means of a fine filament 62–295 μm long before broadening to form the thallus].   *Halodaphnea awabi* (N. Kitancharoen et al.) M. W. Dick

4(3)   Zoosporangial exit tubes 1–3, always unbranched.   6

4'(3)   Zoosporangial exit tubes normally single, sometimes branched.   5

5(4') Colony pigmented, gray to light brown; optimum temperature for growth 30–32°C; [parasitic in eggs of *Scylla serrata* (Forsskål); hyphae (12)26(40) μm diam., zoosporangia 42–1150 × 5–15 μm; zoospores pyriform or slipper-shaped, diplanetic but not dimorphic, (3.8)4.5(5.0) × (5.0)6.3(10.0) μm, cysts (4.5)5.0(7.5) μm diam.].

Halodaphnea hamanaensis (Bian & Egusa) M. W. Dick (type species)

5'(4') Colony hyaline; optimum temperature for growth 25°C (15–30°C) [parasitic in *Panulirus japonicus* (von Siebold); holocarpic, hyphae stout, branched and septate, 10–22(64) μm diam., subthalli transformed into zoosporangia or gemmae; zoospores pyriform or reniform 4.0–5.0 × 7.0–10.0 μm, diplanetic; zoospore cysts 5.0–7.0 μm diam., germinating by means of a fine filament 14–253 μm long before broadening to form the thallus].

Halodaphnea panulirata (N. Kitancharoen & Hatai) M. W. Dick

6(4) Zoosporangia with 1-several broad discharge tubes, (6)8–9(10) × (40)200–300(510) μm containing several ranks of zoospores [parasitic in *Portunus pelagicus* L.; holocarpic, hyphae stout, becoming septate with age, 10–38 μm diam., subthalli transformed into zoosporangia or thick-walled gemmae, 22–190 μm diam.; zoosporogenesis intrasporangial; zoospores pyriform or subglobose (4.0)4.7(6.5) × (5.0)6.3(8.0) μm, diplanetic; zoospore cysts (4.0)5.2(7.0) μm diam., germinating by means of a fine filament 5–190 μm long before broadening to form the thallus; optimum temperature for growth 25°C (20–30°C)].

Halodaphnea okinawaensis (K. Nakam. & Hatai) M. W. Dick

6'(4) Zoosporangia with 1–2 infrequently-branched discharge tubes [discharge tubes straight, wavy or coiled, usually with a cone-like base, 6–14 × 20–780 μm, parasitic in *Brachionus plicatilis* Müller (Rotifera); usually holocarpic (eucarpic with age or in suboptimal growth temperatures), hyphae stout, saccate, becoming septate with age, 15–50 μm diam., subthalli transformed into sporangia or developing into thick-walled gemmae, 40–200 μm diam.; zoosporogenesis intrasporangial; zoospores pyriform (4.0)4.6(5.6) × (4.8)6.0(7.4) μm, monoplanetic, isokont; zoospore cysts (4.8)5.5(6.0) μm diam., germinating by means of a fine filament 8–250 μm long before broadening to form the thallus; optimum temperature for growth 25°C (20–30°C)].

Halodaphnea parasitica (K. Nakam. & Hatai) M. W. Dick

7(1) Thallus septate but not usually with endothallial contraction and new wall formation; zoospore size large (volume equivalent >275 μm³); sexual reproduction present or absent (*Salilagenidium*) 10

7'(1) Thallus elements rounding up to form new endothallial walls; zoospore size variable (volume equivalent <250 μm³); sexual reproduction absent.

(*Haliphthoros*) 8

8(7') Parasitic in crustaceans or mollusks; zoospore size medium-large (volume equivalent >100 μm³); resting bodies not known. 9

Key to the Species of *Salilagenidiaceae* and *Haliphthoraceae*

8'(7') Parasitic in larvae of mollusks (*Venus*); zoospore size small (volume equivalent <75 μm^3); "resting bodies" 40 × 45 μm up to 80 × 90 μm formed as lateral diverticula (oogonia?) [thallus initially filamentous, 10–15 × 82 μm, becoming both inflated and with tapered extremities, resembling rhizoids; intrathallial walled bodies (31–33 × 46–56 μm); zoosporangia intramatrical, 5 × 15–142 μm; zoospores 2.0 × 5.0 μm, anisokont, with a shorter straminipilous flagellum].
Haliphthoros zoophthorum (Vishniac) M. W. Dick

9(8) Parasitic in eggs of mollusks (*Urosalpinx* (*U. cinerea* (Say)), *Haliotis*), and crustaceans (*Penaeus*); sporangia with short exit tubes (7.0)8.0(14.0) μm long; zoospores monoplanetic [holocarpic; hyphae (7.0)14.0–18.8(40.0) μm; zoospores pyriform, subspherical or elongate, (5.0)6.8–7.2(10.6) × (6.7)8.5(12.2) μm, monoplanetic, cysts (6.8)7.8–8.5(8.6–20.0) μm, very slender germ tube 0.5(1.8–2.2) μm diam.].
Haliphthoros milfordensis Vishniac (type species)

9'(8) Parasitic in larvae of prawns (*Penaeus monodon* Fabricius; sporangia with long exit tubes 620 × (7.50)–(12.5) μm; zoospores polyplanetic, dimorphic(?) [hyphae stout, branched, irregular, nonseptate (10.0)21.0(37.5) μm, intrathallial bodies (190 × 100 μm) not disarticulating, remaining connected like beads; first-formed zoospores pyriform, subspherical or elongate, with lateral flagellar insertion, 5.0–7.5 × 7.5–12.5 μm, second-formed zoospores slipper-shaped and slightly shorter (10.0 μm), cysts 5.0–7.5(12.5) μm, very slender germ tube]
Haliphthoros philippinensis. Hatai *et al.*

10(7) Sexual reproduction not known. 11

10'(7') Sexual reproduction known. 13

11(10) Zoospores of medium size (volume equivalent <400 μm^3). 12

11'(10) Zoospores of large size (volume equivalent >500 μm^3) [parasitic in eggs and larvae of crabs (*Scylla serrata*)].
Salilagenidium scyllae (Bian *et al.*) M. W. Dick

Salilagenidium thermophilum (K. Nakam. *et al.*) M. W. Dick

These two species are scarcely separable from the published descriptions:

Salilagenidium scyllae: hyphae thick, irregular, branched, 7.5–17.0(40.0) μm diam., sparingly septate, nonsegmented, thalloid elements becoming sporangia; zoosporangial discharge tubes short or long, 37–500 × 4–10 μm, apex dilating to form a deliquescent vesicle; cytoplasm not filling vesicle prior to cleavage, vesicle not persistent, zoospores reniform, pyriform, ovoid or oblong, (7.0)10.0(15.0) × (8.0)12.5(17.5) μm, released by deliquescence of vesicle or singly through a pore on the vesicle, monoplanetic; cysts (7.5)10.0(15.0) μm diam., cyst wall 1.5 μm thick.

Salilagenidium thermophilum: hyphae thick, irregular, branched, 8.0–24.0(40.0) μm diam., nonseptate, becoming sporangia; zoosporangial discharge tubes 34–440 × 6–14 μm, vesicular membrane not apparent; cytoplasic cleavage completed after discharge, "vesicle" 36–80 μm diam.; zoospores pyriform to subglobose, laterally biflagellate, 8.0–14.0 (mean 10.3) × 10.0–16.0 (mean 13.3) μm; monoplanetic; cysts 6.0–16.04 μm diam.; thermotolerant (15)30–40(45)°C.

12(11) Parasitic in muscles and swimmerets of crustaceans (shrimps) (*Pandalus borealis* Krøyer), holocarpic, culturable; hyphae wide, irregular, branched, 7.0–10.0 μm diam, partial cleavage within the sporangium, discharge tube 86–240 × 7–10 μm, vesicle formed, zoosporogenic protoplasm not filling vesicle, zoospores released by rupture of vesicle, vesicle persistent, 9.6 × 12.9 μm globose, reniform, pyriform or elongate, cysts 5.5–12.0 μm.

    Salilagenidium myophilum (Hatai & Lawhav.) M. W. Dick

12'(11) Parasitic in stomach of crayfish (*Penilia*); mycelium nonseptate, of uniform diameter, 4.2–5.2(7.0) μm, with homogeneous protoplasm; zoosporangia spherical or ellipsoidal, on short side branches, smooth walled; zoosporogenesis intrasporangial, released through a pore over a period of 2 min, swimming away 30 min later; zoospores 30–50 in a zoosporangium, of irregular shape, with an anteriorly inserted flagellum [known only from original locality].

    '*Hyphochytrium peniliae* N. J. Artemczuk & L. M. Zélézinskaja' *nom. illeg.*

13(10') Zoosporogenesis partly extrasporangial in a vesicle; antheridia absent; culturable.   14

13'(10') Zoosporogenesis intrasporangial; zoospore initials in a single row distally; antheridia present, hypogynous [parasitic in eggs of crabs (*Mytilus edulis* L.) and possibly lamellibranchs *Barnea* and *Cardium*; hyphae 7.5–20.0(40.0) μm diam.; zoosporangia undifferentiated from hyphae, occasionally proliferous; zoospores pyriform, 8.0–14.0 μm long, cysts 6.0–11.0 μm diam., diplanetic; oogonia rare, oospores single, nearly plerotic, 17.5–30.0(37.0) μm diam., oospore wall two layered, 7.5 μm thick].

    Salilagenidium marinum (D. Atkins) M. W. Dick

14(13) Parasitic in eggs and larvae of crabs [*Callinectes* (*C. sapidus* Rathbun), *Limulus*], and barnacles (*Chelonibia*); hyphae sparingly septate, extramatrical hyphae (5)8–14(50) μm diam., 1030 μm long, aseptate; zoosporangium with vesicle persistent after discharge; tip of exit tube gelatinizing and contents flowing into the thick gelatinous envelope, never filling it, vesicle persistent, zoospores 9.3 × 12.5 μm; cysts 8.0–10.0(11.3) μm diam., monoplanetic, germ tube 2.5 μm diam., oogonia intercalary, oospores (18)25(36) μm diam., [wall 3 μm thick, subeccentric oil reserve].

    Salilagenidium callinectes (Couch) M. W. Dick (type species)

14'(13) Parasitic in barnacles [*Chthamalus* (*C. fragilis* Darwin)]; hyphae stout, irregular, branched, highly vacuolate 10–18(39) μm diam., becoming segmented, segments behaving as zoosporangia, zoosporogenesis within a vesicle formed from the swelling of the exit tube apex, cytoplasm entering the vesicle only after the latter is completely developed, cytoplasm not filling the vesicle, vesicle disappearing immediately after discharge, zoospores reniform, 6.8–8.5 × 8.5–10.2 μm, oogonia intercalary or terminal, 19–47 μm diam., oospores 1(2), aplerotic, (16)21–25(27) μm diam.

    Salilagenidium chthamalophilum (T. W. Johnson) M. W. Dick. *For taxonomy see Refs. 14, 36, 37.*

Note: Supplementary information in brackets does not form part of the key dichotomy.

necrosis) or EUS (epizootic ulcerative syndrome), caused by *Saprolegnia parasitica*, occasionally reaches epidemic levels, and the disease continues to be under investigation because it has commercial importance in relation to fish farming (25,26). Another disease of vertebrates is Pythiosis insidiosi, caused by *Pythium insidiosum* (27).

Other parasites of invertebrates include the endoparasites of the Aschelminthes (Table 3), such as *Myzocytiopsis, Gonimochaete, Chlamydomyzium* (Myzocytiopsidales), *Nematophthora* (Leptomitales: Leptolegniellaceae), *Catenaria* (Chytridiomycetes: Blastocladiales) *Olpidium, Endochytrium, Rhizophydium* and *Phlyctochytrium* (Chytridiomycetes: Chytridiales), and *Haptoglossa* (Plasmodiophoromycetes), but with the exception of *Nematophthora* (28), their use in biological control has yet to be assessed. Although a few species have an extensive bibliography, the information for most of the species is restricted to the type descriptions, which are therefore cited in Table 3.

The fungal parasites of noncellular (protozoan) Protoctista are even less well known, with reports again limited to (often early) original descriptions. Parasite/host relationships are listed in Table 4.

The gut commensals placed in the Neocallimastigales have been reviewed by Li and Heath (29), Trinci et al. (30), and Ho and Barr (5).

III. METHODS OF DETECTION

A. Gross Features

Parasites on fish, crustacea, and insects are immediately obvious and ultimately destructive. Extramatrical hyphae are usually abundant on the surface of diseased fish and insects and are visible with the unaided eye. Willoughby (31) has provided a summation of the different peronosporomycete infections of fish, but caution is still necessary in determining the distinction between aggressive obligate parasitism and casual wound parasitism of moribund fish. The fungal thalli of the Salilagenidiales may be largely intramatrical, but these can readily be seen in whole mount preparations of the host limbs and gills. Microscopically obvious intramatrical disease is also apparent in aschelminths. In both Aschelminthes and marine Crustacea the diseased animal shows little sign of distress initially, even though the thallus development may be extensive. This is particularly true of rotifers, which continue to move and feed even though the major part of the body cavity may be occupied by the fungus. The gross features of Aphanomycosis of crayfish (Krebspest disease) are somewhat less obvious, but the disease causes rapid population extermination (11,15).

B. Disease Development

In fish, the disease most commonly encountered is an initially superficial ulceration, which rapidly becomes more deep-seated, causing histologically recogniz-

Table 4 Peronosporomycetes and Chytridiomycetes associated with Protozoan Protoctista

Aphanomyces acinetophagus (190, 37: 843)	In *Acineta flava* Claparède & Lachmann (Suctoria)
Aphanomycopsis cryptica (Canter, in Ref. 201)	In *Ceratium hirundinella* (O. F. Müll). Bergh. (Dinomastigota, Gonyaulacales)
Aphanomycopsis peridiniella (Boltovskoy and Aramb, in Ref. 202)	In cysts of *Peridinium willei* Huitf.-Kass (Dinomastigota)
Ciliomyces spectabilis (203)	In cysts of *Kahiella simplex* (Horvath, Ciliat) from air-dried meadow soil, after remoistening)
	From fallen inflorescences of *Cecropia* sp. (Moraceae)
Endemosarca anomala (204)	In *Colpoda* spp. (Ciliata)
	From old fruits of *Annona muricata* L. (Annonaceae)
Endemosorca hypsalysis (205)	In *Colpoda* spp. (Ciliata)
	From fallen flowers of *Althaea* sp. (Malvaceae)
Endemosarca ubatubensis (205)	In *Colpoda* spp. (Ciliata)
Eurychasmopsis multisecunda (Canter, in Ref. 42)	In *Podophrya* sp. (Suctoria) parasitic in
Myzocytiopsis parthenospora (Karling, in Ref. 171)	In eggs of *Chaetonotus iarus* O Müller (Gasterotricha; see also Table 3)
Nucleophaga amoebae (206)	In the hypertrophied nucleus of *Thecamoeba verrucosa* (Gläser; as its synonym, *Amoeba verrucosa* Gläser)
Nucleophaga hypertrophica (207)	Type material not verified
Nucleophaga peranemae (208)	In *Peranema trichophorum* (Ehrenb.) Stein (Sarcomastigophora, Euglenida)
Olpidiomorpha pseudosporae (209, 37: 123)	Within the zoocyst of *Pseudospora leptoderma* Scherff.
Olpidium arcellae (210)	Saprobic (1) on *Arcella* sp.
Olpidium diffugiae (211, 37: 154)	In *Diffugia* sp.
Olpidium leptophrydis (211, 37: 154)	In zoocysts of *Leptophrys vorax* (Cienk.)
Olpidium pseudosporeanum (211, 37: 146)	In zoocysts of *Pseudosporopsis bacillariacearum* Scherff. and *Pseudospora parasitica* Cienk.

Table 4   Continued

Olpidium vampyrellae (211, 37:146)	In zoocysts of *Vampyrella* spp.
Pseudosphaerita drylii (212)	In *Phacus acuminatus* Stokes (Sarcomastigophora, Euglenida)
Pseudosphaerita euglenae (206, 37: 963)	In *Euglena* sp. (Sarcomastigophora, Euglenida)
Pseudosphaerita radiata (P. A. Dang., in Ref. 213, 37: 963)	In *Cryptomonas ovata* Ehrenb. (Cryptophyceae, Cryptomona)
Rhizoblepharis amoebae (214)	In *Amoeba binucleata* Grüb. (Rhizopoda)
Sphaerita amoebae (215)	In *Thecamoeba sphaeronucleolus* [Greeff; as "*Amoeba sphaeronucleolus*" (Rhizopoda)]
	Not determinable
Sphaerita dangeardii (37: 126)	In *Euglena sanguinea* Ehrenb. (Sarcomastigophora, Euglenida)
Sphaerita endogena (*pro parte*) (37: 126)	In *Entamoeba citelli* E. R. Becker (Rhizopoda; from the ground squirrel *Citellus*)
Sphaerita entamoebae (216)	
Sphaerita minor (217)	In *Trichomonas* sp. (Parabasilia)
Sphaerita normeti (218)	In *Entamoeba histolytica* Schaedinn [as "*Entamoeba coli*" (Rhizopoda)]
Sphaerita nucleophaga (215)	In *Thecamoeba sphaeronucleolus* [Greeff; as "*Amoeba sphaeronucleolus*," also in "*Amoeba terricola*" (Rhizopoda)]
Sphaerita phaci (219)	In *Phacus pleuronectes* (O. S. Müller) Dujardin (Sarcomastigophora, Euglenida)
Sphaerita plasmophaga (215)	In *Thecamoeba sphaeronucleolus* [Greeff; as "*Amoeba sphaeronucleolus*" (Rhizopoda)]
Sphaerita trachelomonadis Skvortzov (37: 127)	In *Trachelomonasteres* Maskell var. *glabra* Skvortzov (1927) and *T. swirenkoi* Skvortzov (Sarcomastigophora, Euglenida)
Spirospora paradoxa (209)	In cysts of *Vampyrella* sp. (possibly *Vampyrella spirogyrae* Cienk.)

Note: listed in alphabetic order with nomenclatural citation (expanded in References) with the type host and habitats where noted. (Page numbers in brackets refer to Sparrow, 1960.) Note that the spelling of authorities for taxa may not correspond with that in the References list.

able granulomas composed of host macrophage cells and hyphae (31). It is thought that initial entry is through weakened or less-protected dermis on vulnerable extremities, such as fins. The disease is progressive through muscle tissue, resulting in massive loss or necrosis, eventually causing death after several days and severe mutilation. A less common disease syndrome is caused by initial infection in the gill/pharynx/gut tissues. Asphyxiation and starvation result in more rapid death with reduced superficial symptoms.

Melanization of cuticle tissue in the region of hyphal penetration is a characteristic defense symptom to Aphanomycosis of crayfish (Krebspest disease) (32–34). Melanization will effectively immobilize but not kill the infective hyphae; the extent to which this melanization is effective in constraining the fungus will determine the success or failure of the infection. Söderhäll and Cerenius (35) have related the immune reaction of crayfish to the elicitation of defense processes by β-1, 3-glucans, which are, of course, characteristic wall carbohydrates of peronosporomycetes.

IV. CHARACTERIZATION

A. Methods for Taxonomic Decision Making Including MB

The morphology and the morphogenesis of the Peronosporomycetes are of taxonomic and phylogenetic importance. The most obvious characters relate to thallus form, but for identification to class and lower hierarchical levels zoospore morphology and morphogenesis are normally essential. (Many of the pathogens do not—or do not readily—reproduce sexually.) Transmission electron microscopy (TEM) is diagnostic at class level, particularly with respect to mitochondrial profiles and vesicular inclusions. Several 18S rDNA sequences are now available (21,36) for representatives of the Saprolegniales and Pythiales, but identification probes for animal parasites have yet to be published. Diagnostic features are summarized below under the subheads thallus morphology and protoplasmic features, zoosporangia and zoosporogenesis, zoospore and zoospore cyst morphology, sexual reproduction, biochemistry, and molecular biology.

1. Thallus Morphology and Protoplasmic Features

The thalli of the Peronosporomycetes may be filamentous, composed of hyphae forming a mycelium; or coralloid (eucarpic or holocarpic), allantoid, or ellipsoid (holocarpic) (8,36). Chytridiales (37) are monocentric with or without an assimilative system composed of branched rhizoids. Blastocladiales may be pseudofilamentous or composed of catenulate chains of cells. All hyphae show tip growth. Only with holocarpic thalli (both septate and nonseptate) may there be superficial resemblance between the Peronosporomycetes and the Chytridiomycetes. Obligate parasites may be entirely confined *within* a single host protoplast (endobiotic parasites of protozoa) or intercellular within tissues or in the haemocoel of arthropods and aschelminths.

In the Peronosporomycetes the vegetative thallus is bounded by a wall membrane at maturity, but may be naked initially in some endobiotic parasites. Septa are normally only present to delimit reproductive structures or act as retraction septa (as in old mycelia of *Pythium*). In the Saprolegniaceae there is frequently excessive synthesis of wall material at the septum, resulting in the development of an irregular peg or callus on one or both sides of the septum. Hyphae may be of relatively narrow diameter (10–20 μm in *Aphanomyces*) or broad diameter (20–40 μm in *Saprolegnia*). Generalized intussusception of wall material occurs in older hyphae of *Saprolegnia*, and these hyphae may have diameters up to 120 μm.

The appearance of the protoplasm, using light microscopy, can often provide distinctive diagnostic features to an experienced worker, but it is difficult—and may be misleading—to describe these differences for a novice. Note should be made of the presence and kind of cytoplasmic streaming and the granulation of the cytoplasm, particularly the coarseness of the granulation and the "glassy" character of the groundplasm. (Contrast the gravelly appearance of the protoplasm in hyphae of *Saprolegnia* with the sparce large inclusions suspended in a translucent matrix in *Halodaphnea*.)

The thallus is initiated from a uninucleate zoospore cyst or an aplanospore, from a multinucleate asexual spore or propagule, or from a uninucleate sexually produced oospore. In most of the endoparasitic fungi that have been studied by TEM, an extremely fine penetration tube of approximately 0.1–0.2 μm in diameter enters the host, and subsequent tip expansion enables the formation of the first unit of the thallus.

At the ultrastructural level mitochondria are conspicuous in TEMs. Mitochondrial cristae of Peronosporomycetes appear either as longitudinal cylindrical profiles with unconstricted connections to the inner mitochondrial membrane or as transverse circular sections, while in the Chytridiomycetes (with the exception of the obligately anaerobic Neocallimastigales) the cristae are platelike with various oblique—but never circular—profiles. Dictyosomes (homologues of the Golgi bodies of animals) are also well developed, but with relatively few cisternae in each stack. Of the remaining vesicular systems, the two most abundant are those of the lipid vesicles, and in the Peronosporomycetes, the Dense Body Vesicles (DBVs), which constitute a "family" of vesicles of different appearence or characteristic size. Dense body vesicles are implicated in diverse functions in the life history of the Peronosporomycetes, and while they are often prominent in reproductive morphogenesis, they can be found at all developmental stages. Dense body vesicles are characterized by the possession of an electron-opaque core or inclusion surrounded by a more electron-lucent zone, the boundaries between which may be more or less blurred by myelin-like configurations (1). Dense body vesicles (DBVs) can vary in the proportions of core to matrix; at different phases of the life history and organ development they may show different and developmentally intergrading ultrastructure. The prominence and presumed importance of DBVs may be related to the phosphate/polyphosphate storage differences between Eumycota (including the Chytridiomycetes) and the Peronosporomycetes (38).

2. Zoosporangia and Zoosporogenesis

Asexual reproductive units in eucarpic fungi may develop in terminal, lateral, or intercalary positions. The adaptive diversity of asexual reproduction in the Peronosporomycetes is considerable with respect to the shape and dimensions of the zoosporangium, the determinate or indeterminate zoosporangial renewal, the presence of a discharge vesicle, the site and mechanism of protoplasmic cleavage, and zoospore discharge. Molecular biology does not support a phylogenetic, as opposed to a taxonomic, significance for zoosporangial morphogenesis. (See below.)

For most species, a more or less differentiated zoosporangium is delimited from the vegetative system by a septum, and the vegetative system is eucarpic. Alternatively, the entire thallus may assume the role of a zoosporangium or series of zoosporangia (holocarpic development, irrespective of whether the sporangia mature simultaneously or sequentially), frequently with little morphological modification other than the formation of an exit tube and dehiscence papilla. Cleavage of zoospore initials occurs either within the zoosporangium (*intrasporangial zoosporogenesis*) or after discharge of the sporangial protoplasm (*extrasporangial zoosporogenesis*). Intrasporangial zoospores may be released by enzymatic decay of a papilla or a circumcissile ring, leaving an operculum in both phylogenetic lines.

In the Saprolegniaceae (*Saprolegnia*) zoosporogenesis is intrasporangial, and cleavage furrows developed from dictyosome-derived cisternae become confluent first with the prominent central tonoplast vacuole and eventually breach the plasmamembrane. There is a consequent loss of *volume* (approximately 10%) of the zoosporangium as turgor is lost. Zoospore discharge is achieved by imbibition of water through the zoosporangial cell wall in response to the release of osmotically active β-1,3-glucans within the confines of the zoosporangium (39). In the Leptolegniaceae (*Leptolegnia*, *Aphanomyces*) the zoosporangial contents are released before the zoospores are fully formed and discrete (40,41). In the Pythiales (*Pythium*) the extrusion of uncleaved multinucleate protoplasm takes place into an extra-plasma-membranic, membranous, glucan-polymer vesicle (the *homohylic vesicle*) that is formed simultaneously with discharge and that is confluent with the sporangium wall. In the Myzocytiopsidales a single tonoplast is not present, but a few tonoplastlike vacuoles become apparent early in the presporangial stage and persist while the cleavage cisternae reach an advanced stage of orientation (9). In *Blastulidium* and *Eurychasmopsis* planonts become parietally rearranged prior to *intrasporangial* encystment (42). In the Salilagenidiales most of the cytoplasm is peripheral at the midcleavage stage prior to discharge, and large vacuoles probably merge with the cleavage cisternae so that the development shares similarities with but is not identical to those of the Saprolegniaceae, Leptolegniaceae, or Myzocytiopsidaceae. Coincident with this zoosporogenesis is the presence or absence of a gelatinous matrix surrounding the planonts as they are discharged (10,43). A "vesicle" develops from the exit tube apex as a gelatinous matrix *prior to* the extrusion of the protoplasm. [See also some Myzocytiopsidales (44).] In the Salilagenidiales, in contrast to *Pythium*, the protoplasm *does not fill* this clearly defined vesicle. At maturity the gelatinous matrix (the vesicle) becomes partially inverted and collapses down the outside of the exit tube during zoospore maturation, often remaining as a sleeve after discharge. The boundary of a gelatinous matrix can often be distinct, and in light microscopy it may be difficult to distinguish between such a boundary and the presence of a membranous vesicle. A "membrane" (a *precipitative vesicle*) may thus be formed as a precipitation reaction between such a colloidal matrix and the environment.

If hyphal regrowth and zoosporangial renewal takes place, it may be *through* the zoosporangial septum (*internal renewal*), with the successive zoosporangial septa formed above or below the primary zoosporangial septum. Zoosporangia may also be produced in sequence on the same determinate axis (*basipetal development*) or by a *lateral* branch (*cymose renewal*). Conversely, sporangial development may be arrested to produce resting bodies (*hyphal bodies* in *Pythium* or *gemmae* in *Saprolegnia*), which may either germinate to produce zoospores after the manner of the species or germinate directly, producing a hypha.

3. Zoospore and Zoospore Cyst Morphology

The fungal zoospore is a *normally* uninucleate, motile naked cell or planont. Substantial differences in zoospore volume and concomitant microtubular cytoskeletal array and complexity may influence zoospore shape and ultrastructure. In the great majority of the Peronosporomycetes there is only one zoospore form: This is the *principal zoospore* form (8,45), which is reniform or bean-shaped with flagella *laterally inserted in a groove*. The flagella are inserted on a protuberance, the *kinetosome boss*, which may bridge the flagellar groove. The angle of divergence between the two flagella (and their subtending kinetosomes) is approximately 130–150° (46).

The flagella are of different lengths (*anisokont*) and of different ultrastructural morphology (*heterokont*). The anteriorly directed flagellum (the *straminipilous flagellum*) is ornamented with stiff *tubular tripartite [flagellar] hairs* (TTHs), which are usually in two rows (see Ref. 14). The posteriorly directed flagellum is unornamented or with a *fibrillar surface coat*. In the Peronosporomycetes the flagella may either be subequal or markedly anisokont, and the straminipilous flagellum are shorter. The two rows of stiff TTHs reverse the thrust of quasi-sinusoidal beat so that the straminipilous flagellum pulls the zoospore through the water. (For further references see Ref. 7.)

Sometimes a sequence of two or more zoosporic phases may be interspersed by encysted phases. This phenomenon is known as *polyplanetism*. The principal-form zoospore exhibits polyplanetism in both the Saprolegniomycetidae (*Saprolegnia*) and Peronosporomycetidae (*Pythium*). In *Saprolegnia* the more or less ovoid zoospore formed within the zoosporangium may develop flagella that are subapically inserted (the *auxiliary zoospore*) (45), but the flagellar root system for the straminipilous flagellum of the auxiliary zoospore is deficient. The auxiliary zoospore is not known to be polyplanetic. When the polyplanetism involves the production of an auxiliary-form zoospore followed by one or more planetic phases of the principal-form zoospore, the zoospore production is termed *dimorphic*. The cysts formed from each kind of zoospore are also often morphologically distinct. Zoospore cysts may have a variety of ornamentation on their surfaces. The principal-form zoospores of some species of *Saprolegnia* have distinctive split-ended hairs (the boat-hook hairs), which are derived from preformed structures in cytoplasmic vesicles of the zoospore (47).

The zoospore cysts of auxiliary-form zoospores are formed after retraction of the flagella so that tufts of TTHs are left on the outside of the cyst at the point of flagellar insertion and retraction. In contrast, the flagella of principal-form zoospores are shed and leave no plasma-membranic ornamentation. However, both kinds of cysts for different species of peronosporomycetes show considerable diversity with respect to protoplasmically originating cyst wall ornaments (7,36). The most remarkable of these ornaments are the boat-hook hairs, which have split-ended tips. The length and arrangement of the boat-hook hairs is thought to be diagnostic (very long and in tufts for *Saprolegnia parasitica*) (7,48,49).

The flagellar base is composed of the kinetosome and the attached roots of microtubules. Between the kinetosome and the axoneme there is a transitional zone (1,50,51). The root system is composed of six units; each of the kinetosome bases has two roots. Independent variation in the ultrastructure of each of these four roots occurs between taxa, and sometimes within a species. Variation in the detailed ultrastructure of the transitional zone also occurs, but in each case there is a transitional plate, attached to the kinetosome at the end distal to the nucleus. The transitional plate transects the axoneme core in a plane *above* that of the cell plasma membrane and extends to the flagellar plasma membrane (50,51). Within the axoneme core (i.e., inside the cylinder of the nine doublets) and distal to the transitional plate there is an electron-opaque structure appearing in longitudinal sections like a concertina or double helix of repeating units.

The flagellar root systems of the Chytridiomycetes are quite distinct (51,52), often with prominent striated fiber roots and a proximal mitochondrion. The transitional plate is on a plane corresponding to the cell plasma membrane. Intergeneric diversity of the flagellar root systems also occurs in this class.

Many protists possess extrusomes (peripheral vesicles) that extrude or evert their contents to the exterior but that are not involved with wall synthesis. The Peronosporomycetes have a highly developed complex of extrusomes associated with encystment and germination (53,54). Prominent peripheral vesicles have been figured for zoospores of *Eurychasmopsis* (42) and *Myzocytiopsis* (9,55).

The shedding or retraction of flagella at encystment is an important and possibly fundamental diagnostic criterion, but is rarely observed or recorded (56). There may be total detachment and loss (principal-form zoospores) or a range of retraction mechanisms. The method of contact with the host is a diagnostic characteristic. Contact may be established by the flagellar tip, but in most peronosporomycetes the zoospore comes to rest with the ventral groove adjacent to the substratum or host (42), followed by encystment.

4.   Sexual Reproduction

Peronosporomycetes have oogamous sexual reproduction in which the production of an egg (oosphere or female gamete) in an *oogonium* (receptor gametangium) is generally—but not necessarily—accompanied by an *antheridium* (donor gametangium). Neither flagellate gametes nor flagellate zygotes are formed. Gametangia may be developed terminally, subterminally, in an intercalary position on main or branch hyphae, as terminal or lateral appendages to a nonmycelial thallus, or from the entire thallus. When an antheridium is present and fertilization occurs, there is the injection of a small part of the antheridial protoplasm into the oogonium through a *fertilization tube*. In the Saprolegniaceae the fertilization tube may be branched, although there is no evidence that the number of branches ever equates with the number of apparently mature oospores; it is usually much less. In the holocarpic, nonseptate and therefore heterothallic *Eurychasmopsis* a *fertilization hypha* develops when each cell of a chain of endogenous cells within the donor gametangium "germinates" to cross the space between noncontiguous gametangia. However, the majority of species of most of the genera are homothallic and heterothallism may be secondarily derived. In myceliar species having antheridial production, the subtending antheridial branch grows toward the oogonium under hormonal attraction (57–60) and the antheridium is differentiated after contact with the oogonium. The oogonial initial develops as a swelling without septation or by a transformation of a gametangial segment. The first trigger for this morphogenetic change must be endogenous, either from a nutritional or biochemical stimulus.

Gametangia of the Peronosporomycetes are *meio-gametangia* in which meiosis occurs. *Synchronous meioses* occur in the *coenocytic gametangia*. In paired gametangia the meioses are also either simultaneous, or nearly so, between the two protoplasts. This unique feature makes it possible for karyogamy to take place between two haploid nuclei from *adjacent* meioses in the *same* gametangium (automictic sexual reproduction). Sexual reproduction can thus occur without a separate male gametangium or antheridium (3, Fig. 2). Because of this phenomenon, the absence of a male gametangium does not necessarily indicate parthenogenetic (i.e., no meiosis, no karyogamy) development so that automictic sexual reproduction and parthenogenesis cannot be distinguished without cytological evidence (61). In *Nematophthora* the nonseptate ellipsoid thallus apparently functions as an automictic gametangium, eventually containing several oospores. In the Myzocytiopsidales and a few Pythiales the thallus becomes septate, and adjacent segments assume the function of gametangia. In such cases the thallus segment is the site of meiosis (*thalloid meiosis*). *Gametangial copulation* (Myzocytiopsidales) occurs when the two gametic protoplasts condense to a common *pore* in the contiguous walls of the gametangia; the condensation is dependent on vesicle expansion distal to the point of union in *both* gametangia (3). The fertilization tube, which forces the oosphere away from the point of union, and the conjugation pore, toward which both gametangial protoplasts converge, appear to be quite different morphogenetic processes.

After meiosis the contents of the receptor gametangium (oogonium) become separated as one or several uninucleate and initially unwalled gametes; however, there is no differentiation of the donor gametangial contents into discrete uninucleate gametes (but note the "endocellular" antheridia of *Eurychasmopsis*). The sexual process (meiosis and karyogamy) and the morphogenesis of the two kinds of gametangia should therefore be considered as separate criteria.

Many species have oogonia with a smooth, more or less spherical outline, but many others are ornamented. The pattern of development of wall ornamentation depends on three criteria: initial expansion, secondary primordial initiation, and wall deposition. The diversity of oogonial form depends on the sequential or simultaneous expression of these criteria. The deposition of an internal layer or layers results in the thick oogonial walls, particularly characteristic of the Saprolegniales, but the deposition of this inner layer is often excluded from certain regions to form the simple pits (*Saprolegnia parasitica*).

When the antheridium is of regular occurrence, the shape, origin, and position of the antheridium can be useful taxonomic criteria. The *mode of application* of the antheridium to the oogonium is distinctive for species and sometimes groups of species, but the continuous interspecific cline of variation between fully apical and completely lateral attachments makes reference to *accurate illustrations essential*. Antheridial *origin* can either be characteristic for a species, or highly variable within a species. Antheridia may arise from just below the oogonium (*closely monoclinous*), from the hypha subtending the oogonial branch (*monoclinous*), or from a different hyphal system (*diclinous*).

The morphogenesis of the oosphere is of major phylogenetic significance in the Peronosporomycetes in two ways. First, the protoplasmic organization subsequent to meiosis follows one of two alternatives: *centripetal oosporogenesis* (vacuolar or tonoplast involvement) or *centrifugal oosporogenesis* (protoplasmic rearrangement frequently with residual periplasm). Cleavage in the polyoosporous taxa of the Saprolegniales is essentially centrifugal and similar to cleavage in zoosporangia, with a tonoplast vacuole fusing with cleavage cisternae to form peripheral mounds of presumptive oosphere protoplasts. When the oogonial plasma membrane is finally breached, the oospheres tumble to the center of the oogonium. The extent to which the oogonial cavity is filled by the oosphere is variable and can be taxonomically diagnostic. In the Pythiales the oogonioplasm does not contain a tonoplast vacuole, and the peripheral oogonioplasm gradually loses its organelles to define a centripetal oosphere. Subsequently, the oospore (zygote) wall forms at this boundary, so that the spore is more or less aplerotic within the oogonium. In *Pythium* the concept of an *aplerotic index* has been proposed to aid taxonomic assessment of species differences (62,63).

Second, deposition of the oospore wall and distinctive cytoplasmic reorganization occurs during the development of the oospore. The oospore wall itself is complex; possibly up to seven or more layers can be distinguished, but not all organisms have all layers. Functionally, the most important layer is the deposit of resorbable glucan polymers outside the plasma membrane forming the *endospore*, analagous to the albumin of a bird's egg. The endospore usually shows concentric layers (64). Within the oospore, the protoplasmic contents become reorganized. Two complementary processes of vesicular coalescence proceed simultaneously. The DBVs gradually coalesce to form a large single membrane-bound structure, the *ooplast*, which can be seen clearly in the oospores of almost all Peronosporomycetes. At the same time there may be a variable degree of coalescence of lipid globules. Both contribute to the characteristic bubbly appearance of the partially mature oospore. One or more transparent ellipsoid zones in the mature oospore mark the position(s) of nuclei (*nuclear spots*). At the most extreme state of coalescence, seen in some species of *Achlya*, the mature oospore contains a single ooplast and a single lipid globule. The patterns of lipid coalescence (*centric*, *subcentric*, *subeccentric*, and *eccentric* states) have taxonomic value and possibly significance in both ecology and phylogeny.

Under both light and electron microscopy the ooplast has different characteristics within the class. In most of the Saprolegniales the ooplast is fluid, with Brownian movement of granules, but in the Pythiales the ooplast appears to be homogeneous under light microscopy. Using TEM, the ooplast of the Pythiales is electron-opaque with a dispersed electron-lucent phase. The phases are reversed in the Saprolegniales. In the Salilagenidiales the ooplast descriptions suggest closer similarities to the Saprolegniales.

Parity between estimates of chromosome number as obtained from light microscopical methods and pulsed gel electrophoresis might be difficult to establish because of the possibilities of autopolyploidy, polysomy, and chromosome polymorphy (65).

Many Chytridiomycetes have no known sexual system (37). In the Blastocladiales, however, the resistant sporangium functions as a meiosporangium, producing haploid planonts (gametes). The haploid and diploid phases of the nuclear cycle may occur in different hosts. (See *Coelomomyces*, below.)

5. Biochemistry

Cantino (66) was the first to incorporate a range of biochemical criteria into the systematics of the flagellate fungi, emphasizing that the genetic loss of a biochemical pathway or attribute was unlikely to be restored. Such pathway loss is still considered to be of phylogenetic importance (7,8). Three kinds of organic synthesis have diagnostic value.

The amount of fibrillar wall material is much less in the fungi, and in the Peronosporomycetes cellulose (β-1,4-glucan) tends to be masked by the much larger amounts of β-1,3- and β-1,6-glucans. The fibrillar component of the walls of Chytridiomycetes is chitinous, but glucosamine also occurs in walls of the Peronosporomycetes, and the presence of polymerized chitin has been confirmed for Saprolegniaceae (67).

There are two fundamental lysine synthesis pathways (68). The Peronosporomycetes are characterized by possession of the DAP (α,ϵ-diaminopimelic acid) pathway, while the Chytridiomycetes possess the AAA (α-aminoadipic acid) pathway. The evolution of the DAP pathway is thought to antedate that of the AAA pathway (the occurrence of which correlates with chitinous cell walls) because the DAP pathway interferes with that for chitin biosynthesis (69). Although the AAA lysine synthesis is correlated with the presence of mitochondria with flat, platelike cristae (and chitin synthesis), DAP lysine synthesis may be associated with a range of mitochondrial ultrastructure.

Detailed requirements for exogenous sterols differ between genera, as does the ability to make and utilize sterols with certain substituents. Similarities between *Lagenidium* and *Phytophthora* can be contrasted with the similarities between *Salilagenidium*, *Achlya*, and *Plerogone* (70). Sterol metabolism is considered to be an important factor in the survival of *Lagenidium giganteum*. Polyene antibiotics act on Eumycota but not the Peronosporomycetes; since these antibiotics are thought to function by acting on membrane-bound sterols, the selectivity of the antibiotics suggests that there is a difference between these two groups of heterotrophs with respect to their membrane-bound sterols (71–73).

6. Molecular Biology

Molecular biological investigations of animal parasites have yet to provide a sufficient body of comparative data. However, comparisons of the data for the saprobic and plant pathogenic taxa indicate that there will be considerable scope for developing clear protocols for identification.

The total genomic contents vary widely: the length of the nuclear genome has been shown to vary between approximately 60 and 250 Mb in *Phytophthora* (74); the length of the mitochondrial genome has been shown to vary between 36.4 and 73.0 kb with the presence of an inverted repeat, ranging in length from approximately 10 to 30 kb (75,76). Data for 18S rDNA (21) and mitochondrial phylogeny (77) have confirmed the dichotomy of the subclasses Peronosporomycetidae and Saprolegniomycetidae, but also indicate polyphyly at the genus level. Existing genera have been defined solely by zoosporangial morphology and morphogenesis. Restriction mapping has revealed variability in the intergenic regions of the ribosomal DNA; both the position (within the NTS of the rDNA repeat unit) and the occurrence of inverted copies of 5S rDNA genes show variation *within* the subclass Peronosporomycetidae and *within* the subclass Saprolegniomycetidae.

Molecular biological analyses are accumulating so rapidly that this summary is inevitably "dated." Nevertheless, it is noteworthy that sequences for the ribosomal gene alone may not be sufficient to clarify phylogenetic relationships (21), even though these data will aid identifications.

B. Orders, Families, and Genera Covered

The following more detailed accounts do not follow a common pattern because the amount of information available is most variable. Some species (including *Saprolegnia parasitica*, *Aphanomyces astaci*, *Pythium insidiosum*, and *Lagenidium giganteum*) are treated in more detail, but the remaining pathogens are dealt with in collective groups by means of tabulation or key. Tables 1 and 2 provide the systematic framework, but for most of the pathogens a host/habitat subhead is more convenient.

1. The Freshwater Fish Pathogens

Saprolegnia parasitica Coker (78). Illustrations: (47, 49, 78, oogonia).
Thallus mycelial, extensive, hyphae 30–118 μm diam. *Zoosporangia* cylindrical with internal renewal, (75)150–200(1050) μm × (20)30–45(80)

μm. *Zoosporogenesis* intrasporangial. *Zoospores* dimorphic and polyplanetic. *Zoospore cysts* 9–11 μm diam. with tufts of long, often recurved boat hook hairs. *Oogonia* seldom formed in disease situations and not always capable of induction in culture; when formed, terminal, or intercalary, spherical or pyriform-ellipsoidal, 54–146 × 18–72 μm, wall with or without relatively small simple pits.

Antheridia diclinous. *Oospores* (2)14–23(40) per oogonium, (16)18–24(28) μm diam. The small mean size of the oospore is considered diagnostic.

The disease of salmonid fish was first reported by fly fishermen in the nineteenth century in Irish and Scottish river systems, where it caused periodic but not always annual damage to game-fish stocks (salmon and trout). Accounts of the disease have accumulated over a long period, and the disease has a large bibliography. (See Refs. 31, 79, 80.) *Saprolegnia parasitica* is now a noteworthy disease in fish farms and hatcheries in Europe, North America, and southeast Asia (31). Subsequent to Coker's description in 1923 of the species causing the disease, several other related fish parasites (81–84) and fish hosts have been reported (49,81,82,85–87). Of the additional genera reported, the more common are *Achlya* and *Aphanomyces*. The species of *Achlya* are usually from the lacustrine, eccentric-oospored group (45). *Aphanomyces* is discussed separately below. Other species of *Saprolegnia* (*S. shikotsuensis* Hatai et al.) (88,89) are also associated with the disease. Although the most common disease syndrome is of massive superficial lesions as described in the introductory paragraphs, visceral mycoses have also been reported (90,91). The geographic range is now worldwide (31), with additional reports from India (93) and Australia (94).

One of the most difficult problems in the pathology of this disease is the determination of those organisms that are causal and the dismissal of those organisms that are opportunistic wound-site secondary invaders. Many of the associated saprolegniaceous fungi fall into the latter category or have not been unequivocally shown to be primary pathogens. The greatest confusion has arisen between the disease-causing *Saprolegnia parasitica* and the opportunist saprobe *S. diclina*. The distinctions between these two species have been discussed at the physiological (48,95–98), ultrastructural (7,47,49), and molecular biological (99) levels. The cyst ornamentation of the principal-form zoospore, consisting of tufts of long, flexuous "boat hook" hairs, is distinctive and diagnostic (47,49), although other *Saprolegnia* saprobes may have intermediate ornamentation (7) and not all species of saprobic *Saprolegnia* species have yet been examined. Similarly, the problem with molecular methodology will be the ability to screen the causal agent from opportunists when so few species of the genus have been analysed.

Study of the environmental constraints on disease establishment has only just been touched upon (100). Much work has still to be carried out on the mycotic aspects of the disease when it is epidemic in fish-farm culture (101). For many years the standard treatment for this disease has been the application of the toxic Malachite Green (102), but a more sophisticated approach to fungicide application is now being put forward (103).

Aphanomyces pisci [*nom. illeg.* (no Latin)] (104). Illustrations: (104).

Thallus mycelial with delicate, profusely branched hyphae, 7.2–9.0 μm diam. *Zoosporangia* formed from undifferentiated lengths of mycelium, filamentous, unbranched. Zoosporogenesis with up to 40 spores per zoosporangium. *Zoospores* of the principal form. *Zoospore* cysts 7.0–7.5 μm diam. *Gemmae* abundant, elongate and lobed (reminiscent of lobulate *Pythium* sporangia). *Oogonia* absent. *Antheridia* absent. *Oospores* absent.

Aphanomyces piscicida (105) [no illustrations].

Aphanomyces invadans Willoughby et al. (as "*invaderis*") (106) Illustrations: (106)

Thallus mycelial, hyphae moderately branched, 11.7–16.7 μm diam., young hyphae narrower (8.3 μm diam.). *Zoosporangia* formed from undifferentiated lengths of mycelium, filamentous, often complex, branched systems, approximately 330–930 μm in length. *Zoosporogenesis*-producing spores encysting at lateral orifices, *cysts* 6.7–10 μm diam. *Zoospores* of the principal form. *Oogonia* unknown. *Antheridia* unknown. *Oospores* unknown.

Aphanomyces frigidophilus (107). Illustrations: (107).

Thallus mycelial, hyphae moderately branched, 7–10 μm diam. *Zoosporangia* formed from undifferentiated lengths of mycelium, filamentous, unbranched. *Zoosporogenesis* incomplete at time of discharge, encysting at mouth of zoosporangium. *Zoospores* of the principal form. *Oogonia* abundant, lateral on short hyphae, pyriform or subspherical, 16–25 μm diameter, with crenulate wall. *Antheridia* absent. *Oospores* single, 14–22 μm diameter.

The first record of another fish disease caused by a straminipilous fungus was in 1944 (108). The disease is apparently due to penetration from the gut (compare Refs. 90, 91), presumably caused by ingestion of the pathogen spores. The infected fish develops an abnormal dorsal hump and the abdominal cavity becomes grossly swollen. This is followed by obvious mycelial development in the dorsal musculature; eruption of hyphae to the outside has not been observed, so the source of the infective unit has not been established. Fish-to-fish transfer via fungal propagules apparently does not occur (106). It is possible that zoosporogenesis only occurs after death and initial decay of the host, but such sporulation, which has been sought, has not been found either; the source of infection remains an enigma. More recent studies (109–111) of the disease syndrome have revealed systemic granulomas.

Epizootic ulcerative disease syndrome (EUS) due to *Aphanomyces* may be caused by the same organism (92), and has also been reported (93) in estuarine conditions (compare diseases caused by *Saprolegnia*, reviewed above). There may therefore be some confusion between the identities of the pathogens and the disease symptoms. Enhanced salinity tolerance may be a factor in the disease caused by *Aphanomyces* (112).

The species taxonomy of the causal organism or organisms assigned to the genus *Aphanomyces* is not at all clear. Oospores, which would be diagnostic, have not been described for all binomials, but the fungus is known to produce *Aphanomyces*-like zoosporangia according to some cultural studies (108,110). The growth rates and temperature optima of pathogenic *Aphanomyces* spp. (including *A. astaci*; see below) are different and the southeast Asian fish-pathogenic isolates are thermophilic, growing well at 30°C. Neither isozyme nor molecular biological surveys of numerous isolates have been carried out. Such studies might resolve whether the causal agent is in fact a distinct species or more than one species. It has also been suggested (31) that the causal agent might be a *Leptolegnia*, a genus that is now placed in a separate family together with *Aphanomyces* (21).

2. The Freshwater Crustacean Pathogen

Aphanomyces astaci (113).
Thallus mycelial with hyphae of limited extent, of uniformly narrow width, 7.5–9.5 µm diam. *Zoosporangia* formed from undifferentiated lengths of mycelium, filamentous, unbranched. *Zoosporogenesis* almost completed intramatrically, planonts formed in single file, remaining connected by protoplasmic strands during discharge, up to 40 per zoosporangium. *Zoospores* of the principal form. *Zoospore cysts* 8.0–9.5 µm diam. *Oogonia* not normally formed. [NB: the description by Rennerfelt (114) of minutely spiny-walled, uniovulate oogonia 41–48 µm diam. *Antheridia* monoclinous, 1–2 per oogonium. *Oospores* 22.4–28.83 µm diam. should be treated with caution.]

The disease of crayfish was first reported in 1903 (see Refs. 113, 115, 220) and was apparently the result of the introduction of the American crayfish into European river systems before 1860. At some stage some of the Californian stock is presumed to have been contaminated with its benign parasite, *Aphanomyces astaci*. The European crayfish lacks immunity to this necrotic parasite and the European and Asia Minor commercial production of crayfish has fallen dramatically and probably irreparably (11,15,220). The fungus has an extensive bibliography (see Refs. 115, 116), and considerable work has been published on its physiology (117,118), biochemistry (35) (including an extensive bibliography), and pathobiology (119).

The disease of the European crayfish is noteworthy because of the sudden and total destruction of the population in any given river and lake system, the streambeds being strewn with dead crayfish. There are a few isolated river systems in which the European crayfish is as yet uninfected, but quarantine protocols are liable to be circumvented by the irresponsible introduction of nonnative crayfish. Once stock of moderately resistant crayfish are present, the fungal inoculum will remain, so that any reintroduction of the European crayfish stock will immediately become annihilated. The spread of the disease throughout European river systems has been traced from Italy and central Europe into northern Europe and Russia, Scandinavia, Iberia, and Asia Minor.

In moderately resistant North American crayfish, the disease starts as superficial lesions on the exoskeleton, particularly in thinner areas such as the limb and abdomen joints. However, the hyphal infections which result from encysted zoospores, are arrested, but not killed, by melanization (33,119,120). In the susceptible European crayfish this melanin reaction is weak and develops slowly, so that the fungus is able to penetrate the haemocoel, rapidly causing death. The haemocytes of the haemocoel are thought to be responsible for the defense reaction whereby the activation of the crayfish prophenoloxidase enzyme by the fungal infection (35,121) results in the melanin deposit around, and stasis of, the fungal hyphae.

The range of hosts and their relative susceptibilities have been discussed on a number of occasions (11,15,32,34). In Asia Minor the causal agent of the disease, in at least some cases, appears to be *Saprolegnia parasitica* (49); however, the pathobiology appears to be identical.

3. The Warm Temperate/Tropical Mammalian Pathogen

Pythium insidiosum (27,121,122). Illustrations: (27).

Thallus mycelial, hyphae 4–6 (10) μm diam., branches almost at 90°, sometimes septate, more so in vivo than in vitro; sometimes with club-shaped appressoria. *Zoosporangia* produced only in water cultures, filamentous and undifferentiated from assimilative hyphae, 45–700 μm × 3–4 (distally 5–8) μm. *Zoosporogenesis* extrasporangial in a vesicle 20–60 μm diam. *Zoospores* of the principal form, 12–14 × 6–8 μm. *Zoospore cysts* 8–12 μm diam., ultrastructural morphology undescribed. *Oogonia* seldom formed, subglobose, intercalary with septa at a distance from the swelling, (19)23–30(36) μm diam. *Antheridia* 1(2–3), diclinous. *Oospores* aplerotic, single, (17)20–25(27) μm diam. with a distinct ooplast. *Cardinal temperatures* min 10°C, optimum 34–36°C, max 40–45°C.

The disease was originally reported from India under the name *burusattee* as a disease of horses grazing in or near stagnant water, but the fungal nature of the causal organism was not noted until 1901 (123). The disease has subsequently been described as "phycomycosis of horses" (124), "swamp cancer" (125), "granular dermatitis" (126), "equine pythiosis" (122), finally being named as "pythiosis insidiosi" (127). The causal organism was initially regarded as non-sporulating mycelium, but zoospores were reported in culture in 1974 (125) when the organism was regarded as a species of *Pythium*. The taxonomy of the *Pythium* species is involved, depending on the interpretation of the earlier names using the specific epithet *"destruens"* (62). However, following the description of sexual reproduction (27), the nomenclature has stabilized as given above. References to names such as *"Pythium gracile"* should be regarded as "form names," not taxonomic entities, and much work still needs to be done before there is a fully reliable antigenic (128) or molecular biological framework (21) in which this pathogen can be placed.

The most recent, comprehensive review of the pathobiology and bibliography will be found in Ref. 127, although the mycology is not always accurate. As indicated by the disease names listed above, there is a granuloma with an extensive lesion. This may include the production of coralloid "kunkers" that contain viable hyphae. The disease is not confined to ungulates; a number of other mammals are known to be capable of infection, including man. Mammal-to-mammal transfer has only been achieved by using rabbits.

The fungus is thermophilic, as would be expected for a parasite of mammals. Its survival outside mammals presumably depends on existence in warm, stagnant water, but extensive ecological surveys for its presence have not been carried out. It is therefore regarded as being restricted to the tropics and warm temperate zones, and is particularly prevalent in southeast Asia (including Japan) and Australasia, and Central and South America. Surprisingly, it has not been reported from Africa. It is not regarded as indigenous in Europe.

4. The Tropical Amphibian Pathogen

Olpidium/Rhizophydium new genus (129). Illustrations: (129,130).

A disease of amphibians (rain forest frogs) in Australia and Central America has recently been noted (129–132). The disease, termed chytridiomycosis, is caused by monocentric, rhizoidal chytrids, which are endobiotic within and utilize the keratinized epidermal cells. Thickening of the skin results, and it is suspected that interference with water and gaseous transport results in mortality. There is insufficient evidence as yet, however, to justify a causal relationship between the disease and the epidemics reported (132,133).

These reports have stimulated a resurgence of interest in the infection of amphibian eggs by *Saprolegnia* and *Achlya* (Jeffries, personal communication). It has been known for many decades that frog spawn is often damaged by sapro-legniaceous fungi, but it has not been established whether parasitism rather than wound damage is the cause. *Saprolegnia parasitica* can be isolated from such substrata, a factor which adds to the complexity of the pathobiology of this fungus, discussed in other sections of this chapter. The diversity of the egg-laying behavior of the amphibian may affect the potential for pathogenicity (134).

5. The Mosquito and Other Insect Pathogens

> *Lagenidium giganteum*, *Leptolegnia chapmanii*, *Crypticola* spp., *Couchia circumplexa*, *Aphanomycopsis sexualis*, *Coelomomyces* spp, and *Catenaria spp.*

The fungi grouped under this heading are from diverse flagellate families: *Lagenidium giganteum* (Pythiaceae), *Leptolegnia chapmanii* (Leptolegniaceae), *Couchia circumplexa* (Saprolegniaceae), *Aphanomycopsis sexualis* (Leptolegniellaceae), *Crypticola clavulifera* and *C. entomophaga* (Crypticolaceae), *Catenaria spinosa*, *C. uncinata*, *C. ramosa* (Catenariaceae, Blastocladiales), and *Coelomomyces* spp. (Coelomomycetaceae, Blastocladiales (135)).

6. The Biflagellate Taxa

Lagenidium giganteum is cosmopolitan, having been found in Europe, North and South America, and Australia. *Leptolegnia chapmanii* has so far only been reported from the eastern states of North America and Argentina. *Couchia circumplexa*, *Aphanomycopisis sexualis*, and *Crypticola entomophaga* are known only from eastern North America, while *Crypticola entomophaga* has been recorded only from tropical Australasia.

> *Lagenidium giganteum* (136). Illustrations: (136–138, 140, 141).
> *Thallus* of hyphae, initially nonseptate, becoming septate to give allantoid segments 7–10 μm in diam. and 50–300 μm long. *Zoosporangia* developed from allantoid segments, 20–40 μm × 25–60 μm. *Zoosporogenesis* with a homohylic vesicle, 20–50 μm diam. *Zoospores* of the principal form, 8–9 μm × 9–10 μm. *Zoospore cysts* 10–14 μm diam. *Oogonia* (26)38–42(63) μm diam. *Antheridia* with various origins. *Oospores* single, 18–30 μm diam.
> *Leptolegnia chapmanii* (141). Illustrations: (141).
> *Thallus* extensive, of slender, nonseptate hyphae, gemmae abundant, usually lateral; variable in shape, large swollen, sometimes papillate. *Zoosporangia* 70–240 μm × 15–40 μm. *Zoospores* dimorphic. *Zoospore cysts* (first-formed) 13–15 μm diam., (from principal-form zoospores) 10–14 μm diam. *Oogonia* unpitted, with short papillate projections, (26)38–42(63) μm diam. *Antheridia* with various origins. *Oospores* 1 (3), rarely maturing, subeccentric or eccentric, (18)36–40(52)9 μm diam.
> *Couchia circumplexa* (142). Illustrations: (142).
> *Thallus* of extramatrical hyphae, forming appressoria. *Zoosporangia* termi-

nal, ellipsoidal to clavate, 53–166 (−344) μm × 32–95 μm. *Zoospores* dimorphic. *Zoospore cysts* 14–22 μm diam. *Oogonia* spherical, unpitted, 30–84 μm diam. *Antheridia* diclinous. *Oospores* 1–5, 23–47 μm diam.
Aphanomycopsis sexualis (143). Illustrations: (143).
Thallus endobiotic, richly branched, nonseptate except to delimit gametangia. *Zoosporangia* developed from thallus, 206–280 μm × 88–120 μm, with 1–10 discharge tubes. *Zoospores* dimorphic, first-formed zoospores lacking flagella and encysting immediately, principal-form zoospores 5.4–8.3 μm × 4.2–6.3 μm. *Zoospore cysts* 6–7.3 μm diam., with short spines. *Oogonia* swollen portions of the thallus, 82–240 μm × 44–72 μm. *Antheridia* present, septate. *Oospores* single, ovoid, 52–170 μm × 44–72 μm; oospore wall 2.2–4.3 μm diam.
Crypticola clavulifera (144). Illustrations: (144).
Thallus limited, becoming septate to give allantoid segments 50–60 μm long. *Zoosporangia* developed unilaterally from allantoid segments, 35–40 μm × 12–13 μm. *Zoosporogenesis* intrasporangial. *Zoospores* of the principal form, 8–10 × 5–6 μm. *Zoospore cysts* 10–14 μm diam. *Oogonia* unknown.
Crypticola entomophaga (14, 36). Illustrations: (145).
Thallus ellipsoidal or irregular, nonseptate, up to 215–280 μm × 175–245 μm in caddis fly eggs. *Zoosporangia* developed from thallus, with 1–4 discharge tubes up to 3.7 mm long. *Zoospores* of the principal form (dimorphic), polyplanetic, (8.5−)11(−16) μm × (6−)6.9(−8.5) μm. *Zoospore cysts* (9−)10.5(−13.5) μm diameter. *Oogonia* unknown.

7. The Uniflagellate Taxa

Catenaria spinosa (146). Illustrations: (146).
Thallus simple or branched, with irregular catenulate swellings 3.8–17 μm diam., and a few blunt rhizoids, becoming septate with one-celled isthmuses. *Zoosporangia* of two types: thin walled and resistant: thin walled zoosporangia 25–69 μm × 24–66 μm. *Zoospores* 7–11 × 4.5–7.5 μm. *Resistant sporangia* golden brown, 30–62 μm × 25–61 μm, wall 1.3–2.5 μm thick with short, spinelike appendages.
Catenaria uncinata (147). Illustrations: (147).
Thallus simple or branched, with irregular catenulate swellings 3.8–17 μm diam., and a few blunt rhizoids, becoming septate with equidistant swellings separated by one-celled isthmuses. *Zoosporangia* of two types: thin walled and resistant; thin-walled zoosporangia 22–62 μm × 18–56 μm. *Zoospores* 3.1–5.6 × 2.1–3.4 μm. *Resistant sporangia* golden brown, 34–57 μm × 32–54 μm, with elongate, sometimes curved, spinelike appendages tapering to an uncinate tip.

Catenaria ramosa (147). Illustrations: (147).

Thallus simple or branched, with irregular catenulate swellings 3.8–17 μm diam., and a few blunt rhizoids, becoming septate with one-celled isthmuses. *Zoosporangia* of two types: thin walled and resistant; thinwalled zoosporangia 28–54(−70) × 25–49 μm. *Zoospores* 3.5–6.8 × 2.4–4.8 μm. *Resistant sporangia* golden brown, 20–66(-76) μm × (8-)13–48 μm, appendages sparce, often branched.

Species of *Catenaria* parasitic in animals appear to be confined to insect (Diptera) eggs in unidentified species of *Chironomus*, in *Glyptotendipes lobiferus* Say., and in *Dicrotendipes modestus* Say., respectively. Host and habitat differences between these species have been recorded (147), but further work on their biology and potential for control of the insects is needed.

The genus *Coelomomyces* contains over 60 species and varieties (37,135,148–150), the species often being partly based on host. The species are listed in alphabetic order after the type species in Table 5, with *Index of Fungi* (IF) references.

Generic description for *Coelomomyces* (adapted from 149): *Thallus* simple or branched, without rhizoids or cell walls, giving rise to "hyphal" segments by division or attenuation and fracture by host movement. *Thinwalled zoosporangia*: apparently absent. (See Ref. 151; check alternate host.) *Resistant sporangia*: formed as "hyphal" segments, developing a wall after detachment, golden brown with a very thin outer membrane and a smooth, pitted, striated, or ridged wall. (See Ref. 152.) *Zoospores*: as for the order, except that they lack a well-defined nuclear cap. (See Ref. 153.)

8. Commentary

As a result of the search for methods of biological control of mosquitoes there has been a significant amount of research into the basic biology and biochemistry of these fungi, particularly *Lagenidium giganteum* (16–19), for which over 150 references have been cited (36), and *Coelomomyces* (148). Habitats for agents of biological control have also been screened, and tree-hole and leaf reservoir (phytotelemata) (142,154) mosquito habitats have yielded a number of new taxa. Insect pathogens are reported from dipteran eggs (142–147), but with few corroborative collections and little data on pathobiology.

It is characteristic that most of these pathogens infect the larval stages, rapidly gaining access to the haemocoel, with extensive intramatrical growth resulting in death followed by extramatrical sporogenesis. An extensive review (148) of *Coelomomyces* summarized much of the information on this genus. Its suitability for biological control was severely compromised by four major factors:

Table 5 The Genus *Coelomomyces*

Species	IF References
C. stegomyiae Keilin (1921)	IF3: 5, 213; IF5: 527
C. africanus AJ Walker (1938)	IF5: 526
C. angolensis H Ribeiro (1992)	IF6(6): 329
C. anophelesicus MOP Iyengar (1935)	IF3: 213
C. arcellaneus Couch & Lum (1985)	IF5:526
C. arsenjevii Koval & ES Kuprian. (1981)	IF5: 115
C. ascariformis Van Thiel (1962)	IF3: 5, 213
C. azerbajdzanicus ES Kuprian. & Koval (1981)	IF5: 526
C. beirnei Weiser & McCauley (1971)	IF4: 129, 208
C. bisymmetricus Couch & HR Dodge (1947)	IF3: 213
C. borealis Couch & Service (1985)	IF5: 526
C. cairnsensis Laird (1962)	IF3: 213
C. canadensis (Weiser & McCauley) Nolan (1978) (as canadense)	IF4: 560
C. carolinianus Couch et al. (1985)	IF5: 526
C. celatus Couch & Hembree (1985)	IF5: 526
C. chironomi Răsín (1929)	IF4: 129, 208
C. ciferrii Arêa Leão & Carlota Pedroso (1965)	IF3: 351
C. couchii Nolan & B. Taylor (1979)	IF5: 7
C. cribrosus Couch & HR Dodge (1947)	IF3: 213
C. dentialatus Couch & Rajap. (1985)	IF5: 526
C. dodgei Couch (1945)	IF3: 213
C. dubitskii Couch & Bland (1985)	IF5: 526
C. elegans Couch & Rajap. (1985)	IF5: 526
C. fasciatus Couch & MOP Iyengar (1985)	IF5: 526
C. finlayae Laird (1962)	IF3: 5, 213
C. grassei Rioux & Pech (1962)	IF3: 213
C. iliensis Dubitskií et al. (1973)	IF4: 208; IF5: 334, 384, 526
C. indianus MOP Iyengar (1935)	IF3: 213
C. iranii Weiser, Zaim & Saebi (1991)	IF6(4): 200
C. neotropicus Lichtw. & LD Gómez (1993)	IF6(10): 540
C. iyengarii Couch (1985)	IF5: 526
C. keilinii Couch & HR Dodge (1947)	IF3: 213
C. lacunosus Couch & OE Sousa (1985)	IF5: 526
C. lairdii Maffi & Nolan (1977) (as lairdi)	IF4: 623
C. lativittatus Couch & HR Dodge (1947)	IF3: 213
C. maclayae Laird (1962)	IF3: 5, 213
C. madagascaricus Couch & Grjebine (1985)	IF5: 526
C. milkoi Dudka & Koval (1973)	IF4: 242
C. musprattii Couch (1985)	IF5: 526
C. notonectae Bogoyavl. (1922)	IF5: 873

Table 5 The Genus *Coelomomyces*

Species	IF References
C. omorii Laird et al. (1975)	IF4: 404
C. opifexi Pillai & JMB Sm. (1968)	IF4: 208
C. orbicularis Couch & Muspratt (1985)	IF5: 526
C. orbiculostriatus Couch & Pras. (1985)	IF5: 526
C. pentangulatus Couch (1945)	IF3: 213
C. ponticulus Nolan & Mogi (1980)	IF5: 526
C. psorophorae Couch (1945)	IF3: 213; IF5: 526
C. punctatus Couch & HR Dodge (1947)	IF3: 213
C. quadrangulatus Couch (1945)	IF3: 213
C. rafaelei Coluzzi & Rioux (1962)	IF3: 299; IF4: 404
C. reticulatus Couch & AJ Walker (1985)	IF5: 527
C. rugosus Couch & Service (1985)	IF5: 527
C. sculptosporus Couch & HR Dodge (1947)	IF3: 213
C. seriostriatus Couch & JB Davies (1985)	IF5: 527
C. solomonis Laird (1962)	IF3: 213
C. sulcatus Couch & MOP Iyengar (1985)	IF5: 527
C. tasmaniensis Laird (1962)	IF3: 214
C. thailandensis Couch et al. (1985)	IF5: 527
C. triangulatus JN Couch & WW Martin (1985)	IF5: 527
C. tuberculatus Couch & Rodhain (1985)	IF5: 527
C. tuzetiae Manier et al. (1970)	IF4: 30
C. utahensis Romney et al. (1985)	IF5: 527
C. uranotaeniae Couch (1945)	IF3: 214
C. walkeri Van Thiel (1962)	IF3: 5, 314

Note: Species, with authorities in recommended form, listed in alphabetic order after the type species, with *Index of Fungi* (IF) references (up to vol. 6, part 17).
Source: Refs. 37, 135, 148–150.

the species appeared to be strongly host-specific (149,150); the life history of at least some species was discovered to be heteroceous, involving resistant sporangial production confined to the alternate host, usually a microcrustacean (155); it was observed (156) that the accumulation of fungal material in the mosquito ovaries did not always result in the death of the host; and the economic aspects of mass fungal inoculum preparation and its failure to regenerate were noted— heavy inoculum resulted in early instar death, while lighter inoculum loads failed to provide sufficient decrease in the population levels (157).

Lagenidium, in contrast, is much more amenable to manipulation and has a much wider host range. There remain problems in formulating protocols for biological control, but a full discussion is not appropriate here.

9. The Parasites of Marine Crustacea and Mollusks: (Blastulidium Salilagenidium *spp.*, Haliphthoros *spp.*, Atkinsiella dubia, *and* Halodaphnea *spp.*)

The fungal parasites of marine littoral crabs and prawns have been described from the northern hemisphere Atlantic and Pacific shores, starting with the descriptions by Atkins (158–161) and Couch (43).

Egg masses are often the prime sites of parasitism. In most cases of juvenile and adult infections, the fungus develops in the haemocoel and can completely fill the swimmerets of shrimps and prawns. Epidemics have been reported (13) from the mariculture centers in the Asian Pacific where the crustacean hosts form an important part of the protein diet. The occurrence of these fungi on crabs and barnacles does not appear to have the same economic impact (11).

Since all of these species are placed in a single order, the Salilagenidiales (previously part of the polyphyletic Lagenidiales) (36), the distinctions between the species, including hosts and any cultural information, are most concisely presented in the form of a fully descriptive key. Note that the Haliphthoraceae was specifically erected for species without sexual reproduction (162) and that a new generic name (*Halodaphnea*) has become necessary for *Halicrusticida*, to which a number of *Atkinsiella* species were assigned. (For further taxonomic references, see Refs. 14.)

Comparative molecular biological data are not yet available, and relationships within the order and with other orders are still conjectural.

Another biflagellate fungal parasite of crabs was described and named as *Pythium thallasium* (161), but Plaats-Niterink (163) regarded the generic placement as doubtful because of the mode of sporangial renewal. Until this parasite is rediscovered and redescribed its classification must be regarded as *incerta sedis*.

Three species of the Labyrinthista could also be mentioned here. *Labyrinthuloides haliotidis* is the only member of this genus reported as a parasite of marine animals (164,165)—in this case the abalone of the north-east Pacific littoral. The molecular biology of this organism has been documented (166), and although straminipilous, its fungal status and generic attribution are in doubt. *Labyrinthula thaidis* B. A. Cox & J. G. Mackin (as *L. thaisi*) (167) and *L. jeremarina* L. Rolf (*nomen nudum*) (167) have been described from the gill tissue of gateropod mollusks (*Thais* and *Littorina*, respectively). Again the descriptions lack salient features for satisfactory placement in a classification. Both *Labyrinthuloides* (Thraustochytriales) and *Labyrinthula* (Labyrinthista) are genera primarily inhabitants of plant material and detritus.

> *Blastulidium paedophthorum* (Synonym: *Blastulidiopsis chattonii*). A. Sigot (169). Illustrations (168).

The parasite of *Daphnia* from freshwater lagoons could be mentioned here.

It has rarely been found, and its affinities are unclear (36) in spite of ultrastructural and taxonomic work that has been carried out (170).

10. *The Parasites of Aschelminthes* (Myzocytiopsis *spp.*, Gonimochaete *spp.*, Chlamydomyzium *spp., and Other Flagellate Fungi*)

Table 3 lists the flagellate fungal parasites of the Aschelminthes (nematodes and rotifers). Many of these taxa are known only from the original descriptions (cited in Table 3) and a few subsequent recorded collections. The most commonly encountered flagellate fungal parasites of the Aschelminthes are placed in the Myzocytiopsidales (171), and a key to these species is given in Refs. 36 and 55. However, the one flagellate fungal pathogen of nematodes that has received much attention is *Nematophthora gynophila* (28), placed in the Leptolegniellaceae because of its distinctive oospore wall construction. Chytrid parasites, particularly of nematode eggs, also occur where nematodes are abundant. Three species of *Catenaria* (Blastocladiales) have been described as parasites of nematodes; *C. allomycis* is also reported as occurring in nematodes (172).

Much of the research effort in the biological control of nematodes has been directed to the facultatively parasitic ascomycetes because these fungi are more easily cultured. Control of nematode infestation by *Nematophthora* is effective eventually, but may cause commercial loss in the interim (28).

Unfortunately, little attempt has been made to understand the ecology of any of these parasites in the complex soil ecosystem. Given the number of species already described, it is probable that there is considerable niche specificity. Host specificity of these obligate parasites is probably not critical since transfer in laboratory conditions between nematodes, and nematodes and rotifers, is easily achieved. The same situation appears to be true for the Haptoglossales. Future studies may show that the critical features are the habitat of the nematode in relation to air–water interfaces and the precise mechanism of fungal spore attachment to the host; thus in different species of *Myzocytiopsis* the zoospores encyst either at body orifices or at random over the cuticle. In *Haptoglossa* the shape of the cyst containing the injection apparatus and the prevalence of a zoosporic phase differ between species. Multiple infections by one species are common, but the size of each thallus may be reduced with multiple infections. It is much rarer for a host to contain more than one fungal species, but this situation has been recorded (9). Usually only one species of a given genus is found in an individual. Nematodes and rotifers infected by Myzocytiopsidales or Haptoglossales remain active for a long time after much of the body space has been taken up by the parasites; fungal spore release may be by thallus conversion into sporangia or zoosporangia, which develop exit tubes through the skin (36,55), or by sporangial disintegration within the body cavity and release by the total disintegration of the host (44).

Nothing is known of the biochemical progress of disease in aschelminths caused by flagellate fungi. No molecular biological data have yet been published for any of these flagellate fungi, although such studies are underway (G. W. Beakes and S. L. Glockling, personal communication).

Endosphaerium funiculatum (173). Illustrations: (173).

Endosphaerium is a monotypic genus (species *E. funiculatum*) erected for a parasite on Aschelminthes in the mantle cavity of a freshwater pelecypod bivalve mollusk. It is known only from the original report. Unfortunately, the description of the fungus lacks almost all the diagnostic criteria necessary for classification. The author placed the fungus in the Pythiaceae, but it is difficult to see why. The fungus must be regarded as doubtful until rediscovered and redescribed.

11. Parasites of Noncellular Protoctista

The records for fungal pathogens of protozoan parasites are meager. Table 4 lists the hosts and cites the original descriptions, most of which are rather old records. Few subsequent supportive collections have been reported for any of these species. Diagnostic criteria that would now be considered essential are mostly lacking for species placed in *Sphaerita* and *Pseudosphaerita* (174), and so the taxonomy itself must be considered highly dubious. The comparable algal/fungal and algal/protozoan relationships have been reviewed (42), and more details from the original descriptions are available. (See Ref. 175.) Very little ecological research has been carried out on the occurrence, virulence, and distribution of these parasites, and further generalizations would be premature.

REFERENCES

1. MW Dick. Fungi, flagella and phylogeny. Mycol Res 101:385–394, 1997.
2. JD Silberman, ML Sogin, DD Leipe, CG Clark. Human parasite finds taxonomic home. Nature (London) 380:398, 1996.
3. MW Dick. Sexual reproduction in the Peronosporomycetes. Can J Bot, suppl. 1, secs. E-H, 73:S712–S724, 1995.
4. CJ Umphlett. Morphological and cytological observations on the mycelium of Coelomomyces. Mycologia 54:540–554, 1962.
5. YW Ho, DJS Barr. Classification of anaerobic gut fungi from herbivores with emphasis on rumen fungi from Malaysia. Mycologia 87:655–677, 1995.
6. MW Dick. Sexual reproduction: Nuclear cycles and life-histories with particular reference to lower eukaryotes. Bio J Linn Soc 30:181–192, 1987.
7. MW Dick. Phylum oomycota. In: L Margulis, JO Corliss, M Melkonian, D Chapman, eds. Handbook of Protoctista. Boston: Jones and Bartlett, 1990, pp. 661–685.

8. MW Dick. Peronosporomycetes. In: McLaughlin and McLaughlin, eds. The Mycota. Germany: Springer Verlag, 2001, pp. 39–72.

9. SL Glockling. Predacious and parasitoidal fungi in association with herbivore dung in deciduous woodlands. Ph.D. thesis, University of Reading, Reading, England 1994.

10. DJ Alderman. Fungal diseases of marine animals. In: EBG Jones, ed. Recent Advances in Aquatic Mycology. London: Paul Elek, 1976, pp. 223–261.

11. DJ Alderman. Fungi as pathogens of non insect invertebrates. In: RA Samson, JM Vlak, D Peters, eds. Fundamental and applied aspects of invertebrate pathology. Foundation of the Fourth International Colloquium of Invertebrate Pathology, 1986, Netherlands, Wageningen, 1986, pp. 354–355.

12. N Kitancharoen, K Hatai. A marine oomycete Atkinsiella panulirata sp. nov. from philozoma of spiny lobster, Panuliratus japonicus. Mycoscience 36:97–104, 1995.

13. K Nakamura, M Nakamura, K Hatai, Zafran. Lagenidium infection in eggs and larvae of mangrove crab (Scilla serrata) produced in Indonesia. Mycoscience 36: 399–404, 1995.

14. MW Dick. The species and systematic position of Crypticola in the Peronosporomycetes, and new names for the genus Halocrusticida and species therein. Mycol Res 102:1062–1066, 1998.

15. DJ Alderman, D Holdich, I Reeve. Signal crayfish as vectors in crayfish plague in Britain. Aquaculture 86:3–6, 1990.

16. RC Axtell, ST Jaronski, TL Merriam, CD Grant, RK Washino, EE Lusk, RL Coykendall. Efficacy of the mosquito fungal pathogen, Lagenidium giganteum (Oomycetes: Lagenidiales). In: CD Grant, RK Washino, EE Lusk, RL Coykendall, eds. Proceedings and Papers of the Fiftieth Annual Conference of the California Mosquito and Vector Control Association. California Mosquito and Vector Control Association, Sacramento, 1983, pp. 41–42.

17. JL Kerwin, RK Washino. Efficacy of Romanomeris culicvorax and Lagenidium giganteum for mosquito control: Strategies for use of biological control agents in rice fields of the Central Valley of California. In: CD Grant, JC Combs, RL Coykendall, EE Lusk, RK Washino, eds. Proceedings and Papers of the Fifty-Second Annual Conference of the California Mosquito and Vector Control Association. California Mosquito and Vector Control Association, Sacramento, 1985, pp. 86–92.

18. JL Kerwin, DA Dritz, RK Washino. Pilot scale production and application in wildlife ponds of Lagenidium giganteum (Oomycetes: Lagenidiales) to mammals. J Amer Mosquito Cont Assoc 10:451–455, 1994.

19. RK Washino, M Laird. Biocontrol of Mosquitoes Associated with California Rice Fields with Special Reference to the Recycling of Lagenidium Giganteum Couch and Other Microbial Agents: Biocontrol of Medical and Veterinary Pests. New York: Praeger, 1983, pp. 122–139.

20. CC Lopez Lastra, MM Steciow, JJ Garcia. Registro más austral del hongo Leptolegnia chapmanii (Oomycetes: Saprolegniales) como patógeno de larvas de mosquitos (Diptera: Culicidae). Revista Iberoamericana de Micologia (Espaqa) 1999.

21. MW Dick, CM Vick, TAJ Hedderson, G Gibbings, C Lopez Lastra. 18S rDNA for species of Leptolegnia and other Peronosporomycetes: Justification for the subclass taxa Saprolegniomycetidae and Peronosporomycetidae and division of the

Saprolegniaceae sensu lato into the families Leptolegniaceae and Saprolegniaceae. Mycol Res 103: 1119–1125, 1999.

22. JN Couch. Sporangial germination of Coelomomyces punctatus and the conditions favoring the infection of Anopheles quadrimaculatus under laboratory conditions. Proceedings Joint US–Japan Seminar on Microbial Control of Insect Pests, Fukuoka, Japan, April 21–23, 1967; pp. 93–105.

23. JN Couch. Mass production of Coelomomyces, a fungus that kills mosquitoes. Proceed Natl Acad Sci, USA 69:2043–2047, 1972.

24. CJ Umphlett. Infection levels of Coelomomyces punctatus, an aquatic fungus parasite, in a natural population of the common malaria mosquito Anopheles quadrimaculatus. J Invert Path 15:299–305, 1969.

25. K Hatai, G Hoshiai. Mass mortality in cultured coho salmon (Oncorhnchus kisutch) due to Saprolegnia parasitica Coker. J Wildl Dis 28:532–536, 1992.

26. K Hatai, W Rhoobunjongde, S Wada. Haliphthoros milfordensis isolated from gills of juvenile kuruma prawn (Penaeus japonicus). Nihon Kin Gakkai Kaiho 33:185–192, 1992.

27. AWAM de Cock, L Mendoza, AA Padhye, L Ajello, HH Prell. Pythium insidiosum sp. nov., the etiologic agent of pythiosis. J Clin Microbio 25:344–349, 1987.

28. BR Kerry. Fungal parasites of cyst nematodes. Agric Ecosyst Environ 24:293–306, 1988.

29. J Li, IB Heath. Chytridiomycetous gut fungi, oft overlooked contributors to herbivore digestion. Can J Microbio 39:1003–1013, 1993.

30. APJ Trinci, DR Davies, K Gull, MI Lawrence, BB Nielsen, A Rickers, MK Thodorou. Anaerobic fungi in herbivorous animals. Mycol Res 98:129–152, 1994.

31. LG Willoughby. Fungi and Fish Diseases. Pisces Press, Stirling, Scotland, 1994.

32. T Unestam, DW Weiss. The host–parasite relationship between freshwater crayfish and the crayfish disease fungus, Aphanomyces astaci: Responses to infection by a susceptible and a resistant species. J Gen Microbio 60:77–90, 1970.

33. L Nyhlén, T Unestam. Wound reactions and Aphanomyces astaci growth in crayfish cuticle. J Invert Path 36:187–197, 1980.

34. M Persson, L Cerenius, K Söderhäll. The influence of haemocyte number on the resistance of the freshwater crayfish, Pacifastacus leniusculus Dana, to the parasitic fungus, Aphanomyces astaci. J Fish Dis 10:471–477, 1987.

35. K Söderhäll, L Cerenius. Crustacean immunity. Ann Rev Fish Dis 3–23, 1992.

36. MW Dick. Straminipilous Fungi, Systematics of the Peronosporomycetes including accounts of the Marine Straminipilous Protists, the Plasmodiophorids and Similar Organisms. Dordecht, The Netherlands: Kluwer Academic Publishers, 2001.

37. FK Sparrow. Aquatic Phycomycetes. 2nd rev. ed. Ann Arbor, MI: University of Michigan Press, 1960.

38. GA Chilvers, FF Lapeyrie, PA Douglass. A contrast between Oomycetes and other taxa of mycelial fungi in regard to metachromatic granule formation. New Phytol 99:203–210, 1985.

39. NP Money, J Webster, R Ennos. Dynamics of sporangial emptying in Achlya intricata. Exp Mycol 12:13–27, 1988.

40. WC Coker. Leptolegnia from North Carolina. Mycologia 1:262–264, 1909.

41. HC Hoch, JE Mitchell. The ultrastructure of zoospores of Aphanomyces euteiches and their encystment and subsequent germination. Protoplasma 75:113–138, 1972.

42. HM Canter, MW Dick. Eurychasmopsis multisecunda gen. nov. sp. nov., a parasite of the suctorian Podophrya sp. (Ciliata). Mycol Res 98:105–117, 1994.

43. JN Couch. A new fungus on crab eggs. J Elisha Mitchell Sci Soc 58:158–162, 1942.

44. SL Glockling, MW Dick. New species of Chlamydomyzium from Japan and pure culture of Myzocytiopsis species. Mycol Res 101:883–896, 1997.

45. MW Dick. Saprolegniales. In: GC Ainsworth, FK Sparrow, AL Sussman, eds. The Fungi: An Advanced Treatise, IVB. New York: Academic, 1973; pp. 113–144.

46. DJS Barr. The zoosporic grouping of plant pathogens. Entity or non-entity? In: ST Buczacki, ed. Zoosporic Plant Pathogens: A Modern Perspective. London: Academic, 1983; pp. 43–83.

47. GW Beakes. A comparative account of cyst coat ontogeny in saprophytic and fish-lesion (pathogenic) isolates of the Saprolegnia diclina–parasitica complex. Can J Bot 61:603–625, 1983.

48. K Hatai, LG Willoughby, GW Beakes. Some characteristics of Saprolegnia obtained from fish hatcheries in Japan. Mycol Res 94:182–190, 1990.

49. K Söderhäll, MW Dick, G Clark, M Fürst, O Constantinescu. Isolation of Saprolegnia parasitica from the crayfish Astacus leptodactylus. Aquaculture 92:121–125, 1991.

50. DJS Barr. The phylogenetic and taxonomic implications of flagellar rootlet morphology among zoosporic fungi. BioSystems 14:359–370, 1981.

51. DJS Barr. Evolution and kingdoms of organisms from the perspective of a mycologist. Mycologia 84:1–11, 1992.

52. L Lange, L Olson. The uniflagellate phycomycete zoospore. Dansk Bot Arkiv 33: 7–95, 1979.

53. AR Hardham, F Gubler, J Duniec. Ultrastructural and immunological studies of zoospores of Phytophthora. In: Lucas JA, Shattock RC, Shaw DS, Cooke LR, eds. Phytophthora. Cambridge, UK: Cambridge University Press, 1990; pp. 326–336.

54. AR Hardham, HJ Mitchell. Use of molecular cytology to study the structure and biology of phytopathogenic and mycorrhizal fungi. Fung Gen Bio 24:252–284, 1998.

55. MW Dick, SL Glockling. Three new species of the Myzocytiopsidaceae (Peronosporomycetes), with a key to the nematophagous species of the family. Bot J Linn Soc. Submitted.

56. WJ Koch. Studies on the motile cells of chytrids. V. Flagellar retraction in posteriorly uniflagellate fungi. Amer J Bot 55:841–859, 1968.

57. JR Raper. Sexual hormones in Achlya. Amer Scientist 39:110–120, 130, 1951.

58. JA Galindo, ME Gallegly. The nature of sexuality in Phytophthora infestans. Phytopathology 50:123–128, 1960.

59. JW Hendrix. Sterols in growth and reproduction of fungi. Ann Rev Plant Path 8: 111–130, 1970.

60. TC McMorris. Antheridiol and the oogoniols, steroid hormones which control sexual reproduction in Achlya. Phil Trans Roy Soc, London 284:459–470, 1978.

61. MW Dick. Morphology and taxonomy of the Oomycetes, with special reference to Saprolegniaceae, Leptomitaceae and Pythiaceae. II. Cytogenetic systems. New Phytol 71:1151–1159, 1972.

62. MW Dick. Keys to Pythium. published by the author, Reading, UK, 1990.
63. S Shahzad, R Coe, MW Dick. Biometry of oospores and oogonia of Pythium (Oomycetes): The independent taxonomic value of calculated ratios. Bot J Linn Soc 108:143–165, 1992.
64. GW Beakes. Ultrastructural aspects of oospore differentiation. In: HR Hohl, G Turian, eds. The Fungal Spore: Morphogenetic Controls. London: Academic, 1981; pp. 71–94.
65. FN Martin. Electrophoretic karyotype polymorphisms in the genus Pythium. Mycologia 87:333–353, 1995.
66. EC Cantino. Physiology and phylogeny in the water molds—A re-evaluation. Q Rev Bio 30:138–149, 1955.
67. V Bulone, H Chanzy, L Gay, V Girard, M Fèvre. Characterization of chitin and chitin synthase from the cellulosic cell wall fungus Saprolegnia monoica. Exp Mycol 16:8–21, 1992.
68. HJ Vogel. Distribution of lysine pathways among fungi: Evolutionary implication. Amer Nat 98:435–446, 1964.
69. HB LéJohn. Enzyme regulation, lysine pathways and cell wall structures as indicators of major lines of evolution in fungi. Nature (London) 231:164–168, 1972.
70. LR Berg, GW Patterson. Phylogenetic implications of sterol biosynthesis in the oomycetes. Exp Mycol 10:175–183, 1986.
71. RA Fletcher. Plant growth regulating properties of sterol-inhibiting fungicides. In: SS Purchit, ed. Hormonal Regulation of Plant Growth and Development, Dordrecht, Netherlands: Martinus Nijhoff, 1987; pp. 103–114.
72. JM Griffith, AJ Davis, BR Grant. Target sites of fungicides to control Oomycetes. In: W Köllered, ed. Target Sites of Fungicide Action, Boca Raton, FL: Chemical Rubber Company Press, 1992, pp. 69–100.
73. Y Cohen, MD Coffey. Systemic fungicides and the control of Oomycetes, Ann Rev Phytopath 24:311–338, 1986.
74. T Van der Lee, I De Witte, A Drenth, C Alfonso, F Govers. AFLP linkage map of the oomycete Phytophthora infestans. Fung Gen Bio 21:278–291, 1997.
75. GR Klassen, SA McNabb, MW Dick. Comparison of physical maps of ribosomal DNA repeating units in Pythium, Phytophthora and Apodachlya. J Gen Microbio 133:2953–2959, 1987.
76. SA McNabb, GR Klassen. Uniformity of mitochondrial DNA complexity in Oomycetes and the evolution of the inverted repeat. Exp Mycol 12:233–242, 1988.
77. DSS Hudspeth, SA Nadler, MES Hudspeth. A cytochrome c oxidase II molecular phylogeny of the Peronosporomycetes (Oomycetes). J 1999.
78. WC Coker. The Saprolegniaceae, with notes on other water molds. Chapel Hill, NC: University of North Carolina Press, 1923.
79. GA Neish, GC Hughes. Diseases of Fishes, Book 6: Fungal Diseases of Fishes. Reigate, UK: T.F.H. Publications, 1980.
80. RJ Roberts. Fish Pathology. 2nd ed. London: Baillierè Tindall, 1989.
81. WN Tiffney, FT Wolf. Achlya flagellata as a fish parasite. J Elisha Mitchell Sci Soc 53:298–300, 1937.
82. WN Tiffney. The identity of certain species of the Saprolegniaceae parasitic to fish. J Elisha Mitchell Sci Soc 55:134–151, 1939.

83. WW Scott. Fungi associated with fish diseases. Dev Indus Microbio 5:109–123, 1964.

84. N Kitancharoen, K Hatai, R Ogihara, DNN Aye. A new record of Achlya klebsiana from snakehead, Channa striatus, with fungal infection in Myanmar. Mycoscience 36:235–238, 1995.

85. WN Tiffney. The host range of Saprolegnia parasitica. Mycologia 31:310–321, 1939.

86. WW Scott, CO Warren. Studies of the host range and chemical control of fungi associated with diseased tropical fish. technical bulletin. vol. 171. Blacksburg, VA: Virginia Agricultural Experiment Station, 1964, pp. 1–24.

87. JE Bly, LA Lawson, DJ Dale, AJ Szalai, RM Durborow, LW Clem. Winter saprolegniosis in channel catfish. Dis Aquat Organ 13:155–164, 1992.

88. K Hatai, S Egusa, T Awakura. Saprolegnia shikotsuensis sp. nov., isolated from kokanee salmon associated with fish saprolegniasis. Fish Path 12:105–110, 1977.

89. LG Willoughby. Saprolegnia polymorpha sp. nov., a fungal parasite on Koi carp, in the U.K. Nova Hedwigia 66:507–511, 1998.

90. HS Davis, EC Lazar. A new fungus disease of trout. Trans Amer Fish Soc 70:264–271, 1940.

91. K Hatai, S Egusa. Studies on visceral mycosis of salmonid fry. II. Characteristics of fungi isolated from the abdominal cavity of amago salmon fry. Fish Path 11:187–193, 1977.

92. JH Lilley, RJ Roberts. Pathogenicity and culture studies comparing the Aphanomyces involved in epizootic ulcerative syndrome (EUS) with other similar fungi. J Fish Dis 20:135–144, 1997.

93. RC Srivastava. Fungal parasites of certain freshwater fishes in India. Aquaculture 21:387–392, 1980.

94. GC Fraser, RB Callinan, LM Calder. Aphanomyces species associated with red spot disease, an ulcerative disease of estuarine fish from eastern Australia. J Fish Dis 15:173–182, 1992.

95. G Beakes, H Ford. Esterase isoenzyme variation in the genus Saprolegnia, with particular reference to the fish pathogenic S. diclina–parasitica complex. J Gen Microbio 129:2605–2619, 1983.

96. SE Wood, LG Willoughby, GW Beakes. Experimental studies on uptake and interaction of spores of the Saprolegnia diclina–parasitica complex with external mucus of brown trout (Salmo trutta). Trans Brit Mycol Soc 90:63–73, 1988.

97. K Hatai, G-I Hoshiai. Characteristics of two Saprolegnia species isolated from Coho salmon with saprolegnosis. J Aquat Animal Health 5:115–118, 1993.

98. K Yuasa, K Hatai. Relationship between pathogenicity of Saprolegnia spp. isolates to Rainbow Trout and their biological characteristics. Fish Path 30:101–106, 1995.

99. FI Molina, S-C Jong, M Guozhong. Molecular characterization and identification of Saprolegnia by restriction analysis of genes coding for rRNA. Anthonie van Leeuwenhoek 68:65–74, 1995.

100. JE Bly, LA Lawson, AJ Szalai, LW Clem. Environmental factors affecting outbreaks of winter saprolegniosis in channel catfish Ictalurus punctatus (Rafinesque). J Fish Dis 16:541–549, 1993.

101. RJ Roberts, LG Willoughby, S Chinabut. Mycotic aspects of epizootic ulcerative disease (EUS) of Asian fishes. J Fish Dis 16:169–183, 1993.

102. DJ Alderman. Malachite green: A review. J Fish Dis 8:289–298, 1985.

103. LG Willoughby, RJ Roberts. Towards strategic use of fungicides against Saprolegnia parasitica in salmonid fish hatcheries. J Fish Dis 15, 1–13, 1992.

104. RC Srivastava. Aphanomycosis—A new threat to fish population. Mykosen 22: 25–29, 1979.

105. K Hatai. Special Report of Nagasaki Prefectural Institute of Fisheries, 8:1980.

106. LG Willoughby, RJ Roberts, S Chinabut. Aphanomyces invaderis sp. nov., the fungal pathogen of freshwater tropical fish affected by epizootic ulcerative syndrome. J Fish Dis 18:273–275, 1995.

107. N Kitancharoen, K Hatai. Aphanomyces frigidophilus sp. nov. from eggs of Japanese char, Salvelinus leucomaenis. Mycoscience 38:135–140, 1997.

108. L Shanor, HB Saslow. Aphanomyces as a fish parasite. Mycologia 36:413–415, 1944.

109. K Hatai, S Egusa, S Takahashi, K Ooe. Study on the pathogenic fungus of mycotic granulomatosis I. Isolation and pathogenicity of the fungus from cultured-ayu infected with the disease. Fish Path 12:129–133, 1977 (in Japanese).

110. K Hatai, K Nakamura, SA Rha, K Yuasa, S Wada. Aphanomyces infection in Dwarf Gourami (Colisa lalia). Fish Path 29:95–99, 1994.

111. S Wada, K Yuasa, S-A Rha, K Nakamura, K Hatai. Histopathology of Aphanomyces infection in Dwarf Gourami (Colisa lalia). Fish Path 29:229–237, 1994.

112. TH Shafer, DE Padgett, DA Celio. Evidence for enhanced salinity tolerance of a suspected fungal pathogen of Atlantic menhaden, Brevoortia tyrannus Latrobe. J Fish Dis 13:335–344, 1990.

113. F Schikora. Die Krebspest. Fischerei Zeitung 9:529–532, 549–555, 561–566, 581–583, 1906.

114. E Rennerfelt. Untersuchungen uber die Entwicklung und Biologie der Krebspestpilze, Aphanomyces astaci Schikora. Mitt Anst f Binnerfischerei bei Drottningholm, Stockholm, 10:1–21, 1936.

115. WW Scott. A monograph of the genus Aphanomyces. technical bulletin. Blacksburg, VA, Virginia Agricultural Experiment Station, 151:1–95, 1961.

116. DJ Alderman, JL Polglase. Aphanomyces astaci: Isolation and culture. J Fish Dis 9:367–379, 1986.

117. T Unestam. Chitinolytic, cellulolitic, and pectinolytic activity in vitro of some parasitic and saprophytic Oomycetes. Physiologia Plantarum 19:15–30, 1966.

118. T Unestam, FH Gleason. Comparative physiology of respiration in aquatic fungi. II. The Saprolegniales, especially Aphanomyces astaci. Physiologia Plantarum 21: 573–588, 1968.

119. L Cerenius, K Söderhäll, M Persson, R Ajaxon. The crayfish plague fungus Aphanomyces astaci–Diagnosis, isolation and pathobiology. Freshwater Crayfish 7:131–144, 1988.

120. K Söderhäll, R Ajaxon. Effect of quinones and melanin on mycelial growth of Aphanomyces spp. and extracellular protease of Aphanomyces astaci, a parasite on crayfish. J Invert Path 39:105–109, 1982.

121. J de Haan, LJ Hoogkammer. Hyphomycosis destruens equi. Archiv fuer Wis-
      senschaftliche und Practische Tierheilkunde 13:395–410, 1903.
122. WA Shipton. Pythium destruens sp. nov., an agent of equine pythiosis. J Med Vet
      Mycol 25:137–151, 1987.
123. J de Haan, LJ Hoogkammer. Hyphomycosis destruens. Veeartsenijk Bl v Ned Indie
      13:350–374, 1901.
124. CH Bridges, CW Emmons. A phycomycosis of horses caused by Hyphomyces des-
      truens. J Am Vet Med Assoc 138:579–589, 1961.
125. PKC Austwick, JW Copeland. Swamp cancer. Nature (London) 250:84, 1974.
126. T Ichitani, J Amemiya. Pythium gracile isolated from the foci of granular dermatitis
      in the horse (Equus cabalus). Trans Mycol Soc Japan 21:263–265, 1980.
127. L Mendoza, L Ajello, MR McGinnis. Infections caused by the oomycetous patho-
      gen Pythium insidiosum. J Mycol Méd 6:151–164, 1996.
128. L Mendoza, L Kaufman, P Standard. Antigenic relationship between the animal
      and human pathogen Pythium insidiosum and nonpathogenic Pythium species. J
      Clin Microbio 25:2159–2162, 1987.
129. JE Longcore, AP Pessier, DK Nichols. Batrachochytrium dendrobatidis gen et sp.
      nov., a chytrid pathogenic to amphibians. Myucol 91:219–227, 1999.
130. AP Pessier, DK Nichols, JE Longcore, MS Fuller. Cutaneous chytridiomycosis in
      poison dart frogs (Dendrobates sp.) and White's tree frogs (Litoria caerulea). J Vet
      Diagnost Invest 11:194–199, 1999.
131. L Berger, R Speare, P Daszak, DE Green, AA Cunningham, CL Goggin, R Slo-
      combe, MA Ragan, AD Hyatt, KR McDonald, HB Hines, KR Lips, G Marantelli,
      H Parkes. Chytridiomycosis causes amphibian mortality associated with population
      declines in the rain forests of Australia and Central America. Proceed Natl Acad
      Sci USA 95:9031–9036, 1998.
132. J-M Hero, GR Gillespie. Epidemic disease and amphibian declines in Australia.
      Cons Bio 11:1023–1025, 1997.
133. RA Alford, SJ Richards. Lack of evidence for epidemic disease as an agent in the
      catastrophic decline of Australian forest frogs. Cons Bio 11:1026–1029, 1997.
134. JM Kiesecker, AR Blaustein. Influences of egg laying behavior on pathogenic in-
      fection of amphibian eggs. Cons Bio 11:214–220, 1997.
135. JE Longcore. Chytridiomycete taxonomy since 1960. Mycotáxon 60:149–174,
      1996.
136. JN Couch. A new saprophytic species of Lagenidium, with notes on other species.
      Mycologia 27:376–387, 1935.
137. LG Willoughby. Pure culture studies of the aquatic phycomycete, Lagenidium gi-
      ganteum. Trans Brit Mycol Soc 52:393–410, 1969.
138. LG Willoughby. Aquatic fungi from an antarctic island and a tropical lake. Nova
      Hedwigia 22:469–488, 1971.
139. PT Brey. Observations of in-vitro gametangial copulation and oosporogenesis in
      Lagenidium giganteum. J Invert Path 45:276–281, 1985.
140. A Domnas, S Jaronski, WK Hanton. The zoospores and flagellar mastigonemes of
      Lagenidium giganteum (Oomycetes, Lagenidiales). Mycologia 78:810–817, 1986.
141. RL Seymour. Leptolegnia chapmanii, an oomycete pathogen of mosquito larvae.
      Mycologia 76:670–674, 1984.

142. WW Martin. Couchia circumplexa, a water mold parasitic in midge eggs. Mycologia 73:1143–1157, 1981.

143. WW Martin. Aphanomycopsis sexualis, a new parasite of midge eggs. Mycologia 67:923–933, 1975.

144. SP Frances, AW Sweeney, RA Humber. Crypticola clavulifera gen. et sp. nov. and Lagenidium giganteum: Oomycetes pathogenic for dipterans infesting leaf axils in an Australian rain forest. J Invert Path 54:103–110, 1989.

145. WW Martin. The development and possible relationships of a new Atkinsiella parasitic in insect eggs. Amer J Bot 64:760–769, 1977.

146. WW Martin. A new species of Catenaria parasitic in midge eggs. Mycologia 67: 264–272, 1975.

147. WW Martin. Two additional species of Catenaria (Chytridiomycetes, Blastocladiales) parasitic in midge eggs. Mycologia 70:461–467, 1978.

148. JN Couch, CE Bland, eds. The genus Coelomomyces. New York: Academic, 1985.

149. JN Couch. Revision of the genus Coelomomyces, parasitic in insect larvae. J Elisha Mitchell Sci Soc 61:124–136, 1945.

150. JN Couch, HR Dodge. Further observations on Coelomomyces, parasitic on mosquito larvae. J Elisha Mitchell Sci Soc 63:69–79, 1947.

151. MP Madelin, A Beckett. The production of planonts by thin-walled sporangia of the fungus Coelomomyces indicus, a parasite of mosquitoes. J Gen Microbiol 72: 185–200, 1972.

152. CE Bland, JN Couch. Scanning electron microscopy of sporangia of Coelomomyces. Can J Bot 51:1325–1330, 1973.

153. WW Martin. The ultrastructure of Coelomomyces punctatus zoospores. J Elisha Mitchell Sci Soc 87:209–221, 1971.

154. JO Washburn, DE Egerter, JR Anderson, GA Saunders. Density reduction in arval mosquito (Diptera: Culicidae) populations by interactions between a parasitic ciliate (Ciliophora: Tetrahymenidae) and an opportunistic fungal (Oomycetes: Pythiaceae) parasite. J Med Entomol 25:307–314, 1988.

155. HC Whisler. Life history of species of Coelomomyces. In: JN Couch, CE Bland, eds. The Genus Coelomomyces. New York: Academic, 1985, pp. 9–22.

156. CJ Lucarotti. Invasion of Aedes aegypti ovaries by Coelomomyces stegomyiae. J Invert Path 60:176–184, 1992.

157. CJ Umphlett. Ecology of Coelomomyces infections of mosquito larvae. J Elisha Mitchell Sci Soc 84:108–114, 1968.

158. D Atkins. On a fungus allied to the Saprolegniaceae found in the pea crab Pinnotheres. J Mar Bio Assoc UK 16:203–219, 1929.

159. D Atkins. Further notes on a marine member of the Saprolegniaceae, Leptolegnia marina n. sp., infecting certain invertebrates. J Mar Bio Assoc UK 33:613–625, 1954.

160. D Atkins. A marine fungus, Plectospira dubia n. sp. (Saprolegniaceae) infecting crustacean eggs and small Crustacea. J Mar Biol Assoc UK 33:721–732, 1954.

161. D Atkins. Pythium thalassium sp. nov. infecting the egg mass of the pea crab, Pinnotheres pisum. Trans Brit Mycol Soc 38:31–46, 1955.

162. HS Visniac. A new marine phycomycete. Mycologia 50:66–79, 1958.

163. AJ van der Plaats-Niterink. Monograph of the genus Pythium. Studies in Mycology, Centraalbureau voor Schimmelcultures, Baarn 21:1–242, 1981.
164. SM Bower. Labyrinthuloides haliotidis n. sp. (Protozoa: Labyrinthomorpha), a parasite of juvenile abalone. Can J Zool 65:1996–2007, 1987.
165. SM Bower, DJ Whitaker, RA Elston. Detection of the abalone parasite Labyrinthuloides haliotidis by a direct fluorescent antibody technique. J Invert Path 53:281–283, 1989.
166. DD Leipe, SM Tong, CL Goggin, SB Slemenda, NJ Pieniazek, ML Sogin. 16S-like rDNA sequences from Developayella elegans, Labyrinthuloides haliotidis, and Proteromonas lacertae confirm that stramenopiles are a primarily heterotrophic group. Eur J Protist 32:449–458, 1996.
167. BA Cox, JG Mackin. Studies on a new species of Labyrinthula (Labyrinthulales isolated from the marine gasteropod Thais haemastoma). Trans Amer Microscopical Soc 93:62–70, 1974.
168. C Pérez. Sur un organisme nouveau, Blastulidium paedophthorum, parasite des embryons de Daphnies. Compte rendu des Séances de la Société de Biologie 55: 715–716, 1903.
169. A Sigo. Une chytridiacée nouvelle, parasite des oeufs de Cyclops: Blastulidiopsis chattoni n. g., n. sp. Comptes Rendus de la Société de Biologie de Strasburg 108: 34–37, 1931.
170. J-F Manier. Cycle et ultrastructure de Blastulidium paedophthorum Pérez 1903 (phycomycète lagénidiale) parasite des oeufs de Simocephalus vetulinus (Mull.) Schoedler (crustacé, Cladocère). Parasitologica 12:225–238, 1976.
171. MW Dick. The Myzocytiopsidaceae. Mycol Res 101:878–882, 1997.
172. JS Karling. Chytridiomycetarum Iconographia. J. Cramer, Vaduz, Lichtenstein, 1977. Catenaria in nematodes.
173. PN D'Eliscu. Endosphacrium funiculatum gcn. nov., sp. nov., a new predaceous fungus mutualistic in the gills of freshwater pelecypods. J Invert Path 30:418–421, 1977.
174. JS Karling. The present status of Sphaerita, Pseudophaerita, Morella, and Nucleophaga. Bull Torrey Bot Club 99:223–228, 1972.
175. JN Couch. Observations on the genus Catenaria. Mycologia 37:163–193, 1945.
176. N Sorokin. Les végétaux parasites des Angullulae. Annales des Sciences Naturelles, Botanique, Série VI 4:62–71, 1876.
177. HT Tribe. A parasite of white cysts of Heterodera: Catenaria auxiliaris. Trans Brit Mycol Soc 69:367–376, 1977.
178. W Birchfield. A new species of Catenaria parasitic on nematodes of sugarcane. Mycopathologia 13:331–338, 1960.
179. JS Karling. The structure, development, identity, and relationship of Endochytrium. Amer J Bot 24:352–364, 1937.
180. C Drechsler. A nematode-destroying phycomycete forming immotile spores in aerial evacuation tubes. Bull Torrey Bot Club 73:1–17, 1946.
181. SY Newell, R Cefalu, JW Fell. Myzocytium, Haptoglossa, and Gonimochaete (fungi) in littoral marine nematodes. Bull Marine Sci 27:177–207, 1977.
182. GL Barron. A new Gonimochaete with an oospore state. Mycologia 77:17–23, 1985.

183. GL Barron. Nematophagous fungi: A new Gonimochaete. Can J Bot 51:2451–2453, 1973.
184. GL Barron. A new and unusual species of Haptoglossa. Can J Bot 68:435–438, 1990.
185. C Drechsler. Three fungi destructive to free-living terricolous nematodes. J Wash Acad Sci 30:240–254, 1940.
186. GL Barron. Two new fungal parasites of bdelloid rotifers. Can J Bot 59:1449–1455, 1981.
187. GL Barron. Host range studies for Haptoglossa and a new species, Haptoglossa intermedia. Can J Bot 67:1645–1648, 1989.
188. GL Barron. A new Haptoglossa attacking rotifers by rapid injection of an infective sporidium. Mycologia 72:1186–1194, 1980.
189. JGN Davidson, GL Barron. Nematophagous fungi: Haptoglossa. Can J Bot 51:1317–1323, 1973.
190. AF Bartsch, FT Wolf. Two new saprolegniaceous fungi. Amer J Bot 25:392–395, 1938.
191. A Valkanov. Über Morphologie und Systematik der rotatorienbefallenden Oomyceten. Godishnik na Sofiiskiya Universitet (Jahrbuch der Sofiana Universität) 27:215–233, 1931.
192. B Kerry, DH Crump. Two fungi parasitic on females of cyst-nematodes (Heterodera spp.). Trans Brit Mycol Soc 74:119–125, 1980.
193. SN Dasgupta, R John. A contribution to our knowledge of the zoosporic fungi. Bull Bot Surv India 30:1–82, 1988.
194. SL Glockling. Isolation of a new species of rotifer-attacking Olpidium. Mycol Res 102:206–208, 1998.
195. U Kiran, R Dayal. Chytrids from leaf litter in ponds of Varanasi VIII. Genus Olpidium (Braun) Rabenhorst—New species and records. Proceedings of the National Academy of Sciences, India, 1992, Vol. 62 pp. 295–298.
196. LJ Dogma. Philippine zoosporic fungi: Olpidium sparrowii, a new chytridiomycete parasite of rotifer eggs. Kalikasen, Philip Bio 6:9–20, 1977.
197. GL Barron, E Szijarto. A new species of Olpidium in nematode eggs. Mycologia 78:972–975, 1986.
198. A Scherffel. Endophytische Phycomyceten-Parasiten der Bacillariaceen und einige neue Monadinen. Ein Beitrag zur Physiologie der Oomyceten (Schröter). Arch Protistenk 52:1–141, 1925.
199. N Arnaudow. Ein neuer Räderthiere (Rotatoria) fangender Pilz. Flora, Jena 116:109–113, 1923.
200. C Apstein. Synchaetophagus balticus, ein in Synchaeta lebender Pilz. Wissenschaftliche Meeresuntersuchungen der Kommission zur Wissenschaftlichen Untersuchungen der Deutschen Meere, Abteilung Kiel 12:163–166, 1910.
201. HM Canter, SI Heaney. Observations on zoosporic fungi of Ceratium spp. in lakes of the English Lake District: Importance for phytoplankton population dynamics. New Phytol 97:601–612, 1984.
202. A Boltovskoy. Relacion huesped-parasito entre el quiste de Peridinium willei y el oomicete Aphanomycopsis peridiniella n. sp. Limnobios 2:635–645, 1984.
203. I Foissner, W Foissner. Ciliomyces spectabilis, nov. gen., nov. sp., a zoosporic

fungus which parasitizes cysts of the ciliate Khaliella simplex. I. Infection, vegetative growth and sexual reproduction. Zeitschrift für Parasitenkunde 72:29–41, 1986.

204.   GW Erdos. A new species of Endemosarca from the Seychelles. Mycologia 65: 229–232, 1973.

205.   GW Erdos, LS Olive. Endemosarca: A new genus with proteomyxid affinities. Mycologia 63:877–883, 1971.

206.   P-A Dangeard. Mémoire sur les parasites du noyau et du protoplasma. Le Botaniste 4:199–248, 1895.

207.   H Epstein. Über parasitische Infektion bei Darmamöben. (in Russian). Archives de la Société russe de Protistologie 1:46–81, 1922.

208.   A Hollande, HH de Balsac. Parasitism du Peranema trichophorum par une chytridinée du genre Nucleophaga. Archives de Zoologie Expérimentale et Générale 82: 37–46, 1942.

209.   A Scherffel. Beiträge zur Kenntnis der Chytridineen, Teil III. Arch Protistenk 54: 510–528, 1926.

210.   N Sorokine. Apeçu systématique des Chytridiacées récoltées en Russie et dans l'Asie Centrale. Archives Botaniques du Nord de la France 2:1–42, 1883.

211.   A Scherffel. Einiges über neue ungenügend bekannte Chytridineen (Der Beiträge zur Kenntnis der Chytridineen, Teil II). Arch Protistenk 54:167–260, 1926.

212.   R Pérez-Reyes, M Madrazo-Garibay, EL Ochoterena. Pseudosphaerita dryli sp. nov. parásito de Phacus acuminata Stokes (Sarcomastigophora, Euglenidae). Revista Latino-americana de Microbiologia (Mexico) 27:89–92, 1985.

213.   JS Karling. Simple Holocarpic Biflagellate Phycomycetes. published by the author, New York, 1942.

214.   P-A Dangeard. Note sur un nouveau parasite des amibes. Le Botaniste 7:85–87, 1900.

215.   O Mattes. Über Chytridineen in Plasma und Kern von Amoeba terricola. Arch Protistenk 47:413–430, 1924.

216.   ER Becker. Endamoeba citelli sp. nov. from striped ground squirrel Citelli tridecemlineatus, and the life-history of the parasite, Sphaerita endamoebae sp. nov. Bio Bull Marine Bio Lab, Woods Hole 50:444–454, 1926.

217.   A da Cunha, J Muniz. Parasitismo de Trichomonas por Chytridaceae do genero Sphaerita Dang. Brazil-Medico 36:386, 1923.

218.   A Lwoff. Chytrininées parasites des amibes de l'homme. Possibilitéde leur utilisation comme moyen biologique de lutte contre la dysentrie amibienne. Bulletin de la Société de Pathologie Exotique 18:18, 1925.

219.   TL Jahn. On certain parasites of Phacus and Euglena: Sphaerita phaci, sp. nova. Arch Protistenk 79:349–355, 1933.

220.   DJ Alderman. Geographical spread of bacterial and fungal diseases of crustaceans. Rev Sci Tech Off Int Epiz 15:603–632, 1996.

3

Zygomycetes

The Order Mucorales

M. A. A. Schipper and J. A. Stalpers
*Centraalbureau voor Schimmelcultures, Institute of the Royal
Netherlands Academy of Arts and Sciences, Utrecht, The Netherlands*

I. INTRODUCTION

This chapter deals with the members of the Mucorales that cause mycotic diseases in mammals and birds, with emphasis on identification of the species involved. As the species are generally only identifiable when cultured, special attention has been given to cultural characters, growth conditions, and mating, while the number of clinical data is limited, as these are treated in detail in various modern handbooks (1–4).

Mycotic diseases caused by Mucorales are referred to as mucormycoses. Two other terms occur in recent literature: zygomycoses, a broader term, which also covers diseases caused by Entomophthorales (e.g., *Basidiobolus, Conidiobolus*), the only other order of the Zygomycetes causing diseases in warm-blooded vertebrates (see Chap. 4), and phycomycoses, which even include diseases caused by *Pythium* and related species, which are currently classified in a different kingdom, the Chromista (See Chap. 2.) Since the diseases caused by Mucorales are fairly uniform with respect to epidemiology, clinical aspects, and pathology, and are with very few exceptions clearly distinct from those caused by Entomophthorales, the term mucormycoses is preferred here.

Mucormycoses cover a wide range of diseases, including cutaneous, gastrointestinal, pulmonary, rhinocerebral, and disseminated types. The causal agents belong to the genera *Absidia, Apophysomyces, Cokeromyces, Cunninghamella, Mortierella, Mucor, Rhizomucor, Rhizopus, Saksenaea, Syncephalastrum,* and *Thermomucor*.

Prompt identification is essential, because mucormycoses belong to the most rapidly progressing mycoses. Identification at the genus and species level is usually only possible in culture. A correct identification also requires some knowledge of related species as well as of the more generally occurring saprophytes.

The current keys are not restricted to accepted pathogens only, but build according to the "positive key" principle. The first key starts at class level and only those orders containing pathogenic organisms are considered further. This is repeated at family, genus, and species levels. The pathogenic species are described in some detail.

II. HISTORY

Fürbringer (5) was the first to report on the pathogenic behavior of Mucorales. Lichtheim (6) isolated two species of Mucorineae, which he suspected to be pathogenic, and he inoculated both strains into rabbits to test this hypothesis. The strains were identified as *Mucor rhizopodiformis* (= *Rhizopus microsporus* var. *rhizopodiformis*) and *Mucor (Absidia) corymbifera*. Barthelat (7) published a review of all known pathogenic Mucorineae, including clinical reports, and after testing the reaction of various animals, such as rabbits, guinea pigs, and chickens, accepted the following species as pathogenic: *Mucor corymbifer, M. ramosus, M. truchisi, M. regnieri* (all *Absidia corymbifera*), *Mucor pusillus, Rhizomucor parasiticus* (both *Rhizomucor pusillus*), and *Rhizopus cohnii* (= *Rhizopus microsporus*). No pathogenicity could be reported for *Mucor mucedo, M. racemosus, M. alternans* (*M. circinelloides*), and *Rhizopus nigricans* (= *Rh. stolonifer*).

Ainsworth and Austwick (8) reviewed the fungal diseases of animals, while Ader and Dodd (9) reviewed the literature on mucormycoses up to 1978.

III. METHODS

A. Direct Examination

Material obtained from sputum, skin scraping, tissue preparation, or sinus aspirates may give a fast clue to the presence of a mucormycosis, because in such cases hyphae are often—but not necessarily—abundantly present, and these broad (5–15 μm), irregularly branched, aseptate, or rarely septate hyphae are quite diagnostic. They are only characteristic for mucormycosis, however; identification at family, genus, or species level from human material is generally impossible, as sporogenous structures are typically absent. Several methods can be applied, and these are discussed below.

1. Direct Smear

Direct observation can be made of the material mounted in 10% KOH. The slide is gently heated. This method is fast, but not always sufficient (10).

2. Methenamine Silver Reaction

The methenamine silver reaction of Grocott and Gomori is based on a mixture of chromic acid, methamine-silvernitrate, gold chloride, and sodium thiosulphate. The method takes about 1 hr.

3. Fluorescence

Uvitex 2B (Fungiqual, CIBA-Geigy, Basal, Switzerland). Uvitex 2B is described by Kuyper et al. (11). It binds specifically to chitin of the cell walls of fungi. The preparation takes about 30 min. The slide has to be examined at 400 nm.

Blankophor P (Bayer, Germany). Blankophor P is described by Monod et al. (12). The preparation takes about 20 min, and observation is at 420 nm. A similar stain is Fungi-Fluor (Polysciences).

B. Cultivation of the Organisms

Mucorales generally grow well on artificial media and without special effort isolates from more than 90% of the material diagnosed as mucormycosis by direct examination were successful (11). In particular, malt agar (MA), oatmeal agar (OA), or cornmeal agar (CMA) are generally applicable, while the popular Sabouraud agar ranks among the less recommendable media. The optimal conditions for the production of sporangia or zygospores vary for the different species; they are summarized in Table 1. Especially in *Apophysomyces*, *Mortierella*, and *Saksenaea* sporulation may be poor or absent when the conditions are not optimal.

Mixed or contaminated cultures can be purified by monosporangium isolates. Dip a hot inoculation needle in agar, touch a sporangium with the now coated needle, and inoculate a new petri dish.

Because of the invasive growth of Mucorales it is advisable to examine and isolate from attacked but not already necrotic tissue or from sinus aspirates. In case of isolation from superficial smears false positives are more likely to occur, as the spores of zygomycetes occur abundantly in the air. Grinding of

Table 1 Summary of Culture Conditions for the Pathogenic Taxa of the Mucorales

Name	Media	Temp. (°C)	Storage (°C)	Remark
Absidia	MA4, PDA, OA, CMA	20, 36	5	
Apophysomyces	YPPS, MA4	33	16	
Cokeromyces	MA4, PDA, OA, YPPS	20–22	5	
Cunninghamella	MA2, PDA, PCA, OA	20–22	5	
Mortierella	PDA, CMA, MA2	22	5	See also under *M. wolfii*
M. wolfii		35	16	
Mucor	MA4, CMA, PDA	20	5	
M. amphibiorum			16	
M. indicus		25–35	16	
Rhizomucor	MA4, OA, PDA	36	5	
Rhizopus	MA4, CMA, OA		16	
Rh. oryzae		27		
Rh. microsporus		36		
Saksenaea	Hay infusion agar (double layer)	25		See also under *Saksenaea*
Syncephalastrum	MA4, PDA, OA	36	5	
Thermomucor	MA4		5	

Note: For medium formulae see Section III.C.

material can give negative results, because the mycelium is coenocytic and the methods may result in only damaged (dead) cells.

The viability of vegetative spores is often limited; subculturing should include transfer of pieces of substrate mycelium. Many species sporulate poorly when oxygen concentrations are low. (Beware of closely fitting petri dishes.)

Mucorales are generally heterothallic and the occurrence of zygospores is quite rare. Matings can assist in confirming a supposed identity, however. (See Sec. V.B.)

The morphological study includes both macroscopical cultural characters and microscopic morphology. Water mounts are preferred over other mounting fluids, which may produce structural changes (swelling or shrinkage). The characteristics of sporangiophores, sporangia (size), stolons, and rhizoids should be studied in situ in undisturbed colonies with a stereomicroscope.

To prevent delay in identification cultures should be incubated on at least two different media at various temperature; for example, 25, 30, 36, and 40°C. For the suppression of bacterial growth, see Sec. III.C.

C. Preparation of Media

Media are sterilized by autoclaving at 121°C for 15 min, unless stated otherwise. Formulae for extracts and additives (when complex) are given separately. All media are for 1 liter and contain 15 g agar.

1. *CMA (Cornmeal Agar).* Add 60 g freshly ground cornmeal agar to 1 liter water, heat to boiling, and simmer gently for 1 hr. Squeeze and filter through cloth. Fill up to 1 liter and sterilize.

2. *Hay (hay-infusion agar).* Sterilize 50 g hay for 30 min at 121°C, filter, fill bottles, adjust to 1 liter and adjust pH to 6.2 with KH_2PO_4. Sterilize.

3. *MA2 (Malt Agar).* Dilute brewery malt with water to 10% sugar solution (level 10 on Brix saccharose meter). Sterilize. Fill 200 ml (2%) or 400 ml (4%) up to 1 liter. Sterilize. There are also good commercial MAs available.

4. *OA (Oatmeal Agar).* Wrap 30 g oatmeal flakes in cloth and hang in pan. Bring to the boil and simmer gently for 2 hr. Squeeze and filter through cloth. Sterilize.

5. *PCA (potato–carrot agar).* Add 20 g pealed and chopped carrots and 20 g scrubbed and diced potatoes to 1 liter water and boil and simmer for 1 hr. Boil again for 5 min and filter. Sterilize.

6. *PDA (potato-dextrose agar).* Add 200 g scrubbed and diced potatoes to 1 liter water and boil for 1 hr. Let it pass through a fine sieve, add 20 g glucose, and boil until dissolved. Sterilize. *Avoid new potatoes.*

7. *SNA (synthetischer nährstoffarmer agar, Synthetic poor medium).* Add 1 g K_2HPO_4, 1 g KNO_3, 0.5 g $MgSO_4.7H_2O$, 0.5 g KCl, 0.2 g glucose, and 0.2 g saccharose to 1 liter distilled water. Sterilize. Pieces of filter paper may be added as carbon source.

8. *YPSS (yeast powder-soluble starch agar).* Boil 1 g K_2HPO_4, 0.5 g $MgSO_4.7H_2O$, 15 g soluble starch, and 4 g yeast extract (Difco) in 1 liter of water until the ingredients are dissolved. Fill up to 1 liter.

For the suppression of bacterial growth, 1 ml of one of the following antibiotic solutions is added to the petri dishes before pouring out the agar, to give final concentrations (in parentheses) of: penicillin-G (50 ppm); streptomycin (30–50 ppm); aureomycin (10–50 ppm) neomycin (100 ppm); and novobiocin (100 ppm) or vanomycin (10 ppm). Chloramphenicol (50 ppm) is resistant to autoclaving and can be added before sterilization.

IV. MUCORMYCOSES

Generally, the immune system prevents infections with the Mucorales, but in immunocompromised patients (e.g., AIDS, diabetes, hepatitis, after transplantations and immunosuppression, intravenous drug abuse, leukemia), infections can be acute and serious (13,14). Characteristic is vascular invasion by hyphae, leading to thrombosis, infarction, and necrosis of tissue.

Invasion by the fungus generally occurs through

Inhalation of the spores
Ingestion
Traumatic implantation
Surgery
Contamination of burn wounds
Traumatized skin
Ears, nose, nails, and eyes

The frequent occurrence of the spores in the air (especially *Mucor* and *Rhizopus*) explains reported laboratory infection; for example, from the use of contaminated ectoplast bandages, postoperative wound infection, and infected protheses.

In particular the use of deferoxamine incurs an increased risk for mucormycosis. Dialysis patients having an iron overload based on a history of frequent transfusions and treated with deferoxamine frequently developed severe mucormycoses (15). Rhizoferrin, a siderophore isolated from *Rh. microsporus*, may mediate the infection (16), while Seeverens et al. (17) stressed the role of oxygen radicals. Boelaert et al. (18) advised against the prolonged use of deferoxamine therapy because of the increased risk of mucormycosis.

The following main types of mucormycoses are recognized (3,19):

Cutaneous mucormycosis: Infection generally occurs by injection or by implantation in wounds, especially burns, but it may also result from ingrowing mycelium from other organs. Most known mucoralean pathogens have been reported from cutaneous or subcutaneous infections: *Apophysomyces*, *Cunninghamella*, *Rhizopus*, *Saksenaea*, and *Syncephalastrum*.

Rhinocerebral mucormycosis: Infection is generally by inhalation and germination of the sporangiospores in the nose or paranasal sinuses. Development can be fast, especially when arterial walls are infected. Most reports concern *Rhizopus* spp., but cases caused by *Saksenaea* may also occur.

Pulmonary mucormycosis: Infection originates generally from inhaled sporangiospores or from disseminated mucormycosis. It may be acute, resulting in relatively large amounts of mycelium, or subacute, by spreading through the airways. Because of the oxygen-rich environment, growth can be very fast, often with lethal consequences. Most cases are caused by *Absidia* or *Cunninghamella*, some by *Mucor* spp.

Gastrointestinal mucormycosis: The infection is generally caused by indigestion or following surgery in the abdomen. Cases have been reported involving *Absidia*, *Corymbifera*, and *Rhizopus* spp.

Disseminated mucormycosis: Infection generally starts in the lungs or paranasal cavities, and may spread into the central nervous system, the vascular system, or the brain. Severely immunocompromised patients may develop mucormycosis in virtually any organ. Causal agents are: *Rhizomucor* spp. and *Saksenaea. vasiforme*.

V. IDENTIFICATION

A. Morphological Characters

Characters observed in microscopical slides include the following.

1. Vegetative Characters

Mycelium: total of aerial and submerged hyphae produced. In the Zygomycetes the hyphae are in principle aseptate, except in two cases: to separate reproductive structures and to separate dead or damaged parts from the active mycelium.

Stolons: creeping aerial hyphae from which rhizoids *and* sporangiophores are produced (Fig. 1).

Rhizoids: short, rootlike hyphae, which can be simple or branched (Fig. 1).

2. Reproductive Characters

Sporangiophores: branched (Fig. 2) or unbranched structures bearing sporangia. When branched, the development is sequential. Sporangiophores can be hyphoid or specialized, and then often terminating with a swollen structure on which the sporangia are more or less simultaneously produced (Figs. 3, 4). The branches can be monopodial (a persistent axis produces more fertile branches) or sympodial (continued growth after the main axis has produced a terminal sporangium; Fig. 2). Sporulation is indeterminate when the conidiophore length increases with continuing sporulation, or determinate when new sporangia are produced below the older ones.

Sporangia: collective term for a reproductive cell whose contents become transformed into asexual spore(s). An axial part of the mature sporangium can remain sterile and is called a *columella* (Fig. 5). The apophysis

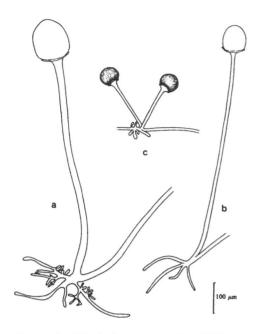

Figure 1   Schematic representation of *Rhizopus* groups; (a) *Rh. stolonifer* group, (b) *Rh. oryzae* group, (c) *Rh. microsporus* group.

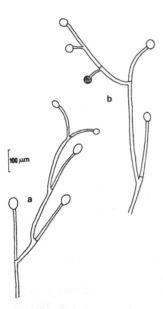

Figure 2   Branching patterns; (a) sympodial, (b) monopodial.

Figure 3 *Syncephalastrum racemosum*—sporangiophore and merosporangia.

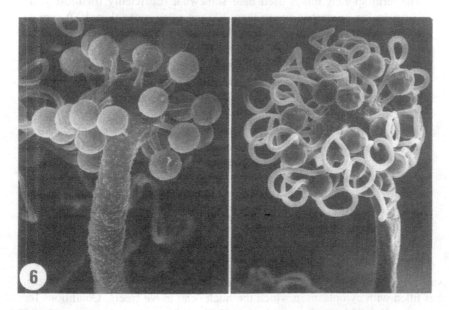

Figure 4 *Cokeromyces recurvatus*—young and mature sporangiophore.

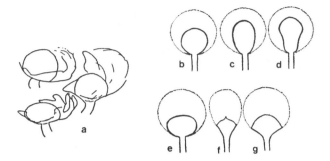

Figure 5 Columella types (from Ref. (84): (a) with distinct collar (*M. ramosissimus*), (b) globose (*M. hiemalis*, (c) ovoid (*M. silvaticus*), (d) pyriform (*M. mucedo*), (e) hemisphaerical (*Zygorhynchus vuilleminii*), (f) hemisphaerical with apophysis and apical projection, (g) hemisphaerical with broad apophysis (*Rhizopus* sp.).

is the basal section of the columella, which is not surrounded by the sporangium wall (Fig. 6). The sporangial wall can be smooth, spiny, transparent, or dark. At maturity the wall can be deliquescent (dissolving in a drop of liquid) or persistent (breaking). After spore liberation remnants of the original sporangial wall may form a *collar* (Fig. 7).

The term sporangium is used here somewhat restrictedly for those cells producing numerous (more than 30, and often more than 100) spores. Two additional terms are used for special forms: sporangioles and merosporangia.

Sporangioles: small, usually globose sporangia, containing few (1–30) spores. The wall is usually persistent. The monospored sporangioles (sometimes called *conidia*) are formed on pedicels that are often borne on vesicles (Fig. 8).

Merosporangia: cylindrical sporangia in which the spores are produced in a single row (Figs. 8).

Chlamydospores: thick-walled, terminal, or intercalary cells that are functional resting spores.

Zygospores: diploid cells resulting from the fusion of two haploid cells (Figs. 9, 10).

B. Mating

As a rule Zygomycetes produce hyphae without regular septation, resulting in tubes filled with cytoplasm in which the nuclei can move freely. Conditions for mating species of Mucorales are given in Table 2. Septa are only produced near

Figure 6 Sporangiophore of *Absidia*; (a) sporangium, (b) columella, (c) apophysis.

reproductive structures and to separate damaged or "aged" parts of the mycelium. Zygomycetes are mainly characterized by their sexual reproduction: fusion
of coenocytic gametangia resulting in a thick-walled zygospore (Figs. 9, 10).
Three types are distinguished: heterothallic, with gametangia originating from
different mycelia; homothallic, gametangia produced by a single mycelium; and
azygosporic, when the "zygospore" is produced by a single gametangium only
(parthenogenic zygospores).

   Blakeslee (20) showed that the majority of the Zygomycetes are heterothallic and unable to produce zygospores unless two genetically different mycelia
(mating types) were present. He designated such strains + and −. Burgeff (21)
and Plempel (22) discovered that + and − mycelia produced mating type-specific
substances (sex hormones), that act complementarily to produce pheromones.

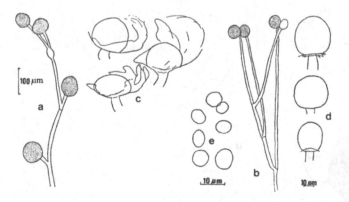

Figure 7 *Mucor ramosissimus*: (a–b) sporangiophores, (c–d) columellae with collar,
(e) sporangiospores.

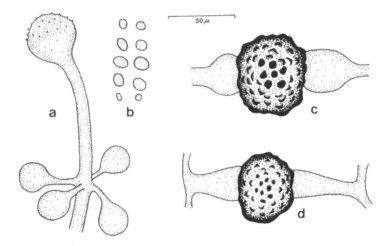

Figure 8 *Cunninghamella bertholletiae* (from Ref. 71): (a) conidiophore, (b) conidia, (c–d) zygospores.

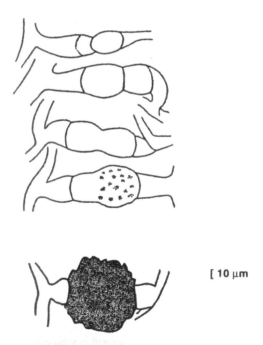

Figure 9 Various stages in the development of zygospores in *Mucor hiemalis*.

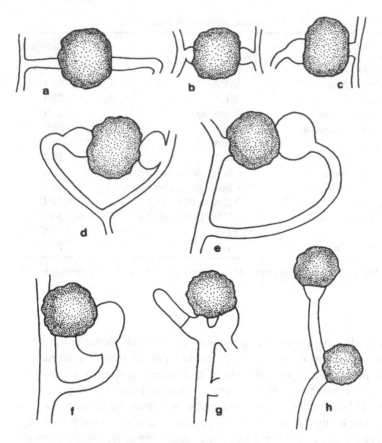

Figure 10 Morphological variation in zygospore formation from heterothallic to homo-thallic azygospores; a-c heterothallic with different types of suspensors; (a) *Rhizomucor pusillus*, (b) *Mucor hiemalis*, (c) *Mucor genevensis*; d-g homothallic with variously shaped suspensors and decreasing distances of the origin of participating hyphae: (d) *Rhizopus sexualis*, (e) *Rh. homothallicus*, (f) *Zygorhynchus moelleri*, (g) *Z. japonicus*, (h) azygo-spore, *Mucor bainieri*.

These pheromones induce zygophore formation and act as chemotropic agents between mating partners. The initial stages of the mating process depend on the mating type only, and may occur between + and − strains of quite different taxa, thus allowing an identification and a consistent use of the designations + and −. The completion of the process, starting with the lysis of the fusion wall, is considered species-specific.

A confusing element is the production of azygospores in both legitimate

Table 2 Mating Conditions for Pathogenic Species of *Mucorales*

Taxa	Medium[a]	Temperature
Absidia corymbifera	MYA	33°C
Cunninghamella bertholletiae	MYA	35°C
Mucor amphibiorum	MEA	25°C
M. circinelloides f. circinnelloides	Whey agar	25°C
M. circinelloides f. lusitanicus	MEA	25°C
M. hiemalis f. hiemalis	MEA	20°C
M. hiemalis f. luteus	MEA	20°C
M. racemosus f. racemosus	Cherry decoction agar	10°C
M. racemosus f. sphaerosporus	Cherry decoction agar	10°C
Rhizomucor pusillus	MEA (YPSS, YEA)	27°C
Rhizopus microsporus group	YEA (YPSS, MEA)	30°C
Rh. oryzae	YEA	30°C
Syncephalastrum racemosum	YEA	25°C

[a] Abbreviations: MYA, malt yeast extract agar; MEA, malt extract agar; YPSS, yeast powder-soluble starch agar; YEA, yeast extract agar.

matings, which also yield normal zygospores, and in interspecific pairings, in which normal zygospores do not occur. Zygospores result from the interaction between different hyphae or parts of hyphae, each secreting its own sex-specific substance. In heterothallic species the + and − sites are located on hyphae from different thalli. In homothallic species the copulating hyphae are connected either in the substrate or on the aerial mycelium, but the + and − sites are distinctly separate. Azygospores are produced from one single site only, which therefore might be +/− in genetic constitution or where agglomerations of + and − are close together. Following a very careful study of the mating processes of Mucorineae, Ling Young (23) suggested that azygospore formation in intraspecific matings may result from weak sexual potency of the partners leading to anomalies of the fusion processes and the prevention of actual zygospore formation. Blakeslee and others accepted the occurrence of such azygospores in interspecific contrasts as anomalies caused by a ''strong sexual ability of the partners.''

It seems that azygospores, whether formed in intraspecific or interspecific crosses, are produced through similar processes. They are certainly not rare. In mating experiments with strains of 19 different species, Schipper (24) found zygospore-like bodies in 150 interspecific pairings. These were often pale, small, sometimes slightly misshapen, and were produced rather late in comparison with most intraspecific matings. They were never really abundant and were generally produced within a wall of interwoven hyphae and incomplete conjugations. After careful observation of their development it was found that the fusion wall did not disintegrate and that only one suspensor was fully developed. Under the scanning electron microscope the azygospores in interspecific matings resembled normal zygospores arrested at an early stage of development (25).

Although zygospores are a characteristic of Zygomycetes, they are not known from many species. There are several reasons for this situation: the circumstances for zygospore production are more critical than those for sporangiospore production, homothallic strains are relatively rare, and the mating types of heterothallic species are only rarely isolated from the same locality. The germination of zygospores is slow (often after a resting period), and zygospores are thus an unlikely source of contamination.

C. Zygospore Production in Pathogenic Species

All types of zygospores mentioned above occur in the spectrum of pathogenic *Mucorales*, but for a number of species zygospores are unknown. As the conditions for the production of zygospores are usually critical, Table 2 specifies them for the species concerned.

Isolates of *Rhizopus azygosporus* show a striking morphological resemblance to azygosporic strains in the progeny of *Rh. microsporus* x *Rh. rhizopodiformis*. Schipper et al. (26) found that they produced apparently normal zygospores. Germinating zygospores formed germsporangia. A part of the progeny were azygosporic strains, with morphological characteristics intermediate between those of the parents. In *Rh. microsporus* var. *chinensis*, which also has features intermediate between var. *microsporus* and var. *rhizopodiformis*, the mating ability was found to be rather poor or absent. In old slants azygospores may occur.

The close relationship between taxa of the *Rh. microsporus* group was also confirmed by antigenic studies on vars. *microsporus*, *rhizopodiformis*, progeny of *microsporus* x *rhizopodiformis*, and var. *chinensis*. *Rh. microsporus* var. *chinensis*, *Rh. azygosporus*, and azygosporic isolates obtained from matings between *microsporus* x *rhizopodiformis* show similarity in general morphology, which is intermediate between those of the parents. Azygosporic strains resulting from *Rhizopus microsporus* x *rhizopodiformis* may lose the ability to produce azygospores or produce them in reduced numbers. *Rhizopus azygosporus* retains azygospores through careful subculturing with azygosporic material. From these observations it could be concluded that *Rh. microsporus* var. *chinensis* and *Rh. azygosporus* may be the results of spontaneous matings of *Rh. microsporus* var. *microsporus* with var. *rhizopodiformis*.

Absidia blakesleeana, *A. corymbifera*, *Rhizomucor pusillus*, and *Rh. microsporus* not only share a number of physiological characters, but also morphological ones; they are all thermotolerant, potentially pathogenic, have an optimal zygospore production at 30–33°C on malt yeast extract agar (MYA), are capable of good growth at high sugar concentrations (20–40%), share antigens (27), and produce azygospores in interspecific matings (*A. blakesleeana* with *A. corymbifera* and *Rh. pusillus*, *Rh. microsporus* with *Rh. pusillus*).

VI. TAXONOMY

A. Zygomycetes

The division Zygomycota contains two classes: the Trichomycetes with obligate parasites on arthropods (not considered here) and the Zygomycetes. They are mainly characterized by their sexual reproduction; the fusion of coenocytic gametangia (zygogamy) results in a thick-walled zygote, the zygospore.

The Zygomycetes include two orders that are of interest to medical mycology, Mucorales and Entomophthorales, which differ mainly in asexual, but to some extent also sexual reproduction. In the Mucorales, nonmotile sporangiospores are formed in sporangia, or else in the subdivisions of sporangia usually called merosporangia when they are elongate and contain sporangiospores in one row, and sporangioles when they are globose and contain clusters of a few (or a single) sporangiospores: Single-spored sporangioles occurring in some families are recognized as conidia when they are discharged in toto and when their membrane is coalescent with the wall of the spore. The zygospores develop directly from the fused gametangia. The ornamentation and color of the zygospore wall, the equality or inequality of the size of the suspensors, and the presence or absence of projections originating from the suspensors may contribute to the definition of families and genera. The pathogenic Entomophthorales are treated in Chap. 4.

Relevant literature includes refs. 28 and 29.

Key to the Orders of the Zygomycetes

1a.	Mycelium more or less regularly septate	**2**
1b.	Mycelium aseptate, except near propagative structures	**3**
2a.	Merosporangia with two spores	*Dimargaritales*
2b.	Merosporangia with a single spore	*Kickxellales*
3a.	Spores forcibly discharged (see Chap. 4)	
		Entomophthorales
3b.	Spores not forcibly discharged (but compare *Pilobolus*, where the sporangium is forcibly discharged)	**4**
4a.	Endomycorrhizal	*Glomales*
4b.	Not endomycorrhizal	**5**
5a.	Asexual spores absent	*Endogonales*
5b.	Asexual spores present	**6**
6a.	Obligate parasites on invertebrates (except *Piptocephalidaceae*)	*Zoopagales*
6b.	Not parasitic on invertebrates	**Mucorales**

B. Mucorales

The hyphae are thin-walled, nonseptate (septa only occurring near reproductive organs or in old, necrotic hyphae), coenocytic, and 3 to 12 μm wide. The sporangiospores are produced in sporangia or sporangioles.

The order Mucorales contains 13 families. Many species are not known to form zygospores, but their systematic position is easily established by their similar asexual characteristics.

Most of the Mucorales grow rapidly on culture media. Characteristics are generally optimally produced on MA, OA, or potato-dextrose agar; they may be less expressed on common medical media such as Sabouraud dextrose agar (SDA), maltose agar (MA), or synthetic mucor agar (SMA). The members of most families are free-living saprophytes in soil, on decaying vegetable matter such as stale bread and other food, manure from various mammals, and so on, whereas those of only a few families can be facultative parasites on other fungi or higher plants. They are not aquatic. Many species, including those capable of producing disease in humans or mammals, are widely distributed, and their asexual spores are prevalent in the air. Pathogenicity to vertebrates is an exceptional feature in the life of these fungi and is determined essentially by predispositions of the host (diabetes, malignant hemopathia, immunosuppression, etc.). In the host tissues, only a vegetative mycelium is usually formed, the morphology of which is characteristic of the Mucorales and indistinguishable among all the species of the order.

Key to the Families of the *Mucorales*

1a.	Cylindrical multispored merosporangia simultaneously produced on terminal swellings; spores in a single row	*Syncephalastraceae*
1b.	Cylindrical multispored merosporangia absent	**2**
2a.	Sporangia present, sporangioles absent	**3**
2b.	Sporangioles present, sporangia present or absent	**8**
3a.	Sporangiophores with greenish metallic luster, tall (up to 15 cm), unbranched	*Phycomycetaceae*
3b.	Sporangiophores without metallic luster	**4**
4a.	Sporangia becoming more or less intactly separated from sporangiophore (either violently or passively); sporangial walls apically cutinized	*Pilobolaceae*
4b.	Sporangial wall deliquescent or breaking on the sporangiophore	**5**
5a.	Sporangial wall persistent, fracturing into 2–4 segments along preformed suture(s)	*Gilbertella*
5b.	Sporangial wall deliquescent or breaking into numerous small parts, leaving a collar; preformed sutures absent	**6**

6a.	Sporangia lacking a columella (rudiments sometimes present)	**Mortierellaceae**
6b.	Sporangia with a distinct columella	**7**
7a.	Sporangia flask-shaped, with a long neck	**Saksenaeaceae**
7b.	Sporangia globose to pyriform	**Mucoraceae**
8a.	Sporangia and sporangioles present	**9**
8b.	Sporangia absent	**10**
9a.	Sporangia and sporangioles borne on separate and distinct sporangiophores; sporangiospores striate	*Choanephoraceae*
9b.	Sporangia and sporangioles borne on the same or morphologically similar sporangiophores; sporangiospores smooth	**Thamnidiaceae**
10a.	Sporangioles on sporangiophores	**Thamnidiaceae**
10b.	Sporangioles produced on vesicles	**11**
11a.	Sporangioles always monospored	**12**
11b.	At least a number of the sporangioles multispored	**13**
12a.	Sporangiophores without sterile apical branch; sporangioles on distinct vesicle	**Cunninghamellaceae**
12b.	Sporangiophores terminating with a sterile spine, sporangioles sympodial, or produced on small vesicles.	*Chaetocladiaceae*
13a.	Sporangioles borne on secondary vesicles, which are borne on branches arising from large, primary vesicles; sporangioles with persistent appendaged walls	*Radiomycetaceae*
13b.	Sporangioles borne on primary vesicles; sporangioles dehiscing by fracture from a preformed zone	*Mycotyphaceae*

1. Mucoraceae

All sporangia are globose or pyriform, rarely dumbbell-shaped, contain few to many sporangiospores, and are provided with a columella. The columella may or may not show an apophysis. The sporangium membrane is usually not cutinized and is persistent or diffluent; in the latter case it may leave a "collar." Merosporangia, sporangioles, or conidia are always absent. The morphology of the sporangium is important for the definition of genera.

Key to the Genera of the *Mucoraceae*

1a.	Homothallic zygospores with extremely unequal suspensors formed on the same aerial hypha.	*Zygorhynchus*
1b.	Zygospores when present not formed on a single aerial hypha. (Be aware of azygospores with one suspensor only.)	**2**
2a.	Parasitic on fungi.	**3**
2b.	Not growing on fungi.	**5**
3a.	On Zygomycetes, forming galls (facultative parasite). In	

pure culture narrow sporangiophores are produced on
the apices of substrate hyphae, on the natural substrate
on aerial hyphae. (Compare *Mucor hiemalis*.)   *Parasitella*

3b.   On Hymenomycetes.   **4**

4a.   On Agaricales, preferably *Mycena*. Aerial hyphae with
short, lateral spinelike branchlets. Sporangia
multispored.   *Spinellus*

4b.   On Boletales, preferably *Boletus*. Sporangiophores dichotomously branched, producing few-spored sporangioles.   *Syzygites*

5a.   Sporangiophores always on distinct stolons and rhizoids,
hardly ever on aerial or substrate hyphae.   **6**

5b.   Sporangiophores originating from substrate hyphae or
aerial hyphae. Stolons or rhizoids may also be present.   **12**

6a.   Sporangia globose.   **7**

6b.   Sporangia pyriform or dumbbell-shaped.   **8**

7a.   Apophysis absent. Dark-colored sporangia on unbranched sporangiophores arising from well-developed
stolons opposite distinct rhizoids.
(Note: Poorly sporulating strains with abundant chlam-   **Rhizopus**
ydospores in the mycelium are sometimes recognized as
Amylomyces.)

7b.   Apophysis present, hemispherical.   *Gongronella*

8a.   Sporangia dumbbell-shaped.   *Halteromyces*

8b.   Sporangia pyriform.   **9**

9a.   Sporangia arising from substrate hyphae.   *Protomycocladus*

9b.   Sporangia borne on aerial hyphae.   **10**

10a.   Besides sporangiospores also dark angular spores (conidia) borne singly on branches of aerial mycelium.   *Chlamydoabsidia*

10b.   Spores never borne singly.   **11**

11a.   Sporangiophores may arise opposite rhizoids; columella
funnel- or bell-shaped.   **Apophysomyces**

11b.   Sporangiophores never opposite rhizoids. Columella globose to conical, sometimes with projection.   **Absidia**

12a.   Species thermophilic, growing at 36°C. Sporangia dark.   **13**

12b.   Species not growing at 36°C.   **14**

13a.   Apophysis present, conspicuous.   **Thermomucor**

13b.   Apophysis absent. Sporangia globose, dark, with distinct
columella. Sporangiophores branched, originating from
short aerial hyphae or from stolons with simple or
sparsely branched rhizoids.   **Rhizomucor**

14a.   Sporangia hyaline.   *Actinomucor*

14b.   Sporangia dark.   **15**

15a.   Sporangiophores arising from aerial hyphae. Apophysis
present or absent.   *Hyphomucor*

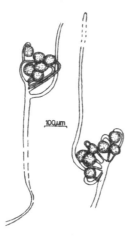

Figure 11   Indeterminate circinate sporangiophores of *Circinella umbellata*.

15b.	Sporangiophores arising from the substrate.	**16**
16a.	Sporangiophores circinate, branching below sporangia (Fig. 11). Apophysis present, short. Sporangia globose, with persistent wall, breaking at maturity. Collars present, long. Columella cylindrical to conical	*Circinella*
16b.	Sporangiophores branched or unbranched. Main pattern not circinate, although short, circinate branches occasionally may occur (Fig. 12). Apophysis absent	**Mucor** (Table 3).

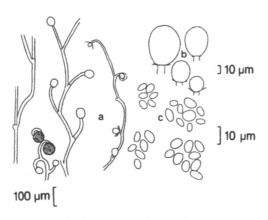

Figure 12   *Mucor circinelloides*: (a) sporangiophores, (b) columellae, (c) sporangiospores.

Table 3 Differential Characteristics of the Genera of Mucoraceae That Include Pathogenic Species

Genus	Rhizoids and stolons	Sporangiophores	Sporangia	Columellae	Apophysis	Sporangiospores	Zygospores when present	Figure
Rhizopus	Present	Single or in tufts; usually unbranched; mostly brown	Globose; gray, or brown	Subglobose to slightly elongated or pyriform	Present	Angular to globose ornamented; striate to globose	Rough; mostly between unequal suspensors	1, 22
Absidia	Present	Branched; often in corymbs; almost hyaline	Pyriform	Hemispherical; often with projection(s)	Present; conspicuous; conical	Globose to cylindrical; smooth	Smooth to slightly roughened; with equatorial ridges; suspensors almost equal	3, 9
Rhizomucor	Present	Monopodially or sympodially branched; dark brown	Globose; gray; opaque and glittering	Subglobose to slightly pyriform; brown	Absent	Globose to subglobose; small; smooth	Rough; between equal suspensors	17, 18
Mucor	Absent	Branched or unbranched; mostly hyaline	Globose	Various forms (e.g., globose, depressed, pyriform, elongated)	Absent	Globose to cylindrical	Rough; between equal suspensors	4a–d, 7, 13–16
Thermomucor	Present	Monopodially or sympodially branched; dark brown	Globose; gray; opaque and glittering	Subglobose to slightly pyriform; brown	Present	Globose to subglobose; small; smooth	Smooth; suspensors unequal	20, 21

Absidia. All sporangia are pyriform. The columella, which is always provided with a conspicuous, conical apophysis, is usually hemispherical on its top and often bears one or more projections. The sporangiophores may originate from stolons in the intervals between rhizoids or from a finely branched aerial mycelium. (The latter is the case for the pathogenic species.) In the zygosporogenic species the suspensors may carry projections enveloping the zygospore, but these are not found in the pathogenic species, *A. corymbifera.*

Type species: *A. repens*
Relevant literature: Hesseltine and Ellis (30–32); Ellis and Hesseltine. (33,34); Scholer and Müller (35); and Nottebrock et al. (36).

Key to the Species of *Absidia* [After Schipper (37)]

1a.	Colonies 40 mm or less in diameter, in a month at 25°C; suckerlike branches in the substrate mycelium (species of uncertain position)	2
1b.	Colonies usually filling petri dish (90 mm diameter) in a few days; suckerlike substrate hyphae absent	3
2a.	Homothallic, zygospores present at 25°C	*A. parricida*
2b.	Zygospores absent (unknown)	*A. zychae*
3a.	Determinate growth of the fertile aerial hyphae, generally ending in a large pyriform sporangium; good growth at 36°C	4
3b.	Indeterminate growth of the fertile aerial hyphae; typically no growth at 36°C (subgenus *Absidia*)	6
4a.	Stolons and rhizoids absent; sporangiophores arising from the substrate (excluded from *Absidia*)	
4b.	Stolons and rhizoids present; sporangiophores arranged in random fashion on the stolons; whorls or verticils not obvious (subgenus *Mycocladus*)	5
5a.	Sporangiospores subglobose, partly with roughened walls; at 45°C growth insignificant to absent	*A. blakesleeana*
5b.	Sporangiospores subglobose to ellipsoid or ellipsoid-cylindrical, smooth; at 45°C rather good growth	**A. corymbifera**
6a.	Sporangiospores globose or short ellipsoidal	7
6b.	Sporangiospores cylindrical or lacrymoid-cuneate	11
7a.	Sporangiospores globose-short ellipsoidal or slightly angular; sporangiophores both of the usual *Absidia*-type and single, short sporangiophores in series along stolons	*A. repens*
7b.	Sporangiospores globose; sporangiophores of one type only	8

8a.	Columellae with projections of intricate shape	*A. macrospora*
8b.	Columellae with a single apical projection	**9**
9a.	Sporangia up to 35 μm diam	*A. californica*
9b.	Sporangia up to 50 μm diam	**10**
10a.	Young colonies bluish	*A. coerulea*
10b.	Young colonies greenish	*A. glauca*
11a.	Sporangiospores lacrymoid-cuneate	*A. cuneospora*
11b.	Sporangiospores cylindrical with rounded ends	**12**
12a.	Sporangiospores variable in size, 3–6 × 2–3.5 μm	*A. heterospora*
12b.	Sporangiospores up to 5 × 2.5 μm	**13**
13a.	At 30°C no growth; optimal at 15°C	*A. psychrophila*
13b.	At 30°C growth; optimal at 20–24°C	
14a.	Homothallic zygospores present (Note: *A. anomala* may differ in unequal, monoappendiculate suspensors *and* sporangia larger than 70 μm.)	*A. spinosa*
14b.	Zygospores absent	**15**
15a.	Whorls of 5 or more sporangiophores quite common	*A. cylindrospora*
15b.	Whorls of more than 3 sporangiophores rare	**16**
16a.	Sporangiophores unequal in length	*A. fusca*
16b.	Sporangiophores equal in length	*A. pseudocylindrospora*

The pathogenic species of *Absidia* are:

Absidia corymbifera (Cohn) Saccardo & Trotter (Fig. 13)
Syn.: *Absidia ramosa* (Lindt) Lendner

Mycelium is rapid-growing, very high, and light dirty gray to almost white in some strains, but comparatively low (3–4 mm), whitish to dark grayish-brown, depending on the spore production. The complicated pattern with many superimposed ramifications bearing sporangia is basically the same in all strains. Stolons (distinct horizontal hyphae) are narrow, producing radially diffusing hyphae (rhizoids) for adherence to the substratum. Sporangiophores are finely branched, botryose (grapelike) to racemose (corymb), usually without a septum beneath the sporangium, and subterminally slightly brownish. Sporangia pyriform, which are 10–120(−150) μm in diameter, are transparent and almost colorless when young, becoming opaque (sometimes glittering) and grayish-brown to greenish-beige; the sporangium membrane is diffluent. Apophysis is long, and conical, often bearing a collar. Columella are hemispherical to short-ovoidal above the apophysis and 5–70(−85) μm in diameter, often with one to two mamilliform projections. Sporangiospores are smooth, slightly yellow to greenish, and globose to long-ellipsoid. The mean diameters in typical "globose strains" are 3.2–4.0 × 3–3.5

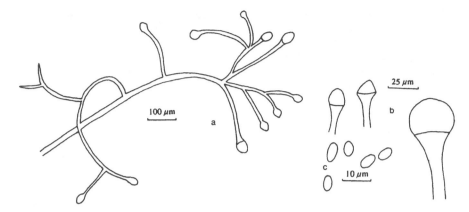

Figure 13 *Absidia corymbifera*: (a) branching pattern of sporangiophores on stolon, (b) columellae, (c) sporangiospores.

μm, in "intermediate strains" 3.5–5.0 × 3.0–3.8 μm, and in "elongate strains" 4.0–5.0 × 2.3–3 μm. Heterothallic. Zygospores are short, ellipsoid, 60–100 × 45–80 μm, thick-walled with very flat projections approximately 10 μm in diameter and one to three equatorial ridges, and bright reddish-brown. The suspensors are devoid of projections, equal or slightly unequal, the thicker suspensors or both often with a slightly roughened wall. The maximum temperature of growth is 48–52°C.

Absidia corymbifera was probably responsible for the first reported case of human mucor-mycosis. It has been considered the third most frequent agent after *Rh. oryzae* and *Rh. rhizopodiformis*, but Kwon-Chung and Bennett (3) consider it to be very infrequent. The species has been isolated from brain, pulmonary, kidney, and cutaneous infections (e.g., Ref. 38). It has been reported to be the most frequent agent of mammalian mucormycosis (including bovine mycotic abortion caused by Mucorales) and was also repeatedly isolated from mucormycoses in birds.

A. hyalospora (Saito) Lendner has been shown to be pathogenic to mice in axenic experiments (39). Schipper (37) treated this species as a synonym of *A. blakesleeana*, but in the neotype (CBS 173.67) the sporangiospores are up to 7(−10) μm in diameter, while in typical *A. blakesleeana* the sporangiospores are 4 to 6 μm (average 5 μm) in diameter.

Apophysomyces. The genus is closely related to *Absidia*. The sporangiophores may arise opposite the rhizoids, however. The pyriform sporangia show well-defined funnel- to bell-shaped apophysis. There is only one species, *A. elegans* Misra et al. (Fig. 14).

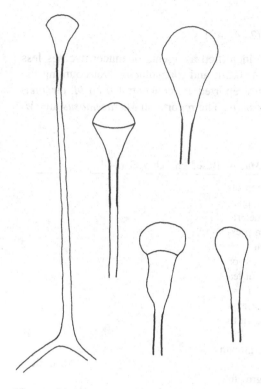

Figure 14 Funnel- to bell-shaped apophyses of *Apophysomyces elegans*.

The colony is brownish, and the aerial mycelium is sparse. Sporangiophores develop on the stolons with rhizoids below, straight or slightly curved, unbranched, brown, 150–250(−500) × 3.5–6 μm. The sporangia are terminal, 20–60 μm in diameter. The apophysis is funnel- to bell-shaped. The columella are hemispherical. The sporangiospores are ellipsoid to oblong, (3−)5.5–8 × 4–5 μm. Optimal development is at 36°C. For relevant literature; see Misra et al. (40).

A. elegans has been reported to cause mucormucosis after contamination of burn wounds.

Mucor. Sporangiophores are simple or more often branched, originating from substrate hyphae (stolons absent), and the rhizoids are absent. Sporangia are globose, and without apophysis. The zygospores are thick-walled and reddish-brown to dark brown or black, with warty to stellate ornamentation. The suspensors are as a rule opposed and without projections.

Type species: *M. caninus*
Relevant literature: Schipper (24,41,42)

Members of the genus *Mucor* were identified as agents of mucormycoses less frequently than those of *Rhizopus*, *Absidia*, and *Rhizomucor*. Not counting the obvious errors in diagnosis the names reported are concentrated on *M. circinelloides*, *M. hiemalis*, and *M. ramosissimus*. The reports on *M. racemosus* and *M. indicus* are uncertain.

Key to the Species of *Mucor* [Based on Schipper (24)]

1a.	Tallest sporangiophores usually repeatedly sympodially branched; sporangia usually less than 100 μm in diameter; zygospores or azygospores, when present, reddish-brown to dark brown	**2**
1b.	Tallest sporangiophores unbranched or weakly sympodially branched or at least some sporangia with a diameter of more than 100 μm; zygospores or azygospores, when present, dark brown to black	**14**
2a.	Growth and sporulation at 37°C, growth at 40°C	**M. indicus**
2b.	Poor or no growth at 37°C, no growth at 40°C	**3**
3a.	Reddish-brown azygospores borne terminally and subterminally on branched azygophores present in monosporangial cultures	*M. bainieri*
3b.	Reddish-brown azygospores on branched azygophores absent in monosporangial cultures	**4**
4a. μm	Spores hyaline, ellipsoid, 1.5–2 × 1	*M. nanus*
4b.	Spores at least 4 μm long	**5**
5a.	Sporangiospores ellipsoid ($1.3 < Q < 1.6$) or subglobose and approximately 4 to 6 μm in diameter, smooth; chlamydospores in sporangia uncommon	**6**
5b.	Sporangiospores broadly ellipsoid ($1.15 < Q < 1.3$) or globose and up to 8 to 10 μm in diameter, smooth or verrucose; chlamydospores in sporangiophores abundant or rare	**11**

6a. Colonies low, velvety; recurved sporan-
     giophores absent   7
6b. Colonies otherwise; recurved sporangio-
     phores present or absent   8
7a. Sporangia blackish, with extremely per-
     sistent walls; columellae applanate; spo-
     rangiophores with swollen regions   **M. ramosissimus**
7b. Sporangia grayish-brown, with diffluent
     walls; columellae mostly subglobose;
     sporangiophores of regular width   *M. zonatus*
8a. Colonies grayish; sporangiophores up to
     7(−10) µm in diameter; sporangia
     blackish   9
8b. Colonies brownish; sporangiophores up
     to 14(−17) µm in diameter; sporangia
     brownish to brownish-gray, rarely dark
     brown   10
9a. Sporangiospores subglobose   **M. circinelloides f. janssenii**
9b. Sporangiospores ellipsoid   **M. circinelloides f. griseocyanus**
10a. Columellae globose   **M. circinelloides f. lusitanicus**
10b. Columellae obovoid   **M. circinelloides f. circinelloides**
11a. Sporangiophores repeatedly branched in
     a sympodial fashion, with long branches
     originating just below the preceding spo-
     rangium and at acute angles (except oc-
     casionally at the top)   *M. fuscus*
11b. Sporangiophores branched in a mixed
     sympodial and monopodial fashion, not
     as described above   12
12a. Columellae usually with apical projec-
     tions; sporangiospores globose to subglo-
     bose, brownish, punctate; chlamydo-
     spores in sporangiophores rare   *M. plumbeus*
12b. Columellae usually smooth; sporangio-
     spores subglobose or broadly ellipsoid,
     grayish, generally smooth; chlamydo-
     spores in sporangiophores usually abun-
     dant   13
13a. Sporangiospores mainly subglobose   **M. racemosus f. sphaerosporus**
13b. Sporangiospores broadly ellipsoid   14
14a. Columellae obovoid to broadly pyri-
     form, often 50 µm and more in height;
     monopodial branches short   *M. racemosus f. racemosus*

14b.   Columellae subglobose, obovoid or ellip-
       soid, mostly 20 μm or less in height,
       rarely up to 40 μm; monopodial
       branches long   *M. racemosus f. chibinensis*
15a.   Tall sporangiophores slightly sympodi-
       ally branched, branches occurring rather
       late; sporangia with a diameter of less
       than 80 μm, sporangiospores ellipsoid   **16**
15b.   Tall sporangiophores unbranched except
       for an infrequent single sympodial
       branch or at least some sporangia with
       a diameter of more than 80 μm or spor-
       angiospores not ellipsoid   **20**
16a.   Zygospores present in monosporangial
       cultures   *M. genevensis*
16b.   Zygospores absent in monosporangial
       cultures   **17**
17a.   Branches arising just below the black-
       ish-brown sporangia, often swollen at
       the base; columellae globose; sporangio-
       spores cylindrical   *M. hiemalis f. silvaticus*
17b.   Branches arising at a longer distance be-
       low the yellowish to dark-brown spo-
       rangia   **18**
18a.   Sporangia yellowish; columellae glo-
       bose; sporangiospores narrowly ellip-
       soid   **M. hiemalis f. luteus**
18b.   Sporangia brownish to dark brown; colu-
       mellae globose and ellipsoid; sporangio-
       spores ellipsoid to cylindrical-ellipsoid   **19**
19a.   Sporangiospores mainly cylindrical ellip-
       soid   *M. hiemalis f. corticola*
19b.   Sporangiospores ellipsoid, sometimes
       flattened at one side   *M. hiemalis f. hiemalis*
20a.   Young sporangiophores erect   **21**
20b.   Young sporangiophores transitorily re-
       curved   **46**
21a.   Tall sporangiophores sympodially
       branched   **23**
21b.   Tall sporangiophores unbranched or in-
       frequently with a single branch   **22**
22a.   Sporangia often more than 150 μm in di-
       ameter   **27**
22b.   Sporangia usually not exceeding 150
       μm in diameter   **30**

23a. Colonies approximately 1 mm in height; at least some sporangia over 100 µm in diameter and having persistent walls; sporangiospores approximately 8 × 6 µm   *M. algariensis*

23b. Colonies more than 1 mm in height   **24**

24a. Sporangiophores of two types: tall and short, sympodially branched; sporangia on tall sporangiophores with deliquescent walls, sporangia on short sporangiophores with persistent walls; sporangiospores ellipsoid mostly 7 × 4 mm   *M. saturninus*

24b. Sporangiophores either of one type only or of two types; sporangiospores ellipsoid, over 7 µm in length or subglobose   **25**

25a. Sporangiospores ellipsoid   *M. flavus*

25b. Sporangiospores globose or subglobose   **26**

26a. Sporangiospores globose or subglobose, 4 to 5 µm in diameter, both at 5°C and 20°C   *M. minutus*

26b. Sporangiospores globose or subglobose, 3.5 to 7 µm in diameter, when grown at 20°C, ellipsoid, 5.5–8 × 3–4 µm when grown at 5°C   *M. strictus*

27a. Columellae ovoid or conical   **28**

27b. Columellae obovoid, ellipsoid or pyriform   **29**

28a. Mature sporangia dorsiventrally flattened, often 400 µm and over in diameter; sporangiospores 25 µm and over in average length; mesophilic; aerial mycelium abundant   *M. plasmaticus*

28b. Mature sporangia slightly dorsiventrally flattened, rarely exceeding 200 µm in diameter; sporangiospores usually 7.5 to 10 µm in length; psychrophilic: poor sporulation at 20°C; slow superficial growth at all temperatures   *M. psychrophilus*

29a. Sporangiospores usually ellipsoid, up to 10 µm in length; rapid growth on cherry agar or acid beerwort agar   *M. piriformis*

29b. Sporangiospores typically cylindrical-ellipsoid, often over 10 µm in length; restricted growth on cherry agar or acid beerwort agar   *M. mucedo*

30a. Zygospores or azygospores present in
     monosporangial cultures   31
30b. Zygospores or azygospores absent in
     monosporangial cultures   32
31a. Sporangiospores bacilliform   *M. bacilliformis*
31b. Sporangiospores globose or broadly el-
     lipsoid   *M. azygosporus*
32a. Sporangiospores ellipsoid or cylindrical-
     ellipsoid, 8 μm or less in length   33
32b. Sporangiospores ellipsoid, over 12 μm
     in length and/or globose   37
33a. Sporangiospores narrow ellipsoid, with
     a granule (globule) at each end   34
33b. Sporangiospores ellipsoid, without gran-
     ules at the ends   35
34a. Columellae mostly obclavate   *M. guilliermondii*
34b. Columellae globose to subglobose   *M. subtilissimus*
35a. Sporangia less than 100 μm in diameter   *M. microsporus*
35b. Sporangia over 100 μm in diameter
     present   36
36a. Colony very dark gray; sporangia blu-
     ish-black; columellae mouse-gray or
     brownish   *M. mousanensis*
36b. Colony gray; sporangia pale brown or
     dark gray; columellae hyaline, with or
     without brownish contents   *M. ucrainicus*
37a. Sporangia 75 μm or less in diameter   38
37b. Sporangia often 100 μm or more in di-
     ameter   43
38a. Sporangia up to 75 μm in diameter; spo-
     rangiophores often 15 μm or more in
     width   39
38b. Sporangia not exceeding 60 μm in diam-
     eter; sporangiophores not exceeding 10
     μm in width   40
39a. Sporangiospores globose, 3.5 to 5.5 μm
     in diameter, nonagglutinate   *M. amphibiorum*
39b. Sporangiospores ellipsoid or plano-
     convex-ellipsoid, up to approximately
     18 × 10 μm, agglutinate   *M. odoratus*
40a. Sporangiophores 8 μm or less in width;
     globose giant cells present   41
40b. Sporangiophores up to 10(−15) μm in
     width; giant cells absent   42

41a.	Sporangiospores ellipsoid	*M. zychae var. zychae*
41b.	Sporangiospores (sub)globose	*M. zychae var. linnemanniae*
42a.	Sporangiospores ellipsoid or flattened at one side	*M. prayagensis*
42b.	Sporangiospores subglobose	*M. sinensis*
43a.	Sporangiospores extremely variable in shape and size: ellipsoid, less than 10 µm in length, mixed with much larger, ellipsoid to globose sporangiospores up to 30 µm in length	*M. inaequisporus*
43b.	Sporangiospores less variable, either ellipsoid or (sub)globose	**44**
44a.	Sporangiospores (sub)globose; sporangiophores with a sterile, sickle-shaped branch at the base	*M. falcatus*
44b.	Sporangiospores ellipsoid; sporangiophores unbranched	**45**
45a.	Columellae ellipsoid-obovoid or globose; no growth at 37°C	*M. variosporus*
45b.	Columellae applanate to elongated applanate or conical; growth at 37°C	*M. variabilis*
46a.	Sporangia up to 250 µm or more in diameter	**47**
46b.	Sporangia 200 µm or less in diameter	**49**
47a.	Sporangiospores oblong ellipsoid, up to 40 µm in length	*M. oblongiellipticus*
47b.	Sporangiospores ellipsoid, up to 21 µm long	**48**
48a.	Spores $11.5-16.5 \times 6.8-8.4$ µm	*M. oblongisporus*
48b.	Spores $12.5-21 \times 7.5-12$ µm	*M. grandis*
49a.	Sporangia 60 µm or less in diameter	*M. guilliermondii*
49b.	Sporangia up to 125 µm and over in diameter	**50**
50a.	Sporangiospores irregularly polyhedral	*M. tuberculisporus*
50b.	Sporangiospores ellipsoid or oblong-ellipsoid	**51**
51a.	Sporangiola absent	*M. recurvus var. recurvus*
51b.	A few short, permanently recurved branches bearing sporangiola with persistent spiny walls present in aging cultures	*M. recurvus var. indicus*

The pathogenic species of *Mucor*:

Mucor circinelloides Tieghem (with the formae *circinelloides*, *lusitanicus*, *griseocyanus*, and *janssenii*; Fig. 12. Growth fast, reaching 7 cm within 2 days. Sporangiophores repeatedly sympodially branched, sometimes recurved, up to 6 mm tall and 20 μm wide. Sporangia brown or blackish, usually less than 100 μm in diameter. Sporangiospores subglobose, 4–6 μm diameter, or broadly ellipsoid to ellipsoid, 4.3–6.8 × 3.7–4.7 μm, mostly 5.4 × 4 μm. Temperature: from 5–30°C good growth and sporulation, at 37°C poor growth and sporulation. *M. circinelloides* has been reported from subcutaneous skin lesions and from urine without any evidence of infection. The species was isolated from peritoneal granulomata of cattle and swine, in fowl, and in ganders. Occasionally identification may be handicapped by atypical features. The distance between the sporangium and the next lateral branch can be so reduced that the sporangia seem to be sessile, and furthermore the sporangia often abort and development fails. Thin-walled sterile sporangia, terminal swellings, and swellings in the sporangiophore frequently occur. The sporangia in such cultures are small, the columellae subglobose to applanate, and the sporangiospores tend to be subglobose and variable in size.

M. ramosissimus Samutsevitsch (Fig. 7). Growth relatively slow, reaching 7 cm in 7 days. Sporangiophores repeatedly sympodially branched, up to 2 mm tall and 18 μm wide. Sporangia yellowish to blackish, globose to dorsiventrally flattened, up to 75 μm in diameter, with extremely persistent walls. Sporangiospores subglobose to broadly ellipsoid, 4–7(−8) μm in diameter or 5–8 × 4.5–6 μm. Zygospores unknown. Temperature: from 5–30°C good growth and sporulation, at 36°C growth extremely restricted. *M. ramosissimus* has been isolated from cerebral mucormucosis and chronic mucormucosis of the face (43).

M. indicus Lendner (Fig. 15): Growth fast, reaching 7 cm within 2 days. Sporangiophores repeatedly sympodially branched, up to 10 mm tall and 14 μm wide. Sporangia brown, up to 75 μm wide. Sporangiospores subglobose to ellipsoid, 5.4–6.4 × 3.8–4.8 μm (at 30°). Zygospores black, up to 100 μm in diameter, with pointed-stellate ornamentation and unequal suspensors. Temperature: from 10–15°C poor growth and no sporulation, from 25–37°C good growth and sporulation, at 40°C growth but no sporulation. Occasionally strains of *M. indicus* and *M. circinelloides* are rather similar. A second region of profusely branched short sporangiophores was never observed in *M. indicus*, whereas it is often present in *M. circinelloides*. Globose columellae predominate in *M. indicus*, while in *M. circinelloides* obovoid columellae are most common. In case of doubt, the temperature relations can generally be decisive. *M. indicus* has been reported from gastric mucormycosis.

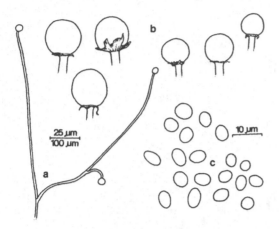

Figure 15 *Mucor indicus*: (a) sporangiophore, (b) columellae, (c) sporangiospores.

M. hiemalis Wehmer f. *luteus* (Fig. 16): Growth fast, reaching 7 cm within 2 days. Sporangiophores unbranched at first, then with few branches, up to 15 mm tall and 14 μm wide. Sporangia yellowish to dark brown, up to 70 μm wide, walls deliquescent. Sporangiospores narrowly ellipsoid to cylindrical, 5.7–8.7 × 2.7–5.4 μm. Zygospores blackish-brown, up to 70 μm in diameter, with low warted ornamentation and equal suspensors. Temperature: from 5–25°C good growth and sporulation, at 30°C stunted growth, at 37°C no growth. *Mucor hiemalis* has been isolated from subcutaneous infections.

M. racemosus Fres. f. *racemosus* (Fig. 17): Growth fast, reaching 7 cm within 2 days. Sporangiophores monopodially or sympodially branched, up to 20 mm tall and 17 μm wide. Chlamydospores abundant to rare in the sporangiophores. Sporangia hyaline, becoming brownish to brownish gray, globose, up to 80(−90) μm in diameter. Sporangiospores subglobose to broadly ellipsoid 5.5–8.5(−10) × 4–7 μm in diameter or 5.5 to 8 μm in diameter. Zygospores reddish-brown to bright brown, with low warts and equal suspensors. Temperature: from 5–30°C good growth and sporulation; maximum temperature 32°C. Considered doubtful by the maximum temperature of 32°C. Some reports on natural infections are nevertheless quite convincing: mucormycose in horse, chicken, and even human infections (pulmonary) in a renal transplant recipient (Table 4).

Rhizomucor.   Aerial mycelium is relatively low (up to 2 mm), and is brownish. Sporangiophores are branched, with simple or weakly branched rhi-

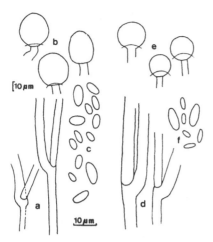

Figure 16 *Mucor hiemalis f. hiemalis*: (a) sporangiophores, (b) columellae, (c) sporan-giospores *M. hiemalis f. luteus*, (d) sporangiophores, (e) columellae, (f) sporangiospores.

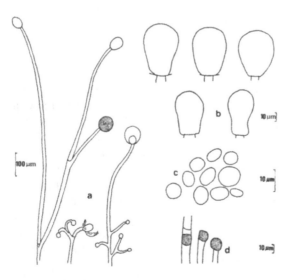

Figure 17 *Mucor racemosus*; (a) sporangiophores, (b) columellae, (c) sporangiospores, (d) chlamydospores.

Table 4 Differential Characteristics of the Pathogenic Species of *Mucor*

Species	Maximum temperature of growth (°C)	Sporangiophores	Sporangia	Columellae	Sporangiospores
M. circinelloides	36	Sympodially branched; long branches erect, and short branches sometimes circinate	Diameter varying from 20–80 μm; membrane diffluent in larger but rupturing or persistent in small sporangia	Globose, in larger sporangia ellipsoid	Ellipsoid; length up to 6–7 μm or subglobose; 4–6 μm diameter
M. ramosissimus	36	Sympodially branched; erect with swollen regions	Up to 75 μm in diameter but mostly smaller; membrane extremely persistent	Alpanate; absent in in small sporangia	Ellipsoid; length up to 7–8 μm
M. indicus	42	Sympodially branched; erect	up to 75 μm in diameter; membrane diffluent	Subglobose to aplanate, rarely elongate	Subglobose to ellipsoid ± 5.5 × 4.5 μm
M. racemosus	32	Sympodially or monopodially branched; (mixed) typically with abundant chlamydospores	Up to 80–90 μm in diameter; membrane mostly diffluent but persistent in smaller sporangia	Subglobose to pyriform, often with truncate basis	Subglobose or broadly ellipsoid; 8–10 μm long; smooth or verruculose
M. hiemalis f. luteus	30	Sympodially erect	Up to 60(-70) μm in diameter; diffluent	Globose	Narrow ellipsoid to nearly fusiform; 3.3–9.5 × 1.4–4 μm

zoids, with comparatively small sporangia (diameter up to 100 μm) with opaque, glittering walls. Columellae are small, subglobose to slightly pyriform, without or with a minute apophysis. Sporangiospores are subglobose. Zygospores are globose, covered with blunt projections, formed in the aerial mycelium between smooth, isogamous opposite suspensors. Growth starts at 22–24°C, optimal temperature is 37–40°C, and the maximum temperature >50°C.

Type species: *Rhizomucor parasiticus* (Lucet & Cost.) Schipper
Relevant literature: Schipper (44)

[Note: not included are: *Rh. nanitalensis* Joshi (45) with deformed sporangiospores and *Rh. variabilis* R.Y. Zheng & G.Q. Chen (46), which does not belong in *Rhizomucor*, because it is not thermophilic, has pale colored sporangia, and the sporangiophores have a different branching pattern.]

Key to the Species of *Rhizomucor* [on MEA, 36°C, After Schipper (44)]

1a.	Poor growth, hardly any sporulation, sporangiophores up to 35 μm wide; osmophilic	*Rh. tauricus*
1b.	Good growth and sporulation; sporangiophores up to 15 μm wide; osmotolerant	2
2a.	Zygospores absent or when present more than 50 μm in diameter	*Rh. pusillus*
2b.	Zygospores present, less than 50 μm in diameter	*Rh. miehei*

The pathogenic species of *Rhizomucor* are:

Rhizomucor pusillus (Lindt) Schipper (Fig. 18)
Syn.: *Rhizomucor parasiticus* (Lucet & Costantin) Schipper

The colony is 1(−2) mm high, and brownish. Stolons are up to 17 μm in diameter, with roughened walls, and are brown. The rhizoids are weakly developed. The sporangiophores branch in a mixed monopodial-sympodial fashion, are mostly subterminal, and are up to 11(−15) μm in diameter. They are brownish, with slightly encrusted walls. The sporangia are gray and glittering. They are up to 80(−100) μm in diameter, with encrusted walls that rupture at maturity. The columellae are ovoid to pyriform, up to 45 × 38 μm, light brown to mouse-gray, smooth, and rarely with very small protrusions. The collars are poorly defined to undefined. The sporangiospores are subglobose. A few are ellipsoid, typically 3 to 4 μm in diameter. The zygospores are formed in aerial mycelium. They are globose to slightly compressed, up to 70 × 63 μm (including blunt spines). They are dark brown to black. The suspensors are equal, elongated, and conical. They are heterothallic and rarely homothallic.

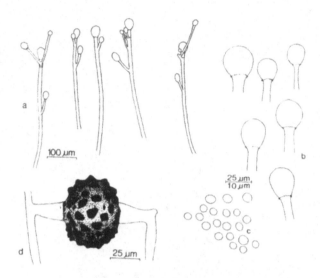

Figure 18 *Rhizomucor pusillus*: (a) sporangiophores, (b) columellae, (c) sporangio-spores, (d) zygospore between suspensors.

Rhizomucor pusillus is a common causative agent of animal mycoses (39). Artificially invoked mortal mycoses were described as early as 1886 by Lindt (47). Scholer (39) observed macroscopical changes in the kidneys. The species was frequently isolated from diseased mammals such as cattle, horses, swine, and seals (8), causing, for example, bovine mucormycotic abortion. There are only a few reports on human mucormycosis [lung (48), endocarditis (49), cutaneous infections (50,51)].

> *Rhizomucor miehei* (Cooney & Emerson) Schipper (Fig. 19): Differentiating characters from *Rh. pusillus*: colony color dirty gray to deep olive, branching of sporangiophores mostly sympodially, homothallic, zygospores up to 50 μm in diameter. Assimilation of sucrose is negative. Reports of animal mucormycosis caused by "zygosporic *M. pusillus*" may actually be *Rh. miehei*. All available information points to a similar pathogenic behavior as reported under *Rh. pusillus*.
> C. *Rhizomucor tauricus* (Milko & Schkurenko) Schipper (Fig. 20): The species differs from *Rh. pusillus* in the olivaceous color and weak branching (due to poor sporulation) and the requirement a of a high (10–50%) concentration of sucrose (osmophilic). Zygospores are unknown. In view of its growth at 37°C and above (up to 55°C), the species may be considered potentially pathogenic (Table 5).

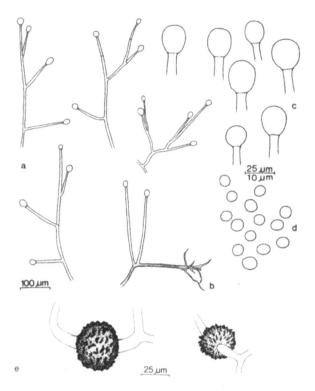

Figure 19 *Rhizomucor miehei*: (a) sporangiophores, (b) rhizoids, (c) columellae, (d) sporangiospores, (e) zygospore between suspensors.

Thermomucor Subrahamanyam, Mehrotra & Thirumalachar (Figs. 21, 22). Colonies are up to 2 mm high. They are grayish-brown to brown, with stolons up to 20 μm in diameter. The rhizoids are generally simple, up to 200 μm long and 10 μm wide. The sporangiophores are brownish, up to 40 μm wide, and with roughened walls. They are repeatedly sympodially branched. The sporangia are dark brown or gray and glittering, and are up to 125 μm in diameter. Collars are absent. Columellae are dark brown, hemispherical, up to 85 × 100(−150) μm, and subtended by a dark brown apophysis. The sporangiospores are subglobose, hyaline to slightly brown, and smooth. They are 3.5–6(−6.5) μm. The zygospores are borne near the substrate. They are reddishbrown to dark brown, smooth, and up to 57 × 51 μm. The suspensors are equal. They are homothallic, with a growth range of 24–53°C.

> Type (and only) species: *Thermomucor indicae-seudaticae* Subrahama-
> nyam, Mehrotra & Thirumalachar

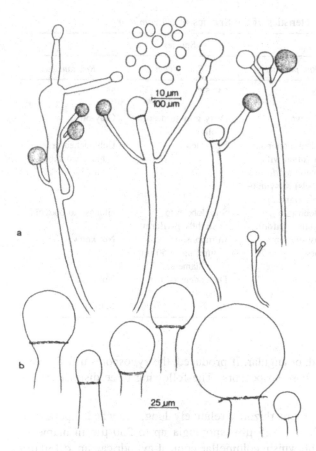

Figure 20 *Rhizomucor tauricus*: (a) sporangiophores, (b) columellae, (c) sporangiospores.

Relevant literature: Subrahamanyan et al. (52), Schipper (53)

Thermomucor resembles *Rhizomucor* in regard to the stolons, rhizoids, repeated branching of the sporangiophores, dark pigment, and high maximum temperature of growth (55°C), but differs by the presence of a conspicuous apophysis and by smooth zygospores. *Thermomucor* is pathogenic to mice, but less so than *Rhizomucor* spp. (H. J. Scholer, unpublished data). The species is considered a potential pathogen.

    Rhizopus.   The sporangiophores are unbranched, single or usually in clusters, and formed on stolons opposite rhizoids. The sporangia are globose, distinctly columellate, apophysate, and grayish to brownish. The sporangiospores

Table 5 Differential Characteristics of the Species of *Rhizomucor*

	Species		
	Rm. pusillus	*Rm. miehei*	*Rm. tauricus*
Maximum temperature of growth (°C)	54–58	54–58	55
Colony color	Hair-brown	Dirty gray to deep olive	Gray-olive
Sporangiophores	Branched monopodially (often with subterminal false umbels) or sympodially (mixed)	Branched	Unbranched or branched subterminally
Columellae	Subglobose to slightly pyriform	Subglobose to slightly pyriform	Globose to obovoid
Zygospores	Almost only in crosses; up to 70 μm in diameter	In monosporic culture; up to 50 μm in diameter	Not known
Thiamine requirement	Independent	Dependent (stimulated)	Not tested
Osmophily	No	No	Marked

are (sub)globose, ellipsoid, or angular. If produced, the zygospores usually have unequal, more or less globose suspensors. The following three distinct groups are distinguished (Fig. 1):

1. *Rh. stolonifer* group: rhizoids relatively long, branched. Sporangiophores up to 2000 × 25 μm; sporangia up to 250 μm in diameter, mouse-gray to brownish; columellae conical-cylindrical, up to 130 μm high; sporangiospores angular-subglobose to ellipsoid, with distinct ridges. No growth at 33°C.
2. *Rh. oryzae* group: rhizoids relatively long, unbranched. Sporangiophores up to 1500(−2000) × 15(−20) μm; sporangia up to 200 μm in diameter, mouse-gray to brownish; columellae conical-cylindrical; sporangiospores angular-subglobose to ellipsoid, with distinct ridges, not typical at higher temperatures. Growth at 40°C; no growth at 45°C.
3. *Rh. microsporus* group: rhizoids small, unbranched. Sporangiophores up to 500(−800) long; sporangia up to 80(−100) μm in diameter, mouse-gray to brownish; columellae variable (Fig. 23); sporangiospores angular-subglobose to ellipsoid, with or without ridges, warted to echinulate. Growth at 45°C.

Type species: *Mucor stolonifer* Ehrenb.
Relevant literature: Schipper (54); Schipper and Stalpers (55)

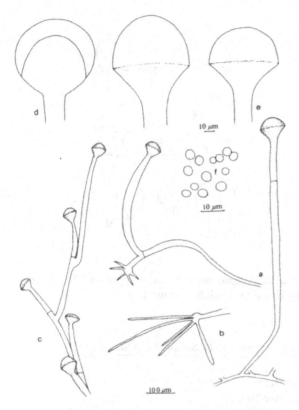

Figure 21 *Thermomucor indicae-seudaticae*: (a) young sporangiophore, (1) on short aerial hypha, (2) on stolon with rhizoids; (b) rhizoids; (c) branched sporangiophore; (d) sporangium with columella; (e) columellae; (f) sporangiospores.

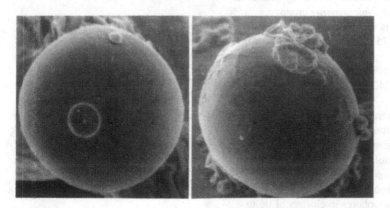

Figure 22 *Thermomucor indicae-seudaticae*: smooth zygospores.

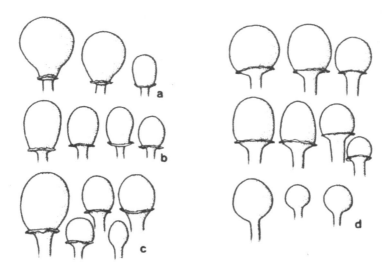

Figure 23 Shapes of columellae in *Rhizopus microsporus*: (a) pyriform, (b) pyriform-ellipsoid, (c) ellipsoid, (d) (sub)globose to subglobose-conical.

<table>
<tbody>
<tr><td colspan="3" align="center">Key to the Species [after Schipper (54); Schipper and Stalpers (55)]</td></tr>
</tbody>
</table>

1a.	Sporangiophores often more than 1 mm long; rhizoids relatively long, branched or not; sporangia over 100 μm in diameter; no growth at 45°C	2
1b.	Sporangiophores up to 0.5(−0.8) mm; rhizoids short, simple; sporangia up to 100 μm in diameter; usually growth at 45°C; *Rh. microsporus* group	3
2a.	Rhizoids usually branched; sporangia up to 275 μm in diameter; no growth at 36°C	*Rh. stolonifer*
2b.	Rhizoids simple; sporangia up to 200(−240) μm in diameter; growth at 36°C	**Rh. oryzae**
3a.	Zygospores or azygospores present in single-spore isolates	4
3b.	No zygospores produced in single-spore isolates	5
4a.	Zygospores present	*Rh. homothallicus*
4b.	Azygospores present	**Rh. azygosporus**
5a.	Sporangiophores in clusters of up to 9(−10)	6

5b.	Sporangiophores in groups of 1–3(−4). *Rh. microsporus* group	7
6a.	Sporangiospores smooth, angular to broadly ellipsoid, up to 8 μm long	*Rh. caespitosus*
6b.	Sporangiospores striate, subglobose to ovoid, 5.8 to 6.8 μm long	**Rh. schipperae**
7a.	Sporangiospores striate, distinctly angular-ellipsoid	**Rh. microsporus var. microsporus**
7b.	Sporangiospores not distinctly striate, (sub)globose to broadly ellipsoid, but sometimes angular	8
8a.	Sporangiospores at least on average smaller than 5 μm; radial growth at 50°C (120 hr) at least 35 mm	**Rh. microsporus var. rhizopodiformis**
8b.	Sporangiospores larger; radial growth at 50°C (120 hr) up to 10 mm	9
9a.	Sporangiospores of two kinds: (sub)globose, up to 9 μm in diameter, and irregular, larger ones	*Rh. microsporus var. oligosporus*
9b.	Sporangiospores globose to ellipsoid or slightly angular, up to 7.5 μm in diameter	**Rh. microsporus var. chinensis**

The pathogenic species of *Rhizopus* are:

> *Rhizopus oryzae* Went & Prinsen Geerligs
> Syn.: *Rhizopus arrhizus* Fischer ss auct.

Mycelium is high grayish-yellow to grayish-brown or slightly olive. The rhizoids are almost hyaline to dark brown, often with a reddish shade. The sporangiophores are single or in tufts. They are essentially unbranched, (500−)1000–1500(−2000) × 15–25 μm, yellowish-brown to dark brown, occasionally with local swellings up to 50 μm wide or bifurcate near the apex. The sporangia are 100–200 μm in diameter. The apophysis is 3–12 μm high. The sporangiospores are variable, roughly rhomboid, sometimes nearly ellipsoid or subglobose, longitudinally striate, (4−)6–8(−12) × (3−)4.5–6(−8) μm. The zygospores are globose or laterally flattened, brown to slightly red-brown, (60−)80–100(−140) μm, with stellate ornamentation, with very unequal suspensors. Sporangia produced in the higher temperature range are often atypical.

Rhizopus oryzae is the most common agent of human mucormycosis, and is responsible for nearly 90% of the cranial (rhinocerebral) form (56) (Table 6).

> *Rhizopus microsporus* van Tieghem: Mycelium yellowish-brown to gray to dark grayish-brown; rhizoids almost hyaline to pale brown; sporangio-

Table 6 Summary of Diagnostic Characters of *Rhizopus* Groups

	Stolonifer-group	*oryzae*	*microsporus*-group
Rhizoids	Complex, well developed	Medium	Simple
Sporangiophores (length)	1–3(-4) mm	Up to 1(-2.5) mm	Up to 0.5(-1) mm
Sporangia (diameter)	Up to 250(-300) μm	Up to 160(-240) μm	Up to 100 μm
Zygospores (diameter)	Black, up to 225 μm	Brown, up to 140 μm	Reddish-brown, up to 90(-100) μm
Suspensors	Approximately equal	Unequal	Unequal
Maximum growth temperature	33°C	45°C	Over 45°C

phores single or groups of 2 to 4, unbranched, up to 500(-800) × 8–15 μm, brownish; sporangia up to 80(-100) μm in diameter; apophysis indistinct; sporangiospores variable, roughly rhomboid, and longitudinally striate *or* irregular *or* regularly subglobose to ellipsoid, minutely spinulose, 3 to 9 μm long; zygospores globose or laterally flattened, reddish-brown, up to 90(-100) μm in diameter, with stellate ornamentation, with very unequal suspensors.

The species is treated after Schipper (54), thus including *Rh. rhizopodiformis* (Cohn) Zopf, *Rh. oligosporus* Saito, and *Rh. chinensis* Saito as varieties, which are often recognized at the species level in the medical literature. For differences see Table 7. *Rhizopus azygosporus* Yuan & Jong, isolated from tempeh in Indonesia, is considered an azygospore-producing intermediate between *Rh. microsporus* var. *rhizopodiformis* and var. *microsporus*.

Rhizopus microsporus ranks second after *Rh. oryzae* as an agent of human mucormycosis, being especially responsible for the cutaneous and gastrointestinal mycoses forms of the disease (14,57). It is only exceptionally found in cranial (rhinocerebral) mycoses (58). It is also known to be hospital-acquired (59). Polonelli et al. (60) conducted antigenic studies with the *Rh. microsporus* complex. The azygosporic *Rh. azygosporus* has been reported from lethal mucormycosis in premature babies (61). *Rh. microsporus* also occurs often in mammals, particularly pigs and cattle.

Rh. schipperae Weitzman et al.: The species is very close to *Rh. microsporus* var. *microsporus*, mainly differing in the larger groups (up to 10) of sporangiophores per stolon, the abundant production of chlamydospores (globose to ellipsoid, thin- to thick-walled, up to 20 μm in diameter) on all media, the poor production of sporangia (and thus a paler color) and negative mating results with the mating types of *Rh. microsporus*. The species was isolated from bronchial wash and lung specimens from a patient with myeloma (62) (Table 7).

2. Mortierellaceae

The sporangia are from one- to many-spored and globose. The columella are absent or very small, and never protrude into the sporangium. The mycelium is often unusually fine.

Key to the Genera of the *Mortierellaceae*		
1a.	Sporangiospores with two vermiform appendages	*Aquamortierella*
1b.	Sporangiospores without vermiform appendages	2
2a.	Thin-walled sporangia borne in sporocarps	*Modicella*
2b.	Sporangia not in sporocarps	3
3a.	Sporangia globose	4
3b.	Sporangia not globose	6
4a.	Sporangiophores arising progressively on specialized fertile hyphae of indeterminate growth	*Dissophora*
4b.	Sporangiospores arising from normal vegetative hyphae	5
5a.	Colonies velvety, not exceeding 3 mm in height; sporangia mostly ochraceous or vinaceous; odor absent or indistinct	*Micromucor*
5b.	Colonies cottony to arachnoid with longer aerial mycelium, ascendent or prostrate; sporangia not pigmented; odor often garliclike	**Mortierella**
6a.	Sporangia transversely elongate, cylindrical or sausage-shaped, with 1–5 apical spines; sporangiophores dichotomously branched	*Echinosporangium*
6b.	Sporangia ovoid to fusiform, without apical spines. Sporangiophores with terminal swelling on which a number of sporangioles develop simultaneously	*Umbellopsis*

Mortierella species rank among the commonest of soil fungi. They are fast growing and freely sporulating in nature. On 2% MA they often produce abundant aerial mycelium but few sporangia. On poorer media, such as Soil Extract Agar or SNA, sporulation is enhanced and much more differentiated. These media, however, are not suitable to isolate the fungus.

Type species:
Relevant literature: Gams (63), Domsch et al. (64)

Table 7 Summary of the Characteristics of the *Rhizopus microsporus* Group

Rhizopus microsporus group	Colony	Sporangiophores	Sporangia	Collumellae	Sporangiospores	Temperature
Rhizopus microsporus var. *microsporus*	Pale brownish-gray	Up to 400 × 10 μm (mostly swollen); brownish; often in pairs	Grayish-black; up to 80 μm	(Sub)globose to conical	Angular to broadly ellipsoid, up to 6–5 (7–5) μm; distinctly striate	46°C: good growth; 50°C: no growth
var. *rhizopodiformis*	Dark grayish-brown,	Up to 500 × 8 μm; brownish; 1–4 together	Bluish to grayish-black; up to 100 μm	Distinctly pyriform columella seem indicative for rhizopodiformis but absence is not conclusive	(Sub)globose; up to 5(–6) μm in diameter; minutely spinulose	50°C: good growth
var. *oligosporus*	Pale yellowish-brown to gray	Up to 300 × 15 μm; brownish; 1–3 together	Blackish, up to 80–100 μm (av. 50–60 μm)	(Sub)globose to conical	(Sub)globose; up to 9 μm in diameter	At 40°C: good growth and sporulating; at 45°C: restricted growth
var. *chinensis*	Pale brownish-gray	Up to 500(–700) × 10(–12) μm; 1–3 together	Blackish; up to 75(–80) μm	Subglobose-conical	Globose-ellipsoid; some slightly angular; up to 7–5 μm diam	At 45°C good growth, but immature; at 50°C: no growth

Rhizopus homothallicus	Dark brownish-gray; zygospores present	Up to 850 × 15 μm; brownish; 1–3 together	Blackish-gray; up to 100(–125) μm	Enlarged conical	Angular globose; up to 7–5(–8) μm	46°C: good growth; 50°C: no growth
Rhizopus azygosporus	Morphologically intermediate between "microsporus" and "rhizopodiformis"; azygospores present in single sporangium isolates					
Rhizopus caespitosus	Gray	Up to 200–500 μm × 10 μm; brownish; 1–9 together	Brownish-black up to 60(–75) μm in diameter	(Sub)globose to conical; brownish	Angular to broadly ellipsoid; up to 6–8 μm	15–36°C: sporulation; 45–50°C: growth, no sporulation
Rhizopus schipperae	Grayish-white	100–460 × 5–15 μm; brown on stolons; single or in pairs on rhizoids; up to 10	Grayish-black; up to 80 μm	Subglobose to conical; hyaline to pale tan	Subglobose to ovoid; 5.8–6.8 × 4.8–5.8 μm; (faintly) striate	23–45°C: good growth

Key to the Sections and Relevant Species of *Mortierella* [After Gams (63)]

1a.	Colonies velvety, not exceeding 3 mm in height; sporangia mostly ochraceous or vinaceous, often with a small columella; without distinctive odor.	subgen. *Micromucor*
1b.	Aerial mycelium consisting of longer, ascendent, or prostrate hyphae, white, cottony, or arachnoid; sporangia usually not pigmented, without or with a rudimentary columella; mostly with a garliclike odor. subgen. *Mortierella*	**2**
2a.	Only chlamydospores present (recognizable as *Mortierella* by colony habit and odor)	sect. *Stylospora*
2b.	Sporangiophores and sporangia (sporangioles) present	**3**
3a.	Sporangiophores always unbranched	**4**
3b.	Sporangiophores branched (at least sometimes)	**6**
4a.	Sporangiophores usually exceeding 200 µm in length	Sect. *Simplex*
4b.	Sporangiophores less than 150 µm in length	**5**
5a.	Sporangia, at least partly, many-spored; sporangiophores with distinctly widening base	Sect. *Alpina*
5b.	Sporangia one-spored; very slender sporangiophores arising in dense rows from the aerial hyphae	Sect. *Schmuckeri*
6a.	Branches arising mainly from the lower part of the sporangiophore (basitonous)	**7**
6b.	Branches arising from the middle or upper part of the sporangiophore (mesotonous or acrotonous)	**10**
7a.	Sporangia containing many, or at least several, smooth or ornamented spores	Sect. *Hygrophila*
7b.	Sporangia one-spored, often ornamented	Sect. *Stylospora*
8a.	Sporangia with 1 to 2 spores; sporangiophores short, with broad base, strongly tapered in the middle part and arising in dense rows from the aerial hyphae	Sect. *Haplosporangium*
8b.	Sporangia many-spored	**12**
9a.	Branches arising from the uppermost part of the sporangiophore in clusters from an inflated region, chlamydospores absent (when chlamydospores present; Compare **M. wolfii** under 24a)	Sect. *Actinomortierella*
9b.	Branching in another way	**13**
10a.	Sporangiophores racemosely branched with a thick main stem and thin, short branches; Sect. *Mortierella*	**11**
10b.	Sporangiophores often bent upwards above an ascendent basal part and with a minute columella; Sect. *Spinosa*	**14**

11a.	Spores with reticulate walls	*M. reticulata*
11b.	Spores with smooth or granulate walls	**12**
12a.	Sporangia up to five-spored; spores finely warty, 11 to 16 µm in diameter; irregularly lobate chlamydospores present in the agar	*M. oligospora*
12b.	Sporangia many-spored; aerial chlamydospores regularly spinulose	**13**
13a.	Spores smooth-walled, 10 to 12 µm in diameter	*M. polycephala*
13b.	Spores finely echinulate, 12 to 15 µm in diameter	*M. echinulata*
14a.	Sporangiophores generally not exceeding 200 µm in length	**15**
14b.	Sporangiophores typically much longer	**17**
15a.	Spores globose, 18 to 25 µm in diameter, with a double membrane	*M. acrotona*
15b.	Spores much smaller, with a single membrane	**16**
16a.	Chlamydospores constantly absent; spores 3 to 4 µm in diameter	*M. pulchella*
16b.	Chlamydospores scarcely produced, lemon-shaped, about 6 µm in diameter; spores 4–7(−10) µm in diameter	*M. epicladia*
17a.	Sporangiophores with umbellate branches arising from the same level; spores planoconvex, 12–14 × 5–7 µm	*M. umbellata*
17b.	Sporangiophores with branches inserted at different levels; spores different	**18**
18a.	Spores globose to subglobose	**19**
18b.	Spores ellipsoidal to cylindrical	**21**
19a.	Spores not exceeding 4 µm in diameter	*M. parvispora*
19b.	Spores larger	**20**
20a.	Sporangiophores often exceeding 1 cm in length; spores more than 10 µm in diameter	*M. nantahalensis*
20b.	Sporangiophores 200 to 800 µm long; spores less than 10 µm in diameter	*M. gamsii*
21a.	Chlamydospores densely covered with blunt spines; spores 4.0–5.5 × 2.0–3.0 µm	*M. fimbricystis*
21b.	If present, chlamydospores smooth-walled or irregularly lobate	**22**
22a.	Spores not exceeding 5.5 µm in length; chlamydospores regularly globose, lemon-shaped, or absent	**23**
22b.	Spores larger; chlamydospores bearing irregularly lobate appendages	**24**
23a.	Chlamydospores absent or tardily produced and lemon-shaped; sporangiophores 400 to more than 1500 µm long; spores 3.5–4.0(−5.0) × 2.0–2.5 µm	*M. jenkinii*

23b.	Chlamydospores abundantly produced, globose, 20 to 60 μm in diameter, thick-walled; sporangiophores usually not exceeding 300 μm; spores 3–4 × 1.2–2.0 μm	*M. cystojenkinii*
24a.	Sporangia leaving a pronounced collarette; spores with a double membrane	*M. wolfii*
24b.	Sporangia not leaving a collarette but with a trace of a columella; spores with a single membrane	*M. exigua*

Pathogenic Species of Mortierella. *Mortierella wolfii* Mehrotra & Baijal (Fig. 24) is characterized by sporangiophores 80 to 250 μm long, branched immediately below the apex, leaving a conspicuous collar after the dehiscence of the sporangia. The sporangiospores are short-cylindrical, 6–10 × 3–5 μm. Characteristic chlamydospores are up to 35 μm in diameter, with short, dichotomously branched outgrowths (amoeboid, stylospores) may be present; The odor is garlic-like. According to Crisan (65), the maximum temperature of growth is 48°C. Sporulation is fostered on soil extract agar incubated at temperatures above 25°C, but some strains are still difficult to identify because of the paucity of spore production.

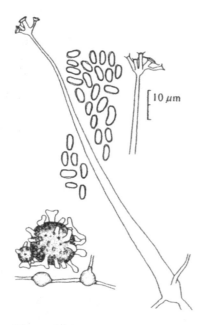

Figure 24 *Mortierella wolfii* (from Gams in Ref. 64): sporangiophores with acrotonous branching, sporangiospores, and chlamydospores, smooth or with pseudopodiform ornamentation.

M. wolfii is not known with certainty as a human pathogen but was proved to be a common agent of bovine mycotic abortion in New Zealand (66–70). A significant number of the cows were also affected by pneumonia caused by the same fungus. Previously reported cases from New Zealand, which were first attributed to *M. alpina* and *M. zychae*, are probably also due to *M. wolfii*. More recently, the latter species was also found as an agent of bovine mycotic abortion in Great Britain and the United States.

M. polycephala has been mentioned as the causal agent of lung mycosis and mycotic abortion of cattle (64), but as the maximum temperature of growth of this species is 26°C, this is probably a misidentification; it most likely is *M. wolfii*.

3. Cunninghamellaceae

The family contains only one genus, *Cunninghamella* Matr. It is characterized by monosporous sporangioles on short pedicels, arising simultaneously on swollen globose to obovoid vesicles, borne apically on branched or unbranched sporangiophores.

Key to the Species of *Cunninghamella* (71)

1a.	Homothallic; branching of conidiophores never verticillate	*C. homothallica*
1b.	Heterothallic; branching of conidiophores simple or verticillate	2
2a.	Mature colonies grayish due to dark conidia	3
2b.	Colonies whitish to yellowish	6
3a.	Conidia always globose, echinulate, brown	*C. phaeospora*
3b.	At least some conidia not globose, smooth, or echinulate	4
4a.	Conidia smooth, globose to ovoid or lacrymoid	*C. polymorpha*
4b.	Conidia echinulate, globose to ellipsoid or ovoid	5
5a.	Good growth at 42°C	**C. bertholettiae**
5b.	No growth at 40°C	*C. elegans*
6a.	Colonies powdery because of abundant sporulation; conidiophores irregular, pseudoverticillately branched	*C. echinulata*
6b.	Colonies not powdery	7
7a.	Colonies remaining white; conidia with long echinulae; conidiophores irregularly to cymosely branched	*C. vesiculosa*
7b.	Colonies becoming yellowish; conidia with short echinulae; conidiophores simple or verticillately branched	*C. blakesleeana*

Pathogenic Species of Cunninghamella. There has been an ongoing debate if *C. elegans* Lendner is the correct name for the pathogenic species, which in a more restricted sense is named *C. bertholletiae* Stadel (72). Samson (71) considered them identical (based on morphology only). Weitzman and Crist (73) reported that clinical isolates and saprophytic strains were morphologically similar, but incompatible mating partners. Moreover, the clinical isolates grew well at 42°C, while the other strains did not. They concluded that *C. elegans* and *C. bertholletiae* are distinct species. Elders (unpublished, 1986) confirmed the morphological similarity. She also found close serological relationships. She discovered that saprophytic isolates did not grow at 40°C, while the clinical isolates displayed good growth. Their maximum growth temperature was 46°C. Mating incompatibility was found as indicated by Weitzman and Crist, with one exception: *C. elegans* CBS 657.85 (origin unknown), which could be successfully mated with various strains of both *C. bertholletiae* and *C. elegans*.

Cunninghamella bertholletiae Stadel (Fig. 28). Growth is rapid. Mycelium is high, light gray to brownish. Hyphae are up to 20 μm wide. Conidiophores have verticillate or solitary branches. The vesicles are subglobose to pyriform. The apical ones are up to 40 μm in diameter, and those on the lateral branches are 10 to 30 μm in diameter. Conidia are globose, ovoid, or ellipsoid, 7 to 11 μm in diameter, or 6–10 × 9–13 μm, short echinulate, subhyaline, and brownish in mass. The zygospores, are brown, tuberculate, 18–28 × 25–35 μm, and heterothallic. *C. bertholletiae* has been isolated from several cases of mucormycosis in man.

4. Saksenaeaceae

The family contains only a single genus, *Saksenaea* Saksena, with only one species, *Saksenaea vasiformis* Saksena (Fig. 25).

The Mycelium is well developed and fast-growing. The sporangia are single or more rarely in groups of two, with basal, dichotomously branched rhizoids. The sporangiophores are erect, 24–65 × 6.5–9.5 μm. The sporangia are terminal. They are flask-shaped with a spherical venter, 22–52 × 16–44 μm, with a long neck, 54–200 × 6.5–11 μm, with expands apically, 8–14.5 μm in diameter. The columellae are distinct, and hemispherical. The sporangiospores are subcylindrical, smooth, and 2.8–4.2 × 1.5–2 μm.

Maximum temperature: 44°C
Relevant literature: Saksena (74)

When isolated, *S. vasiformis* does not sporulate on routine mycological media and thus is likely to be discarded as a contaminant. Sporulation can be obtained by growing the species on hay infusion agar (20 ml per petri dish) and 28–30°C in the dark. Padhye and Ajello (75) placed blocks of SDA–agar containing the fungus on a plate with 20 ml sterile distilled water with 0.2 ml of filter-sterilized 10% yeast extract solution in the dark at 37°C. After 7 days abundant sporulation was obtained.

The species was recorded to cause cutaneous and subcutaneous infections, necrotizing cellulitis, osteomyelitis, cranial and rhinocerebral infections, and dis-

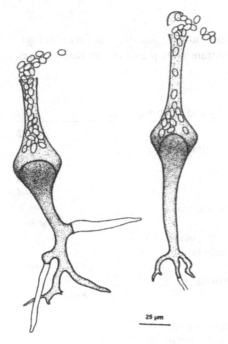

Figure 25 *Saksenaea vasiformis* (from Ref. 4): sporangia liberating sporangiospores.

seminated infections (3,76,77). In most cases the infection was acquired through traumatized skin (78).

5. Syncephalastraceae

The family contains one genus with one species and is characterized by the formation of numerous cylindrical merosporangia with 3 to 18 spores in a single row on a globose vesicle. The sporangial wall is deliquescent.

Syncephalastrum racemosum Cohn (Fig. 3). The colony is up to 15 mm high, and is olive gray. The sporangiophores are typically racemosely branched. They are up to 15 μm in diameter; with terminal vesicles up to 40 μm in diameter. They are brownish, often subtended by septa and bearing numerous merosporangia and are up to 80 μm in diameter. The merosporangia are cylindrical-clavate, about 25 × 5 μm, with 3 to 18 spores (average 7). The sporangiospores are globose to subglobose, verruculose, and 3 to 5 μm in diameter. The maximum temperature is 40°C. For relevant literature see Schipper and Stalpers (79).

S. racemosum has once been reported from a cutaneous infection (80) and as causal agent of bovine mycotic abortion (81).

6.   Thamnidiaceae

Sporangia and sporangioles are borne on the same or separate (but morphologically similar) sporangiospores (or only sporangioles present). Sporangiospores
are thin-walled and smooth.

Key to the Genera [after Benny and Kimbrough (82)]

1a.	Sporangia and sporangioles present	**2**
1b.	Sporangia absent, only sporangioles (containing up to 10 spores) present	**9**
2a.	Sporangioles pyriform to ellipsoid	**3**
2b.	Sporangioles globose to subglobose	**4**
3a.	Rhizoids and stolons present; columella of sporangioles subglobose, ovoid, or ellipsoid	*Thamnostylum*
3b.	Rhizoids and stolons absent; columella of sporangioles often slightly constricted or apically broadened (obpyriform)	*Pirella*
4a.	Sporangia subtended by a clavate to obovoid swelling	*Fennellomyces*
4b.	Subsporangial swelling absent	**5**
5a.	Both unispored and multispored sporangioles present; sporangiole wall verrucose to minutely spinulose	*Backusella*
5b.	Unispored sporangioles absent	**6**
6a.	Sporangiole wall distinctly spinulose	*Kirkomyces*
6b.	Sporangiole wall smooth to verrucose	**7**
7a.	Sporangioles borne on well-developed vesicles arising from main axis of sporangiophore	*Thamnostylum nigricans*
7b.	Sporangioles not borne on vesicles, arising directly from the sporangiophore or on lateral branchlets	**8**
8a.	Sporangioles arising directly from sporangiophores or on verticillate branchlets	*Helicostylum*
8b.	Sporangioles borne terminally on dichotomously branched lateral branchlets	*Thamnidium*
9a.	Sporangioles obpyriform	*Thamnostylum*
9b.	Sporangioles globose to subglobose	**10**
10a.	Sporangioles arising from the tips of the ultimate branches of the sporangiophores, not subtended by vesicles	*Ellisomyces*
10b.	Sporangioles arising from vesicles	**11**
11a.	Sporangiophores terminally condensed dichotomously branched (appearing lobate); sporangiole pedicels up to 60 µm long	*Zychaea*

11b. Sporangiophores terminally unbranched; vesicles
subglobose to broadly clavate; sporangiole pedi-
cels up to 120 μm long   **Cokeromyces**

Pathogenic species. Cokeromyces contains only one species, *C. recurvatus* Poitras (Fig. 4). Growth is rather slow. Sporangioles are produced at the tip of long, recurved stalks arising from a terminal vesicle of usually unbranched sporangiophores. Vesicles are 12 to 32 μm wide, with 5 to 30 sporangioles. The sporangiole pedicel is elongated, becoming up to 120 μm long. The sporangioles are globose, 8.5 to 12.5 μm, with 12 to 20 spores. The spores are ovoid to irregular, smooth, and about 4.5 × 2.5 μm. The zygospores are abundant, globose, brown, ornamented, and 33 to 55 μm in diameter. They are homothallic.

The species occurs commonly on the dung of small animals, but was once reported from a case of chronic cystisis (83).

REFERENCES

1. RD Baker. Mucormycosis (opportunistic phycomycosis). In: RD Baker, ed. Human Infection with Fungi, Actinomycetes and Algae, 1971; pp. 832–918.
2. A Balows, WJ Hausler Jr, KL Herrmann, HD Isenberg, HJ Shadomy. Manual of Clinical Microbiology. 5th ed. American Society for Microbiology, 1991.
3. KJ Kwon-Chung, JE Bennett. Medical Mycology. Philadelphia: Lea & Febiger, 1992.
4. GS de Hoog, J Guarro, eds. Atlas of Clinical Fungi. Baarn: CBS, 1995.
5. P Fürbringer. Beobachtungen über Lungenmycose beim Menschen. Virchows Arch Pathol Anat Physiol 66:330–365, 1876.
6. L Lichtheim. Ueber pathogene Mucorineen und die durch sie erzeugten Mykosen des Kaninchens. Zentralbl Klin Med 7:140–177, 1884.
7. GJ Barthelat. Les mucorinées pathogènes et les mucormycosis chez les animaux et chez l'homme. Arch Parasitol 7:1–116, 1903.
8. GC Ainsworth, PKC Austwick. Fungal Diseases of Animals. Farnham Royal: Commonwealth Agricultural Bureaux, 1973
9. P Ader, JK Dodd. Mucormycosis and entomophthoromycosis—A bibliography. Mycopathologia 68:67–99, 1979.
10. R Siebenmann, T Wegmann. Generalisierte Mucormykose. Schweiz Med Wochenschr 98:537–543, 1968.
11. PJ Kuyper, J Bruins, J Dankert. Mucormycose: Een zeldzame schimmelinfectie? Analyse 262–266, 1991.
12. M Monod, F Baudraz-Rosselet, AA Ramelet, E Frenk. Direct mycological examination in dermatology: A comparison of different methods. Dermatologia 179:183–186, 1989.
13. RVP Hutter. Phycomycetous infection (mucormycosis) in cancer patients: A complication of therapy. Cancer 12:330–350, 1959.

14. BC West, AD Oberle, KJ Kwon-Chung. Mucormycosis caused by Rhizopus microsporus var. microsporus: Cellulitis in the leg of a diabetic patient cured by amputation. J Clin Microbiol 33:3341–3344, 1995.
15. JR Boelaert, AZ Fenves, JW Coburn. Amer J Kidney Dis 18:660–667, 1991.
16. H Drechsel, J Metzger, S Freund, G Jung, JR Boelaert, G Winkelmann. Rhizoferrin—A novel siderophore from the fungus Rhizopus microsporus var. rhizopodiformis. Bio Metals 4:238–243, 1991.
17. HJJ Seeverens, GJ Tijhuis, GJHM Ruijs, BA Kazzaz, RH Kauffmann. Dialysis associated mucormycosis and desferroxiamine treatment: A case report with review of the role of oxygen radicals. Netherlands J Med 41:275–279, 1992.
18. JR Boelaert, M de Locht, J van Cutsem, V Kerrels, B Cantiniaux, A Verdonck, HW van Landuyt, Y-J Schneider. Mucormycosis during deferoxamine therapy is a siderophore-mediated infection. Amer Soc Clin Invest 91, 1993.
19. NL Goodman, MG Rinaldi. Agents of zygomycosis. In: A Balows et al, eds. Manual of Clinical Microbiology. 5th ed. Washington, DC: American Society for Microbiology, 1991.
20. AF Blakeslee. Sexual reproduction in the Mucorineae. Proc Amer Acad Arts Sci 40:205–319, 1904.
21. H Burgeff. Untersuchungen über Sexualität und Parasitismus bei Mucorineen. I Bot Abhandl 4:1–135, 1924.
22. M Plempel. Die Sexualstoffe der Mucoraceae. Ihre Abtrennung und die Erklärung ihrer Funktion. Arch Mikrobiol 26:151–174, 1957.
23. Ling Young. Étude des phénomènes de la sexualité chez les Mucorinées. Revue gén Bot 42:567–768, 1930.
24. MAA Schipper. On certain species of Mucor with a key to all accepted species. Stud Mycol 17:1–52, 1978.
25. JA Stalpers, MAA Schipper. Comparison of zygospore ornamentation in intra- and interspecific matings in some related species of Mucor and Backusella. Persoonia 11:53–63, 1980.
26. MAA Schipper, W Gauger, H van den Ende. Hybridization of Rhizopus species. J Gen Microbio 131:2359–2365, 1985.
27. PA Hessian, JMB Smith. Antigenic characterization of some potentially pathogenic mucoraceous fungi. Sabouraudia 20:209–216, 1982.
28. CW Hesseltine, JJ Ellis. Mucorales. In: GC Ainsworth, FK Sparrow, AS Sussman, eds. The Fungi. vol. 4B. Basidiomycetes and Lower Fungi. Academic, New York–London, 1973; pp. 187–217.
29. DL Hawksworth, PM Kirk, BC Sutton, DN Pegler. Ainsworth and Bisby's Dictionary of the Fungi. 8th ed. Wallingford: CAB International, 1995.
30. CW Hesseltine, JJ Ellis. Notes on Mucorales, especially Absidia. Mycologia 53:406–426, 1961.
31. CW Hesseltine, JJ Ellis. The genus Absidia: Gongronella and cylindrical-spored species of Absidia. Mycologia 56:568–601, 1964.
32. CW Hesseltine, JJ Ellis. Species of Absidia with ovoid sporangiospores I. Mycologia 58:761–785, 1966.
33. JJ Ellis, CW Hesseltine. The genus Absidia: Globose-spored species. Mycologia 57:222–235, 1965.

34. JJ Ellis, CW Hesseltine. Species of Absidia with ovoid sporangiospores II. Sabouraudia 5:59–77, 1966.
35. HJ Scholer, E Müller. Beziehungen zwischen bio-chemischer Leistung und Morphologie bei Pilzen aus der Familie der Mucoraceen. Path Microbio 29:729–741, 1966.
36. H Nottebrock, HJ Scholer, M Wall. Taxonomy and identification of mucor-mycosis-causing fungi. 1. Synonymity of Absidia ramosa with A. corymbifera. Sabouraudia 12:64–74, 1974.
37. MAA Schipper. Notes on Mucorales. 1. Observations on Absidia. Persoonia 14: 133–148, 1990.
38. M Darja, MI Davy. Pulmonary mucormycosis with cultural identification. Can Med Assoc J 89:1235–1238, 1963.
39. HJ Scholer. Mucormykosen bei Mensch und Tier. Taxonomie der Erreger, Chemotherapie im Tierexperimenten in der Klinik. Habilitationsschrift, Medical Fac. Basel: University of Basel, 1970.
40. PC Misra, KJ Srivastava, K Lata. Apophysomyces, a new genus of Mucorales. Mycotaxon 8:377–382, 1979.
41. MAA Schipper. A study on variability in Mucor hiemalis and related species. Stud Mycol 4:1–40, 1973.
42. MAA Schipper. On Mucor circinelloides, M. racemosus and related species. Stud Mycol 12:1–40, 1976.
43. JD Bullock, LM Jampol, AJ Fezza. Two cases of orbital phycomycosis with recovery. Amer J Ophthal 78:811–815, 1974.
44. MAA Schipper. On the genera Rhizomucor and Parasitella. Stud Mycol 17:53–71, 1978.
45. MC Joshi. A new species of Rhizomucor from India. Sydowia 35:100–103, 1982.
46. R-Y Zheng, G-Q Chen. A non-thermophilic Rhizomucor causing human primary cutaneous mucormycosis. Mycosystema 4:45–57, 1991.
47. W Lindt. Mitteilungen über einige neue pathogene Schimmelpilze. Arch Exp Pathol Pharmakol 21:269–298, 1886.
48. JL Nicod, C Fleury, J Schlegel. Mycose pulmonaire double à Aspergillus fumigatus Fres. et à Mucor pusillus Lindt. Schweiz Z Allg Pathol Bakteriol 15:307–321, 1952.
49. MS Erdos, K Butt, L Weinstein. Mucormycotic endocarditis of the pulmonary valve. JAMA 222:951–953, 1972.
50. RD Meyer, MH Kaplan, M Ong, D Armstrong. Cutaneous lesions in disseminated mycomycosis. JAMA 225:737–738, 1973.
51. BS Kramer, AD Hernandez, RL Reddick, AS Levine. Cutaneous infection, manifestation of disseminated mucormycosis. Arch Derm 113:1075–1076, 1977.
52. A Subrahamanyam, BS Mehrotra, MJ Thirulamachar. Thermomucor, a new genus of Mucorales. Ga J Sci 35:1–4, 1977.
53. MAA Schipper. Thermomucor (Mucorales). Antonie van Leeuwenhoek 45:275–280, 1979.
54. MAA Schipper. A revision of Rhizopus. I. The Rh. stolonifer group and Rhizopus. Stud Mycol 25:20–34, 1984.
55. MAA Schipper, JA Stalpers. A revision of Rhizopus. II. The Rh. microsporus group. Stud Mycol 25:20–34, 1984.

56. HJ Scholer, L Peter. Serology of mucormycosis causing fungi and of mucormycosis. In: HJ Preusser, ed. Medical Mycology. Zentralbl Bakteriol Parasitenkd Infektionskr Hyg Abt 1. suppl. 8:183–191, 1980.

57. P Neame, D Rayner. Mucormycosis: Report on twenty-one cases. Arch Pathol 70: 261–268, 1960.

58. EJ Bottone, I Weitzman, BA Hanna. Rhizopus rhizopodiformis: Emerging etiological agent of mucormycosis. J Clin Microbiol 9:530–537, 1979.

59. G Gartenberg, EJ Bottone, GT Kensch, I Weitzman. Hospital-acquired mucormycosis (Rhizopus rhizopodiformis) of skin and subcutaneous tissue. N Eng J Med 299:1115–1118, 1978.

60. L Polonelli, G Dettori, G Morace, R Rosa, M Castagnola, MAA Schipper. Antigenic studies on Rhizopus microsporus, Rh. rhizopodiformis, progeny and intermediates (Rh. chinensis). Antonie van Leeuwenhoek 54:5–18, 1988.

61. MAA Schipper, MM Maslen, CG Hogg, CW Chow, RA Samson. Human infection by Rhizopus azygosporus and the occurrence of azygospores in Zygomycetes. J Med Vet Mycol 34:199–203, 1996.

62. I Weitzman, DA McGough, MG Rinaldi, P Della-Latta. Rhizopus schipperae, sp. nov., a new agent of zygomycosis. Mycotaxon 54:217–225, 1996.

63. W Gams. A key to the species of Mortierella. Persoonia 9:381–391, 1977.

64. KH Domsch, W Gams, T-H Anderson. Compendium of soil fungi I. Eching, IHW Verlag, 1993.

65. EV Crisan. Current concepts of thermophilism and thermophilic fungi. Mycologia 65:1171–1198, 1973.

66. JMB Smith. An interesting bovine mycotic complex in New Zealand. NZ Vet J 14: 226, 1966.

67. ME di Menna, ME Carter, DO Cordes. The identification of Mortierella wolfii isolated from cases of abortion and pneumonia in cattle and a search for its infection source. Res Vet Sci 13:439–442, 1972.

68. ME Carter, DO Cordes, ME di Menna, R Hunter. Fungi isolated from bovine mycotic abortion and pneumonia with special reference to Mortierella wolfii. Res Vet Sci 14:201–206, 1973.

69. MJ Corbel, SM Eades. Cerebral mucormycosis following experimental inoculation with Mortierella wolfii. Mycopathologia 60:129–133, 1977.

70. K Wohlgemuth, WV Knudtson. Abortion associated with Mortierella wolfii in cattle. J Amer Vet Med Assoc 171:437–439, 1977.

71. RA Samson. Revision of the genus Cunninghamella. K Ned Akad Wet Versl Afd Natuurk C 72:322–325, 1969.

72. O Stadel. Über einen neuen Pilz, Cunninghamella bertholletiae. Diss. Kiel, 1911.

73. I Weitzman, MY Crist. Studies with clinical isolates of Cunninghamella. I. Mating behaviour. Mycologia 61:1024–1033, 1979.

74. SB Saksena. A new genus of the Mucorales. Mycologia 45:426–436, 1953.

75. AA Padhye, L Ajello. Simple method of inducing sporulation by Apophysomyces elegans and Saksenaea vasiformis. J Clin Microbiol 26:1861–1863, 1988.

76. L Ajello, DF Dean, RS Irwin. The zygomycete Saksenaea vasiformis as a pathogen of humans with a critical review of the etiology of zygomycosis. Mycologia 68:52–62, 1976.

77.	L Kaufman, AA Padhye, S Parker. Rhinocerebral zygomycosis caused by Saksenaea vasiformis. J Med Vet Mycol 26:237–246, 1988.
78.	MS Mathews, U Mukundan, MK Lalitha, S Aggarwal, SM Chandy, AA Padhye, EP Ewing. Subcutaneous zygomycosis caused by Saksenaea vasiformis in India: A case report and review of the literature. J Mycol Méd 3:95–98, 1993.
79.	MAA Schipper, JA Stalpers. Spore ornamentation and species concept in Syncephalastrum. Persoonia 12:81–85, 1983.
80.	A Kamalam, AS Thambiah. Cutaneous infection by Syncephalastrum. Sabouraudia 18:19–20, 1980.
81.	PD Turner. Syncephalastrum associated with bovine mycotic abortion. Nature (London) 204:399, 1964.
82.	GL Benny, JW Kimbrough. The Zygomycetes. Gainsville, FL: University of Florida, 1977.
83.	P Axelrod, KJ Kwon-Chung, P Frawley, H Rubin. Chronic cystitis due to Cokeromyces recurvatus. J Infec Dis 155:1062–1064, 1987.
84.	H Zycha, II Pilze. Mucorineae in Kryptogamenflora der Mark Brandenburg, Band VIa. Leipzig: Gebr. Borntraeger, 1935.

4

Zygomycetes

The Order Entomophthorales

Shung-Chang Jong
American Type Culture Collection, Manassas, Virginia, U.S.A.

Frank M. Dugan
USDA-ARS Western Regional Plant Introduction Station, USDA, Pullman, Washington, U.S.A.

I. INTRODUCTION

Numerous species of Entomophthorales cause disease in animals, including insects and other lower animals, but few are documented as causing disease in humans. *Basidiobolus ranarum* Eidam (= *B. haptosporus* Drechsler), *B. incongruus* Drechsler, and *Conidiobolus coronatus* (Costanin) Batko are the recorded agents of disease in humans (1–3).

Consensus on nomenclature and identity of these fungi is not universal. De Hoog and Guarro utilize *Delacroxia coronata* as the preferred name for *C. coronatus* and imply a distinction between *B. ranarum* and *B. haptosporus* (4). King avoided applying the name *B. ranarum* to human-derived isolates (5,6), and Evans and Richardson applied the name *B. meristosporus* to human-derived strains (7).

Diseases induced by these agents have received various names, the most accepted of which are entomophthoromycosis basidiobolae and entomophthoromycosis conidiobolae, or simply basidiobolomycosis and conidiobolomycosis, for the agents in *Basidiobolus* and *Conidiobolus*, respectively (8,1). Infection by *Conidiobolus incongruus* is rare (6,9), but is now known from immunocompromised patients (10). *Basidiobolus ranarum* occasionally infects humans, and very rarely other mammals; most cases are confined to tropical regions (11). Children are the principal victims; infection typically occurs in the body trunk or extremi-

This chapter is based on that by Douglas S. King in the first edition.

ties (12). *Conidiobolus coronatus* afflicts humans and other higher mammals, especially in the tropics. Infection of the nasal passages (rhinoentomophthoromycosis) is typical (4). The majority of case reports are from tropical rainforest areas in west Africa. Adult agricultural and outdoor workers are the usual victims (13). *Conidiobolus coronatus* typically inhabits decaying vegetation. *C. incongruus* has also been recovered from that substrate. *Basidiobolus ranarum* is most commonly isolated from gastrointestinal tracts or dung of amphibians or other animals, but also from soil or decayed vegetation (1,10).

II. MORPHOLOGY AND CYTOLOGY OF THE ENTOMOPHTHORALES

A. Propagative and Vegetative Structures

Conidia (asexual spores) are produced singly at the apices of simple or branched conidiophores. Most species produce two types of conidia, although *C. coronatus* can produce four types. Each type of conidium can germinate to produce vegetative mycelium, or germination may result in production of another conidium. Except in the genus *Massospora*, which lacks forcibly discharged conidia, primary conidia are forcibly discharged from conidiophores arising from vegetative elements. In *Conidiobolus*, such discharge is typically effected by pressure developed within the conidium; release of the conidium is accompanied by eversion of the conidial wall at the point of attachment to the conidiophore, resulting in a prominent basal papillum. In *Basidiobolus*, discharge is effected via pressure developed within the conidiophore, which is ruptured in the discharge process. Most genera are uniform with respect to conidial shape, nuclear number, and mode of discharge of the primary conidia. Primary conidia are phototropic; they frequently germinate to produce smaller, secondary conidia, which are also phototropic. Some species of *Entomophthora* and *Conidiobolus* produce two types of secondary conidia: One type resembles the primary conidia, while the other has a different shape (6).

In several species of *Entomophthora* and *Conidiobolus* and most species of *Basidiobolus*, primary and secondary conidia may also produce capillospores. These latter conidia, borne atop very long, thin conidiophores, are elongated and passively discharged. Multiplicative conidia (microconidia) of various sorts typify some species (e.g., *B. microsporous* and *C. incongruus*); villous conidia are produced by *C. coronatus*.

Conidia function primarily as disseminating propagules. Chlamydospores and zygospores serve as resting spores for surviving unfavorable environmental conditions. Chlamydospores develop from pre-existing vegetative structures by formation of a thick wall. Zygospores are the result of sexual reproduction; they are formed after the conjugation of equal or unequal gametangia. Homothallism (both gametangia from the same individual) is typical of the order. Azygospores are formed in some species of *Entomophthora*; these spores are formed without the prior production of gametangia (6).

The mycelium (vegetative hyphae) of entomophthorales may be composed of short segments, which Thaxter termed "hyphal bodies" (14); or true hyphae may be present. In some species of *Conidiobolus*, vegetative growth is restricted to coenocytic mycelium (15).

B. Characters of Modern Families

Classic treatments of the Entomophthorales utilized only one family (Entomophthoraceae), with up to 10 genera (6,16). Humber presented a revised classification with six families: Entomophthoraceae, Completoriaceae, Ancylistaceae, Meristacraceae, Neozygitaceae, and Basidiobolaceae (17). Characters of the primary conidia were foundational for assignment to family: morphology of the conidium and its papillar region, the number of nuclei and number of wall layers in the conidium, its mode of discharge, and morphology of conidiogenous cells and conidiophores. Table 1 provides a synopsis of the principal characters and genera for each family. Humber provides a more extensive synopsis (17); Hawksworth et al. provide a key to families (18).

III. PATHOGENIC GENERA AND SPECIES

A. Basidiobolus

1. Biology

Members of the genus *Basidiobolus* (Fig. 1) are distinguished primarily by the conidiophores and manner of discharge of the conidia and the development of the gametangia for production of zygospores. Globose conidia (Fig. 1a) frequently produce replicative conidia (Fig. 1b). Forcible discharge results when the swollen subconidial vesicle of the conidiophore ruptures and contracts. The conidiophore apex, discharged along with the conidium, usually separates in flight (Fig. 1c). This discharge of the apex of the vesicle, attached (albeit temporarily) to the conidium, is among the entomophthorales unique to *Basidiobolus* (5,6).

Another distinctive character of *Basidiobolus* is the "beaked" zygospore (Fig. 1d). Conjugation typically occurs between contiguous segments of the same hypha (even between cells of a recently divided conidium), but may also occur between different hyphae. During conjugation, each gametangium produces a protuberance; after the contents of one gametangium flow into the other, the zygospore is formed, leaving the beak-shaped protuberances attached to the side of the spore.

Table 1 Families of the Entomophthorales

Entomophthoraceae
 Conidiophores simple to dichotomously or digitately branched, with conidiogenous
  cells apical on each branch.
 Primary conidia unitunicate or bitunicate, forcibly discharged (except *Massospora*)
  by papillar eversion, or by propulsion of conidiophore contents.
 Nuclei typically 5–12 μm, staining readily with aceto-orcein or bismark brown; no
  prominent nucleolus.
 Pathogens of insects or other arthropods.
 *Batkoa, Entomophaga, Entomophthora, Erynia, Eryniopsis, Furia, Massospora, Pan-
  dora, Strongwellsea, Tarichium, Zoophthora.*
Completoriaceae
 Conidiophores simple, short, with no separate conidiogenous cells.
 Conidia unitunicate, forcibly discharged by papillar eversion.
 Nuclei large, with condensed, "granular" chromatin at interphase.
 Obligate intercellular parasites of fern gametophytes.
 Completoria
Ancylistaceae
 Conidiophores simple or infrequently branched, bearing a single terminal conidium
  per branch.
 Primary conidia unitunicate, forcibly discharged by papillar eversion.
 Nuclei small (3–5 μm), not condensed ("granular") at interphase, not staining
  strongly with aceto-orcein or bismark brown.
 Saprobes in soil or pathogens of insects or other invertebrates, or facultative para-
  sites of vertebrates.
 Ancylistes, Conidiobolus, Macrobiotophthora
Meristcraceae
 Conidiophores unbranched, each bearing several conidia.
 Primary conidia unitunicate, forcibly discharged by papillar eversion or passively dis-
  charged.
 Nuclei small (3–5 μm), with prominent nucleolus, no significant "granular" nucleo-
  plasm in interphase, not staining strongly with aceto-orcein or Bismark brown.
 Obligate pathogens of soil invertebrates, especially nematodes and tardigrades.
 Ballocephala, Meristacrum, Zygonemomyces.
Neozygitaceae
 Conidiophores typically simple, cylindrical to clavate, with apical conidiogenous
  cell.
 Conidia unitunicate, melanized, multinucleate, with small or truncate papilla, dis-
  charged by papillar eversion.
 Nuclei small (3–5 μm), condensed chromatin inconspicuous, staining poorly with
  aceto-orcein or other nuclear stains.
 Obligate pathogens of mites and insects, especially homopterans.
 Neozygites, Thaxterosporium.

Table 1   Continued

Basidiobolaceae
  Conidiophores simple or seldom branched, with a swollen apex.
  Primary conidia unitunicate and uninucleate forcibly discharged with circumsessile
    rupture of conidiophore.
  Nuclei large (typically >10 µm), with no condensed "granular" chromatin during
    interphase, not staining strongly with aceto-orcein or Bismark brown.
  Saprobes in soil, colonizers of the guts of amphibians or reptiles, or facultative para-
    sites in vertebrates.
  Basidiobolus.

With the exception of *B. microsporus*, species of *Basidiobolus* produce capillospores. Capillospores of *Basidiobolus* are distinguished by a mucilaginous apex (Fig. 1e). The microspores of *B. microsporus* are morphologically similar to capillospores, but differ both developmentally and in size, and they lack an adhesive apex.

In nature, species of the genus are characteristically associated with the gut contents of reptiles and amphibians, but most have also been isolated from plant detritus and can grow on common mycological media (19,20–21). *B. microsporus* is associated with rodent dung; *B. magnus* with decayed plant material (22).

Strains of *Basidiobolus* exhibit a tendency for loss of diagnostic characters (conidia and zygospores) when the strains are maintained in culture. Strains should be preserved in liquid nitrogen vapor as quickly as possible after isolation.

2. Taxonomy

Modern literature on basidiobolomycocis most frequently uses the names *Basidiobolus ranarum* or *B. haptosporus*. We prefer *B. ranarum*, with synonyms as follows (23):

> *Basidiobolus ranarum* Eidam, 1887
> = *B. meristosporus* Drechser, 1955
> = *B. heterosporus* Srinivasan & Thirumalachar, 1965
> = *B. haptosporus* Drechsler, 1947
> = *B. haptosporus* var. *minor* Srinivasan & Thirumalachar, 1965

Basidiobolus is the sole genus of the Basidiobolaceae (Table 1). In addition to characters listed in Table 1, the production of "beaked" zygospores (Fig. 1d) is diagnostic for the genus. The large nuclei, with prominent nucleoli, are also distinctive (17,24,25,26). King (6) found the traditional morphological and other criteria used for species separation less than satisfactory: odor and the texture of the zygospore wall, both primary characters for species separation, were not

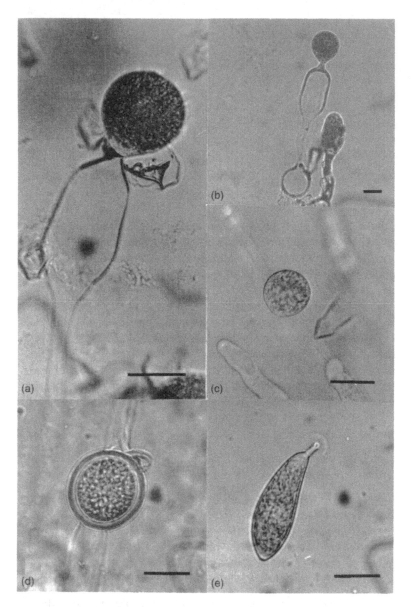

Figure 1 *Basidiobolus ranarum*. Bars = 20 μm. (a) Primary conidium on conidiophore; (b) replicative conidium on conidiophore arising from another conidium; (c) discharged conidium and conidiophore apex; (d) mature zygospore with "beak"; (e) capillospore with adhesive material at apex. (Source: Ref. 6.)

consistently reliable, nor was optimum temperature for growth. Kwon-Chung and Bennett reviewed attempts at species separation via antigenic studies, restriction analysis of rDNA, and isozyme banding (1). They concluded that "the problem of speciation of *Basidiobolus* is still unresolved," but that "it is reasonable to accept *B. ranarum* as the pathogen that infects man." Species other than those in the above synonymy are *B. magnus*, *B. philippinensis*, and *B. microsporus*. King (6) considered *B. microsporus*—but not *B. philippinensis*—distinct from *B. haptosporus*. He did not place *B. ranarum* or *B. magnus* into synonymy with *B. haptosporus* (6). Natural infection of mammals other than humans is rare (6).

3. Infection, Diagnosis, and Treatment

Basidiobolus ranarum most probably gains entrance to human tissues via insect bites and minor trauma (1). Basidiobolomycosis occurs almost exclusively in children. An enlarged subcutaneous mass is the primary symptom, with the upper limbs and buttocks most frequently involved (7). Tissue for biopsy should be recovered from the growing nodules on the border of the subcutaneous masses. Tissue for biopsy should be inoculated onto culture media without delay; it should not be kept for any extended time in a refrigerator (27). Biopsy tissue can be mounted in potassium hydroxide, teased apart, and gently heated prior to microscopic examination; an eosinophilic, sleevelike coating typically surrounds broad hyphae embedded in the tissue. Sabouraud glucose agar with antibacterial antibiotics but without cycloheximide is normally used for isolation. Fast-growing colonies are initially whitish, later becoming buff or gray, with a waxy, radiate surface. Some colonies have a *Streptomyces*-like odor (1,3,7). Immunofluorescence techniques are of growing importance in diagnosis (28,29). Treatment is with saturated solutions of potassium iodide, or more rarely with ketoconazole (1,30).

B. Conidiobolus

1. Biology

King extensively described the taxonomy of the genus *Conidiobolus* (15,31,32). All species display forcible discharge of conidia via pressure developed within the conidium. Conidia are produced atop unbranched conidiophores (Figs. 2a, 2b). After discharge, each conidium displays a papillum (Figs. 3c, 3b) because of eversion of the wall adjacent to the condiophore. Multiplicative conidia (Fig. 3c) are common to several species, including *C. incongruus*. As many as 40 are produced from a single conidium. They are forcibly discharged but are not phototropic. Only *Conidiobolus coronatus* produces villous conidia (Fig. 2d). These conidia may be produced from primary or replicative conidia by the production of the thick wall and villous appendages. Chlamydospores (Fig. 2e) are

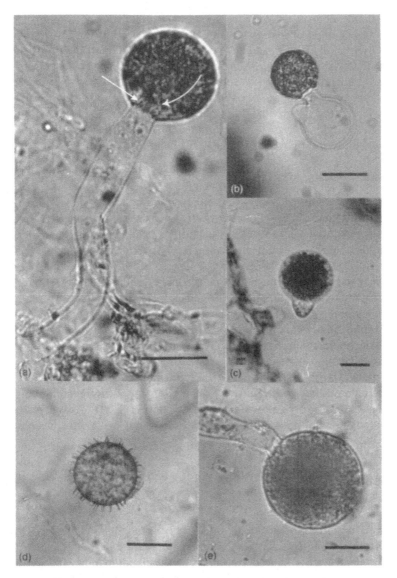

Figure 2 *Conidiobolus coronatus.* Bars = 20 μm. (a) Primary conidium on conidiophore arising from vegetative mycelium. Arrows indicate dome-shaped septum separating
conidium and conidiophore that splits during discharge of the conidium; (b) replicative
conidium on conidiophore arising from a globose (either primary or replicative) conidium;
(c) discharged globose conidium; note rounded shape of papillum (see Fig. 3b); (d) relatively thin-walled villose conidium; (e) chlamydospore arising from hyphae in agar.
(Source: Ref. 6.)

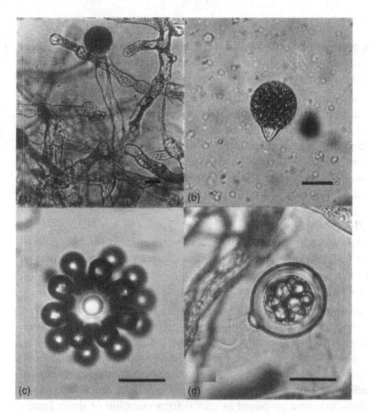

Figure 3 *Conidiobolus incongruus.* Bars = 20 μm. (a) Almost mature primary conidium on conidiophore arising from vegetative mycelium; (b) globose discharged conidium;
note pointed shaped of papillum (see Fig. 2c); (c) Multiplicative conidia on radially proj-
ecting sterigmata (= microconidiophores) arising from a discharged primary conidium;
(d) mature zygospore. The zygospore lies within the persistent wall of one of the gametangia. The other gametangium is not visible. (Source: Ref. 6.)

reported from several species, including some strains of *C. coronatus.* Zygospores (Fig. 3d) are known in several species, including *C. incongruus,* but not
C. coronatus (6,7)

   C. coronatus and other members of the genus are consistently isolated from
plant debris, although some, including *C. coronatus,* are also isolated from the
fruiting bodies of higher fungi or from insects. *C. coronatus* is cosmopolitan and
frequently isolated, but *C. incongruus* is infrequently isolated. The tendency to
loss of sporulation ability in culture is less than that of *Basidiobolus* strains, but
is still a problem: Some human-derived isolates fail to produce villose conidia
(5).

2. Taxonomy

There is general consensus on the identity of the agents of conidiobolomycosis in humans: *Conidiobolus incongruus* and the more common *C. coronatus*. Our nomenclator for these agents follows King (5):

> *Conidiobolus incongruus* Drechsler, 1960
> = *Conidiobolus gonimodes* Drechsler, 1961
> *Conidiobolus coronatus* (Costantin) Bakto, 1964
> = *Boudierella coronata* Costantin, 1897
> = *Delacroixia coronata* (Costantin) Saccardo & Sydow, 1899
> = *Entomophthora coronata* (Costantin) Kevorkian, 1937
> = *Conidiobolus coronatus* (Costantin) Srinivasan & Thirumalachar, 1964
> = *C. villosus* Martin, 1925
> Conidiobolomycosis in horses usually results from infection by *C. coronata*, but *C. lamprauges* has also been implicated in infection of horses:
> *Conidiobolus lamprauges* Drechsler, 1953
> = *C. nanodes* Drechler, 1955

In classic taxonomy, *Conidiobolus* was separated from *Entomophthora* by the saprophytic habit of the former and the insect-pathogenic habit of the latter. *C. coronata*, being at least occasionally isolated from insects, was placed in *Entomophthora*, but more species of *Entomophthora* have been cultured, and more species of *Conidiobolus* have been isolated from insects. Although characters of the primary conidiophore (macronematous for *Entomophthora* versus micronematous for *Conidiobolus*) were proposed as useful for separation of these genera (33), King regarded the criteria as unreliable (26). Humber proposed cytological critera (Table 1) (17). Recent investigations of sterol content suggest separation of *C. coronatus* (as *Delacroixia coronatus*) from other fungi in *Entomophthora*, *Basidiobolus*, and *Conidiobolus* (34).

Conidiobolus coronatus is a common saprophyte in plant debris, an occasional pathogen of insects, and a well-known pathogen of horses and humans, especially in the tropics (1,35,36). Although most reports address infections of horses or humans, *C. coronatus* has also caused infection in chimpanzees (37,38). There is evidence that *C. coronatus* can sporulate in host tissue (39). Recent research has focused on proteases and lipases, which may be of importance with regard to pathogenesis (13,40,41).

Conidiobolus incongruus, also a saprophyte in plant debris, is a pathogen of humans, and has now been diagnosed as causing rhinocerebral and nasal zygomycosis in sheep (42). Infections in humans are very rare (9,43,44). *C. lamprauges* Dreschler, found in plant debris and occasionally isolated from insects, has been isolated from a nasopharyngeal nodule of a horse (45). King provided a key to 26 species of *Conidiobolus* (32).

3. Infection, Diagnosis, and Treatment

Infection of nasal mucosa and sinuses is typical for conidiobolomycosis; disease is frequently fatal in immunocompromised individuals (3). The method of entry is uncertain, although traumatized nasal mucosa is a possibility. Obstruction of nasal passages, rhinorrhea, and nosebleeds are common symptoms. Edema may produce conspicuously swollen areas in or adjacent to sinuses: "Relentless progression over many years is the rule" (1). Culture of tissue from biopsy should be inoculated onto Sabouraud glucose agar without cycloheximide. Colonies are fast-growing, glabrous or powdery, white becoming beige to brown, flat, with a pale reverse (2,7). Immunodiffusion has been successfully employed for serodiagnosis (46). Although *C. coronatus* is susceptible to amphotericin B, and although iodides, ketoconazole, fluconazole, or surgery have resulted in instances of improvement in individual patients, there is no consistently satisfactory treatment for conidiobolomycosis (1,47,48). Occasional instances of successful therapy include amphoteracin B plus terbinafine (49), fluconazole (50), and ketoconazole plus potassium iodide (51).

REFERENCES

1. KJ Kwon-Chung, JE Bennett. Medical Mycology. Philadelphia: Lea & Febiger, 1992.
2. G St-Germain, R Summerbell. Identifying Filamentous Fungi: A Clinical Laboratory Handbook. Belmont, CA: Star, 1996.
3. DA Sutton, AW Fothergill, MC Rinaldi. Guide to Clinically Significant Fungi. Baltimore: Williams & Wilkins, 1998.
4. GS de Hoog, J Guarro. Atlas of Clinical Fungi. Baarn, The Netherlands: Centraalbureau voor Schimmelcultures, and Reus, Spain: Universitat Rovira I Virgili, 1995.
5. DS King. Systematics of fungi causing entomophthoromycosis. Mycologia 71:731–745, 1979.
6. DS King. Entomophthorales. In: DH Howard, ed. Fungi Pathogenic for Humans and Animals. Part A: Biology. New York: Marcel Dekker, 1983, pp. 61–72.
7. EGV Evans, MD Richardson. Medical Mycology: A Practical Approach. Oxford, UK: Oxford University Press, 1989.
8. CW Emmons, CH Binford, JP Utz, KJ Kwon-Chung. Medical Mycology. 3rd ed. Philadelphia: Lea & Febiger, 1977.
9. DS King, SC Jong. Identity of the etiological agent of the first deep entomophthoraceous infection of man in the United States. Mycologia 68:181–183, 1976.
10. TJ Walsh, G Renshaw, J Andrews, J Kwon-Chung, RC Currion, HI Pass, J Taubenberger, W Wilson, PA Pizzo. Invasive zygomycosis due to *Conidiobolus incongruus*. Clin Infec Dis 19:423–430, 1994.
11. SR Davis, DH Ellis, P Goldwater, S Dimitriou, R Byard. First culture-proven Austra-

lian case of entomophthoromycosis caused by *Basidiobolus ranarum*. J Med Vet
Mycol 32:225–230, 1994.

12. S Sood, S Sethi, U Banarjee. Entomophthoromycosis due to *Basidiobolus hap-tosporus*. Mycoses 40:345–346, 1997.

13. HC Gugnani. Entomophthoromycosis due to *Conidiobolus*. Eur J Epidem 8:391–396, 1992.

14. R Thaxter. The Entomophthoraceae of the United States. Mem Boston Soc Nat Hist 4:133–201, 1888.

15. DS King. Systematics of Conidiobolus (Enthomophthorales) using numerical taxonomy. III. Descriptions of recognized species. Can J Bot 55:718–729, 1977.

16. DS King, RA Humber. Identification: Entomophthorales. In: HD Burges, ed. Microbial Control of Insects, Mites, and Plant Diseases. vol. 2. New York: Academic, 1982.

17. RA Humber. Synopsis of a revised classification for the Entomophthorales (Zygomycotina). Mycotaxon 34:441–460, 1989.

18. DL Hawksworth, PM Kirk, BC Sutton, DN Pegler. Ainsworth & Bisby's Dictionary of the Fungi. 8th ed. Wallingford, Oxon, UK: CAB International, 1995.

19. J Coremans-Pelseneer. Biologie des champignons du genre *Basidiobolus* Eidam 1886. Saprophytisme et pouvoir pathogène. Acta Zool Pathol 60:1–143, 1974.

20. JA Hutchison, MA Nickerson. Comments on the distribution of *Basidiobolus ranarum*. Mycologia 62:585–587, 1970.

21. MA Nickerson, JA Hutchison. The distribution of the fungus *Basidiobolus ranarum* Eidem in fish, amphibians and reptiles. Am Midl Naturalist 86:500–502, 1971.

22. JA Hutchison, DS King, MA Nickerson. Studies on temperature requirements, odor production and zygospore wall undulation of the genus *Basidiobolus*. Mycologia 64:467–474, 1972.

23. SC Jong, JM Birmingham, G Ma. Stedman's ATCC Fungus Names. Baltimore: Williams & Wilkins, 1993.

24. K Gull, APJ Trinci. Nuclear division in *Basidiobolus ranarum*. Trans Brit Mycol Soc 63:457–460, 1974.

25. CF Robinow. Observations on cell growth, mitosis, and division in the fungus *Basidiobolus ranarum*. J Cell Bio 17:123–152, 1963.

26. NC Sun, CC Bowen. Ultrastructural studies of nuclear division in *Basidiobolus ranarum* Eidam. Caryologia 25:471–494, 1972.

27. DP Burkitt, AMM Wilson, DB Felliffe. Subcutaneous phycomycosis. Brit Med J 1: 1669–1672, 1964.

28. G Michel, P Ravisse, J Lohoue-Petmy, JP Steinmetz, C Winter, A Mbakop, P Ave, MA Ruffaud. Five new cases of entomophthoromycosis observed in Cameroon: Role of immunofluorescence in their diagnosis. (in French). Bull Soc Path Exot 85:10–16, 1992.

29. JL Pecarrere, M Huerre, P Lafond, P Esterre, C Raharisolo, P De Rotalier. Entomophthoromycoses in Madagascar. (in French). Arch Inst Pasteur Madagascar 61: 99–102, 1994.

30. Z Nazir, R Hasan, S Pervaiz, M Alam, F Moazam. Invasive retroperitoneal infection due to *Basidiobolus ranarum* with response to potassium iodide–Case report and review of the literature. Ann Trop Paediatr 17:161–164, 1997.

31. DS King. Systematics of Conidiobolus (Enthomophthorales) using numerical taxonomy. I. Biology and cluster analysis. Can J Bot 54:45–65, 1976.

32. DS King. Systematics of Conidiobolus (Enthomophthorales) using numerical taxonomy. II. Taxonomic considerations. Can J Bot 54:1285–1296, 1976.

33. MC Srinivasan, MJ Narasimhan, MJ Thirumalachar. Artificial culture of *Entomophthora muscae* and morphological aspects for differentiation of the genera *Entomophthora* and *Conidiobolus*. Mycologia 56:683–691, 1964.

34. JD Weete, SR Gandhi. Sterols of the phylum Zygomycota: Phylogenetic implications. Lipids 32:1309–1316, 1997.

35. PF Jungerman, RM Schwartzman. Veterinary Medical Mycology. Philadelphia: Lea & Febiger, 1972.

36. DT Zamos, J Schumacher, JK Loy. Nasopharyngeal conidiobolomycosis in a horse. J Amer Vet Med Assoc 208:100–101, 1996.

37. AD Roy, HM Cameron. Rhinophycomycosis entomophthorae occurring in chimpanzee in the wild in East Africa. Amer J Trop Med Hyg 21:234–237, 1972.

38. R Vanbreuseghem, Ch de Vroey, M Takashio. Practical Guide to Medical and Veterinary Mycology. 2nd ed. New York: Masson, 1978.

39. A Kamalam, AS Thambiah. Lymph node invasion by *Conidiobolus coronatus* and its spore formation in vivo. Sabouraudia 16:175–184, 1978.

40. JI Okafor. Purification and characterization of protease enzymes of *Basidiobolus* and *Conidiobolus* species. Mycoses 37:265–269, 1994.

41. S Phadtare, M Rao, V Deshpande. A serine alkaline protease from the fungus *Conidiobolus coronatus* with a distinctly different structure than the serine protease subtilisin Carlsberg. Arch Microbiol 166:414–417, 1996.

42. PJ Ketterer, MA Kelly, MD Connole, L Ajello. Rhinocerebral and nasal zygomycosis in sheep caused by *Conidiobolus incongruus*. Austr Vet J 69:85–87, 1992.

43. R Busapakum, U Youngchaiyud, S Sriumpal, G Segretain, H Fromentin. Disseminated infection with *Conidiobolus incongruus*. Sabouraudia 21:323–330, 1983.

44. EF Gilbert, GH Khoury, RS Pore. Histopathological identification of *Entomophthora* phycomycosis. Arch Pathol 90:583–587, 1970.

45. RA Humber, CC Brown, RW Kornegay. Equine zygomycosis caused by *Conidiobolus lamprauges*. J Clin Microbiol 27:573–576, 1989.

46. L Kaufman, L Mendoza, PG Standard. Immunodiffusion test for serodiagnosing subcutaneous zygomycosis. J Clin Microbiol 28:1887–1890, 1990.

47. AR Costa, E Porto, JRP Pegas, VMS dos Reis, MC Pires, C da S. Lacaz, MC Rodrigues, H Muller, LC Cuce. Rhinofacial zygomycosis caused by Conidiobolus coronatus: A case report. Mycopathologia 115:1–8, 1991.

48. A Restrepo. Treatment of tropical mycoses. J Am Acad Derm 31:S91–102, 1994.

49. NT Foss, MR Rocha, VT Lima, MA Velludo, AM Roselino. Entomophthoromycosis: Therapeutic success by using amphotericin B and terbinafine. Dermatology 193:258–260, 1996.

50. HC Gugnani, BC Ezeanolue, M Khalil, CD Amoah, EU Ajuiu, EA Oyewo. Fluconazole in the therapy of tropical deep mycosis. Mycoses 38:485–488, 1995.

51. D Mukhopadhyay, LM Ghosh, A Thammayya, M Sanyal. Entomophthoromycosis caused by *Conidiobolus coronatus*: Clinicomycological study of a case. Auris Nasus Larynx 22:139–142, 1995.

5

Onygenales

Arthrodermataceae

Dexter H. Howard
UCLA School of Medicine, Los Angeles, California, U.S.A.

Irene Weitzman
*Columbia University College of Physicians and Surgeons, New York,
New York, and Arizona State University, Tempe, Arizona, U.S.A.*

Arvind A. Padhye
*National Center for Infectious Disease, Centers for Disease Control
and Prevention, Atlanta, Georgia, U.S.A.*

I. INTRODUCTION

The family Arthrodermataceae of the order Onygenales comprises human and animal pathogens and a number of nonpathogenic forms isolated from soil and various keratin-containing substrates such as birds' nests and animal fur that are not known to cause infections. The pathogens are referred to as dermatophytes, a term that designates a large, closely related group of keratinophilic fungi that cause infections of the skin, hair, and nails. Most of these fungi were originally described as Hyphomycetes (anamorphs), but several are now known to have a perfect state (teleomorph) placed in the genus *Arthroderma* in the family Arthrodermataceae of the order Onygenales.

The infections produced by the dermatophytes, called dermatophytoses (syn. tinea, ringworm), are generally restricted to the nonliving, cornified layers of the skin and its appendages (1, 1a, 2), but occasionally the dermis and subcutaneous tissue may be involved in a form of the disease known as Majocchi's granuloma, and pseudomycetomas have been described (1–5). Although the pen-

141

etration of tissue by the dermatophytes is usually quite superficial, adsorption of products from the fungi leads to sensitization of the host, an event that is manifested by specific delayed hypersensitivity and other sorts of allergic responses. Generalized cutaneous infections (6, 7), primary invasive cutaneous infections (8, 9), and more rarely disseminated systemic infections have been reported in immunocompromised patients (10).

II. HISTORICAL ANTECEDENTS

The development of knowledge about the ringworm fungi parallels that of medical mycology in general, for these microorganisms were among the first to be identified as human and animal pathogens (11). The story, with variations, has been told on several occasions (11–13). Briefly, recognition of the dermatophytes began in 1837 with the work of Robert Remak, who found arthroconidia and hyphae in the hair shafts of a child with favus. Remak did not publish his findings at this time but allowed them to be incorporated into the thesis of a close friend, Xavier Hube (11, 12). It has been said that Remak did not recognize the importance of his discovery (12) and that it remained for Lucas Schoenlein to claim favus as a mycotic disease (14). In doing this Schoenlein was strongly influenced by Bassi's classic work on the fungal nature of the muscardine disease of silkworms (12). In 1840, Remak published his observations on the "plant-like nature" of the structures he had observed in the hair shafts of patients with favus (15), and he subsequently showed that the infection was contagious by successfully inoculating himself (16). In 1845, he described the fungus of favus more completely and named it in honor of Schoenlein (12, 17). Some have interpreted this as an effort to curry favor with Schoenlein in order to obtain a university chair, an appointment which at that time was generally denied to Jews (18). If this were the motivation for the somewhat unusual behavior of Robert Remak, the effort was not totally successful, for he was appointed "extraordinary professor" (but not full professor, which he wanted) at the University of Berlin. To his credit he eschewed the obvious alternative of a certificate of baptism to achieve his goal (12).

David Gruby also refused to buy a university chair with a certificate of baptism (12). Gruby published his thesis in Vienna and then settled down in Paris. From 1841 to 1844 he published a series of papers that are generally acknowledged to have established him as the founder of medical mycology (11, 13, 19). In the first two papers of this series Gruby described, independently of Remak and Schoenlein, the fungal nature of favus (20, 21). The next year he gave an account of an ectothrix dermatophytosis of the beard (22), and although he did not name the fungus, it was subsequently described as *Microsporon (Trichophyton) mentagrophtes* by Charles Robin (23). In 1843, Gruby (24) published his account of classic epidemic ringworm of the scalp and named the etiologic agent *Microsporum audouinii* in honor of Victor Audouin, director of the Paris Natural History Museum. His final published work on dermatophytes was one in which he established the etiologic relation of *T. tonsurans* Malmsten, 1848 to endothrix dermatophytosis (25). Although extraneous to the subject of this chapter, it should be noted that Gruby also described the fungal nature of thrush (26). Interestingly, Gruby made no further mycological observations after 1844, but concentrated on his medical practice in Paris and attended many notables, including George Sand, Alexander Dumas (both father and son), Frederic Chopin, Franz Liszt, and Heinrich Heine (27).

In the middle of the nineteenth century there arose a concept that caused considerable confusion for a number of years. The work of Anton De Bary (28) and that of the Tulasne brothers (29) had established the fact of polymorphism among certain fungi, notably the parasites of cereals. The term polymorphism denoted that one fungus could appear in different forms in the various stages of its life cycle. This concept came to be widely and often inappropriately applied with the disastrous consequences that all fungi and bacteria were thought to be much the same and simply developmental stages of one another (28).

The result of this extension of the concept of polymorphism to those fungi in which it played no role for a time obscured the specific etiology of infectious diseases (30). The one person who was important in rescuing the idea of the specificity of cause in the mycoses was Raimond Sabouraud, who insisted that there were different species of dermatophytes, which he placed in four genera: *Achorion, Epidermophyton, Microsporum,* and *Trichophyton* (31). In spite of the firm foundation laid by Sabouraud, however, the literature on dermatophytes subsequently became confused by investigators who went too far in the opposite direction and based species identification on small variations in the clinical appearance of lesions or on slight differences in colonial morphology. The situation became so muddled that in 1935 some 118 species of dermatophytes were described by C. W. Dodge in his monographic consideration of medical mycology (32).

Emmons proposed the basis for the current taxonomic treatment of the dermatophytes (33). He adopted a rigorous approach, based on accepted rules of nomenclature, to the classification of the ringworm fungi and established three genera (*Microsporum, Trichophyton,* and *Epidermophyton*) to contain all the species considered legitimate. Emmons's proposal is widely accepted as valid (1, 2), although alternative schemes are suggested from time to time.

The existence of a sexual phase of growth among certain dermatophytes was indicated by the observations of a number of early workers. (See Ref. 34 for a review of the literature.) Nannizzi (35) described cleistothecia in a strain of *Sabouraudites (Microsporum) gypseum.* He named the isolate *Gymnoascus gypseus.* Nannizzi's work was generally discredited because it was thought that he dealt with contaminated substrates. His observations were proven to be correct by Griffin (36) and Stockdale (34), however, and the existence of a sexual form of growth was also demonstrated for species in the genus *Trichophyton* by the results of studies by Dawson and Gentles (37, 38).

The rediscovery of a sexual phase of growth among the dermatophytes led to a flurry of taxonomic and genetic studies. Few other groups of medically important fungi have been subjected to review as often as have the dermatophytes (39, 40). Accordingly, in this chapter the major emphasis is placed on directing the reader to appropriate reviews in which more comprehensive coverage of a given topic is presented.

III. DERMATOPHYTE MEMBERS OF THE FAMILY ARTHRODERMATACEAE

A. General Considerations

1. Definitions

The dermatophytes comprise a large, closely related group of fungi that cause infections of the hair, skin, and nails (1, 2). The word tinea is used to designate these infections, known colloquially as ringworm, and an adjective is appended to that word to indicate the primary area of involvement. For example, tinea capitis is ringworm of the scalp, tinea barbae is ringworm of the beard, tinea corporis is ringworm of the general body surfaces, tinea cruris is ringworm of the groin, tinea pedis is ringworm of the feet, and tinea unguium is ringworm of the nails (1, 2, 40, 41). The diseases may also be termed collectively dermatophytoses (1, 2, 40).

The species of dermatophytes covered in this chapter are listed in Table 1.

2. Identification

The dermatophytes are identified primarily on the basis of their appearance in materials from the host and in culture (1, 40, 45–47).

The Appearance of Dermatophytes in the Tissues of the Host.   Direct examination of materials from a host infected with a dermatophyte not infrequently reveals the fungus. Specimens of skin, nails, and hair from suspicious lesions are placed on a slide in a drop of 10–20% potassium hydroxide (KOH) or in KOH with Calcofluor white (47). A coverslip is placed over the drop. The preparation is heated gently over a low flame and flattened before being examined microscopically. The various species of dermatophytes look alike in skin scales and nail scrapings (1). The appearance is that of septate, branching hyphae that may break up into barrel-shaped or rounded arthroconidia (1). Invasion of the hair may be

Table 1 Anamorph-Teleomorph States of the Dermatophytes

Anamorphs	Teleomorphs
Epidermophyton	Unknown
E. floccosum	Unknown
Microsporum	*Arthroderma*
M. audouinii	Unknown
M. canis var. *canis*	*A. otae*
M. canis var. *distortum*	*A. otae*
M. cookei	*A. cajetani*
M. equinum	Unknown
M. ferrugineum	Unknown
M. fulvum	*A. fulva*
M. gallinae	Unknown
M. gypseum	*A. gypseum*
M. gypseum	*A. incurvatum*
M. nanum	*A. obtusum*
M. persicolor	*A. persicolor*
M. praecox	Unknown
M. racemosum	*A. racemosum*
M. vanbreuseghemii	*A. grubyi*
Trichophyton[a]	*Arthroderma*
T. concentricum	Unknown
T. equinum var. *equinum*	Unknown
T. equinum var. *autotrophicum*	Unknown
T. gourvilii	Unknown
T. megninii	Unknown
T. mentagrophytes var. *erinaceae*	*A. benhamiae*
T. mentagrophytes var. *interdigitale*	*A. benhamiae*, *A. vanbreuseghemii*
T. mentagrophytes var. *mentagrophytes*	*A. benhamiae*
T. mentagrophytes var. *quinckeanum*	*A. benhamiae*
T. rubrum	Unknown
T. schoenleinii	Unknown
T. simii	*A. simii*
T. soudanense	Unknown
T. tonsurans var. *tonsurans*	Unknown
T. tonsurans var. *sulfureum*	Unknown
T. verrucosum var. *album*	Unknown
T. verrucosum var. *ochraceum*	Unknown
T. verrucosum var. *discoides*	Unknown
T. violaceum	Unknown
T. yaoundei[b]	Unknown

[a] *T. raubitschekii* (42) and *T. kanei* (43) would appear to be variants of *T. rubrum* which "is a highly variable species" (1). *T. krajdenii* (44) is excluded because some consider it to be *T. mentagrophytes* var. *nodulare*. This variety of *T. mentagrophytes* has not been validated, but isolates of *T. krajdenii* mate with *A. benhamiae* (A. A. Padhye, unpublished data).

[b] Not validly described.

either of the ectothrix or endothrix sort (1). Details of the appearance of different species of dermatophytes in and on the surface of hair will be given with the description of individual species. (For further details on direct microscopic examinations see Refs. 40, 45–47.)

Wood's Light. A Wood's lamp emits ultraviolet light at approximately 365 nm. Hairs invaded by some species of *Microsporum* emit a greenish-yellow fluorescence. Hairs involved by most species of *Trichophyton* do not fluoresce, but *T. schoenleinii*-infected hairs emit a dull bluish-white fluorescence. *Epidermophyton* spp. ordinarily do not invade hair. (See discussion of *E. floccosum*, Section IB1a.) Infected skin scales and nails do not fluoresce.

Cultures. Specimens from lesions suspected of being tinea are cultured. Several recipes for suitable media are available (40). Some variation of glucose peptone agar with inhibitors is generally employed. One such recipe that is commonly used is Sabouraud peptone glucose agar (Emmons's modification) with cycloheximide and chloramphenicol (40). The colonial appearance will form a part of the description to be given for each species considered. (For other useful culture media see Refs. 40, 45–48.)

Generally the dermatophytes are identified solely on the basis of their colonial and microscopic morphology, and this will be the approach taken in the description of taxonomy to follow. Physiological tests may also be of great value in their identification, however. For example, it was recognized early on that the inability of *M. audouinii* to grow on rice grains served to distinguish it from *M. canis* and *M. gypseum* (49). The classic work of Lucille Georg and her colleagues led eventually to the development of a series of nutritional tests that could be used to augment species recognition among the *Trichophyton* spp. (40, 41, 46, 47). A series of *Trichophyton* agars for distinguishing among *Trichophyton* spp. is available. (For a discussion see Refs. 40, 47.) Ancillary diagnostic approaches include (1) the ability to perforate hair in vitro (40), (2) urease production (46), and (3) growth at various temperatures (e.g., *T. verrucosum* grows optimally at 37°C). More recently the techniques of molecular biology have been applied to taxonomic decisions. (For a review see ref. 40.)

3. Epidemiology

The natural occurrence of dermatophytes varies among the species. Typical associations are generally recognized, although it is not always easy to describe a species as belonging to only one category (1, 2). *Anthropophilic* species are obligate parasites of humans. *Zoophilic* species infect humans but are predominantly found on other animals (cattle, cats, dogs, etc). *Geophilic* species are commonly found in soil (1, 2). Among the geophilic species are those of *Microsporum*, *Epidermophyton*, and *Trichophyton*, which have not been found to cause disease.

Table 2 Anamorph-Teleomorph States
of the Saprophytic Arthrodermataceae

Anamorphs	Teleomorphs
Epidermophyton	Unknown
E. stockdaleae	Unknown
Microsporum	*Arthroderma*
M. amazonicus	*A. borellii*
M. boullardii	*A. corniculatum*
M. magellanicum	Unknown
M. ripariae	Unknown
Microsporum sp.	*A. cookiellum*
Trichophyton	*Arthroderma*
T. ajelloi[a]	*A. uncinatum*
T. fischeri	Unknown
T. flavescens	*A. flavescens*
T. georgiae	*A. ciferrii*
T. gloriae	*A. gloriae*
T. longifusum	Unknown
T. mariatii	Unknown
T. phaseoliforme	Unknown
T. terrestre	*A. insingulare*,
	A. lenticularum,
	A. quadrifidum
T. vanbreuseghemii	*A. gertleri*

[a] There are reports of infections caused by *T. ajelloi*, but these
isolations remain doubtful.

Because of their morphological similarity to pathogens, they are members of the
Arthrodermataceae and are included in this chapter (Tables 1, 2).

Some of the dermatophytes have a very restricted geographic distribution,
whereas others are cosmopolitan. Examples of each are recorded in Table 3. An
excellent review of this topic has appeared recently (40).

4. Taxonomy

Prior to 1961 the taxonomy of the dermatophytes was based on the anamorphic
forms. The practical needs of the clinical laboratory also involve observations
on this stage for identification. The customary arrangement of a discussion of
the dermatophytes is thus based on the conidia of the asexual phase of growth
even when comprehensive coverage is also given the teleomorphic forms (1, 2,
40, 41, 45).

Table 3 Geographic Distribution of Dermatophytes

Anthropophilic	Zoophilic	Geophilic
Cosmopolitan		
E. floccosum	M. canis var. canis	M. cookei
M. audouinii	M. equinum	M. fulvum
T. mentagrophytes	M. gallinae	M. gypseum
var. interdigitale	T. equinum	M. nanum
T. rubrum	T. mentagrophytes	M. persicolor
T. tonsurans	var. mentagrophytes	T. ajelloi
var. tonsurans	T. verrucosum	
var. sulfureum	var. album	
	var. ochraceum	
	var. discoides	
Limited		
T. concentricum	M. canis var. distortum	M. praecox
M. ferrugineum	T. equinum	M. racemosum
T. gourvilii	var. autrophicum	M. vanbreuseghemii
T. megninii	T. mentagrophytes	T. simii
T. schoenleinii	var. erinacei	
T. soudanense	var. quinckeanum	
T. violaceum		
T. yaoundei		

In 1934, Emmons proposed that the dermatophytes be classified in three genera: *Epidermophyton*, *Microsporum*, and *Trichophyton* (33). These genera were distinguished by the morphology of the large, multiseptate macroconidia that each produced. A second type of conidia, the small unicellular, spherical, oval, or pyriform microconidia, was used to some degree for differentiation of species. The scheme was widely adopted and remains the dominant one in the field today (1, 2, 40, 41, 44–46).

The morphological features that distinguish the three genera are shown in Figure 1. The macroconidia of *Epidermophyton* are smooth-walled, clavate, and borne in clusters of two or three; those of *Microsporum* are rough-walled, mostly spindle-shaped, and usually borne singly; and those of *Trichophyton* are smooth-walled, mostly cylindrical, and borne singly.

B. Introduction to the Arthrodermataceae

The morphology of the anamorphic (asexual) states of the dermatophytes suggested to earlier investigators that these fungi were closely related to the Gymnoascaceae. The discovery of the teleomorphic (sexual) state of the der-

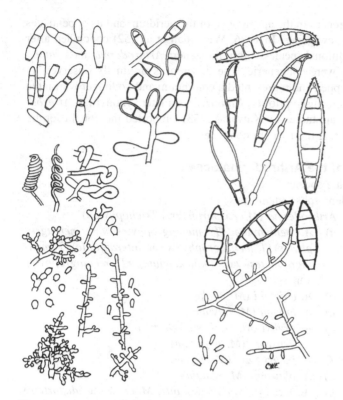

Figure 1 Spore types in the three genera of dermatophytes. Source: From Ref. 33, p. 337.

matophytes revealed them to be Ascomycetes, and the appearance of their peridial hyphae (hyphae making up the outer covering wall of the fruiting body) reinforced their affiliation with the Gymnoascaceae. Currah's thorough study of the taxonomy of the Onygenales (51), however, resulted in his revision of this order by creating four families—Onygenaceae, Arthrodermataceae, Myxotrichaceae, and Gymnoascaceae—on the basis of expanded characters rather than earlier classification schemes that placed primary emphasis on the characteristics of the peridium and its appendages. Currah's major distinction between the families in the Onygenales was based on ascospore morphology, the peridium and its appendages, the nature of occurrence of the anamorphs, and the substrate of habitat. The teleomorphic genera of the dermatophytes, *Arthroderma* and *Nannizzia*, were placed in the Arthrodermataceae rather than the Gymnoascaceae.

The genus *Nannizzia* was created by Stockdale in 1961 (34) to accommodate *N. incurvata*, the newly described teleomorph of *Microsporum gypseum*. Although she considered *N. incurvata* to most closely resemble *Arthroderma curreyi*, enough differences in the appearance of the peridium and its appendages warranted creating a new genus. In 1986, Weitzman et al. (52) concluded after a careful study of additional species within the genera *Arthroderma* and *Nannizzia* that the two taxa were congeneric. The differences originally observed by Stockdale now overlapped and represented a continuum and no longer warranted two separate genera. Owing to priority, *Nannizzia* is a later synonym of *Arthroderma*. Recent molecular studies by Kawasaki (53) supported the conclusion by Weitzman et al. (52) that the two are congeneric.

Index to Species of the Arthrodermataceae

I.   Pathogenic species
  A.   Telemorphs (anamorphs)
    1.   *Arthroderma* (*Microsporum* and *Trichophyton*)
      a)   *A. benhamiae* (*T. mentagrophytes* var. *mentagrophytes*) (*T. mentagrophytes* var. *interdigitale*) (*T. mentagrophytes* var. *quinckeanum*) (*T. mentagrophytes* var. *erinacei*)
      b)   *A. cajetani* (*M. cookei*)
      c)   *A. fulvum* (*M. fulvum*)
      d)   *A. grubyi* (*M. vanbreuseghemii*)
      e)   *A. gypseum* (*M. gypseum*)
      f)   *A. incurvatum* (*M. gypseum*)
      g)   *A. obtusum* (*M. nanum*)
      h)   *A. otae* (*M. canis* var. *canis; M. canis* var. *distortum*)
      i)   *A. persicolor* (*M. persicolor*)
      j)   *A. racemosum* (*M. racemosum*)
      k)   *A. simii* (*T. simii*)
      l)   *A. vanbreuseghamii* (*T. mentagrophytes* var. *interdigitale*)
  B.   Anamorphs (Teleomorph unknown)
    1.   *Epidermophyton*
      a)   *E. floccosum*
    2.   *Microsporum*
      b)   *M. audouinii*
      c)   *M. equinum*
      d)   *M. ferrugineum*
      e)   *M. gallinae*
      f)   *M. praecox*
    3.   *Trichophyton*
      a)   *T. concentricum*
      b)   *T. equinum* var. *equinum; T. equinum* var. *autotrophicum*

    c) *T. gourvilii*
    d) *T. megninii*
    e) *T. rubrum*
    f) *T. schoenleinii*
    g) *T. soudanense*
    h) *T. tonsurans* var. *tonsurans; T. tonsurans* var. *sulfureum*
    i) *T. verrucosum* var. *ochraceum; T. verrucosum* var. *album; T. verrucosum* var. *discoides*
    j) *T. violaceum*
    k) *T. yaoundei*

II. Saprophytic nonpathogens
    A. Teleomorphs (anamorphs)
        1. *Arthroderma* (*Microsporum* and *Trichophyton*)
            a) *A. borellii (M. amazonicum)*
            b) *A. ciferrii (T. georgiae)*
            c) *A. cookiellum (Microsporum* sp.)
            d) *A. corniculatum (M. boullardii)*
            e) *A. flavescens (T. flavescens)*
            f) *A. gertleri (T. vanbreuseghamii)*
            g) *A. gloriae (T. gloriae)*
            h) *A. insingulare (T. terrestre)*
            i) *A. lenticularum (T. terrestre)*
            j) *A. quadrifidum (T. terrestre)*
            k) *A. uncinatum (T. ajelloi)*
    B. Anamorphs (teleomorph unknown)
        a) *E. stockdaleae*
        b) *M. magellanicum*
        c) *M. ripariae*
        d) *T. fischeri*
        e) *T. longifusum*
        f) *T. mariatii*
        g) *T. phaseoliforme*

C. Pathogenic Species: Teleomorphs

Arthroderma Currey ex Berkeley emend. Weitzman, McGinnis, Padhye et Ajello. Mycotaxon 25:505, 1986. Type species: *A. curreyi* Berkeley. Outlines of British Fungology, p. 357, 1860 = *Nannizzia* Stockdale, Sabouraudia 1:45, 1961.

Ascocarp globose (Fig. 2a), whitish to pale yellow or buff; peridium (Fig. 2b) consisting of a densely packed network of interwoven hyphae; perid-

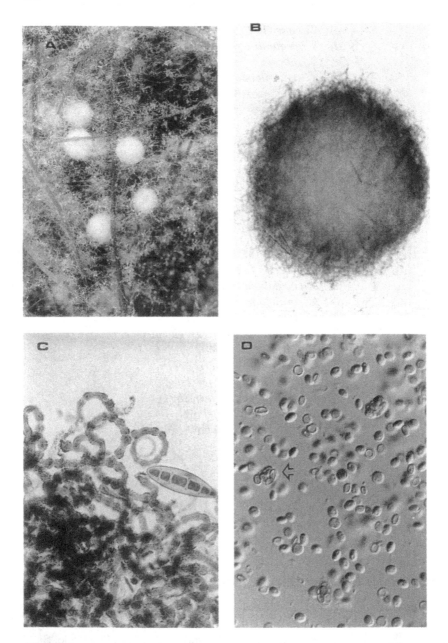

Figure 2 Morphology of the telemorphic state of the Arthrodermataceae. The examples shown are those formed by *Arthroderma uncinatum* (anamorph = *Trichophyton ajelloi*). (a) Ascocarps on hair, ×25. (b) Single ascocarp, ×240. (c) Peridial hyphae, ×450. (d) Asci (arrow) and ascospores, ×1400.

ial hyphae hyaline (Fig. 2c), pale yellow or buff, septate, usually uncinately, verticillately, or dichotomously branched; outer peridial hyphal cell walls echinulate, densely asperulate, or verruculose, with 1–3 slight to moderate constrictions, or deeply constricted in the middle, symmetrically or asymmetrically dumbbell-shaped; peridial appendages consist of tightly to loosely coiled spiral hyphae, which may be terminal or lateral, some species producing additional terminal appendages consisting of elongate, slender, tapered hyphae, or macroconidia; gymnothecial initials composed of a clavate antheridium surrounded by a coiled ascogonium; asci globose, subglobose, or oval, evanescent, eight-spored (Fig. 2d), 3.9–8 × 3.5–7.5 μm; ascospores oval (Fig. 2d), lenticular, oblate, smooth, hyaline, yellow in mass, 1.5–6 × 1.4–4 μm; homothallic or heterothallic; *Chrysosporium*, *Microsporum*, or *Trichophyton* anamorphs.

I. *Arthroderma benhamiae* Ajello & Cheng, Sabouraudia 5:232, 1967

   A. Conidial state

      Trichophyton mentagrophytes (Robin) Blanchard, *sensu lato*, Traite de pathologie general, vol. 2, Masson et Cie, Paris, 1896, p. 811.

      Distinctive varieties:

      Trichophyton interdigitale Priestley, Med. J. Aust. 4: 417–475, 1917.

      Trichophyton quinckeanum (Zopf) MacLeod & Muende, Practical Handbook of the Pathology of the Skin. 2nd ed. London: H. K. Lewis & Co. Ltd., 1940, p. 383.

      Trichophyton erinacei (Smith & Marples) Padhye & Carmichael, Sabouraudia 7: 178–181, 1969.

   B. Description

      Heterothallic. Ascocarp globose, pale white, 250–450 μm in diameter, excluding appendages. Peridial hyphae hyaline, septate, dichotomously branched, thin-walled. Cells dumbbell-shaped, echinulate, asymmetrically constricted, 8–12 × 4.5–5.2 μm. Appendages of two sorts: (1) elongated, smooth-walled hyphae tapering at the apex, 60–200 μm long; (2) smooth-walled, spiral hyphae. Asci globose to ovate, 4.2–7.2 × 3.6–6.6 μm, thin-walled, evanescent, eight-spored. Ascospores hyaline, smooth, oblate, 1.5–1.8 × 2.5–2.8 μm, yellow in mass.

      Colonies develop rapidly on glucose peptone agar; floccose powdery to granular, cream, yellowish to peach-colored. Reverse pale yellow to brown to reddish-brown. Microconidia (Fig. 3a) clavate, borne laterally on undifferentiated hyphae in

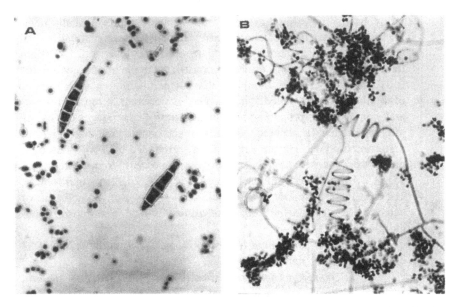

Figure 3 Micro- and macroconidia (a, ×850) and spiral hyphae (b, ×640) of *Trichophyton mentagrophytes*.

floccose strains or nearly spherical on conidiophores, forming grapelike clusters in granular strains. Macroconidia (Fig. 3a) rare or abundant in granular forms, clavate, smooth-walled, 20–50 × 6–8 μm, with 3–5 septa. Spiral hyphae (Fig. 3b) and nodular organs may be present.

C. Discussion

This species is a common cause of tinea pedis, tinea corporis, and tinea unguium. It may also cause tinea barbae and tinea capitis. The hairs show small-spored ectothrix involvement. *Arthroderma benhamiae* also infects cattle, horses, dogs, cats, sheep, pigs, rabbits, squirrels, monkeys, chinchillas, silver foxes, laboratory rats, mice, and other animals (1, 2, 41). The fungus, which is worldwide in distribution, has also been isolated from soil (1, 2, 41). The microconidia of highly granular cultures are infectious, and laboratory-acquired infections may occur in humans and animals (54).

The discovery of the sexual phase of growth of *T. mentagrophytes* has helped to sort out some of the confusion about the status of species and varieties of the genus that had been based on morphological variations. For example, the organism causing mouse favus, *T. quinckeanum*, was shown to be conspecific with *T. mentagrophytes* by Ajello et al. (55), and many isolates of *T. interdigitale* are compatible with tester strains of *A. benhamiae* (56). Weitzman (57) has summarized the work that eventually led to recognition of *T. erinacei* as a variety of *T. mentagrophytes*, and she emphasized that a large number of tester strains of diverse origin may be necessary in order to reveal true relationships. Weitzman et al. (58) made a detailed cytological study of *A. benhamiae* and have determined the chromosome number to be 4 (59).

Among the dermatophytes *A. benhamiae* is well suited to genetic studies because: (1) under properly controlled conditions it abundantly produces uninucleate microconidia that are hardy enough for easy manipulation in the laboratory, (2) it is an important human pathogen, (3) it is easily mated, and (4) the ascospores are reasonably easy to manipulate and give a good percentage of viability. Kwon-Chung has critically reviewed early reports of reversion of pleomorphic mutants of dermatophytes and of a putative suppressor mutation in *T. mentagrophytes* (39). Rippon (60) and Rippon and Garber (61) reported a difference in extracellular elastase production and virulence between one strain each of the (+) and (−) mating types of *A. benhamiae*. Chu-Cheung and Maniotis showed that the locus controlling elastase segregated independently of the mating type locus but was closely linked to that governing colonial morphology (62). Regrettably, the authors did not assess virulence in the segregants. In a recent study the usefulness of restriction enzyme analysis of mitDNA was assessed in taxonomy of *T. interdigitale* (63).

II. *Arthroderma cajetani* (Ajello) Ajello, Weitzman, McGinnis et Padhye, comb. nov., Mycotaxon 25:505, 1986; basionym: *Nannizzia cajetani* (as "*cajetana*") Ajello, Sabouraudia 1:175, 1961
   A. Conidial state
      Microsporum cookei Ajello, Mycologia 51:69–70, 1959.
   B. Description
      Heterothallic. Ascocarp globose, pale yellow, 368–686 μm in diameter. Peridial hyphae hyaline, septate, verticillately branched. Cells echinulate, thin-walled, and slightly constricted at the site of septations. Appendages of two kinds: (1) elongate, slender, tapered, smooth hyphae up to 480 μm in length, 3.6 μm in diameter at the base, 2.4 μm at midlength, and 1.2 μm at the tip; (2) elongate, smooth-walled, slender hyphae coiled

into spirals. Asci globose to ovate, 6–9 μm in diameter. Asco-spores ovate, smooth-walled, and golden, 3–3.6 × 1.8 μm.

Colonies on glucose peptone agar grow rapidly and are flat and spreading, with a rather powdery surface that is yellowish, buff, or dark tan. The reverse is deep purplish-red. Macroconidia are oval to ellipsoidal, thick-walled, and echinulate, 31–50 × 10–15 μm with six septa. Microconidia obovate, are produced abundantly.

C. Discussion

This species is geophilic and has a worldwide distribution. The fungus may be isolated from dogs and monkeys (1, 2). Tinea corporis has been reported rarely in humans (1, 2). Genetic analysis of the incompatibility system in *N. cajetani* has been accomplished by Kwon-Chung (64).

III. *Arthroderma fulvum* (Stockdale) Weitzman, McGinnis, Padhye et Ajello, comb. nov., Mycotaxon 25:505, 1986; basionym: *Nannizzia fulva* Stockdale, Sabouraudia 3:120, 1963; syn: *N. gypsea* (Nannizzi) Stockdale var. *fulva* (Stockdale) Apinis, Mycol. Paper 96:33, 1964

A. Conidial state

Microsporum fulvum Uriburu, Argent Med I: 563–582, 1909

B. Description

Heterothallic. Ascocarp globose, pale buff, 500–1250 μm in diameter, excluding appendages. Peridial hyphae are hyaline, septate, branched hyphae with thin, densely verrucose walls. Up to four branches arising in succession at the apex of the same cell; the branches are straight or curved over toward the ascocarp. Appendages of three kinds: (1) straight, slender, smooth-walled, septate hyphae up to 250 μm long, 3–4.5 μm in diameter at the base, tapering to 1.5–2 μm at the tip; (2) slender, smooth-walled, septate, spiral hyphae, not branched or much branched, each branch loosely to tightly coiled with up to 15 turns; (3) moderately thick-walled, verrucose macroconidia, usually cylindrical, tapering toward the apex and the base, or clavate, sometimes ellipsoid or fusiform, 30–50 × 9–12 μm, with up to five septa. Asci subglobose, evanescent, thin-walled, 5–7 μm in diameter, with eight ascospores. Ascospores smooth-walled, lenticular, 1.5–2 × 2–5.4 μm, yellow in mass.

Colonies on glucose peptone agar are dense, downy to granular, pale buff to rosy-buff. The reverse is rosy-buff to amber. Macroconidia are predominantly cylindrical, slightly tapering toward both ends, and with a rounded apex or clavate, occasionally ellipsoid or fusiform, 25–28 × 5–12 µm, with up to five septa and moderately thick, verrucose walls. Whiplike appendages (see *N. gypsea*) are rare. Macroconidia borne singly or on branched conidiophores. Microconidia 1.7–3.3 × 3.3–8.3 µm, clavate, smooth-walled or slightly rough-walled, sessile or on short pedicels, borne laterally along the hyphae. Spiral hyphae similar to those seen on ascocarp are formed abundantly on the vegetative mycelium.

C. Discussion

This species produces tinea corporis and tinea capitis in humans (1). Hair involvement is of the ectothrix sort but there is no fluorescence under a Wood's lamp. The fungus is also reported from dogs and occasionally from other animals (1, 2). *Arthroderma fulvum* is a geophilic dermatophyte found throughout the world.

   Genetic studies on the relationship between the enzyme elastase and the mating type of *N. fulva* have been reviewed by Kwon-Chung (39).

IV. *Arthroderma grubyi* (Georg, Ajello, Friedman et Brinkman) Ajello, Weitzman, McGinnis et Padhye, comb. nov., Mycotaxon 25:505, 1986; basionym: *Nannizzia grubyi* (as "*grubyia*") Georg, Ajello, Friedman et Brinkman, Sabouraudia 1:194, 1962

A. Conidial state

Microsporum vanbreuseghemii Georg, Ajello, Friedman et Brinkman, Sabouraudia 1:189–196, 1962.

B. Description

Heterothallic. Ascocarp globose, white to pale buff, 150–600 µm in diameter, exclusive of appendages. Peridial hyphae are hyaline, septate, branching dichotomously and uncinately, mostly curved to the outside of the main hypha. Cells are moderately thick-walled, densely echinulate, and moderately constricted in the center. Appendages terminate bluntly or taper into smooth-walled, loosely coiled hyphae with two to three turns or into elongate, thin, smooth-walled, loosely coiled, septate hyphae tapering from 3.0 µm in diameter at the base to 1.5 µm at the tip, with 30–50 tight coils. Macroconidia thick-walled, cylindrofusiform, multiseptate, and densely echinulate, borne laterally and terminally on the peridial hyphae. Asci are globose, thin-walled, evanescent, 4.8–6.0 µm in diameter, with eight ascospores. Ascospores hyaline, pale yellow, smooth-walled, ovate, 2.4 × 3.0 µm, yellow in mass.

Colonies on glucose peptone agar are rapid growing, flat with powdery to fluffy surface. Surface color white to yellowish; pink to deep rose colonies have also been described (65). The reverse is light yellow to bright lemon yellow. (Some strains are colorless.) Macroconidia are numerous, cylindrofusiform with thick, echinulated walls, and with 5–12 septa. The size range is 58.8–61.7 × 10.4–10.6 µm. Macroconidia are numerous, pyriform to obovate, 9.2 × 4.0 µm, borne laterally on the hyphae, sessile, or on pedicels.

C. Discussion

This species was originally described from tinea lesions in a Malabar squirrel, a dog, and two humans. The fungus has been isolated from soil (1, 2). Experimental infections in guinea pigs resulted in hyphal elements within the hair shafts and large numbers of spores on the surface of the hair. The hairs did not fluoresce under a Wood's lamp. In a natural human infection, however, the hairs did fluoresce (65).

V. *Arthroderma gypseum* (Nannizzi) Weitzman, McGinnis, Padhye et Ajello, comb. nov., Mycotaxon 25:505, 1986; basionym: *Gymnoascus gypseus* Nannizzi, Atti. Accad. Fisioscr. Siena Med.-fis. 2:93, 1927; syn: *Nannizzia gypsea* (Nannizzi) Stockdale, Sabouraudia 3: 119, 1963. *N. gypsea* (Nannizzi) Stockdale var. *gypsea*, Mycol. Paper 96:32, 1964

A. Conidial state

Microsporum gypseum (Bodin) Guiart & Grigorakis, sensu lato Lyon Med. 141:369–378, 1928.

B. Description

Heterothallic. Ascocarp globose, pale buff, 500–1250 µm in diameter, excluding appendages. Peridial hyphae hyaline, pale buff, septate, branched with thin, densely verruculose walls. Up to four branches arising in succession at the apex of the same cells. Appendages are of three kinds: (1) straight, slender, smooth-walled, septate hyphae, up to 250 µm long, 2.5–4.0 µm at the base tapering to 1.5–2.0 µm; (2) slender, smooth-walled, septate, spiral hyphae, rarely branched, 2.5–3.5 µm at the base tapering to 1.5–2 µm, loosely to tightly coiled with a variable number of turns; (3) moderately thick-walled, verrucose, ellipsoid or fusiform macroconidia, 35–55 × 10–13.5 µm, with up to five septa. Asci are subglobose, thin-walled, evanescent, 5–7 µm in diameter, with eight ascospores. Ascospores smooth, lenticular, 1.5–2 × 2.5–4 µm, yellow in mass.

Colonies on glucose peptone agar are coarsely granular,

Figure 4 Macroconidia of *Microsporum gypseum*, ×875.

radiating, with arachnoid edge, surface rosy-buff, while the reverse is buff to cinnamon. Macroconidia ellipsoid to fusiform, 25–58 × 8.5–15 μ, with moderately thick, verrucose walls (Fig. 4). Slender whiplike appendages, up to 30 μm long, borne on the apical cells of the conidia, are seen in some isolates but not in others.

Microconidia are clavate, 1.7–3.5 × 3.3–8.3 μm, unicellular, smooth-walled, or slightly roughened, sessile or on pedicels, borne laterally on the hyphae. Spiral hyphae similar to those borne on the ascocarp are also formed on the vegetative mycelium.

C. Discussion

This species is worldwide in distribution. It is found in the soil and causes an inflammatory tinea corporis and tinea capitis in humans (66). Small-spore ectothrix involvement of hairs is seen but generally no fluorescence. The fungus also infects a variety of animals (1, 2, 41). A genetic study of a phenomenon of pleomorphism has been made by means of induced mutants of this fungus and of *A. incurvatum* (67–69). The results of such studies have been comprehensively reviewed (39, 57).

VI. *Arthroderma incurvatum* (Stockdale) Weitzman, McGinnis, Padhye et Ajello, comb. nov., Mycotaxon 25:505, 1986; basionym: *Nannizzia incurvata* Stockdale, Sabouraudia 1:46, 1961

A. Conidial state

Microsporum gypseum (Bodin) Guiart & Grigoraksi, sensu lato Lyon Med. 141:369–378, 1928.

B. Description

Heterothallic. Ascocarp globose, pale buff to yellow, 350–650 µm in diameter, excluding appendages. Peridial hyphae pale buff, hyaline, septate, verticillately branched with up to five branches, which usually curve in toward their main axis. Cells moderately thick-walled, densely asperulate, with one to three more or less symmetrical constrictions, usually up to 7–8 µm in diameter but occasionally reaching 11 µm. Free ends numerous and appendages of three kinds: (1) elongate, slender, smooth-walled, septate, occasionally branched hyphae, up to 300 µm long, 3.0–4.5 µm in diameter at the base tapering to 1.5–2.0 µm in diameter, straight or loosely coiled; (2) elongate, slender, smooth-walled, septate, occasionally branched hyphae, 2.5–3.5 µm in diameter at the base tapering to 1.5–2.0 µm in diameter, tightly coiled with up to 30 spirals; (3) macroconidia thick-walled, asperulate spindle-shaped, 40–57 × 10–12.5 µm, with 1–5 septa. Asci globose to ovate, 5–7 µm in diameter, eight-spored. Ascospores yellow, smooth-walled, lenticular, 2.8–3.5 × 1.5–2.0 µ.

Colonies grow rapidly on most laboratory media, becoming powdery and buff to cinnamon-brown in color. Reverse pale yellow, tan or reddish-brown, occasionally red in some isolates. Microconidia clavate, thin-walled, and sessile on the hyphae. Macroconidia numerous, large, echinulate, thick-walled, ellipsoid, with 3–9, commonly 4–6 septa, 8–12 × 30–50 µm.

C. Discussion

This species produces tinea corporis or tinea capitis. Hair involvement is of the small-spored ectothrix type and the hairs usually do not fluoresce. The fungus, which also infects animals, is found in soil throughout the world.

A description of the morphogenesis of the sexual phase of growth was included with the original description (34) and has been re-examined in detail by Kwon-Chung (71).

VII.   *Arthroderma obtusum* (Dawson et Gentles) Weitzman, McGinnis, Padhye et Ajello, comb. nov., Mycotaxon 25:505, 1986; basionym: *Nannizzia obtusa* Dawson et Gentles, Sabouraudia 1:56, 1961

    A. Conidial state

        Microsporum nanum (Fuentes, Aboulafia & Vidal) Fuentes, Mycologia 48:613–614, 1956.

    B. Description

        Heterothallic. Ascocarp globose, pale buff, 250–450 μm in diameter, excluding appendages. Peridium about 50 μm thick. Peridial hyphae pale yellow, hyaline, septate, branching mostly dichotomous, occasionally verticillate, angle between branch and main hypha usually obtuse. Cells thick-walled, echinulate, cylindrical, 8–20 μm. Average 13 × 4–7 μm, may have one or two slight constrictions. Appendages of two sorts: (1) septate, smooth-walled, tightly coiled, lateral or terminal; (2) elongate, slender, septate, hyphae up to 450 μm long, terminal. Asci subglobose, thin-walled, evanescent, 5.5–6.5 × 5–6 μm, eight-spored. Ascospores hyaline, smooth or finely roughened, lenticular, 2.7–3.2 × 1.5–2 μm, yellow in mass.

        Colonies white, cottony, and spreading, becoming granular and buff-colored. Reverse red to brown. Macroconidia (Fig. 5) small, clavate, thick-walled, echinulate, generally with one septum but a few spores may have two or even three septa, 12–18 × 4–7.5 μm.

    C. Discussion

        This species is worldwide in distribution. The fungus causes tinea capitis, in which the hairs display the small-spored ectothrix type of involvement but do not fluoresce. *Arthroderma obtusum* is primarily a pathogen of pigs (1, 2).

VIII. *Arthroderma otae* (Hasegawa et Usui) McGinnis, Weitzman, Padhye et Ajello, comb. nov., Mycotaxon 25:505, 1986; basionym: *Nannizzia otae* Hasegawa et Usui, Jpn J Med Mycol 16:151, 1975

    A. Conidial state

        Microsporum canis Bodin, Arch Parasitol 5:5–30, 1902. Distinctive variety: *Microsporum distortum* di Menna & Marples, Trans Br Mycol Soc 37:372–374, 1954.

    B. Description

        Heterothallic. Ascocarps are globose, white becoming buff with age, 280–700 μm in diameter, exclusive of appendages. Peridial hyphae hyaline, septate, echinulate, constricted at the cell junction, usually dichotomously but occasionally verticillately branched. The tips of peridial hyphae are blunt and

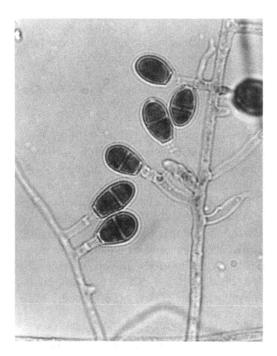

Figure 5   Macroconidia of *Microsporum nanum*, ×675.

curved toward the ascocarp. Appendages of three kinds: (1) long (up to 150 μm), slender, straight, smooth-walled, and septate hyphae; (2) spiral or long, coiled hyphae with 10–15 turns; and (3) hyphal appendages with thick-walled, echinulate macroconidia (Fig. 6a). Asci are subglobose, thin-walled, evanescent, 5–7 μm in diameter, with eight ascospores. Ascospores are smooth, lenticular, 2–2.5 × 2.5–4.8 μm.

Colonies on glucose peptone agar grow fairly rapidly, forming a cottony or woolly mycelium, white to buff in color. The reverse is yellow to orange-brown. (The *distortum* variety may be colorless.) Macroconidia are numerous, large, thick-walled, spindle-shaped, echinulate, 8–20 × 40–150 μm, with 6–15 septa. The macroconidia of the *distortum* (Fig. 6b) variety are basically spindle-shaped, but all bend and are grossly distorted. Their size is roughly similar to the *canis* variety but difficult to measure because of the distortions. Microconidia are usually scarce, clavate to elongate, sessile, or on short pedicels borne laterally on the hyphae. Racquet hyphae, pec-

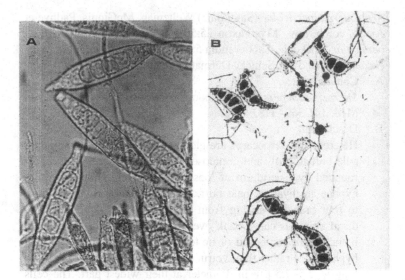

Figure 6 Macroconidia of *Microsporum canis* var. *canis* (a, ×875) and *M. canis* var. *distortum* (b, ×450).

tinate hyphae, nodular bodies, and chlamydospores may be seen.

C. Discussion

The fact that *A. otae* (*M. canis* var. *canis*) grows on polished rice grains whereas *M. audouinii* does not was one of the earliest uses of nutritional tests in identification of dermatophytes (49). *Arthroderma otae* (*M. canis* var. *canis*) causes tinea capitis, tinea corporis, including primary invasive infections (70), and rarely, tinea unguium (72). A mixed infection with *T. mentagrophytes* was reported in an AIDS patient (73). The source of infection is almost invariably an animal. The infected hairs fluoresce under a Wood's lamp and the small-spores ectothrix type of hair shaft involvement is seen. The fungus is zoophilic, being frequently isolated from dogs and cats and occasionally from monkeys, horses, rabbits, and pigs. The mating behavior of *A. otae* has been studied (74).

A. otae (*M. canis* var. *distortum*) species produces tinea capitis. As with the *canis* variety, the hairs show ectothrix involvement and fluoresce under a Wood's lamp. Cases have been reported in monkeys, dogs, horses, and pigs (41). The fungus occurs in the United States, Australia, and New Zealand (41).

IX. *Arthroderma persicolor* (Stockdale) Weitzman, McGinnis, Padhye et
     Ajello, comb. nov., Mycotaxon 25:505, 1986; basionym: *Nannizzia
     persicolor* Stockdale, Sabouraudia 5:357, 1967; syn: *Nannizzia quin-
     ckeani* Balabanov et Schick, Dermatol Venerol 9:35–36, 1970
     A. Conidial state
         Microsporum persicolor (Sabouraud) Guiart & Grigorakis, Lyon
         Med. 141:369–378, 1928.
     B. Description
         Heterothallic. Ascocarps are globose, 350–900 μm in diameter,
         pale buff to buff, and composed of a central mass of asci sur-
         rounded by a peridium of loosely interwoven hyphae. Peridial
         hyphae are hyaline, pale buff, septate, and branching, with up
         to four branches arising from the apex of the same cell. The
         distal branches curve back over the ascocarp, and the ultimate
         branches are composed of up to five cells. Cells of the peridial
         hyphae have rather thin, verruculose walls and are 5.5–27.5 μm
         long × 2.5–7.5 μm in diameter at their widest part. The cells
         of the distal branches are dumbbell-shaped, with one slight con-
         striction in the middle; they are symmetrical in both face and
         side views. Appendages are numerous, branched, spirally coiled
         hyphae. Asci subglobose, thin-walled, evanescent, 4.5–6.0 ×
         5.0–7.0 μm in diameter, and contain eight ascospores. Asco-
         spores are hyaline, smooth-walled, lenticular, 2.5–3.3 × 1.6–
         2.1 μm, pale yellow.
             Colonies on glucose peptone agar are fast-growing, flat at
         first, becoming fluffy and folded, yellowish-buff to pink. The
         reverse is variable from peach, rose, to deep ochre. Macroconidia
         are not abundantly produced, rather thin-walled, clavate to fusi-
         form, echinulate at the tip, usually with six septa. Microconidia
         produced in abundance, clusters, clavate, fusiform, or globose.
         Stalked, elongate, clavate conidia are distinctive.
     C. Discussion
         This species is worldwide in distribution, producing tinea infec-
         tions in wolves, bats, shrews, mice, horses, and more rarely, in-
         fection of humans (2, 75). It has also been isolated from soil (79).
         At one time the fungus was considered to be a *Trichophyton*, *T.
         persicolor*, closely related to—or in some individuals' views,
         synonymous with—*T. mentagrophytes*. (See Ref. 45.) Stockdale
         established that *A. persicolor* was a distinct taxon and expressed
         surprise that the conidial state should be *Trichophyton* (80). She
         recorded the slightly verruculose walls of the macroconidia, a
         feature used subsequently to support inclusion of the species in
         Microsporum as *M. persicolor*.

X. *Arthroderma racemosum* (Rush-Munro, Smith et Borelli) Weitzman, McGinnis, Padhye et Ajello, comb. nov., Mycotaxon 25:505,
   1986; basionym: *Nannizzia racemosa* Rush-Munro, Smith et Borelli, Mycologia 62:858, 1970
   A. Conidial state
      Microsporum racemosum, Borelli, Acta Med. Venez. 12:148–
      151, 1965.
   B. Description
      Heterothallic. Ascocarp globose, pale buff, 300–700 μm in di-
      ameter, exclusive of appendages. Peridial hyphae hyaline, sep-
      tate, verticillately and dichotomously branched. Up to four
      branches arising in succession at the apex of the same cell,
      many branches curve back over the ascocarp. Cells are thin-
      walled, echinulate, often more or less symmetrically con-
      stricted with one to three constrictions, up to 8 μm in diameter.
      Appendages of three kinds: (1) slender, smooth-walled, septate
      hyphae, 2.3–3.0 μm in diameter at the base tapering to 1.5–
      2.0 μm at the tip, coiled with up to 15 turns; (2) straight, slen-
      der, smooth-walled hyphae, 3.0–4.0 μm in diameter at the base,
      tapering to 1.7–2.0 μm at the tip, up to 380 μm long; (3) deli-
      cate, smooth-walled hyphae, 0.7–0.9 μm in diameter, which
      are up to 200 μm long. Asci globose or oval, 4.4–5.5 μm in
      diameter, containing eight spores. Ascospores hyaline, smooth-
      walled, oval, 2.5–3.0 × 1.2–1.5 μm, yellow in mass. The dis-
      tinctive feature of the ascocarps of this fungus is the flagel-
      lumlike appendages (45).
      Colonies are very fast-growing on glucose peptone agar.
      Said to be the fastest-growing dermatophyte (45). The surface
      of the colony is powdery cream-white. The reverse is grape-
      red. Macroconidia spindle-shaped, echinulate, rather thick-
      walled, up to 60 μm long, 5–10 septa. The macroconidia
      frequently have a terminal filament. Microconidia are quite
      distinctive. They are club-shaped, mostly stalked, and are borne
      in large, wandlike clusters (racemes) (81).
   C. Discussion
      This fungus has been reported from soils of South America and
      Romania (82) but is probably worldwide. Infections in humans
      have been reported in Illinois (83, 84).
XI. *Arthroderma simii* Stockdale, Mackenzie & Austwick, Sabouraudia
    4:113, 1965
   A. Conidial state
      Trichophyton simii (Pinoy) Stockdale, Mackenzie & Austwick,
      Sabouraudia 4:114, 1965; synonyms: *Epidermophyton simii* Pi-

noy, C R Soc Biol (Paris) 72:59, 1912; *Pinoyella simii* (Pinoy) Castellani & Chalmers, Manual of Tropical Medicine. 3rd ed. London: Bailliere, Tindall & Cox, 1919, p. 1023.

B. Description

Heterothallic. Ascocarp globose, pale buff to buff, 200–700 μm, thin, excluding appendages. Peridial hyphae pale buff, septate, with somewhat thin, verruculose walls. Up to three secondary branches may arise from the apex of the same cell of a peridial hypha. Distal branches curve over the ascocarp, and the ultimate branches are rarely composed of more than two or three cells. Cells of the inner parts of the peridial hyphae are 12.5–21 × 3–4.5 μm, not constricted but swelling toward the apex from which the branches arise, 5–7 μm in diameter. Cells of the outer branches are asymmetrically constricted, 6.5–12.5 × 4.2–6.7 μm at their widest part. Appendages slender, smooth-walled, septate, spiral hyphae. Smooth-walled, cylindrical to fusiform macroconidia are produced by the peridial hyphae. Asci subglobose, thin-walled, evanescent, 5–6.7 μm in diameter, eight-spored. Ascospores hyaline, smooth, oblate, 2.5–3.5 × 1.5–2.5 μm, yellow in mass.

Colonies velvety with a finely granular uniform surface and a fluffy asteroid margin. The surface is pale buff and the reverse straw to salmon. Macroconidia numerous, borne terminally on complex branched hyphae, hyaline, smooth-walled, fusiform, occasionally cylindrofusiform, 30–80 × 6–11 μm, with 4–7 (up to 10) septa. In older cultures the macroconidia become constricted at the septa as a result of an increase in the diameter of individual cells forming intercalary, thick-walled chlamydospores. Microconidia rare in young cultures, more abundant in old cultures, clavate to pyriform, 2–6.5 × 1.5–4 μm, sessile or on short pedicels, and borne singly along the sides of the hyphae. Spiral hyphae are moderately abundant on the vegetative mycelium of old cultures.

C. Discussion

Arthroderma simii is of restricted geographic distribution, as most of the cases have been reported in or from India (41, 45); also reported from Brazil and Guinea (45). The fungus produces tinea in monkeys, poultry, and dogs (85). Several human cases have been reported, and the type of hair involvement is endothrix (1, 2). The organism also occurs in the soil (1). Some workers have considered *A. simlii* and A. persicolor to be geophilic (1, 87, 88) while others consider them to be zoophilic

(2, 86). Both species have been isolated from soil and from the hair of animals that did not display lesions (79, 87).

A fascinating attribute of *A. simii* is that it induces a form of stimulated growth of the mycelium of several different species of dermatophytes (89). This stimulation occurs in strains of mating type opposite that of the tester *A. simii* type. The interaction, of course, never goes on to formation of fertile ascospores within asci except in opposite mating types of *A. simii* itself, but the observation of stimulated growth has been used to presume the juxtaposition of opposite mating types. In some instances isolates of species not known to engage in a sexual form of growth have been identified as being of a single mating type. For example, all isolates of *T. rubrum* were shown in this manner to be the (−) mating type (90). Such observations strengthen the argument that the reason certain species cannot be induced to reveal a sexual phase of growth is that they are all of one mating type. Thus, if virulence were associated with mating type, one could imagine the evolutionary disappearance of a mating type. This would be especially likely to happen if the environment were highly restricted. It is interesting to reflect on the fact that only one of the anthropophilic species (*A. vanbreuseghemii*) has been shown to have a perfect state.

The chromosome number of *A. simii* is 4 (59), and the incompatibility system has been subjected to a comprehensive study (91). At one time it was suggested that *A. simii* was one of the "three" teleomorphs of the *T. mentagrophytes* complex (92). Although this concept has been reiterated on one recent occasion (93), it would appear that the results of careful studies affirm the distinctness of *T. simii* as the single anamorph of *A. simii* (57, 94, 95). The relationships are, however, close (57, 95).

XII. *Arthroderma vanbreuseghemii* Takashio, Ann Soc Belge Med Trop 53:427, 1973

   A. Conidial state
   Trichophyton mentagrophytes (Robin) Blanchard, Traite de pathologie generale, vol. 2. Paris: Masson et Cie, 1896, p. 811.

   B. Description
   Heterothallic. Ascocarps are globose, 300–650 μm in diameter, straw- or buff-colored. Peridial hyphae are hyaline or slightly buff, septate, interwoven, dichotomously branched, and thin-walled. Cells dumbbell-shaped, echinulate, asymmetrically constricted, 8–12 × 4.5–5.2 μm. Appendages of two sorts: (1) elon-

gate, smooth-walled hyphae tapering at the apex, 60–200 μm long; (2) smooth-walled spiral hyphae with up to 15 turns. Asci globose to ovate, 4.2–7.2 × 3.6–6 μm, thin-walled, evanescent, with eight ascospores. Ascospores are hyaline, smooth, oblate, 1.5–1.8 × 2.5–2.8 μm, yellow in mass.

Colonies described under *A. benhamiae.*

C. Discussion

This species is the second perfect state of the *T. mentagrophytes* complex. (See *A. benhamiae.*) The differences are the larger ascospores of *A. vanbreuseghemii* and no interspecific crosses. Mating studies in the complex have been comprehensively recorded by Takashio (94). Variations in the colonial morphology have been presented (94). The disease pattern and distribution have been described under *A. benhamiae.*

D. Pathogenic Species: Teleomorphs Unknown

Epidermophyton Sabouraud, Arch Med Exp Anat Pathol 19:754–762, 1907; This genus is characterized on the basis of the macroconidia, which are 6–10 × 8–15 μm in size, clavate, smooth-walled, with 0–4 (usually 2–3) septa. No microconidia are found. The type species is *Epidermophyton floccosum* (Harz) Langeron & Milochevitch, 1930.

XIII.   *Epidermophyton floccosum* (Harz) Langeron & Milochevitch, Ann Parasitol Hum Comp 8:495–497, 1930

A. Description

Colonies. The fungus grows slowly on glucose peptone agar. The colonies have a velvety to powdery surface that is gently folded in a number of radiating furrows and is khaki-yellow in color. The reverse is yellow to tan. Usually, isolates mutate early and almost invariably, with the production tufts of white, sterile hyphae that soon overgrow the entire colony. Macroconidia (Fig. 7) clavate, smooth, fairly thick-walled, with 0–4 (usually 2–3) septa; the macroconidia are borne in characteristic clusters of twos and threes. Microconidia are not produced. Chlamydospores, racquet hyphae, and nodular bodies are often abundantly produced.

B. Discussion

This species is worldwide in distribution. The fungus produces tinea pedis and tinea cruris. There are rare reports of tinea unguium (41) and one report of tinea capitis (96). A dog has been reported to have yielded *E. floccosum*, but animals are generally not infected (41).

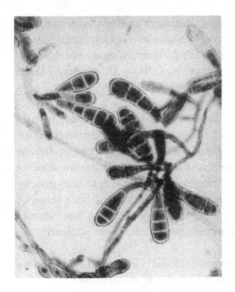

Figure 7   Macroconidia of *Epidermophyton floccosum*, ×475.

Microsporum Gruby, C. R. Acad. Sci. (Paris) 17:301–303, 1843. The genus is characterized on the basis of the macroconidia, which are large, thick-walled, echinulate, fusiform to obovate conidia with 1–15 septa, generally borne singly. Smaller one-celled microconidia are produced on short pedicels or are sessile on the hyphae. Pectinate hyphae, racquet hyphae, nodular bodies, coils, and chlamydospores may be present. Type species is *Microsporum audouinii* Gruby, 1843.

XIV.   *Microsporum audouinii* Gruby, C R Acad Sci (Paris) 17:301–303, 1843; synonyms: *Sabouraudites (Microsporum) langeronii* Vanbreuseghem, Ann Parasitol Hum Comp 25:409–417, 1950; *Microsporum rivalieri* Vanbreuseghm, Sabouraudia 2:215–224, 1965

A.   Description

Colonies on glucose peptone agar grow rather slowly, form a matted, velvety surface with straggly edges, and are light tan to grayish-white in color. The reverse is buff-salmon to orange-brown. Macroconidia are rarely found in cultures. When observed they are distorted and very irregular with thick, echinulate, or smooth walls. Some may be nearly spindle-shaped. The number of septa is from 2 to 8. Microconidia, when present, are sessile or borne on short pedicels along the hyphae, clavate, and single-celled. Racquet hyphae, pectinate hyphae, nodular bodies, and chlamydospores may be present and are not infrequently the only microscopic structures visible. As pointed out earlier, the poor growth of *M. audouinii* on rice grains serves to distinguish the species from *M. canis (A. otae)* and *M. gypseum (A. gypsea* and *A. incurvata)*. Also, *M. audouinii* does not perforate hair whereas *M. canis* does.

B. Discussion

This species produces classic, epidemic tinea capitis in children. The hairs show ectothrix involvement and fluoresce under a Wood's lamp. Rare in the United States and Europe, mainly present in Africa, Rumania, and Haiti (2). *Microsporum audouinii* is one of the forms of ringworm most susceptible to griseofulvin therapy, and the incidence of disease has been remarkably reduced wherever the drug is used. A few animal infections have been reported, but *M. audouinii* is clearly anthropophilic (41).

XV. *Microsporum equinum* (Delacroix & Bodin) Gueguen, Les champignons parasites de l'homme et des animaux domestiques, Josephy Van In & Cie, Lierre, 1904, p. 144

A. Description

Colonies on glucose peptone agar are downy to powdery, pale buff to pinkish-buff. The reverse is yellow to buff-amber. Macroconidia elliptical to fusiform, thick-walled, echinulate, 18–62 × 5–13.5 μm, with 4–8 septa and borne terminally in clusters (97, 98). Microconidia sessile or on short pedicels, pyriform to clavate, 3–8 × 2–3.5 μm. Spiral hyphae and chlamydospores have been seen (97). The description of this species in the early literature (97) varies in detail from that more recently published (98).

B. Discussion

This species is now included in most standard lists of *Microsporum* spp. (1, 2, 40). It was originally reported in Europe, Java, and Uruguay (99), and is said to cause tinea of horses. Small-spored ectothrix involvement is described (97). The fungus is said to infect humans (98). The results of a recent study support the view that *M. equinum* is a distinct species and not a synonym of *M. canis* (98).

XVI. *Microsporum ferrugineum* (Ota), Bull Soc Pathol Exot 15:588–596, 1922; synonym: *Trichophyton ferrugineum* (Ota) Langeron & Milochevitch, Ann Parasitol Hum Comp 8:422–436, 1930

A. Description

Colonies on glucose peptone agar are slow-growing, heaped with many deep furrows, glabrous and waxy, deep yellow to orange in color. Reverse is buff to brownish. A white velvety cover may form over the colony. Sectoring occurs with great variations in color intensities. Macroconidia are rarely observed. Such structures have been reported when isolates are cultured on a potato dextrose charcoal agar or diluted Sabouraud agar (41, 91, 100), however. The macroconidia so generated were typical for the genus *Microsporum* (i.e., spindle-shaped, echinulate, thick-walled, 40.5–47 × 8.5–10.5 μm, with 2–8 septa). The mycelium comprises hyphae, some of which have prominent cross walls ("bamboo" hyphae). Chlamydospores may be seen.

B. Discussion

The fungus causes tinea capitis in children. Hair involvement is of the small-spored ectothrix sort. The hairs fluoresce under a Wood's lamp. The organism is strictly anthropophilic (41) and is found in middle Europe, Asia, the former USSR, and Africa (41).

XVII. *Microsporum gallinae* (Megnin) Grigorakis, Ann Dermatol Syphilol 10:18–53, 1929; synonym: *Trichophyton gallinae* Silva & Benham, J Invest Dermatol 18:432–472, 1952

A. Description

Colonies on glucose peptone agar grow moderately rapidly, are flat at first with radial folds. Edges of the colonies may be irregular. The reverse of the colony is deep strawberry-red. The pigment diffuses into the medium. Macroconidia relatively thin-walled, echinulate at the tips, 6–8 × 15–50 μm, frequently curved, with 2–10 septa. Microconidia pyriform to clavate, smooth-walled, borne laterally on the hyphae.

B. Discussion

This species is a very rare incitant of tinea capitis, tinea corporis, and tinea cruris in humans. The ectothrix type of hair involvement is seen. More typically seen as a cause of tinea in chickens and other fowl (1, 41). The distribution is worldwide.

XVIII. *Microsporum praecox* Rivalier, Ann Inst Pasteur 86:276–284, 1953

A. Description

Colonies on glucose peptone agar are moderately fast-growing, cream to yellowish-tan; powdery, with a yellowish-orange reverse. There is a report of an isolate from a case of tinea capitis whose macroscopic appearance differed from that of the original description by Rivalier (77). Macroconidia long, narrow, lanceolate, with fairly thick, roughened walls, 60–65 × 8–10 μm, with 6–9 septa. Microconidia are absent.

B. Discussion.

This species was isolated from a pustular, vesicular tinea lesion on the wrist from an adult (45), from tinea capitis (77), from saprophytic sites associated with horses (76), and from a tinea capitis in a patient with sickle cell anemia (78).

Trichophyton Malmsten, *Trichophyton tonsurans* harskararde Mogel, 1845 (transl.), Arch Anat Physiol Wiss Med 1–19, 1848.

The genus is characterized on the basis of the macroconidia, which are elongate, clavate, to fusiform, generally thin-walled, smooth, and have 0–10 septa. The size ranges from 8–50 × 4–8 μm. Smaller one-celled microconidia are usually produced. These cells are globose, 2.5–4 μm, or clavate, to pyriform, and 2–3 × 2–4 μ. Spiral hyphae and chlamydospores may be produced. The type species is *Trichophyton tonsurans* Malmsten, 1845.

XIX. *Trichophyton concentricum* Blanchard, Traite de pathologie generale. vol. 2. Paris; Masson et Cie, 1895, pp. 811–926

A. Description

Colonies on glucose peptone agar are very slow-growing and are raised deeply folded, smooth and white, becoming cream to amber or brown, and covered with short, gray hyphae. The reverse is cream to brown. Macroconidia not observed. Microscopically similar to *T. schoenleinii*. The hyphae are swollen, bearing chlamydospores and aborted branches. Microconidia have been reported (45).

B. Discussion

The fungus causes tinea imbricata [i.e., a form of tinea corporis in which concentric rings of scales occur on the skin (1, 2)]. The disease has been reported in the South Pacific islands, Guatemala, southern Mexico, and central Brazil (1).

XX. *Trichophyton equinum* (Matruchot & Dassonville) Gedoelst, Les champignons parasites de l'homme et des animaux domestiques. Joseph Van In & Cie, Lierre, 1902, p. 88

A. Description

Colonies grow rapidly on glucose peptone agar and are flat, developing folds with age, surface white, cottony with yellow color in the edge around the new growth. The reverse is bright yellow, becoming dark pink to brown with age. Macroconidia are rarely found in cultures grown on glucose peptone agar but may be produced on wort agar (101). They are slightly clavate, thin-walled, smooth, with 3–4 septa. Microconidia are nearly spherical to slightly pyriform, borne laterally along the hyphae, sessile or on short pedicels and in clusters.

B. Discussion

This species—considered by some as being synonymous with or

a variety of *T. mentagrophytes*—was re-evaluated as a separate taxon (101). The fungus is rare in humans and commonly causes ringworm in horses (102). The large-spored ectothrix type of hair involvement is seen. There is a variety recorded from New Zealand and Australia with less exacting nutritional requirements than is customary. This form has been referred to as *T. equinum* var. *autotrophicum* (103).

XXI.   *Trichophyton gourvilii* Cantanei, Bull Soc Pathol Exp 26:377–381, 1933

    A.   Description

Colonies on glucose peptone agar are waxy with a heaped and folded surface, becoming velvety with age. The color is lavender to deep garnet-red. The reverse is reddish-purple. The colonies are said to resemble those of *T. violaceum* and *T. soudanense* (41). Macroconidia smooth-walled, cylindrical. Microconidia pyriform, smooth-walled, borne laterally on the hyphae. Not all isolates produce micro- and macroconidia.

    B.   Discussion

This fungus is found in Africa (1), where it produces tinea capitis and tinea corporis. Endothrix type of hair involvement is observed.

XXII.   *Trichophyton megninii* Blanchard, Traie de pathologic generale. vol. 2. Paris: Masson et Cie, 1896, pp. 811–926; synonym: *Trichophyton kuryangei* Vanbreuseghem & Rosenthal, Ann Parasitol Hum Comp 36:797–802, 1961 (see Ref. 46)

    A.   Description

Colonies grow slowly on glucose peptone agar and are cottony to velvet, white at first, becoming pink with age. A nondiffusible rose to red pigment develops on the reverse side. Macroconidia are scarce, clavate, thin and smooth-walled, with 2–10 septa, said to resemble those of *T. rubrum* (45). Microconidia are numerous, small, pyriform to clavate, borne singly or in clusters along the hyphae.

    B.   Discussion

Produces tinea barbae predominantly and to a lesser extent tinea corporis and tinea capitis. Hair involvement of the large-spored ectothrix type. Predominantly anthropophilic, it has been reported on dogs (41). The fungus is found in Europe and Africa.

XXIII.   *Trichophyton rubrum* (Castellani) Sabouraud, Br J Dermatol 23:389–390, 1911; synonym: *Trichophyton fluviomuniense* Pereiro Miguens, Sabouraudia 6:312–317, 1968 (see Ref. 71)

A. Description

Colonies on glucose peptone agar are cottony and white, later becoming velvety. The reverse side is reddish to rose-purple. Macroconidia elongate, cylindrical, thin and smooth-walled, with 3–8 septa. Generally, these are scarce on glucose peptone agar but are formed in cultures on heart infusion tryptose agar (15, 57). Microconidia (Fig. 8) numerous, clavate, 2–3 × 3–5 µm, borne singly along the hyphae. Chlamydospores, racquet hyphae, and nodular bodies may be seen in primary cultures. A wide variety of colonial types has been described (1).

B. Discussion

This species is worldwide in distribution and clinically is among the most important dermatophyte pathogens. It produces tinea corporis (see Ref. 104 for a recent report on invasive infections), tinea pedis, tinea cruris, and tinea unguium. Nail lesions have been refractory to treatment, but newer drugs have succeeded in producing a clinical and mycological cure. The fungus rarely causes disease in animals, but instances are recorded (41). *Trichophyton rubrum* does not perforate autoclaved hair, a test useful in distinguishing isolates of it from *T. mentagrophytes* (105).

Mating studies with *A. simii* (90) have revealed that all

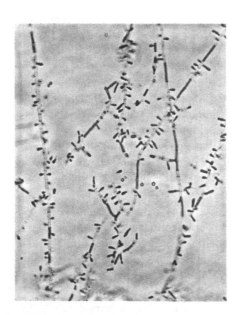

Figure 8   Microconidia of *Trichophyton rubrum*, ×560.

tested isolates of *T. rubrum* were the same mating type (−). It may be that virulence or some other selective factor has eliminated the (+) mating type of *T. rubrum*.

XXIV. *Trichophyton schoenleinii* (Lebert) Langeron & Milochevitch, Ann Parasitol Hum Comp 8:465–508, 1930

   A. Description

   Colonies are very slow-growing on glucose peptone agar, heaped with many irregular folds, waxy smooth, later becoming velvety white. Macroconidia have never been reported. Microconidia are virtually never seen in primary isolation; old cultures become velvety and develop microconidia. Antlerlike terminal branching ("favic chandeliers") is the single most obvious morphological feature. Chlamydospores may be seen.

   B. Discussion

   The fungus is widespread in Eurasia and North Africa (1), where it produces favus or tinea favosa, which is a chronic, cicatricial form of tinea capitis distinguished by the formation of scutula, crusts, scarring, atrophy, and permanent hair loss. Hair involvement is quite distinctive. Within the hair, hyphae and air spaces accompanied by degenerate hyphae are seen. A dull bluish-white fluorescence of the hairs under a Wood's lamp has been described. Tinea corporis has been observed, as have infections in animals (41).

XXV. *Trichophyton soudanense* Joyeux, C R Soc Biol (Paris) 73:15–16, 1912; synonym: *Langeronia soudanensis* Vanbreuseghem, Ann Parasitol Hum Comp 25:493–508, 1950

   A. Description

   Colonies are very slow-growing, flat to folded, suede-like, yellow, dried-apricot colored, often with a fringed border (106). Sterile rapidly growing hyphae develop readily in culture. Macroconidia not seen. Microconidia are seen in cultures grown on potato dextrose agar and are pyriform, smooth-walled, borne laterally on hyphae, sessile or on short pedicels (106). The mycelium is characterized by hyphae that break up into segments or arthroconidia and by reflexive branching. Produces dark brown colonies on Lowenstein Jensen medium.

   B. Discussion

   This species produces tinea capitis and tinea corporis. It is found principally in Africa, but cases have been reported in England, Brazil, and the United States (41). The differentiation between *M. ferrugineum* and *T. soudanense* has been studied (107).

XXVI. *Trichophyton tonsurans* (Malmsten, 1845). *Trichophyton tonsurans* harskarande Mogel, Stockholm (transl.) Arch Anat Physiol Wiss Med 1–99, 1848

    A. Description

        Colonies grow fairly slowly on glucose peptone agar, flat, somewhat powdery or velvety, white, yellowish to pinkish surface. The reverse is mahogany-red. The sulfureum variety produces pale to deep yellow colonies. Macroconidia are not frequently seen, but when found are thin-walled, club-shaped, or sinuous, with 3–5 septa. Some more abundantly sporulating strains have been described (26). Microconidia are numerous, clavate, 2–5 × 3–7 μm, and are borne laterally along the hyphae on pedicels of various lengths (Fig. 9). Chlamydospores and racquet hyphae may be seen. Growth stimulated by thiamine.

    B. Discussion

        This species produces black-dot tinea capitis and tinea corporis [caused institutional and family outbreaks—role of families also—tinea corporis gladiatorum (contact spreads outbreaks)]

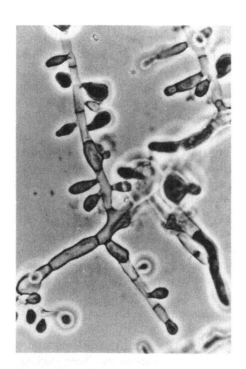

Figure 9 Microconidia of *Trichophyton tonsurans*, ×875.

and tinea pedis. Hair shaft involvement is of the endothrix sort. Animal infections have been rarely reported. Two varieties have been described: *T. tonsurans* var. *tonsurans* and *T. tonsurans* var. *sulfureum* (108, 109).

XXVII. *Trichophyton verrucosum* Bodin, Les champignons parasites de l'homme. Paris: Masson et Cie, 1902, p. 121

  A. Description

Colonies are very slow growing on glucose peptone agar, heaped, deeply folded, glabrous and waxy, sometimes with a fine white velvety surface. Colonial variants have been described and named [var. *ochraceum*, var. *album*, and var. *discoides* (45)]. These variants were originally described as species (1, 105, 106), but are not currently so considered (40). On glucose peptone agar generally only chlamydospores and hyphae are observed microscopically. On enriched media microconidia are produced. They are clavate to pyriform, sessile, borne laterally along the hyphae. Macroconidia are rare and are "rat-tailed" or "string bean"-shaped with 3–5 septa, smooth, thin-walled. All strains require thiamine; most require thiamine and inositol. Growth stimulated at 35–37°C.

  B. Discussion

This species produces highly inflammatory tinea capitis, tinea barbae, and tinea corporis. The hairs show ectothrix type of involvement. The fungus is worldwide in distribution. Characteristically, it is a zoophilic species found primarily in cattle, and has also been recorded in donkeys, dogs, goats, sheep, and horses (41).

XXVIII. *Trichophyton violaceum* Bodin, Les champignons parasites de l'homme. Paris: Masson et Cie, 1902, p. 113

  A. Description

Colonies are slow-growing on glucose peptone agar, waxy, wrinkled, heaped up or verrucous, with a deep purplish-red pigmentation. Colonies tend to produce sector variants. Generally, only torturous, tangled hyphae are observed. Pyriform microconidia borne laterally may be seen, but only rarely. Clavate macroconidia (45) with 3–5 septa, and smooth, thin walls have been reported. Growth stimulated by thiamine.

  B. Discussion

This species produces tinea corporis and tinea capitis of the black-dot variety. Hair invasion is of the endothrix sort. Worldwide in distribution, although unusual in the United States (1). Reported to infect animals other than humans (41).

XXIX. *Trichophyton yaoundei*, Cochet & Doby-Dubois, Sem Hop Paris 33: 26–30, 1957.

A. Description

Colonies are very slow growing on glucose peptone agar, glabrous, raised, folded, white to cream in color at first, becoming tan to brown with age. The pigment diffuses into the medium. Sectoring variants devoid of diffusible pigment on the reverse occur regularly. Macroconidia are not formed. Microconidia rare, but when seen are pyriform and borne laterally on the hyphae. Chlamydospores are present.

B. Discussion

This species produces endothrix-type tinea capitis and is found in equatorial Africa (1).

E. Saprophytic Species: Teleomorphs

XXX. *Arthroderma borellii* (Moraes, Padhye et Ajello) Padhye, Weitzman, McGinnis et Ajello, comb. nov., Mycotaxon 25:505, 1986; basionym: *Nannizzia borellii* Moraes, Padhye et Ajello, Mycologia 67: 1112, 1975

A. Conidial state

Microsporum amazonicum Moraes, Borelli & Feo, Med. Cutanea 2:201–205, 1967.

B. Description

Heterothallic. Ascocarp globose, dull white or pale yellow, 300–500 μm in diameter, excluding appendages. Peridial hyphae are hyaline, septate, and dichotomously branched. Cells are thin-walled, finely echinulate, slightly constricted in the middle, and swollen near the septa. Appendages are of two kinds: (1) elongate, septate, smooth-walled, tapering hyphae up to 150 μm in length, 2–3 μm in diameter at the base and 1.2– 1.5 μm at the tip; (2) elongate, smooth-walled, slender hyphae coiled into spirals; macroconidia may be at the end of some hyphal appendages. Asci are hyaline, subglobose, 4–5 μm in diameter with eight ascospores. Ascospores hyaline, oblate, 2– 3 × 2.5 μm in diameter, yellow in mass.

Colonies on glucose peptone agar are fluffy to powdery with a distinctive gray-olive-buff color (45). Macroconidia are symmetrical, spindle-shaped, echinulate, thick-walled, with 1– 4 septa. Microconidia not observed.

C. Discussion

The fungus was isolated from hair of spiny rats (*Proechimys guyannensis*) and from hair of a rat belonging to the genus *Ory-zymys*. Although the organism is associated with animals, it is a saprophyte since naturally occurring lesions have not been described.

XXXI. *Arthroderma ciferrii* Varsavsky & Ajello, Riv Patol Veg 4:358–359, 1964

    A. Conidial state

       Trichophyton georgia Varsavsky & Ajello, Riv Patol Veg 4: 357–358, 1964.

    B. Description

       Homothallic. Ascocarp globose, pale buff to pale brownish vinaceous, 450–700 μm in diameter, excluding appendages. Peridial hyphae hyaline, septate, uncinately branched with curled end. Cells of the peridial hyphae dumbbell-shaped, symmetrically or asymmetrically constricted, densely asperulate, 8–12 μm long, 7–8.5 μm at enlarged ends in width. Appendages of two kinds: (1) slender, smooth-walled spiral hyphae, borne terminally, and (2) slender, smooth-walled, septate, tapering hyphae extending outward, 60–96 μm long. Asci subglobose, thin-walled, evanescent, 4.8 × 6 μm, 8-spored. Ascospores hyaline, smooth, oblate, 2.5–3 × 1.4–2.8 μm, yellow in mass.

       Colonies on glucose peptone agar are flat and granular with an umbonate center. Center pale vinaceous and downy. Flat area surrounding umbonate center powdery to granular, vinaceous brown. Reverse irregularly spotted with dark red or dark vinaceous brown. Macroconidia thought not to occur, but microconidia have septa and might be regarded as macroconidia (45). Microconida variable in size and shape, elongate, clavate, occasionally pyriform to subglobose, borne laterally or terminally, singly on short pedicels, or sessile on the hyphae. Microconidia smooth-walled, usually nonseptate, occasionally with two or three septa, 4–6.5 × 2–2.5 μm.

    C. Discussion

       This fungus is found in soil, on hair and feathers, or in birds' nests or excreta (111). The organism has been reported from the United States, Venezuela, Australia, Czechoslovakia, and Yugoslavia (111).

XXXII. *Arthroderma cookiellum* (de Clercq) Weitzman, McGinnis, Padhye et Ajello, comb. nov., Mycotaxon 25:505, 1986; basionym: *Nannizzia cookiella* de Clercq, Mycotaxon 18:23, 1983

    A. Conidial state *Microsporum de Clercq*, Mycotaxon 18:23–28, 1983.

Heterothallic. Ascocarp globose, pale buff to yellow, 400–800 μm in diameter. Peridial hyphae pale buff, verrucose, septate, dividing in two or more branches. Distal branches straight or curved in "running legs." Peridial cells 10–20 × 2–5 μm, asperulate, with appendages that are either slender, smooth-walled, 64–400 × 2 μm, or as 10–14 μm wide spirals with 8–22 turns. Asci subglobose, 4–6 × 4 μm, 8-spored, evanescent. Ascospores golden-yellow in mass, smooth-walled, 2–3 μm. Macroconidia numerous, oval-shaped, dumpy, 18–34 × 17 μm, predominantly four-celled, thick-walled, 2–4 μm, very densely verruculose, the "warts" resembling sometimes "pseudopodes." Microconidia numerous, pyriform, sometimes very elongate, 1–2 × 2–8 μm.

On glucose peptone agar, colonies rapidly growing. Aerial mycelium, ochraceous powdery, at the margin white and fluffy; the down becomes violet after 7 days. Reverse brownish-purple with a nondiffusing pigment and a violet border by transparency. There is no growth at 37°C. Inoculation on the scarified skin of guina pigs is negative. Habitat in soil of Ivory Coast. Keratinophilic.

XXXIII. *Arthroderma corniculatum* (Takashio et de Vroey) Weitzman, McGinnis, Padhye et Ajello, comb. nov., Mycotaxon 25:505, 1986; Basionym: *Nannizzia corniculata* Takashio et de Vroey, Mycotaxon 14:384, 1982

A. Conidial state
Microsporum boullardii Dominik & Majchrowicz, Ecol Pol Ser A 13:415–447, 1965.

B. Description
Heterothallic. Ascocarp globose, pale-buff to yellow-buff, 370–870 μm diameter. Peridial hyphae pale-buff, hyaline, septate, verticillately branched at the distal ends of hyphal cells, with up to four branches, straight or curved. Distal branches composed of one to five cells curve moderately or coil to form semicircle, or a complete circle or even more. Cells 10–28 (−34) μm long, with maximal breadth of 3–7 μm and with minimal breadth of 3–5 μm, moderately thick-walled, densely asperulate, slightly thickened toward their ends, except the terminal cells, which are slightly tapering and have blunt ends. These terminal cells may have an appendage: elongate, slender, smooth-walled hyphae, 2–2.5 μm in diameter, more or less tightly coiled with up to 11 spirals of (10-) 12–20 μm in diameter. Asci globose to ovate 4–5 × 5–5.5 μm diameter, eight-spored. Ascospores yellow, smooth-walled, lenticular, 1.7 × 2.7 μm.

Colonies on glucose peptone agar are rapid growing, flat with a dense downy to floccose chamois-like surface, tawny-buff to pink-buff in color. Reverse red to purplish, resembles some isolates of *M. fulvum.* (See *A. fulva.*) Macroconidia are fairly thick-walled, echinulate, clavate to bullet-shaped, 24–58 × 7.5–12 µm, 4–6 septa, borne singly. Microconidia single-celled, clavate, and abundant.

C. Discussion

This fungus has been isolated from soil of Guinea, Africa (4, 45). Seemingly, this is the only report of its isolation (99). Its ability to produce disease is unproven.

XXXIV. *Arthroderma flavescens* Rees, Sabouraudia 5:206–207, 1967

A. Conidial state

Trichophyton flavescens Padhye & Carmichael, Can J Bot 9: 1535, 1971.

B. Description

Heterothallic. Ascocarp globose, pale buff, 450–650 µm in diameter, excluding appendages. Peridial hyphae hyaline, septate, with two inwardly curving branches developed from apex of some peridial cells. Cells dumbbell-shaped, thick-walled, asymmetrically constricted, echinulate, 11–16.5 × 3–4.8 µm. Appendages smooth-walled, septate; spiral hyphae usually terminal. Smooth-walled, septate macroconidia are rarely attached to peridial hyphae. Asci subglobose, thin-walled, evanescent, 5.9–8.3 × 4.8–7.5 µm, eight-spored. Ascospores hyaline, smooth, oblate, with equatorial rim, 3.8–4 × 1.8–2 µm, yellow in mass.

Colonies on glucose peptone agar white to pale yellow with buff center. Reverse bright yellow, darkening to yellow-brown. Microconidia borne laterally or terminally smooth, thin-walled, sessile or on short pedicels, unicellular, occasionally two-celled, pyriform, ovate, 5–16 × 4–8 µm. Macroconidia numerous, hyaline, usually 2–6 septa, smooth, thin-walled, cylindrical with rounded apices, 26–86 × 8–14 µ.

C. Discussion

This species is found in Australia, where it was isolated from feathers (111).

XXXV. *Arthroderma gertleri* Bohme, Mykosen 10:251, 1967

A. Conidial state

Trichophyton vanbreuseghemii Rioux, Jarry & Jiminer, Nat Monspel Ser Bot 16:153–162, 1964.

B. Description

Heterothallic. Ascocarp globose, pale yellow, 200–600 µm in

diameter, excluding appendages. Peridial hyphae hyaline, septate, uncinately branched, curving over the ascocarp. Cells dumbbell-shaped, echinulate, usually asymmetrically, but occasionally symmetrically constricted. Appendages septate spirals varying in length and number of turns, borne terminally. Multiseptate, thin-walled, smooth, cylindrical macroconidia are sometimes attached to the pyridial hyphae. Asci subglobose, thin-walled, evanescent, 4 × 5 μm, eight-spored. Ascospores hyaline smooth, oblate, 1.5–2.8 × 2–2.5 μm; yellow in mass.

Colonies flat or with radial grooves, floccose at first, becoming velvety [described as the texture of "fine glove leather" (45)] with a finely granular center, white to buff. Reverse creamy white to pale buff. Microconidia pyriform, subglobose, ovoid, thin-walled, unicellular, sessile or on short pedicels, borne laterally or terminally, 2–7 × 1.5–2.5 μm. Macroconidia numerous, thin-walled, smooth, cylindrical, multiseptate, borne singly, 30–55 × 6–8 μm.

C.   Discussion

This dermatophyte is probably worldwide in distribution and has been isolated from a tinea of the hand (45). Large-spored ectothrix involvement of guinea pig hair is reported (41). The fungus has been isolated from soil (45, 111).

XXXVI.   *Arthroderma gloriae* Ajello, Mycologia 59:257, 1967.

A.   Conidial state

Trichophyton gloriae Ajello, Mycologia 59:257, 1967.

B.   Description

Heterothallic. Ascocarp globose, pale yellow, 250–450 μm in diameter, excluding appendages. Peridial hyphae thin-walled, septate, hyaline, uncinately branched. Cells dumbbell-shaped, echinulate, symmetrically constricted, 7.5 × 5 μm. Appendages moderately slender, thin-walled hyphae pointed at the end, 150 μm long. Asci subglobose, thin-walled, evanescent, 3.5 × 4.6 μm, eight-spored. Ascospores hyaline, thin-walled, smooth, oblate, 2–2.5 ×1.5–2 μm, yellow in mass.

Colonies downy, slightly folded, pale cream with a brownish center. Reverse yellow with yellow-brown center. Microconidia pear-shaped, borne singly on short pedicels or sessile, 1.5–6 × 1.5–2.5 μm. Macroconidia numerous, borne singly, more often in clusters of 4–30, cylindrical, multiseptate, with 1–10 septa, smooth walls up to 10 μm thick, 9–60 × 3–7 μm.

C. Discussion
This fungus has been reported from the soil and on animal hair from several states of the United States (45).

XXXVII. *Arthroderma insingulare* Padhye & Carmichael, Sabouraudia 10:47, 1972

A. Conidial state
Trichophyton terrestre, sensu lato Durie and Frey, Mycologia 49:401, 1957.

B. Description
Heterothallic. Ascocarp globose, white to pale yellow, 250–500 µm in diameter, excluding appendages. Peridial hyphae pale yellow, hyaline, septate, uncinately branched, usually on the outside of the main hypha. Cells thick-walled, echinulate, asymmetrically dumbbell-shaped, 8–12 µm long, 5–6 µm wide at enlarged ends and 3–4 µm wide at the internode. Appendages spiral, septate, smooth-walled, terminal hyphae, which vary considerably in length and number of turns. Asci subglobose, thin-walled, evanescent, 4–6 × 3.5–5 µm, eight-spored. Ascospores hyaline, smooth, oblate, 2.5–3 × 2.2–2.5 µm, yellow in mass.

Colonies on glucose peptone agar are white and fluffy, becoming pale yellow, and downy to granular. Reverse yellowish-brown. Some strains develop a red pigment on the reverse. Microconidia elongate, pyriform, borne singly or in groups, 3–6.5 × 1.5–3.5 µm. Macroconidia cylindrical or rarely clavate, smooth, thin-walled, sessile, 8–52 × 4–5 µm, with 2–6 septa. Intermediate forms present. Terminal or intercalary chlamydospores, spirals, racquet hyphae, antler-like hyphae, and nodular bodies occasionally seen. The colonial description of *T. terrestre* given at this point for convenience should more properly have been introduced under *Arthroderma quadrifidum*, where it was first used.

C. Discussion
This species has been isolated from soil, hair, and feathers. The fungus is known to occur in Canada, the United States, Hungary, and Czechoslovakia.

XXXVIII. *Arthroderma lenticularum* Pore, Tsao & Plunkett, Mycologia 57: 970–971, 1965

A. Conidial state
Trichophyton terrestre, sensu lato Durie & Frey, Mycologia 49:401, 1957.

B. Description
Heterothallic. Ascocarp globose, pale buff, 300–600 µm in diameter, excluding appendages. Peridial hyphae pale yellow, hy-

aline septate, uncinately branched usually to one side, the outside of the main hyphae. Cells thick-walled, echinulate, dumbbell-shaped, symmetrically constricted, 5.5–8.5 × 7–10 μm. Appendages septate, spiral hyphae borne terminally. Asci subglobose, thin-walled, evanescent, 4–4.8 × 5–5.6 μm, eight-spored. Ascospores hyaline, smooth, oblate, 2.2–3 × 1.5–7.5 μm, yellow in mass.

    For colonies see description of *A. insingulare*. This fungus is found in soil in the United States (41).

XXXIX. *Arthroderma quadrifidum* Dawson & Gentles, Sabouraudia 1:56–57, 1961

    A. Conidial state

        Trichophyton terrestre, sensu lato Durie & Frey, Mycologia 49:401, 1957. Synonym: *Trichophyton thuringiense* Koch, Mykosen 12:287–290, 1969. (See Ref. 63.)

    B. Description

        Heterothallic. Ascocarp globose, pale buff, 400–700 μm in diameter, excluding appendages. Peridial hyphae pale yellow, hyaline, septate, uncinately branched usually to one side, the outside of the main hypha. Cells thick-walled, strongly echinulate, dumbbell-shaped when young but when mature resembling a short humerus bone with condyles much accentuated and formed on one face only, 8–13 × 5–9 μm. Appendages septate spirals varying considerably in length and number of turns, borne terminally. Asci subglobose thin-walled, evanescent, 4–6 × 3.5–5 μm, eight-spored. Ascospores hyaline, smooth, oblate, 2.5–3 × 2.2–3 μm, yellow in mass.

        Colony (see description of *A. singulare*).

    C. Discussion

        This fungus is of worldwide occurrence and is found in soil, on feathers, and on animal hair (41).

        Pore and Plunkett (112) reported three more teleomorphs in the *T. terrestre* complex, but these new species were not named or described by the authors and so have not been included in compilations of the *T. terrestre* complex (38, 41, 45, 106).

XL. *Arthroderma uncinatum* Dawson & Gentles, Sabouraudia 1:55, 1961

    A. Conidial state

        Trichophyton ajelloi (Vanbreuseghem) Ajello, Sabouraudia 6:148, 1968. Synonym: *Keratinomyces ajelloi* Vanbreuseghem, Bull Acad R Med Belg. 38:1076, 1952.

    B. Description

        Heterothallic. Ascocarp globose, pale buff, 300–900 μm in diameter, excluding appendages. Peridial hyphae pale yellow, hyaline, septate, uncinately branched, usually to the outside of the main hyphae. Cells thick-walled, strongly echinulate, symmetrically dumbbell-shaped, 7–11 × 4–7 μm. Appendages septate, smooth-walled, spirals that vary considerably in length and number of turns, borne terminally. Sometimes smooth-walled, multiseptate, fusiform macroconidia are produced by peridial hyphae. Asci subglobose, thin-walled, evanescent, 5.4–7.2 × 5–6.5 μm, eight-spored. Ascospores hyaline, smooth, oblate, 2.5–3.2 × 1.8–3 μm, yellow in mass.

Colonies flat to folded, with a downy to powdery surface, cream to orange-tan. Reverse with or without diffusable, vinaceous red pigment that becomes bluish-black with age. Macroconidia produced in abundance, cylindrical to fusiform, multiseptate (8–12 septa), thick-walled, smooth, 18–60 × 8–12 μm. Microconidia present in some strains, pyriform to obovate, 3–9 × 2–5 μm (56).

C. Discussion

It is to be noted that there are two colonial forms of *T. ajelloi*: one producing a diffusible bluish-black pigment, the other not. In the original description of *A. uncinatum* it was recorded that nonpigmented varieties did not mate with pigmented ones (38). Moreover, the size of the macroconidia of the two forms has been reported to differ (113). Thus, the suggestion was made that the nonpigmented isolates might represent a distinct taxon (113). Weiztman et al. (114), however, and Padhye and Carmichael (111, 115) examined a larger number of strains, observed fertile crosses between pigmented and nonpigmented isolates, and concluded that the lack of pigment did not denote a separate taxon.

Weitzman and Silva-Hutner (116) studied the genetics of pigmentation in *A. uncinatum* and concluded from the segregation pattern that there was a single gene difference between pigmented and nonpigmented parents. *Arthroderma uncinatum* has a haploid complement of four chromosomes (59). A variety of *T. ajelloi* with very small macroconidia has been described at *T. ajelloi* var. *nanum* (117, 118)

F. Saprophytic Species: Teleomorphs Unknown

XLI.   *Epidermophyton stockdaleae* Prochacki & Englehardt-Zasada, Mycopathol. Mycol. Appl. 54:341–345, 1974

A. Description

Colonies grow fairly rapidly on glucose peptone agar, flat, with a slightly umbonate center, powdery to granular, and buff cream-colored. The reverse is yellow, becoming ferruginous to blush ferruginous. Macroconidia numerous, smooth-walled, clavate, or cylindrical, borne laterally or terminally, 20–50 × 3.8–12.9 μ, with 2–9 septa. Microconidia not formed. Degenerate macroconidia such as those seen in *A. simii* are observed.

B. Discussion

This species was isolated from soil in Szezecin, Poland.

XLII. *Microsporum magellanicum* Coretta & Piontelli, Sabouraudia 15: 1–10, 1977

A. Description

Colonies on glucose peptone agar are finely granular, powdery, cream to yellowish in color; reverse yellow to ochre. Macroconidia are hyaline, ovate or clavate, rarely cylindrical, measuring 14.4–21.6 × 4.8–7.2 μm, verrucous and pedicillate-walled also verruculose, with 1–2 or rarely 4–6 septa, borne singly. Microconidia not seen, but young macroconidia are pyriform or clavate with a truncate base, lightly verruculose, measuring 2.4–3 ×4.8–5.8 μm, sessile or pedunculate often with a septum. Chlamydospores observed rarely.

B. Discussion

This keratinophilic fungus was isolated from soil collected in the extremity of Chile, in southern Shetland, and the Antarctic continent. It has only been described in this saprophytic situation.

XLIII. *Microsporum ripariae* Hubalek & Rush-Munro, Sabouraudia 11: 287–289, 1973

A. Description

Colonies on glucose peptone agar are densely powdery to velvety and floccose with some zonate growth. Reverse is light yellow or lemon yellow and brownish in the center. Macroconidia are produced in limited numbers, borne singly, hyaline, relatively thin-walled, verrucose to verruculose, fusiform to cigar-shaped or almost cylindrical, 27–47 × 6.3–11 μm, with 4–6 septa. Microconidia are numerous, unicellular, hyaline, smooth, thin-walled, and sessile or borne on short, lateral branches, often in bunches. They are pyriform and on the average 3.2 × 2.1 μm. Arthroconidia and terminal chlamydospores are occasionally observed.

B. Discussion

This fungus has only been reported one time. It was isolated from feathers and nests of the sand martin (*Riparia riparia*) in central Europe. Although isolated from a seemingly saprophytic source, it is capable of causing experimental infections in guinea pigs and humans.

XLIV. *Trichophyton fischeri* Kane, Sabouraudia 15:231–241, 1977 (50)

    A. Description

Colonies develop moderately slowly, are velvety to cottony, and white. The reverse is blood red. Macroconidia are sinuous, few in number, and approximately 50 μm long by 3–5 μm wide. Filaments bear club-shaped lateral projections that are 9–12 μm long by 6–8 μm wide. These projections have a broad attachment to the hyphae and have 0–3 septa. Two types of microconidia are seen: slender, club-shaped conidia 4–5 μm long by 2 μm wide, and subglobose microconidia 3–4 μm long by 2–3 μm wide. Both types are borne singly on the hyphae (46, 50).

    B. Discussion

T. fischeri appears to be a saprophyte (46). A recent report of the first isolation of the fungus in the United States again affirms its saprophytic occurrence (119). *T. fischeri* resembles closely *T. rubrum* and must be distinguished from the pathogen (114).

XLV. *Trichophyton longifusum* (Florian & Galgoczy) Ajello, Sabouraudia 6:148, 1968

    A. Description

Colonies on glucose peptone agar are fluffy to powdery, and white to pale yellow in color. The reverse is yellowish-brown. Macroconidia are long and cylindrical, 7–9 μm in diameter, and up to 300 μm long, occurring in clusters, and characteristically may arise from one another and merge into terminal hyphae. Microconidia not reported in original description (117).

    B. Discussion

Isolated from soil in Hungary (120). The species was transferred to the genus *Trichophyton* by Ajello (117) and otherwise has seemingly received scant attention.

XLVI. *Trichophyton mariatii* Tapia de Fossaeri, Mizrachi, Padhye et Ajello, Proc. 5th Int'l. Conf. on the Mycoses, Sci. Publ. no. 396, 154–158, 1980

    A. Description

Colonies grow fairly rapidly on glucose peptone agar and are flat and velvety, pale yellow; reverse dark yellow to apricot. Macroconidia are scanty, smooth, thin-walled, carrot-shaped,

3 to 8 single cells on short stalks or sessile, 35–56 × 6–18
μm. Microconidia sessile or with short stalks; varying sizes and
shapes: elongated, clove- or pear-shaped; mostly one-celled,
rarely two- or three-celled, 3–6.5 × 1.3–2.0 μm.

B. Discussion

A saprophyte isolated from soil in Caracas, Venezuela. The
basis for differentiating *T. mariatii* from other geophilic species
is the shape of the macroconidium: broad at the base (8–10
μm) and narrower at the tip (4–5 μm), resembling a carrot.
Attempts to cross *T. mariatii* with strains of *A. benhamiae*, *A.
lenticularum*, and *A. quadrifidum* failed.

XLVII. *Trichophyton phaseoliforme* Borelli & Feo, Acta Med Venez 13:
176–177, 1966

A. Description

Colonies on glucose peptone agar are powdery, white to bright
cinnamon colored. Microconidia are cylindrical and produced
in terminal clusters. Microconidia are curved and "cashew-
nut" shaped, borne laterally on the hyphae and in large numbers on thickened and enlarged hyphae in the ascocarp-like
(pycnidia?) structures.

B. Discussion

This species deserves further work to establish various morphological features that might alter its current taxonomic treatment.
It was isolated from soil in Romania, from an apparently normal rodent in Venezuela, and is perhaps worldwide in distribution (45).

NOTE ADDED IN PROOF

During the last revision of the manuscript for this chapter, a review article on
tinea capitis appeared in the literature. Since the review contained 354 references
and dealt predominantly with therapy, it is included here as an addendum which
will interest some readers of this chapter (121).

REFERENCES

1.   KJ Kwon-Chung, JE Bennett. Medical Mycology. Philadelphia: Lea & Febiger,
     1992.
1a. AA Padhye, I Weitzman. In: L Ajello, R Hay, volume 4 eds. Mycology, Topley &
     Wilson's Microbiology and Microbial Infections, ninth edition. London, Sidney,
     Auckland: Arnold Publishers, 1998. New York: co-published in the USA by Oxford
     University Press, Inc, 1998.

2. JW Rippon. Medical Mycology. Philadelphia: Saunders, 1988.

3. BC West, KJ Kwon-Chung. Mycetoma caused by Microsporum audouinii. Amer J Clin Pathol 73:447–454, 1980.

4. AWJ Chen, JWL Kuo, J-S Chen, C-C Sun, S-F Huang. Dermatophyte pseudomycetoma: A case report. Brit J Derma 129:729–732, 1993.

5. MG Rinaldi, EA Lamazor, EH Roeser, CJ Wegner. Mycetoma or pseudomycetoma? A distinctive mycosis caused by dermatophytes. Mycopathologia 81:41–48, 1983.

6. DE Allen, R Snyderman, L Meadows, SR Pinnell. Generalized Microsporum audouinii infection and depressed cellular immunity associated with a missing plasma factor required for lymphocyte blastogenesis. Amer J Med 63:991, 1977.

7. M Lowinger-Seoane, JM Torres-Rodriguez, N Madrenys-Brunet, S Aregall-Fusté, P Saballs. Extensive dermatophytoses caused by Trichophyton mentagrophytes and Microsporum canis in a patient with AIDS. Mycopathologia 120:143–146, 1992.

8. MC Grossman, AS Pappert, MC Garzon, DN Silvers. Invasive Trichophyton rubrum infection in the immunocompromised host: Report of three cases. J Amer Acad Derm 33:315–318, 1995.

9. D King, LW Cheever, A Hood, TD Horn, MG Rinaldi, WG Merz. Primary invasive cutaneous Microsporum canis infections in immunocompromised patients. J Clin Microbio 34:460–462, 1996.

10. AN Araviysky, RA Araviysky, GA Eschkov. Deep generalized trichophytosis. Mycopathologia 56:47, 1975.

11. L Ajello. Milestones in the history of medical mycology: The dermatophytes. In: K Iwata, ed. Recent Advances in Medical and Veterinary Mycology. Proceedings of the Sixth Congress of the International Society for Human and Animal Mycology. Tokyo: University of Tokyo Press, 1977, pp. 3–11.

12. GC Ainsworth. Introduction to the History of Mycology. Cambridge: Cambridge University Press, 1976.

13. FM Keddie. Medical mycology, 1841–1870. In: Medicine and Science in the 1860s. London: Wellcome Institute of the History of Medicine, 1969, pp. 137–140.

14. JL Schoenlein. Zur Pathogenie der Impetigines. Arch Anat Physiol 82, 1839.

15. R Remak. Zur kenntniss von der pflanzlichen Natur der Porrigo lupinosa. Med Ztg 9:73–74, 1940.

16. R Remak. Gelungene Impfung des Favus. Med Ztg 11:137, 1842.

17. R Remak. Diagnostische und pathogenetische Untersuchungen in der Klinik des Hern Geh. D Raths. Schoenlein auf dessen Veranlassung angestellt und mit Benutzung andersweitiger Beobachtungen veroffentlecht. Berlin: Hirschwald, 1845.

18. JA Alkiewicz. On the discovery of Trichophyton schoenleinii (Achorion schoenleinii). Mycopathol Mycol Appl 33:28–32, 1967.

19. W Bullock. The History of Bacteriology. London: Oxford University Press, 1938 (reprint, New York: Dover, 1960).

20. D Gruby. Mémoire sur une végétation qui constitue la vraie teigne. C R Acad Sci (Paris) 13:72–75, 1841.

21. D Gruby. Sur les mycodermes qui constituent la teigne faveuse. C R Acad Sci (Paris) 13:309–311, 1841.

22. D Gruby. Sur une espèce de mentagre contagieuse résultante du développement d'un nouveau cryptogame dans la racine depoils de la barbe de l'homme. C R Acad Sci (Paris) 15:512–513, 1841.

23. C Robin. Histoire naturelle des végétaux parasites qui croissent sur l'homme et sur les animaux vivants. Paris, 1853.

24. D Gruby. Recherches sur la nature, la siège et le développement du porrigo décalvans ou phytoalopécie. C R Acad Sci (Paris) 17:301–303, 1843.

25. D Gruby. Recherches sur les cryptogames qui constituent la maladie contagieuse du cuir chevelu décret sous le nom de teigne tondante (Maton), Herpes tonsurancs (Cazenave). C R Acad Sci (Paris) 18:583–585, 1844.

26. D Gruby. Recherches anatomiques sur une plante cryptogame qui constitue le vrai muguet des enfants. C R Acad Sci (Paris) 14:634–636, 1842.

27. L Le leu. Le Dr Gruby, P-V Stock, ed. Paris, 1908.

28. A De Bary. Vergleichende Morphologie und Biologie die Pilze, Mycetozoen, und Bacterien, Leipzig, 1884. (English transl. Oxford, at the Clarendon Press 1887).

29. LR Tulasne, C Tulasne. Selecta fungorum carpologia. 3 vols. Paris, 1861–1865. (English transl. by WB Grove, ed. by AHR Buller, Oxford, 1931).

30. F Loeffler. Vorlesungen uber die geschichtliche Entwickelung der Lehre von den Bakterien. Leipzig: FCW Vogel, 1887. (English transl. by DH Howard).

31. RJA Sabouraud. Les Teignes. Paris: Masson et Cie, 1910.

32. CW Dodge. Medical Mycology. St. Louis: Mosby, 1935.

33. CW Emmons. Dermatophytes: Natural grouping based on the form of the spores and accessory organs. Arch Derm Syphilol 30:337–362, 1934.

34. PM Stockdale. Nannizzia incurvata gen nov sp nov a perfect state of Microsporum gypseum (Bodin) Guiart et Grigorakis. Sabouraudia 1:41–48, 1961.

35. A Nannizzi. Richerche sull'origine saprofitica del funghi delle tigne. 2. Gymnoascus gypseum sp n forma ascofora del Sabouraudites (Achorion) gypseum (Bodin) Ota et Langeron. Atti Accad Fisiocr Siena 10:89–97, 1927.

36. DM Griffin. A perfect stage of Microsporum gypseum. Nature (London) 186:94–95, 1960.

37. CO Dawson, JO Gentles. Perfect stages of Keratinomyces ajelloi. Nature (London) 183:1345–1346, 1959.

38. CO Dawson, JO Gentles. The perfect states of Keratinomyces ajelloi Vanbreuseghem, Trichophyton terrestre Durie and Frey, and Microsporum nanum Fuentes. Sabouraudia 1:49–57, 1961.

39. KJ Kwon-Chung. Genetics of fungi pathogenic for man. Crit Rev Microbio 3:115–133, 1974.

40. I Weitzman, RC Summerbell. The Dermatophytes. Clin Microbio Rev 8:240–259, 1995.

41. ES Beneke, AL Rogers. Medical Mycology and Human Mycoses. Belmont, CA: Star, 1996.

42. J Kane, IF Salkin, I Weitzman, CM Smitka. Trichophyton raubitschekii sp nov. Mycotaxon 13:259–266, 1981.

43. RC Summerbell. Trichophyton kanei, sp nov, a new anthropophilic dermatophyte. Mycotaxon xxvii(2):509–523, 1987.

44. J Kane, JA Scott, RC Summerbell, B Diena. Trichophyton krajdenii, sp nov, an anthropophilic dermatophyte. Mycotaxon 45:307–316, 1993.

45. G Rebell, D Taplin. Dermatophytes: Their Recognition and Identification. Coral Gables, FL: University of Miami Press. (See 4th printing, 1979, for citations in this chapter.)
46. J Kane, R Summerbell, L Sigler, S Krajden, G Land. Laboratory Handbook of Dermatophytes: A Clinical Guide and Laboratory Handbook of Dermatophytes and Other Filamentous Fungi from Skin, Hair, and Nails. Belmont, CA: Star, 1997.
47. DH Larone. Medically Important Fungi. Washington, DC: ASM, 1995.
48. R Summerbell. Identifying Filamentous Fungi. Belmont, CA: Star, 1996.
49. LK Georg, LB Camp. Routine nutritional tests for the identification of dermatophytes. J Bacteriol 74:113–121, 1957.
50. J Kane. Trichophyton fischeri sp nov: A saprophyte resembling Trichophyton rubrum. Sabouraudia 15:231–241, 1977.
51. RS Currah. Taxonomy of the Onygenales; Arthrodermataceae, Gymnoascaceae, Myxotrichaceae, and Onygenaceae. Mycotaxon 24:1–216, 1985.
52. I Weitzman, MR McGinnis, AA Padhye, L Ajello. The genus Arthroderma and its later synonym Nannizzia. Mycotaxon 25:505–518, 1986.
53. M Kawasaki, M Aoki, H Ishizaki, N Nishio, T Mochizuki, S Watanabe. Phylogenetic relationships of the genera Arthroderma and Nannizzia inferred from mitochondrial DNA analysis. Mycopathologia 118:95–102, 1992.
54. M Hironaga, T Fujisaki, S Watanabe. Trichophyton mentagrophytes skin infections in laboratory animals and a cause of zoonosis. Mycopathologia 73:101, 1981.
55. L Ajello, L Bostick, SL Cheng. The relationship of Trichophyton quinckeanum to Trichophyton mentagrophytes. Mycologia 60:1185–1189, 1968.
56. AA Padhye, JW Carmichael. Mating behavior of Trichophyton mentagrophytes varieties paired with Arthroderma benhamiae mating types. Sabouraudia 7:178–181, 1969.
57. I Weitzman. Genetic studies of mating reactions of zoopathogenic fungi. In: ES Kuttin and GL Gaum, eds. Human and Animal Mycology. Amsterdam: Excerpta Medica, 1980, pp. 251–254.
58. I Weitzman, PW Allderdice, M Silva-Hunter, OJ Miller. Meiosis in Arthroderma benhamiae (-Trichophyton mentagrophytes). Sabouraudia 6:232–237, 1968.
59. I Weitzman, PW Allderdice, M Silva-Hutner. Chromosome numbers in species of Nannizzia and Arthroderma. Mycologia 62:89–97, 1970.
60. JW Rippon. Elastase: Production by ringworm fungi. Science 157:947, 1967.
61. JW Rippon, ED Garber. Dermatophyte pathogenicity as a function of mating type and associated enzymes. J Invest Derma 53:445, 1969.
62. S Chu-Cheung, J Maniotis. A genetic study of extracellular elastinohydrolysing protease in the ringworm fungus Arthroderma benhamiae. J Gen Microbio 47:299–304, 1973.
63. T Mochizuki, K Takada, S Watanabe, M Kawasaki, H Ishizak. Taxonomy of Trichophyton interdigitale (Trichophyton mentagrophytes var interdigitale) by restriction enzyme analysis of mitochondrial DNA. J Med Vet Mycol 28:191–196, 1990.
64. KI Kwon-Chung. Studies on the sexuality of Nannizzia. I. Heterothallism vs fertile isolates. Sabouraudia 6:5–13, 1967.
65. LK Georg, L Ajello, L Friedman, SA Brinkman. A new species of Microsporum pathogenic to man and animals. Sabouraudia 1:189–196, 1962.

66. J Alsop, AP Prior. Ringworm infection in a cucumber greenhouse. Brit Med J 1: 1081, 1961.

67. I Weitzman. Variation in Microsporum gypseum. I. A genetic study of pleomorphism. Sabouraudia 3:195–204, 1964.

68. I Weitzman, M Silva. Linkage group I of Nannizzia incurvata. Mycologia 58:580, 1966.

69. I Weitzman, M Silva. Variation in the Microsporum gypseum complex II. A genetic study of spontaneous mutation in Nannizzia incurvata. Mycologia 58:570–579, 1966.

70. WJ Barson. Granuloma and pseudogranuloma of the skin due to Microsporum canis. Arch Derm 121:895, 1985.

71. KJ Kwon-Chung. Studies on the sexuality of Nannizzia. II. Morphogenesis of gametangia in N. incurvata. Mycologia 61:593–605, 1969.

72. I Bournerias, M Feuihade De Chauvin, A Datry, I Chambrette, J Carriere, A Devidas. Unusual Microsporum canis infections in adult HIV patients. J Amer Acad Derm 35:808, 1996.

73. M Lowinger-Seoane, JM Torres-Rodriguez, N Madrenys-Brunet, S Aregall-Fuste, P Saballs. Extensive dermatophytoses caused by Trichophyton mentagrophytes and Microsporum canis in a patient with AIDS. Mycopathologia 120:143–146, 1992.

74. I Weitzman, AA Padhye. Mating behavior of Nannizzia otae (Microsporum canis). Mycopathologia 64:17–22, 1978.

75. AA Padhye, F Blank, PJ Koblenzer, S Spatz, L Ajello. Microsporum persicolor infection in the United States. Arch Derm 108:561–562, 1973.

76. C DeVroey, C Wuytack-Raes, F Fossoul. Isolation of saprophytic Microsporum praecox Rivalier from sites associated with horses. Sabouraudia 21:255–257, 1983.

77. I Weitzman, S McMillen. Isolation in the United States of a culture resembling Microsporum praecox. Mycopathologia 70:181–186, 1980.

78. AA Padhye, JG Detweiler, A Frumkin, GS Bulmer, I Ajello, MR McGinnis. Tinea capitis caused by Microsporum praecox in a patient with sickle cell anaemia. J Med Vet Mycol 27:313–317, 1989.

79. N Contet-Audonneau, G Percebois. Microsporum persicolor isolement du sol. Bull Soc Fr Mycol 15:193–196, 1986.

80. PM Stockdale. Nannizzia persicolor sp nov, the perfect state of Trichophyton persicolor. Sabouraudia 5:355–359, 1967.

81. D Borelli. Microsporum racemosum nova species. Acta Med Venez 12:148–151, 1965.

82. I Alteras, R Evolceanu. First isolation of Microsporum racemosum-Dante Borelli, 1965 from Romanian soil: New data on its pathogenic properties. Mykosen 12: 223–230, 1969.

83. V Daum, DJ McCloud. Microsporum racemosum: First isolation in the United States. Mycopathologia 59:183, 1976.

84. JW Rippon, TW Andrews. Case report—Microsporum racemosum: Second clinical isolation from the United States and the Chicago area. Mycopathologia 64:187, 1978.

85. D Gruby. Recherches sur les cryptoganes qui constituent la maladie contagieuse

du cuir chevelu decrite sous le nom de tiegne tondante (Mahon), Herpes tonsurans (Cazenave). C R Acad Sci 18:583–585, 1844.

86. S Tanaka, RC Summerbell, R Tsuboi, T Kaaman, T Matsumoto, TL Ray. Advances in dermatophytes and dermatophytosis. J Med Vet Mycol 30 suppl 1:29–39, 1992.

87. AA Padhye, MJ Thirumalachar. Isolation of Trichophyton simii and Cryptococcus neoformans from soil in India. Hindustan Antibiot Bull 9:155–157, 1967.

88. T Matsumoto, L Ajello. Current taxonomic concepts pertaining to the dermatophytes and related fungi. Int J Dermatol 26:491–499, 1987.

89. PM Stockdale. Sexual stimulation between Arthroderma simii Stockdale, Mackenzie and Austick and related species. Sabouraudia 6:176–181, 1968.

90. CN Young. Pseudocleistothecia in Trichophyton rubrum. Sabouraudia 6:160–162, 1968.

91. KJ Kwon-Chung. Genetic study of the incompatibility system in Arthroderma simii. Sabouraudia 10:74–78, 1972.

92. M Takashio. The Trichophyton mentagrophytes complex. In: K Iwata, ed. Recent Advances in Medical and Veterinary Mycology. Proceedings of the Sixth Congress of the International Society for Human and Animal Mycology. Tokyo: University of Tokyo Press, 1977, pp. 271–276.

93. M Heronga, S Watanabe. Mating behavior of 334 Japanese isolates of Trichophyton mentagrophytes. Mycologia 72:1159–1170, 1980.

94. M Takashio. Taxonomy of dermatophytes based on their sexual states. Mycologia 71:968–976, 1979.

95. I Weitzman, AA Padhye. Is Arthroderma simii the perfect state of Trichophyton quinckeanum? Sabouraudia 14:65–74, 1976.

96. M Flammia, P Vannini, EM Difonzo. Tinea capitis in the Florence area between 1985 and 1993. Mycoses 38:325, 1995.

97. NF Conant. Studies in the genus Microsporum. III. Taxonomic studies. Arch Dermat Syphilol 36:781–808, 1937.

98. AA Padhye, I Weitzman, L Ajello. Mating behavior of Microsporum equinum with Nannizzia otae. Mycopathologia 69:87–90, 1979.

99. L Ajello. Natural history of the dermatophytes and related fungi. Mycopath Mycol Appl 53:93–110, 1974.

100. R Vanbreuseghem, C DeVroey, M Takashio. Production of macroconidia by Microsporum ferrugineum Ota 1922. Sabouraudia 7:252–256, 1970.

101. LK Georg, W Kaplan, LB Camp. Trichophyton equinum—A reevaluation of its taxonomic status. J Invest Derm 29:27–37, 1957.

102. LK Georg, W Kaplan, LB Camp. Equine ringworm with special reference to Trichophyton equinum. Amer J Vet Res 18:798–810, 1957.

103. JMB Smith, RD Jolly, LK Georg, MD Connole. Trichophyton equinum var autotrophicum: Its characteristics and geographical distribution. Sabouraudia 6:296–304, 1968.

104. ME Grossman, AS Pappert, MC Garzon, DN Silvers. Invasive Trichophyton rubrum infection in the immunocompromised host: Report of three cases. J Amer Acad Derm 33 Number (2, part 1):315, 1995.

105. AA Padhye, CN Young, L Ajello. Hair perforation as a diagnostic criterion in the identification of Epidermophyton, Microsporum, and Trichophyton species. Pro-

ceedings of the Fifth International Conference on the Mycoses. Pan American Health Organ Sci publ. no. 396, Washington, DC, 1980, pp. 115–120.

106. JW Rippon, M Medenica. Isolation of Trichophyton soudanense in the United States. Sabouraudia 3:301–302, 1964.

107. I Weitzman, S Rosenthal. Studies in the differentiation between Microsporum ferrugineum Ota and Trichophyton soudanense Joyeux. Mycopathologia 84:95–101, 1984.

108. AA Padhye, I Weitzman, E Domenech. An unusual variant of T. tonsurans var sulfureum. J Med Vet Mycol 32:147–150, 1994.

109. GS de Hoog, J Guarro. Atlas of Clinical Mycology. Centralbureau voor Schimmelcultures. Baarn, The Netherlands, 1997.

110. GC Ainsworth, LK Georg. Nomenclature of the faviform trichophytons. Mycologia 46:9–11, 1954.

111. AA Padhye, JW Carmichael. The genus Arthroderma Berkeley. Can J Bot 49: 1525–1540, 1971.

112. PS Pore, OA Plunkett. Biological species and variations in Arthroderma. Mycopath Mycol Appl 31:225–241, 1967.

113. PM Stockdale. Personal observations on the production of sexual forms of dermatophytes. Ann Soc Belge Med Trop 44:821–820, 1964.

114. I Weitzman, I Kozma, M Silva-Hutner. Some observations on Arthroderma uncinatum. Sabouraudia 7:216–218, 1969.

115. AA Padhye, JW Carmichael. The mating reaction in the Trichophyton terrestre complex. Sabouraudia 11:64–69, 1973.

116. I Weitzman, M Silva-Hutner. Genetic studies on the segregation of pigmentation in Arthroderma uncinatum (=Trichophyton ajelloi). Abstr Ann Meet Amer Soc Microbio MM 12:120, 1970.

117. L Ajello. A taxonomic review of the dermatophytes and related species. Sabouraudia 6:147–159, 1968.

118. L Ajello, AA Padhye. An orthographic correction. Mycotaxon 8:383–384, 1979.

119. SA Rosenthal, JS Scott, RC Summerbell, J Kane. First Isolation of *Trichophyton fischeri* in the United States. J Clin Microbiol 36:3389–3391, 1998.

120. E Florian, J Galgoczy. Keratinomyces longifusus sp nov from Hungary. Mycopathol Mycol Appl 24:73–80, 1964.

121. AK Gupta, RC Summerbell. Tinea capitis. Med Mycol 38:255–287, 2000.

6

Ascomycetes

The Onygenaceae and Other Fungi from the Order Onygenales

Lynne Sigler
University of Alberta, Edmonton, Alberta, Canada

I. INTRODUCTION

The ascomycete order Onygenales is important from the medical perspective because it includes the sexual stages of the true fungal pathogens of humans and animals (i.e., the dermatophytes and the dimorphic fungi capable of causing disease in an otherwise healthy host). The Onygenales includes three families: the Arthrodermataceae, including dermatophytes (treated in this volume, Chap. 5); the Onygenaceae, including the dimorphic fungi; and the Gymnoascaceae. Although a prior version of this volume (1) treated the Onygenaceae as part of the Eurotiales, the most recent treatment of the ascomycetes (2) maintains it within the order Onygenales separately from the Eurotiales. This chapter describes the pathogenic members of the Onygenaceae as well as some nonpathogenic species that may resemble them and that are rare to common contaminants in clinical specimens. A few members of the Gymnoascaceae are also treated. A fourth family, the Myxotrichaceae, formerly included within the Onygenales (3, 4), appears instead to have affinities to the inoperculate discomycetes (5). Notwithstanding its probable disparate relationship to onygenalean fungi, some members of the Myxotrichaceae are included here for convenience.

Although phylogenetic relationships of many of the true pathogenic fungi are known, most medical mycologists continue to use the name of the anamorph

(asexual or mitotic stage) rather than the name of the teleomorph (sexual or meiotic stage). The reasons are because a vast body of literature has been published
in which these names are used and because the majority of these fungi are heterothallic in compatibility. The anamorph is the stage commonly encountered in
primary isolation from the specimen, and the teleomorph is seldom seen.

The species described in this chapter are ones with known or inferred placement within the order Onygenales. A few nonpathogenic "look-alikes" of uncertain affinity are included for convenience. Species are described under the name
in common use (i.e., usually the name for the anamorph). Table 1 lists the species
in the order in which they are covered and provides comments on their known
or inferred teleomorphs and occurrence as pathogens.

II. TAXONOMIC CONCEPTS

Heterothallic compatibility is common among onygenalean fungi and the teleomorph can be obtained only through mating trials. These tests are of great value
in confirming identity and in establishing biological relationships, but are rarely
employed in diagnostic laboratories because of the need for appropriate test isolates, the length of time required for development and maturity of the sexual
structures, and—at least for the dimorphic fungi—the risks of handling the fungi.
For this reason, medical mycologists have paid greater attention to the characteristics of the anamorph (asexual or mitotic stage) because it is the stage encountered in primary isolation from the specimen. Even though connections between
anamorphs and teleomorphs have been firmly established for many fungi, the
physical separation between the anamorph and teleomorph means that relationships often have been difficult to discern. The correct phylogenetic position of
Cooccidioides immitis, for example, a fungus known for 100 years and for which
there is a vast body of literature, remains uncertain, but modern approaches place
it clearly within the Onygenaceae. A suggestion of a close relationship among
Histoplasma capsulatum, *Blastomyces dermatitidis*, and *Emmonsia* species,
made first in the late 1940s, has been borne out by evidence that their teleomorphs occur in the same genus of the Onygenaceae and by molecular phylogenetic studies.

III. CHARACTERISTICS OF THE ORDER AND FAMILIES

Members of the Onygenales are united by their formation of prototunicate asci
with eight ascospores and by ascospores that are small (usually less than 8 µm
in length), single-celled, and light colored (never dark brown or black). Prototunicate asci are usually globose or subglobose; they are irregularly arranged within

Table 1 Overview of the Species in Order of Their Coverage and Comments on Their Known or Inferred Teleomorphs and Pathogenic Potential

Anamorph	Teleomorph	Family affinity	Occurrence as a pathogen	Figure number
Histoplasma capsulatum	*Ajellomyces capsulatus*	Onygenaceae	Dimorphic pathogen; common but limited in endemic areas; may be associated with outbreaks	1, 2
Histoplasma capsulatum var. *duboisii*	*Ajellomyces capsulatus*	Onygenaceae	Dimorphic pathogen; sporadic even in endemic areas	
Histoplasma capsulatum var. *farciminosum*	Not known; relationship to *Ajellomyces* inferred	Onygenaceae	Dimorphic pathogen; sporadic, mainly causing infection in horses in endemic areas	
Blastomyces dermatitidis	*Ajellomyces dermatitidis*	Onygenaceae	Dimorphic pathogen; sporadic even in endemic areas; may be associated with outbreaks	1, 3
Emmonsia crescens	*Ajellomyces crescens*	Onygenaceae	Dimorphic pathogen; very rarely pathogenic to humans; common in rodents and small mammals as shown by surveys	1, 4
Emmonsia parva	Not known; relationship to *Ajellomyces* inferred	Onygenaceae	Dimorphic pathogen; very rarely pathogenic to humans; sporadic in rodents and small mammals as shown by surveys	
Emmonsia pasteuriensis	Not known; relationship to *Ajellomyces* inferred	Onygenaceae	Dimorphic pathogen; known from a single case of infection	

Table 1 Continued

Anamorph	Teleomorph	Family affinity	Occurrence as a pathogen	Figure number
Paracoccidioides brasiliensis	Not known; relationship to *Ajellomyces* inferred	Onygenaceae	Dimorphic pathogen; common but limited in endemic areas	5
Coccidioides immitis	Not known; relationship to *Uncinocarpus* inferred	Onygenaceae	Dimorphic pathogen; common in endemic areas; may be associated with outbreaks	6
Chrysosporium	*Aphanoascus fulvescens*	Onygenaceae	Rarely pathogenic; common cutaneous contaminant and rare agent of cutaneous infection.	7
Chrysosporium zonatum	*Uncinocarpus orissi*	Onygenaceae	Pathogenic; soil fungus known from a single case of deep infection and two cases of pulmonary colonization	8
Chrysosporium	*Nannizziopsis vriesii*	Onygenaceae	Pathogenic; uncommon agent of cutaneous infection in reptiles	
Chrysosporium	*Renispora flavissima*	Onygenaceae	Nonpathogenic; soil fungus shown for comparison with *H. capsulatum*	9
Chrysosporium keratinophilum	*Aphanoascus keratinophilus*	Onygenaceae	Nonpathogenic; uncommon contaminant of cutaneous specimens; shown for comparison	10
Chrysosporium articulatum	Not known	Onygenaceae	Nonpathogenic; uncommon but regular contaminant of cutaneous specimens; compare *C. anamorph* of *Aphanoascus fulvescens*	11

Chrysosporium carmichaelii	Not known	Not known	Nonpathogenic; uncommon but regular contaminant of cutaneous specimens; compare *C. undulatum*	12
Chrysosporium lobatum	Not known	Not known	Nonpathogenic; uncommon but regular contaminant of cutaneous specimens	13
Malbranchea	*Uncinocarpus reesii*	Onygenaceae	Nonpathogenic; uncommon contaminant; shown for comparison with *C. immitis*	14
Malbranchea gypsea	Not known	Not known	?Pathogenic; status as agent of onychomycosis not confirmed	15
Onychocola canadensis	*Arachnomyces nodosetosus*	Gymnoascaceae	Pathogenic; uncommon agent of onychomycosis	16
anamorph absent	*Gymnascella dankaliensis*	Gymnoascaceae	Pathogenic; uncommon agent of onychomycosis	17
anamorph absent	*Gymnascella hyalinospora*	Gymnoascaceae	Pathogenic; uncommon contaminant; known from single case of pulmonary infection	18
Geomyces pannorum	Not known	Myxotrichaceae	Nonpathogenic; common contaminant of cutaneous specimens	19
anamorph absent or arthroconidia	*Myxotrichum deflexum*	Myxotrichaceae	Nonpathogenic; uncommon contaminant of cutaneous specimens	20, 21
Ovadendron sulphureo-ochraceum	Not known	?Myxotrichaceae	Pathogenic; rare agent of eye infection	22

the ascomata (ascocarps, sexual fruiting bodies) and their cell walls lyse at or near maturity, allowing for passive discharge of the ascospores. Ascomata have varied morphologies ranging from nonostiolate, usually globose, pseudoparenchymatous structures (i.e., cleistothecia) to a loose network of differentiated hyphal cells or of branched hyphae (i.e., often called gymnothecia) that sometimes extend into elaborate hooked, spiralled, or branched appendages; to clusters of more or less naked asci and ascospores. Under cultural conditions, the ascomatal appendages are sometimes found associated with the anamorph. Anamorphs are solitary, single-, or multicelled aleurioconidia or alternate arthroconidia in which part or all of the subtending or intervening cell lyses to release the conidium. Characters that have been used to separate families include features of the ascospore wall ornamentation, correlated to some extent with ascospore shape, habitat, capacity to degrade keratinous or cellulosic substrates or lacking these enzymatic capacities, and features of the anamorph. Members of the Arthrodermataceae and Onygenaceae (Table 2) are often keratinolytic, and these families are distinguished by their ascospore wall morphologies (smooth in the former and punctate or punctate-reticulate in the latter) and conidial types. The key characters for the Onygenaceae are described below.

A. Family Onygenaceae

Members of the Onygenaceae have an ability to degrade keratinaceous substrates, and are often associated with mammals. Ascospore walls are ornamented with small pits or depressions (also called puncta), with netlike ridges (reticulate) or a combination of both, or with minute spiny projections (muriculate). Anamorphs are aleurioconidia that are predominantly single-celled and placed in the genera *Histoplasma*, *Blastomyces*, *Emmonsia*, *Paracoccidioides*, and *Chrysosporium*, or are alternate arthroconidia and placed in the genus *Malbranchea* (Table 2). It should be noted, however, that the latter two anamorph genera are not monophyletic; some species are of uncertain affinity.

Species of Medical Relevance

1. *Ajellomyces* McDonough & Lewis (Fig. 1).
    Synonym: *Emmonsiella* Kwon-Chung.
    Ascomata (gymnothecia) solitary, discrete, globose, becoming irregularly stellate by formation of coiled appendages, buff, small, 80 to 350 μm in diameter. *Peridial hyphae* pale brown, smooth, branched, individual cells obtusely diamond-shaped (swollen near the center and constricted at the septa) or unswollen. *Appendages* helically coiled, thick-walled, yellowish-brown. *Asci* subglobose, clavate or pyriform, evanescent, eight-spored. *Ascospores* small, globose, hyaline, measuring 1 to 1.5 μm in diameter,

muriculate (having short, sharp outgrowths) by scanning electron microscopy, but appearing almost smooth by light microscopy.

Comments. Similarities in growth habit and conidial type have long been noted among the dimorphic fungi *Histoplasma*, *Blastomyces*, *Emmonsia*, and *Paracoccidioides*, but the species have been retained in separate genera largely based on differences in their parasitic forms, in their virulence, and in the clinical syndromes that they elicit. Recent studies have confirmed a close relationship among them. Cultural mating experiments proved that *Blastomyces dermatitidis*, *Histoplasma capsulatum*, and *Emmonsia crescens* are heterothallic ascomycetes and that their ascomata and ascospores are characteristic of the genus *Ajellomyces* (6–8). Phylogenetic trees derived from comparison of large subunit ribosomal and ITS region DNA sequences place sexual and asexual species of all four genera as members of the *Ajellomyces* clade (9, 10) and show that *Emmonsia* and *Blastomyces* species are more closely related to each other than to *Histoplasma* species. These findings lend some support to arguments that have been made in favor of maintaining *Emmonsiella* as a separate genus, but molecular phylogenetic studies all show the *Ajellomyces* clade to be well supported (9–11), even though the relationships within it are not clearly resolved. Ambiguities are found in ubiquinone data that show that the principal ubiquinones are Q-10(H_2) for *H. capsulatum* and *Emmonsia* species but Q-10 for *B. dermatitidis* (12, 13). *Paracoccidioides* has been placed within the *Ajellomyces* clade, suggesting that its teleomorph (if found) will belong there, but its closest relative is not yet clearly elucidated (9, 10, 14). This evidence of close relationship opens the possibility of combining the anamorphic genera. This could be useful from the taxonomic point of view because relationships become more obvious, but is disadvantageous because the anamorphic names are clinically relevant and in widespread use.

Molecular phylogenetic studies show the *Ajellomyces* clade to be separated from other members of the family Onygenaceae. It may belong in a separate family of the Onygenales. The minute spiny projections on the ascospore wall are unusual within the family, and similar features are found only in one other monotypic genus (8).

a. *Ajellomyces capsulatus* (Kwon-Chung) McGinnis & Katz, Mycotaxon 8:158, 1979
Synonym: *Emmonsiella capsulata* Kwon-Chung, Science 177:368, 1972.
Description as for genus. Ascomata 80 to 250 µm in diameter. *Peridial hyphae* uniform in diameter (unswollen) and not constricted at the septa. *Appendages* helically coiled, smooth, originating near the center

Table 2 Comparison of Arthrodermataceae and Onygenaceae (Onygenales)

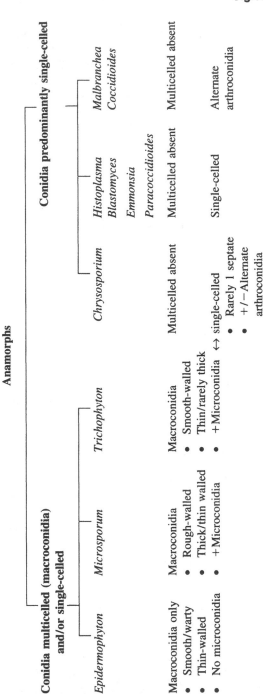

General features of Arthrodermataceae and Onygenaceae

- Often keratinolytic
- Conidia having lytic dehiscence (solitary, single-, or multicelled aleurioconidia and/or alternate arthroconidia)
- Prototunicate asci
- Ascomata with varied morphologies (naked asci, gymnothecia, cleistothecia)
- Ascospores light or brightly colored, single-celled
- Often heterothallic

Anamorphs

Conidia multicelled (macroconidia) and/or single-celled				Conidia predominantly single-celled	
Epidermophyton	*Microsporum*	*Trichophyton*	*Chrysosporium*	*Histoplasma* *Blastomyces* *Emmonsia* *Paracoccidioides*	*Malbranchea* *Coccidioides*
Macroconidia only	Macroconidia	Macroconidia	Multicelled absent	Multicelled absent	Multicelled absent
• Smooth/warty	• Rough-walled	• Smooth-walled			
• Thin-walled	• Thick/thin walled	• Thin/rarely thick	Single-celled ↔ single-celled	Single-celled	Single-celled
• No microconidia	• +Microconidia	• +Microconidia ↔ single-celled	• Rarely 1 septate		Alternate arthroconidia
			• +/– Alternate arthroconidia		

Teleomorph(s)

| Unknown | Arthroderma or unknown | Arthroderma or unknown | Arthroderma | Aphanoascus Uncinocarpus Nannizziopsis other genera or unknown | Ajellomyces or unknown | Uncinocarpus Auxarthron other genera or unknown |

Arthrodermataceae

Onygenaceae

Onygenales
Conidial and ascomatal types in the Onygenales

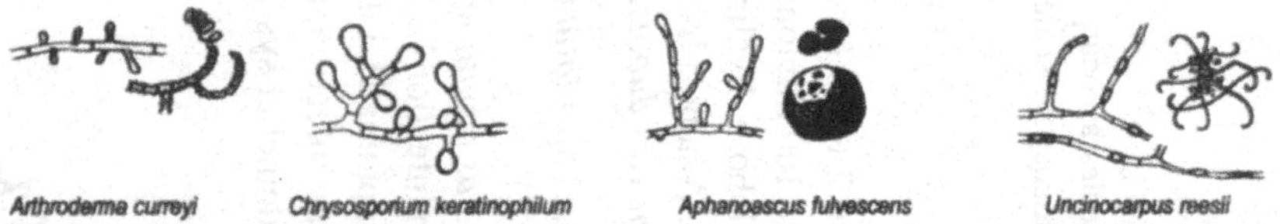

Arthroderma curreyi Chrysosporium keratinophilum Aphanoascus fulvescens Uncinocarpus reesii

Source: illustrations of teleomorphs courtesy of © R. S. Currah.

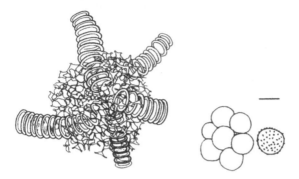

Figure 1 *Ajellomyces* species ascocarp and ascospores (Note that these structures are drawn to different scales; ascospore scale bar approx 1 µm.) Source: ascocarp © R. S. Currah (with permission).

of the ascoma, 2 to 10 (commonly 3–5) per ascoma, 1.7 to 3 µm in diameter, 30 to 100 µm long. *Asci* pyriform or clavate, to 10 µm wide. *Ascospores* globose, hyaline, muriculate, 1.2 to 1.5 µm in diameter. *Heterothallic*. Anamorphs *Histoplasma capsulatum* var. *capsulatum*, *H. capsulatum* var. *duboisii*.

i.   *Histoplasma capsulatum* var. *capsulatum* Darling, JAMA 46:1285, 1906 (Fig. 2).
     Occurrence as a pathogen. Etiologic agent of histoplasmosis capsulati, an infection having several clinical manifestations, including asymptomatic infection (most common), acute or chronic pulmonary infection, and disseminated disease involving the reticuloendothelial system (15–18).

Figure 2 *Histoplasma capsulatum* in vitro and in vivo. Bar approximately 10 µm.

Description (See also Table 3.) Detailed descriptions of the tissue and colonial and microscopic morphologies have been published in medical mycology texts (e.g., 15, 16, 18, 19). In culture, single-celled conidia of two sizes are produced and are commonly referred to as macro- and microconidia. Conidia are sessile (borne directly on the sides of the hypha) or borne at the ends of short or long stalks that have parallel sides (i.e., are unswollen) or are slightly swollen. Macroconidia are globose or broadly pyriform, sometimes smooth, but typically tuberculate; that is, having tuberous or fingerlike projections that vary in length (2–6 μm) and width (0.5–2 μm). Microconidia are globose, smooth to slightly roughened, and predominantly sessile. As shown by Berliner (20) in a detailed study of phenotypic variation, primary subculture may yield albino (also called A type) or brown (B type) colonies or both. Albino colonies are white and cottony and have slightly broader aerial hyphae. They are more likely to have smooth to slightly warty macroconidia and abundant microconidia, but sporulation may be lost in serial subculture. Additionally, they are difficult to convert to the yeast form. Brown colonies are buff to pale brown with a diffusible tan pigment, velvety to almost granular in age, and have narrow hyphae (≤2 μm wide). Sporulation is more stable and consists predominantly of the typical tuberculate macroconidia.

Comments. Use of a commercially available DNA probe (Accu-Probe; Gen-Probe Inc., San Diego, CA) (18, 21) and/or exoantigen test are the methods now most commonly used to confirm identification of an isolate suspected of being *H. capsulatum*. The DNA probe proved useful in delineating an unusual case of dual infection in which culture yielded a mixed growth of *H. capsulatum* and *B. dermatitidis* (22). Conversion to the yeast phase may be achieved by subculture onto enriched media containing cysteine and incubation at 35–37°C, but conversion can be difficult to achieve and may require multiple subcultures. Histopathologic staining of biopsy specimens or smears of fluids demonstrating presence of intracellular small oval yeast cells and history of residence or travel in endemic areas are useful confirmatory findings.

Only a few soil fungi produce large, echinulate to tuberculate conidia with morphologies resembling the macroconidia of *H. capsulatum*, and these can be distinguished by the absence of the microconidial state and differences in colonial morphology. These species are rarely isolated as contaminants in clinical

Table 3 Characteristics of the Dimorphic Fungi[1]

	Ecology & Distribution	Mycelial form	Teleomorph	Parasitic form in vivo	Confirmatory tests	Selected refs
Histoplasma capsulatum var. *capsulatum*	Soil associated with dung of birds & bats. North America in areas along Mississippi, Missouri & Ohio River valleys, rarely in Central & South America, Asia, Europe	Narrow hyphae forming solitary sessile microconidia and tuberculate macroconidia at the ends of unswollen stalks. Microconidia subglobose or broadly pyriform, 4–5 μm long & 3–4 μm wide. Macroconidia broadly pyriform, 8–17 μm long & 8–13 μm wide, or globose, 8–13 μm in diameter, bearing tuberous projections 2–6 μm long and 0.5–2 μm wide.	*Ajellomyces capsulatus*	Intracellular, rarely extracellular, small yeast cells, 3 to 5 μm in length, with single narrow based bud	Convert to yeast form using enriched media (brain heart infusion or blood agar with added cysteine) at 37°C; multiple subculture may be required for complete conversion. Exoantigen shows specific m, or h and m, precipitin bands DNA probe test.	15, 16, 17, 18, 19, 20, 21, 23, 26
Histoplasma capsulatum var. *duboisii*	Rarely isolated from soil, intestinal contents of bats, aardvarks, baboons; habitat poorly known. Central Africa, Madagascar	Similar to var. *capsulatum*	*Ajellomyces capsulatus*	Extracellular, thick-walled yeast cells, 12–15 μm in length, with single narrow based bud, occasionally in short chains	Convert to yeast form using enriched media) at 37°C. Differentiate from var. *capsulatum* by larger size of yeast cells.	15, 16, 17, 18, 19, 26, 27

	Habitat/hosts	Morphology	Teleomorph relationship	Tissue form	Laboratory characteristics	References
Histoplasma capsulatum var. *farciminosum*	Isolated from equines. Africa, Asia, Eastern Europe	Similar to var. *capsulatum* but very slow growing and not sporulating on rich media.	Not known. Relationship to *Ajellomyces* inferred.	Intracellular, rarely extracellular, small yeast cells, 3 to 5 μm in length, with single narrow based bud	Convert to yeast phase on brain heart infusion agar (with or without blood) at 37°C; several subcultures may be required. No specific exoantigen test available.	15, 16, 17, 26, 28, 29
Blastomyces dermatitidis	River & forest soil, beaver dams, habitat poorly known. North America in areas along Mississippi & Ohio River valleys, eastern & central Canada, rarely Africa & Asia	Narrow hyphae forming solitary sessile conidia and 1, rarely 2, conidia at the ends of narrow, usually unswollen stalks. Conidia mostly pyriform, 4–5 μm long & 3–4 μm wide	*Ajellomyces dermatitidis*	Large yeast cells, 10 to 12 μm in diameter, with thick, double contoured wall and single broad-based bud.	Convert to yeast form on enriched (brain heart infusion agar) at 37°C or on conversion media (Kane) at 37°C or lower; serial subculture may be required. Exoantigen shows specific A precipitin band. DNA probe test.	6, 8, 9, 15, 16, 18, 19, 21, 23, 30, 32, 33, 34, 36
Emmonsia crescens	Many animal hosts (about 118 species of rodents, lagomorphs, carnivores, insectivores, marsupials), rarely recovered from soil or other habitats. Worldwide except Africa, Australia	Narrow hyphae forming solitary sessile conidia and 1–3 conidia at the ends of narrow, often slightly swollen stalks. Conidia subglobose, often slightly broader than long, 2.5–4 μm long & 3–5 μm wide, smooth to finely roughened.	*Ajellomyces crescens*	Oval or globose adiaspores, 50 to 500 μm, commonly 300 μm in diameter, walls 10–70 μm thick	Maximum growth temperature 37°C (lower for some strains); forming adiaspores (enlarged chlamydospore-like structures) measuring 20–140 μm. Exoantigen tests may show non-specific precipitin lines against *B. dermatitidis* and *H. capsulatum* antisera	8, 9, 10, 15, 23, 38, 42, 43

Table 3 Continued

	Ecology & Distribution	Mycelial form	Teleomorph	Parasitic form in vivo	Confirmatory tests	Selected refs
Emmonsia parva	Few animal hosts (about 21 species of rodents, lagomorphs, carnivores & marsupials), rarely recovered from soil or other habitats. Geographically restricted (parts of US, Kenya, Europe, Australia)	Narrow hyphae forming solitary sessile conidia and 1–3 conidia at the ends of narrow, often distally swollen stalks. Conidia subglobose, often slightly broader than long, 2.5–4 µm long & 3–5 µm wide; smooth to finely roughened.	Not known Relationship to *Ajellomyces* inferred	Oval or globose adiaspores, 10 to 40 µm in diameter, walls thinner	Maximum growth temperature 40°C, forming adiaspores measuring 8–20 µm diameter. Exoantigen tests may show non-specific precipitin lines against *B. dermatitidis* and *H. capsulatum* antisera	8, 9, 10, 15, 23, 38, 42, 43a
Emmonsia pasteuriana	Known only from Italy.	Narrow hyphae forming solitary sessile conidia and 1–8 conidia at the ends of narrow, distally swollen stalks. Conidia subglobose to oval, often slightly broader than long, 2 µm long & 3–4 µm wide; smooth to finely roughened.	Not known Relationship to *Ajellomyces* inferred	Small and large yeast-like cells; oval to globose with narrow based buds.	Conversion to yeast phase at 37°C on enriched media; no adiaspores formed.	8, 39, 44

					References
Paracoccidioides brasiliensis	Rarely isolated from soils (coffee plantation, rain forest), armadillos, habitat considered to be forest areas of moderate climate, medium to High annual rainfall. Central and South America.	Narrow hyphae, often remaining sterile; occasionally forming lateral or terminal pyriform conidia and solitary intercalary arthroconidia that are often slightly swollen (2–5.5 × 2–3.5 μm).	Not known. Relationship to *Ajellomyces* inferred.	Thick walled central cell up to 30 μm in diameter, bearing 1 to many daughter cells of varying size and joined to mother cell by narrow pedicel, occasionally occurring in short chains.	On enriched media (brain heart infusion agar or blood agar), converts to yeast stage at 37°C. Exoantigen test shows specific 1, 2, 3 precipitin bands.	9, 10, 14, 15, 16, 18, 19, 45, 46, 47, 48
Coccidioides immitis	Soil of arid regions, animal burrows, has been found in mixed infections with *E. parva* in lungs of rodens & other burrowing animals. Geographically restricted to arid regions of southwest U.S. & parts of Central & South America	Hyphae forming alternate arthroconidia. Arthroconidia cylindrical 3–8 μm long and 3–5 μm wide.	Not known Relationship to *Uncinocarpus* (Onygenaceae) inferred	Spherules 30 to 60 μm with walls 2 μm thick and containing endospores 2–5 μm in diameter.	In vitro conversion not recommended and rarely done. Exoantigen test shows specific F precipitin band. DNA probe test.	15, 16, 18, 19, 33, 49, 50, 51, 52, 53, 54, 55, 56, 57

[1] Modified from Sigler (38).

specimens, but laboratorians may encounter them in proficiency testing programs. They include the *Chrysosporium* anamorph of *Arthroderma tuberculatum* (Arthrodermataceae) (23), the *Chrysosporium* anamorph of *Renispora flavissima* (Onygenaceae) (23), the *Botryotrichum* anamorph of *Chaetomium histoplasmoides* (Chaetomiaceae) (24), and some *Sepedonium* species (anamorphs of Hypocreaceae) (23). None is closely related to *H. capsulatum*. Indeed, as mentioned above, the closest relatives to species of *Histoplasma* are *Blastomyces dermatitidis* and species of *Emmonsia*.

Studies of compatibility among >1000 soil and human isolates of *H. capsulatum* of North American origin showed that mating type was significantly associated with pathogenicity, with a ratio of 7(−):1(+) among human isolates compared with 1:1 for soil isolates, and with conversion to the yeast form, with a higher percentage of conversion among − than + mating type strains (25). Recent studies have revealed genetic differences among populations of *H. capsulatum*. Based on DNA thermal denaturation, Guého et al. (9) could distinguish three geographic populations of *H. capsulatum*, but the LSU rDNA sequences were identical or showed only one base difference. Peterson and Sigler (10) observed significant genetic difference between two North American isolates, including one from a human source and another derived from germination of a single ascospore of *A. capsulatus*. Kasuga et al. (26) have suggested that the differences are sufficient to recognize six groups that may be considered as phylogenetic species.

ii. *Histoplasma capsulatum* var. *duboisii* (Vanbreuseghem) Ciferri, Manual di Micologia Medica, Tomo II:342, 1960.

Synonym: *Histoplasma duboisii* Vanbreuseghem, Ann. Soc. Belge Med. Trop. 32: 578, 1952.

Occurrence as a pathogen. Etiologic agent of histoplasmosis duboisii, a systemic infection endemic in parts of central Africa, in which lesions occur most commonly in the bone, skin, and subcutaneous tissues (15–18).

Description (See also Table 3.) The colonial and microscopic features correspond with those described for the variety *capsulatum* (15–19).

Comments. Initially recognized as a distinct species because of the larger size of yeast cells in tissue and different clinical manifestations, *H. duboisii* was later given variety status. Evidence supporting this decision came from mating studies (27) that demonstrated compatibility between *H. duboisii* × *H. capsulatum*. Ascospores produced in these crosses failed to germinate, however, leaving open a possibility that these may be separate species, except that failure to germinate occurred also among some *H. capsulatum* × *H. capsulatum* crosses. As with the latter, − mating types predominated. Comparison of partial rRNA gene sequences showed a level of variation within the genus not exceeding 1% and supported the varietal status of *duboisii* (9, 14), but comparison of DNA sequences of protein-coding genes supported recognition of *H. duboisii* as a distinct species (26).

iii. *H. capsulatum* var. *farciminosum* (Rivolta) Weeks, Padhye & Ajello, Mycologia 77:969, 1985.
Synonym: *H. farciminosum* (Rivolta) Redaelli & Ciferri, Boll. Sez. Ital. Soc. Int. Microbiol. 6:377, 1934.
Occurrence as a pathogen. Etiologic agent of histoplasmosis farciminosi or epizootic lymphangitis, a systemic infection of equines that affects mainly the subcutaneous lymph nodes and lymphatics of the neck (15–17, 28, 29).
Description (See also Table 3.) Weeks et al. (29) obtained sporulation only when isolates were grown on a weak medium, such as soil extract agar. Colonies were very slow growing, attaining diameters of only 5 to 6 mm after 12 weeks incubation at 30°C. Sporulating areas of the colony were buff in color and demonstrated abundant micro- and macroconidia having similar morphologies to those of var. *capsulatum*.
Comments. Varietal status has been proposed based on antigenic and morphologic similarities in the yeast cells and macroconidia (29). No sexual stage is known. Groupings based on comparison of partial LSU rRNA sequences support the distinction of *farciminosum* at the variety level except that genetic differences are slightly greater between it and the other varieties (9, 14). Analysis of different genes suggested a close relationship between equine isolates and some var. *capsulatum* isolates from South America (26).

b. *Ajellomyces dermatitidis* McDonough & Lewis, Mycologia 60:77, 1968
Description as for genus. Ascomata 200 to 350 μm in diameter. *Peridium* is composed of hyphae in which individual cells are obtusely diamond-shaped (swollen near the center and constricted at the septa). *Appendages* helically coiled, generally 3 to 5 per ascoma, breadth of helix 12 to 40 μm. *Asci* globose or subglobose, 3.5 to 7.5 μm. *Heterothallic.* Anamorph *Blastomyces dermatitidis*.

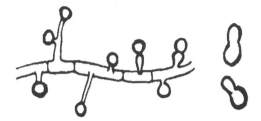

Figure 3   *Blastomyces dermatitidis* in vitro and in vivo.

i.   *Blastomyces dermatitidis* Gilchrist & Stokes, J Exp Med 3:76, 1898
     (Fig. 3).
     Nomenclature. This name remains in common use for the agent of
        blastomycosis even though it has long been recognized as invalid
        under the International Code of Botanical Nomenclature. See
        Carmichael (30), van Oorschot (31), or de Vries (1), for detailed
        lists of synonyms and explanation of the problem. Recent
        changes to the code suggest that a proposal for conservation
        might be successful and ensure retention of this important name.
     Occurrence as a pathogen. Etiologic agent of blastomycosis
        (Gilchrist's disease or North American blastomycosis), a sys-
        temic infection involving the lung but frequently progressing to
        other sites, including the skin and bone (15, 16, 18, 32).
     Description (See also Table 3.) The tissue, colonial, and micro-
        scopic morphologies have been described extensively (e.g., 15,
        16, 18, 19, 23, 30). Aleurioconidia are sessile (borne directly on
        the sides of the hypha) or borne at the ends of short stalks that
        are unswollen (having parallel sides) or slightly swollen at the
        apical end (nearest to the conidium). Conidia are smooth to echi-
        nulate (spiny walled), broadly pyriform or subglobose, and often
        flattened apically, so that they are slightly broader than long, and
        have a narrow truncate base. Occasional isolates show conidia of
        larger dimensions that are smooth or ornamented with irregular
        protrusions (8), or helical coils and peridial hyphae characteristic
        of the sexual stage. Colonies are initially white, becoming buff
        to pale brown, sometimes with a diffusible tan pigment, velvety
        to almost granular in older cultures. Sporulation is heavy in buff-
        pigmented colonies, but may be absent or sparse in isolates that
        are glabrous (having little aerial hyphae).

Comments. The presence of large, thick-walled yeast cells with a single broad-based bud in cutaneous material or biopsy specimens provides presumptive evidence of infection. The commercial probe (AccuProbe) has been evaluated for specificity in confirming *B. dermatitidis* in culture (18, 21, 33). The specificity was 100% when the nontarget isolates were morphologically similar but not biologically related (21), but dropped to 59% when isolates of the biologically related *Paracoccidioides brasiliensis* (i.e., a member of the *Ajellomyces* clade) were included (33). Conversely, the exoantigen test was found to be 100% specific for isolates of both species (33). These species occur in different endemic areas and express different morphologies in their yeast stages. Isolates may be converted to the yeast phase by subculture onto enriched or conversion media (e.g., Blasto "D" medium; see Refs. 34, 35) and incubating at 35–37°C. Although temperature is considered the primary determinant for dimorphism in this species (36), medium constituents also play a role in transition (34). Glabrous and poorly sporulating strains are more difficult to convert and may require multiple subcultures.

Fungi that form conidia resembling those of *B. dermatitidis* have been reviewed recently in detail (23). The most similar conidial morphologies are found among species of *Emmonsia*. (See Table 3.) As has long been hypothesized (30, 37), *Emmonsia* species are the closest biological relatives to *B. dermatitidis* (see also discussion under *H. capsulatum* and *Emmonsia*) (8, 9, 10, 38), but they express different in vivo morphologies, forming nonreplicating thick-walled adiaspores rather than budding cells. Isolates demonstrating transitional features have recently been reported from human cases, however (9, 10, 39, 40); thus the features defining the genera are not as clear as previously supposed. Arguments in favor of combining the genera have been made, but to retain use of the name *Blastomyces* rather than *Emmonsia*, it must first be conserved. (See Ref. 10.)

Kwon-Chung (41) studied compatibility among >40 human isolates of *B. dermatitidis* from North America and Africa. She demonstrated that incompatibility was of the bipolar type; that mating competence was greater in buff-pigmented colony types; and that ascospores showed a 1:1 ratio of + to − mating type. She also confirmed previous findings of incompatibility between North American and African isolates. As reviewed by di

Salvo (32), antigenic and morphologic differences have been reported for African strains. In comparing sequences of a 592 nucleotide region covering the two most divergent regions of the LSU rRNA, Guého et al. (9) found nine base pair differences between African and North American isolates, almost the same level of difference as found between the latter and *E. parva* (eight bases).

c. *Ajellomyces crescens* Sigler, J Med Vet Mycol 34:305, 1996
    Description as for genus. Ascomata 80 to 250 µm in diameter. *Peridium* is composed of hyphae in which individual cells are obtusely diamond-shaped (swollen near the center and constricted at the septa). *Heterothallic.* Anamorph *Emmonsia crescens.*

    i. *Emmonsia crescens* Emmons & Jellison, Ann NY Acad Sci 89:96, 1960 (Fig. 4).
        Synonym: *Chrysosporium parvum* var. *crescens* (Emmons & Jellison) Can J Bot 40:1164, 1962.
        Occurrence as a pathogen. Etiologic agent of adiaspiromycosis, a fairly common pulmonary infection of rodents and small burrowing animals. Infection in humans is rare, usually confined to the lung, with rare progression to other sites (15, 38, 42).
        Description (See also Table 3.) Detailed descriptions of the tissue and cultural and microscopic morphologies have been published recently (8, 15, 23, 38). The colonial and microscopic morphologies resemble those of *Blastomyces dermatitidis*, but *Emmonsia* species differ in their propensity to form one to three conidia at the ends of narrow, often slightly swollen stalks. Growth of *E. crescens* isolates is moderately to strongly inhibited by addition of cycloheximide (400 µg/ml) to culture media and almost completely inhibited at 35–37°C. At this temperature, most isolates will develop thick-walled adiaspores (globose or subglobose

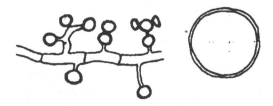

Figure 4   *Emmonsia* species in vitro and in vivo.

chlamydosporelike structures) that measure 20 to 140 μm in diameter.

Comment. Adiaspores in vivo may be much larger and reach dimensions of 50 to 500 μm with walls from 10 to 70 μm thick. Adiaspores are unusual parasitic forms since they arise from swelling or enlargement of inhaled conidia rather than by multiplication of the conidia within the tissues. Available evidence suggests that *Emmonsia* species are weak pathogens and that the functional disability and elicitation of symptoms is dependent upon the number of conidia inhaled (43). The presence of these large and thick-walled structures in tissue provides presumptive evidence of infection by *E. crescens*. Discovery of their presence has often been made accidentally during histopathological examination of tissues when other syndromes were suspected. Although *E. crescens* has been obtained in culture from several animal species by dissection of lung tissue, so far there has been no documented recovery of an authentic isolate from human tissue. Several isolates from human sources that were examined by Peterson and Sigler (10) demonstrated genetic differences that placed them outside the *E. crescens* clade.

d. Species Within the *Ajellomyces* Clade Lacking Known Teleomorphs

  i. *Emmonsia parva* (Emmons & Ashburn) Ciferri & Montemartini, Mycopath Mycol Appl 10:314, 1959.

Synonym: *Chrysosporium parvum* var. *parvum* (Emmons & Ashburn) Carmichael, Can J Bot 40:1164, 1962.

Occurrence as a pathogen. Etiologic agent of adiaspiromycosis in animals, but less widely distributed and in fewer animal hosts than *E. crescens*. There is a single report of disseminated infection in a human with AIDS (15, 38, 42, 43a).

Description (See also Table 3.) Detailed descriptions of the tissue and cultural and microscopic morphologies have been published recently (8, 15, 23, 38). *E. parva* isolates are slightly inhibited by cycloheximide (at 400 μg/ml) in culture media and moderately inhibited at 37°C. At 40°C, growth is almost completely inhibited, and at this temperature most isolates develop thin to slightly thick-walled adiaspores (globose or subglobose chlamydosporelike structures) measuring 8 to 20 μm in diameter.

Comment. Molecular phylogenetic analysis demonstrated a close relationship between *E. parva* and *B. dermatitidis* and also revealed considerable genetic variation among isolates originally thought to represent *E. parva* (10). Isolates from human skin lesions appeared to be distinct (10, 40). Similarly an isolate causing cutaneous disseminated mycosis in a patient with AIDS was found to be molecularly and morphologically distinct and was described as a new species, *E. pasteuriana* (9, 39, 44).

ii. *Emmonsia pasteuriana* Drouhet, Guého & Gori, J Med Mycol 8: 70, 1998.

Occurrence as a pathogen. There is a single report of disseminated cutaneous infection in a human with AIDS.

Description (See also Table 3.) The following is based on the published description of the ex-type culture and only known isolate (39, 44). Colonies are moderately slow growing, yellowish-white, densely woolly, furrowed or zonate with powdery or glabrous sectors. The fungus is tolerant of cycloheximide. Hyphae are narrow. Conidia are solitary and formed on the sides of the hyphae (sessile), or one to eight conidia are borne at the tips of narrow stalks that are distally swollen to form a vesiclelike structure. Conidia appear smooth to finely roughened by light microscopy, tuberculate by scanning electron microscopy, and are slightly broader than long. Conversion to the yeast phase may be obtained by subculture onto enriched media and incubation at 37°C. Colonies on brain heart infusion agar appear smooth, light greyish-brown, and consist of oval to lemon-shaped, small (2–4 μm) yeastlike cells having narrow-based buds. In vivo, yeast cells are more variable in size and shape.

Comments. Variation in colony texture appears similar to that reported for other species of *Emmonsia* (8). *E. pasteuriana* differs in forming clusters of up to eight conidia on a swollen stalk, by lacking expression of adiaspores, and by conversion to a yeastlike stage in vitro. Notwithstanding the unusual expression of a yeast phase, the microscopic morphology and comparison of partial LSU rDNA sequences suggested that the Italian isolate from a patient with AIDS fit within the concept of the genus *Emmonsia* and was a close relative of *E. crescens*.

iii. *Paracoccidioides brasiliensis* (Splendore) Almeida, C R Soc Biol Paris 106:316, 1930 (Fig. 5).

Occurrence as a pathogen. Etiologic agent of paracoccidioidomycosis, a systemic infection endemic in parts of Central and South America that manifests as benign or progressive lung infection but often progresses to form chronic granulomatous lesions of the skin or mucous membranes, especially of the mouth, nose, and gastrointestinal tract (15, 16, 18, 45).

Description (See also Table 3.) Detailed descriptions of the tissue

Figure 5   *Paracoccidioides brasillensis* in vitro and in vivo.

and colonial and microscopic morphologies have been published in medical mycology texts (15, 16, 18, 19, 45). Colonies are very slow growing and vary in texture from glabrous to densely velvety. They may be furrowed and are initially white, but darken to buff. Sporulation is often absent on richer media, such as Sabouraud dextrose agar, but may be induced by growing an isolate on nutritionally poor media such as tap water, glucose salts, or yeast extract agars (46).

Comments. Diagnosis and identification may be confirmed by observation of the distinctive yeast stage, either in vivo or in vitro by subculture onto enriched media and incubation at 37°C. Dimorphism in *P. brasiliensis* has been reviewed (47) and stated to be strictly thermal. In a case of imported paracoccidioidomycosis known to the author, however, the primary isolate was recovered in the yeast form on phytone yeast extract agar at 28–30°C, and could be maintained for several subcultures under these conditions without becoming hyphal. As discussed earlier, the DNA probe test may fail to differentiate isolates of *P. brasiliensis* from *B. dermatitidis*, but these species occur in different endemic regions and express different yeast morphologies. The ecology and substrate preferences of *P. brasiliensis* are still rather poorly understood, as the fungus has rarely been isolated from soil or other habitats (48).

Recent molecular phylogenetic studies (9, 10, 14) have confirmed that the closest biological relatives of *P. brasiliensis* are members of the *Ajellomyces* clade; that is, *Blastomyces*, *Emmonsia*, and *Histoplasma*, as suggested by Dowding (37), but the relationships among them require further study.

2. *Coccidioides immitis* Stiles in Rixford & Gilchrist, Johns Hopkins Hosp Rep 1:243, 1896 (Fig. 6).

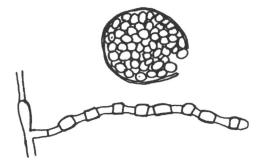

Figure 6   *Coccidioides immitis* in vitro and in vivo.

Teleomorph: unknown but relationship to *Uncinocarpus* inferred.

Occurrence as a pathogen. Etiologic agent of coccidioidomycosis, a systemic infecton with several clinical manifestations, including asymptomatic benign infection (most common), acute or chronic progressive lung infection, and disseminated disease, especially involving the skin, subcutaneous tissues, and bone (15, 16, 18, 49).

Description (See also Table 3.) Detailed descriptions of the tissue and colonial and microscopic morphologies have been published in medical mycology texts (15, 16, 18, 19, 49). Conidia are formed in an intercalary position by concentration of cytoplasm and organelles in some cells while other cells become devoid of contents and the walls autolyse. Conidia formed in this way have been called alternate arthroconidia, arthroaleuries, or enteroarthric conidia. In *C. immitis*, arthroconidia are formed in branched fertile hyphae and are closely spaced, with intervening cells often (but not always) shorter than the arthroconidia (50). They are cylindrical to barrel-shaped or cuneiform if terminal or lateral, and measure from 3 to 5 μm in width. Vegetative hyphae are septate and often show swellings at one end of the cell (i.e., racquet hyphae). Colonies are moderately fast growing, white to yellowish-white or pale buff, velvety to powdery, occasionally glabrous. Variation in colonial texture and pigmentation has been reported in the literature (e.g., see Ref. 15), but some isolates having unusual features have been reidentified as species of *Malbranchea* (50).

Comments. The demonstrated presence of characteristic spherules containing endospores in fluid or tissue specimens or in animal tissue following inoculation is confirmatory evidence for diagnosis of coccidioidomycosis; however, spherules are sometimes lacking or structures are not definitive (51). The exoantigen test and DNA probe are commercially available tests commonly used to confirm identification of an isolate (33). In vitro conversion is not recommended because of the potential for misidentification when arthroconidia of similar *Malbranchea* species swell and round up but lack endospores.

The endosporulating spherules and lack of a sexual cycle has made the taxonomic position of *C. immitis* the subject of considerable speculation. The alternate arthroconidia of its saprobic phase are typical of some *Malbranchea* species (50), and it has been hypothesized that its closest relatives would be found among some keratinophilic species of *Malbranchea* having meiotic stages in the Onygenaceae (52). Depending upon the fungi included in the analysis, molecular phylogenetic studies have placed *C. immitis* within the order Onygenales (9, 53) or within the family Onygenaceae with *Uncinocarpus reesii*, an ascomycete with a *Malbranchea* anamorph, as a close relative (11, 54). The sexual stage of *C. immitis* remains undiscovered, but Burt et al. (55) report molecular evidence of recombination. Some isolates produce helically coiled hyphae (56, 57). Helical, spiral, or uncinate hyphae often occur as ascomatal appendages in onygenalean fungi, and their presence in cultures of anamorphs suggests a potential for sexual reproduction.

3. *Chrysosporium* Corda
    Type species: *C. merdarium* (Link) Carmichael.
    Teleomorphs are placed in *Aphanoascus*, *Nannizziopsis*, *Renispora*, *Uncinocarpus*, *Arthroderma*, and some other genera (as on page 303).
    Comments. Many species of *Chrysosporium* are anamorphs of onygenalean fungi. The genus as circumscribed by Carmichael (30) was heterogeneous, but has been useful for the placement of anamorphic fungi that produce solitary, usually nonseptate aleurioconidia (i.e., conidia with lytic dehiscence), and occasional arthroconidia. He broadened the concept of the genus to include *Blastomyces* and *Emmonsia*, but their exclusion is now supported by genetic data. Some *Chrysosporium* species are encountered as rare to common contaminants from cutaneous specimens, and a few species have the potential to cause infection (23). Some reports concerning *Chrysosporium* species as etiologic agents must be viewed with caution, however, since the isolated organism has neither been identified to species nor documented well enough to confirm the etiology, and a representative culture has not been maintained (58, 59).
    Species of Medical Relevance.

    a. *Aphanoascus fulvescens* (Cooke) Apinis, Mycopathol Mycol Appl 35: 101, 1968 (Fig. 7).
        Occurrence as a pathogen. Rare cause of nail, skin, and scalp infection (35).
        Description. Detailed descriptions of the colonial and microscopic morphologies have been published recently (35). Colonies grow moder-

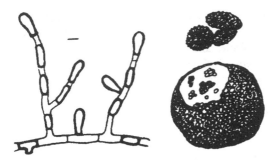

Figure 7 *Aphanoascus fulvescens* (note conidia and ascospores are not drawn to the same scale). Bar approx 5 μm. Source: ascomata© R. S. Currah (used with permission).

ately rapidly, are tolerant of cycloheximide, and are somewhat inhibited at 37°C. They are yellowish-white, becoming buff to greyish as ascomata develop, velvety to granular. The *Chrysosporium* anamorph is distinguished by the development of sessile or terminal pyriform or clavate aleurioconidia on short or long branches and numerous intercalary cylindrical or barrel-shaped arthroconidia (i.e., alternate arthroconidia; see Fig. 7). Conidia are fairly large, and measure 6 to 20 by 3 to 5(8)μm. Ascomata usually develop within 3 weeks and are globose yellowish-brown cleistothecia containing yellowish-brown, ovoid, irregularly reticulate ascospores measuring 4 to 5 by 2.5 to 3.5 μm.

    Comments. Species of *Aphanoascus* occur in soil. They are strong keratinophiles and commonly isolated by hair bait techniques. Isolates are often obtained in the anamorphic stage, and with the exception of *A. fulvescens*, it can be difficult to obtain or maintain the teleomorph in culture. The anamorph of *A. fulvescens* has been suggested to be *C. keratinophilum*, but the teleomorph of the latter is a different species. The data of Leclerc et al. (14) place the two species on one branch of the phylogenetic tree of the Onygenales.

  b.   *Chrysosporium zonatum* Al Musallam & Tan, Persoonia 14:69, 1989 (Fig. 8).

    Synonym: *C. gourii* P.C. Jain, Deshmukh & S.C. Agrawal, Mycoses 36: 77, 1993.

    Teleomorph: *Uncinocarpus orissi* (B. Sur & G.R. Ghosh) Sigler & Flis, Can J Bot 76:1627, 1998.

    Synonyms: *Pseudoarachniotus orissi* B. Sur & G.R. Ghosh, Kavaka 12: 67, 1985; *Gymnoascus arxii* Cano & Guarro, Stud Mycol 31:61, 1989.

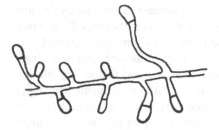

Figure 8   *Chrysosporium zonatum* (*Uncinocarpus orissi*).

Occurrence as a pathogen. Known from two cases of pulmonary colonization and a single case of disseminated infection involving the lung
and bone in a patient with chronic granulomatous disease (59, 60).

Description. Detailed descriptions of the colonial and microscopic morphologies have been published recently (57, 59). *C. zonatum* is
thermotolerant and cycloheximide tolerant. Colonies grow faster at
37°C than at 25°C, reaching diameters of 7.5 to 8 cm in 14 days. They
are initially yellowish-white but darken to buff, especially on potato
dextrose agar at the higher temperature. Aleurioconidia are borne at
the ends of short or long stalks that are characteristically curved, or
are sessile. They are smooth or warty, clavate or broadly obovoid,
rounded at the tip, and measure (3.5) 4 to 8 (13) by (2.5) 3 to 5 μm.
Intercalary arthroconidia are rare. *C. zonatum* is heterothallic, and the
teleomorph is obtained by mating compatible isolates. Ascomata are
discrete, more or less globose gymnothecia composed of reddish-
brown ascospores surrounded by loose wefts of undifferentiated racquet hyphae that may form conidia on side branches. Ascospores are
oblate, appearing round in face view and flattened with truncate ends
in side view, and measure 4.5 to 7 by 3 to 4.5 μm. They appear smooth
to minutely roughened under light microscopy and pitted under *SEM*
(57).

Comments. This strongly keratinolytic and thermotolerant species appears to have a broad geographic distribution in subtropical and
warmer temperate regions. The anamorph and teleomorph have been
described independently and under more than one name; the connections among them have been established through matings (57).

c.   *Chrysosporium* anamorph of *Nannizziopsis vriesii* (Apinis) Currah, Mycotaxon 24:164, 1985.

Occurrence as a pathogen. Reported as the cause of cutaneous mycoses
in lizards and snakes with rare invasion into deep tissues (61, 62).

Description. Detailed descriptions of the tissue and colonial and microscopic morphologies have been published recently (61, 62). Asexual
isolates of this fungus have come from several infected reptiles. Colonies are moderately fast growing, flat or umbonate, and sometimes
zonate (showing concentric zones of denser and thinner mycelium),
yellowish-white, and powdery. Growth is similar on media containing
cycloheximide but is highly restricted at 37°C. Aleurioconidia are pyriform or clavate, single-celled or rarely one-septate, and commonly
sessile or borne at the ends or sides of short stalks that arise at 90°
angles. The conidia are commonly 3 to 6 by 1.5 to 2.5 µm, but may
be up to 12.5 µm in length. Arthroconidia measuring 1.5 to 3.5 by 4
to 9 µm long form either in an alternate position (i.e., separated by
cells that ultimately lyse) or in chains, and these are separated by
fission at the septum (i.e., schizolytic dehiscence). A characteristic
feature is the development of solitary lateral branches that are undulate
(wavy) and sparsely septate. Rarely these fragment to form arthroconidia. The teleomorph has been found in cultural conditions only in
two soil isolates. Ascomata are discrete, yellowish-white, and composed of branched asperulate hyphae that are slightly constricted at the
septa. Ascospores are globose, punctate-reticulate, and measure 2.5 to
3 µm in diameter. Development of ascomata occurs optimally at 30°C.

Comments. Histopathologic sections of the skin lesions on affected reptiles show presence of hyaline hyphae that in some instances erupt
through the surface of the epidermis to form aerial terminal arthroconidia (61, 62). The sessile or short stalked aleurioconidia borne at right
angles to the hyphae are reminiscent of some *Trichophyton* species,
including *T. mentagrophytes* and *T. terrestre*, but this dermato-
phytelike fungus differs in its propensity to form schizolytic arthroconidia and undulate hyphae, in its failure to form macroconidia, and
in comparison with *T. mentagrophytes*, in its restricted growth at
37°C. Preliminary results from molecular genetic testing suggest that
the pathogenic anamorphic isolates are closely related to teleomorphic
soil isolates (63), but further testing is being done to assess the degree
of genetic diversity among them. Currah (3) placed *Nannizziopsis* in
the Onygenaceae, but Guarro et al. (64) treated it within the Eurotiales
and transferred some other species to the genus.

d. Notes on Saprophytic Species

    i. *Chrysosporium* anamorph of *Renispora flavissima* Sigler, Gaur,
        Lichtwardt & Carmichael, Mycotaxon 10:134, 1979 (Fig. 9). Colo-
        nies are moderately slow growing, lemon yellow, and powdery.
        The conidia are large, reaching a size similar to those of the macro-

Figure 9 *Chrysosporium* an. of *Renispora flavissima*.

conidia of *Histoplasma capsulatum* (Table 3) and demonstrating tubercles on the surface. This is a soil fungus that is known only from its original habitat in Kansas. Growth is strongly inhibited at 37°C and there is no conversion to a yeast phase. It is heterothallic, forming gymnothecial ascomata in mated isolates.

ii.   *Chrysosporium keratinophilum* (Frey) Carmichael, Can J Bot 40: 1157, 1962 (Fig. 10). Colonies are moderately rapid growing, flat, dense, velvety to granular, yellowish-white. Large (mostly 10 to 12 by 6 to 8 μm), smooth, broadly pyriform conidia are formed in clusters at the ends of acutely branched, unswollen stalks or sessile (23). The teleomorph, *Aphanoascus keratinophilus* Punsola & Cano, is rarely recovered, and the conditions for its inducement in strictly anamorph strains are not yet known. As shown by Leclerc et al. (14) and expected by their known teleomorphs, *C. keratinophilum* and *Aphanoascus fulvescens*, cluster together in the phylogenetic tree of the Onygenales.

iii. *Chrysosporium articulatum* Scharapov, Nov. Syst. niz. Rast. 15: 146, 1978 (Fig. 11). The microscopic morphology is very similar to the *C.* anamorph of *Aphanoascus fulvescens* (Fig. 7), but *C. arti-*

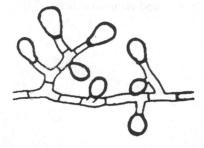

Figure 10 *Chrysosporium keratinophilum.*

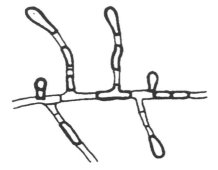

Figure 11   *Chrysosporium articulatum.*

culatum is not known to form a teleomorph, and colonies are more
rapid growing and usually cottony to woolly (23).

iv.   *Chrysosporium carmichaelii* van Oorschot, Stud Mycol 20:15,
1980 (Fig. 12). Colonies are moderately slow growing, yellowish-
white, umbonate, velvety to woolly. Growth is enhanced in the
presence of the vitamin thiamine but there is no growth at 37°C.
Fertile hyphae bear small aleurioconidia (mostly 3.5 to 4 by 2.5 to
3 µm) at the ends or on the sides of short, often slightly curved
branches that often occur in dense clusters associated with sterile
undulate or curved hyphae (23).

v.   *Chrysosporium lobatum* Scharapov Nov. Syst. niz. Rast. 15:144,
1978 (Fig. 13). Colonies are slow growing, velvety or felty, with
an irregular lobate margin, initially pale grayish green and often
developing pinkish-grey to violet tints in the center and diffusible
brown pigment. Colonies are stimulated by the vitamin thiamine.
Small (mostly 2.5 to 3.5 by 2 to 2.5 µm), smooth to rough conidia
form at the ends of short, very narrow, peglike stalks that arise at
a 90° angle from the vegetative hyphae and occur in clusters (23).

Figure 12   *Chrysosporium carmichaelii.*

Figure 13   *Chrysosporium lobatum.*

4. *Malbranchea* Saccardo

Type species: *M. pulchella* Saccardo & Penzig.

Teleomorphs. Teleomorphs of keratinolytic species of *Malbranchea* are placed in the genera *Aphanoascus*, *Auxarthron*, and *Uncinocarpus* and some other genera of the Onygenaceae. Some cellulolytic species have known or inferred affinities within the genus *Myxotrichum*.

Comments. There is no evidence to confirm a pathogenic role for any species of *Malbranchea*. A few species are seen occasionally as clinical contaminants. It has been shown that the arthroconidia of some species survive in and may be recovered from animal tissue after inoculation (50). The genus encompasses species producing alternate arthroconidia, (i.e., arthroconidia that are separated from each other by one or more cells that undergo lytic dehiscence). Sigler and Carmichael (50) divided the species into two groups: one containing species in which the fertile hyphae are strongly curved or arcuate and the other in which they are straight. Species among the latter group show closest morphologic similarity to *Coccidioides immitis*, and two of the species described were based on isolates that had previously been thought to represent atypical isolates of *C. immitis*. In a key to the species, the *Malbranchea* anamorph of *Uncinocarpus reesii* was placed in a couplet with *C. immitis* as the most similar species, and a close phylogenetic relationship between these species has subsequently been demonstrated by molecular methods (11, 54).

Notes on Saprophytic Species.

a. *Malbranchea* anamorph of *Uncinocarpus reesii* Sigler & Orr, Mycotaxon 4:462, 1976 (Fig. 14). Colonies are moderately fast growing, flat, dense, velvety to powdery, yellowish-white to buff. Growth is strongly to moderately inhibited at 37°C. Arthroconidia are separated by one or more cells of irregular length, and the intervening cells often show signs of collapse (23, 50, 57). Arthroconidia are cylindrical to slightly barrel-shaped, club or slightly wedge-shaped if terminal, and measure 3.5 to 6 (8) by 2.5 to 3.5 (4) µm. Thick-walled, brown uncinate (curved at the tip) appendages develop from the vegetative hyphae in many isolates. *U. reesii* is heterothallic and when compatible strains are

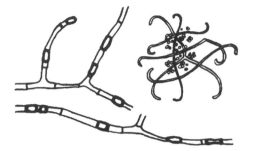

Figure 14 *Uncinocarpus reesii* showing ascocarp and *Malbranchea* anamorph (not drawn to same scale). Source: ascomata © R.S. Currah (used with permission).

mated, the resultant gymnothecia are composed of a loose net-work of these uncinate appendages (23, 50, 57). Ascospores are oblate with truncate or slightly rounded apices and are ornamented on the surface with small pits that are hardly discernible by light micro-scopy. They are pale yellowish-brown individually and reddish-brown in mass.

Comments. The closely spaced, fairly broad arthroconidia of *Coccoidi-oides immitis* are unlikely to be confused with those of *Malbranchea* species, and isolates suspected to be *C. immitis* may be confirmed by the use of probe or exoantigen tests.

b. *Malbranchea gypsea* Sigler & Carmichael, Mycotaxon 4:455, 1976 (Fig. 15). Colonies are slow growing, white, often furrowed and velvety. Cy-lindrical to slightly barrel-shaped, narrow arthroconidia measuring (2.5) 3 to 6(9) by 2 to 2.5 μm are formed in straight, branched fertile hyphae and separated by one or more cells (23, 50).

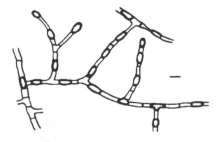

Figure 15 *Malbranchea gypsea*. Bar approximately 5 μm.

Comments. This slow-growing fungus is unlikely to be confused with *C. immitis.* Its recovery from a dystrophic nail on three occasions is suggestive of a possible role in onychomycosis, but its role has not been substantiated (23).

B. Family Gymnoascaceae

Members of the Gymnoascaceae are a diverse group (3), and Currah (4) has suggested that some taxa may belong outside the family. They are not keratinolytic and not or only weakly cellulolytic as determined by in vitro methods. Ascospores are smooth or ornamented with thickened knobs and are oblate or discoid, sometimes with one or two longitudinal ridges. Anamorphs are lacking for many species or are arthroconidia or aleurioconidia. (See also Table 1.)

Species of Medical Relevance

1. *Onychocola canadensis* Sigler & Congly, J. Med. Vet. Mycol. 28:409, 1990 (Fig. 16). Teleomorph: *Arachnomyces nodosetosus* Sigler & Abbott, J Med Vet Mycol 32:280, 1994.

   Colonies are very slow growing, often digging into the agar and eventually cracking it. They are initially glabrous, yellowish-white to pale grey, gradually developing floccose tufts of white to grey aerial mycelium and often developing diffusible tan pigments. Isolates are cycloheximide-tolerant. Arthroconidia are formed in the aerial mycelium and are one- or two-celled, initially cylindrical but commonly rounding up, and remaining connected in adherent chains (35, 65). Dehiscence occurs by splitting (schizolysis) of adjacent conidia or by lysis of thin-walled segments of the hyphae. Detached conidia measure 4 to 8 by 2 to 5 μm if 0-septate and 8 to 17 by 2.5 to 5.5 μm if 1-septate. Brown, knobbed, uncinate, or spiralled appendages form in older cultures and especially on media such as phytone

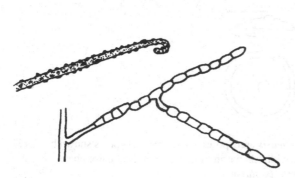

Figure 16   *Onychocola canadensis* arthroconidia and seta.

yeast extract or blood agar. *O. canadensis* is heterothallic, but the teleomorph has been obtained among matings of only a few isolates after prolonged incubation. Ascomata are globose cleistothecia, and bear knobbed appendages; ascospores are oblate, pale brown, and smooth (65, 66).

Comments. As mycologists have gained familiarity with this slowgrowing and often poorly sporulating fungus, they have recorded its isolation from nails, and less commonly from the skin of feet or hands in geographically separated areas (Canada, Australia, New Zealand, the United Kingdom, and Europe). Gupta et al. (67) have suggested that *O. canadensis* may be isolated occasionally from dystrophic nails without evidence of pathology. In one unusual finding, it was incidentally isolated from a bronchial washing (Sigler, unpublished data).

2. *Gymnascella dankaliensis* (Castellani) Currah, Mycotaxon 24:77, 1985 (Fig. 17). Colonies are moderately slow growing and appear yellowish-white and sterile on routine isolation media. Ascosporulation occurs on media such as oatmeal salts and Takashio or Leonian's agars (35), and as ascospores mature, colonies darken to yellowish- or brownish-orange. Asci develop as naked clusters sometimes associated with slightly differentiated hyphae. Mature ascospores are oblate and pale orange-brown, and measure 5.5 to 7 by 3 to 4.5 μm (3, 68). They appear round with irregular thickenings in face view and rhomboidal (broader in the center than on the ends) with a longitudinal rim and polar thickenings in side view. No anamorph is formed.

Comments. This weakly cellulolytic species is mainly recorded from the soil of warmer climates. In the one reported case of nail infection, hyphal elements in tissue were irregularly swollen (68).

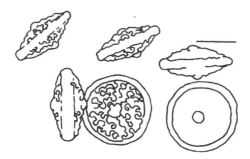

Figure 17 *Gymnascella dankaliensis* ascospores. (Note that ascospores shown here are drawn to a different scale than conidia shown in other figures.) Bar approximately 3 μm. Source: © R. S. Currah (used with permission).

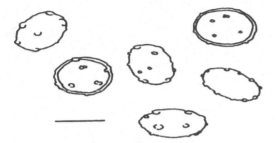

Figure 18 *Gymnascella hyalinospora* ascospores (Note that ascospores shown here are drawn to a different scale than conidia shown in other figures.) Bar approximately 3 μm. Source: © R.S. Currah (used with permission).

3. *Gymnascella hyalinospora* (Kuehn, Orr & Ghosh) Currah, Mycotaxon 24: 84, 1985 (Fig. 18). Colonies are moderately slow growing, initially white and glabrous or with fine nap of aerial mycelium, and gradually form sectors, patches, or tufts of yellowish-orange to yellowish-green aerial hyphae on potato dextrose or oatmeal salts agar. *G. hyalinospora* is thermotolerant and slightly cellulolytic. Asci develop in the colored areas of the colonies and are naked or loosely associated with yellow, slightly thick walled hyphae (3). Ascospores are oblate, yellow, smooth-walled under light microscopy, and measure 2 to 3.5 by 2 to 2.5 μm. No anamorph is formed.

    Comments. G. hyalinospora is mainly recorded from soil, and has been re-covered rarely from lesions of humans and animals without confirmation of a pathogenic role. A recent report documenting invasive pulmonary infection in a immunosuppressed patient confirms its pathogenic potential (69).

C. Family Myxotrichaceae

Members of the Myxotrichaceae have fusiform or ellipsoidal ascospores that are striate or smooth; they degrade cellulosic substrates, and some have *Geomyces* or *Malbranchea* anamorphs (3, 50). (See also Table 1.)

Species of Medical Relevance

1. *Geomyces pannorum* (Link) Sigler & Carmichael, Mycotaxon 4:377, 1976 (Fig. 19). Colonies are slow growing, smooth or furrowed, white, tan, pale yellow or pale grey, and glabrous, fasciculate, powdery or cottony in texture. *G. pannorum* is tolerant of cycloheximide and is psychrophilic, growing bet-

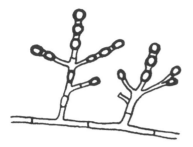

Figure 19   *Geomyces pannorum.*

ter at 18°C than at 25°C and failing to grow at 37°C. The species is distinguished by short, slender conidiophores that branch acutely at the tip to form
three to four verticillate branches (23, 30, 50). The septate fertile branches
develop basipetally into short chains of slightly swollen conidia that are separated from each other by a short cell. Conidia may also develop on the sides
of the branch (sessile). Conidia are cuneiform (wedge-shaped) or barrel-
shaped, smooth or roughened, and measure 2 to 5 by 2 to 4 µm. *G. pannorum*
is a common fungus found in temperate soils worldwide. It is a regular contaminant, especially of nails, but it has not been substantiated as a cause of
onychomycosis (23). No teleomorph is known for isolates recognized as *G.
pannorum. Geomyces vinaceus* encompasses isolates with similar microscopic features, but having colonies that are purplish-red or vinaceous; the
teleomorph is *Pseudogymnoascus roseus* (23, 50).

2.   *Myxotrichum deflexum* Berkeley, Ann. Nat. Hist. 1:260, 1838 (Figs. 20, 21).
Colonies are slow growing, yellowish-white to light grey, often developing
patches of pink or wine red, and expressing a pink diffusible pigment on
some media. With development of ascomata on media such as oatmeal salts
and Takashio or Leonian's agars (35), colonies darken to grey or black. Ascomata are discrete or confluent and are composed of a meshlike network of
branched dark brown to black hyphae (3). Commonly the lateral branches
of the ascomatal hyphae are bent downwards (deflexed) and terminate in
hyaline filaments that eventually disintegrate. Ascospores are ovoid to fusiform, slightly striate, and measure 4 to 5.5 by 2.5 to 3.5 µm. Alternate arthroconidia are formed by some isolates. *M. deflexum* is uncommonly isolated
from nails, but has not been substantiated as a cause of onychomycosis.

3.   *Ovadendron sulphureo-ochraceum* (van Beyma) Sigler & Carmichael, Mycotaxon 4:392, 1976 (Fig. 22). Colonies are moderately slow growing, yellowish-white to yellowish-green, and velvety. Narrow curved or loosely heli-

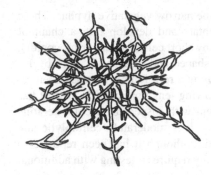

Figure 20 *Myxotrichum deflexum* ascocarp. Source: © R. S. Currah (used with permission).

Figure 21 *Myxotrichum deflexum* ascospores. Bar approximately 4 μm. Source: © R. S. Currah (used with permission).

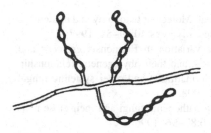

Figure 22 *Ovadendron sulphureo-ochraceum.*

cally coiled lateral branches arise from the narrow vegetative hyphae. These fertile branches become basipetally septate and develop into a chain of slightly swollen arthroconidia separated by short cells. Mature arthroconidia, released by lytic dehiscence, are barrel-shaped and measure 2.5 to 4 by 1.5 to 2.5 µm (50). *O. sulphureo-ochraceum* is a rare fungus that has been reported as an agent of eye infection following lens implantation (70). It has been isolated several times from a patient with suspected mycotic keratitis and incidentally recovered from sputum. It is moderately cellulolytic and has been recovered from decayed wood. Although it has been reported to be slightly keratinolytic (50), this capability requires retesting with additional available isolates.

REFERENCES

1. GA De Vries. Ascomycetes: Eurotiales, Sphaeriales, and Dothidiales. In: DH Howard, ed. Fungi Pathogenic for Humans and Animals. New York: Marcel Dekker, 1983, pp. 81–111.
2. OE Eriksson et al. (Eds). 2001. Outline of Ascomycota. Myconet 7:1–88, 2001.
3. RS Currah. Taxonomy of the Onygenales: Arthrodermataceae, Gymnoascaceae, Myxotrichaceae and Onygenaceae. Mycotaxon 24:1–216, 1985.
4. RS Currah. Peridial morphology and evolution in the prototunicate ascomycetes. In: DL Hawksworth ed. Ascomycete Systematics: Problems and Perspectives in the Nineties. New York: Plenum, 1994, pp. 281–293.
5. S Hambleton. Mycorrhizas of the Ericaceae: Diversity and systematics of the mycobionts. Ph.D. dissertation. Edmonton, Alberta, Canada: University of Alberta, 1998, 145 pp.
6. ES McDonough, AL Lewis. Blastomyces dermatitidis: Production of the sexual stage. Science 156:528–529, 1967.
7. KJ Kwon-Chung. Studies on Emmonsiella capsulata. I. Heterothallism and development of the ascocarp. Mycologia 45:109–121, 1973.
8. L Sigler. Ajellomyces crescens sp. nov., taxonomy of Emmonsia species, and relatedness with Blastomyces dermatitidis (teleomorph Ajellomyces dermatitidis). J Med Vet Mycol 34:303–314, 1996.
9. E Guého, MC Leclerc, GS de Hoog, B Dupont. Molecular taxonomy and epidemiology of Blastomyces and Histoplasma species. Mycoses 40:69–81, 1997.
10. SW Peterson, L Sigler. Molecular genetic variation in Emmonsia crescens and E. parva, etiologic agents of adiaspiromycosis, and their phylogenetic relationship to Blastomyces dermatitidis (Ajellomyces dermatitidis) and other systemic fungal pathogens. J Clin Microbio 36:2918–2925, 1998.
11. BH Bowman, TJ White, JW Taylor. Human pathogenic fungi and their close non-pathogenic relatives. Molec Phylogen Evol 6:89–96, 1996.
12. K Fukushima, K Takeo, K Takizawa, K Nishimura, M Miyaji. Reevaluation of the teleomorph of the genus Histoplasma by ubiquinone systems. Mycopathologia 116:151–154, 1991.

13. K Takizawa, K Okada, Y Maebayashi, K Nishimura, M Miyaji, K Fukushima. Ubi-quinone systems of the form-genus Chrysosporium. Mycoscience 35:327–330, 1994.
14. MC Leclerc, H Philippe, E Guého. Phylogeny of dermatophytes and dimorphic fungi based on large subunit ribosomal RNA sequence comparison. J Med Vet Mycol 32: 331–341, 1994.
15. KJ Kwon-Chung, JW Bennett. Medical Mycology. Philadelphia: Lea & Febiger, 1992.
16. JW Rippon. Medical Mycology: The Pathogenic Fungi and the Pathogenic Actino-mycetes. Philadelphia: Saunders, 1988.
17. R Tewari, LJ Wheat, L Ajello. Agents of histoplasmosis. In: L Ajello, R Hay, eds. Topley & Wilson's Microbiology and Microbial Infections. vol. 4. London, UK: Edward Arnold, 1998, pp. 373–393.
18. DH Larone, TG Mitchell, TJ Walsh. Histoplasma, Blastomyces, Coccidioides, and other dimorphic fungi causing systemic mycoses. In: PR Murray, EJ Baron, MA Pfaller, FC Tenover, RH Yolken, eds. Manual of Clinical Microbiology. 7th ed. Washington, DC: American Society for Microbiology, 1999, pp. 1259–1274.
19. GS de Hoog, J Guarro, J Gene, MJ Figueras. Atlas of Clinical Fungi. Utrecht, Neth-erlands: Centraalbureau voor Schimmelcultures, 2000.
20. MD Berliner. Primary subcultures of Histoplasma capsulatum. I. Macro and micro-morphology of the mycelial phase. Sabouraudia 6:111–118, 1968.
21. L Stockman, KA Clark, JM Hunt, GD Roberts. Evaluation of commercially available aridinium ester-labeled chemiluminescent DNA probes for culture identification of Blastomyces dermatitidis, Coccidioides immitis, Cryptococcus neoformans and His-toplasma capsulatum. J Clin Microbio 31:845–850, 1993.
22. RC Summerbell, A Li, M Kuhn, L Turgeon, DT Janigan, AR Syed, M Decastro. Mixed Histoplasma and Blastomyces infection: Facilitation of diagnosis by rRNA hybridization. Amer Soc for Microbio Abstracts F-115, 1995.
23. L Sigler. Chrysosporium and molds resembling dermatophytes. In: J Kane, RC Sum-merbell, L Sigler, S Krajden, G Land. Laboratory Handbook of Dermatophytes. A Clinical Guide and Laboratory Manual of Dermatophytes and Other Filamentous Fungi from Skin, Hair and Nails. Belmont, CA: Star, 1997, pp. 261–311.
24. LM Carris, DA Glawe. Chaetomium histoplasmoides: A new species isolated from cysts of Heterodera glycines in Illinois. Mycotaxon 29:383–391, 1987.
25. KJ Kwon-Chung, RJ Weeks, HW Larsh. Studies on Emmonsiella capsulata (His-toplasma capsulatum) II. Distribution of the two mating types in 13 endemic states of the United States. Amer J Epidemi 99:44–49, 1974.
26. T Kasuga, JW Taylor, TJ White. Phylogenetic relationships of varieties and geo-graphical groups of the human pathogenic fungus, Histoplasma capsulatum Darling. J Clin Microbio 37:653–663, 1999.
27. KJ Kwon-Chung. Perfect state (Emmonsiella capsulata) of the fungus causing large-form African histoplasmosis. Mycologia 67:980–990, 1975.
28. FW Chandler, W Kaplan, L Ajello. Color Atlas of the Histopathology of Mycotic Diseases. Chicago: Year Book Medical Publishers, 1980, pp. 70–72.
29. RJ Weeks, AA Padhye, L Ajello. Histoplasma capsulatum variety farciminosum: A new combination for Histoplasma farciminosum. Mycologia 77:964–970, 1985.
30. JW Carmichael. Chrysosporium and some other aleuriosporic hyphomycetes. Can J Bot 40:1137–1173, 1962.

31. CAN Van Oorschot. A revision of Chrysosporium and allied genera. Stud Mycol 20:1–89, 1980.

32. AF Di Salvo. Blastomyces dermatitidis. In: L Ajello, R Hay, eds. Topley & Wilson's Microbiology and Microbial Infections. vol. 4. London, UK: Edward Arnold, 1998, pp. 337–355.

33. AA Padhye, G Smith, PG Standard, D McLaughlin, L Kaufman. Comparative evaluation of chemiluminescent DNA probe assays and exoantigen tests for rapid identification of Blastomyces dermatitidis and Coccidioides immitis. J Clin Microbio 32: 867–870, 1994.

34. J Kane. Conversion of Blastomyces dermatitidis to the yeast form at 37°C and 26°C. J Clin Microbio 20:594–596, 1984.

35. J Kane, RC Summerbell, L Sigler, S Krajden, G Land. Laboratory Handbook of Dermatophytes: A Clinical Guide and Laboratory Manual of Dermatophytes and Other Filamentous Fungi from Skin, Hair and Nails. Belmont, CA: Star, 1997.

36. J Domer. Blastomyces dermatitidis. In: P Szaniszlo, ed. Fungal Dimorphism. New York: Plenum 1985, pp. 51–67.

37. ES Dowding. The pulmonary fungus Haplosporangium parvum, and its relationship with some human pathogens. Can J Res E, 25:195–206, 1947.

38. L Sigler. Agents of adiaspiromycosis. In: L Ajello, R Hay, eds. Topley & Wilson's Microbiology and Microbial Infections. vol. 4. London, UK: Edward Arnold, 1998, pp. 571–583.

39. E Drouhet, E Guého, S Gori, M Huerre, F Provost, M Borgers, B Dupont. Mycological, ultrastructural and experimental aspects of a new dimorphic fungus Emmonsia pasteuriana sp. nov. isolated from a cutaneous disseminated mycosis in AIDS. J Mycol Med 8:64–77, 1998.

40. ME Kemna, M Weinberger, L Sigler, R Zeltser, I Polachek, IF Salkin. A primary oral blastomycosis-like infection in Israel. Amer Soc Microbio Annu Mtg Abstr F75, 1994.

41. KJ Kwon-Chung. Genetic analysis on the incompatibility system of Ajellomyces dermatitidis. Sabouraudia 9:231–238, 1971.

42. DM England, L Hochholzer. Adiaspiromycosis: An unusual fungal infection of the lung. Amer J Surg Path 17:876–886, 1993.

43. A de Almeida Barbosa, AC Moreira Lemos, LC Severo. Acute pulmonary adiaspiromycosis: Report of three cases and a review of 16 other cases collected from the literature. Rev Iberoam Micol 14:177–180, 1997.

43a. E Echaverria, EL Cano, A Restrepo. Disseminated adiaspiromycosis in a patient with AIDS. J Med Vet Mycol 31:91–97, 1993.

44. S Gori, E Drouhet, E Guého, M Huerre, A Lofaro, M Parenti, B Dupont. Cutaneous disseminated mycosis in a patient with AIDS due to a new dimorphic fungus. J Mycol Med 8:57–63, 1998.

45. B Wanke, AT Londero. Paracoccidioides brasiliensis. In: L Ajello, R Hay, eds. Topley & Wilson's Microbiology and Microbial Infections. vol. 4. London, UK: Edward Arnold, 1998, pp. 395–407.

46. B Bustamante-Simon, JG McEwen, AM Tabares, M Arango, A Restrepo. Characteristics of the conidia produced by the mycelial form of Paracoccidioides brasiliensis. Sabouraudia: J Med Vet Mycol 23:407–414, 1985.

47. F San-Blas, G San-Blas. Paracoccidioides brasiliensis. In: P Szaniszlo, ed. Fungal Dimorphism. New York: Plenum, 1985, pp. 93–120.
48. ML Silva-Vergara, R Martinez, A Chadu, M Madeira, G Freitas-Silva, CM Leite Maffei. Isolation of a Paracoccidioides brasiliensis strain from the soil of a coffee plantation in Ibia, State of Minas Gerais, Brazil. Med Mycol 36:37–42, 1998.
49. D Pappagianis. Coccidioides immitis. In: L Ajello, R Hay, eds. Topley & Wilson's Microbiology and Microbial Infections. vol. 4. London: Edward Arnold, 1998, pp. 357–371.
50. L Sigler, JW Carmichael. Taxonomy of Malbranchea and some other hyphomycetes with arthroconidia. Mycotaxon 4:349–488, 1976.
51. L Kaufman, G Valero, AA Padhye. Misleading manifestations of Coccidioides immitis in vivo. J Clin Microbio 36:3721–3723, 1998.
52. L Sigler. Perspectives on Onygenales and their anamorphs by a traditional taxonomist. In: DR Reynolds, JW Taylor, eds. The Fungal Holomorph: A Consideration of Mitotic, Meiotic and Pleomorphic Speciation, Wallingford, UK: CAB International, 1993, pp. 161–168.
53. BH Bowman, JW Taylor, TJ White. Molecular evolution of the fungi: Human pathogens. Molec Bio Evol 9:893–904, 1992.
54. S Pan, L Sigler, GT Cole. Evidence for a phylogenetic connection between Coccidioides immitis and Uncinocarpus reesii (Onygenaceae). Microbio 140:1481–1494, 1994.
55. A Burt, DA Carter, GL Koenig, TJ White, JW Taylor. Molecular markers reveal cryptic sex in the human pathogen Coccidioides immitis. Proceed Natl Acad Sci USA 93:770–773, 1996.
56. GF Orr. The use of bait in isolating Coccidioides immitis from soil: A preliminary study. Mycopath Mycol Appl 36:28–32, 1968.
57. L Sigler, AL Flis, JW Carmichael. The genus Uncinocarpus (Onygenaceae) and its synonym Brunneospora: New concepts, combinations and connections to anamorphs in Chrysosporium, and further evidence of its relationship with Coccidioides immitis. Can J Bot 76:1624–1636, 1998.
58. L Sigler, MJ Kennedy. Aspergillus, Fusarium, and other opportunistic moniliaceous fungi. In: PR Murray, EJ Baron, MA Pfaller, FC Tenover, RH Yolken, eds. Manual of Clinical Microbiology. 7th ed. Washington, DC: American Society for Microbiology, 1999, pp. 1212–1241.
59. E Roilides, L Sigler, E Bibashi, H Katsifa, N Flaris, C Panteliadis. Disseminated infection due to Chrysosporium zonatum in a patient with chronic granulomatous disease and review of non-Aspergillus infections in patients with this disease. J Clin Microbio 37:18–25, 1999.
60. S. Hayashi et al. Pulmonary colonization by Chrysosporium zonatum associated with allergic inflammation in an immunocompetent subject. J Clin Microbio 40:1113–1115, 2002.
61. JA Pare, L Sigler, DB Hunter, RC Summerbell, DA Smith, KL Machin. Cutaneous mycoses in chameleons caused by the Chrysosporium anamorph of Nannizziopsis vriesii (Apinis) Currah. J Zoo Wildl Med 28:443–453, 1997.
62. DK Nichols, RS Weyant, EW Lamirandel, RT Mason, L Sigler. Fatal mycotic der-

matitis in captive brown tree snakes (Boiga irregularis). J Zoo Wildl Med 30:111–118, 1999.

63. RC Summerbell, L Sigler, A Li, JA Pare. Chrysosporium anamorph of Nannizziopsis vriesii: An agent of mycotic infection in reptile and human hosts. Atlanta, Amer Soc Microbio Annual Mtg F-107, 1998.

64. J Guarro, J Cano, Ch de Vroey. Nannizziopsis (Ascomycotina) and related genera. Mycotaxon 42:193–200, 1991.

65. L Sigler, SP Abbott, A Woodgyer. New records of nail and skin infection due to Onychocola canadensis and description of its teleomorph Arachnomyces nodosetosus sp. nov. J Med Vet Mycol 32:275–285, 1994.

66. SP Abbott, L Sigler, RC Currah. Delimitation, typification and taxonomic placement of the genus Arachnomyces. Systema Ascomycetum 14:79–85, 1996.

67. AK Gupta, CB Horgan-Bell, RC Summerbell. Onychomycosis associated with Onychocola canadensis: Ten case reports and a review of the literature. J Amer Acad Derm 39:410–417, 1998.

68. RC Summerbell. Nondermatophytic molds causing dermatophytosis-like nail and skin infection. In: J Kane, RC Summerbell, L Sigler, S Krajden, G Land. Laboratory Handhook of Dermatophytes: A Clinical Guide and Laboratory Manual of Dermatophytes and Other Filamentous Fungi from Skin Hair and Nails. Belmont, CA: Star, 1997, pp. 213–259.

69. PC Iwen et al. Pulmonary infection caused by Gymnascella hyalinospora in a patient with acute myelogenous leukemia. J Clin Microbio 38:375–381, 2000.

70. BL Lee et al. Ovadendron sulphureo-ochraceum endophthalmitis after cataract surgery. Am J Ophthalmol 119:307–312, 1995.

7

Ascomycetes

Aspergillus, Fusarium, Sporothrix, Piedraia, and Their Relatives

Richard Summerbell
Centraalbureau voor Schimmelcultures, Utrecht, The Netherlands

This chapter also includes pathogenic and opportunistic members of the *Eurotiales*, *Hypocreales*, *Ophiostomatales*, and *Pseudeurotiaceae ss. str.*, plus the families *Didymosphaeriaceae* and *Piedraiaceae* of the order *Dothideales* and mycetoma-causing members of the genus *Leptosphaeria* (*Pleosporales*).

I. INTRODUCTION

The hierarchical classification of organisms into taxa begins with the species, and then advances to the genus, family, order, class, phylum or division, and kingdom. Each of these major levels may also contain some recognized subtaxa (e.g., species may contain varieties). In the years immediately prior to the advent of molecular phylogeny, the families, orders, and classes of fungi in the phylum *Ascomycota* were in disrepute. It was realized that they were partially based on relatively uncomplicated form characteristics that could easily have evolved convergently, so that to some extent they represented artificial groups rather than naturally biologically related groups. Fortunately, studies based on sequencing genes such as the 18S and 26S ribosomal subunits have allowed more natural groupings of ascomycetous fungi to be clarified. To a large extent, the traditional taxonomy of these organisms is retained, while the correct assignment of species in groups long suspected of heterogeneity, such as the *Pseudeurotiaceae* (1), is facilitated. Currently, the orders *Eurotiales* and *Hypocreales* appear to be supported by molecular phylogenetic studies, although the latter is paraphyletic if its offshoot, the *Clavicipitales* or *Clavicipitaceae*, is considered a separate order.

In the past, formally delineated biological groups based on sexual species were kept separate from anamorphic species described based on asexual states. The reason was that the asexual states usually did not display sufficient information to allow them to be meaningfully biologically grouped. Recently, however, ongoing biological studies, particularly those involving sequencing, have disclosed the biological affinities of the majority of anamorphs of medical and economic interest, if these were not already known. In the presentation below, therefore, anamorphic species are represented along with teleomorphic species in the orders and families named. It should be noted, however, that anamorphs are still accorded separate generic names; therefore, even though *Sporothrix schenckii* is very closely related to the teleomorph genus *Ophiostoma*, it remains in the separate anamorph genus *Sporothrix* under our current nomenclatural rules and does not become "*Ophiostoma schenckii*" (as one recent publication erroneously named it). The close teleomorphic affinities of anamorphs, where they are known, will be emphasized in this chapter through the device of inserting the name of a sufficiently closely related teleomorph genus in braces between anamorph genus name and epithet in the primary entry for the species (e.g., *Sporothrix {aff. Ophiostoma} schenckii*). This is an extension of the traditional device of inserting additional information about superseded relationships into the middle of binomials [e.g., *Candida (Torulopsis) glabrata*]. As with its prototype, the device used here should not be taken to imply a new nomenclatural combination. Again, as with the prototype, it is not a formal nomenclatural usage, but rather a literary one. The standard taxonomic abbreviation "*aff.*" (*affine*) is used both to obstruct the appearance of a new coinage and also to indicate that in current taxonomy, where Hennigian phylogenetic principles are not always strictly adhered to, a highly phenotypically divergent teleomorph might still be placed in a separate genus, even if phylogenetic indications seemed to contradict this. The prototype for this decision is the genus *Homo*, which is generally held apart from the chimpanzee/bonobo genus *Pan*, even though cladistic analysis tends to amalgamate the genera into a revised genus *Homo* (2).

In this chapter, emphasis is given to the form in which the fungus is most likely to be seen in the clinical laboratory. In cases in which one morph is very unlikely to be seen in the clinical lab, it is generally mentioned but not described.

It should be noted that some phylogenetic studies [e.g., that of Suh and Blackwell (1)] have found the affinity between the Eurotiales (classically considered to contain the holomorphs of *Aspergillus*, *Penicillium*, and relatives) and the Onygenales (containing the holomorphs of dermatophytes, *Blastomyces*, *Coccidioides*, and relatives) to be so great that the two orders are combined into an expanded concept of the Eurotiales. In fact, this concept coincides with some older concepts that unified these two groups, such as the concept used by de Vries (3). This expanded concept is not followed here, but the close relationship between these two medically important groups is noted with interest. Possibly in the future additional phylogenetic studies will generate sufficient supporting evidence that the two groups can comfortably be combined.

A. Notes on the Descriptions

Unless mentioned otherwise, descriptions are based on growth on modified Leonian's agar (4) at 7 days at 25°C in the dark. For most fungi such descriptions differ minimally from those obtained on other common fungal sporulation media with moderate nutrient levels, such as malt extract agar and potato glucose agar. Colonial appearances and growth rates may differ substantially on media low in nutrients, such as soil extract agar, or high in nutrients, such as any formulation of Sabouraud's peptone-glucose agar.

B. A Note on the List of Included Species

It is axiomatic that in a book entitled *Pathogenic Fungi of Humans and Animals* the described species should be pathogenic. On the other hand, medical literature concerning less common opportunistic fungi contains a nearly overwhelming number of spurious case reports based on false attributions of pathogenicity to contaminating organisms, as well as similarly misleading reports based on misidentified organisms. Also, many likely credible reports have been rendered "irreproducible results" from the technical point of view because the authors did not describe unusual organisms well enough that identifications could be verified, and/or did not deposit voucher cultures in professional culture collections. Medical mycology is an unusual mycological subfield to review, in that a high proportion of the published work is the product of professionals in intersecting fields who are nonetheless mycological amateurs, whether physicians or laboratory technologists, and the result, as with all work done by amateurs, covers the whole range from thoroughly professional to tragicomic. Considerable effort, however, has been made to ensure that the present chapter is minimally influenced by erroneous literature, including that extensively cited in other books and reviews. In order to inform readers why some apparent records of pathogenicity have been rejected and to correct the record in general, some notes are included on rejected, reinterpreted, or questionable case reports. These appear in the general paragraph discussing the pathogenicity of described organisms and occasionally in notes on species previously reported as pathogenic for which no cases can be verified. In general, every effort has been made to avoid making or recapitulating ambiguous statements such as "this species has been isolated from infected skin" or "this fungus has been reported from peritonitis," which, despite their apparent promise of etiologic relevance, may well contain no medical information whatsoever and in some cases may further the accumulation of a pseudo-literature supporting and promoting diagnostic error.

Mycology is entering an era in which molecular identification will greatly increase the accuracy of case reports involving unusual organisms. It is therefore useful to know which elements of the past 100 years of medical literature are suitably reliable, based on standard criteria of taxonomic documentation and reproducibility, in order to be meaningfully compared with the results that will be generated in upcoming decades. Also, since more accurate identification methods can in some cases simply generate better quality, spurious reports imputing etiologic status to incidental organisms, the basic gold standards of medical fungal ecology that are used in the reliable confirmation of cases are rather rigorously adhered to in the text below. For the sake of accuracy, no charity is given to ambiguity.

II. *EUROTIALES*

The order *Eurotiales*, as roughly defined here, contains three families: *Eremomycetaceae*, *Thermoascaceae*, and *Trichocomaceae*. By far the greatest number of fungal pathogens is in the last-named family. *Eremomycetaceae* are represented in medical mycology only by *Arthrographis kalrae*, the anamorph of *Eremomyces langeronii*. It is dealt with in Chap. 11. The family Thermoascaceae contains *Thermoascus* and *Byssochlamys* species, as well as some related *Paecilomyces* anamorphs, by far the most biomedically prominent of which is *P. variotii*. Another medically important *Paecilomyces* species, *P. lilacinus*, is completely unrelated, with affinities in the order *Hypocreales*, family *Clavicipitaceae*, rather than *Eurotiales*. It is therefore included in the section of this chapter devoted to Hypocrealean fungi rather than being retained with *P. variotii*. The major human and animal pathogens in the Eurotiales consist of medically important aspergilli, as well as *Penicillium marneffei* and *P. variotii*.

A. *Aspergillus* Fr: Fr.

The medically important members of the genus *Aspergillus* comprehend anamorphs related to seven teleomorph genera: *Emericella, Eurotium, Fennellia, Hemicarpenteles, Neopetromyces, Neosartorya,* and *Petromyces*. Teleomorphs are described where species forming them have been recorded in connection with human and animal disease. In addition, the recently described *Neopetromyces* is added in order to distinguish it from the similar *Petromyces*. Species are described in biological groups rather than in pure alphabetical order. The reason for this organization is that it highlights the biological patterns that emerge when *Aspergillus* species are considered in the context of their natural relationships. These are as outlined by Samson (5), with some modifications to reflect recent molecular studies (6). Within each natural subgroup, teleomorphic species are considered before anamorphic species, and within formal taxonomic sections of the genus, the namesake and type species of the section is treated first, then related species are listed in alphabetical order beneath it (e.g., *A. versicolor* is listed first in the section *Versicolores*). In the section *Aspergillus*, where there is no such namesake, species are just listed alphabetically.

To make species easy to find, Table 1 displays the species in the order in which they appear in this chapter. At the same time, it summarizes the teleomorphs most closely corresponding to common anamorphs seen in the clinical lab. See also the rapid "*Aspergillus* finder" (Table 2).

An overview of the clinically important aspergilli and their characteristic features is presented in Table 3.

In nature, aspergilli are diversely competent, mostly generalist saprobes, usually possessing most (but not all) of the following capabilities: thermotolerance, osmotolerance, cellulose degradation, opportunistic human/animal pathogenesis, plant seed pathogenesis/degradation, insect pathogenesis/degradation, and hydrocarbon degradation. Some species are particularly strongly specialized for osmotolerance; that is, growth on materials of low water activity. Although aspergilli are commonly called "soil fungi," in biomedical literature they are common in soil only in tropical or subtropical areas, deserts, some grasslands, and indoor plantings. In any case, soil is only one of their many characteristic niches.

In general the anamorph genus *Aspergillus* consists of fungi that have no melanized elements, excepting black conidia in some species. They generally form erect, aseptate, thick-walled conidiophores with an expanded vesicle at the apex and a thick-walled "foot cell" integrating more or less at a 90° angle into the subtending hypha in the growth medium at the base. On the apical vesicle, fertile elements are formed synchronously (not sequentially, as in members of the genus *Penicillium*); that is, primordia of fertile elements in the initial stages of formation can be seen to be all the same size on each individual vesicle. Fertile elements may consist of phialides alone or of short branches (metulae) bearing clusters of phialides apically. Conidia are formed in interconnected chains and are strongly hydrophobic. Conidia en masse may be black, white, or "brightly colored" (e.g., green, blue, yellow, sandy-brown, ochraceous).

1. *Eurotium* Link:Fr.

Ascomata are globose cleistothecia, 100 to 150 µm in diameter, and usually bright yellow to orange-red. *Asci* are eight-spored, globose, and evanescent. *Ascospores*

Table 1 Natural (Teleomorph-Based) Groups of *Aspergillus* Species as Arranged in This Chapter

Teleomorph	*Aspergillus* subtaxon		Raper/Fennell group	Species: teleomorph, (anamorph)
	Subgenus	Section		
Eurotium	*Aspergillus*	*Aspergillus*	*A. glaucus* grp.	*Eurotium amstelodami*/(*Aspergillus vitis*) *E. chevalieri*/(*A. chevalieri*) *E. herbariorum*/(*A. glaucus*) *E. repens*/(*A. reptans*) *E. rubrum*/(*A. rubrobrunneus*)
		Restricti	*A. restrictus* grp.	(*A. restrictus*) (*A. caesiellus*) (*A. conicus*) (*A. penicillioides*)
Emericella	*Nidulantes*	*Nidulantes*	*A. nidulans* grp.	*Emericella nidulans*/(*A. nidulans*) *E. dentata*/(*A. nidulans* var. *dentatus*) *E. echinulata*/(*A. nidulans* var. *echinulatus*) *E. quadrilineata*/(*A. tetrazonus*) *E. rugulosa*/(*A. rugulovalvus*) *E. unguis*/(*A. unguis*)
		Versicolores	*A. versicolor* grp.	(*A. versicolor*) (*A. granulosus*) (*A. janus*) (*A. sydowii*) (*A. varians*)
		Usti	*A. ustus* grp.	(*A. ustus*) (*A. deflectus*)

Fennellia	*Nidulantes*	*Flavipedes*	*A. flavipes* grp.	*Fennellia flavipes*/(*A. flavipes*)
				F. nivea/(*A. niveus*)
				(*A. carneus*)
		Terrei	*A. terreus* grp.	(*A. terreus*)
Neopetromyces	"*Circumdati*"[a]	*Candidi*	*A. candidus* grp.	(*A. candidus*)
	Circumdati	*Circumdati*	*A. ochraceus* grp.	(*A. ochraceus*)
				(*A. sclerotiorum*)
Petromyces		*Flavi*	*A. flavus* grp.	*Petromyces alliaceus*/(*A. alliaceus*)
				(*A. flavus*)
				(*A. oryzae*)
				(*A. avenaceus*)
				(*A. tamarii*)
		Nigri	*A. niger* grp.	(*A. niger*)
				(*A. japonicus*)
Neosartorya	*Fumigati*	*Fumigati*	*A. fumigatus* grp.	*Neosartorya pseudofischeri*/(*A. thermomutatus*)
				N. fischeri/(*A. fischerianus*)
				(*A. fumigatus*)
Hemicarpenteles		*Clavati*	*A. clavatus* grp.	(*A. clavatus*)
				(*A. clavato-nanica*)

[a] Recent molecular studies have shown that this group was particularly strongly displaced from its natural affinities by phenotypic taxonomy, and it is arranged with its related groups. The related subgeneric nomenclature, however, has not yet been changed.

Table 2 *Aspergillus* Finder [Look up the *Aspergillus* or Teleomorph Name (or Common Synonym) Alphabetically and Locate It in the Biological Arrangement]

Name	Related teleomorph group	Page number
Aspergillus alliaceus	*Petromyces*	298
A. amstelodami	*Eurotium*	252
A. avenaceus	*Petromyces*	303
A. caesiellus	*Eurotium*	264
A. candidus	*Fennellia*	290
A. carneus	*Fennellia*	288
A. chevalieri	*Eurotium*	256
A. clavatonanica	*Hemicarpenteles*	315
A. clavatus	*Hemicarpenteles*	313
A. conicus	*Eurotium*	266
A. deflectus	*Emericella*	283
A. fischeri	*Neosartorya*	308
A. flavipes	*Fennellia*	285
A. flavus	*Petromyces*	298
A. fumigatus	*Neosartorya*	310
A. glaucus	*Eurotium*	257
A. granulosus	*Emericella*	277
A. janus	*Fennellia*	278
A. japonicus	*?Petromyces*	306
A. nidulans	*Emericella*	267
A. niger	*?Petromyces*	304
A. niveus	*Fennellia*	286
A. ochraceus	*Neopetromyces*	294
A. oryzae	*Petromyces*	301
A. penicillioides	*Eurotium*	266
A. quadrilineatus	*Emericella*	271
A. reptans	*Eurotium*	262
A. restrictus	*Eurotium*	263
A. rubrobrunneus	*Eurotium*	263
A. rugulosus	*Emericella*	272
A. sclerotiorum	*Neopetromyces*	295
A. sydowii	*Emericella*	279
A. tamarii	*Petromyces*	303
A. terreus	*Fennellia*	290
A. tetrazonus	*Emericella*	271
A. unguis	*Emericella*	274
A. ustus	*Emericella*	281
A. varians	*Emericella*	280

Table 2 Continued

Name	Related teleomorph group	Page number
A. versicolor	Emericella	275
A. vitis	Eurotium	252
Emericella dentata (T)		270
E. echinulata (T)		270
E. nidulans (T)		267
E. quadrilineata (T)		271
E. rugulosum (T)		272
E. unguis (T)		274
Eurotium amstelodami (T)		252
E. chevalieri (T)		256
E. herbariorum (T)		257
E. repens (T)		262
E. rubrum (T)		263
Fennellia flavipes (T)		285
F. nivea (T)		286
Neosartorya fischeri (T)		308
N. pseudofischeri (T)		308
N. spinosa (T)		308
Petromyces alliaceus (T)		298

Note: (T) indicates teleomorph names.

are rough or smooth, oblate, usually with at least one of the following: equatorial crests, an equatorial furrow, or a distinct flattened equatorial band. The anamorphs are members of *Aspergillus* subgenus *Aspergillus* section *Aspergillus*, formerly considered the *A. glaucus* series (7). They have uniseriate conidiophores.

Species with a Known Teleomorph and with Anamorphs in A. *Subgenus* Aspergillus *Section* Aspergillus *(Formerly* Aspergillus glaucus *Group).* This is a highly unified and distinct group of species, in which most isolates readily form the teleomorph homothallically in culture on common media, and do so even more reliably on concentrated media used for osmotolerant fungi. In ecology, members of this group are osmotolerant colonizers of dry substrates or other substrates with low water activity due to high solute concentrations (e.g., salted materials, high-sugar substrates). They are thus often encountered in food mycology, and are also major colonizers of household dust, where they may grow on relatively dry shed-skin scales. Growth on leather stored in slightly humid conditions, as well as other semidry surfaces, is common. Their isolation

Table 3 Laboratory Characters of the Medically Important Aspergilli and Species Frequently Confused with Them

Species	Aspergillus head structure	Typical colony surface color	Stipe	Special features
Uniseriate group				
Eurotium amstelodami (*Aspergillus vitis*)	Uni[a]	Dull green with yellow tufts (ascomata)	Smooth	Ascospores with central groove and rough valves
E. chevalieri (*A. chevalieri*)	Uni	Dull green with yellow tufts (ascomata)	Smooth	Ascospores with central groove and two broad sur-rounding ridges; valves smooth
E. herbariorum (*A. glaucus*)	Uni	Dull green with yellow, red, and orange tufts (ascomata and sterile hyphae)	Smooth	Ascospores with shallow groove and sometimes small ridges; valves smooth
E. rubrum	Uni	Dull green with yellow, red, and orange tufts (ascomata and sterile hyphae)	Smooth	Ascospores with shallow groove, valves smooth; growth rate on special me-dia differs from *E. herbari-orum*
E. repens	Uni	Dull green with yellow to orange tufts (ascomata and sterile hyphae)	Smooth	Ascospores with flat equato-rial band; valves smooth
A. restrictus	Uni	Dark green	Smooth	Small colonies; small vesicles with persistent chains of long-ellipsoidal conidia; no growth at 42°C
A. fumigatus	Uni	Blue-green to smoky blue-green	Smooth, greenish	Conidial masses in upright col-umns; good growth at 42°C
A. fumigatus (*dysgonic*)	Uni or irregular	Dirty white to pale green	Poorly formed	No conidiation, or aberrant, small heads; conidia also sometimes aberrant; good growth at 42°C

Species	Seriation	Color	Stipe	Distinguishing features
Neosartorya pseudofischeri	Uni	Creamy (white ascomata)	Smooth	Ascospores with erect triangular flaps on valves, a central groove, and broad surrounding ridges
N. fischeri	Uni	Creamy (white ascomata)	Smooth	Ascospores with netlike reticulum on valves, plus central groove and surrounding ridges
N. spinosa	Uni	Creamy (white ascomata)	Smooth	Ascospores with small conical spines on valves, plus central groove and surrounding ridges
A. clavatus	Uni	Blue-green	Smooth	Elongate club-shaped vesicles
A. japonicus	Uni	Purple-black	Smooth	Large, round vesicle; conidia spiny
Uni- or biseriate group (all with rough, apically expanding stipes)				
A. flavus	Uni or bi[a]	Yellow-green	Roughened, expanded toward apex	Metulae extend around whole vesicle surface; conidia smooth to finely roughened
A. parasiticus	Mostly uni, few bi	Dark olive green	Roughened, expanded toward apex	Metulae if present extend around whole vesicle surface; conidia strongly roughened
A. tamarii	Uni or bi	Olive brown	Roughened, expanded toward apex	Metulae extend around whole vesicle surface; conidia strongly roughened
Biseriate group with metulae surrounding vesicle				
A. avenaceus	Bi	Dull olive green	Almost smooth (very fine roughening); parallel sides	Metulae extend around whole vesicle surface; black sclerotia present

Table 3 Continued

Species	*Aspergillus* head structure	Typical colony surface color	Stipe	Special features
A. ochraceus	Bi	Pale yellow-brown	Roughened, parallel sides	Metulae extend around whole vesicle surface; conidia smooth or finely rough; sclerotia absent or pink/purple
A. sclerotiorum	Bi	Pale yellow	Roughened, with parallel sides	Metulae extend around whole vesicle surface; conidia smooth or finely rough; sclerotia abundant, cream-colored
A. niger	Bi	Dirty black	Smooth	Metulae extend around whole vesicle surface; conidia roughening irregular ridges and bars
A. japonicus[b]	Uni	Purple-black	Smooth	No metulae; conidia roughening small conical spines
A. candidus	Bi	White	Smooth or finely roughened, hyaline	Metulae extend around whole vesicle surface; no aleurioconidia in substrate mycelium
Biseriate group with metulae on upper portion of vesicle				
A. flavipes	Bi	Pinkish-buff	Smooth or finely rough; brownish	Metulae from upper 3/4 vesicle; aleurioconidia present in substrate mycelium
A. niveus	Bi	White	Smooth, hyaline	Metulae from upper 2/3 of vesicle; aleurioconidia may be detected

A. carneus	Bi	Pinkish to pinkish brown	Smooth, hyaline or faint brown	Metulae from upper 2/3 of vesicle; irregularly disposed phialides may be seen; aleurioconidia in substrate mycelium
A. terreus	Bi	Sandy brown	Smooth	Metulae from upper 2/3 of vesicle in nearly parallel arrangement; conidial masses columnar; aleurioconidia in substrate mycelium
Emericella nidulans (*A. nidulans*)	Bi	Forest green, later with yellow tufts (ascomata)	Smooth, brown	Metulae from upper 3/4 of vesicle; ascospores purple-red with two crests, valves smooth; round hülle cells present
E. echinulata	Bi	Forest green, later with yellow tufts (ascomata)	Smooth, brown	Metulae from upper 3/4 of vesicle; ascospores purple-red with two crests; valves with small spines; round hülle cells present
E. quadrilineata	Bi	Dull green, later with yellow tufts (ascomata)	Smooth, brown	Metulae from upper 3/4 of vesicle; ascospores purple-red with four crests, two large ones flanked by two smaller ones; valves smooth; round hülle cells present
E. rugulosa	Bi	Dull green, soon yellow to purple brown with heavy ascoma formation	Smooth, pale brown	Metulae from upper 3/4 of vesicle; ascospores purple-red with two crests, valves rugose (irregular folded roughening); round hülle cells present

Table 3 Continued

Species	*Aspergillus* head structure	Typical colony surface color	Stipe	Special features
A. unguis	Bi	Dull green	Smooth, brown	Metulae from upper 2/3 of vesicle; ascomata usually not formed; colony surface bears erect spinelike hyphae ("setae")
A. versicolor	Bi	Grey-green, emerald green, pinkish or tawny	Smooth, hyaline	Metulae from upper 3/4 of vesicle; round hülle cells sometimes present; small penicillate conidiophores may be present
A. sydowii	Bi	Turquoise grey to dull blue-green	Smooth, hyaline	Metulae from upper 3/4 of vesicle; small penicillate conidiophores may be present
A. granulosus	Bi	Dull green to purple-brown	Smooth, hyaline	Metulae from upper 3/4 of vesicle; colony surface with clumps of hülle cells
A. ustus	Bi	Grey-brown	Smooth, brown	Metulae from upper 3/4 of vesicle; colony reverse yellow; long, sinuous hülle cells may be present
A. deflectus	Bi	Grey-brown	Smooth, red-brown, vesicle deflected abruptly downward	Metulae from upper 3/4 of vesicle; colony reverse yellow to reddish, sinuous hülle cells may be present

[a] uni = uniseriate; bi = biseriate.
[b] *A. japonicus* is given a second entry to facilitate comparison with *A. niger.*

and growth are favored by high-solute media not used in medical mycology, such as Czapek's + 20% sucrose agar and dichloran 18% glycerol agar, but most isolates will grow and sporulate, at least to some degree, on common medical mycology sporulation media such as potato dextrose agar. In secondary metabolism, the members of the group investigated so far form a similar spectrum of compounds, including physcion and echinulin, which are not considered mycotoxins in the strict sense of Frisvad and Thrane (8).

The species in this group have often not been distinguished from one another, and the group as a whole—or *A. glaucus* used as a general term—has a pathogenic record that cannot be attributed to individual species. In some cases authors have attributed responsibility to *A. glaucus*, but lack of detail—and in larger surveys, failure of more common *Eurotium* species to appear—leads one to suspect that all *Eurotium* anamorphs were lumped under this name. Nonetheless, such attributions will be discussed under *E. herbariorum* (anamorph *A. glaucus*) below. All species in *Aspergillus* subgenus *Aspergillus* produce small aspergilli with uniseriate heads, leading to possible confusion with dysgonic (poorly and aberrantly sporulating), host-adapted isolates of *A. fumigatus*, as is illustrated in several definite examples of mistaken identity mentioned under *A.* subgenus *Aspergillus*, section *Restricti* below. Only a small number of records contain sufficient detail to allow verification that the identification was correct.

Young et al. (9) ascribed three cases of pathologically verified systemic aspergillosis in immunocompromised patients to members of the *A. glaucus* group, not further identified. These cases made up 5% of total aspergilloses seen in a 39-patient survey (7.7% of the patients seen were attributed *A. glaucus* infection; the difference between the two percentages arises from the occurrence of mixed aspergilloses in the patient population). One attributed *A. glaucus* group infection was pulmonary, one disseminated, and one affected the spleen concomitant with an *A. fumigatus* pulmonary infection. No identification details were given. This report came from a prestigious institution (the National Institutes of Health) and may well be legitimate, despite its absence of taxonomic documentation and its indefinite identifications. It raises, however, the question of why *A. glaucus* group members did not continue to cause a similar proportion of invasive aspergilloses through the swell of well-documented opportunistic mycoses of the later 1970s and onward. In fact, as can be seen in the present writing, a high proportion of the reports of invasive opportunistic mycoses caused by *Aspergillus* species with small, green aspergilli, potentially confused with host-adapted *A. fumigatus* (i.e., *Aspergillus* subgenus *Aspergillus*, and for workers experiencing the initial difficulties of distinguishing biseriate and uniseriate structures, the section *Versicolores*) are indeed from more than 25 years ago. This finding is difficult to explain except by postulating at least a degree of influence arising from a shift from less accurate to more accurate identification. The inclusion of the "*A. glaucus* group" in the bibliographically unsupported list of invasive aspergilloses originated by Rinaldi (10) and reproduced and extended by Rippon (11) appears to derive from the report of Young et al. (9), based on a comparison of the body sites listed. Well-verified examples of comparable cases must be documented before these records can be accepted.

Table 4 gives statistics on the number of reports obtained in the online Medline database with keywords designed to roughly determine the proportion of aspergilloses caused by different species in immunocompromised patients as of August 8, 2000. (See the footnote in Table 4 for the keywords used.) The records obtained were scanned to eliminate inappropriate entries, but no attempt was made to enumerate multiple cases recorded in individual reports or to detect overlap. None of the 435 accepted Medline reports on aspergillosis in immunocompromised patients featured *A. glaucus* or other members of section *Aspergillus* in their searchable text. In comparison, the roughly predicted number of reports based on a hypothetical 5% ratio of cases [as per Young et al. (9)] is 22; that is, approximately the level of reporting found for *A. niger*. This crude prediction ignores the possible tendencies for interestingly rare causal agents to be disproportionately frequently reported and for highly prevalent opportunists to be treated as too quotidian to report except in novel cases or in series. Such tendencies would cause the expected number of reported section *Aspergillus* cases to rise. Although species are certainly reported in some omnibus papers without their names entering the text searchable in Medline, it is presumed that in most if not all cases in which species identifications are given more than desultory attention, the names will appear in summary material. There is no obvious reason why an identification of *A. glaucus* from invasive aspergillosis should elude abstracting more frequently than one of, for example, *A. niger*. The Fischer exact probability that the Young et al. (9) data and the above Medline data on section *Aspergillus* case frequencies are drawn by chance from the same population is <0.005. On the other hand, the p value derived from χ^2 testing for *A. fumigatus* in a comparison of both data sets is 0.466, indicative of high homogeneity. This suggests that some or all of the "*A. glaucus* group" cases recorded by Young et al. (9) were based on misidentifications.

***Eurotium amstelodami* Mangin, Anamorph *Aspergillus vitis* Novobr.**
[Synonyms: *Aspergillus amstelodami* (Mangin) Thom & Church, *Aspergillus hollandicus* Samson & W. Gams].   This species has been conclusively demonstrated as an agent of a cerebral abscess in an otherwise healthy female (12) and of a leg tumor in a Galápagos tortoise, *Testudo elephantopus* (13) (Figs. 1 and 2). An Argentinian case of black grain mycetoma consistently grew a rugose-ascospored *Eurotium* from both native and stringently surface-disinfected grains (14). Although identified according to the taxonomy of the day (1939) as *E. chevalieri*, now regarded as a smooth-ascospored species (minute roughening is visible in electron microscopy), the isolate illustrated would most probably be

Table 4 Aspergillosis Case Reports Involving
Immunocompromised Patients, as Determined by Species
Name Search on Online Medline (August 2000)

Species named	Number of reports	Percent
A. fumigatus	299	68.7
A. flavus	74	17.0
A. niger	25	5.7
A. terreus	17	3.9
A. nidulans	5	1.1
A. ustus	5	1.1
Neosartorya fischeri	2	0.5
A. granulosus	1	0.2
A. flavipes "group"	1	0.2
A. ochraceus	1	0.2
A. oryzae[a]	1	0.2
A. versicolor	1	0.2
Emericella echinulata[b]	1	0.2
E. quadrilineata[c]	1	0.2
Neosartorya pseudofischeri[d]	1	0.2
Total species records[e]	435	

Note: Search term string used: case + [species epithet] + (Aspergillus or [teleomorph genus name]) + (immunocompromis* or immunodefic* or immunosuppress* or compromised or lymphoma or leukem* or HIV or AIDS or steroid* or corticosteroid* or transplant*) not (mouse or mice or rat or rats or alveolitis or hypersensitivity). Items in brackets were filled in appropriately for each organism. Various permutations of this string gave similar results.
[a] Byard et al. (159): not actually A. oryzae. (See text for that species.)
[b] Originally identified as A. nidulans var. echinulatus.
[c] Originally identified as A. quadrilineatus. (Anamorph name now changed to A. tetrazonus.)
[d] Originally identified as N. fischeri var. spinosa.
[e] Searched epithets giving nil records: amstelodami, avenaceus, candidus, carneus, chevalieri, clavatus, deflectus, glaucus, herbariorum, hollandicus, niveus, parasiticus, penicillioides, restrictus, rugulosa, rugulosus, sclerotiorum, sydowii, tamarii, unguis.

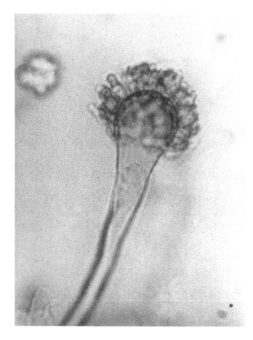

Figure 1 Conidiophore of *Aspergillus vitis*, anamorph of *Eurotium amstelodami*.

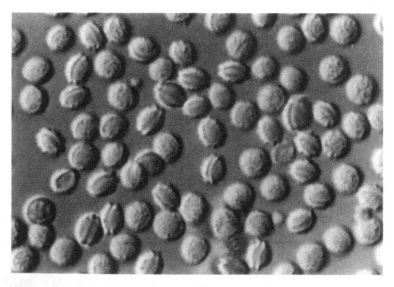

Figure 2 Ascospores of *Eurotium amstelodami*.

identified as *E. amstelodami* today. The rough ascospores also have very minute equatorial crests or ridges, consistent with this species. (See the discussion of *E. chevalieri* below.)

A report of *A. amstelodami* causing human hypopyon corneal ulcer subsequent to puncture by a bamboo stick (15) contains no identification details, but since this specific identification cannot be attempted without at least seeing ascospores, this case, in which corneal scrapings were positive for fungal elements, is tentatively accepted. The original description of *Aspergillus montevideensis* Tal. & Mack., a name long considered and now molecularly supported (6) as a synonym of *E. amstelodami*, includes a convincing case report describing the postsurgical eardrum colonization from which the isolate was obtained (16). The isolate is available in several culture collections [e.g., Centraalbureau voor Schimmelcultures (CBS)], American Type Culture Collection (ATCC)]. A later report of causation of human otitis by Wadhwani and Srivastava (17) is not accepted for reasons mentioned under *Fennellia nivea* below. A report by Grigoriu and Grigoriu (18) that the fungus was among the most common agents of nondermatophytic onychomycosis in Switzerland must be regarded as dubious. (See the discussion of that report under *Emericella unguis*, below, along with the discussion of *A. glaucus* onychomycosis records.)

An unusual case report by Janke (19) records *A. vitis* (as *A. amstelodami*) as the purported cause of a large, psoriasislike scalp lesion causing diffuse hair loss in an otherwise healthy woman. The paper's histologic report states "in epithelium, especially of the hair follicle, fungal spores were visible by gram stain" (transl. W. Gams). This species, however, would be expected to form typical *Aspergillus* tissue filaments if it were causing infection. If conidia were formed, vesiculate conidiophores should also be seen. It is possible dermatophyte substrate arthroconidia or artifacts were seen in this study rather than *A. vitis* conidia, especially as the structures seen are not described. *Aspergillus* species, apart from *A. terreus*, do not produce structures resembling conidia within infected tissue. (Conidia formed in colonizing infections such as aspergilloma are not formed within tissue, but rather after mycelium breaks through a substrate/air interface.) The reason this seemingly misinterpreted case of "aspergillosis capitis" (19) is mentioned is that the present author was also once sent three successive scalp specimens from a similar case in which a pronounced scalp lesion on a child gave rise to *E. amstelodami* each time it was cultured, even though definitive direct microscopic corroboration of infectious status could not be found. The unresolved etiology of the case precluded publication. It might be proposed that *E. amstelodami*, which saprobically colonizes shed-skin flakes in house dust, may do the same to a limited extent on lesions in which sufficient dead keratinous material has accumulated. If so, however, filamentation and conidiophore formation should again be seen, as mentioned above for infection. Comparable cases may provide clues as to the origin of such "scalp aspergilloses." A case involving a heavily outgrowing *Penicillium* species also repeatedly showed only "brown, double-walled spores" in direct microscopy of recurring "light brown crusts" on the scalp of a 3-year-old boy (20). The family cat in veterinary examination had similar crusts, also apparently composed of *Penicillium* conidia. No dermatophyte was isolated from either the boy or the cat, nor were filaments seen in microscopy or, in the boy's case, in a punch biopsy performed after 6 weeks of oral ketoconazole therapy appeared ineffective. A further 4 weeks of ketoconazole at doubled dosage was considered to have cured the infection. This conidial crust phenomenon, also mentioned in a previous *Penicillium* case involving two brothers (21), as well as in additional unreported cases mentioned in that paper, may derive from adhesion of environmentally formed conidia from an unknown source to scalp oils or serous exudate. Also, growth of the fungus in hair products or skin medicaments, with subsequent transfer of conidia to the scalp by the patient or caregiver, is possible, but no candidate materials have been identified in case reports to date. The variants of Munchausen syndrome (attention-seeking construction of false medical symptoms in oneself or a proxy, usually a child) provide perhaps the most parsimonious potential explanation, and certainly isolates from such cases should be fully identified to rule out organisms such as *Penicillium roquefortii*, the blue cheese fungus, and species not growing at body temperature. It is not clear, however, how Munchausen practitioners would acquire pure *Eurotium*. The conspicuous presence of such organisms indoors on moldy cloth, leathers, papers, or particleboard may provide an opportunity in some cases.

The ecology of *A. amstelodami* is as discussed above for *A.* subgenus *Aspergillus* section *Aspergillus* in general.

Description. Colonies grow 7 to 10 mm in diameter in 7 days at 25°C, at first whitish, but rapidly becoming deeply dusty grey-green to dark green with conidiation of the *Aspergillus* anamorph. *Ascomata* arise as small, yellowish, spheroidal tufts scattered on surfaces of mature (usually 7–20-day-old) colonies, sometimes massing densely and turning the colony center yellowish overall. The colony reverse is pale to yellow-brown.

Ascomata under the microscope are mostly 75 to 150 μm in diameter, globose, nonostiolate, with thin pseudoparenchymatous peridium. *Asci* contain eight lens-shaped *ascospores*, 4.5–6 × 3.5–4 μm, rough-walled, each with an equatorial furrow flanked by two rough, irregular ridges. *Conidiophores* are mostly 250 to 350 μm high, with thick, hyaline walls, aseptate, uniseriate (i.e., phialides directly attached to vesicle), with vesicles globose or nearly so. *Conidia* are in dry, upright chains, grey-greenish to brownish in transmitted light, subglobose to globose, sometimes distinctly barrel-shaped with two conspicuously flattened ends, finely to distinctly roughened, 3.5–5.5 × 3.5–5.0 μm.

***Eurotium chevalieri* Mangin, Anamorph *Aspergillus chevalieri* Mangin** [Synonym: *Aspergillus chevalieri* (Mangin) Thom & Church]. Few human or animal case reports concern this species, and the majority are not definitive. The most likely genuine human case examined so far is that of da Fonseca (22), who grew a pure culture of a *Eurotium* species from eight separate biopsy-derived mycetoma grains from a foot infection. The fungus was originally identified as *E. chevalieri* according to the concepts of Thom and Church (23), and the detailed descriptions and illustrations accord well; however, according to da Fonseca, the isolate was then examined by prominent French mycologist Paul Vuillemin, who, following M. L. Mangin's original diagnosis of *A. amstelodami* as a species with smooth-valved ascospores, reidentified the smooth-ascospored case isolate as *A. amstelodami*. It has been carried forward in subsequent review literature under this name. In more recent times, *E. amstelodami* has been neotypified as a rough-ascospored species [see discussion by Pitt (24)], and da Fonseca's fungus now clearly best fits *E. chevalieri*. Unfortunately, the isolate seems not to have been preserved for confirmation, but it appears likely that *E. chevalieri* or a very similar *Eurotium* may rarely cause mycetoma.

Most other case literature mentioning this species appears unreliable. Three cases of cutaneous infection attributed by Naidu and Singh (25) are based only on the results of outgrowth from a single dermatologic mycology scrape in each case. These appear to be typical cases of tinea manuum, in which the causal dermatophyte failed to grow out in an initial evaluation and a surface contaminant was taken to be an etiologic agent. Confirmation of this interpretation is obtained from a related paper in which *E. chevalieri* and a large number of additional contaminating species are interpreted as skin pathogens by Naidu (26). Implication of *E. chevalieri* in otitis by Wadhwani and Srivastava (17) is not accepted for reasons given under *Fennellia nivea* below.

E. chevalieri is similar in ecology and morphology to *E. amstelodami*, but differs by having strongly pulley-shaped, smooth-valved ascospores with two prominent equatorial crests (Fig. 3).

***Eurotium herbariorum* (Wiggers: Fr) Link, Anamorph *Aspergillus glaucus* Link.** This species has a fragmentary and often dubious record as a human and animal pathogen. As indicated in the general remarks on *Aspergillus* subgenus *Aspergillus* above (pages 245–252), the anamorph name may have frequently been used for any incompletely identified *Eurotium* anamorph. There is no record of human or animal infection by this species that combines adequate establishment of causality with adequate verifiability of identification (e.g., any of: clinching description, diagnostic illustration, collection deposition, or referral to a recognized *Aspergillus* expert).

The classic study on otomycosis by Gregson and La Touche (27) clearly links a case of this affliction to an isolate identified as "*Aspergillus herbariorum*" (a nonexistent name combination). Although the development of yellow ascomata (incorrectly called "perithecia") is mentioned, and in the context of other descriptions, confirms that a *Eurotium* was examined, no microscopic description

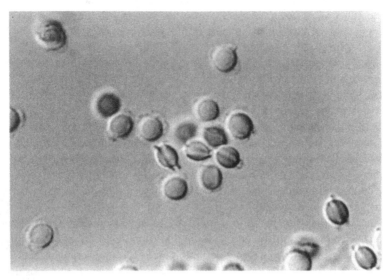

Figure 3   Ascospores of *Eurotium chevalieri.*

is given that could confirm this species or rule out other members of the genus. This appears to be the most strongly confirmed case involving *E. herbariorum* published to date. An authoritative study on keratitis in the region of Mumbai, India, reported 23 cases involving *A. glaucus* among 327 cases in which fungal etiology was confirmed by direct microscopy (28). Three hundred of these cases were also reconfirmed by reculture from scrapings taken the day after the initial examination. Unfortunately, no identification information was given, although differential organisms such as *A. fumigatus* (43 cases) and *A. flavus* (36 cases) were significantly more common than *A. glaucus*, as would be expected. This general indication of accuracy suggests that at least group-level identification for the reported *A. glaucus* isolates is likely to be correct, although there is no means of determining whether *A. glaucus* itself was encountered. An analogous report from climatologically distinct north India mentions three mycologically unverifiable *A. glaucus* cases etiologically confirmed either by direct microscopy or repeated culture (29).

In south Florida, *A. glaucus* has also been included without mycological elaboration in a list of fungi isolated from keratitis by Rosa et al. (30). Direct microscopy was done in this series, but results were not reported, except in general statements indicating that fungal elements were seen in connection with about one-third of the tabulated cases.

One of 13 well-established invasive orofacial aspergilloses in leukemia patients is attributed to *A. glaucus* by Dreizen et al. (31); the lack of case and confirmatory identification detail, plus the failure of the laboratory to identify 4/12 aspergilli cultured from the remaining cases (possibly indicating a low level of familiarity with the genus, as is also suggested by the lack of a name in *Eurotium* for the case isolate) makes this report difficult to accept. (In the present author's experience, the great majority of clinically encountered aspergilli are among the easiest fungi for a relatively experienced laboratorian to identify; therefore, publication of numerous genus-level identifications for etiologic *Aspergillus* isolates strongly suggests reliance on a mycologically unspecialized laboratory.) An *A. glaucus* strain isolated from cellulitis in an HIV-infected child by Shetty et al. (32) is not etiologically connected by direct microscopy and is reported as an unsupported identification in a table in which three of seven *Aspergillus* isolates from cases are not identified to species. A similar pattern is again found in a report on aspergillosis in children with cancer; an unsubstantiated *A. glaucus* isolation from a traumatized skin site (armboard or intravenous site) of a 16-year-old leukemia patient is reported without explicit evidence of causality in a survey in which 15 of 66 reported aspergilli remained unidentified (33). The authors stated that only five of the six skin infections reported were histopathologically examined, but four of these five showed fungal filaments. Whether or not the *A. glaucus* case drew one of the two short straws is not discernible. The report by Weingarten et al. (34) that two of three successive cases of fulminant nasal and paranasal sinus aspergillosis in immunocompromised patients were caused by *A. glaucus* (the third did not yield a culture) appears highly unlikely to be correct, given the statistical anomaly posed by the coincidence of rare events and absence of common events, the incomplete (no *Eurotium* name) and undocumented identifications, and the highly aggressive nature of the pathogen, which disseminated and caused death in one case. Since sinuses, like bronchi, pulmonary cavities, and outer ear canals, are classic sites for ongoing *A. fumigatus* and *A. flavus* establishment and for generation of atypical host-adapted isolates, it seems highly probable that such isolates were misidentified in this report.

A well-established case of a fungal frontal bone lesion in an Indian farmer (35) gives no written details for identification of *A. glaucus* as the causal agent, but includes a nondescript photograph that could represent numerous *Aspergillus* species, including host-adapted *A. fumigatus*, as well as members of *Penicillium* subgenus *Aspergilloides*. No structures resembling the characteristic ascomata of *E. herbariorum* are described or depicted. A report of mycotic keratitis by Venugopal et al. (36) is not confirmed by demonstration of fungal filaments in corneal scrapings; indeed, the authors state that such elements were only seen in 19.6% of cases attributed fungal etiology. Such diagnostic leaps of faith, justifiable clinically in rapidly progressing eye infections, unfortunately do not provide mycological documentation.

Gage et al. (37) diagnosed *A. glaucus* endocarditis after a man recently given an artificial cardiac prosthesis developed a fever and yielded *A. glaucus* in one blood culture, although "many subsequent blood cultures during this febrile period showed no growth." When the patient recovered without treatment, they concluded that "some patients are able to successfully combat *Aspergillus* infection on their own." Studies of blood cultures [e.g., Bille et al. (38)], however, generally reveal, among species isolated only rarely, a number of fungi with no known pathogenic record that are most likely contaminants. Since ascospores are extremely resistant, ascosporic fungi seem particularly likely to survive disinfection procedures and to contaminate the small plug of skin taken up in all percutaneous syringe samples, as well as any chemically surface-disinfected or otherwise inadequately sterilized equipment. Such events may have led to confusion in the past (e.g., a hypothesis that the *Thermoascus crustaceus* metabolite cyclosporin was the cause of AIDS, based on several isolations of this ascosporic fungus from monocyte cultures of AIDS patients) (39). The most probable interpretation of Gage et al.'s case (37), then, is not spontaneous recovery from cardiac aspergillosis, but rather, single contaminated blood culture plus fever of undiagnosed origin. In any event, the identification of *A. glaucus* in this case is not substantiated. An earlier, etiologically well demonstrated *Aspergillus* endocarditis case gave no substantiation for identification of the causal agent as *A. glaucus* (40).

An attributed case of *A. glaucus* otomycosis by Bambule et al. (41) lacks direct microscopic verification; the authors state that such verification was obtained in some cases, but do not state which. Well-demonstrated mycotic middle ear infection is attributed to *A. glaucus* by Wulf (42); the identification is substantiated by a microphotograph that shows globose-vesiculate aspergilla compatible with *A. glaucus* or another *Eurotium* anamorph, although *A. fumigatus* cannot be completely ruled out. Vennewald et al. (43) list *A. glaucus* among 31 cultures from paranasal sinusitis, but state that only 12 of these cultures connected with positive direct microscopy and fail to indicate which these were. A connection by Ponikau et al. (44) with causation of allergic fungal sinusitis cannot be accepted for reasons given below under *A. versicolor*. More problematic is the attribution of five cases of *Aspergillus* onychomycosis to *A. glaucus* by Bereston and Waring (45). These authors confirmed the nondermatophyte onychomycoses impeccably; however, the assertion that *A. glaucus* made up 83.3% (five of six) of sequential nail aspergilloses seen is a profound statistical anomaly compared to all later surveys (e.g., Refs. 46, 47), in which this species is rare (and poorly etiologically linked) or absent. The present author's laboratory evaluated at least 75,000 nail specimens from 1985 to 2000 without elucidating a confirmed or strongly suspected *Eurotium* onychomycosis (45; unpublished data). At the same time, the normally relatively common *A.* subgenus *Nidulantes*, section *Versicolores* (=*A. versicolor* group) is absent from Bereston and Waring's study, suggesting that the Army Medical School credited for identifications may have made a systematic error. Since no vouchers were deposited and no identification criteria were given, this matter cannot be investigated.

A nail biopsy figured by Contet-Audonneau et al. (48: Fig. 10) purports to show *A. glaucus* filaments in nail. Cases in this study, however, were not verified by later repeat isolation. Furthermore, no fungal identification characters were given. The authors stated that they were able to distinguish dermatophyte from nondermatophyte filaments in biopsies. They characterized dermatophyte filaments, however, as being "regular in diameter and sometimes associated with a mass of arthrospores." This describes dermatophyte filaments from distal-subungual onychomycosis, but not those from superficial white onychomycosis, in which irregular and frondose filaments are common (49, 50). The figured biopsy depicted a superficial infection. Very little research has been done to date around the distinction of dermatophyte and nondermatophyte onychomycoses in biopsy slides stained with nonspecific fungal stains, and repeat isolation studies are required as a gold standard to allow the validation of results. Identification criteria should be given for such unusual fungi as *A. glaucus*.

In summary, *E. herbariorum* is a possible opportunistic pathogen badly in need of an authoritative case report. Based on the "where there's smoke there's fire" principle, it is likely a small but significant number of opportunistic mycoses, especially keratitis cases, are indeed caused by *Eurotium* species, although clearly the probability of involvement of the thermotolerant and previously etiologically substantiated *E. amstelodami* is much greater than that of the relatively uncommon and less thermotolerant *E. herbariorum/A. glaucus*, for which no fully credible report of pathogenicity appears to exist.

Like other members of its genus (see general discussion of *A.* subgenus *Aspergillus* section *Aspergillus* above), *E. herbariorum* is adapted to grow in conditions of low water activity. Carefully identified isolates have mainly been from cereals, spices, dried fruits, and other low-water-activity food products (51).

Description. Colonies grow 5 to 10 mm in diameter in 7 days at 25°C, at first whitish, but rapidly becoming deeply dusty grey-green to dark green with conidiation of the *Aspergillus* anamorph. *Ascomata* arise as small, yellowish, spheroidal tufts on surfaces of mature (usually 14 or more days old) colonies, often massing densely amid orange to reddish sterile hyphae. The colony center then becomes yellow-orange to reddish, and progressively more reddish with age. The colony reverse is pale orange or deep reddish-brown.

Ascomata under the microscope are mostly 100 to 150 μm in diameter, globose, nonostiolate, with thin pseudoparenchymatous peridium. *Asci* contain eight lens-shaped *ascospores*, 6–7 × 5–6 μm, smooth-walled, each with a shallow equatorial furrow, sometimes with small ridges or flattened crests (Fig. 4). *Conidiophores* are mostly 500 to 700 μm high, with thick, hyaline walls, aseptate, uniseriate (i.e., phialides directly attached to vesicle), with vesicles globose or nearly so. *Conidia* in dry upright chains, grey-greenish to brownish in transmitted

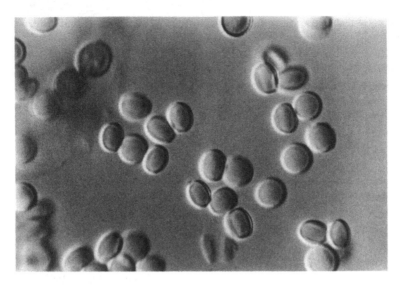

Figure 4   Ascospores of *Eurotium herbariorum.*

light, subglobose to globose, sometimes distinctly barrel-shaped with two con-spicuously flattened ends, spinose, 5.0–6.5 × 3.0–5.0 μm.

The very similar *Eurotium rubrum* König et al. is distinguished by its slightly smaller ascospores, 5.0 to 6.0 μm wide, and by its different growth rates on specialized diagnostic media, as outlined by Pitt and Hocking (51). The only slightly less similar *E. repens* de Bary has ascospores with barely discernible equatorial furrows, and also has more yellow or orange and less red colony pig-mentation deriving from colored sterile hyphae. These identifications should be confirmed by comparison of reference isolates, by experts, or by molecular analy-sis, and should not be published on the basis of comparison with descriptive literature alone. Voucher isolates should be deposited in professional culture col-lections (i.e., collections operated by dedicated, full-time staff) in connection with case reports.

***Eurotium repens* de Bary, Anamorph *Aspergillus reptans* Samson & W. Gams.**   This species has not been documented as an agent of human infection. It co-occurred with *Microascus cinereus* in a case of sinusitis, but pathogenic etiology was established only for *M. cinereus* (52). Another record from sinusitis by Vennewald et al. (43) cannot be accepted for reasons given above under *E. herbariorum.* For identification, see notes under the description of *E. herbari-orum* above.

Eurotium rubrum König et al., Anamorph *Aspergillus rubrobrunneus* **Samson & W. Gams.** This species has not been conclusively documented as an agent of human infection. It was reported under the invalid name *Aspergillus sejunctus* Bain. & Sart. from keratitis by Shukla et al. (15), but corneal scrapings were negative for fungal elements. For identification, see the notes under the description of *E. herbariorum* above.

Anamorphic Species in A. *Subgenus* Aspergillus, *Section* Restricti *Gams et al.* (*Formerly* Aspergillus restrictus *Group*). The three recognized, closely related species in this section are particularly specialized osmotolerant organisms, very common in dry environments such as household dust, but frequently missed because of their poor growth on ordinary environmental fungal isolation media. Accurate environmental numbers are obtained with high-osmoticum media such as dichloran 18% glycerol agar. These species are seldom seen in the medical mycology laboratory as contaminants, but are occasionally fortuitously isolated from exposed materials such as skin or respiratory secretions. As outlined below, they have never been convincingly recorded as opportunistic pathogens. Indeed, the general ecological picture of an opportunistic mold pathogen is of a plurivorous, enzymatically versatile, thermotolerant, generalist decomposer isolated from a wide variety of habitats, at least some of which are osmotically similar to animal tissue (i.e., high water activity). The specialized *Restricti* do not fit this picture well. No group of fungi can be ruled out a priori as opportunists, but in reference to such specialized fungi, adequately mycologically and etiologically documented case reports, preferably backed up by deposition of case isolates in a recognized culture collection, must be especially strongly recommended.

Aspergillus {aff. Eurotium} restrictus **G. Smith.** Many if not all reports attributing pathogenicity to this slow-growing osmophilic species are specious or unconfirmed. An attribution of endocarditis by Mencl et al. (53) in a case also mentioned by Pospisil et al. (54) and Resl et al. (55) is based on an isolate that is clearly recognizable from the description and photograph of Mencl et al. as a typical dysgonic *A. fumigatus* with reduced conidiophores. The authors admirably confirmed this by showing the isolate doubled its growth rate at 37°C on Czapek–Dox medium, whereas the growth rate of *A. restrictus* itself at 35°C is only about 25% of its optimal growth at 25° (56). [An optimum of 30° and maximum of 40.5° was shown by Smith and Hill (57) in a medium of higher water activity; *A. restrictus* does not grow at all at 37° in standard taxonomic studies on Czapek's agar (51).] Similarly, a report of *A. restrictus* aspergilloma by Estrader et al. (58) appears to be based on host-adapted *A. fumigatus*; smooth, spherical conidia 2 to 3 μm in diameter were seen rather than the elongated 4 to 6 μm conidia of *A. restrictus*. The patient had previously been precipitin-positive against *A. fumigatus* but had a single negative test at the time the purported *A. restrictus* was isolated. Possibly a false-negative test was involved. A report of ostensibly sero-logically confirmed *A. restrictus* aspergilloma by Tamaki et al. (59) was based on using the case isolate, not a standard *A. restrictus* strain, as representative of that species for antigen extraction. The authors appear merely to have shown that the case isolate, perhaps itself an *A. fumigatus*, had a stronger precipitin reaction with the patient's serum than did an extract of a heterogeneous *A. fumigatus* isolate (which gave a negative result). No identification characters were adduced for the isolate involved. A report of *A. restrictus* onychomycosis in Schönborn and Schmoranzer (60) cannot be confirmed. Although these authors state they tried to obtain confirmatory repeat samples for onychomycoses when possible, there is no mention of which cases were confirmed in this way. Furthermore, the authors reported a number of nondermatophyte onychomycosis "cases" in which direct nail microscopy was negative, suggesting that they were not always obser-vant of standard confirmatory criteria. (See also the discussion under *A. conicus* below.)

It may be noted that *A. fumigatus* produces a well-known ribotoxin, restric-tocin (61, 62), which received its name because it was extracted from a medically isolated strain misidentified as *A. restrictus* and so deposited in the ATCC (ATCC 34475 = NRRL 2869). This illustrates the frequency and strong scientific influ-ence of this common identification error. The identification of ATCC 34475 has since been corrected by ATCC.

Description. The colonies are slow-growing (6 to 12 mm in 7 days), dense, often heaped, powdery with sporulation, and dark green, with the reverse pale or dark green.

Conidiophores are mostly 75 to 200 μm high, with smooth, hyaline walls. The vesicles are narrowly hemispherical, 6 to 12 μm in diameter [10 to 18 μm on the high-osmoticum Czapek-yeast extract 20% sucrose (51)], bearing uniseri-ate (i.e., with phialides directly attached to vesicles) fertile elements (Fig. 5). *Conidia* are in dry, strongly coherent chains, often remaining in discernible paral-lel columns in wet mounts, at first nearly cylindrical, then barrel-shaped to ellip-soidal at maturity. They are rough-walled, deep green in color, mostly 4.0 to 5.0 (sometimes to 6.0) μm × 3.0 to 3.5 μm.

There is no growth at 37°C on most media, and no growth at 42°C on any medium.

This fungus may easily be distinguished from all variants of atypical, host-adapted *A. fumigatus* by temperature testing. *A. fumigatus* almost always grows at 45°C; its optimum temperature is 40–42°C (51).

***Aspergillus {aff. Eurotium} caesiellus* Saito.** This species strongly re-sembles *A. restrictus*, but grows more rapidly on certain specialized media, such as Czapek's agar (7) and Czapek-yeast extract 20% sucrose agar (63). Unlike *A. restrictus*, it grows weakly at 37°C on Czapek-yeast extract agar. An isolate from

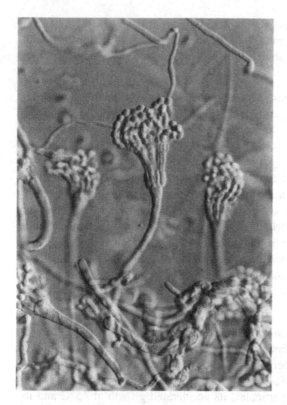

Figure 5 *Aspergillus restrictus.*

an aspergilloma was identified by Otčenášek et al. (64) as *A. caesiellus*. It is described as nonsporulating on Sabouraud agar and with very rare production of mostly aberrant conidiophores on Czapek–Dox agar. The few *A. caesiellus* isolates known in collections, however, sporulate on all media tested, and do so heavily on malt extract agar (7, 63), which is nutritionally similar to many general growth and sporulation media used in medical mycology laboratories. The case isolate also had a growth rate around half that of *A. caesiellus*, and a ferruginous brown to brown-black reverse, in contrast to the pale to deep green reverse typical of *A. caesiellus*. As the principal author was also involved in the demonstrable misidentification of dysgonic *A. fumigatus* as *A. restrictus* (see above), it appears possible that the identification of this case isolate as *A. caesiellus* was a similar error, despite the exact coincidence of the isolate's conidial size range with that given by Raper and Fennell (7) for *A. caesiellus*. Because of prolonged host/fungus contact, aspergilloma is one of the most predictable sources of aberrant and poorly and/or atypically sporulating *A. fumigatus* isolates. Host-adapted *A. fumigatus* isolates strongly resembling the isolate of Otčenášek et al. (64), with restricted colonial growth and aberrantly long conidia, are described by Raper and Fennell (7, p. 244).

Aspergillus {aff. Eurotium} conicus **Blochwitz.** Although considered a dubious taxon and a probable synonym of *A. restrictus* by Pitt and Samson (63), this species has recently been revalidated by molecular study (6). It was reported without identification details or credit by Jones (65) from a well-substantiated case of postsurgical fungal endophthalmitis. As the fungus was extracted from the eye 23 days after its implantation, it is unlikely to have been a dysgonic form of *A. fumigatus*; this form generally results from prolonged growth in the host. The assigned identity of the fungus, however, cannot be verified, and the absence of comment on the unusual attribution, plus the lack of acknowledged expert confirmation for the highly esoteric identification, suggests a low degree of mycological awareness. Microphotographs of *A. conicus* in Raper and Fennell (7, p. 230), could easily engender misidentification of an *A. fumigatus* with smaller than usual aspergilli, as well as other Eurotialean fungi with small, vesiculate conidiophores.

Aspergillus {aff. Eurotium} penicillioides **Speg.** Maršálek et al. (66) recorded a case of aspergillosis, first pulmonary and then disseminated, which they stated was caused by a member of the *A. restrictus* series, most probably *A. penicillioides*. Their excellent description of the isolate, however, reveals it as a host-adapted *A. fumigatus*. In particular, its accelerated growth at 45°C and its prolonged absence of conidiation on all media rule out the section *Restricti* entirely. The isolate was clearly quite aberrant, with small, cerebriform colonies on Sabouraud agar at 20° and 37°C, and its identification [by G.A. de Vries of Centraalbureau voor Schimmelcultures (CBS)] was not unreasonable, given the paucity of descriptions of such *A. fumigatus* isolates at the time. Regrettably the isolate was not preserved in CBS. A case from a lung lesion in a roe deer was attributed to *A. penicillioides* by Fragner et al. (67), but was also based on an *A. fumigatus* isolate, as evidenced by a growth rate at 24°C, approximately tenfold that of the former species. Figures depict conidial heads typical of *A. fumigatus*. The conidial structures of *A. penicillioides* are shown in Fig. 6.

2. *Emericella* Berk. & Br.

Ascomata are globose cleistothecia, yellow, reddish, or brown, usually formed after more than 1 week of colonial growth on fungal sporulation media [e.g., potato dextrose agar, Leonian's agar (4)] and seen as characteristically colored tufts on the colony surface, often more dense near the colony center. *Peridium* is surrounded by masses of hülle cells, which are seen as subglobose to globose

Figure 6 *Aspergillus penicillioides.*

cells up to 25 μm in diameter with thick, hyaline cell walls occupying most of the cellular volume. *Asci* are eight-spored, round, and evanescent. *Ascospores* are oblate, with equatorial crests, and often with a distinct species-specific color. Corresponding anamorphs and evolutionarily radiating anamorphic species are in *Aspergillus* subgenus *Nidulantes*, sections *Nidulantes*, *Versicolores*, and *Usti*, with biseriate conidial heads. Formerly the species included below were considered members of the *Aspergillus nidulans*, *A. versicolor*, and *A. ustus* series (7).

Species with a Known Teleomorph (which may not form in every culture) Anamorphs in Aspergillus *Subgenus* Nidulantes, *Section* Nidulantes *Gams et al. (Formerly* Aspergillus nidulans *Group)*

Emericella nidulans (Eidam) Vuill.; Anamorph *Aspergillus nidulans* (Eidam) Winter, nom. cons. This species is a well-known agent of serious opportunistic infection (11, 68, 69) (Figs. 7 and 8). Although much less common

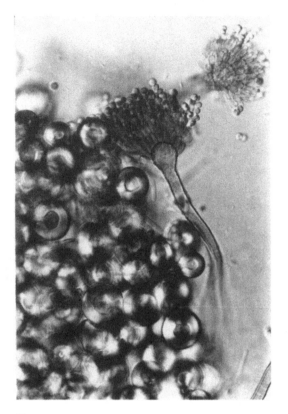

Figure 7 *Emericella nidulans*, conidiophore, conidia and sheath of hülle cells at margin of an ascoma.

as an infectious agent than *A. fumigatus*, *A. flavus*, and *A. terreus*, it may potentially cause a similar range of infections. A relatively large number of reports are in connection with chronic granulomatous disease (70), where it was determined by case review to be more virulent and more difficult to treat than *A. fumigatus*. Causation of mycetoma in tropical areas is also regularly seen (71). Another specific association is with guttural pouch disease of the horse (72). In human nasal sinusitis, this fungus is relatively rare, but may be confirmed in direct microscopy if present when cleistothecia are produced in situ (73). It also causes a variety of marginal opportunistic mycoses such as otomycosis (27). In ecology, it is a widely distributed soil fungus, particularly in the tropics and subtropics, and is also a prominent colonizer of decomposing plant debris, including seeds and grains (74). In secondary metabolism it produces the poorly water-soluble hepatic carcinogen sterigmatocystin, as well as penicillin (8).

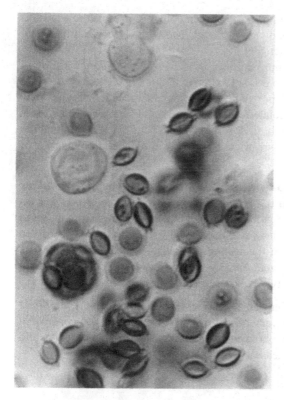

Figure 8   *Emericella nidulans*, ascospores.

Description.   Colonies grow 35 to 65 mm in diameter in 7 days at 25°C, at first white, but rapidly becoming deeply dusty bluish-grey-green with conidiation of the *Aspergillus* anamorph. *Ascomata* in macroscopic examination arises as small, yellowish, spheroidal tufts scattered on surfaces of mature (usually 7- to 20-day-old) colonies, sometimes massing densely and turning the colony center yellowish overall.

Ascomata in microscopy are 100 to 150 μm in diameter, globose, nonostiolate and surrounded by a thick layer of subglobose to globose, hyaline, thick-walled hülle cells. (See genus description.) *Asci* contain eight vinaceous red, lens-shaped *ascospores*, 3.8–6 × 3.5–4 μm, each with two parallel, flat, equatorial ridges approximately 1 μm wide, giving the spore a Saturn-shaped appearance in side view and a fried-egg appearance in face view.

Conidiophores are 75 to 100 μm high, with thick brown walls. They are aseptate, and biseriate (i.e., bearing metulae and phialides). *Conidia* are in dry, upright chains, usually massed into short, upright columns, greenish in transmitted light, subglobose to globose, distinctly roughened with verrucose ornamentation, and 3 to 3.5 μm in diameter.

Although homothallic in pure culture, in *subspecific population structure* this species appears to outcross to a small extent in nature (75). There is therefore no evidence of extended clonal populations, and diverse genetic subtypes may occur among relatively closely related isolates. There are a number of heterokaryon incompatibility subgroups within *A. nidulans*, and parasexual recombination is largely—but not completely—restricted to within the genetic boundaries of these groups (76).

Emericella dentata **(Sandhu & Sandhu) Horie, Anamorph** *A. nidulans* **var.** *dentatus* **Sandhu & Sandhu.**   The original ex-type isolate of this species was, according to its collectors Sandhu and Sandhu (77), "isolated from diseased human fingernails." No argument was made, however, nor any evidence given for a causal connection between the isolated organism and the infection. The species is distinguished from *E. nidulans* by the conspicuously dentate (toothed or starlike) morphology of the two equatorial crests on its ascospores (Fig. 9).

Emericella echinulata **(Fennell & Raper) Horie, Anamorph** *A. nidulans* **var.** *echinulatus* **Fennell & Raper.**   Until recently this species was considered a variant of *E. nidulans*. It was obtained from disseminated infection of a child with chronic granulomatous disease (78).

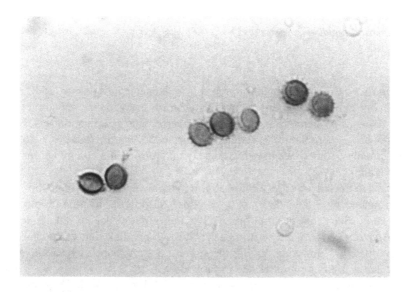

Figure 9   *Emericella dentata*, ascospores.

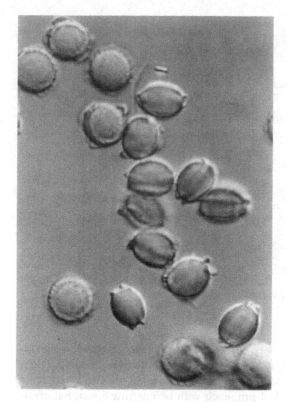

Figure 10   *Emericella echinulata*, ascospores.

Description.   The description is the same as for *E. nidulans*, except that the valves of the ascospores are distinctly ornamented with small spines (Fig. 10).

Emericella quadrilineata **(Thom & Raper) C.R. Benj., Anamorph** *Aspergillus tetrazonus* **Samson & W. Gams** (Synonym: *Aspergillus quadrilineatus* Thom & Raper).   This species has been well documented from fungal sinusitis in an immunocompromised patient (79). A report of skin infection in two Indian sheep (80) is unsubstantiated and appears to be based on fortuitous isolation. In nature *E. quadrilineata* is a soil fungus.

Description.   Colonies grow 35 to 65 mm in diameter in 7 days at 25°C. They are at first white, but rapidly become dull greyish-green with conidiation of the *Aspergillus* anamorph. The reverse is yellow-brown. *Ascomata* in macroscopic examination arise as small, yellowish, spheroidal tufts scattered on surfaces of mature (usually 7–20-day-old) colonies, sometimes massing densely and turning the colony center dull yellowish.

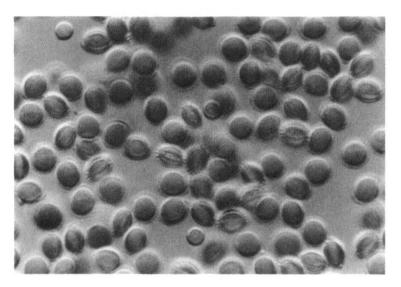

Figure 11   *Emericella quadrilineata*, ascospores.

Ascomata in microscopy are 125 to 150 μm in diameter, globose, nonostio-late, and surrounded by a thick layer of subglobose to globose, hyaline, thick-walled hülle cells (see genus description). *Asci* contain eight vinaceous red, lens-shaped *ascospores*, 4.5–6 × 3.5–4 μm, each with two narrow equatorial crests flanked by two smaller ridges, giving the spore a distinctive appearance of pos-sessing four crests (Fig. 11).

Conidiophores are 40 to 200 μm high, with thick, smooth, brown walls. They are aseptate and biseriate (i.e., bearing metulae and phialides). *Conidia* radiate en masse to columnar, in wet mount appearing greenish, globose, finely roughened, and 2.5 to 3.5 μm in diameter.

***Emericella rugulosa* (Thom & Raper) C.R. Benj., Anamorph *Aspergillus rugulovalvus* (Thom & Raper) Samson & W. Gams *apud* Samson & Pitt** (Common Synonym: *Aspergillus rugulosus* Thom & Raper).   This species has been definitively linked to mycotic abortion in cattle in the Midwest by Knudtson and Kirkbride (81).

In nature, *E. rugulosa* is a soil fungus of warm soils, especially tropical and grassland, and composts.

Colonies grow 10 to 15 mm in diameter in 7 days at 25°C. They are at first white, but rapidly become grey-green with conidiation of the *Aspergillus* anamorph, then yellowish to purple-brown with accumulating hülle cells. *As-*

comata arise as small, spheroidal tufts, at first yellowish to brownish, then scattered on surfaces of mature (usually 7–20-day-old) colonies, sometimes massing densely and turning the colony center yellowish to purple-brown overall. The reverse is usually pale to intense orange-brown to red-brown.

Ascomata in microscopy are seen as cleistothecia, 100 to 200 μm in diameter, globose, with thick reddish walls, surrounded by subglobose to globose yellowish, thick-walled hülle cells. (See genus description.) *Asci* contain eight red-brown, lens-shaped *ascospores*, 5–6.5 × 3.0–4.5 μm, with heavily roughened walls and two parallel, flat, equatorial ridges approximately 1 μm wide (Fig. 12).

Conidiophores are 60 to 100 μm high, with thick, brown, smooth walls. They are aseptate and biseriate (i.e., bearing metulae and phialides). *Conidia* are in dry, upright chains, usually massed into short, upright columns, subglobose to globose, distinctly roughened, and 3 to 4 μm in diameter.

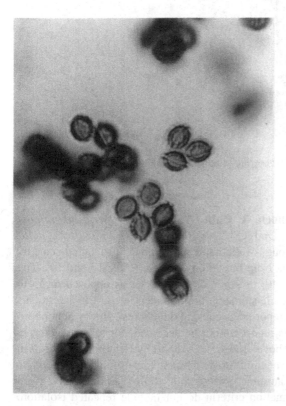

Figure 12 *Emericella rugulosa,* ascospores.

Figure 13 *Emericella unguis, Aspergillus unguis* conidiophores.

Emericella unguis Malloch & Cain, Anamorph *Aspergillus unguis*
(Émile-Weil & Gaudin) C.W. Dodge.   This species is very similar to *A. nidulans* except that spiny, erect, rough-walled spicular hyphae, also called setae by
some authors, generally emerge from the colony surface (Figs. 13 and 14). The
species also grows only 20 to 40 mm in 7 days at 25°C, as opposed to 35 to
65 mm for *A. nidulans*. Fertile cleistothecia are not seen in most isolates. This
uncommon organism has been reported as causing onychomycosis by Schönborn
and Schmoranzer (60), but this report cannot be accepted, for reasons given in
the discussion of *A. restrictus* above. Likewise, a report by Grigoriu and Grigoriu
(18) stating that *A. unguis* was the third most common agent of nondermatophyte
onychomycosis obtained in a Lausanne, Switzerland, clinic seems unlikely, given
that the authors' stated confirmation criteria do not include repeated isolation.
Also, no identification characters are given, the statistical anomaly of this rare
organism appearing prominently is not discussed, and some of the other fungi

Figure 14 *Emericella unguis*, spicular hyphae.

listed as prominent agents of onychomycosis are common contaminants that rarely cause this disease. The original report of *A. unguis* onychomycosis by Émile-Weil and Gaudin (82), however, is unequivocal, as the fungus was illustrated sporulating in direct microscopy of clinical material and the isolate was preserved as an ex-type of the species.

Anamorphic Species in A. *Subgenus* Nidulantes, *Section* Versicolores *Gams et al.* (Formerly *A. versicolor* Group).

Aspergillus {aff. Emericella} versicolor **(Vuill.) Tirab.** This fungus appears to cause opportunistic onychomycosis, usually in elderly patients (83, 84). It may, however, be found to do so less commonly than the closely related *A. sydowii* when the two species are methodically distinguished (47). This distinction may be rendered difficult in a few cases by an intermediate blue-green color form that may grow from infected nails. Because of the red reverse pigmentation and exudate associated with this form, it is tentatively included under *A. sydowii* in the present work pending molecular resolution.

Like *Aspergillus glaucus*, *A. versicolor* is a name that may sometimes be loosely used to signify the former Raper and Fennell (7) group concept rather than the species as recognized currently. Therefore, unless colony characters, especially conidial mass color, are specified, it may be difficult to know that this species per se has been identified. For example, a review of agents of onychomycosis by Torres-Rodríguez and López-Jodra (85, p. 128, Fig. 9) depicts a nail-infecting "*A. versicolor*" isolate that is clearly "greyish-turquoise" on Sabouraud agar [colors 24 D-E 3-4 in the Methuen color monograph (86)]. This is well within the published Methuen color range for typical *A. sydowii* isolates (87), but is not normally consistent with *A. versicolor*, whether live or in photos. Compare similar published color photos of *A. sydowii* on Sabouraud agar in St. Germain and Summerbell (88, p. 77) and Summerbell (89, p. 233), in contrast to quite different photos of *A. versicolor* on pp. 76 and 236, respectively, of the same works. Conidial mass color is the main feature distinguishing these species; Pitt (87) recommends Czapek yeast extract (CYA) medium as best for clear separation.

Non-onychomycosis case reports involving *A. versicolor* are very rare, and some are inadequately evidenced. An isolate from an intracranial lesion in an otherwise healthy male (90) appears etiologically well connected and well identified (by R. Vanbreuseghem), despite atypical descriptive notes mentioning slightly roughened conidiophores and smooth conidia (features which, if taken at face value, would indicate *A. flavus*). A record from an etiologically well-documented mycotic osteomyelitis of the sacrum by Liu et al. (91) does not substantiate the identification of *A. versicolor* in any way; a microphotograph claiming to represent the culture appears to show only filaments. The identification is uncredited. Fungal names are repeatedly misspelled in the paper, giving an impression of unfamiliarity with mycology. A record from a localized lung infection of a 16-year-old leukemia patient contains no verification either of species identification or causality. In particular, no direct microscopic result is mentioned for lung lobe tissue resected in handling the case, and the patient, who had been treated with amphotericin B and itraconazole, was negative for fungus at autopsy (33). A case of keratitis was well documented by Anderson et al. (92). No identification characters were given, but Duke University, the site of the study, was a major mycology center at the time and a likely source of complex identifications that were given in connection with other cases in the same paper. A recent case report connecting *A. versicolor* to nodular dermal lesions in an Argentinian patient treated with corticosteroids was likewise etiologically well attested, but included no details supporting the species identification (93). A somewhat similar case, however, involving a single subcutaneous granuloma in the lip of a horse was accompanied by extensive descriptive notes confirming *A. versicolor* (94), probably not contradicted by an uncharacteristic microphotograph used as an illustration. A recent record by Ponikau et al. (44) connecting *A. versicolor* with allergic fungal sinusitis (AFS) cannot be accepted. In that study, all fungi growing from loosened mucus in nasal washings were interpreted as "colonizing the mucus" and "associated with AFS." It was noted, however, that all healthy control volunteers were also "colonized" by similar organisms. Clearly, the majority of "colonizers" were trapped propagules from air. The list of species given is recognizably a list of common air spora in the area in which the study was done— including species not growing at or near body temperature—mixed with a small proportion of isolates probably originating in genuine infections. "*Aspergillus versiforme*," also mentioned in the same publication, is a name that cannot be traced to a known *Aspergillus* species and may be based on a verbal error. An isolation from otomycosis is mentioned by Yassin et al. (95). Unfortunately, even though the authors did direct microscopy on each specimen, they did not publish the results or mention an interpretation policy (e.g., "isolation of an organism was considered significant if compatible elements were seen in direct microscopy"). Their results are therefore uninterpretable, particularly for infrequently isolated species not elsewhere well confirmed as agents of otomycosis.

Natural and anthropogenic habitats of *A. versicolor* appear to be similar to those of *A. sydowii*, q.v. Its principal mycotoxin is the hepatotoxin and carcinogen sterigmatocystin.

Description. Colonies are moderately slow-growing, 1 to 1.5 cm after 7 days, beginning whitish but soon characteristically dusty grey-green, emerald green, orange-brown, or pinkish-brown (the first two colors based on masses of conidia, the last two strongly influenced by mycelial color), typically with a pale colony reverse.

Conidiophores are mostly up to 500 (occasionally to 700) μm high, with thick, hyaline (clear) walls. They are aseptate and biseriate (i.e., bearing metulae and phialides) (Fig. 15). Reduced uniseriate conidiophores resembling structures of the genus *Penicillium* are also frequently seen, and may predominate in some cultures. These structures may be shorter than 25 μm. *Conidia* are in dry, upright chains, usually massed into short, upright columns, greenish in transmitted light, subglobose to globose, distinctly roughened with verrucose ornamentation, and 2 to 3.5 μm in diameter. Uncommonly, colonies form subglobose, thick-walled hülle cells similar to those seen in *E. nidulans*.

Aspergillus {aff. Emericella} granulosus Raper & Thom. This species was well causally connected to a case of disseminated infection in a heart transplant patient (96). It is otherwise mainly known as a soil fungus.

Description. The colonies are moderately slowly growing (2.0 to 3.5 cm in 7 days). They are very irregularly granular and tufted in texture, beginning whitish but soon characteristically pale olive to purple-brown, with the reverse dull yellow to red-brown.

Conidiophores are up to 300 μm high, with thick, hyaline (clear) walls. They are aseptate and biseriate (i.e., bearing metulae and phialides), with ovoidal

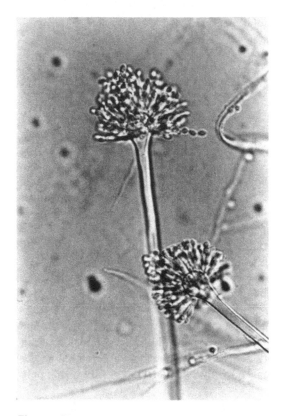

Figure 15   *Aspergillus versicolor.*

vesicles (Fig. 16). *Conidia* are blue-greenish, subglobose to globose, rough-walled (verruculose), and 3.5 to 5.5 μm in diameter. Globose to subglobose hülle cells are abundant, often scattered across the colony surface in conspicuous clumps.

This species is distinguished from the far more common *A. versicolor* mainly by its formation of conspicuous clumps of hülle cells in a faster-growing colony and by its larger conidia.

Aspergillus {aff. Fennellia} janus **Raper & Thom.**   This species was reported from a patient who had keratitis subsequent to ocular puncture by a twig (97). No direct microscopy confirming fungal filaments in the cornea was done. Moreover, no characters confirming the fungal identification were given. The infection responded to antifungals. This suboptimal evidence is all that links this fungus to human and animal pathogenicity so far. *A. janus* is similar to *A. versicolor* but is distinguished by: (1) the production, especially on Czapek's agar,

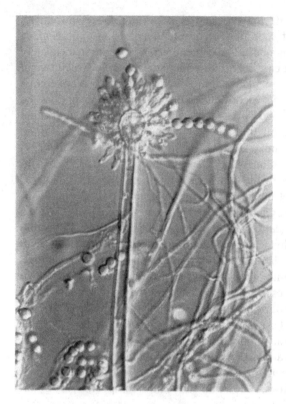

Figure 16 *Aspergillus granulosus.*

of two distinct types of aspergilli, one tall (2000 to 2500 μm long), with clavate vesicles and white conidia, and the other shorter (300 to 400 μm long), with ovoid vesicles and dark green conidia, as well as some heads of mixed character, and (2) regular production on Czapek's agar of masses of cream to yellow hülle cells at 24°C. According to Raper and Fennell (7), it varies considerably according to the exact media and temperatures used in identification, and is best identified under the conditions outlined in their monograph. Whether any of its distinctive features might be produced on any media commonly used in medical mycology has not been investigated. A recent molecular study indicates that it is not in fact closely related to *A. versicolor*, despite its phenotypic similarities (6).

Aspergillus {aff. Emericella} sydowii (**Bain. & Sart.**) **Thom & Church.** This fungus is primarily significant in medical mycology as an agent of opportunistic onychomycosis (46, 47), usually in the elderly patient. It is an abundant airborne contaminant and colonizer of poorly stored medical specimens, so attributions of other opportunistic infectious capacities to it need to be reviewed with caution. Some early reports of causation of onychomycosis [e.g., those of da Alecrim and Vital (98) and Schönborn and Schmoranzer (60)] are inadequately documented or dubious (e.g., a case reported in the latter paper in which fungal elements were not detected in direct microscopy). Well-confirmed ocular keratitis was reported by Shukla et al. (15). Identification details were not given, but the identity of the fungus was confirmed by the Commonwealth Mycological Institute (now Commonwealth Agricultural Bureau International; CABI) and the strain was conserved. On the other hand, a keratitis record by Prasad and Nema (99) lacks direct microscopic verification. Also, the mycology in this paper is supported only by its figures, none of which purports to show *A. sydowii*. The figures presented include a nondescript image of aspergillus heads that appears identically twice, once labeled *A. flavus* and the second time *Curvularia lunata*, while a photo of *C. lunata* is labeled *A. fumigatus*. A pulmonary and pericardial infection concurrent with systemic staphylococcal infection in a moribund neonate with adrenal cortical hypofunction was reported by Zimmerman (100). No mycological details were given. The identity of the fungus, however, was confirmed by two general bacteriology laboratories. This unverifiable record or unacknowledged references to it have appeared in many reviews over the past 45 years [e.g., Rippon (11)], firmly establishing *A. sydowii* as a systemic opportunist.

The species' most common indoor habitat may be damp structural material, especially papered walls and wallboard, but it also occurs in soils, on seeds, and on many types of decaying litter, foodstuffs, fabrics, and insect cadavers (74). Its principal secondary metabolites, sydowinin and related compounds, are of uncertain toxigenic significance (8).

Description. Colonies are moderately slow-growing, attaining 1.5 to 2 cm in 7 days. They begin whitish but are soon characteristically navy blue to blue-green, typically with beads of red-brown exudate on the colony surface and with a similarly reddish-brown colony reverse.

Conidiophores are up to 500 μm high, with thick, hyaline (clear) walls. They are aseptate and biseriate (i.e., bearing metulae and phialides) (Fig. 17). Reduced uniseriate conidiophores resembling structures of the genus *Penicillium* are also frequently seen, as in *A. versicolor* (Fig. 18). *Conidia* are in dry, upright chains, usually massed into short, upright columns, blue-green in transmitted light, subglobose to globose, distinctly roughened with verrucose ornamentation, and 2 to 3.5 μm in diameter.

***Aspergillus {aff. Emericella} varians* Wehmer.** *A. varians* is very similar to *A. versicolor* (q.v.) except that it has significantly larger conidia (3.8 to 5 μm in diameter rather than 2 to 3.5 as in *A. versicolor*). It also has conidiophores on malt extract agar distinguishably formed in two size classes, tall and short. *A. versicolor* may have variability in its conidiophores but no such regular strati-

Figure 17 *Aspergillus sydowii*, full-sized conidiophore.

fication. An isolate linked with some hesitation to *A. varians* has been a validly attributed causality of a case of onychomycosis by Torres-Rodríguez et al. (101). A later publication by the same research group (84) refers to this isolate as *A. versicolor*.

    Anamorphic Species in A. *Subgenus* Nidulantes, *Section* Usti *Gams et al.* (Formerly *A. ustus* Group).

    Aspergillus {aff. Emericella} ustus **(Bain.) Thom & Church.**   This species rarely causes human or animal disease, but some cases of pulmonary, disseminated, or primary cutaneous infection in immunocompromised patients have recently been well documented (102–107). *A. ustus* formed part of a mixed fungal community growing in grossly visible patches on burn eschar that had been coated with therapeutic emulsion. The depth of penetration of the organism was not determined, but there was no evidence of penetration into living tissues (108).

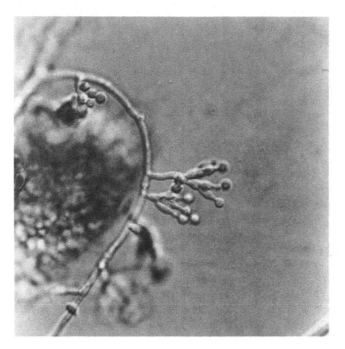

Figure 18   *Aspergillus sydowii*, reduced conidiophore.

Endocarditis connected with a prosthetic cardiac valve has been well demonstrated (109). Causation of onychomycosis has been suggested by Walshe and
English (46), and probably occurs but has not yet been well confirmed [i.e., by
successive repeat isolation, rather than the relatively unreliable (110) counting
of positive inoculum fragments employed by Walshe and English (46)]. A report
by Wadhwani and Srivastava (17) alleging causation of otitis must be discounted
for reasons mentioned in the discussion of *Fennellia nivea* below.

In nature, *A. ustus* is most common in soils and on seeds, especially grains
and peanuts (111). Its characteristic secondary metabolites, austamide, austdiol,
austins, and austocystins, are considered significant mycotoxins (8).

Description.   The colonies are moderately rapidly growing (30 to 50 mm
in 7 days). They are velvety to slightly floccose, beginning whitish but soon
characteristically greyish-brown, typically with a yellow to yellow-brown colony
reverse and yellowish soluble pigment.

Conidiophores are 75 to 350 μm high, with thick brown walls. They are
aseptate and biseriate (i.e., bearing metulae and phialides) (Fig. 19). *Conidia* are
in dry, upright chains, usually seen as radiating or forming short columns, brown-

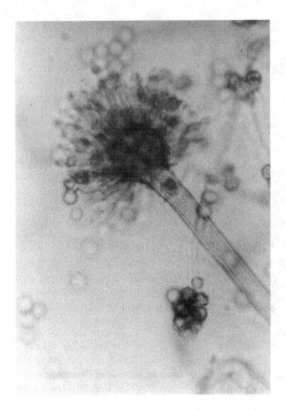

Figure 19 *Aspergillus ustus*, conidiophore.

ish in transmitted light, globose, strongly roughened, and 3 to 4.5 μm in diameter. A minority of colonies will form elongate, irregular, thick-walled "squashed banana" hülle cells (Fig. 20).

Aspergillus {aff. Emericella} deflectus **Fennell & Raper.**   This species has been linked in two publications (five cases) to disseminated infection in dogs, mostly in the notoriously *Aspergillus*-susceptible German shepherd breed (112), but also in one case in a springer spaniel (113).

Isolates from nature have mainly derived from soils in areas of tropical to mediterranean climate.

Description.   Colonies grow restrictedly (1.0 to 1.5 cm in 7 days), beginning whitish, but soon characteristically greyish drab, reverse yellow to dull orange-red.

Conidiophores are mostly 40 to 50 μm long, with thick, red-brown walls. They are aseptate and biseriate (i.e., bearing metulae and phialides), with small

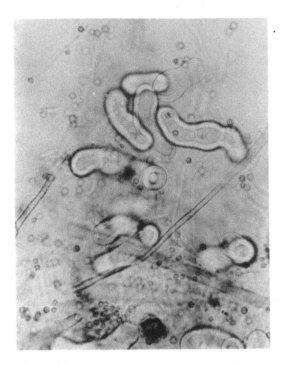

Figure 20   *Aspergillus ustus*, hülle cells. From Laboratory Handbook of Dermatophytes, by J. Kane et al., 1997, Star Publishing Company, Belmont, CA (see Ref. 89). Used with permission.

vesicles characteristically abruptly bent downward, giving a "briar pipe appearance" (Fig. 21). *Conidia* are in dry chains, more or less columnar, brownish in transmitted light, globose, rough-walled, and 3 to 3.5 μm in diameter. A minority of colonies will form elongate, irregular, thick-walled "squashed banana" hülle cells.

3.   Fennellia Wiley & Simmons

Ascomata are globose cleistothecia, yellow and surrounded by masses of rounded to elongate hülle cells. *Asci* are eight-spored, globose, and evanescent. *Ascospores* are hyaline to pale yellow, lens-shaped, smooth or rough-walled, with a sometimes inconspicuous equatorial groove, and sometimes also with inconspicuous crests. Anamorphs and evolutionarily radiating anamorphic species are members of *Aspergillus* subgenus *Nidulantes* sections *Flavipedes* and *Terrei*, formerly considered the *A. terreus* and *A. flavipedes* series (7). They have biseriate conidiophores with fertile elements arising from the upper portions of the vesicle.

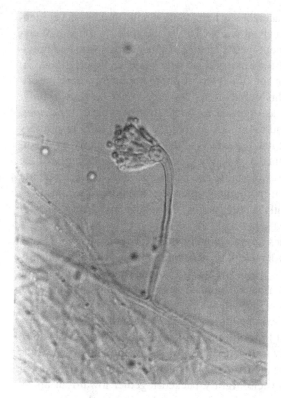

Figure 21 *Aspergillus deflectus.*

Species with a Known Teleomorph (Which May Not Form in Every Culture) Anamorphs in Aspergillus *Subgenus* Nidulantes *Section* Flavipedes *Gams et al. (Formerly* A. flavipes *Group)*

Fennellia flavipes Wiley & Simm., Anamorph *Aspergillus flavipes* (Bain. & Sart.) Thom & Church.   Barson and Ruymann (114) reported a fatal *A. flavipes* infection in a teenaged leukemic bone marrow transplant patient. It began as a palmar skin infection after brief contact with an armboard during catheter placement. No confirmatory identification details or mycological credit were noted. Apart from this reference to *A. flavipes* per se causing disease, there are also a small number of references to members of the "*A. flavipes* group" doing so. Roselle and Baird (115) received an identification of "*A. flavipes* group" from the Centers for Disease Control (CDC) in Atlanta, for a fungus well demonstrated as causing osteomyelitis in the lumbar vertebrae subsequent to surgical evacuation of an intracerebellar hematoma. Such "group"-level identifications were occasionally received from CDC at the time. [Compare Weiss and Thiemke (102).] Also, a patient with a history of tuberculosis who had previously been successfully treated for aspergillosis developed an aspergilloma; *A. flavipes* group was isolated from sputum (116). A photograph of a clearly biseriate *Aspergillus* is depicted in this publication. The conidiophores shown, however, are not over 50 μm long, as opposed to >150 μm for typical *A. flavipes*, so an atypical isolate may be indicated. An *Aspergillus* isolate from respiratory secretions may always be incidental, regardless of the patient's underlying condition (117), and even when obtained as a heavy growth, so the significance of the isolation in this case is not entirely certain. The patient gave a positive antibody response to *A. fumigatus* antigen in counterimmunoelectrophoresis studies.

F. flavipes is predominantly a soil fungus, especially in warmer parts of the globe.

Description. The colonies are moderately slowly growing (15 to 35 mm in 7 days). They are deeply powdery with dense conidiophores, characteristically orange-grey to dull pinkish-buff, "with the overall coloration resulting from the off-white spore masses against a background of brownish conidiophores" (7). The reverse is pale to pale yellow-brown.

Ascomata are seldom formed in culture and are not described in detail here. [See Wiley and Simmons (118).] *Conidiophores* are mostly 150 to 450 (−800) μm high, with smooth to slightly roughened, mostly yellow to light brown walls, biseriate (i.e., bearing metulae and phialides), with metulae arising from the upper 1/2 to 3/4 of the subglobose vesicle, tightly packed together toward an apical concentration (Fig. 22). *Conidia* are in dry, short, upright columns, hyaline, whitish en masse, globose, smooth, and 2 to 3 μm in diameter. In the submerged mycelium of some isolates, conidia of a second type, aleurioconidia, are found attached to hyphae by short pegs. These conidia, which may be sparsely or densely produced, are refractile, globose to ovoid, and 4 to 6 μm long. *Hülle cells*, if present, are elongate, sometimes branched and swollen, with heavy walls.

Fennellia nivea **(Wiley & Simm.) Samson, Anamorph** *Aspergillus niveus* **Blochw.** A male with poorly controlled diabetes who had had a pulmonary lesion caused by *Aspergillus flavus* surgically removed 5 years earlier was reported by Seabury and Samuels (119) to have developed a second lesion caused by *A. niveus*, which was also surgically treated. No mycological characters were given to allow verification of the unusual identification, nor was a mycologist credited. *A. niveus* was also reported as an agent of otitis by Wadhwani and Srivastava (17). The cases in this paper are poorly attested, and otitis appears to be ascribed to all fungi grown or seen in direct microscopy, including smut teliospores that could only have formed on infected plants. An *Aspergillus* conidiophore shown in direct microscopy of ear material, seemingly in connection with

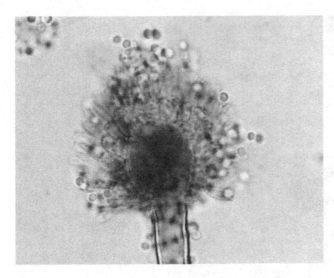

Figure 22 *Fennellia flavipes* CBS 260.73, conidiophore apex. Apparent dark color of vesicle is an artifact of staining.

the report of *A. niveus*, has a subglobose vesicle with metulae all around and an untapered stipe attachment point. It may be *A. candidus* or *A. ochraceus*. *A. niveus* has spathulate to hemispheral vesicles with only the upper 1/2 to 2/3 bearing metulae. The vesicle tapers gradually into the stipe in many of its conidiophores. A record from sinusitis by Vennewald et al. (43) cannot be accepted for reasons given above under *E. herbariorum*.

   F. nivea is predominantly a soil fungus, especially in tropical to midtemperate parts of the globe (74).

   Description.   Colonies are moderately slow-growing (20 to 30 mm in 7 days). They are deeply powdery with dense conidiophores, characteristically white, sometimes at least in part pale yellow or cream, with or without beads of yellowish or reddish exudate, and with yellow-brown, red-brown, dark green, or nearly black reverse colors.

   Ascomata are seldom formed in culture and are not described in detail here. [See Wiley and Fennell (120).] *Conidiophores* are mostly 100 to 600 (−1000) µm high, with smooth hyaline walls. They are biseriate (i.e., bearing metulae and phialides), with metulae arising from the upper 1/2 to 2/3 of the subglobose vesicle, and tightly packed together toward an apical concentration (Fig. 23). *Conidia* are in dry, short, upright columns, and hyaline. They are white to dull whitish-buff en masse, globose or subglobose, smooth, and 2 to 3.5 µm in diameter. In the submerged mycelium of some isolates, conidia of a second type, aleuri-

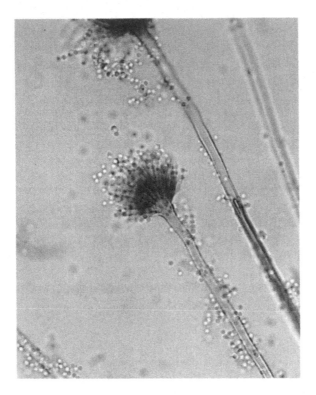

Figure 23 *Fennellia nivea*, conidiophores.

oconidia, are found attached to hyphae by short pegs. These conidia, which may be sparsely or densely produced, are refractile, globose to ovoid, and 3 to 5 μm long. *Hülle cells*, if present, are present as tufted, yellow masses of globose to elongate, thick-walled cells.

Anamorphic Species in A. *subgenus* Nidulantes, *Section* Flavipedes *Gams et al. (Formerly the* Aspergillus flavipes *group).*

***Aspergillus {aff. Fennellia} carneus* Blochw.** This fungus was isolated on several successive occasions from a patient described only as having pneumonia (121). No direct microscopic examination of either fluids or tissues was reported. Possibly an allergic bronchopulmonary colonization was involved, although incidental sources (e.g., contamination due to continuous exposure to an environmental source) cannot be completely ruled out. In nature, this fungus mostly derives from tropical soils (122).

Figure 24 *Aspergillus carneus. conidiophore.*

Description. The colonies are moderately slow-growing (15 to 30 mm in 7 days). They are powdery, beginning whitish, but soon characteristically pinkish to pinkish-brown, sometimes orange-brown. The reverse is pale or tinged with yellow or brown.

Conidiophores are mostly 80 to 200 μm high, with smooth walls that are hyaline or faintly brownish. The vesicles are clavate to nearly spherical, bearing biseriate (i.e., with metulae and phialides) fertile elements (Fig. 24). Fertile elements may also occasionally be irregularly disposed along the conidiophore (74). *Conidia* are in dry chains, spherical, smooth, and mostly 2.5 to 3 μm in diameter. In the submerged mycelium, a conidia of a second type, aleurioconidia, may be found attached to hyphae by short pegs. These conidia, which may be sparsely produced, are refractile, globose to ovoid, and 4 to 6 μm long.

Anamorphic Species in A. *Subgenus* Nidulantes, *Section* Terrei *Gams et al. (Formerly* Aspergillus terreus *Group).*

Aspergillus {aff. *Fennellia*} terreus Thom. This species is one of the five prominent and regularly encountered opportunistic fungi in the genus *Aspergillus*, along with *A. fumigatus, A. flavus, A. nidulans,* and *A. niger.* It and *A. nidulans,* however, are generally much less commonly encountered than the other three species listed. *A. terreus* causes the full range of known aspergilloses, including pulmonary and disseminated aspergillosis in the immunocompromised patient (69, 123), allergic bronchopulmonary aspergillosis (124), otomycosis (27, 125), onychomycosis (47, 49, 126), and various other infections (69). In the immunocompromised patient, it appears to cause higher morbidity and to be significantly less responsive to amphotericin B therapy than other aspergilli (127). The occurrence of aleurioconidia in histopathology may facilitate rapid diagnosis when the pathologist is aware of their significance (128). *A. terreus* may cause serious infections in both birds (69) and dogs (129, 130). In nature and the anthroposphere, *A. terreus* is a fungus of composts, soils, seeds, and foodstuffs (74, 124). It produces as major mycotoxins the mutagen patulin, the nephrotoxin citrinin, and the neurotoxin citreoviridin (8).

Description. The colonies are moderately rapidly growing (25 to 65 mm in 7 days). They are deeply powdery with dense conidiophores, and characteristically medium sandy brown. The reverse is pale to pale brownish.

Conidiophores are 100 to 250 μm high, with smooth, hyaline walls that are straight or gently undulate (Fig. 25). They are aseptate and biseriate (i.e., bearing metulae and phialides), with metulae arising from the upper one-half to two-thirds of the subglobose vesicle and conspicuously appressed against one another in a nearly parallel orientation. *Conidia* are in dry, upright columns, in wet microscopic mounts hyaline to slightly yellow in transmitted light, globose to ellipsoidal, smooth, and 2 to 2.5 μm in diameter. In the submerged mycelium, conidia of a second type, aleurioconidia, are found attached to hyphae by short pegs. These conidia, which may be sparsely or densely produced, are refractile, globose to ovoid, and 4 to 6 μm long.

Host-adapted (dysgonic) isolates of *A. terreus* may have few or no aspergillary heads and produce only aleurioconidia, yielding a colony reminiscent of the filamentous morph of *Blastomyces dermatitidis* (89, 131). Colonies will be whitish, dense, and often radially furrowed, as with dysgonic *A. fumigatus.*

Anamorphic Species Currently in A. *subgenus* Circumdati, *Section* Candidi *Gams et al., but with Greater Biological Affinities to Teleomorph Genus* Fennellia *(Formerly the* Aspergillus candidus *Group).*

Aspergillus {aff. *Fennellia*} candidus Link. Recent molecular studies (6) have shown that this fungus and its close relatives are more closely related

Figure 25 *Aspergillus terreus* conidiophore.

to the teleomorph genus *Fennellia* than to *Petromyces* and *Neopetromyces*, as was long thought.

This species was well documented as an agent of brain granuloma in a man who had injured his head in a fall from a horse (132). A direct microscopically confirmed case of sphenoid sinus infection (133) contains no confirmatory identification information; however, the identity of the organism is emphasized as a key feature of the study. As relatively few hyphomycetes are easier to identify than *A. candidus*, this identification is accepted as likely correct. A necrotizing coinfection by *A. candidus* and *A. niger* beginning in the ear of a leukemia patient was documented by Falser (134). Five case records from otomycosis by Yassin et al. (95) are regarded as unsubstantiated for reasons mentioned under *A. versicolor* above. On the other hand, an otomycosis case registered by Gregson and La Touche (27) is well attested, even though an unusually dark colony reverse made the authors cautiously label the isolate "*A. candidus* sp." Well-confirmed cases of onychomycosis have been documented (135, 136), as well as cases problematically attested from direct-microscopic-positive single specimens yielding heavy

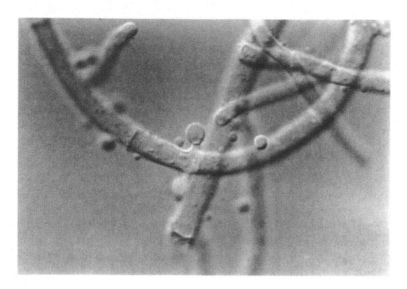

Figure 26   *Aspergillus terreus* aleurioconidium.

outgrowth (46, 137). A fungus identified as *A. candidus* was repeatedly cultured from the sputum and needle biopsy of a patient with a pulmonary cavity by Iwasaki et al. (138), but could not be traced in pathology examinations after the affected portion of lung was resected. Photographs of the fungus isolated are not typical of *A. candidus*, but may show diminutive heads known from this species, as illustrated by Raper and Fennell (7, p. 348).

A. *candidus* is mainly known from warmer soils and from seeds, especially grains (74). Major secondary metabolites are terphenyllin and derivatives, as well as chlorflavonin (139).

Description.   The colonies are moderately slow-growing (10–25 mm in 7 days). They are densely powdery and characteristically white to pale cream. The reverse is pale to moderate yellow.

Conidiophores are often 200 to 500 (but up to 1000) µm high, with smooth to finely roughened, hyaline walls. They are aseptate, attaching abruptly to a globose vesicle, which is biseriate (i.e., bearing metulae and phialides), with metulae all around the perimeter in a radial orientation (Figs. 27 and 28). *Conidia* are found in dry, upright chains. They often radiate or mass into two or more short columns per head, in wet microscopic mounts. They are grey-brown, globose, strongly roughened, and 3.5 to 4.5 µm in diameter. *Sclerotia*, if formed, are red-purple to black.

A. *candidus* can be distinguished from the similarly whitish *A. niveus* by

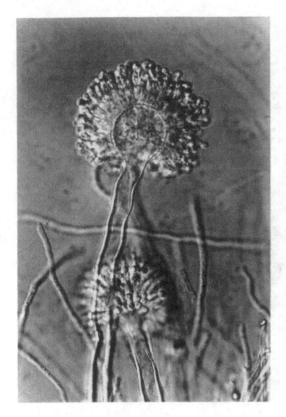

Figure 27 *Aspergillus candidus*, full-sized conidiophores.

its fertile elements covering the whole vesicular surface, as opposed to only the upper half to two-thirds in *A. niveus*. Subsurface aleurioconidia can also often be found to confirm an identification of *A. niveus*.

4. *Neopetromyces* Frisvad & Samson

The teleomorph is formed as hard sclerotioid bodies that mature as ascostromata containing one to two ascomata. *Ascostromata* are globose, subglobose, or elongate, 400 to 600 μm in diameter, and pale yellow, with an outer layer of thick-walled angular to subglobose cells and an inner layer of thin-walled cells. *Ascomata* develop as subglobose to globose chambers within the ascostromata, and are 155 to 300 μm in diameter. *Asci* are eight-spored, globose, and evanescent. *Ascospores* are oblate, subglobose to broadly ellipsoidal, with spiny walls and a small equatorial ridge, 4.5–5.5 × 3.5–5.8 μm. The description is based on *N.*

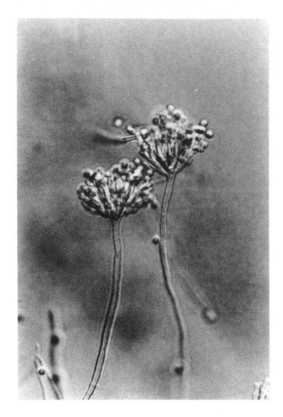

Figure 28   *Aspergillus candidus*, reduced conidiophores.

(Petromyces) muricatus (Udagawa et al.) Frisvad & Samson (140, 141). Ana-morphs are members of *Aspergillus* subgenus *Circumdati* section *Circumdati*, formerly considered the *A. ochraceus* series (7) (excluding the anamorph of *Petromyces alliaceus* Malloch & Cain). They have biseriate conidiophores. The sole species forming the teleomorph has no record as a human or animal pathogen.

   Anamorphic Species in A. *subgenus* Circumdati, *Section* Circumdati *Gams et al. (Formerly the* Aspergillus ochraceus *Group).*

   ***Aspergillus {aff. Neopetromyces} ochraceus* Wilhelm.**   This species has been well documented as an agent of one case of allergic bronchopulmonary aspergillosis (142) and in one case of suppurative otitis media (142a). A "sulfur yellow" *Aspergillus* described as "probably *A. ochraceus*" was well documented from keratitis in rural Bangladesh by Williams et al. (143). Wierzbicka et al. (144) isolated and correctly identified *A. ochraceus* on a single occasion from the sputum of a neutropenic leukemia patient with a cavitating pulmonary lesion that responded to antifungal drugs. As was shown by Staib et al. (117), who recorded numerous fortuitous isolations of aspergilli from uninfected AIDS patients, such isolations are not necessarily significant. Wierzbicka et al. (144) stated that serum antibodies precipitated in vitro against antigens of both *A. fumigatus* and *A. ochraceus*, a finding more likely consistent with a cross-reaction than with a dual infection. (No cross-absorption steps were mentioned.) The case report thus neither fully substantiated *A. ochraceus* infection nor ruled out *A. fumigatus*. Bovine abortion was attributed to *A. ochraceus* by Muñoz et al. (145); however, the well-described and illustrated organism is clearly *A. terreus*. Paranasal sinusitis caused by *A. ochraceus* was alleged by Bassiouny et al. (146); however, no direct microscopic examination was recorded in this case, and fungi such as *Penicillium melinii* with no record of human or animal pathogenicity were recorded from other inadequately documented cases.

In nature and the anthroposphere, *A. ochraceus* is a soil fungus, especially in warm regions, and is also commonly isolated from seeds, coffee beans, decaying insects, and indoor carpets. It produces a highly economically significant, nephrotoxic grain-contaminating mycotoxin, ochratoxin A (8).

Description. The colonies are moderately rapid-growing (45 to 55 mm in 7 days). They are deeply powdery with dense conidiophores and characteristically chalky yellow to pale ochraceous yellow-brown, with the reverse pale to brownish.

Conidiophores are up to 1500 μm high, with granular, pale yellow-brown walls, attaching abruptly to a globose to subglobose vesicle, which is biseriate (i.e., bearing metulae and phialides), with metulae all around the perimeter in a radial orientation (Fig. 29). *Conidia* are in dry, upright chains, often massing into two or more short columns per head, in wet microscopic mounts hyaline. They are globose to subglobose, smooth or finely roughened, and 2.5 to 3.5 μm in diameter. Some colonies form pinkish to purple, irregular, rock-hard sclerotia up to 1 mm in diameter.

Aspergillus {aff. Neopetromyces} sclerotiorum **G.A. Huber.** This species was well documented as an agent of an onychomycosis by Feuilhade de Chauvin and de Bièvre (147). A record from sinusitis by Vennewald et al. (43) cannot be accepted for reasons given above under *E. herbariorum*. *A. sclerotiorum* is mainly a tropical soil fungus.

Description. The colonies are moderately rapid-growing (40 to 55 mm in 7 days), with velvety to floccose white mycelium mostly covered with light yellow conidial masses and white to buff sclerotia. The colony reverse is usually light yellow.

Conidiophores are mostly 400 to 1200 and sometimes as tall as 2000 μm high, with rough yellow or brownish walls, attaching relatively abruptly to a

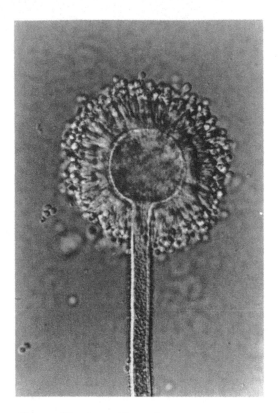

Figure 29 *Aspergillus ochraceus.*

globose vesicle, which bears biseriate fertile elements (i.e., metulae and phialides) in larger heads, with these elements all around the perimeter in a radial orientation (Fig. 30). *Conidia* are in dry, radiating chains, smooth to finely roughened, globose, and mostly 2.5 to 3.0 μm in diameter. The sclerotia are white to buff, mostly 1 to 1.5 mm in diameter.

This species must be distinguished from *A. ochraceus*, which often has no sclerotia but may have pale pink to purple sclerotia, distinct from the creamy sclerotia of *A. sclerotiorum. A. sclerotiorum* is also distinct in conidial color from *A. ochraceus*. The former is pale yellow, whereas the latter is a deeper color, described as ranging through wheat, ochraceous, buff, or amber-yellow (122).

5. *Petromyces* Malloch & Cain

The teleomorph is formed as hard sclerotioid bodies that mature as ascostromata containing 1 to 8 ascomata. *Ascostromata* are ovate to elliptical, 1000–3000 ×

Figure 30 *Aspergillus sclerotiorum.*

500–700 μm, at first whitish, but soon maturing grey and then black, with an outer sclerenchymatous layer composed of thick-walled angular to subglobose cells. *Ascomata* develop as globose chambers within the ascostromata. *Asci* are eight-spored, globose, and evanescent. *Ascospores* are ellipsoidal, with smooth walls and an equatorial furrow scarcely visible in young spores only, 5.5–9 × 5.0–7.0 μm. The description is based on *P. alliaceus.* (See below.) Anamorphs are now considered members of *Aspergillus* subgenus *Circumdati* section *Flavi* (8), based on biochemical and genetic evidence; they were previously placed in the *A. ochraceus* group by Raper and Fennell (7).

Species with a Known Teleomorph, and with Anamorphs in Aspergillus *Subgenus* Circumdati *Section* Flavi *Gams et al. (Formerly* Aspergillus flavus *Group).*

Petromyces alliaceus **Malloch & Cain, Anamorph** *Aspergillus alliaceus* **Thom & Church.** This species was reported as the agent of a chronic otitis externa subsequent to surgery (148). It was isolated on two separate occasions, but direct microscopy was not recorded, leaving the organism's role in the disease process in question. A record from sinusitis by Vennewald et al. (43) cannot be accepted for reasons given above under *E. herbariorum.*

The species is known from nature in soils, especially those of deserts and grasslands, and may also grow from root crops in the onion and garlic family, Alliaceae. It produces ochratoxin A and B and kojic acid as major secondary metabolites (141).

Description. The colonies are rapid-growing (60 to 70 mm in 7 days), with velvety to floccose white mycelium covered sparsely at maturity with cream to yellow-gold conidial masses and dark grey to black sclerotia, the latter often formed in concentric accumulations. The colony reverse is pale tan to yellow-brown.

Conidiophores are a mixture of intergrading large and small types, stipes as short as 40 and as long as 2000 μm high, with smooth, hyaline walls, attaching relatively abruptly to a globose to pyriform vesicle, which bears biseriate fertile elements (i.e., metulae and phialides) in larger heads and uniseriate elements (phialides only) in smaller heads, with these elements all around the perimeter in a radial orientation (Fig. 31). *Conidia* are in dry, radiating chains. They are hyaline, smooth, subglobose to ovoid, and mostly 3.0 to 3.5 μm in diameter. *Sclerotia/ascostromata* are ovate to elliptical, 1000–3000 × 500–700 μm, white, then grey with white tips, then black. Production of ascospores is seldom seen and requires 3 to 10 months of incubation (7, 149).

This species must be distinguished from members of section *Circumdati,* such as *A. ochraceus* and *A. sclerotiorum,* which have pale to pink or purple colored—not black—sclerotia, and conidia in paler shades of yellow.

Anamorphic Species in A. *subgenus* Circumdati, *Section* Flavi *Gams et al.* *(Formerly the* Aspergillus flavus *Group).*

Aspergillus {aff. Petromyces} flavus **Link.** *A. flavus* has essentially caused every known variant of aspergillosis, in particular: (1) pulmonary, disseminated, and other systemic infections in immunocompromised, especially neutropenic patients; (2) rare, idiopathic, systemic, localized, and occasionally progressive infections in apparently immunocompetent patients; (3) chronic colonizing infections, mainly of predisposed hosts (e.g., allergic bronchopulmonary aspergillosis), mostly in long-term asthmatic and cystic fibrosis patients (150), pulmonary aspergilloma (fungus ball) in lungs of persons with pre-existing cavitation, chronic mycotic sinusitis, and otomycosis (outer ear canal surface infestation); and (4) opportunistic mycoses of especially vulnerable body sites (e.g., ocular infections subsequent to traumatic introduction, otomycosis, onychomycosis,

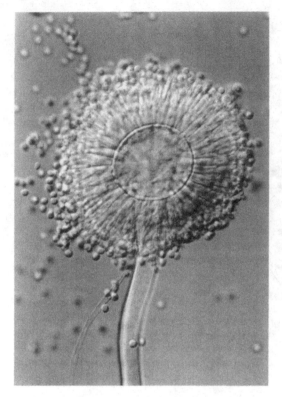

Figure 31 *Petromyces alliaceus*, conidiophore.

dialysis-related peritonitis, and endocarditis). In comparison with *A. fumigatus*, however, it generally causes a higher proportion of infections of sinuses and ear canals and a lower proportion of pulmonary infections in proportion to the overall frequency of significant isolation of both fungi. This trend arises possibly because its larger conidia are not as effective as the very small conidia of *A. fumigatus* at penetrating deeply into the lungs, but may impact more effectively in more exposed airways. In places such as Sudan, in which *A. flavus* is notoriously prevalent, however, this trend is scarcely noticeable, and the species overwhelmingly causes most aspergillosis.

Aspergillus flavus is a cosmopolitan compost-degrading fungus as well as a general colonizer of warm soils. Some of its variants, as well as closely related species such as *Aspergillus parasiticus*, have particular relationships as seed pathogens and degraders of plants such as *Zea mays*, maize corn. Extensive notes on sites of isolation are given by Domsch et al. (74). *A. flavus* produces the

Figure 32   *Aspergillus flavus.*

notorious hepatic mutagens in the aflatoxin B series as well as cyclopiazonic acid. As a toxin producer it may be a major cause of hepatic cancer as well as occasional outbreaks of acute liver toxicosis, especially in countries in which its levels in peanut and maize products are not stringently regulated.

Description.   The colonies are rapidly growing (mostly 50 to 70 mm in 7 days), deeply powdery-granular, yellow-green to green, sometimes with visible white marginal or central mycelium, and the reverse is pale to brown. Atypical, host-adapted colonies, degenerated after prolonged growth in the patient, may be deeply floccose and white with a greenish cast imparted by scattered conidiophores in the aerial mycelium.

Conidiophores are often around 400 to 800 (but up to 2500) μm high. The stipes are hyaline, with characteristic granular wall roughening (must be distinguished from adherent bacteria or fat droplets on smooth-stalked species)

(Fig. 32). The walls are generally parallel, but with a slight funnel-like expansion at the apex supporting the globose to subglobose vesicle. The aspergillary heads of both biseriate (i.e., bearing metulae and phialides) and uniseriate (phialides only) character are often found, with larger heads generally more complex. Metulae or phialides are attached around the whole vesicle perimeter or at least the apical three-quarters in a radial orientation. *Conidia* are in dry chains. They are radiate, green, finely roughened or rarely smooth, globose to subglobose, and 3.5 to 5.0 µm in diameter. Although common in isolates from the environment, *sclerotia* are rare in clinical isolates except from nail and skin, but when seen are stony-hard, usually red-brown to black, roundish, mostly 400 to 800 µm in diameter.

Specific polymerase chain reaction (PCR) primers for identification or direct detection of *A. flavus* have been published in recent years. For example, primer pairs based on genes related to aflatoxin biosynthesis (151) detect only *A. flavus* and *A. parasiticus*. Sandhu et al. (152) devised a specific 28S rDNA-based probe for *A. flavus* to be used after PCR amplification of clinical specimens or unknown fungal cultures using universal fungal 28S primers. Primers based on alkaline protease genes for detection of *A. flavus* were published by Tang et al. (153). Walsh et al. (154) found that single-strand conformational polymorphism analysis could distinguish among 18S rDNA-based PCR products from *A. flavus*, *A. fumigatus*, and selected other medically important fungi.

In *subspecific population structure A. flavus* consists of two genetically isolated groups, one (group I) with considerably more genetic variation than the other (group II) (155). Although the species is largely clonal, there is some evidence of past or ongoing recombination in population genetics statistical tests.

A. flavus should be distinguished from the closely related *A. parasiticus*, which differs by having mostly or entirely uniseriate aspergilla, strongly roughened conidia, and a more somber color described as "dark yellow-green" by Samson et al. (111) as opposed to "green" for *A. flavus* (111, p. 54, Table 3), and "dark olive or deep dark green" by Klich and Pitt (122) as opposed to "olive green, olive or parrot green" [color names from Kornerup and Wanscher (86)]. The distinction is best made after having seen authentic cultures of each species. *A. parasiticus* produces aflatoxins of the B and G series—the latter are never produced by *A. flavus*—and does not produce cyclopiazonic acid. *A. parasiticus* has little record as a human and animal pathogen, but this may be due to inattention; that is, failure to distinguish it from *A. flavus*.

***Aspergillus* {*aff. Petromyces*} *oryzae* (Ahlb.) Cohn.**   The name *A. oryzae* was originally based on domesticated *A. flavus*-like strains used in soy fermentation processes. In morphological characters, however, these domesticated isolates intergraded with some wild isolates, and descriptions of *A. oryzae* such as that of Domsch et al. (74) tended to include both domesticated and similar wild isolates. In contrast to *A. flavus* as classically defined, *A. oryzae* tended to have more brownish-green conidial masses, a more floccose colony, longer conidiophores (up to 4000 to 5000 μm in length), and larger conidia, mostly 5 to 6 μm in diameter. Genuine *A. oryzae* isolates do not produce aflatoxins (111). Recent population genetics studies have shown that they are closely similar to one of 16 multilocus genotypes discovered in a preliminary survey of *A. flavus* genetic diversity (155). Some isolates considered *A. flavus* are highly genetically similar, and in terms of normal biological taxonomic criteria, *A. oryzae* is not regarded as a separate species. It is maintained as a regulatory category for aflatoxin-free koji molds and as an acknowledgment of the differentiating effects of domesticating selection (155). At the time of this writing, however, it is not at all clear that any wild isolate, no matter how superficially similar, can be included under the name *A. oryzae*. Certainly any isolate given this name today should at the very least be tested using sensitive techniques such as high-powered liquid chromatography to ensure no aflatoxin is produced, regardless of apparent morphological similarity. Ideally, genetic similarity with the *A. oryzae* characters demonstrated by Geiser et al. (155) should be shown. *A. oryzae* has a relatively extensive track record of medical cases that are clinically indistinguishable from cases attributed to *A. flavus* (69). Except possibly for the small number of papers documenting cases connected to soy fermentation processes (e.g., 156, 157), these case reports should all be considered to apply to *A. flavus*. It should be noted that the genetic diversity now revealed within *A. flavus* makes many previously used defining features for *A. oryzae* invalid or in need of reevaluation. For example, Gordon et al. (158) "confirmed" an alleged *A. oryzae* isolate from meningitis in an injection drug addict by showing that patient serum reacted with "*A. oryzae* antigen" but not with *A. flavus* antigen in immunodiffusion tests. (It reacted with both antigens in the more sensitive counterimmunoelectrophoresis technique used in the same study.) The authors did not state if the *A. oryzae* antigen was from a known domesticated reference isolate or from the case isolate. In any event, even if a reference isolate was correctly used, the case isolate to which antibodies were directed may have simply been from an *A. flavus* group I isolate (see *A. flavus* subspecific population structure above) genetically close to *A. oryzae*, whereas the reference *A. flavus* isolate may have been from group II or another genetically relatively distant subgroup of *A. flavus*. In hindsight, it can be firmly stated that identity with *A. oryzae* was not demonstrated.

Still less, then, was such an identity demonstrated in case reports such as that of Byard et al. (159), where a "greenish-yellow" *A. flavus*–like isolate from sinusitis in Canada was called *A. oryzae* because it was seen to form only uniseriate conidiophores! Although *A. oryzae* as an *A. flavus* subtype is not expected to differ significantly in pathogenicity from the species as a whole, documentation of this pathogenicity in a way that conforms to current standards will require sophisticated research methodologies.

Aspergillus {aff. Petromyces} avenaceus **G. Smith.** This species was well demonstrated from chronic invasive sinusitis in an otherwise healthy Sudanese immigrant to the United States by Washburn et al. (160). No identification characters were given, but the identity of the fungus was confirmed by K. B. Raper. Curiously, despite the rarity of the organism, the highly interesting source of isolation, and the examination by one of history's foremost *Aspergillus* authorities, the isolate appears not to have been conserved. This naturally raises a question about whether or not it was sufficiently typical to be identified with full confidence and treated as an authentic isolate.

In nature the uncommon fungus has been isolated from legume seeds, cornmeal, and soil. It produces the distinctive secondary metabolite avenaciolide, and unlike other members of the section *Flavi*, makes no ochratoxins, aflatoxins, cyclopiazonic acid, or kojic acid (141).

Description. The colonies are rapid-growing (35 to 55 mm in 7 days). They are deeply granular, with conspicuous black sclerotia. Conidia are yellowish to dull olive green. The reverse is pale to dirty pink.

Conidiophores are often around 400 to 600 (but up to 1000) μm high, with smooth-appearing [finely roughened when seen dry, *fide* Raper and Fennell (7)] hyaline walls, attaching relatively abruptly to a globose or oblate vesicle without the funnel-like apical stalk expansion seen in *A. flavus* and close relatives (Fig. 33). The vesicle bears biseriate fertile elements (i.e., metulae and phialides), with metulae all around the perimeter in a radial orientation. *Conidia* are in dry radiating chains. They are hyaline, smooth, ellipsoid, mostly 4.0–5.0 × 3.2–4.0 μm, and occasionally as long as 6.5 μm. *Sclerotia* are black.

Aspergillus {aff. Petromyces} tamarii **Kita.** This fungus was reliably documented from an ulcerating cutaneous infection on an eyelid subsequent to implantation in an otherwise healthy woman via the bristle of a toothbrush being used to brush the eyebrows (161).

This fungus in nature colonizes seeds, soils, and various other substrata (74). Like the closely related *A. flavus*, it is an important food spoilage organism, especially in warm areas (51). Its major mycotoxin is cyclopiazonic acid. Unlike *A. flavus*, it produces no aflatoxin.

Description. The colonies are rapid-growing (55 to 70 mm in 7 days), deeply granular, and olive-brown to yellowish-brown. The reverse is pale to yellow-grey.

Conidiophores are often around 600 to 1500 (but up to 2500) μm high, with strongly roughened, hyaline walls, attaching with a slightly flaring, funnel-like expansion to globose vesicle (Fig. 34). Heads with biseriate fertile elements

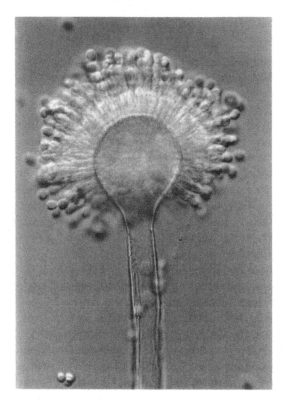

Figure 33 *Aspergillus avenaceus.*

(i.e., metulae and phialides) and uniseriate heads (phialides only) may both be found, in both cases with fertile elements all around the perimeter in a radial orientation. *Conidia* are in dry, radiating chains, globose to subglobose, with thick, heavily roughened walls, mostly 5.0 to 8.0 μm in diameter. *Sclerotia*, if formed, are red-purple, brown, or black.

This fungus is very similar to *A. flavus* in morphology but is distinguished by its considerably browner conidial masses and the thick, heavily roughened walls of its conidia.

Anamorphic Species in A. *Subgenus* Circumdati, *Section* Nigri *Gams et al.* (*Formerly the* Aspergillus niger *Group*).

***Aspergillus* {*aff. Petromyces/Neopetromyces*} *niger* van Tieghem.**   This species is arguably the least virulent of the regularly seen opportunistic aspergilli, and is also an abundant contaminant both of body surfaces and laboratories. It

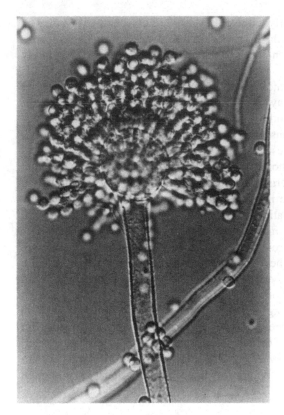

Figure 34   *Aspergillus tamarii.*

is most commonly significantly isolated as an agent of otomycosis (27, 69, 162), but also as an agent of aspergilloma and allergic bronchopulmonary aspergillosis (69, 163). In immunocompromised patients, pulmonary infections or colonizations may occur, and are often characterized by oxalosis, which is extensive production of microscopically conspicuous oxalic acid crystals in sputum (164, 165). Oxalosis may also distinguish *A. niger* in some cases of invasive otomycosis in compromised patients (166). Other infections occurring in immunocompromised patients include disseminated and primary cutaneous infections. A bizarre leprosy-like cutaneous infection of an Egyptian farmer with no known predisposing conditions was immaculately documented in 1967 (167), and a similar primary cutaneous infection was reported in a Swedish metalworker in 1965 (168). To our knowledge no similar cases have been reported since. As with many relatively weak opportunists, *A. niger* may cause dialysis-related peritonitis, endocarditis, and other invasions of particularly vulnerable sites (69). Onychomycosis may be caused, although this is rare and the fungus is a constantly occurring nail contaminant (89, 169). The rare condition proximal subungual onychomycosis can rarely be caused by *A. niger* (170).

In nature and anthropogenic habitats, *A. niger* colonizes soils, seeds, root crops, spices, composts, and many different foodstuffs (51, 74, 124). Its characteristic secondary metabolites, naphtho-T-pyrones and malformins, were not considered to be significant mycotoxins by Frisvad and Thrane (8).

Description.   The colonies are rapid-growing (45 to 70 mm in 7 days) and deeply granular, with conidiophores dense at the colony center and scattered at the margins. They are characteristically fuscous black (i.e., black with a dirty brownish cast), and the reverse is pale to yellow. Atypical, poorly sporulating colonies or colonies on sporulation-suppressive media such as bacteriological brain-heart-infusion-blood agar may have conspicuously yellow mycelium.

Conidiophores are often around 500 (but up to 3000 μm) high, with smooth, hyaline to pale yellow walls. They are aseptate, attaching abruptly to a globose vesicle, which is biseriate (i.e., bearing metulae and phialides), with metulae all around the perimeter in a radial orientation (Fig. 35). *Conidia* are in dry, upright chains, often radiating or massing into two or more short columns per head, in wet microscopic mounts. They are grey-brown, globose, strongly roughened with small, irregular ridges and bars of dark material, and are 3.5 to 4.5 μm in diameter.

In *subspecific population structure* this species appears to be largely or entirely clonal, and can be genetically typed using techniques similar to those used for *A. fumigatus* (171).

Aspergillus {aff. Petromyces/Neopetromyces} japonicus **Saito** (Synonym: *A. aculeatus* Iizuka).   This species has black conidia and superficially resembles *A. niger*. Its aspergilla, however, are uniseriate, and its conidia are purple-black en masse, a color quite distinct from the dirty brownish-black of *A. niger*. Microscopically, the roughening on the conidia consists of evenly spaced spines, not irregular ridges and bars as in *A. niger*. *A. japonicus* appears to have no track record as far as being a pathogen of humans and animals, but has been taken into some review literature because causation of otitis was attributed by Wadhwani and Srivastava (17). These records are not accepted. (See the discussion under *Fennellia nivea* above.) *A. japonicus* was also mentioned under the synonym *A. aculeatus* as a possible agent of a mysterious, diphtheria-like tongue and throat condition in Nigeria (172). The proposed etiology appears to be based on isolation of the species in culture from tongue scrapings and vomitus of a single patient examined only once, without direct microscopic demonstration of any fungal elements in apparently infected, darkened areas of the tongue. It appears likely that the fungus had simply contaminated some food recently eaten

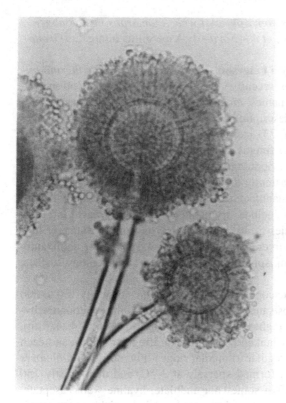

Figure 35 *Aspergillus niger.*

by the patient. *A. japonicus* in nature is a soil and leaf litter decay fungus, mostly from the tropics, but occurring in the temperate zone occasionally as an invader of grapes and rarely as a medical specimen contaminant. [For a description see Klich and Pitt (122) and de Hoog et al. (69).]

6. *Neosartorya* Malloch & Cain

Ascomata are globose cleistothecia, mostly 150 to 500 μm in diameter, and white, with a thin, fragile peridium composed of two to three layers of pseudoparenchymatous cells. *Asci* are eight-spored, globose, and evanescent. *Ascospores* are lens-shaped to nearly spherical, hyaline, with two equatorial crests and no or various types of ornamentation on the valves. Anamorphs are members of *Aspergillus* subgenus *Fumigati* section *Fumigati*, formerly considered the *A. fumigatus* series (7). They have uniseriate conidiophores.

Species with a Known Teleomorph, and with Anamorphs in A. *Subgenus* Fumigati *Section* Fumigati *Gams et al. (Formerly* Aspergillus fumigatus *Group).*

Neosartorya pseudofischeri **Peterson, Anamorph** *Aspergillus thermomutatus* **(Paden) Peterson.** This recently recognized fungus was mentioned in case reports under a variety of names, mostly names of former varieties of *Neosartorya fischeri*, prior to its description. Fortunately, isolates from most cases were obtained by Padhye et al. (173) and preserved in ATCC after being correctly reidentified as this species. The fungus is known, then, to cause osteomyelitis, endocarditis, keratitis, and bronchopulmonary colonization (173). In nature and anthropogenic habitats, it is apparently uncommon, but when found typically occurs on seeds and on cellulosic manufactured materials such as papers and matches (124). Secondary metabolites include tryptoquivaline and trypacidin (124).

Description. The colonies grow approximately 60 mm in diameter in 7 days at 25°C. They are whitish to cream, and granular. *Ascomata* are abundantly formed as small, white to cream, spheroidal tufts massed on the colony surface. The colony reverse is pale.

Ascomata under the microscope are hyaline cleistothecia mostly 150 to 500 μm in diameter. They are globose and non-ostiolate, with a thin pseudoparenchymatous peridium. *Asci* contain eight lens-shaped *ascospores*, 5–6 × 4–4.5 μm, each with an equatorial furrow flanked by two broad, irregular ridges, and each with valves ornamented with irregular triangular ridges (Fig. 36). *Conidiophores* are seldom seen at 25°C, often formed sparsely at 37°C, and then mostly 150 to 500 μm high, with thick, often somewhat undulate, hyaline walls, aseptate, uniseriate (i.e., phialides directly attached to vesicle), with vesicles mostly pyriform. *Conidia* are in dry, upright chains. They are pale blue-green, ellipsoidal, apparently smooth to finely roughened, and 2.5–3 × 2–2.5 μm.

The otherwise similar *N. fischeri* has an anastomosing reticulum and lacks the protruding triangular vertices of *N. pseudofischeri* (Fig. 37). *N. spinosa* has numerous isolated short spines, while *N. glabra* has smooth valves. The characteristic ascospores of these species are illustrated in connection with previously used *N. fischeri* varietal names by Samson et al. (174). The complex relationships of the many *Neosartorya* species now described as well as their relatives in the section *Fumigati* is analyzed using β-tubulin sequences by Varga et al. (175).

A case of disseminated infection in a bone marrow transplant patient was attributed to *N. fischeri* by Lonial et al. (176). According to D. Sutton (personal communication, March 2001), who provided the identification in that case, the organism was provisionally identified as *N. fischeri* var. *spinosa*, currently called *N. spinosa*, but the possibility of the recently described *N. pseudofischeri* was not ruled out. Many authors, including the present one, have identified *N. pseudofischeri* as *N. spinosa* prior to the description of the former as a separate species (177). While this manuscript was in press, Sutton sent the Lonial et al. isolate

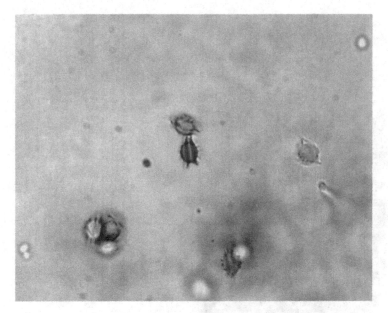

Figure 36 *Neosartorya pseudofischeri* CBS 109512, ascospores.

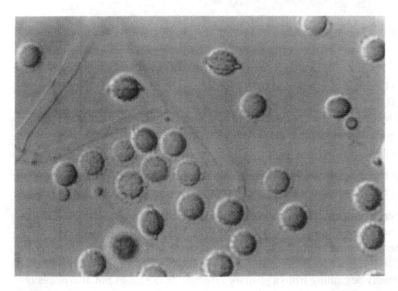

Figure 37 *Neosartorya fischeri*, ascospores.

Figure 38   *Aspergillus fumigatus.*

to CBS, where it was confirmed as *N. spinosa* by R. A. Samson and deposited as CBS 109511. A myeloma patient with pneumonia attributed to *N. fischeri* is described by Chim et al. (178). The identification is unsubstantiated at the species level, but well verified at the genus level, including demonstration of recognizable cleistothecia in infected tissue.

Anamorphic Species in A. *Subgenus* Fumigati *Section* Fumigati *Gams et al. (Formerly* Aspergillus fumigatus *Group).*

***Aspergillus {aff. Neosartorya} fumigatus* Fres.** (Synonym: *Aspergillus phialoseptus* Kwon-Chung).   *A. fumigatus* is the best known and most commonly seen of the filamentous fungal opportunistic pathogens in most parts of the world (11, 68, 69). It causes opportunistic disease in the severely immunocompromised patient, usually beginning from a primary infection in the lungs but also occasionally in other sites, such as the nasal sinuses. Systemic and deep tissue infections, often limited to a single site such as a lung lobe, may be caused in people who are apparently immunocompetent or only weakly immunocompromised. AIDS patients, although relatively rarely affected unless secondarily neutropenic due to cancer or immunosuppressive therapy (e.g., ganciclovir), may develop unusual aspergillosis presentations, such as primary cutaneous lesions (179). *A. fumigatus* also causes various chronic colonizing infections of people with certain predispositions. For example, it is the major agent of chronic bronchopulmonary aspergillosis, a disease featuring an exacerbated allergic response to aspergilli growing on upper respiratory surfaces, mostly in long-term asthmatic and cystic fibrosis patients (150). Other colonizing infections include pulmonary aspergilloma (fungus ball) in the lungs of persons with pre-existing cavitation, chronic mycotic sinusitis, and otomycosis (outer ear canal surface infestation) (11). The species also causes the full spectrum of opportunistic mycoses of specially vulnerable body sites (e.g., ocular infections subsequent to traumatic introduction, otomycosis, onychomycosis, dialysis-related peritonitis, and endocarditis). It causes many animal infections, most notably pulmonary infections of birds—it is, for example, a well-known risk for penguins in zoos (180)—as well as potentially fatal infections of dogs, especially the notably susceptible German shepherd breed, in which it often becomes established as an invasive mycotic sinusitis.

In nature and anthropogenic habitats, it is a thermotolerant compost organism, growing in self-heating decaying vegetation, warm soils, and warm building-related habitats, especially organically enriched (e.g., with bird dung) warm ventilation ducts or ledges, as well as indoor composters and potted plants (74, 124). It also commonly invades seeds, especially grains, and warmth-dried spice and smoking materials. Its principal mycotoxins include gliotoxin, verruculogen, fumitremorgens, fumitoxins, and tryptoquivalins (8).

Description. The colonies are fast-growing (70 to 85 mm in 7 days at 25°C). They are powdery and dull blue-green with a whitish margin. The reverse is pale to slightly greenish.

Conidiophores have phialides directly attached (uniseriate) over the upper two-thirds of the surface. They are greenish, with clavate vesicles and smooth stipes. *Conidia* are borne in interconnected dry chains cohering above conidiophores as compact upright columnar masses. When seen individually, they are subglobose, finely roughened, and 2.5 to 3.0 μm in diameter.

Most isolates grow well at 45°C; a few have maximum growth temperatures around 43°C.

A common host-adapted "dysgonic" (conidiation-impaired) variant usually lacks conidiation on initial outgrowth. It is typically isolated from long-term colonization habitats such as aspergillomas, or bronchopulmonary, nasal sinus, or outer ear colonizations. It can be distinguished preliminarily by good growth at 45°C, and dense, matted, whitish colonies with radial folds. Conidiation can often be reinitiated by growing colonies on modified Leonian's agar (4) at 37°C and making subcultures from the small conidial tufts or faint haze of surface conidiation that may develop. When formed, conidiophores, are often aberrant and strongly reduced in size, sometimes with only one to five phialides. Completely nonsporulating isolates may be identified by exoantigen (181) or specific molecular studies. (See below.)

The difficulty of detecting invasive aspergillosis by culture has led to extensive development of molecular direct detection methods. Many of these involve amplifications using primers detecting all aspergilli or a wider range of fungi followed by posttreatment with nested PCR, specific probes, restriction digests, sequencing, or single-strand conformational studies to yield species-level identifications. These techniques have predominantly utilized ribosomal DNA amplifications, especially the 18S subunit (154, 182–186), the large subunit (152, 187), and the internal transcribed spacer region (188). In a few cases, ribosomal primers specific to *A. fumigatus* have been used directly (189, 190). In other cases, they are used in tandem with general primers (186, 187). Random primers have been designed specifically for *A. fumigatus* (191) or for *A. fumigatus* and closely related *Neosartorya* species (192). The strategy of relatively broad-based amplification for various aspergilli followed up by additional analysis for species determination has also been pursued using alkaline protease (153) and mitochondrial (193) primers. *A. fumigatus*-specific and *Aspergillus* genus-specific primers based on the aspergillopepsin *PEP* gene (194) have been published. In addition, primers based on an 18-KD immunoglobulin E-binding protein (195) specific to *A. fumigatus* are available. In the latter study, the authors' claim that their primers also amplify *A. restrictus* DNA appear to be based on *A. fumigatus* isolates misidentified under that name. (See discussion under *A. restrictus* above.) Their technique is highly specific to *A. fumigatus*, not amplifying even the closely related *Neosartorya fischeri*. Recently a plate-hybridization technique (196) and a quantitative Light Cycler PCR protocol (197) specific for *A. fumigatus* have been reported.

A. fumigatus is a predominantly or entirely clonal organism (198) that may be biotyped using various molecular techniques (171). Patients with colonizing infections may carry several genotypes, while patients with invasive aspergillosis generally are infected by a single genotype (124).

Investigators carrying out epidemiological studies on *A. fumigatus* need to keep certain ecological possibilities in mind that have not always been considered in published studies.

1. Even in a small focus, *A. fumigatus* growing in a hospital or other environment need not consist of a single genotype. It is very common for conducive fungal habitats to support mosaics of multiple genotypes of the same species. Therefore if a small sample size of environmental isolates contains only genotypes different from the one found in an infected patient, this does not mean the patient was not infected by a local

(e.g., in-hospital) source. Typical hospital habitats such as accumulations of bird dung around air intakes are stable habitats of the kind most likely to support a well-developed community of various strain types. The presence in a hospital of *any* genotype of *A. fumigatus*, whether the same as that obtained from the patient or different, suggests that the possibility of a local source for the patient's isolate requires full investigation. A genotype that constitutes only a small fraction of the local *A. fumigatus* population may—through greater virulence or by chance alone—be the one that infects one or more patients.

2. As a largely clonal organism, *A. fumigatus* may manifest as the same genotype in two different sites by coincidence. Elucidating the same genotype from two sources (e.g., a patient and a nearby hospital air duct) does not necessarily mean that the two isolates are connected. More complete sampling might show that local outdoor air, the patient's house, or the clothing or hair of visitors might be just as probable a source of the same genotype as the air duct.

These possibilities indicate that epidemiological testing with *A. fumigatus* is fundamentally a probabilistic problem. For investigating whether or not an environmental source is linked to a patient isolate, local environmental isolates must be studied in sufficient detail to ensure that the full biodiversity has been covered in order to prevent false-negative associations. As in all ecological studies, the diversity of different types present should be studied until adequate coverage has been attained. The percent–coverage statistic outlined by Gochenaur (199) is one means of addressing this question. To prevent false-positive associations, more distally obtained environmental data need to be obtained or generated through de novo studies. This constitutes an essential control study. It needs to be shown that the probability of the patient being exposed to a given genotype in the local environment (e.g., in the hospital) is significantly greater than the risk of exposure in the miscellaneous exterior environments in which the patient or his or her contacts (not to mention any unsterile environmental materials; e.g., foods, tap water, or flowers) may have been present. Of course, a patient who is infected by a genotype making up only 3% of hospital isolates but 57% of local outdoor air isolates may still in reality have been infected from the in-hospital source. This link cannot be conclusively demonstrated by epidemiological genotyping studies, however.

7. *Hemicarpenteles* Sarbhoy & Elphick

Anamorphic Species in A. *Subgenus* Clavati *Section* Clavati *Gams et al. (Formerly* Aspergillus clavatus *Group).*

***Aspergillus** {aff. **Hemicarpenteles**} clavatus* **Desm.** This species is apparently of relatively low virulence. It has been reliably reported from a case of endocarditis by Opal et al. (200). Two case records from otomycosis by Yassin et al. (95) are regarded as unsubstantiated for reasons mentioned above under *A. versicolor*. *A. clavatus* is widely distributed in nature, especially in the tropics, and is common in these areas on decaying foodstuffs, especially grains (122). It is also common in settings connected with the brewing industry (111).

Description.   Colonies are moderately fast-growing (30 to 45 mm in 7 days at 25°C), and are granular or deeply and evenly powdery with a dense felt of long conidiophores. They are dull blue-green with a whitish margin. The reverse is uncolored to pale yellow-brown.

Conidiophores are mostly 500 to 1000 µm long, with phialides directly attached (uniseriate) over the surface of dramatically elongated vesicles up to 250 µm long, and mostly 40 to 60 µm wide (Figs. 39 and 40). Stipes are smooth. *Conidia* are ellipsoidal, smooth, and 3.0–4.5 × 2.5–4.5 µm.

Figure 39   *Aspergillus clavatus*, 10X objective.

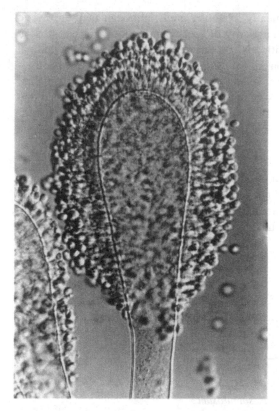

Figure 40 *Aspergillus clavatus*, 40X objective.

Aspergillus {*aff. Hemicarpenteles?*} *clavatonanica* **Batista, da Silva Maia & Alecrim.** This fungus, which resembles *A. clavatus* but has short, re-branching conidiophores, was isolated and described from a painful paronychia-like inflammation in a patient whose work involved washing materials by hand (201). Microscopic examination of purulent exudate revealed "rare and small spherical bodies" considered to be fungal "spores," but the nails themselves contained no discernible fungal elements. The isolation was not repeated and the infection was successfully treated with iodide. Clearly this interesting study did not demonstrate etiology, and no parallel case has been reported since.

B. *Penicillium* Link:Fr.

Penicillium species are among the most common decomposers in nature. They are closely related to *Aspergillus* species but in general are less thermotolerant and are most prominent ecologically in cooler areas, although they are by no means absent in the tropics. The majority of species grow poorly or not at all at 37°C in vitro, and in keeping with this, are not reported from human and animal disease. The prominent exception is *Penicillium marneffei*, which, like other thermally dimorphic fungi probably adapted to infect animals as a means of perennation, has evolved a particulate assimilative phase specialized for growth and dispersal in host tissues and fluids. This particulate phase, analogously with that of *Coccidioides immitis*, is ontogenetically unique in the fungal kingdom, despite its superficial similarity to fission cells of *Schizosaccharomyces*, and indicates an independent evolutionary path to thermally dimorphic facultative pathogenicity.

Penicillium is an amalgam of anamorphs corresponding to two prominent teleomorphic genera, *Eupenicillium* and *Talaromyces*. These will be dealt with separately. The *Penicillium* species in this chapter are presented in a conventional arrangement according to their taxonomic subgenera and teleomorph affinities. The arrangement can be seen at a glance in Table 5.

Table 5 Natural (Teleomorph-Based) Groups of *Penicillium* Species as Arranged in This Chapter

Teleomorph	*Penicillium* subgenus	Species
Eupenicillium	*Aspergilloides*	*Penicillium citreonigrum*
		P. decumbens
		P. spinulosum
	Furcatum	*P. citrinum*
		P. janthinellum
		P. melinii
		P. oxalicum
	Penicillium	*P. brevicompactum*
		P. chrysogenum
		P. commune
		P. cyclopium
		P. expansum
		P. griseofulvum
Talaromyces	*Biverticillium*	*Talaromyces emersonii* (*P. emersonii*)
		T. flavus (*P. dangeardii*)
		T. thermophilus (*P. dupontii*)
		P. marneffei
		P. piceum
		P. purpurogenum
		P. rugulosum
		P. verruculosum

The anamorphs in general are characterized by possessing solitary and undivided, or more commonly brushlike and multiply rebranched, septate and erect or divaricate conidiogenous stipes with a cluster of phialides at the terminus of each branch. Apices may be slightly vesiculate, but not generally more than three times the diameter of the subtending branch. By contrast with *Aspergillus*, *Penicillium* lacks foot cells anchoring the stipes in substrate hyphae, and forms young phialides on an apex sequentially, not synchronously. Tips of phialides formed below the fertile branch apex are bent to bring their alignment closer to that of the uppermost phialides, giving rise to subparallel clusters. By contrast, phialides in most species of *Paecilomyces* tend to splay widely apart, forming divergent clusters. Phialides are typically bottle-shaped (ampulliform) in *Eupenicillium* anamorphs, with an extended "neck" (collula) in a few species. *Talaromyces* anamorphs have distinctive phialides that are lanceolate, meaning "long spearhead-shaped," also called acerose, meaning "sharp." These *Talaromyces*-type phialides have a long basal zone of nearly parallel-sided morphology before narrowing abruptly to a neck at the apex. Conidia in all *Penicillium* species form in dry, interconnected chains. They may be smooth but are often roughened, and are in shades of blue, green, grey-brown, and grey (most typically sage-green, glaucous, or blue). Similar fungi with pink, purple, or white conidia in interconnected chains are in other genera, often *Paecilomyces*. Similar fungi with conidia in slimy heads or unconnected imbricate (overlapping like roof slates, tiles, or shingles) chains are in taxonomically distant genera such as *Clonostachys*.

1. Identification Techniques

Penicillium species are among the most difficult hyphomycetous fungi to identify morphologically, and the means to identify most molecularly are at this writing in a very preliminary state of development. Many species do, however, readily yield to rationalization if the special growth conditions outlined by Pitt (87, 202) are adhered to. The system involves a special inoculation pattern (three points, one at the halfway point of each of three radii dividing the 85-mm plate into three equal 120° sectors) and technique (mashing inoculum in a separately contained small quantity of agar medium prior to use in order to adsorb dry conidia into wet material and prevent the formation of satellite colonies on inoculated plates), use of CYA, malt extract agar (MEA), and 25% glycerol nitrate agar (G25N) at a standard 25°C, and CYA at 37° and 5°. All incubations are carried out for 7 days, and then colony diameters are measured at 90° to the petri plate radius across the original inoculation point, which is now the colony center. Some members of *Penicillium* subgenus *Penicillium* may require additional inoculation onto creatine sucrose agar (111), and especially for some atypical isolates may require the preparation of mycotoxin profiles (111). None of the necessary media or techniques is standard in biomedical laboratories, and apparently etiologically significant *Penicillium* species are best sent for professional expert identification.

Morphological analysis is rendered easier by several procedures. The heavy burden of conidial production, which can completely prevent visualization of fine structures, is best avoided by taking material from areas near the colony margins that are not yet so heavily sporulated. The present author prefers to use a flat-ended inoculating tool or ultrafine forceps to cut a narrow (~1 mm or narrower), parallel-sided radial slice out of the marginal region of heavily sporulating colonies, being careful not to crush the structures. Then the slice is laid on its side in a drop of 95% ethanol (to extract hydrophobic materials and minimize air bubble formation) so that the conidial stipes tend to lie flat along the slide. Regular laboratory slide mounting medium is added (e.g., lactic acid–cotton blue, lactophenol–cotton blue, detergent/water mixtures). The coverslip is angled onto the material to flatten it so that any excess agar is squeezed off to one side rather than over the surface of the fungal material.

Several learned-by-experience morphological characters make preliminary *Penicillium* characterization agonizing for the novice and relatively easy for the developing expert. First, the branching pattern must be analyzed as to whether conidiophores are unbranched, the monoverticillate condition, or whether phialides are supported on a single order of short supporting branches (*metulae*, meaning "between-structures"; singular, *metula*), the biverticillate condition, or on an even more complex branching system. Metulae are often more or less symmetrically arranged, especially in *P.* subgenus *Biverticillium*, the group related to teleomorphs in *Talaromyces*. The next most complex branching condition, terverticillate, features a second order of branches called rami (singular ramus, meaning simply "branch") that support the metulae. Most commonly there is a main ramus and side ramus, which may be curved to more or less parallel the main ramus. Occasionally there are more rami, or rami that are splayed at wider angles. The common household penicilla such as *P. chrysogenum* are terverticillate. Finally, a few *Penicillium* species may have quaterverticillate penicilli, where rami support two intervening orders of branches, ramuli and metulae.

There is an agonizing "grey area" intermediate between some monoverticillate *Penicillium* species placed in *P.* subgenus *Aspergilloides* and a so-called divaricate type of biverticillate *Penicillium* seen in some members of *P.* subgenus *Furcatum*. Monoverticillate species with long unbranched penicilli arising from the agar surface are easy to pigeonhole correctly, but those like *P. decumbens* with short stipes arising from aerial mycelium may be more difficult. Likewise, members of *P.* subgenus *Furcatum* with three or more metulae in a verticil are easy to recognize, but those such as *P. janthinellum* with a mixture of metulae and solitary side branches are more difficult. Here the necessary distinction is between species that regularly make true metulae; that is, groups of two or more short, terminal, phialide-bearing branches arising at an acute angle to one another to form a somewhat compacted cluster (subgenus *Furcatum*), and groups that simply produce short, solitary penicilli, often arising at 90° angles to an aerial branch (subgenus *Aspergilloides*). For the most intractably intermediate cases, the distinction tends to lie in finding the true metulae uncommonly to commonly (subgenus *Furcatum*), as opposed to finding such structures very rarely in slide mounts (generally subgenus *Aspergilloides*).

Similar intermediacy may occur between members of the subgenera *Furcatum* and *Penicillium*, in which structures with metulae may be interspersed with more complex terverticillate structures. This problem is encountered frequently because the common *P. chrysogenum* is one of the most micromorphologically variable members of subgenus *Penicillium*. In dealing with this grey area, the best practice is to look through the microscopic preparation to detect *any* typical terverticillate structures that are present. If even a small number of such structures is formed, the isolate is normally a member of subgenus *Penicillium*, no matter how much the remaining microscopic material may suggest subgenus *Furcatum*.

The initially agonizing decision about whether or not any biverticillate structures seen should be determined as typical of *P.* subgenus *Biverticillium* or *P.* subgenus *Furcatum* has been treated in detail by Pitt (87). The subgenus *Furcatum* tends to feature phialides much shorter than their supporting metulae, whereas in subgenus *Biverticillium* the lengths of these structures are approximately equal. Also, subgenus *Furcatum* species seldom have more than five metulae per stipe, while subgenus *Biverticillium* species usually have more than five. Subgenus *Furcatum* species, as typical osmotolerant penicillia, grow fairly well on the concentrated G25N medium, with colony diameters 10 mm or more in 7 days at 25°C. Subgenus *Biverticillium* members grow less than 10 mm on G25N.

Finally, the initially agonizing decision between ampulliform phialides in *Eupenicillium* anamorphs and lanceolate phialides in *Talaromyces* anamorphs (*P.* subgenus *Biverticillium*) has been conventionally expressed in the generic description above. As an additional aid to the user, the author adds his own observation that if the phialides of *Penicillium* species were scaled up as identically shaped bottles, the phialides of the *Eupenicillium* clade would make plausible beer or wine bottles within the historical range of normal sizes and shapes, apart from being somewhat too curvaceous. *Talaromyces* type phialides, however, would generally appear to be impractically tall, narrow bottles that would be in great danger of toppling over. This imagery expresses a conspicuous regularity in length–width ratio that is sufficiently variable that a more precise, statistical expression would be difficult to formulate in a useful way.

2. Unidentified *Penicillium* in Opportunistic Pathogenesis

The difficulties in identification of *Penicillium*, and the historically weak link between *Penicillium* taxonomists and the biomedical community (still evident in the exclusion of most mycological literature from Medline) have meant that many cases of non-*P. marneffei Penicillium* infection have been attributed to unidentified species. A detailed review is not appropriate here, but the scope of such infections includes keratitis (28), onychomycosis (203, 204), localized pulmonary infection (205, 206), duodenal colonization as a side effect of carbenoxolone and cimetidine therapy (207), dialysis-related peritonitis (208, 209), and endocarditis (210, 211). Two cases of nasal disease with turbinate invasion in dogs were well confirmed (four more were partially evidenced) by Harvey et al. (212); a photograph shows a fungus that could be *Penicillium citrinum*, *P. chrysogenum*, or a close relative of either. Disseminating *Penicillium* invasive sinusitis in a cat has also been established (213). There are a number of older papers linking—in some cases—well-documented mycoses to penicillia identified with now untraceable species concepts [e.g., a well-documented urinary bladder colonization linked in 1911 to a "penicillium glaucum or common mold" (sic)] (214) that produced yellow exudate in culture (suggesting, in light of current knowledge, *P. citrinum* or *P. chrysogenum*, the first of which has been explicitly reported from urinary tract colonization). The present author has not attempted to mine all these papers for currently relevant information.

Many of the reports of *Penicillium* sp. in opportunistic mycosis suffer from the same epistemologic problems as reports with named species [e.g., lack of confirmatory direct microscopy in otomycotic, sinus, pulmonary, and ocular (215) infections and lack of compatible direct microscopy + confirmatory later re-isolation in onychomycosis]. In any case, it is scarcely possible to correlate two successive *Penicillium* isolations from nails when they have not been identified to species, since the isolations may simply be of two different species. Also, many *Penicillium* sp. reports lack published substantiation for the genus-level identification. *Penicillium* sp. identifications without substantiation or with non-specific characterization may well be token misidentifications of host-adapted *A. fumigatus* or *Paecilomyces variotii*, or isolates of *Aspergillus* section *Versicolores* with unusually high proportions of simplified, *Penicillium*-like phialides, especially on inappropriate media such as Sabouraud agar. For example, an uncharacterized "*Penicillium* sp." from olecranon bursitis that was resistant to amphotericin B but susceptible to ketoconazole in vitro (216) would be strongly suspected to be a polyene-resistant Hypocrealean *Paecilomyces* such as *P. lilacinus*. On the other hand, a "*Paecilomyces* sp." depicted by Leigheb et al. (217) from a toe abscess (negative for direct microscopic signs of mycosis) is clearly an etiologically insignificant *Penicillium*. It should be noted that *P. lilacinus* was often called *Penicillium lilacinus* prior to the mid-1970s.

3. Identified *Penicillium* in Cases Suggestive of Pathogenesis

Special mention must be made of the review by Pitt (218), in which 10 non-*P. marneffei Penicillium* isolates from biomedical sources more or less suggestive of clinical significance were identified by a world authority on the taxonomy of the genus. Although none of the cases appears to have been published (meaning that insufficient detail is available to establish or refute etiology), the list of fungi given may provide some indication of possible emerging opportunistic pathogens. Most are members of *P.* subgenus *Biverticillium* and the related teleomorph genus *Talaromyces*. Some of the records, such as a *Penicillium piceum* isolate listed only as having "human nail" as a source, seem unlikely to be significant. All of the isolations but two are from normally contaminated bodily materials, such as respiratory secretions.

4. *Eupenicillium* F. Ludwig

Anamorphic Species in Penicillium Subgenus Aspergilloides *Dierckx.*

Penicillium {*aff. Eupenicillium*} citreonigrum Dierckx. This species may be linked to the teleomorph *Eupenicillium euglaucum*, as proposed by Stolk and Samson (219); however, the matter requires molecular clarification.

In 1929, Talice and Mackinnon (220) meticulously described a case of repeated cultivation of a *Penicillium* species from clumps of mycelium expectorated by a 58-year-old, tuberculosis-free pulmonary disease patient in Uruguay. The species, described as *Penicillium bertai n. sp.*, was synonymized with *P. citreonigrum* by Pitt (202) on the basis of its detailed characterization, even though a representative culture no longer exists. The case was diagnosed as bronchopulmonary mycosis caused by the *Penicillium*, although fungus ball was not excluded. (No X rays were reported.) No similar cases have been reported since, but since few *Penicillium* isolates from human disease are identified to the species level, this absence of reports may not be significant. The description of a species with bright orange-yellow soluble pigment is certainly incompatible with *P. decumbens*, the only other member of *P.* subgenus *Aspergilloides* well linked to human disease. The case isolate was described as growing weakly at 37°C, a property compatible with some isolates of *P. citreonigrum.*

This species in nature is a soil and compost fungus.

Colonies on CYA grow 20 to 27 mm in 7 days. They are mostly velvety in texture, often with some yellow mycelium and with greenish-grey conidiation. The reverse is usually bright yellow, or occasionally yellow-brown, with typically yellow-soluble pigment. Colonies are on MEA 22 to 26 mm, velutinous or floccose, with whitish and/or yellowish mycelium and greenish-grey conidiation and with pale to red-brown reverse, sometimes with yellow to brown soluble pigment. G25N colonies are 11 to 14 mm in diameter. Growth occurs at 5°C but may only be microscopically visible after 7 days. Growth at 37°C is nil or colonies up to 10 mm are formed.

Conidiophores often arise from aerial hyphae. The stipes are mostly 60 to 100 μm long, smooth-walled, and typically monoverticillate, but with a minority with two or three metulae in some isolates, typically lacking an apical vesicle

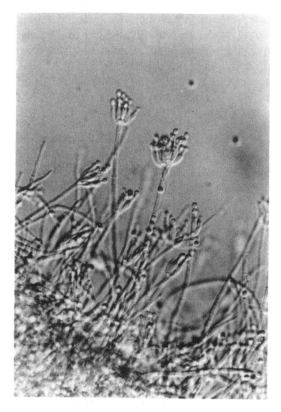

Figure 41   *Penicillium citreonigrum.*

(Fig. 41). The phialides are ampulliform and mostly 5 to 12 μm long. The conidia are globose to subglobose, smooth or with finely roughened walls, and 1.8 to 2.8 μm.

The description of *P. bertai* was based on a purely monoverticillate isolate. *P. citreonigrum* isolates with metulae are most readily distinguished from *P. citrinum* by their significantly faster growth on MEA. The much more blue-tinted conidial color of *P. citrinum* on CYA is also unmistakable for those who have seen each species several times.

Penicillium {aff. Eupenicillium} decumbens Thom.   A case of pulmonary fungus ball caused solely by this *Penicillium* species in an otherwise healthy Japanese farmer was elegantly demonstrated by Yoshida et al. (221). Confirmatory characters for the identification were not given, but a mycological reference center was credited. *P. decumbens* has relatively few look-alikes that grow at

37°C, so this identification is accepted. Alvarez (222) described a case of apparent disseminated infection, evidenced by four positive blood cultures, in a gay man with AIDS and fever. Amphotericin B resolved the fever. The culture was identified by the M. Rinaldi mycology laboratory at the University of Texas Health Science Center.

In nature, this fungus is mainly found in various types of decaying plant material, a niche that extends both to soil and to foodstuffs (51). The majority of records are from soil (74).

Colonies on CYA grow 20 to 30 mm in 7 days. They are mostly velvety in texture. They are dull greyish-green, and the reverse is pale to olivaceous or yellow-brown. Colonies on MEA are similar but 25 to 40 mm. G25N colonies are 11 to 16 mm in diameter. Growth occurs at 5°C but may only be microscopically visible after 7 days. Colonies at 37° may be 5 to 20 mm.

Conidiophores mostly arise from aerial hyphae. The stipes are mostly 20 to 60 μm long, smooth-walled, monoverticillate, and often lacking an apical vesicle but sometimes possessing one approximately twice the diameter of the stipe (Fig. 42). The phialides are ampulliform and mostly 8 to 11 μm long. The conidia are ellipsoidal to subglobose, smooth-walled, and 2.5–3.0 (-4.0) μm.

As mentioned above, this is one of a minority of monoverticillate *Penicil-*

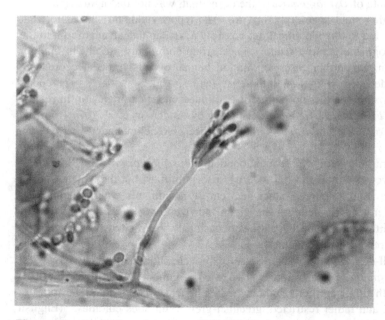

Figure 42 *Penicillium decumbens* CBS 230.81.

lium species that grows at 37°C in vitro, a distinction that may accord with its occasional isolation from opportunistic infection.

***Penicillium* {aff. Eupenicillium}** *spinulosum* **Thom.** This monoverticillate *Penicillium* species was the subject of two bizarre case reports in an otherwise apparently well-worked study on fungal ulcerative keratitis by Anderson et al. (92). Both patients involved were listed in both individual case reports and discussion as being infected by "*Penicillium* sp." that is, "genus *Penicillium*, species undetermined," a common laboratory report. Legends under the figures depicting the infected eyes, however, expanded this to "*Penicillium spinulosum.*" None of the other species names in the paper was contracted to a two-letter abbreviation; furthermore, the absence of similar case reports in the ensuing 40+ years has made the arrival of two independent *P. spinulosum* keratitis patients at the same clinic in a single week in 1957 seem increasingly incredible. Analytic parsimony suggests that *Penicillium* sp. was the correct identification, with *P. spinulosum* perhaps deriving from unwarranted later rationalization of the abbreviation "sp." *P. spinulosum* does not grow at 37°C in vitro, and is therefore not among the more likely *Penicillium* species to have caused a keratitis. As an additional oddity, one of the eyes infected by *Penicillium* sp. was alleged to have had both direct microscopic and cultural evidence of coinfection by a fungus named only as "a corn smut." At the time, however, the culturable budding yeast state of *Ustilago maydis*, the corn smut, was not distinguishable from dozens of other smut anamorphs in vitro. The report yields no clue as to what observations led to this identification; possibly a Ustilaginalean anamorph in the recently recognized genus *Pseudozyma*, a group commonly present in outdoor air, was seen. No further cases of smutted human eyes have been recorded in subsequent decades. In light of all these anomalies, Anderson et al.'s (92) reports of *P. spinulosum* keratitis do not merit continued uncritical inclusion in review literature. "*Fusidium terricola*," coisolated by Anderson et al. (92) from the second case of "*P. spinulosum*" keratitis, is discussed below under *Acremonium implicatum.*

Senturia and Wolf (223) mentioned that *P. spinulosum* was "known to cause otomycosis" in a study of its drug susceptibilities, but the isolate used in testing came from a source "other than ears," presumably environmental. No etiologic or mycologic details were given, nor were any cases described. This study is mentioned only because it has occasionally been cited subsequently as if it were a record of pathogenicity.

A well-demonstrated, severe case of bronchopulmonary mycosis was ascribed to *P. spinulosum* by Delore et al. (224). The authors' description of "fructifications with simple verticils, unbranched; spherical, smooth conidia 3 to 4 μm in diameter; and rather restricted, greenish-grey, velutinous colonies" (English translation by the present author) rules out *P. spinulosum*, which has heavily echinulate conidia and rapid growth. The description is, however, entirely compatible with the expected organism from the infection in question, a host-adapted *A. fumigatus.*

Anamorphic Species in Penicillium *Subgenus* Furcatum *Pitt.*

***Penicillium {aff. Eupenicillium} citrinum* Thom.** A necrotizing *P. citrinum* pulmonary mass in an immunocompromised lymphocytic leukemia patient was thoroughly documented by Mori et al. (225). Another unidentified *Penicillium* species not growing at 37°C was also cultured from the same lesion, but may have been incidental. The case was also summarized in a later review of human *Penicillium* infections (226). A neutropenic leukemia patient was well demonstrated by Mok et al. (227) as having pulmonary and pericardial *P. citrinum* infection. A report by Gilliam and Vest (228) exquisitely documenting a chronic renal infection caused by *P. citrinum* in an otherwise healthy man gives no identification details, but credits an authoritative reference laboratory, the U.S. Department of Agriculture Northern Regional Research Laboratory in Peoria (NRRL). An influential paper by Jones et al. (229) about *Fusarium* keratitis incidentally mentions a severe keratitis case attributed to *P. citrinum*, but gives no confirmatory clinical or identification information. The general quality of the paper and the well-known mycological skill of one coauthor (G. Rebell) make the veracity of this report highly probable, though not formally verifiable. A total of nine corroborating cases of *P. citrinum* keratitis from Nigeria are well documented etiologically and mycologically by Gugnani et al. (230, 231). A well-substantiated case of continuous ambulatory peritoneal dialysis (CAPD)-related peritonitis attributed to *Penicillium* sp. (232) includes a photo of a typical *P. citrinum*-like penicillus. The accompanying description supports a likely identification of *P. citrinum*, despite some anomalies (e.g., a statement that the typical "blue-green" *Penicillium* colony had conidia that under the microscope were "yellow-red"— not a spectrally possible color, except under the familiar name "orange"). The authors state that preserved slides were sent to mycologist M. R. McGinnis for technical description. McGinnis, however, did not originate the expression "yellow-red" (personal communication, September 2000) and a transcription error may be involved.

P. citrinum is an extremely common soil and vegetation decay organism. Its principal mycotoxin, citrinin, is a nephrotoxin.

Colonies on CYA grow 25 to 30 mm in 7 days. They are densely velvety to felt-like, blue-green. The reverse is yellow, yellow-brown, orange-brown, or less commonly red-brown. Clear to pale yellow to reddish-brown exudate or yellow-soluble pigment may be produced. Colonies on MEA are 14 to 18 mm, powdery-granular, and characteristically blue-grey at the margin (but become more greenish-grey centrally). G25N colonies are 13 to 18 mm in diameter. Growth usually occurs at 37°C, but does not occur at 5°C.

Conidiophores are single, biverticillate, characteristically with a V-shaped whorl of 3 to 5 nearly equally long metulae with somewhat vesiculate apices (Fig. 43). Variants with some rami or aerial branches may be seen. The 37°C growth character helps to confirm some of the more usual isolates as *P. citrinum*, provided *P. chrysogenum* has been otherwise ruled out. The stipes are 50 to 200 μm long and smooth-walled. The phialides are ampulliform, and 7 to 8 (−12) μm long. The conidia are globose to subglobose, smooth to finely roughened, and 2.5 to 3.0 μm.

This is usually a very distinctive species, with its 37° growth, its relatively broad CYA colonies often with yellow exudate, reverse or soluble pigment, its significantly smaller MEA colonies with dirty greenish-blue centers and a distinctive marginal zone of faded blue jean blue, and characteristic triads of V-angled, equal metulae.

Figure 43   *Penicillium citrinum.*

Penicillium {aff. Eupenicillium} janthinellum **Biourge.** This species was one of four reported by Sandner and Schönborn (108) as part of a mixed association overgrowing heavily ointment-treated burn eschar, as described above under *A. ustus.* It is not clear that such a record, although medically interesting, constitutes pathogenicity. A case of bronchiolitis obliterans organizing pneumonia (BOOP), "a nonspecific response to pulmonary injury characterized by intraluminal infiltration of the terminal airways," was seen in a man who had apparently inhaled a large quantity of *P. janthinellum* conidia from a fungal mat on contaminated juice (233). Although the patient grew the fungus from bronchial lavage and biopsy, there was no evidence of germination or tissue invasion. This, then, was not an invasive or colonizing disease, but rather an organic dust reaction. *P. janthinellum* is described by Pitt (87, 202).

Penicillium {aff. Eupenicillium} melinii **Thom.** Isolation of this species from a chronic sinusitis patient was reported by Bassiouny et al. (146). The significance of the isolation was not confirmed by direct microscopy. The identification also was not documented except by a nonspecific microphotograph.

Penicillium {aff. Eupenicillium} oxalicum **Currie & Thom.** This species was conclusively demonstrated from posttraumatic keratitis and subsequent endophthalmitis by Rodríguez de Kopp and Vidal (234). E. Piontelli of la Universidad de Valparaiso performed the identification, and a compatible description and photographs were published.

This fungus is a widely distributed vegetation decay and soil organism, with a strong but by no means exclusive affinity for preharvest maize corn (*Zea mays*) (87). Its principal mycotoxin is secalonic acid D.

Colonies on CYA grow 35 to 60 mm in 7 days. They are densely velvety, sometimes wrinkled, usually heavily conidial, and dull green. The colony beneath conidia and the colony reverse are yellow, salmon, orange, orange-brown, or pinkish-brown. Exudate is clear or lacking. Colonies on MEA are 20 to 50 mm and deeply powdery, with a dense layer of conidia colored as on CYA. Heavily conidial masses in this species form continuous sheets of parallel conidial chains, and have a distinctive irideous sheen described by Pitt (87) as "uniquely shiny, even silky." G25N colonies are 12 to 16 mm in diameter. Growth at 37° exceeds 10 mm. At 5°C, conidia may or may not germinate.

Conidiophores are biverticillate, usually with two nonvesiculate metulae closely appressed together (Fig. 44). Verticils of three or four metulae may also be common. Stipes are 200 to 400 µm long and smooth-walled. Phialides are ampulliform to acerose, and mostly 10 to 15 µm long. The conidia are ellipsoidal to ovoid, smooth, and distinctively large (3.5–6.5 [−7.0]) × 2.5–4.0 µm.

Anamorphic Species in Penicillium *Subgenus* Penicillium.

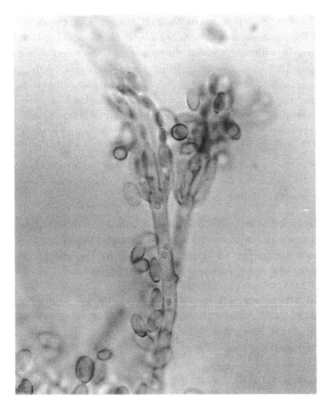

Figure 44   *Penicillium oxalicum.*

Penicillium {aff. Eupenicillium} brevicompactum **Dierckx.**   This fungus was reported from a lung mass in a neutropenic leukemia patient (235). Although the patient had concomitant *Aspergillus fumigatus* cerebral dissemination, the *Penicillium* was consistently isolated first by transbronchial biopsy, then at autopsy, and appeared to be associated with a distinguishable histopathologic presentation. No characterization was given, but the fungus was identified by C. K. Campbell of the Mycology Reference Laboratory, Bristol, England. As with the apparent *Penicillium commune* case described below, however, the apparent causal agent is a fungus never growing at 37°C in vitro, and yet it appeared to cause infection during a time when the patient had a fever. (A figure for degrees of fever was not given in this case.) The widely publicized (51, 74, 87, 202) temperature restrictions on *P. brevicompactum* growth are not discussed by the authors. The general possibilities for reconciling this type of apparent contradiction are outlined under *P. commune.*

It may be noted that the present case, unlike the Huang and Harris (236) *P. commune* case discussed below, featured a contained pulmonary lesion rather than a disseminated *Penicillium* infection. In such a case, it might seem logical that an organism not growing at 37°C or fever temperatures but surviving at those temperatures might be able to grow, at least in the more exposed parts of the respiratory tract, during moments of inhalation when surface temperatures drop, provided no molecules necessary for ongoing metabolism were rendered nonfunctional by the temperature fluctuations. The most recent data and mathematical models indicate, however, that even though exhaled air is only 34.6°C, the lungs have extensive heat recovery mechanisms so that the middle and lower portions of the respiratory system remain close to 37°C (237). In the present case, the author noted that "the fungi invaded blood vessels and bronchioles, forming plugs." Since pulmonary blood vessels are the primary respiratory heat source (238) and are unlikely to drop below body temperature, this suggests that the invader was in fact a thermotolerant organism, not *P. brevicompactum.* The standard of evidence must be raised for an allegation that a fungus not known to grow at body temperature caused an infection. The gold standard would be professional culture collection deposition to allow verification of the identity of the case isolate, plus differential immunohistochemistry or molecular study to certify the same organism was present in the tissue. The case report as it stands is not accepted.

Penicillium {aff. Eupenicillium} chrysogenum **Thom** (Common Synonym: *Penicillium notatum* Westling). This fungus has been reported from a variety of opportunistic infections. A dramatic and well-described case was that of a necrotizing esophagitis in an AIDS patient (239). No identification criteria were given for *P. chrysogenum,* but the identification was credited to reference mycologist D. W. R. Mackenzie. A necrotizing cavitary lung lesion in a squamous cell carcinoma patient was authoritatively documented by D'Antonio et al. (240), and the case isolate was deposited in the ATCC as ATCC 201838. A well-demonstrated report of *P. chrysogenum* colonization of an aortic valve prosthesis, with fungal emboli seeded to the kidney, credited identification to the CDC in Atlanta (241).

Keung et al. (242) attributed fever and generalized follicular and papular rash in a Texas acute myeloid leukemia patient to *P. chrysogenum* on the basis of a single blood culture isolation, and noted resolution of the affliction after treatment with amphotericin B lipid complex and oral itraconazole. The etiologic attribution was supported by a strong positive response to a test for antibody to *Penicillium marneffei* developed by Yuen et al. (243). The case description, however, differs markedly from most infections attributed to aspergilli and other nondimorphic Eurotialean fungi, and was unlike any infection previously reported for *P. chrysogenum.* The observed rapid formation of dispersed skin lesions is more characteristic of organisms producing conidia or particulate assimilative states in vitro, such as yeasts, dimorphic fungi, or hypocrealean fungi, particularly *Fusarium*. Moreover, the serodiagnosis technique of Yuen et al. had been tested for cross-reaction only with *Candida* species and *Cryptococcus neoformans*, at least at the time of its publication (243). Furthermore, *P. marneffei*, as an anamorph in *Penicillium* subgenus *Biverticillium*, which is related to *Talaromyces* teleomorphs, need not have a particularly close antigenic relationship with *P. chrysogenum* (subgenus *Penicillium*, related to *Eupenicillium* teleomorphs), and Keung et al.'s statement that "the *Penicillium* antibody in this patient was highly positive" does not take into account that, for example, anti-*A. fumigatus* antibody is scarcely less likely to react with a *P. marneffei* antigen than is anti-*P. chrysogenum* antibody. The name *Penicillium* as currently used is not a predictor of such specificities. The serological reaction used to support the case attribution is therefore nonspecific and has little evidentiary value in regard to *P. chrysogenum*. The identification of *P. chrysogenum* from the case was confirmed by the laboratory of M. Rinaldi. Although at present the attribution of the case symptomatology to this fungus appears poorly substantiated, similar future case reports or more detail on the specificity of the Yuen et al. (243) serodiagnosis test could tend to confirm it.

A report of posttraumatic endophthalmitis subsequent to perforation of the eye by a metal shard (244) featured credible identification (confirmed by K-J. Kwon-Chung and *Penicillium* expert K. Raper), but only a partial substantiation of causality. *P. chrysogenum* was cultured from the lesion and the patient responded well to topical amphotericin B, but no direct microscopy was reported, nor was there any indication of whether or not the fungus grew out from the cultured surgical swab in a quantity suggestive of a significant source. Seven case records from otomycosis by Yassin et al. (95) are regarded as unsubstantiated for reasons mentioned under *A. versicolor* above. Also, a microphotograph given to document the identification is uncharacteristic of this species. A record by Prasad and Nema (99) from keratitis in India is etiologically and mycologically unsubstantiated. (See the discussion of this paper under *A. sydowii* above.) A case report under the synonymous name *P. notatum* from etiologically well-connected sinusitis (245) mentions "typical rami and phialides" in culture, but otherwise provides no confirmatory characters. The author appears to misinterpret cross and tangential sections of hyphae in a histopathological slide as conidia and phialides. This, plus the use of the arcane synonym and the misspelling of many fungal terms and names make this a dubious report, at least at the species level.

P. chrysogenum is a soil fungus and decomposer of vegetation, especially seeds, and also—because of its growth on crumbs, stored foods, and wall covering papers—an indoor fungus *par excellence*. Roquefortine C is the major toxic secondary metabolite of this species. It also produces meleagrin and penicillin. It is the main industrial source of the latter.

Colonies on CYA grow 35 to 45 mm in 7 days. They are velvety and dull blue-green to dull green. The reverse is typically bright yellow, uncommonly pale or red-brown. Yellow exudate and soluble pigment are usually produced. Colonies on MEA are 25 to 40 mm and velvety, usually with heavy conidiation, colored as on CYA, but often with less exudate and soluble pigment. G25N colonies are 18 to 22 mm in diameter. Growth may or may not occur at 37°, but does occur at 5°C.

Conidiophores are characteristically terverticillate with 1 to 2 rami, but such penicilli are often mixed with biverticillate or less commonly quaterverticillate structures (Figs. 45 and 46). Most penicilli arise from substrate or surface

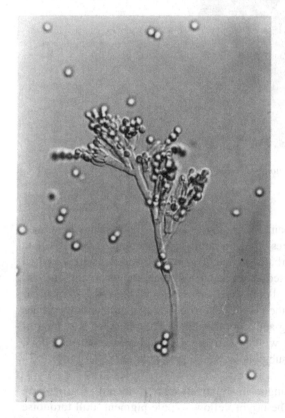

Figure 45 *Penicillium chrysogenum*, variation in conidiophores.

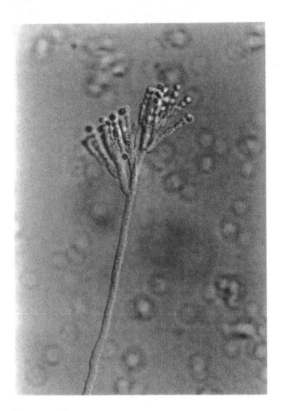

Figure 46 *Penicillium chrysogenum,* variation in conidiophores.

mycelia, but an admixture of penicilli arising as aerial side branches may need to be distinguished from structures borne on rami as part of complex terverticillate penicilli. This problem is exacerbated in heavily squashed slides. There are often structures that could be interpreted either as an additional ramus arising one cell lower on the stipe than the normal rami, or as an extra small biverticillate penicillus diverging one cell beneath the classic terverticillate penicillus above. Stipes are 200 to 300 (−500) μm long and smooth-walled. The phialides are ampulliform and 7 to 8 (−10) μm long, with a relatively wide and green-stained collula. The conidia are ellipsoidal to subglobose, smooth-walled, and 2.5–4.0 × 2.5–3.8 μm.

Typical isolates of this fungus are relatively easy to identify, particularly when they show 37° growth. The bright yellow soluble pigment, dull turquoise conidiation, rapid growth, and terverticillate penicilli with smooth stipes are distinctive. The greenish-tinted phialidic necks provide a subtle character that becomes increasingly useful with habituation. Oil droplets within stipes or adhering to their surfaces must not be confused with roughening. *Penicillium expansum*, which typically has orange-brown soluble pigment and exudate on CYA, must be carefully distinguished in colonies that fail to grow at 37°C. *Penicillium aurantiogriseum*, another similar 37°-negative species, usually has finely roughened stipes, but these may occasionally be smooth. Cultures may have clear or brown exudate and/or brown soluble pigment.

Colonies of *P. chrysogenum* on MEA with closed but unsealed lids may emit an aromatic odor sharply suggestive of pineapple. *P. expansum* smells of apple, one of its characteristic substrates, while *P. aurantiogriseum* mixes the pineapple odor of *P. chrysogenum* with a strong earthy-musty volatile.

Penicillium {aff. Eupenicillium} commune **Thom** [Common Synonym: *Penicillium puberulum* Bain. *sensu* Pitt (202)]. This common food and vegetation decay fungus was apparently well documented by Huang and Harris (236) from a disseminated infection in a leukemic patient who had been treated with prednisone. The fungus was heavily cultured from autopsy specimens heavily invested with septate hyphae. Identification was credited to C. W. Hesseltine of NRRL, and a photograph in the paper is consistent with the current concept of *P. commune*. This species, however, does not grow at 37°C in vitro, and the patient had a fever of 41°C at the time the infection was progressing. No similar cases have been reported since. The reconciliation of these data is scarcely possible, but one of the following is likely to be correct: (1) autopsy materials were secondarily overgrown by *P. commune* so that an original infectious fungus was suppressed, (2) cultures were contaminated in the laboratory, (3) the fungus was misidentified, (4) there is an as yet undocumented thermotolerant variant of *P. commune*, or (5) *P. commune* may become thermotolerant under host conditions. The fourth and fifth possibilities are much less likely than the first three, so *P. commune* cannot yet be considered confirmed as an opportunistic pathogen of humans and animals. The case isolate is not among the 225 NRRL *Penicillium* isolates deposited in the ATCC and is not among the cultures listed on the NRRL website.

Penicillium {aff. Eupenicillium} cyclopium **Westl.** An interesting and well-substantiated case of *Penicillium* invasion of the beak of a macaw, *Ara ararauna*, was attributed to *P. cyclopium* by Bengoa et al. 1994 (246). The identification is unverifiable; the authors stated which characters they looked at, but did not give the results of these examinations. A 45-year-old monograph by Raper and Thom (247) was used for identification, but modern works [e.g., Pitt's 1979 monograph (202)] were not consulted. This species was synonymized with *P. aurantiogriseum* by Pitt (202), and its revival only in context of molecular and mycotoxin studies (248, 249) shows the difficulty of identifying it correctly.

Penicillium {aff. Eupenicillium} expansum **Link.** This species is reported from keratitis in Nigeria by Gugnani et al. (230, 231). Although etiology is well documented in these cases, no identification details are given. Since *P. expansum* does not grow at 37°C and is easily confused with other members of *Penicillium* subgenus *Penicillium*, this identification must be regarded as unconfirmed. *P. expansum* is most frequently isolated as a pathogen of pomaceous fruits (apples, pears) (87) and has been isolated in temperate and subtropical areas, but seldom in the tropics (74).

Penicillium {aff. Eupenicillium} griseofulvum **Dierckx.** This fungus was substantiated as extensively colonizing the body of an unidentified neotropical toucanet (Ramphastidae) that was one of several that died while in the possession of a European bird fancier (250). Identification characters were given, the identification was confirmed by experts, and a culture was deposited in the CDC collection. There must be some doubt about the significance of this isolation, as the fungus was obtained from autopsy specimens taken an unspecified time after the bird's death. *P. griseofulvum* does not grow at 37°C. The present author can find no record of the normal body temperature of a toucanet, but it is noted that many birds have normal body temperatures higher than those of mammals (e.g., 39–41°C in chickens) and that birds respond to illness with fever as mammals do. Finally, although only one bird was autopsied, several had died of an apparent epidemic, and an epidemic connected with a nonthermotolerant common fungus not otherwise known from avian disease seems unlikely.

A record from a cold-blooded animal, the Seychelles giant tortoise, appears to be well demonstrated (251). The fungus caused multiple pericardial and kidney lesions in a zoo tortoise that had previously been injured in a fire.

P. griseofulvum is a very common colonizer of decaying vegetation, including grains, foodstuffs, and animal feeds. Its toxic secondary metabolites include roquefortine C, cyclopiazonic acid, and patulin. It also produces the well-known antifungal drug griseofulvin.

Colonies on CYA grow 20 to 25 mm in 7 days. They are deeply velvety to granular in texture, and greyish-green to yellow-green. The reverse is pale to red-brown or yellow-brown. Clear to pale yellow exudate or red-brown soluble pigment may be produced. Colonies on MEA are 15 to 25 mm and granular. G25N colonies are 16 to 22 mm in diameter. Growth occurs at 5° but not at 37°C.

Conidiophores are single or in loose bundles, terverticillate or quaterverticillate, and very variable in branching pattern, often occurring as aerial side branches diverging acutely from an ascending hypha which also has a terminal penicillus (Fig. 47). The stipes are often undulating. They are greatly variable in length, but up to 500 μm long and smooth-walled. The phialides are ampulliform, unusually short and with a very short neck, and mostly 5.5 to 6.0 (−6.5) μm

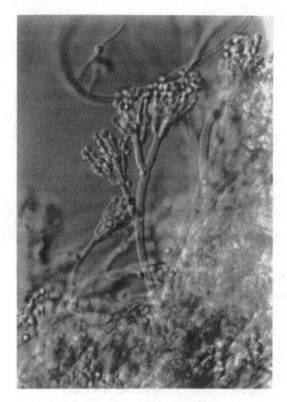

Figure 47 *Penicillium griseofulvum.*

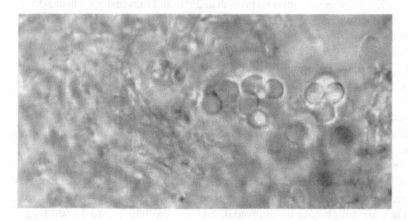

Figure 48 *Talaromyces emersonii* CBS 549.92, ascospores.

long. The conidia are ellipsoidal to subglobose, smooth-walled, and 2.5 to 3.5 ×
2.0 to 2.5 μm.

Colonies on MEA with closed but unsealed lids may emit an aromatic odor
suggestive of mango.

5. *Talaromyces* C.R. Benj.

Ascomata are spheroidal gymnothecia characterized by a peridium of loosely to
densely interwoven, thin hyphae, colored yellow, cream, or white. The ascomatal
initials are often helical or clavate. The asci are usually borne in short chains,
with thin walls deliquescing at maturity. Ascospores are usually eight per ascus.
They are yellow or hyaline, ellipsoidal to spheroidal, and usually ornamented with
fine, short spines, but sometimes bearing ridges or remaining smooth.

Species with a Known Teleomorph and with Anamorphs in P. *Subgenus*
Biverticillium *Dierckx.*

***Talaromyces emersonii* Stolk, Anamorph *Penicillium emersonii* Stolk**
[Common Synonym of the Anamorph: *Geosmithia emersonii* (Stolk) Pitt].   A
purely anamorphic strain of this fungus was repeatedly isolated from a protracted
bronchopulmonary colonization in a 12-year-old cystic fibrosis patient (252). A.
fumigatus colonization preceded this colonization and intermittently co-occurred
with it; however, the itraconazole-susceptible A. *fumigatus* was more readily
eliminated by therapy than was the more resistant P. *emersonii*. On the other
hand, episodes of A. *fumigatus* colonization were symptomatic, whereas P. *emersonii* colonization was benign or nearly so. P. *emersonii* was extensively described and figured, and the identification was confirmed by Eurotialean authority
R. Samson of Centraalbureau voor Schimmelcultures, the Netherlands. According to Pitt (253), the species shows no growth at 25°C and minimal growth at 30°C.
Growth rates below are for 37°C. The species in nature is a compost and soil
thermophile.

Colonies on CYA at 37°C mostly grow 15 to 25 mm in 7 days. They are
velvety in texture, and pale greyish-yellow to drab yellow-brown with conidiation
overlying white mycelium. The reverse is pale to brown. Clear exudate may be
produced. Colonies on MEA are 45 to 70 mm. They are sometimes granular in
appearance due to heavy ascoma formation; otherwise they are velvety, colored
pale yellow or as on CYA. The reverse is yellow or brown. G25N colonies are
0 to 5 mm. Growth does not occur at 5°.

Ascomata are pale yellow or red- to orange-brown, spherical and covered
by a thin layer of reticulate hyphae surrounded by "loose wefts of radiating,
branching, twisted yellowish, encrusted hyphae" (254). *Asci* are formed in coiled
chains. They are subglobose to ellipsoidal, and approximately 8 to 11 μm long.

They contain eight smooth-walled, subglobose to ovoidal ascospores 3.5–4 × 2.7–3.5 μm, sometimes with remnants of a thin, gelatinous coating.

Conidiophores arise from submerged or aerial hyphae. They are biverticillate or with one or two rami, or occasionally with more complex branching patterns, including in extreme cases four to five levels of branching. *Stipes* are 35 to 150 μm long, septate, rough-walled, or rarely smooth (Fig. 49). *Phialides* are acerose with an extended 1 to 2 μm collula, and are 8.5 to 10 μm long overall. *Conidia* are cylindrical, occasionally ellipsoid at maturity, smooth-walled, and 3.5–5 (−10) × 1.5–2.7 μm. The species grows well at 40°C.

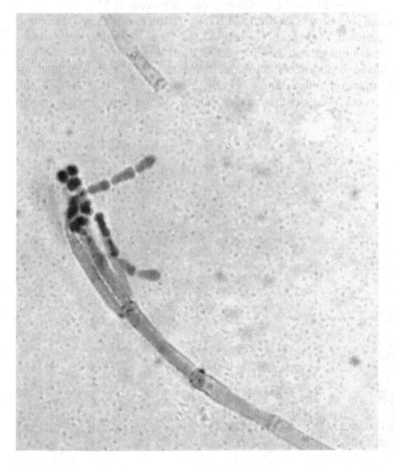

Figure 49 *Talaromyces emersonii*, conidiophore.

Talaromyces flavus **(Klöcker) Stolk & Samson, Anamorph *Penicillium dangeardii* Pitt.**   This species was isolated from direct microscopically verified bovine mycotic abortion by Knudtson and Kirkbride (81) and identified at NRRL. In nature, it is an abundant and widely distributed fungus of soil and organic material.

Colonies on CYA mostly grow 18 to 30 mm in 7 days. They are velvety to floccose in texture, with conspicuous bright yellow mycelium occasionally showing sparse greenish-grey areas of conidiation and usually yellow floccose, round gymnothecia. The reverse is yellow or reddish to brown. Colonies on MEA are 30 to 50 mm, and are sometimes granular in appearance due to heavy ascoma formation, generally similar to CYA colonies. G25N colonies are 2 to 7 mm. Colonies at 37°C are 20 to 45 mm. Growth does not occur at 5°.

Ascomata are yellow, spherical, and 200 to 500 μm in diameter, with tightly interwoven peridium, *Asci* are formed in chains. The *ascospores* are yellow, rough-walled, ellipsoidal, and 3.5–5 × 2.5–3.2 μm (Fig. 50).

Conidiophores arise mainly from aerial hyphae. They are biverticillate or occasionally monoverticillate. *Stipes* are 20 to 80 μm long (Fig. 51) and are septate and smooth. *Phialides* are acerose and 8 to 16 μm long. *Conidia* are el-

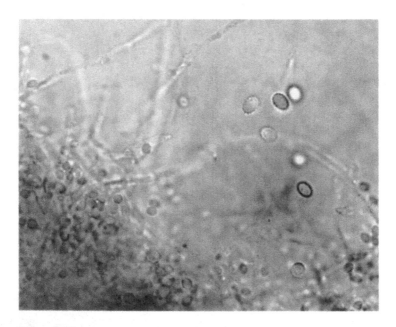

Figure 50   *Talaromyces flavus* CBS 262.78, ascospores.

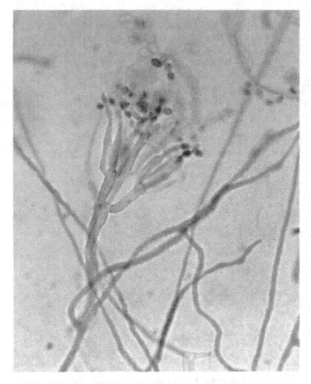

Figure 51 *Talaromyces flavus* CBS 262.78, conidiophore.

lipsoidal to fusiform, smooth or finely roughened, and 2.2–4 × 2.0–2.5 μm. The species grows well at 40°C.

Talaromyces thermophilus **Stolk, Anamorph** *Penicillium dupontii* **Griffon & Maublanc.** This species was isolated from three cases of direct microscopically verified bovine mycotic abortion by Knudtson and Kirkbride (81) and identified at NRRL. In nature, the fungus is primarily a decomposer of composting materials.

There is no growth at 25°C. Colonies on CYA at 37° are 20 to 35 mm in 7 days. They are floccose or tufted, with whitish to pale pinkish-orange mycelium producing greenish-grey areas of conidiation. The reverse is usually deep brown. Colonies on MEA are 15 to 22 mm. The colors are as in CYA colonies except paler or green on the reverse. G25N colonies are 1 to 3 mm. Growth does not occur at 5°.

Ascomata are seldom formed except on sterile oat grains at 45°C (254).

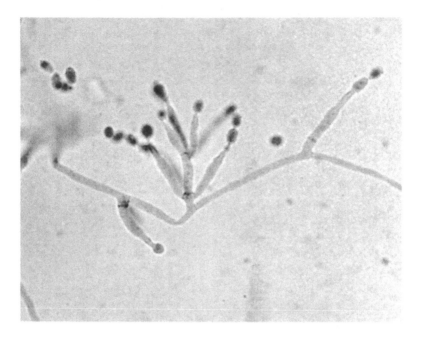

Figure 52   *Talaromyces thermophilus*, conidiophore.

They are cream colored, spherical, and 400 to 1300 μm in diameter, with a thick, tightly interwoven, multilayered peridium. *Asci* are formed in chains. *Ascospores* are ellipsoidal, with irregular ridges more or less longitudinally arranged on the walls, 3.5–4.5 × 2.2–3.5 μm.

Conidiophores arise mainly from aerial hyphae (Fig. 52). They are biverticillate, monoverticillate, or occasionally with a ramus, with *stipes* mostly 5 to 15 μm long and smooth. *Phialides* are acerose and 5 to 10 μm long. *Conidia* are ellipsoidal, smooth or finely roughened, and 3–4 × 1.8–2.5 μm. Conspicuous *chlamydospores* are often present, often on short lateral stalks, and are 4.5 to 6.5 μm.

Anamorphic Species in Penicillium *Subgenus* Biverticillium.

***Penicillium {aff. Talaromyces} marneffei* Segr., Capp. & Sur.**   This species is one of the six virulent, thermally dimorphic, systemic pathogens described in medical mycology thus far (Figs. 53 and 54). The range of diseases it causes and its overall presentation are roughly similar to histoplasmosis. It has, however, many distinctive features, including a distinct endemic range extending eastward from the Indian state of Manipur through mountainous areas of north Myanmar, the south Chinese states of Sichuan, Yunnan, Guangxi, and Guangdong, and most notoriously, northern Thailand and adjacent areas of Laos and Vietnam. It also appears

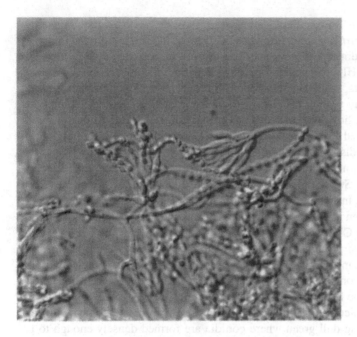

Figure 53 *Penicillium marneffei* conidiophores.

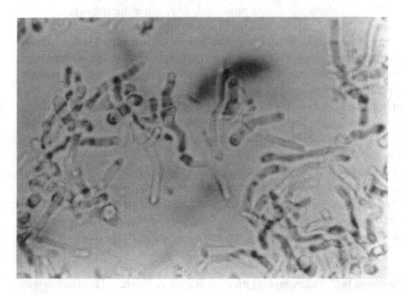

Figure 54 *Penicillium marneffei*, fission-yeast-like state produced at 37°C in vitro.

to be indigenous in Taiwan. Ecologically it is associated with bamboo rats in the genus *Rhizomys* and *Cannomys*. It appears likely to use intermittent animal infection in perennation as dimorphic Onygenalean fungi are suspected to do.

Prior to the HIV pandemic, penicilliosis marneffei was seen in only a small number of patients, mainly immunocompromised, most of whom had acquired the organism many years previously without experiencing significant symptoms (255). Since then, however, an enormous upsurge in HIV-infected persons with penicilliosis has led to hundreds of recorded cases (256, 257). The fungus can be either focal or disseminated, and in the latter case most commonly infects the lung, liver, and skin. Dramatic oral papules and ulcers may be seen in disseminated cases (258). Small, oblong to rounded yeastlike structures are usually seen in histopathology. In some cases, the fission mechanism of progeny production in the host can be seen in close examination of histopathological preparations (255).

Colonies on CYA grow 28 to 32 mm in 7 days. They are velvety in texture, with colonies most prominently showing pale brownish to dull reddish mycelium overlaid by some whitish aerial mycelium. The conidiation is typically light, sometimes patchy on the colony surface. It is drab green. The reverse is red to redbrown. Colonies on MEA are 28 to 30 mm, featuring aerial hyphae aggregated in loose strands (funiculose). They are orange-brown or red, sometimes with few conidia, but turning dull green where conidia are formed densely enough to be apparent. Red exudate may be produced, and bright red soluble pigment is characteristically produced, although atypical isolates may not show this feature. G25N colonies are microscopic. Growth does not occur at 5°. According to Pitt (202), at 37°C on CYA, "colonies 2–5 mm produced, low, sometimes slimy, of white mycelium only." More characteristically, on Sabouraud agar, colonies at 37°C generally convert to a white, fission yeastlike morph in which hyphal-type apical growth produces short hyphae that quickly fragment into individual arthroconidia. Sabouraud colonies at 25°C also characteristically show the red soluble pigment typical of the species.

Conidiophores are biverticillate, arising from surface hyphae or, on MEA, from ropy strands. Stipes are 70 to 190 μm long and smooth-walled, usually bearing only 3 to 5 metulae. Phialides are 6 to 8 μm long. The conidia are ellipsoidal, smooth, and 2.5–4.0 × 2.0–3.0 μm.

An investigation of the *subspecific population structure* showed that two predominant DNA endonuclease restriction patterns were distinguishable (259). Although both types of isolates were found commonly in patients, several isolates from one animal host, *Rhizomys sumatrensis*, appeared to be associated with one type and three isolates from another, *Cannomys badius*, were associated with the other.

***Penicillium {aff. Talaromyces} piceum* Raper & Fenn.**   This fungus was reported without identification characters from two cases of mycotic abortion in an omnibus overview of such infections by Austwick and Venn (260). The text strongly implies but does not state outright that all cases had direct microscopically visible fungal filaments in the placenta, in fetal stomach contents, or in skin. The pathognomonic significance of filaments in stomach contents, according to the authors, had been confirmed in previous published and unpublished studies of experimentally induced mycotic abortions. There is at least a small chance the isolations were fortuitous, in that one case was ascribed to *Polystictus* (currently *Trametes*) *versicolor*, a basidiomycetous wood-decay fungus with no pathogenic record, and according to Austwick and Venn (260) themselves, not growing at 37°C in vitro.

A fungemia case involving *P. piceum* in an immunocompromised human has been summarized (261); the case isolate is deposited as CBS 102383.

Colonies on CYA grow 15 to 25 mm in 7 days. They are velvety in texture. The colonies are dull green in heavily conidial areas and show pale to straw-yellow mycelial areas elsewhere. The reverse is red-brown or brown. Colonies on MEA are 18 to 25 mm, often with clear to yellow exudate. The reverse is yellow to orange. G25N colonies are 4 to 8 mm in diameter. Growth does not occur at 5°, but usually exceeds 25 mm at 37°C.

Little is known about the organism's habitat. Some isolates in collections have come from animal tissues (e.g., the lung of a pig); however, collection records lack etiologic indications.

Conidiophores are biverticillate, distinctively including an expanded, vesiculate stipe apex (Fig. 55). *Stipes* are 15 to 22 μm long, smooth-walled, with vesicle up to 6 μm in diameter. *Phialides* are acerose and 7 to 9 μm long. The *conidia* are mostly broadly ellipsoidal to subglobose, smooth, and 3.0–3.5 × 2.2–2.5 μm, with columns massing in conspicuously conelike clusters when formed on relatively large conidiophores within the species' size range.

Penicillium {aff. Talaromyces} purpurogenum **Stoll.** Two papers by Morin et al. (262) and Breton et al. (263) documented the same case of a neutropenic myeloblastic leukemia patient with a pulmonary infection who yielded this fungus on two successive bronchial lavage specimens, one of which evinced fungal filaments in direct examination. Previous nasal and stool specimens had grown the same fungus. Intravenous inoculation of conidia into normal rabbits resulted in scattered granulomas containing filamentous fungal growth. The causal agent was thoroughly described and illustrated; identification was confirmed by C. de Bièvre.

Colonies on CYA grow 15 to 30 mm in 7 days. They are velvety in texture. The colonies are dark green in heavily conidial areas and show yellow or red mycelial areas elsewhere. The reverse is deep red-purple. Orange or red exudate and bright red soluble pigment are characteristically produced. Colonies on MEA are 20 to 35 mm, usually lacking the red exudate and soluble pigment seen on CYA. The reverse is pale to red-brownish. G25N colonies are 0 to 6 mm in diameter. Growth does not occur at 5°, but usually exceeds 10 mm at 37°C.

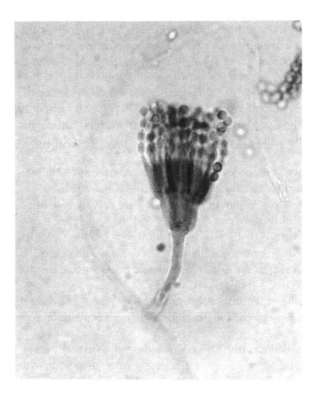

Figure 55   *Penicillium piceum conidiophore.*

Conidiophores are biverticillate, with closely compressed metulae and phialides (Fig. 56). Whole verticils of metulae, if measured across their collective apices, often measure only 8 to 10 μm. *Stipes* are 70 to 300 μm long and smooth-walled. *Phialides* are acerose and 10 to 14 μm long. The *conidia* are mostly ellipsoidal, sometimes subglobose, smooth, or more commonly with finely roughened to heavily warty walls, and are 3.0–3.5 × 2.5–3 μm.

The rather similar *P. verruculosum* lacks red soluble pigment on CYA and has globose, rough conidia. Its conidiophores are not as compressed as those of *P. purpurogenum*, and usually measure around 15 μm or more across the apices. Several other *P.* subgenus *Biverticillium* species with red colony reverse are distinguished by Pitt (87, 202).

Penicillium {aff. Talaromyces} rugulosum Thom.   Neuhann (97) cultured this species from scrapings of a corneal ulcer of a man who had been impaled by a wood fragment. Confirmation of fungal infection by direct microscopy was not recorded, but mycotic keratitis was suggested by positive response to clotrimazole. Characters substantiating the identification were not given (a com-

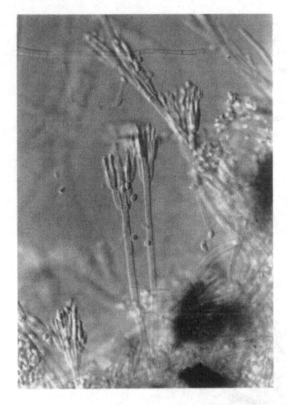

Figure 56   *Penicillium purpurogenum.*

mon shortcoming of ophthalmology literature), but food/water microbiologist H. Barth from Universität Heidelberg was credited with mycological studies. A fungus isolated from any scraping of surface material could be incidental, so this record must be regarded as suggestively but not definitively attested. Most isolates of this species do not grow at 37°C in vitro, but some do (87). An attribution of keratitis to *P. rugulosum* by Świetliczkowa et al. (264) is also etiologically and mycologically unsubstantiated; the record is based purely on growth in culture. A record from sinusitis by Vennewald et al. (43) cannot be accepted for reasons given above under *E. herbariorum.*

Descriptions may be found in works by Pitt (87, 202). The fungus is isolated from soils and rotting vegetal foods.

***Penicillium {aff. Talaromyces} verruculosum* Peyr.**   This species has been definitively documented as an agent of osteomyelitis in a German shepherd dog by Wigney et al. 1990 (265). The case isolate was preserved in the Commonwealth Science and Industrial Research Organization (CSIRO) (North Ryde, NSW, Australia) culture collection. The species in nature is a soil fungus.

Colonies on CYA grow 20 to 30 mm in 7 days. They are velvety or slightly floccose in texture. The colonies are green in heavily conidial areas and show yellow mycelial areas elsewhere, occasionally also with some orange or red coloration. The reverse is orange- to red-brown. Clear or red exudate may be produced, but no soluble pigment. Colonies on MEA are conspicuously broader, 30 to 45 mm, with the reverse usually paler than on CYA. G25N colonies are 2 to 6 mm in diameter. Growth does not occur at 5°, but usually exceeds 20 mm at 37°C.

Conidiophores are biverticillate, with relatively divergent metulae, even occasionally to an angle of 90°, according to Pitt (87), often 15 μm across the apices (Fig. 57). (See description and discussion of *P. purpurogenum*.) *Stipes* are 150 to 250 μm long, smooth-walled, and tend to arise from aerial mycelium. *Phialides* are acerose and 7 to 10 μm long. The *conidia* are mostly globose, sometimes subglobose, heavily roughened, and 3.0 to 3.5 μm in diameter.

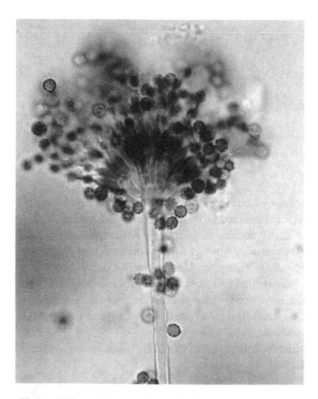

Figure 57 *Penicillium verruculosum* CBS 624.72.

C. *Polypaecilum*

1. *Polypaecilum* {aff. unknown} *insolitum* G. Smith
   [Synonyms: *Scopulariopsis divaricata* Yamashita
   (Invalidly Published)]

This unusual fungus has mainly been isolated from ears of patients with suspected otomycosis (265, 266, 267, 268, 269). No case details were given in any of these studies. More recently, a well-confirmed onychomycosis in an alcoholic patient with dystrophic legs (a sequel of childhood poliomyelitis) and chronic dermatological problems was shown to have *P. insolitum* as an apparent sole agent (270). The identity of the fungus was confirmed by D. Minter at CABI, but the isolate appears not to have been conserved there.

A fungus producing some similar, aberrant structures was isolated by Coutelen et al. (271) from a well-confirmed pulmonary fungus ball and invalidly published as *Scopulariopsis insolita*. No culture was preserved. In his species description of *P. insolitum*, Smith (268) stated that the *S. insolita* isolate, which he had not seen, was "identical" to *P. insolitum*. The isolate studied by Coutelen et al. (271), which was extensively documented with illustrations, produced not only the irregular polyphialidic structures apparently typical of *P. insolitum*, but also numerous inflated structures suggesting aspergilla of the dysgonic, host-adapted form of *Aspergillus fumigatus*. Such structures are not otherwise reported in connection with *P. insolitum*. "*S. insolita*" grew rapidly at 37°C and had compact, cerebriform colony morphology similar to that of atypical *A. fumigatus*. An ad hoc investigation done for the present writing showed that the ex-type isolate of *P. insolitum*, CBS 384.61, grows very slowly at 36°C and not at all at 40°C. Since dysgonic *A. fumigatus* forms various aberrant conidiogenous structures, and since pulmonary fungus ball is the source par excellence of such isolates, the identity of Coutelen et al.'s isolate must be questioned. Smith (268) described *P. insolitum* as producing annellides; however, Yamashita and Yamashita (269) and Piontelli et al. (270) published photographs revealing the apparent annellations as a visual artifact of the conidiation process. Electron microscopy done by Cole and Samson (272) and De Hoog et al. (69) leaves no doubt that the conidiogenous cells of the ex-type are phialides. Smith also described the fungus as grey-brown, in color, but Yamashita and Yamashita (269) showed that on Sabouraud agar it became greenish, similar to their (regrettably unpreserved) 36 isolates from Japanese otomycosis. The nature of this fungus requires further investigation. Regrettably, out of all the published isolates reported by various authors, only the ex-type appears to be available for further studies.

Description. Colonies are 7.5 to 8.5 mm after 7 days at 25°C. They are velvety and radially wrinkled to heaped, cerebriform, or cupulate. They are dirty white to pale brownish-grey to pale grey-green, with the reverse beginning pale

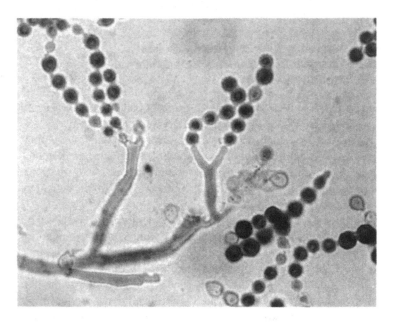

Figure 58   *Polypaecilum insolitum* CBS 384.61.

olive and later becoming coffee brown to brownish-black, with yellow soluble pigment. *Conidiophores* are erect and multiseptate with thin, smooth walls (Fig. 58). They are irregular in width, with occasional irregularly disposed branches, and 6.6–47 × 2–4.2 μm. The main and side branches are terminated by a cluster of 1 to 5, commonly 2, equal, short phialidic protuberances 2.8–9 (−13) × 1.2–2.3 μm. The conidia in interconnected dry chains are ellipsoidal to subglobose, nearly smooth, finely roughened or rugose, and 3.5–5 (−7.2) × 2.4–5 (−6) μm. Chlamydospores are present. They are sometimes abundant, terminal or intercalary, thick-walled, smooth, globose to ellipsoidal or slightly curved, and 8–12 × 6–11 μm. The rough, catenulate conidia with connectives indicate that this fungus is likely of Eurotialean affinity.

D.   *Paecilomyces* Bain. (Eurotialean Part: Olive-Brown and Greenish Species)

This genus has been revealed as heterogeneous by molecular characterizations, although some key studies are as yet unpublished. *P. variotii* is the type species of the genus and will retain the generic name when other species are split off into different genera. The Eurotialean affinity of *P. variotii* has been shown on several occasions (e.g., Refs. 6, 273). The discrepancies among the species are particularly shown in their responses to antifungals (274, 275). The Eurotialean *P. variotii* shows the usual susceptibility of members of its order to the polyenes amphotericin B and natamycin/pimaricin (used in treating ocular infections), while *P. lilacinus* shows the usual tendency of Hypocrealean fungi and their derivatives to be resistant or poorly susceptible to this class of drugs. There are, however, some commonalities that have made the name *Paecilomyces* as currently conceived useful; namely, a tendency for the species in this genus to be relatively weak opportunists, most characteristically causing infections in the presence of implanted avascular devices such as artificial heart valves and corneas, Tenckhoff catheters, and shunts; a tendency for any invasion of tissue to be localized rather than disseminated, except in connection with the most severe neutropenia; a tendency for yeastlike elements, possibly actually conidia in most cases, to be reported in histopathology in disseminated infections of apparently immunocompetent dogs and cats; and a tendency for similar particulate elements recently well verified as conidia (276) to be seen in histopathology of invaded human tissues and bodily fluids. Also, the *Paecilomyces* species seen in medical mycology share a tendency to associate ecologically with manufactured products, especially creams, lotions, cosmetics, plastics, and diagnostic materials containing antifungal inhibitors.

1. Unidentified *Paecilomyces* in Opportunistic Pathogenesis

Clinically important members of the genus *Paecilomyces*, like *Aspergillus* species, are among the easiest opportunistic molds for moderately mycologically specialized laboratory staff to identify, especially since the very distinctive *P. variotii* and *P. lilacinus* are overwhelmingly predominant. On the other hand, mycological trainees invariably find them difficult to distinguish from *Penicillium* species, especially members of *Penicillium* subgenus *Biverticillium* and related species currently classified in *Geosmithia*, as well as *Penicillium janthinellum* and similar species with extended collulas. Therefore, at least since the publication of *Compendium of Soil Fungi* by Domsch et al. in 1980 (74), there has been an increasingly clear split between competent mycology generating species-level reports for *Paecilomyces* and unreliable mycology generating genus-level reports, a phenomenon particularly notable in developed countries with good access to literature and training courses. In *Penicillium*, the chance of an incorrect genus-level identification being given even by an unspecialized laboratory is relatively small; moreover, since individual species are very difficult to identify, a laboratory report of *Penicillium* sp. for a non-*P. marneffei* isolate is to be expected unless species identification has been expressly requested. A report of *Paecilomyces* sp., however, from any situation in which the fungus may have been significant, at the very least indicates a reporter who is unaware of the drug susceptibility differences in this genus, a difference that has been widely written about since the 1960s. Such reports, then, especially in publications, clearly signal suboptimal mycological awareness and must be treated with caution. Review literature should not accept such reports at face value. For example, a recent ''lung infection with paecilomyces [sic] species'' in a child with chronic granulomatous disease gives no mycological substantiation, nor are fungal elements in histopathology mentioned for the biopsy sample that grew the organism (277). Although such a report may be well based, it does not as published rule out: (1) growth of a *Paecilomyces* species as a respiratory contaminant from a cultured biopsy (*P. variotii* in particular is a common contaminant from respiratory sources) or (2) the common misidentification of a *Penicillium* species occurring either as a contaminant or as an etiologic agent. Close reading of a recent report on ''disseminated paecilomycosis'' of a dog reveals isolation of a fungal colony described as ''dark green,'' a color not found in any *Paecilomyces* species (except the colony reverse of *P. carneus*, a species not known as a mammalian pathogen), but common in *Penicillium* (278). Possibly the authors may be unconventionally indicating the vivid olive-brown of *P. variotii*, which has caused similar cases. Serum and bone marrow are stated to have tested positive for anti-*Paecilomyces* immunoglobulin G in an exoantigen procedure, but as is common with such seroidentification reports, the authors do not state if the case isolate or a reference isolate was used as the source of exoantigens. Use of the case isolate tends to confirm that the isolate was etiologic, but does not confirm its identification. In any case, since the genus *Paecilomyces* is heterogeneous, no single reference isolate could legitimately be used to confirm or disconfirm genus identification using such procedures.

As can be seen, the heterogeneity of the genus and the low reliability of most genus-level reports make a compilation of diseases caused by *Paecilomyces* sp. all but uninformative. The present author, therefore, has not attempted such a compilation, and recommends that since there are now over 50 reports in the literature based on identified *Paecilomyces* spp., these more informative reports should be focused on as the best information sources about the pathogenicity of these organisms.

Identification (Table 6).   *Paecilomyces* in its current conception includes fungi forming brushlike clusters of phialides (penicilli) producing conidia in interconnected chains. These structures differ from those of *Penicillium* species by being mostly widely divergent (splayed apart) at the tips rather than tending to bend to form nearly parallel clusters as is typical of *Penicillium*. In addition, *Paecilomyces* phialides typically have an inflated base and an extended, tapered neck (collula, Latin for ''little neck''), whereas *Penicillium* phialides are inflated into a bottlelike shape at the base, midregion, and usually even close to the apex, with just the final 1 to 2 (rarely 4 to 5) μm extended as a significantly thinner

Table 6 Laboratory Characters of the Medically Important *Paecilomyces, Acremonium, Cylindrocarpon,* and *Fusarium* Species

Species	Conidia	Typical colony appearance	Chlamydospores	Special features
Paecilomyces variotii	Fusoid to ellipsoidal, in dry, connected chains	Olive-brown, dusty	+	Thermotolerant
P. lilacinus	Ellipsoidal, in dry, connected chains	Dusty vinaceous pink; dusty or cottony	–	
Acremonium alabamense	Pyriform; in chains	Buff; felty to powdery	–	Obligately thermophilic; growth poor at 25°C, good at 37°C
A. blochii	Subglobose; in long chains or heads	Whitish; moist with bundles of aerial mycelium	–	
A. falciforme	Curved, often two-celled, in heads	Grey-brown surface, purple reverse; wet-looking with sparse aerial hyphae or tufts	+	Long, unbranched, often multicelled conidiophores
A. kiliense	Long-ellipsoidal to cylindroidal, in heads	Pale greyish-brown, sometimes orange-pink; flat or wrinkled; bald or slightly tufted	+	Colony reverse brown after 3 weeks on Sabouraud agar
A. potronii	Ovoidal, in heads	Pale to pale salmon; smooth to slightly granular	–(+)	Thornlike phialides
A. recifei	Cashew-shaped, in heads	Whitish, pale yellow-green or pinkish; smooth to thinly felted or tufted	+/–	
A. spinosum	Subglobose, finely warty to spinulose, in heads or chains	Pale to orange or red-brown; reverse pale to purple-brown; flat to wrinkled or ropy	–	Thornlike phialides

Table 6 Continued

Species	Conidia	Typical colony appearance	Chlamydospores	Special features
A. strictum	Long-ellipsoidal to cylindroidal, in heads	Pale to salmon or pinkish-orange; flat and slimy to very thinly cottony	−	
Cylindrocarpon cyanescens	Small, ovoid, in chains	Pale, with brown to black reverse and blue soluble pigment; clumped; glabrous to velvety	+	Colony small, heaped, mostly composed of chlamydospores
C. destructans	Macroconidia 1- to 3-septate; straight or slightly curved; mostly 20 to 40 μm, microconidia +	Reddish-brown surface and reverse; felty to cottony	+	
C. lichenicola	Macroconidia mostly three- to five-septate; straight with truncate base; mostly 18 to 40 μm; microconidia-, but some underdeveloped one-celled conidia formed	Whitish to pale brown surface, sometimes mixed with red-purple; reverse brown; cottony often with a frothy appearance	+	
Fusarium chlamydosporum	Macroconidia rare; three- to five-septate; microconidia predominant; spindle-shaped, mostly one- to two-celled; "mesoconidia" up to five cells formed (see description)	Ochraceous surface, carmine to red-brown reverse; cottony	+	Many phialides "polyblastic," with multiple openings on separate toothlike protuberances, each giving rise to a single blastoconidium ("mesoconidium")

F. coeruleum	Macroconidia predominant; relatively broad and blunt; mostly 3 (-5) septate; microconidia −, but some underdeveloped one-celled conidia formed	Beige to purple surface; ink-blue reverse; felty	+	Phialides monophialidic; long and thin, to 25 μm
F. dimerum	Macroconidia predominant, short (7 to 11 μm long), pointed, mostly one-septate; microconidia −, but some underdeveloped one-celled conidia formed	Orange, slimy, flat	+	Phialides monophialidic; often relatively short and bottle-shaped
F. incarnatum	Macroconidia predominant; mostly 3 (-5) septate; microconidia sparse; mostly 1 (-2) celled, intergrading with macroconidia, spindle-shaped	Brownish surface; usually ochraceous reverse; cottony	+	Many phialides "polyblastic," with multiple openings on separate toothlike protuberances, each giving rise to a single conidium
F. napiforme	Macroconidia mostly (3-)-5 septate; microconidia in short chains or heads, ovoid to turnip-shaped	Whitish to pale purple surface; pale to deep purplish or vinaceous reverse; cottony	+	Phialides are monophialides, subulate (awl-shaped; i.e., moderately long and tapered), and 14 to 31 μm long
F. nygamai	Macroconidia mostly three-septate; pointed; microconidia in short chains or heads; ovoid to spindle-shaped with truncate bases	Whitish to pale purple surface; pale to deep purplish or vinaceous reverse; cottony	+	Phialides are mostly monophialides, subulate (awl-shaped; i.e., moderately long and tapered), 13 to 40 μm long

Table 6 Continued

Species	Conidia	Typical colony appearance	Chlamydospores	Special features
F. oxysporum	Macroconidia mostly four- to-five-septate; pointed; microconidia in heads; el- lipsoidal to sausage- shaped	Whitish to pale purple sur- face; pale to deep pur- plish or vinaceous re- verse; cottony	+ (or scarce)	Phialides are monophialides, flask-shaped, (i.e., some- what inflated and short), especially in aerial myce- lium; 8 to 14 μm long
F. proliferatum	Macroconidia seldom seen; pointed; three- to five- septate; microconidia in long chains or in heads; club-shaped with truncate bases	Whitish to pale purple sur- face; pale to deep pur- plish or vinaceous re- verse; cottony	–	Phialides are monophialides at first, then proliferate as polyphialides, 11 to 32 μm long
F. sacchari	Macroconidia seldom seen; pointed, three-septate; mi- croconidia in heads; ellip- soidal, ovoid, or sausage- shaped	Whitish to pale purple sur- face; pale to deep pur- plish or vinaceous re- verse; cottony	–	Phialides are monophialides at first, then proliferate as polyphialides; 17 to 30 μm long
F. solani	Macroconidia mostly 3 (-5), septate; relatively broad and blunt; microconidia in heads; ellipsoidal to nearly cylindrical or curved	Whitish to medium brown or red-brown surface; pale to brown to red- brown or blue-greenish reverse; felty	+	Phialides are monophialides, distinctively long and thin, especially in aerial mycelium, 15 to 40 μm long
F. verticillioides	Macroconidia seldom seen; pointed; three-to-five-sep- tate; microconidia in long chains (or in heads on in- appropriate media); club- shaped with truncate bases	Whitish to pale purple sur- face; pale to deep pur- plish or vinaceous re- verse; cottony	–	Phialides are monophialides subulate (awl-shaped; i.e., moderately long and ta- pered), 11 to 32 μm

collula. These morphological subtleties are often greatly clarified by examining the color of conidial masses formed on top of the colony. *Paecilomyces* never has any blue or dark green color in its conidial masses; the closest it comes is the golden olive-brown of *P. variotii* or the bright yellow-green of *Paecilomyces viridis* or *P. leycettanus*. On the other hand, conidial masses of *Penicillium* species are never pink, purple, reddish, or white, as may be seen in *Paecilomyces* species. (Very importantly, however, the colors of the mycelium, exudate, and soluble pigments must be distinguished from the conidial mass colors.) It is important to note that since Sabouraud agar often represses conidiation in some members of both genera, this distinction should be made on potato dextrose agar, modified Leonian's agar, or an equivalent conidiation-promoting medium. There are a few common *Penicillium* species, such as the citrus fruit pathogen, *P. digitatum* (often seen from hospital samples because of consumption of fruit by patients, staff, and visitors), that are similar in color to *P. variotii* and must be distinguished micromorphologically. The clinically isolated *Paecilomyces* species most likely to show some compressed phialidic clusters is *P. lilacinus*, which always shows pink conidiation except in a few strongly host-adapted isolates (in which the pink conidia are present but are too sparse to see en masse) or on richer Sabouraud agar formulations, where overgrowth of white mycelium and secretion of brownish reverse pigments may determine the colony's appearance.

***Paecilomyces {aff. Byssochlamys?} variotii* Bain.** This species is well established as an opportunist, most commonly in connection with implanted artificial devices such as catheters, grafts, shunts, and breast implants. Although rare overall, *P. variotii* infection is regularly reported from continuous ambulatory peritoneal dialysis patients (279, 280). Catheter-associated fungemia may also occur (281). Soft tissue invasions of immunocompromised patients may occur (e.g., an infection in the heel of an 8-year-old boy with chronic granulomatous disease) (282). Another case connected with chronic granulomatous disease manifested as multifocal osteomyelitis (283). Endogenous endophthalmitis has been reported (284). Endocarditis was reported several times in the 1970s and early 1980s (e.g., Ref. 285). Another regular source of etiologically significant isolation is from sinusitis (286). A significant number of cases of disseminated infection from dogs without known immunodeficiencies have been reported (287–290). In some, but not all cases, yeastlike elements (290), sometimes within giant cells (288, 289), have been prominent in histopathology. Mycelial elements may be seen in other parts of the body (288, 290), or in some cases (287) are the only structures seen. In some of these canine cases (288, 290) the etiologic agent was not identified at the species level, but sufficient information was given to allow the identification to be reasonably inferred.

A case of onychomycosis featuring linear melanonychia of both hallux nails showed pigmented conidia in direct microscopy as well as rare filaments, and

grew *P. variotii* heavily in culture (291). No other fungus was grown. The etiologic status of *P. variotii* is thus strongly suggested, although additional isolation
attempts in order to elucidate a possible concomitant dermatophyte would be
ideal in such investigations. *P. variotii* onychomycosis was alleged by Contet-
Audonneau et al. (48), but the record is not accepted for the reasons mentioned
above in connection with *Aspergillus glaucus.*

The fungus is also one of the most common contaminants in medical isolations, and is routinely seen from all nonsterile body sites as an insignificant
growth.

In nature *P. variotii* has been isolated from many types of warm soil and
plant litter (74). The fungus biodegrades many stored and preserved products,
such as foods, ''optical lenses, leather, various chemical solutions, photographic
paper, synthetic rubber, creosoted wood, mouldy cigars and ink'' (74). An affinity
for oils, cosmetic creams, and pharmaceutical emulsions has been noted (292).
The major mycotoxins produced by *P. variotii* are patulin and viriditoxin (111).

Description.   Colonies are variable in growth rate, mostly fast-growing
(30 to 70 mm in 7 days at 25°C, most often >45 mm), heavily powdery and
yellow-brown to olive-brown. The reverse is pale.

Conidiophores are mostly penicillate; that is, consisting of broomlike
branching structures with whorls of two to seven phialides on most branch tips.
The branches themselves may be solitary terminal or side branches, but are often
grouped in whorls (Fig. 59). Subterminal branches may appear in larger conidial
structures as a distinct whorl of *Penicillium*-like metulae. The apex of the branch
bearing these metulae in atypical isolates may become so swollen that conidial measurement may be needed to completely rule out an atypical *Aspergillus
terreus.* [Compare relatively swollen structures depicted by Samson et al. (111,
p. 170)]. Occasional solitary phialides may be seen attached laterally on conidiophore branches, something that would be very unusual in *Penicillium. Phialides*
are mostly divergent, with a basal swelling and a prolonged, thin, tapered to
nearly parallel-sided neck (collula). They are 12 to 20 μm long. *Conidia* are borne
in interconnected, long, dry chains that tend to an unusually great extent to remain
intact even in squash microscopic mounts. They are smooth-walled, ellipsoidal
to fusiform (spindle-shaped), variable in size, and mostly 3.0–5.0 × 2.0–4.0 μm.
Chlamydospores are present in submerged mycelium. They are mostly subglobose, with discernibly thick, brownish walls, solitary or in chains, and 4 to 8 μm
in diameter (Fig. 60). Most isolates grow well at 45°C and may grow up to 50°C.

The species should be distinguished from the much less commonly seen
Thermoascus crustaceus, which begins growth as its *P. variotii*-like *Paecilomyces* anamorph and then forms cleistothecia. It lacks chlamydospores. The presence of these structures can therefore be used to rule *T. crustaceus* out in examinations of young *P. variotii* cultures. The converse should not be inferred (i.e.,

Figure 59   *Paecilomyces variotii,* conidiophore.

immature cultures should not be identified as *T. crustaceus* simply because chlamydospores are not detected).

Paecilomyces {*aff.* unknown} *viridis* Ségr. & Sams.   *P. viridis* has been isolated on several occasions from a systemic infection of *Chamaeleo lateralis,* a chameleon from Madagascar (292, 293). It is noteworthy as a dimorphic infection, in which yeastlike cells are produced in infected tissue, particularly the liver and the spleen. No filaments are seen in the infected host.

A fungus well connected with endophthalmitis by Rodrigues and MacLeod (294) in a boy who had been struck in the eye by a nail was identified at the CDC as *P. viridis.* A description in the case report, however, gives the colony color as yellowish-brown, not the bright yellow-green of *P. viridis.* In addition, the conidia are described as elliptical and appear in a photograph as fusoid, and their stated measurements also agree with *P. variotii,* not with the smaller, glo-

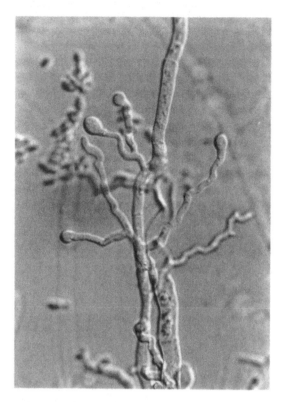

Figure 60   *Paecilomyces variotii*, chlamydospores.

bose to subglobose conidia of *P. viridis. P. viridis* appears only to be known from the chameleon infections mentioned above, while *P. variotii* is a widely distributed fungus well known as an opportunist of mammals. This case report, then, clearly applies to *P. variotii*. This reidentification has been noted previously by Chandler et al. (295) and is mentioned in a review of *Paecilomyces* mycoses by Castro et al. (274).

Description.   Colonies are slow-growing (20 mm in 7 days at 25°C), initially pale, then becoming yellow-green with formation of conidia. The reverse is yellow to yellow-brown.

Conidiophores are not penicillate but rather verticillate; that is, consisting of upright stalks giving rise to successive whorls of conidia at nodes near each septum on the ascending branch. The apex of the conidiophore as well as the tips of occasional side branches generally terminate with a longer than usual phialide, usually emerging from a whorl of laterally arranged, shorter phialides.

Phialides have a globose base and a thin, tapered neck (terminal phialides more evenly tapered) 4 to 9 μm long. *Conidia* are smooth-walled, globose to subglobose, and 2.5 to 3.2 μm in diameter. *Chlamydospores* are absent.

Despite the green conidial color of this species, which led to its inclusion in this chapter subsection along with *P. variotii*, the verticillate conidiophores and the affinity for reptilian disease suggest a closer biological affinity with *P. lilacinus*.

III. *HYPOCREALES*

The best-known *Hypocrealean* fungi in the clinical laboratory are the *Fusarium* and *Acremonium* species (Table 6). They are anamorphs of teleomorph genera that are mainly only seen in agricultural, ecological, or biodiversity studies, such as *Gibberella, Nectria, and Bionectria*. The order *Hypocreales* contains three families that intersect with clinical mycology; namely the *Hypocreaceae* (common anamorph *Trichoderma*, teleomorph *Hypocrea*), the *Nectriaceae* (common anamorphs *Fusarium*, *Cylindrocarpon*, teleomorphs *Nectria*, *Gibberella*), and the family *Bionectriaceae* (common anamorphs *Acremonium*, teleomorph *Emericellopsis*). Although typical Acremonia are Hypocrealean, a number of *Acremonium* species are unrelated; for example, *Acremonium alabamense* is the anamorph of a member of the genus *Thielavia* of the family *Sordariaceae* (296). For convenience, and since the relationships of many *Acremonium* species are not yet well known, all species known from human and animal disease will be treated in this chapter.

Closely related to the *Hypocreales* is the family *Clavicipitaceae*, often treated as the order *Clavicipitales*, within which the best-known genera are the plant pathogen *Claviceps* and the insect pathogen *Cordyceps*. Phylogenetic studies have shown that this group appears to arise as an evolutionary offshoot within the family *Hypocreaceae*, making the latter a heterogeneous evolutionary branch (or to use the technical adjective from phylogenetic systematics, paraphyletic). Taxonomists have not yet reached a consensus about whether or not they must break up such a heterogeneous branch into small branchlets with new names, as the core cladistic philosophy of Willi Hennig (297) argues; prominent evolutionists such as Ernst Mayr (298) and fungal taxonomists such as Seifert et al. (299) have argued against this. Although a few miscellaneous Clavicipitalean anamorphic fungi such as *Beauveria bassiana* have proved to have some medical importance as marginal opportunists, the main intersection of this group with medical mycology appears to be *Paecilomyces lilacinus* and closely related species. (Since this is not a book on toxic fungi, the famous toxic effects of *Claviceps purpurea*, the ergot fungus, and of its artificially derivativized metabolite LSD [lysergic acid diethylamide] will not be discussed, although they are certainly considered medically important.) Since the first part of this chapter ended with Eurotialean *Paecilomyces*, the next part will begin with the Hypocrealean/Clavicipitalean *Paecilomyces*. Note that some Clavicipitalean genera, such as *Beauveria* and *Engyodontium*, are dealt with in Chap. 11 rather than in the current chapter.

Hypocrealean and related medically important fungi have some common factors making the above higher-level taxonomic information useful. Members of this group of fungi have a strong tendency to be resistant to both polyene and azole antifungal drugs, and have also shown resistance to many developmental agents, such as echinocandins and pneumocandins (300, 301). Unlike most Eurotiales, in which conidiation is stimulated by emergence of the fungus through the water–air interface, the conidiation of Hypocrealean fungi typically occurs both in submersion and upon exposure to air. The present author and colleagues pointed out in a 1988 review (302) that this explained the facility of isolating these fungi from blood cultures in disseminated infections, a situation quite different from that found in connection with aspergilloses. It concomitantly explains the rapid seeding of these infections to numerous capillary beds, especially in the skin, a process that gives rise to the widespread ecthyma gangrenosumlike lesions that characterize disseminated *Fusarium* infections. Subsequently, several studies by Wiley Schell, John Perfect, and collaborators have established that this subaqueous conidiation may readily be seen in in vivo materials, potentially allowing rapid identification of Hypocrealean infections by differential histopathology (303–305). The pathologist may profitably learn to recognize the typical phialides and conidia of these fungi in tissues and bodily fluids. Yeastlike states composed of self-replicating phialoconidia may also uncommonly occur (303). It should be noted that, as mentioned above, one Eurotialean fungus, *P. variotii*, may also form copious phialoconidia in tissue (276). Another common feature of many of the medically important Hypocrealean fungi is an ecological association with growth in fluids and with dispersal in aqueous aerosols. Many, though not all, of these fungi bear conidia in sticky heads. This makes them less likely to produce opportunistic infection by contacting the lungs via dry, airborne dust, but may make them more likely to produce opportunistic infection after exposure via tap water, for example, or via contact lens cleaning fluid. An additional commonality that is noteworthy in veterinary mycology is that whereas Eurotialean fungi, especially thermotolerant Aspergilli, are notoriously hazardous opportunistic pathogens of birds, Hypocrealean fungi are notoriously hazardous opportunists of reptiles. (See below.)

Since teleomorphs are rarely seen in connection with Hypocrealean species in the medical laboratory, they are not discussed for most groups below, although an exception is made for *Neocosmospora vasinfecta*, in which the teleomorph does normally occur in cultures.

A. *Paecilomyces* Bain. (Hypocrealean Part—Pink, Purple, and White Conidial Species)

For a general discussion, see Sec. II. C (*Paecilomyces*) above.

1. *Paecilomyces* {aff. *Cordyceps*} *fumosoroseus* (Wize) A.H. Brown & G. Smith

This fungus has been linked to pulmonary infection in a captive giant tortoise by Georg et al. (13). No identification characters were given. Soil mycologist W. B. Cooke was credited with identification. Three *Paecilomyces* species, *P. fumosoroseus*, *P. lilacinus*, and *P. marquandii*, are primarily distinguished, especially in keys (74, 292) by shades of pink, vinaceous, or purple conidial mass color. The normal "greyish vinaceous" color of *P. lilacinus* on many media readily suggests the described pinkish color of *P. fumosoroseus*, so confusion between these organisms is predictable. The characteristic inability of *P. fumosoroseus* to grow at temperatures above 30°C has not been widely publicized since the 1957 monograph of Brown and Smith (305). By contrast, *P. lilacinus* grows to 38°C (74). As *P. lilacinus* has emerged as a pathogen since the 1970s, it has become much more widely known and more likely to be identified correctly; however, the Georg et al. study was published prior to this emergence. Since *P. lilacinus* is a regularly seen pathogen of tortoises and turtles while *P. fumosoroseus* is not, it would be ideal to be able to reconfirm the identification of the isolate Georg et al. (13) examined. The isolate, however, has not been conserved in a collection with a publicly available database.

A disseminated infection attributed to *P. fumosoroseus* in an Indiana cat began as a paw lesion before extending to nasal tissue and the liver (306). Pathology showed many structures described as pseudohyphae and yeast forms, some in macrophages. Histoplasmosis was excluded by repeated cultivation of the *Paecilomyces*. An isolate was deposited in the ATCC and at the CBS, and in both locations has been reidentified as *P. lilacinus* (ATCC 52586; CBS 754.96).

P. fumosoroseus is a generalist insect pathogen also known from leaf litter and other decaying organic substances, including such foods as butter.

Description. Colonies grow 25 to 35 mm in 7 days. They are powdery to cottony, sometimes producing upright, pink synnemata, especially after extended incubation, which are white at first and later pale pink. The reverse is pale to yellow or pinkish.

Conidiophores arise from aerial hyphae or from the substrate. They are often erect, thin (approximately 2 μm), smooth-walled, and scarcely distinguished in aspect from ordinary hyphae. They are of indeterminate length, but are usually up to 100 μm, and are sometimes bundled into erect branched or unbranched synnemata up to 3 cm long and 0.4 cm broad. They are septate, with a terminal whorl of phialides as well as lateral structures that diverge at the septa and that consist either of whorls of four to six phialides or of whorls of short side branches, each bearing a terminal whorl of up to six phialides, or less commonly of single side branches, single phialides, or mixed whorls containing both phialides and side branches. *Phialides* are divergent. They are globose to ellipsoidal at the base, but with a thin, tapered neck (collula). Overall they are 5.7–8 (−18) × 1–2 μm. Very inflated phialides on synnemata may be up to 3.5 μm wide. *Conidia* are borne in interconnected chains. They are smooth-walled, hyaline to pale pink, ellipsoidal to fusiform (spindle-shaped), variable in size, and mostly 3–4 × 1–2 μm (taking into account slight variations occurring on different media).

As noted above, the inability of this fungus to grow above 30°C should be confirmed when an identification is made.

2. Paecilomyces {aff. Cordyceps} javanicus

An endocarditis case involving a porcine aortic valve heterograft and a subsequent cerebral embolism was meticulously documented by Allevato and coinvestigators (307–309). The fungus was identified by medical mycologist John Rippon and a description was given. This description, however, contained phialidic and conidial measurements clearly copied from the published description of Samson (292) and did not give the measurements of the isolate actually studied. Samson's measurements were based on size ranges seen in four isolates studied, and would not be duplicated to the first decimal place by any individual isolate. The isolate does not appear to have been deposited in a public collection. As authentic *P. javanicus* has only been isolated from insects in the Old World tropics, and as the description given by Allevato et al. (307) is a generic account that does not characterize the isolate seen (except in the vague sense that it is presumed to be concordant with the general description), this report must be regarded as unconfirmed. Allevato et al. (307) state that *P. javanicus* "shows poor to no growth at 37°C," whereas the monographic study of Brown and Smith (305) states that growth at this temperature is nil, and even growth at 30°C is restricted. This suggests that Allevato et al. possessed an isolate growing, albeit poorly, at 37°C, deviating significantly from the authentic isolates of *P. javanicus* available in collections.

P. javanicus has been isolated on only a small number of occasions from parasitized insects, mostly lepidopteran pupae, but also the economically important coffee berry borer beetle, *Hypothenemus (Stephanoderes) hampei* Ferr. (Coleoptera: Scolytidae), from Indonesia and Ghana. Although it seems unlikely that *P. javanicus per se* has caused disease in a mammal, the description is given for comparison with similar isolates that may be obtained.

Description. Colonies are 25 to 30 mm in 7 days, deeply matted, and somewhat ropy (funiculose). They are white at first, and later cream or with heavy conidiation pale blue-grey. The reverse is pale to yellow, sometimes with blue-grey zonation on Czapek agar.

Conidiophores arise from aerial hyphae or from the substrate. They are often erect, thin (approximately 2 μm), smooth-walled, and scarcely distinguished in aspect from ordinary hyphae. They are of indeterminate length, but usually up to 50 μm. They are septate, with a terminal phialide or whorl or phialides as well as lateral structures that diverge at the septa and that consist of whorls of two to three phialides, of single short side branches bearing a terminal whorl of up to five phialides, or of single phialides. *Phialides* are divergent, cylindrical at the base but with a thin, tapered neck (collula), and overall 8–14 × 2–2.8 μm. *Conidia* are borne in interconnected chains. They are smooth-walled, hyaline, ellipsoidal to fusiform (spindle-shaped), variable in size, and mostly 4.0–7.4 × 1.0–1.8 μm (taking into account slight variations occurring on different media). Growth is "normal at 24°C, restricted at 30°C, and nil at 37°C" (305). It is very important that case isolates identified as this species and deviating by showing growth at 37°C be deposited with a professional culture collection.

3. *Paecilomyces {aff. Cordyceps} lilacinus* (Thom) Samson
    (Common Synonym in Older Literature:
    Penicillium lilacinum)

This fungus has a very distinctive set of roles in opportunistic disease, mediated by such factors as moderately low virulence (304) and an unusual, multiply resistant drug response pattern (274, 310). In the immunocompromised patient, infections often consist of localized cutaneous or subcutaneous soft tissue lesions, classified by Heinz et al. (304) as "indolent." In a major nosocomial outbreak connected with the use of contaminated skin lotion in the oncology and bone marrow transplant units of a Swiss hospital, nine affected patients had cutaneous lesions, but only one with a very low leukocyte count ($0.13 \times 10^9 \text{ L}^{-1}$) went on to develop dissemination leading to death (310). On rare occasions, soft tissue infections may also occur in immunocompetent hosts (e.g., a reported infection complicated with prepatellar bursitis) (311). Catheter-related fungemia is occasionally reported, sometimes with lung nodules or other indications of disseminated establishment in tissues (312, 313). Leukemia patients have also presented with invasive sinusitis caused by *P. lilacinus* (314), and there has been one case of keratitis progressing to endophthalmitis after surgical intervention in a lymphoma patient (315). A lung abscess in an immunocompetent patient has been demonstrated (316).

Another major association of *P. lilacinus* is with ocular infections in the otherwise healthy patient. Both keratitis (317) and endophthalmitis—usually postsurgical (65, 318)—have been repeatedly reported. Typically, some cases of the latter have been connected with contaminated fluids; for example, a bicarbonate neutralizing solution supplied by a manufacturer of intraocular lens implants (318, 319). *P. lilacinus* cases are unusual in keratitis in responding poorly to pimaricin, the most common polyene drug of choice, which is often effective against related fungi such as *Fusarium solani* (229).

Chronic mycotic sinusitis caused by this organism has been demonstrated (320). In addition, there is one well-demonstrated case of onychomycosis attributed to *P. lilacinus* (321). It should be noted that this cycloheximide-tolerant fungus is one of the most common contaminants from nails in the dermatologic mycology laboratory, and cases of onychomycosis must be stringently confirmed with later consistent repeat isolation. In the published case, this was done on four separate occasions, along with demonstration of compatible fungal elements in direct nail specimen microscopy. A review by Castro et al. (274) lists two earlier studies with inadequately confirmed allegations of onychomycosis caused by *P. lilacinus*. *P. lilacinus* onychomycosis was alleged by Contet-Audonneau et al. (48), but the record is not accepted for reasons mentioned in connection with *Aspergillus glaucus* above.

In veterinary mycology, *P. lilacinus* has a long and extensive history as an agent of opportunistic infections in reptiles, including turtles and tortoises (322, 323) as well as crocodiles (324).

Mammals may also rarely be afflicted (e.g., an armadillo) (325). As mentioned above, a disseminated infection attributed to *P. fumosoroseus* in a cat with no known immunodeficiencies (306) was based on a fungus later reidentified as *P. lilacinus*.

This fungus has been isolated from many soils and decaying vegetative materials, as well as from parasitized insects. In the anthroposphere it colonizes stored nuts and root crops as well as various moist materials, such as the plastics and biomedical solutions mentioned above, with only small amounts of organic material and often with inhibitory additives. In soil it is often remarkably tolerant of exposure to diverse fungicides (74). It is a strong keratin decomposer.

Description. Colonies are 25 to 35 mm in 7 days. They are generally moderately floccose. They are white at first, but soon dusty pinkish to drab vinaceous with conidiation. The reverse is often pale, frequently vinaceous on malt extract agar according to Samson (292), and sometimes brownish on Sabouraud agar.

Conidiophores are variable (Figs. 61–63). The most highly differentiated forms—not always found, especially on Sabouraud and other inappropriate media—arise from the substrate, and are often erect, relatively thick (3 to 4 μm), and 400 to 600 μm long, with yellowish or pale purple-brown, roughened stalks, often bearing several whorls of terminal and subterminal phialides and as well

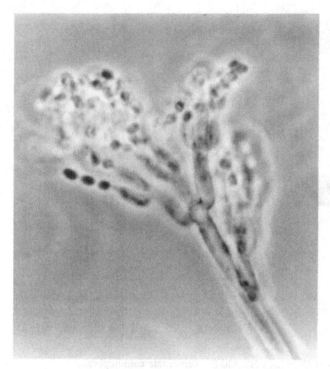

Figure 61 *Paecilomyces lilacinus*, penicillate conidiophore.

as side branches also bearing whorls of phialides. In the most highly developed cases these structures may be arranged in compact, *Penicillium*-like heads composed of a compact whorl of short metulae, sometimes with ramuslike side branches below. In many isolates, however, more typical *Paecilomyces* structures are more commonly seen, with conidiophores bearing several successive whorls of phialides, short, phialide-bearing side branches, and mixtures of phialides and short side branches. Less differentiated conidiophores that are thin and have smooth walls may also be seen and may predominate in some isolates. These conidiophores again tend to have successive whorls of phialides and/or side branches, and may also bear single phialides at some septa. *Phialides* are divergent or with a degree of compacted alignment reminiscent of *Penicillium*, ellipsoidally to cylindrically inflated at the base but tapering to a thin neck (collula). Overall they are 7.5–9 × 2.5–3 µm. *Conidia* are borne in interconnected chains. They are smooth-walled or finely roughened, hyaline, pinkish-purple in mass, ellipsoidal to fusiform (spindle-shaped), and 2.5–3 × 2–2.2 µm. *Chlamydospores* are absent.

Figure 62   *Paecilomyces lilacinus* CBS 430.87, verticillate conidiophore.

This species rather uniformly grows to 38°C (74). *P. marquandii* isolates investigated so far have not grown at 37°C in vitro (74); therefore, temperature tolerance may be helpful in separating these species. As always in biology, however, the possibility of an atypical organism should not be completely ruled out. *P. marquandii*—with its violet color, yellow colony reverse, chlamydospores, and absence of colored or rough-walled conidiophores—is easily distinguished from *P. lilacinus* using other characters where indicated. It should be noted, however, that *P. lilacinus* isolates with little or no formation of large, colored, rough conidiophores are not uncommon from medical isolations. Some degree of in vivo degeneration similar to that seen with *A. fumigatus* may occur, causing an attenuation of the formation of distinctive characteristics.

4.   *Paecilomyces {aff. Cordyceps} marquandii* (Massee) Hughes

Harris et al. (326) reported this fungus from a dermal lesion in the leg of an immunosuppressed renal transplant patient. The tissue invasion with compatible

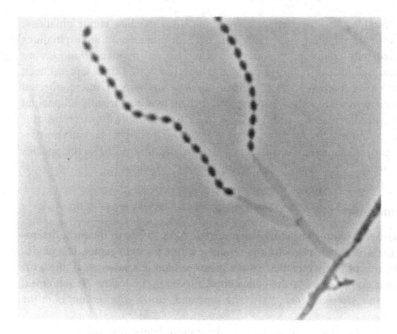

Figure 63 *Paecilomyces lilacinus*, simple conidiophore.

fungal elements was well demonstrated, and the fungus identification was credited to CDC. Since this species does not grow at 37°C in vitro, however, and
since this report comes from a historical period in which the otherwise rather
similar *P. lilacinus* was not yet widely known in medical mycology, this identification must be questioned. No characteristics of the isolate are mentioned by
Harris et al., and the isolate is not deposited in a culture collection with a publicly
accessible database.

This fungus is an insect pathogen and a parasite of mushrooms, especially
Hygrophoraceae, as well as a commonly isolated soil fungus.

Description. Colonies are 18 to 25 mm in 7 days. They are powdery or
floccose, occasionally with small synnemata to 1 cm long. They are white at first
but soon violet to dark vinaceous brown. The reverse is characteristically strongly
yellow, sometimes with yellow soluble pigment.

Conidiophores arise from aerial hyphae or from the substrate. They are
often erect, thin (circa 2.5 to 3 μm), smooth-walled, and 50 to 300 μm. They are
septate, with a terminal whorl of phialides, often also in longer conidiophores
with lateral structures that diverge at the septa and that consist of whorls of either
two to four phialides or short side branches, each bearing a terminal whorl of

up to four phialides, or less commonly of single side branches, single phialides, or mixed whorls containing both phialides and side branches. Single phialides not associated with conidiophores may be seen on aerial hyphae. *Phialides* are divergent and cylindrical to ellipsoidal at the base but with a thin, tapered neck (collula), and overall are 8–15 × 1.5–2 μm. *Conidia* are borne in interconnected chains. They are smooth-walled or finely roughened, hyaline, broadly ellipsoidal to fusiform (spindle-shaped), and 3–3.5 × 2–2.2 μm. *Chlamydospores* are usually present in submerged mycelium. They are globose to ellipsoidal and approximately 3.5 μm in diameter. According to Brown and Smith (305), the species does not grow at 37°C. See the discussion under *P. lilacinus.*

B. *Acremonium*

This group of very simply structured anamorphic fungi is one of the most heterogeneous groups of organisms currently retained within a single genus. The phylogenetic affinities of many members of the genus remain unknown as of this writing, but major components of the genus appear to be in the orders *Hypocreales*, *Sordariales*, and *Microascales* (296). The type species, *A. alternatum*, is in the *Hypocreales*.

Most *Acremonium* species appear to be completely nonpathogenic, and many of those with some opportunistic potential are of very low virulence. They have a relatively strong tendency to cause limited infections even in severely immunocompromised patients (304), in contrast to the many disseminated infections attributed to related *Fusarium* species. Nonetheless, some serious infections do arise. Drug responses may be variable, but many species appear highly resistant to most antifungals, with amphotericin B being the most effective overall (327). As noted above, poor antifungal response is common in Hypocrealean fungi. In vitro–in vivo correlation studies have not been done, and there are several accounts of successful treatment of *A. falciforme* with itraconazole (see below) even though its in vitro minimum inhibitory concentration (MIC) has been measured as 32 μg/ml (i.e., is highly resistant) (69).

1. Unidentified *Acremonium* in Opportunistic Pathogenesis

Clinical cases caused by *Acremonium* species have been thoroughly reviewed by Fincher et al. (328) and Guarro et al. (327). Neither review critically examined documentation of organism identifications, and both replicate some definite or likely errors (e.g., the *Acremonium roseogriseum* identifications discussed below). *Acremonium* isolates from mycetoma are frequently identified to species, especially since the 1970s, and it is now clear that *A. falciforme*, *A. recifei*, and *A. kiliense* are the regularly recurring (though overall very uncommonly seen) agents. In general, other *Acremonium* opportunism, while relatively rare, is regularly seen in connection with keratitis, endophthalmitis, catheter-related perito-nitis, and soft-tissue infection or mycotic arthritis of immunocompromised or physically traumatized patients, and fungemia or disseminated infection of neutropenic and other severely immunocompromised patients. Disseminated in-fections are relatively rare compared to those caused by *Fusarium* species. *Acremonium* spp. are regular causes of superficial white onychomycosis (50). There are a few cases of meningitis or brain lesion, mostly involving known major barrier breaks (e.g., intravenous drug use, spinal anesthetic administration). As with other relatively weak pathogens that are environmentally common, a few cases of infection of high-vulnerability sites, such as prosthetic heart valves, are known (327). Again, however, these infections appear to be less common than those caused by other environmentally common fungi of marginal virulence, such as *Paecilomyces* species. In terms of overall documented incidence, the most significant *Acremonium* infection is keratitis. A brief, useful review of this manifestation has been provided by Kennedy et al. (329).

As the number of medical centers able to identify common opportunistic fungi increases, and as molecular verification techniques become more commonly used, genus-level identifications of *Acremonium* species in connection with case reports become increasingly suspect. A number of recent potentially significant case reports give no genus- or species-level identification information [e.g., recent cases of keratitis after excimer laser photorefractive keratotomy (330) and mixed connective-tissue disease-related infectious esophagitis (331)]. The importance of providing this information may be dramatically seen in cases in which it is present. For example, a recent case report on a severe necrotizing arm tissue infection in a corticosteroid-treated patient (332) ascribes the infection to "*Acremonium* sp." Photographs, however, vividly depict the *Scedosporium* state of *Pseudallescheria boydii*. A parallel "*Acremonium* sp." case from previous literature is discussed under *Fusarium verticillioides* below. As with the identification "*Paecilomyces* sp.," "*Acremonium* sp." in the context of a published case report is currently a confession of inadequate microbiology. Lack of documentation leaves several po-tentially important biomedical assertions of uncertain significance (e.g., an asser-tion that *Acremonium* may cause systemic mycosis in dogs) (333). Of course, how-ever, *Acremonium* sp. will and should be the common laboratory report for acremonia isolated from cases in which they are judged to be insignificant, since species identification by any technique is a nontrivial exercise in this genus.

2. Identification

The distinction of *Acremonium* species from other fungi is considered in detail by Gams (334, 335).

Acremonium species may be readily identified morphologically by their relatively long, narrow phialides (often remarkably long and narrow), which are formed singly or in very simple branching structures, and which bear small, mostly unicellular (often bicellular in *A. falciforme*) conidia either in sticky heads or less commonly in chains. Hyphae are notably thin, usually under 2.5 μm, and hyphae over 4 μm in diameter are generally not found. In *Fusarium*, hyphae are generally relatively broad and diameters over 2.5 μm are common. The colonial growth rate of *Acremonium* species is generally under 35 mm in 7 days at 25°C (often under 20 mm), whereas *Fusarium* species that could potentially be confused generally grow over 50 mm in 7 days, and in most cases exceed 60 mm.

Verticillium species may be superficially *Acremonium*-like. The more complex species have phialides mostly in verticils; that is, in radiating structures arranged like the spokes of a wheel or the supports of a wind-inverted umbrella. Frequently, tall conidiophores may be seen with successive layers of such verticils, reminiscent of the branching pattern of a spruce or of an ornamental Norfolk pine. In the more simply structured *Verticillium* subgenus *Prostrata*, now divided into several genera, phialides tend to be narrow and solitary as in *Acremonium*, but colonies are compact, cushionlike, and deeply, densely woolly. By contrast, *Acremonium* colonies are generally flat to very thinly cottony. *Verticillium* species may sometimes have characters seldom or never found in *Acremonium*, such as crescent-shaped conidia or stalked, often dark-pigmented, multicelled, irregularly rounded chlamydospores (dictyochlamydospores), somewhat reminiscent of *Epicoccum* conidia in shape and associated with substrate mycelium.

Sagenomella species, which are not uncommon as laboratory contaminants, were distinguished from *Acremonium* by Gams (335) because they formed interconnected chains of conidia (conidial chains connecting adjacent conidia with small bridges of cell wall material) in the manner of *Paecilomyces* rather than unconnected chains, as may be seen in some *Acremonium* and *Fusarium* species. Some *Sagenomella* species have teleomorphs in *Sagenoma*, indicating biological relationship with the order *Eurotiales*, not *Hypocreales*.

Some *Acremonium* sequences possibly facilitating identification of some medically important species have been deposited, and ribosomal restriction fragment patterns are given for representative isolates of some species by De Hoog et al. (69). No species, however, has yet been studied molecularly in detail.

3. *Acremonium alabamense* Morgan-Jones Anamorph
   of *Thielavia terrestris* (Apinis) Malloch & Cain

This species was well demonstrated as causing a fatal cerebral infection of an intravenous drug user (336). No identification characters were given, but identification was credited to mycological expert M. R. McGinnis.

The anamorph was described from Alabama pine litter by Morgan-Jones (337). Isolates producing the teleomorph are most frequently isolated from warm soils (e.g., in tropical areas or sun-heated pastures) or from herbivore dung.

Description. This fungus grows poorly at 25°C; the description is based on 37°C growth. Colonies are 76 to 87.5 mm in 7 days. They are floccose to felty, eventually powdery with conidial production, and usually whitish to pale ochraceous on the surface, becoming greyish-buff near the center. The reverse is yellowish-buff near the periphery and reddish-brown near the center.

Conidiophores consist mainly of solitary phialides (Fig. 64). *Phialides* are awl-shaped to nearly cylindrical. They are 8–25 × 1–1.5 (basal width) μm with a thickened ring at the apex and occasionally with percurrent proliferation. *Conidia* are borne in chains or slimy heads. They are smooth-walled, hyaline, obovate, clavate, or pyriform with a truncate base, and 3–6 × 2–3 μm. *Chlamydospores* are not formed.

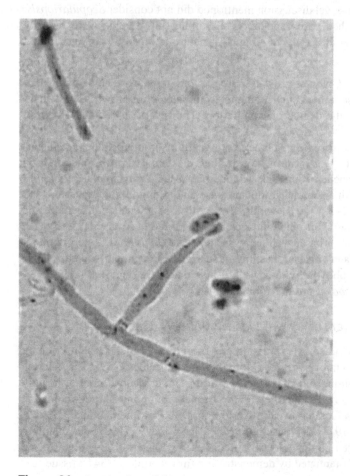

Figure 64 *Acremonium alabamense.*

A. alabamense grows well at 50°C, unlike the rather similar *Acremonium thermophilum* W. Gams & Lacey.

4. *Acremonium {aff.* unknown} *atrogriseum* (Panasenko) W. Gams

This species was reported from keratitis by Read et al. (338). Although etiology was well demonstrated, the identification of the organism was completely unsubstantiated. Discussion ostensibly outlining "differentiation from other, more common fungal organisms" was inappropriately restricted to the genus level without considering species identification. Particularly as this species is readily confused with known opportunistic black *Scopulariopsis* anamorphs (and in fact was originally erroneously described as such an anamorph), this report is not accepted. The genus-level discussion mentioned did not consider *Scopulariopsis*.

Two unpublished cases apparently involving this fungus were recently noted by de Hoog et al. (69). The fungus in nature is mostly from agricultural soils or crop roots.

Description.   Colonies grow 10.5 to 12.5 cm in 7 days. They are powdery and slightly floccose at the center, ranging from pale ochre-brown to brownish-black, according to the degree of conidiation. The reverse is reddish-brown to brown.

Conidiophores arise from submerged mycelium or on prostrate hyphal bundles. They are solitary or in nearly verticillate clusters of up to four, often on short lateral side branches from thick-walled, brown supporting hyphae. *Phialides* are flask-shaped with a somewhat inflated base and a long, thinly tapering apex. They are 8–18 × 2–3.5 (basal width) μm, with a short, pale collarette at the apex. *Adelophialides* (short phialidic necks arising directly from sides of hyphae and not delimited by a basal wall) may be present. *Conidia* are often borne at first in chains that later coalesce as slimy heads. They are smooth-walled, subhyaline to dark grey, obovoid with a weakly apiculate base, and 3.5–4.8 × 1.8–2.1 μm. *Chlamydospores* are not formed.

5. *Acremonium {aff.* unknown} *blochii* (Matr.) W. Gams

This species was obtained by expressing pus from intact nodular lesions seen on extensive areas of skin of two West Bengali water buffaloes (339). The pus was confirmed as containing fungal elements in direct microscopy, and the heavily outgrowing *A. blochii* was identified by the former Commonwealth Mycological Institute (CMI; now Commonwealth Agricultural Bureau International [CABI]).

The original 1911 case description by Bloch and Vischer (340) claiming that *A. blochii* caused gummatous skin ulcers in two relatively severely affected patients was not substantiated by demonstrating fungal etiology (341); tissue sections were negative for invasive fungal elements. The lesions resembled syphilitic ulcers (341). Bloch and Vischer, however, felt that syphilis had been ruled out by a negative Wasserman test.

The habitat of this species is unknown; most isolations have been from medical or veterinary specimens.

Description. Colonies grow 7 to 8.5 mm in 7 days. They are moist, with appressed to tufted aerial mycelium, which are often in ropy bundles, becoming powdery with heavy conidiation on certain sporulation media, such as oatmeal agar. They are usually whitish on both the surface and reverse, whether grown in light or darkness.

Conidiophores consist mainly of solitary phialides (Fig. 65). *Phialides* are spinelike, tapering, and 8–20 (−25) × 1–1.6 (basal width) μm. *Conidia* are borne in long chains or slimy heads. They are smooth-walled, hyaline, subglobose with slightly pointed base, and 2.9–3.3 × 1.9–2.4 μm. *Chlamydospores* are not formed.

6. *Acremonium* {*aff.* unknown} *curvulum* W. Gams.

This fungus was reported without etiologic or mycological substantiation from exogenous endophthalmitis (342). De Hoog et al. (69) pointed out that the similar-

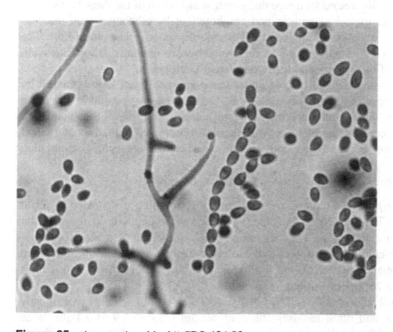

Figure 65 *Acremonium blochii* CBS 424.93.

ity of the well-known opportunist *Acremonium recifei* makes such a record problematical. It is not accepted here.

7.   *Acremonium {aff.* unknown} *falciforme* (Carrión) W. Gams

This fungus uncommonly but regularly causes pale-grain mycetoma in tropical and subtropical areas worldwide (327, 343–345). Successful treatment of such a mycetoma by oral itraconazole alone has been reported in an unusual case involving lesions on the right temporal scalp area of a 70-year-old male (346). Two mycetoma cases involving renal transplant patients have been reported (347, 348), one of which provided sufficient description to allow the identity of the organism to be confirmed (347). In this detailed report, *A. falciforme* was isolated from two arm lesions as well as severely dystrophic fingernails. Direct microscopy to confirm establishment of the organism was done in the skin lesions but not in the nails.

A Kentucky man who had undergone splenectomy and nephrectomy after a diagnosis of renal cell carcinoma developed a disseminated *A. falciforme* infection manifesting as endophthalmitis and diskitis of the lumbar spine (349). No identification characters were given, but the fungus was identified by mycologist Norman Goodman. Itraconazole was again an effective therapy. An endophthalmitis occurring 9 months after penetrating ocular trauma in a Saudi Arabian man was successfully treated with repeated surgical aspiration of the fungal mass, oral ketoconazole, topical natamycin and amphotericin B, subconjunctival miconazole, and an injection of amphotericin B into the anterior chamber (350). The identity of the etiologic agent was confirmed by two unnamed laboratories and substantiated by a morphologically consistent photograph.

A well-documented case of upper gastrointestinal tract lesions caused by *A. falciforme* was reported by Lau et al. (351) in an 11-year-old Hong Kong girl who had undergone bone marrow transplantation to treat severe combined immunodeficiency. The infection was successfully treated with amphotericin B and itraconazole, combined with granulocyte-macrophage colony-stimulating factor.

The habitat of *A. falciforme* in nature is unknown.

Description.   Colonies are slow-growing, reaching a diameter of 30 to 32.5 mm in 7 days at 25°C. They are typically wet-looking with sparse, granular to tufted aerial mycelium. They are initially grey-brown but then becoming grey-violet. The colony reverse becomes violet-purple, most notably on Sabouraud agar. Conidiation may not occur until after 2 to 3 weeks of growth, and is most abundant at 25°C, not forming at all below 20°C (334).

Conidiophores are erect, barely distinguishable from aerial hyphae, mostly unbranched, and often multiseptate, over 50 µm long × 1.9 to 2.1 µm wide (Fig. 66). *Phialides* simply consist of the terminal cells of the conidiophores

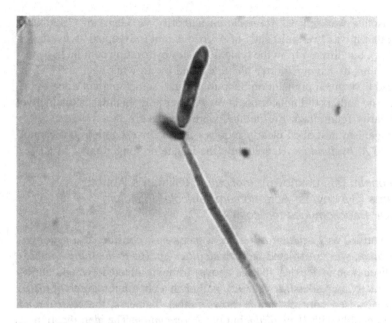

Figure 66 *Acremonium falciforme* CBS 101427.

described, and are not otherwise differentiated. *Conidia* are borne in slimy heads. They are smooth-walled, hyaline, and aseptate or with a single septum (rarely two- or three-septate). They are usually broadly sausage-shaped to broadly crescent-shaped with a rounded apex and a slightly tapered base terminating in a flat, broad detachment scar, mostly 7.8–10 × 2.7–3.3 μm. An isolate from Vanuatu depicted by McCormack et al. (343) has mostly two-celled conidia with hooked and bluntly pointed, *Fusarium*-like apices, in contrast to the more rounded apices depicted and described by Gams (334). Some similar conidia are also depicted by De Hoog et al. (69). *Chlamydospores* are numerous, some intercalary but mostly terminal on short side branches. They are often in chains. They are rounded, brownish at maturity, smooth, thick-walled, and 5 to 10 μm long.

This fungus may appear *Fusarium*-like in culture, particularly in its formation of two-celled, curved conidia and purple colony colors, but was excluded from this genus by Gams (334) on the basis of its slow growth rate and the broad attachment of conidia to the conidiophore.

8. *Acremonium* {*aff.* unknown} *hyalinulum* (Sacc.) W. Gams

A German shepherd dog in South Africa with severe osteomyelitis was found on autopsy to have disseminated mycosis caused by an *Acremonium* species

(352). The isolate was sent to *Acremonium* authority W. Gams of CBS, who commented that it was "reminiscent" of *A. hyalinulum* (miscopied as *hyalinum* by the authors), but differed from the typical isolates by forming conidia in heads instead of chains, by having shorter phialides, and by growing well at 37°C. It also produced dark green pigment on Sabouraud agar, something not done by *A. hyalinulum*. The authors did not comment on whether polyphialides, a distinctive feature ordinarily found in *A. hyalinulum*, were produced.

The organism described clearly requires further investigation and cannot be ascribed to *A. hyalinulum*. Regrettably, the isolate no longer exists at CBS.

9. *Acremonium {aff.* unknown} *implicatum* (Gilman & Abbott) W. Gams (Synonyms: *Acremonium terricola, Fusidium terricola, Paecilomyces terricola*)

A fungus identified as *Fusidium terricola*, a name now considered a synonym of *A. implicatum*, was co-isolated with *Penicillium sp.* (or *Penicillium spinulosum*—see discussion under that species above) from an infected eye and attributed pathogenicity by Anderson et al. (92). Although a photomicrograph of material from the excised corneal button shows fungal filaments, these are broad filaments compatible with *Penicillium* but not *Acremonium*. The identification of *F. terricola* is also unsubstantiated. At the time of the study, members of a number of *Acremonium* species with catenulate conidia may have been identified under this name. *A. implicatum* differs from *A. blochii* and *A. alabamense*—known opportunists producing conidia in chains—by having fusiform (narrowly spindle-shaped, with two pointed ends) conidia rather than conidia with wider shapes, such as subglobose, clavate (club-shaped), or pyriform (pear-shaped). Conidia are pale, not colored like those of *A. atrogriseum*. Polyphialides are not produced, in contrast to *A. hyalinulum*. There are several species closely similar to *A. implicatum*.

10.   *Acremonium {aff. Emericellopsis} kiliense* Grütz

Common synonyms in older literature, and less commonly used synonyms connected with well-known case reports include *Cephalosporium acremonium* (*pro parte*; name also applied to *A. strictum* and other species), *Cephalosporium infestans, Cephalosporium madurae*, and *Hyalopus bogolepofii sensu auct.* (in the sense used by some authors).

This species has long been known as an agent of pale grain mycetoma, mainly in tropical areas. Cases are relatively uncommon (327), but may be quite extensive and severe (345).

Some serious systemic infections have been well documented, including a disseminated infection in a myeloma patient who had undergone bone marrow transplantation (353). The infection was seen as septicemia and catheter colonization combined with papular skin lesions. The fungus was also repeatedly isolated from stool. Mycotic esophagitis causing stenosis was described from an otherwise healthy 11-year-old boy by Simon et al. (354); the identity of *A. kiliense* was well established. Two cases of CAPD-related peritonitis were reported by Lopes et al. (355) from Brazilian patients. The published descriptions and photo of the isolates are nonspecific, not ruling out *A. strictum.*

A case of cranial osteomyelitis subsequent to automobile accident trauma was attributed to *A. kiliense* by Brabender et al. (356). The main affected area, a forehead wound, grew only *Staphylococcus aureus* on culture; however, *A. kiliense* was cultured from three of four subgaleal abscesses remote from the wound. Several direct microscopic and histopathologic examinations disclosed no trace of micro-organisms in any of the samples taken. The patient responded to combined antifungal and antibacterial therapy plus surgery. No diagnostic characters were mentioned for the isolates, but a commercial reference center was credited with identification. Despite impeccable investigation of the case etiology on the part of the attending physicians, this case must still be considered suggestive rather than definitively demonstrated.

A. kiliense was definitively implicated in a fatal case involving a prosthetic heart valve vegetation that embolized and gave rise to a brain abscess in an immunocompetent patient (357). The isolate's identification was confirmed by L. Ajello of CDC and a consistent photograph was published.

A dramatic outbreak of postsurgical endophthalmitis caused by *A. kiliense* was traced by CDC investigators to a building's air system humidifier in an ambulatory center for cataract surgery (358, 359). Patients contracting the infection were operated on early in the day shortly after the air-handling system had been turned on. One apparently cured case from this outbreak recrudesced later as keratitis (360). An earlier case of endophthalmitis secondary to cataract surgery yielded cultures of a fungus identified as *Hyalopus bogolepofii* (Vuill.) Simões Barbosa (361). The relatively detailed description appears to be consistent with *A. kiliense*, the fungus most commonly called by this now disused name, despite a crude line drawing that appears to misrepresent the described and photographed long-ellipsoidal conidia as falcate. A recent case of keratitis progressing to endophthalmitis was attributed to a fungus cited as a "probable *A. kiliense*" by Wang et al. (362). No identification characters were given. Canine keratoconjunctivitis caused by *A. kiliense* was well demonstrated by Mendoza et al. (363). The causal isolate is in ATCC. A well-demonstrated case of keratitis connected to contact lens invasion was described by Lund et al. (364), but the fungus depicted has broadly ovoidal conidia inconsistent with *A. kiliense*. It may be *A. potronii.* Though not depicted, chlamydospores are mentioned in descriptive notes; however, some *A. potronii* isolates may have hyphal swellings, as mentioned in the description below.

An anomalous case of scalp kerion yielding only *A. kiliense* was described by Lopes et al. (365). Direct microscopy showed relatively thin (2 to 4 μm) hyphae and no dermatophyte-like arthroconidia or hair invasion. The patient, a 4-year-old Brazilian boy, had no known immunodeficiency or history of trauma. Unfortunately it is not clear whether the fungus was consistently isolated on two or more successive culture attempts made from the lesion as is generally recommended for purported nondermatophytic cutaneous infections. The authors stated that pus and scrapings were collected from the lesion, and that pus grew *A. kiliense*. They then went on to state that *A. kiliense* was "the unique microorganism isolated from the *rest* of the scrapings on several occasions" (emphasis added). The word rest is emphasized to highlight an ambiguity that may not at first be obvious. Extensive editorial experience has taught that the expression "isolated . . . on several occasions" used by persons other than native English speakers may well simply indicate separate isolation from several inoculum pieces planted at the same time. In the crucial passage quoted, "the rest of the scrapings" may be a mistranslation intended to mean "subsequent samples obtained by scraping the same lesion," in which case *A. kiliense* was consistently isolated in temporally separate examinations and its role in etiology is relatively strongly attested. On the other hand, "the rest of the scrapings" may mean exactly what it appears to mean, while "isolated . . . on several occasions" may be an incorrect rendering of "isolated from several additional inocula." That would make this case weakly evidenced, and possibly a misinterpretation of *A. kiliense* contamination adhering to a purulent dermatophytosis lesion. It is not unknown for dermatophytes such as *T. rubrum* to invade scalp skin without invading hair (366); moreover, sparse hair colonization may be difficult to detect in material from serous, inflamed lesions unless fluorescent dyes or fungal stains are utilized. The etiologic attribution in this case must remain dubious until more explicitly documented later case reports yield consistent findings.

White piedra-like hair concretions repeatedly yielding only a member of the genus *Acremonium* were described by Liao et al. (367). The apparent causal agent, anachronistically determined as *Cephalosporium acremonium*, had *A. kiliense*-like phialides and cylindrical conidia, and a pinkish colony described as becoming pale yellow at maturity on both Sabouraud and potato dextrose agar (PDA) media. Contamination, secondary colonization, or in situ mycoparasitism of poorly viable, classic white piedra concretions (*Trichosporon ovoides*) would be difficult to rule out in such a case without sophisticated techniques such as immunohistochemistry.

An anomalous subcutaneous hyalohyphomycosis affecting the hip and leg of an otherwise healthy French stonecutter was ascribed to *C. acremonium* by Lahourcade and Texier (368, 369). The extensively described and illustrated isolate appears to have been *A. kiliense*. Characteristic chlamydospores were noted in culture.

Well-confirmed *A. kiliense* bovine mycotic abortion was demonstrated by Dion and Dukes (370). The isolate's identity was confirmed by L. Brady of CMI (now CABI).

Description. Colonies grow 12.5 to 16 mm in 7 days. They are flat to slightly wrinkled with a moist surface, often somewhat ropy or slightly cottony, and occasionally spiralling across the medium surface. They are dirty white to pale orange or very pale brown. The reverse is pale brownish, and on Sabouraud agar characteristically rich ochraceous to grey-brown after 14 to 21 days of growth.

Conidiophores consist of solitary phialides or short side branches bearing up to four phialides, sometimes rebranched once or twice, arising from medium surface or more often from ropy strands consisting of several hyphae fascicled together (Fig. 67). *Phialides* are acicular, tapering in outline, often flexuose, and mostly 25–50 × 1.5–2 (basal width) μm. Also, submerged in growth medium, short *adelophialides* (reduced phialidic necks) up to 6 μm long protrude at right angles from vegetative hyphae. *Conidia* are borne in slimy heads. They are smooth-walled, hyaline, ellipsoidal to cylindrical, and 3.1–5.8 × 1–1.6 μm. Conidia formed from submerged adelophialides may be curved. *Chlamydospores* are usually present on submerged mycelium after more than 1 week growth. They are globose to ellipsoidal, thick-walled, and 4 to 8 μm in diameter (Fig. 68). According to Gams (334), chlamydospores form most reliably on oatmeal agar.

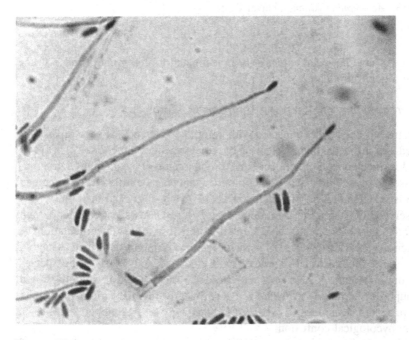

Figure 67   *Acremonium kiliense*, phialides soumis.

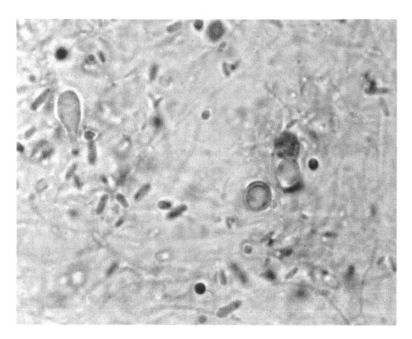

Figure 68 *Acremonium kiliense*, chlamydospores.

Other media have not been extensively investigated to determine if these structures are reliably produced.

11. *Acremonium* {*aff.* unknown} *potronii* W. Gams

This fungus has been well connected with keratitis (371) and has been repeatedly
identified as a suspected but not rigorously proven agent of onychomycosis (334).
A well-documented case of an ulcer beneath the tongue of a cat featured granule-
like masses of mycelium (372). The etiologic agent was identified by an unnamed
mycologist at CBS, but no identification details were given and the culture was
not preserved. This identification, done long before Gams's critical monograph
of *Acremonium* (334), cannot be accepted in the absence of documentation. A
recent keratitis case attributed to *A. potronii* featured a fungus described as having
pale olive-green colonies and spherical conidia (373). A connection to this species
is therefore uncertain. A statement that conidia were ''spherical, smooth-walled
and grouped in verticils,'' with the last term clearly intended to mean ''sticky
heads'' in context (phialides were described separately as ''developed laterally''),
suggests mycological confusion.

The original isolates described as this species were isolated by Potron and

Noisette (374) from an apparent septicemia of a French winegrower, whose course of disease began as a cyst at the site of a horsefly bite. After some weeks in which the patient experienced multiple systemic symptoms, *A. potronii* was isolated both from septic arthritis of the right knee and in association with *Candida albicans* from oral thrush. Concomitant skin lesions were culture-negative. Consistent thin, nonbudding filaments were seen along with *C. albicans* elements in direct smears from the oral lesions; examinations of knee and skin lesion fluid gave negative results.

The knee fluid was only sparsely culture-positive, and the negative direct microscopy was noted as not necessarily contradictory. The aggressive infection responded dramatically to therapy with potassium iodide, the antifungal drug of choice at the time (1911), then recrudesced after the patient ceased therapy prematurely, and was finally eliminated after prolonged additional therapy. Although now lost, the isolate was described and illustrated in detail by Vuillemin (375) and Pollacci and Nannizzi (376), and its connection with the current concept of the species was upheld unequivocally by Gams (334).

In the years just after the description of *A. potronii* two etiologically well-attested cases of aggravated tonsillitis were attributed to this fungus (142a, 376a). In one case (142a) the identification was documented with illustrations as well as confirmation from G. Pollacci, who had studied the original *A. potronii* isolate. It is not clear if the illustrations were from the case strain or from the Vuillemin reference strain. In the other case (376a), the identification was simply attributed to mycologist R. Motta, who worked under the supervision of Pollacci. Since the taxonomy of *Acremonium* was meagerly known in those times, it is difficult to be certain that the same species was actually studied in each case.

An Algerian mycetoma attributed by Montpellier and Catanei (377) to an *Acremonium* "presenting the principal characters of *A. potronii*" was caused by an organism with off-white to pale pinkish, wrinkled colonies producing cylindrical to subglobose but mostly ovoid conidia averaging 5.5 × 3 μm but sometimes elongate, 6.5 × 2.5 μm. Some conidia were formed on adelophialides. Chlamydospores were formed on glucose-free Sabouraud agar. The organism grew more rapidly than Potron and Noisette's isolate (actual rates were not given) and sporulated poorly except on sterilized rye grains. This may be *A. potronii*, but an *A. kiliense* isolate in which some atypically shaped conidia were observed cannot be excluded. Atypically swollen conidia in *Acremonium* species may be formed from subsurface phialides or especially when older conidia begin to swell prior to germination. Montpellier and Catanei based their descriptions on cultures observed over several weeks. Modern students of *Acremonium* would avoid such senescent material.

Kinnas (378) illustrated *A. potronii* in 1965 in connection with a keratitis case from Greece. The case isolate, however, was not identified to species, and the illustrations are an uncredited, redrawn pastiche of Vuillemin's 1910 (375)

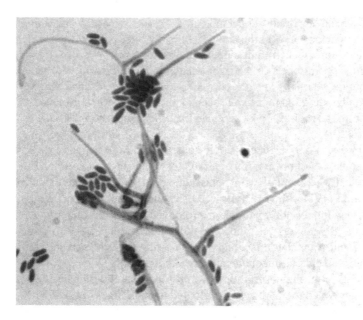

Figure 69   *Acremonium potronii* CBS 251.95.

descriptive drawings of *A. potronii*. Kinnas's text indicates that they are intended only to illustrate the genus *Acremonium*, not *A. potronii* per se. Some later authors interpreted this paper as an *A. potronii* record.

   Description.   Colonies grow 2 to 5 mm in 7 days. They are typically almost smooth to slightly granular, usually pale, but may be pale salmon if exposed to light during growth. The reverse is concolorous with the surface.

   Conidiophores consist mainly of solitary phialides, occasionally arising from small ropy hyphal bundles (Fig. 69). *Phialides* are spinelike, tapering, often gently curving, and mostly (7−) 11–27 × 1–2 (basal width) μm. *Conidia* are borne in slimy heads. They are smooth-walled, hyaline, mostly obovate (egg-shaped, with thick end formed first), and 2.1–4 (−5) × 1.3–2.5 (−3) μm. *Chlamydospores* are not formed; one nail isolate in CBS evinces irregular hyphal swellings (334).

12.   *Acremonium {aff.* unknown} *recifei* (Arêa Leão & Lobo) W. Gams

A. recifei is most commonly isolated from white-grain eumycetoma in tropical areas worldwide (379–381). A case of hyalohyphomycosis not organized as my-

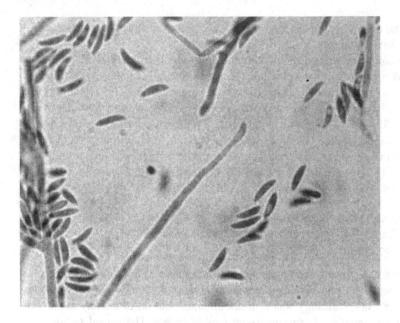

Figure 70 *Acremonium recifei* CBS 485.77.

cetoma was reported by Zaitz et al. (381); however, this Brazilian case concerned a hand dorsum injured only 3 months prior to diagnosis, so it may well have been an infection that would develop into mycetoma if left untreated. Oral itraconazole was effective therapy; no debridement was done. A man in the Netherlands with an eye injury caused by a fragment of coconut shell developed well-confirmed *A. recifei* keratitis (382). A patient with fungemia and respiratory infection after autologous bone marrow transplantation in a multiple myeloma grew multiple cultures identified as *A. recifei* (383). Direct microscopy of respiratory material was positive for consistent filaments. The fungus was identified by mycologist C. de Bièvre, and a consistent description was published. Photographs, however, showed adelophialides, a structure not mentioned by Gams (334) in his description of this species, but possibly occurring under some conditions.

The fungus in nature grows on decaying parts of tropical plants and has been isolated several times from decaying coconut and brazil nut shells (334).

Description. Colonies grow 8.5 to 21 mm in 7 days. They are typically dusty-looking, thinly felted, or—especially in fresh isolates—sometimes raggedly tufted, but also commonly smooth and wet-looking. They are whitish to yellow-greenish if kept in darkness, but pale pink if exposed to light during growth. The reverse is the same as the surface color on most media, but ochre-brown on Sabouraud.

Conidiophores arise directly from the substrate or from bundled aerial my-celium and consist of solitary phialides or, predominantly, of short side branches bearing up to 1 to 3 (or occasionally more) phialides (Fig. 70). *Phialides* are acicular, and tapering in outline, often gently curving or undulate. They are mostly 15–55 × 1.7–2.5 (basal width) μm. *Conidia* are borne in slimy heads. They are smooth-walled, hyaline, strongly curving, and sausage- or cashew-shaped with an apiculate base and sometimes with a thickened apical end, 4–6 (−7.5) × 1.3–2 μm. *Chlamydospores* may be formed in small quantities in older cultures, and are thin-walled and 3.5 μm in diameter.

13. *Acremonium {aff.* unknown} *roseogriseum* (S.B. Saksena) W. Gams (Common Synonym: *Cephalosporium roseogriseum* S. B. Saksena)

This species was reported in connection with mycotic arthritis of the knee of a Florida patient who had experienced puncture wounds from spiny palm fronds in his garden (384). Etiology was well documented, but unfortunately the identification of the agent as *A. roseogriseum* was incorrect. The causal fungus was described as having conidia 20–30 × 4–8 μm, incompatible with any *Acremonium* species, and no mention was made of dark conidial pigmentation. The description of a colony with purple reverse naturally suggests *Fusarium* or possibly *Cylindrocarpon lichenicola*. The isolate was not deposited and its identity cannot be traced.

Zaias (49) gave *Cephalosporium roseogriseum* as the name of a common etiologic agent from superficial white onychomycosis. The description and illustrations do not correspond to this species but rather to an amalgam of *Acremonium potronii* and *A. strictum*, the two species later confirmed by Gams (334) from cultures sent by Zaias to CBS.

A. roseogriseum is a tropical soil fungus with mostly dark grey, clavate to pyriform conidia (334). It has no pathogenic record in humans or animals.

14. *Acremonium {aff.* unknown} *spinosum* (Negroni) W. Gams

This fungus has been well documented as an agent of superficial white onycho-mycosis by Negroni (385, 386) in conjunction with its original species description. Confirmatory repeat isolations are reported in only one (386) of the two strongly overlapping papers published on this case.

Description. Colonies grow 7.7 to 8.4 mm in 7 days. They are tough in texture, moist to slightly scurfy or ropy, and flat, wrinkled, or heaped. They are whitish, ochraceous, orange, or sordid red-brown on the surface. The reverse is pale to purple-brown.

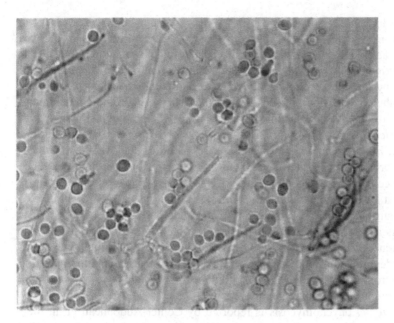

Figure 71 *Acremonium spinosum* CBS 136.33.

Conidiophores consist mainly of solitary phialides arising from single hyphae or hphal bundles (Fig. 71). Sometimes side branches bearing 2 to 3 phialides are seen. *Phialides* are spinelike, tapering, often gently curving, and 8–20 (−30) × 1–1.5 (basal width) μm. *Conidia* are borne in sticky heads or occasionally in chains, mostly with fine warts or spines. They are hyaline, subglobose to broadly ellipsoidal, and 2.7–3.3 × 2.2–2.6 μm. *Chlamydospores* are not formed.

15. *Acremonium {aff. Emericellopsis} strictum* W. Gams
    [Common Synonym in Older Literature: *Cephalosporium acremonium* (Also Applied to *A. kiliense* and Other Species)]

This species has been well etiologically connected and well identified from two cases of disseminated mycosis in severely immunocompromised patients, in one case beginning as a gastrointestinal colonization (303) and in another beginning as onycholysis (387). A culture from the former case was deposited in ATCC. In limited infections of immunocompromised patients, a fatal brain abscess in an immunosuppressed carcinoma patient proved to contain both *A. strictum* and *Fusarium oxysporum* (388), while *A. strictum* was causally connected with a pulmonary infection in a 15-year-old chronic granulomatous disease patient

(389). The latter case isolate is also in ATCC. Well-demonstrated *A. strictum* meningitis in a steroid-treated child with Landry-Guillain-Barré syndrome was tentatively attributed to introduction via repeated lumbar punctures (390). Soft tissue infections by *A. strictum* in the gluteal and femoral regions of an immunocompetent but multiply traumatized, comatose automobile accident patient may have contributed to the patient's death a few days later (388), although the immediate cause of death was a pneumonia that was not microbiologically characterized. *A. strictum* peritonitis in a CAPD patient was recently reported (391).

In the brain and automobile accident cases mentioned above, the identification of the organism is not verifiable in the case report, but one or more reference mycology centers were consulted. In the CAPD case, a partial description is given, but it lacks sufficient detail to confirm the identification. A recent case of fungemia attributed to *A. strictum* in a neutropenic child (392) both describes and depicts a case isolate with ovoid conidia, which is incompatible with this species. Induction of aberrant growth by use of an inappropriate medium or conditions in identification studies cannot be ruled out in this case; a photograph from a slide culture shows vesiculate and distorted forms not found in any *Acremonium* species growing in favorable conditions.

A. strictum onychomycosis is reported by Contet-Audonneau et al. (48), and the pattern of its infection is illustrated in a biopsy study. No identification characters are given, but the distinctive, not at all dermatophyte-like morphology of *Acremonium* filaments in biopsy is here accepted as sufficient evidence of confirmed infection. As detailed above under *A. roseogriseum*, several onychomycosis isolates obtained in classic studies by Nardo Zaias were later identified as *A. strictum.*

In nature this species is commonly isolated from soils and decaying mushrooms and plant parts.

Description.   Colonies grow 11 to 17.5 mm in 7 days. They are variable from isolate to isolate, typically almost smooth to slightly velvety, and less commonly felty or cottony. They are sometimes whitish, especially if kept in darkness, but more typically intensely orange or salmon, especially if exposed to light during growth. The reverse is pale to orange.

Conidiophores consist of solitary phialides or short side branches bearing up to 1 to 3 or occasionally more phialides (Fig. 72). Side branches are sometimes rebranched once or twice, with at least the first second-order side branch typically basitonous (arising from near the base of the main side branch). *Phialides* are acicular, tapering in outline, often gently curving or flexuose, and mostly 25–50 × 1.5–2 (basal width) μm. Also, submerged in growth medium, short *adelophialides* (reduced phialidic necks), mostly 1–5 (−10) μm long, protrude at right angles from vegetative hyphae. *Conidia* are borne in slimy heads. They are smooth-walled, hyaline, long-ellipsoidal to cylindrical, and 3.3–5.5 (−7) ×

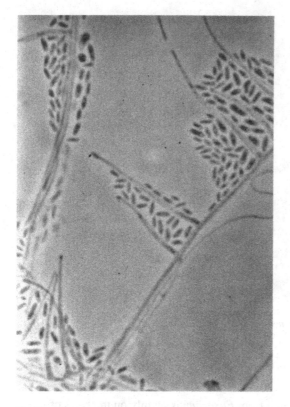

Figure 72 *Acremonium strictum.*

1–2 μm. Conidia formed from submerged adelophialides may be curved. *Chlamydospores* are not formed.

C. *Cylindrocarpon*

The genus *Cylindrocarpon* consists mainly of fungi related to the teleomorph genus *Neonectria*, and is thus closely related to some *Fusarium* species, such as *F. solani*. Most human infections, however, are ascribed to *Cylindrocarpon lichenicola*, a species without a known teleomorph. The genus as a whole is primarily associated with plant pathogenesis. There is a smattering of reports giving genus-level identifications of *Cylindrocarpon* strains from human (393) and animal (394, 395) keratitis, including one human case of keratitis progressing to endophthalmitis (30), but the majority of reported cases are for identified species, as noted below.

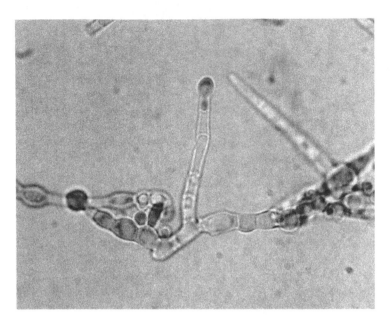

Figure 73 *Cylindrocarpon cyanescens* CBS 518.82.

1. Identification

The distinction of *Cylindrocarpon* from *Fusarium* is mainly on the basis of mac-
roconidial shape; *Cylindrocarpon* macroconidia lack the distinctive (although
sometimes subtle) foot cell possessed by *Fusarium* macroconidia, and also have
a rounded apical end, unlike the pointed apices seen in most *Fusarium* species.
Also, the elongate, cylindrical phialides that are restricted to *F. solani* and related
members of *Fusarium* subgenus *Martiella* (see also *F. coeruleum* below) in *Fusarium* are the norm in *Cylindrocarpon*. *Cylindrocarpon destructans* may be especially likely to cause confusion with *F. solani* (See the discussion under the
former name.) *Cylindrocarpon* species should be grown for identification on fungal sporulation media such as PDA, oatmeal agar, modified Leonian's agar, or
pablum cereal agar (396).

2. *Cylindrocarpon {aff.* unknown} *cyanescens* (de Vries et al.) Sigler

This fungus is known from a single case of pale-grain mycetoma originally contracted from a puncture of the foot dorsum in Indonesia (397). It was originally
described as a *Phialophora* species (398), and was moved out of this notoriously

heterogeneous genus by Sigler (399) because the culture, chlamydospores, and phialides suggested *Cylindrocarpon*, even though no macroconidia were produced. The hypothesis proposed by the current name requires molecular testing.

Description. Colonies grow very slowly, 1 to 1.5 mm after 7 days at 20°C. They are velvety to heaped [described as formed of "cauliflowerlike pustules" by de Vries et al. (398)] and often consist mainly of chains of chlamydospores. They are a pale cream color at first on the surface and reverse. The reverse becomes brownish to black, secreting blue pigment into the agar when freshly isolated. *Conidiophores* are often extended branches bearing a small number of phialides as side branches as well as a terminal phialide. *Phialides* are cylindrical with tapered ends, often rather short and curved when arising as a side branch. They are sometimes aseptate, but often with up to three septa, which are 10 to 40 μm long with a small collarette. *Conidia* are smooth, hyaline, ovoid to ellipsoidal, and 4–7.5 × 2–3 μm. *Chlamydospores* are predominant, mostly in chains or clusters, nearly hyaline at first, then becoming olivaceous brown and partially encrusted with a thin layer of loosely adhering brown material.

This species does not strongly resemble any other arising in medical mycology. It is very slow growing and secretes blue pigment when fresh, and in microscopy of cultures is composed mainly of chlamydospores, while producing a small number of septate phialides giving rise to small, ellipsoidal conidia. Unlike the *Phialophora* species it was once included with, it lacks melanized (dark pigmented) hyphae or conidiophores, although its chlamydospores tend to be pigmented.

3. *Cylindrocarpon destructans* (Zins.) Scholten Anamorph of *Nectria radicicola* Gerlach & L. Nilsson

This species has been well confirmed by Zoutman and Sigler (399) as an agent of mycetoma contracted as the result of a nail puncture of the foot in Antigua.

Four cases of keratitis in horses from Florida, including two cases confirmed with direct microscopy, were reported by Hendrix et al. (400). Identification criteria were given, and the identification of the fungus was confirmed by phytopathological mycologist J. Kimbrough. Despite this, the photographs presented—a view of a primary isolation plate showing loosely textured fungal colonies and a light micrograph showing a preponderance of uncurved macroconidia with three evenly spaced septa and an abruptly tapered, truncate base—suggest that further study of the identity of at least one of the isolated strains may be warranted. The authors state that they ruled out *C. lichenicola* because their strains produced microconidia, unlike that species. In the micrograph (their Fig. 2), however, the few reduced conidia present, referred to as both "microconidia" and "microconidiospores" in the figure legend, are similar to those formed in low proportion in many *Fusarium* and *Cylindrocarpon* species conventionally said not to produce microconidia. Similar 0- or 1-septate conidia are shown in micrographs of *C. lichenicola* in papers by Booth et al. (401), Laverde et al. (402), and Iwen et al. (403), as well as in line drawings in the monographic study (as *C. tonkinense*) of Booth (404). Also, the deeply floccose, effuse, frothy-looking *C. lichenicola* colonies depicted by Laverde et al. (402) are strikingly similar to the colonies depicted by Hendrix et al. (400), and contrast strongly with the compact, felted, and centrally tufted *C. destructans* colony depicted by Zoutman and Sigler (399). In nature, *C. destructans* grows on many types of plant roots in both cultivated and uncultivated soils.

Description.   Colonies grow 7 to 12 mm in 7 days at 25°C. They are densely felty to floccose, usually with raised centers and somewhat folded marginal zones. They are pale at the margin and in floccose areas but otherwise reddish-brown on the surface, sometimes spotted with slimy cream to beige sporodochia. The reverse is light brown or more commonly reddish-brown, with a brownish diffusing pigment. *Conidiophores* are simple or repeatedly branched, with 1 to 3 branches at branch points (Fig. 74). *Phialides* are long and slender, cylindrical to awl-shaped, and 18–35 × 2.5–3 μm, with a small collarette. *Macroconidia* are uncommon in some cultures. They are cylindrical to slightly

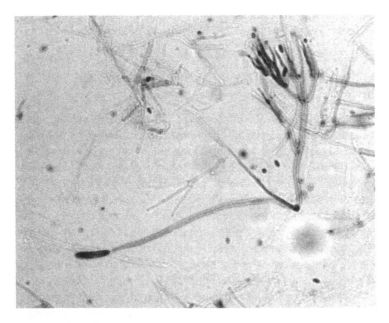

Figure 74   *Cylindrocarpon destructans* CBS 301.93, showing formation of both microconidia and macroconidia.

curved, with a rounded apex and rounded to slightly truncate base. They are mostly 1 to 3-septate, sometimes composed of cells of markedly unequal length, and 20–40 (-50) × 5–6.5 (-7.5 µm). *Microconidia* intergrade in size and shape with macroconidia. They are long-ovoid to long-elliptical and sometimes slightly curved, with no or one septa. Aseptate conidia are 6–14 × 3–4.5 µm. *Chlamydospores* are abundant, terminal or intercalary, solitary, in chains, or in clusters. They are pale and smooth, becoming brownish and sometimes warted at maturity. They are more or less rounded and 8 to 16 µm long. The teleomorph is found only on infected plants and is not described here.

This fungus bears a slight resemblance to *F. solani*, but grows much more slowly. Its macroconidia lack the asymmetrical basal foot cell that *F. solani* macroconidia have. In the *C. destructans*, microconidia are mostly under 10 µm long (some larger ones occur), while *F. solani* has microconidia mostly over 10 µm, with some smaller ones also seen.

4. *Cylindrocarpon {aff. Nectria} lichenicola* (C. Massal.) Hawksw. (Common Synonym: *Cylindrocarpon tonkinense* Bugn.)

This species has primarily been reported in well-substantiated cases of keratitis from Colombia (402), Japan (405), and Argentina (406). There is also an endophthalmitis record from Florida, with etiology attested by growth in multiple cultures and the identification evidenced only by the appearance of the name *C. tonkinense* in a table (407). (See also the discussion of Florida horse keratitis cases attributed to *C. destructans* above.) Other opportunistic infections are also rarely seen. A Nebraska leukemia patient experienced a localized hand ulceration caused by *C. lichenicola*; since the patient was undergoing high-dose chemotherapy, partial amputation of the hand was used in addition to amphotericin B to cure the infection (403). A case of CAPD-related peritonitis is reported by Sharma et al. (408) in a British patient who had had a prolonged Jamaican vacation. The etiology in this case was well verified, but the identification of the organism was not documented or credited. A disseminated *C. lichenicola* infection in an acute myeloid leukemia patient was documented by James et al. (409). The infection was treated successfully with liposomal amphotericin B. The authors speculated that the patient's "past history of athlete's foot" was the source of the infection, even though the patient had "no particular skin or nail lesions" at the time he presented for leukemia chemotherapy. This hypothesis seems highly far-fetched. (See below.) In addition, the question of whether fungal etiology was confirmed in the previous foot condition is not addressed. The *C. lichenicola* isolate involved in dissemination, however, was authoritatively identified by two expert laboratories and collected at the second, the International Mycological Institute, in Egham, UK (now CABI).

The notion of a *C. lichenicola* from athlete's foot in the James et al. (409) study was supported by the citation of a case report by Lamey et al. (410), in which an unidentified *Cylindrocarpon* was reported from intertrigo involving all the toe webs of a male Beninese patient examined in France. In this case, a fungus disclosed by its photograph and description as compatible with *C. lichenicola* was isolated repeatedly and consistently over a 2-month period from scrapings that were positive in direct examination for compatible fungal filaments. No dermatophyte was isolated in four or more consecutive attempts, all positive for the *Cylindrocarpon*. The case appears comparable to intertrigo cases involving African patients discussed under *Fusarium solani* and *F. oxysporum* below. The reason this emerging pattern of hypocrealean intertrigo in otherwise healthy patients appears mainly to affect Africans may be at least partially adumbrated by the case report of Comparot et al. in 1995 (411), discussed below under *F. oxysporum*, in which a possible connection to foot washing multiple times per day for religious reasons is mentioned. (This in itself could not be a sufficient cause, since this religious practice is carried out worldwide, but may combine with other factors such as use of contaminated water.) The probability that a case of typical athlete's foot seen in a non-African patient could be caused by a hypocrealean fungus appears to be extremely small. With regard to the Lamey et al. (410) case report, it should be noted that the conidial measurements given for the fungus seen are compatible with *C. lichenicola* in length (20 to 30 μm), but are too narrow in width (3 to 4 μm). The measurements given, however, conflict with the paper's Fig. 3; according to the measurements, conidia should have a length–width (l/w) ratio of (5-) 6.6–7.5 (-10), whereas the mature conidia shown in the photo, measured with a ruler, have l/w ratios of around 3.5. A conidium 20 μm long at this ratio would have a width of 5.7 μm, while one 30 μm long would be 8.5 μm wide. If the published lengths are correct, therefore, the conidia are within the width range of *C. lichenicola* or even slightly too wide, rather than too narrow. It appears that the conidial measurements given by Lamey et al. (410) could not possibly have been accurate and therefore do not contradict the probability that *C. lichenicola* was involved in this case. Alternatively, the photo may be completely unrepresentative of the isolate seen; in fact, however, it appears typical of this most common *Cylindrocarpon* from human disease.

Description. Colonies grow approximately 41 mm in 7 days at 25°C on PDA. They are velvety to moderately floccose with minute, upright, heavily conidium-producing strands giving a frothy appearance. They are white initially, then yellow to pale brown, sometimes becoming red-brown to purplish-red at the center, especially when the reverse is strongly pigmented. The reverse is buff at first but is soon dark brown, sometimes with a brown diffusing pigment. *Conidiophores* are long or simple, or with a small number of mostly basitonously arising (i.e., formed on the lower half of the conidiophore) branches (Fig. 75). *Phialides* are long-cylindrical, subulate (awl-shaped; i.e., long and tapered) to filiform

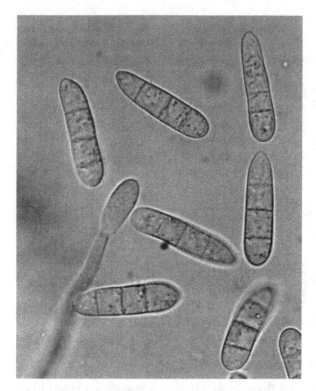

Figure 75 *Cylindrocarpon lichenicola* CBS 109048.

(threadlike, sometimes with a degree of irregular sinuous curvature), and 38–50 × 3–5 (basal diameter) μm. *Macroconidia* are formed in sticky heads. They are smooth-walled, hyaline, cylindrical, long-ellipsoidal, or slightly clavate (i.e., with slightly expanded apical end), with a rounded apex and a distinct, protruding, often slightly laterally displaced truncate basal hilum, usually with three evenly spaced septa, occasionally to five-septate, and mostly 18–40 × 5–7 μm. Reduced forms with 0 or 1 septum but more or less consistent in diameter with typical macroconidia may be present in low proportions. *Chlamydospores* are abundant, especially after 14 days. They are formed terminally, laterally on short branches, sessile along hyphae, intercalated into macroconidia, in dyads or clumps or in intercalary chains, and are hyaline to brownish, smooth to finely roughened, globose, and 6 to 15 μm in diameter.

This very distinctive species can be distinguished from many other *Cylindrocarpon* species by its mostly straight macroconidia with a truncate hilum, and by its lack of differentiated microconidia. (See comments about reduced conidia in discussion of *C. destructans* above.) The same characters distinguish it from the only slightly similar *Fusarium* species, *F. solani*.

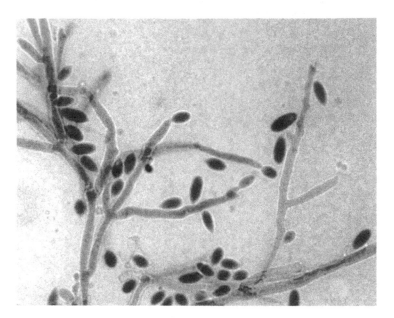

Figure 76   *Fusarium chlamydosporum* CBS 615.87 polyphialides.

5.   *Cylindrocarpon {aff.* unknown} *vaginae* Booth

This species was described by Booth et al. (401) in a publication stated to be about *Cylindrocarpon* species associated with keratitis. No details were given, however, about the isolate obtained from ''the eye of a British farmer following an injury.'' This is thus an unsubstantiated record. *C. vaginae* has predominantly 0 to 1-septate conidia, with a few mainly curved, larger two- or three-septate conidia to 24 μm long, and is primarily distinguished by producing hyaline clumps of up to six chlamydospores from globose vesicles held on short lateral stalks. Colonies are pale with a reverse that gradually becomes grey-brown, and grow rapidly, covering a 90-mm petri plate of PDA medium in 7 days at 30°C (25°C measurement not given).

D.   *Fusarium*

The anamorph genus *Fusarium* contains the asexual states of numerous ascomycetous fungi in the genera *Gibberella*, *Nectria*, and *Cosmospora*, as well as related purely anamorphic species for which no teleomorph is known. At the present moment, a few phylogenetically outlying but morphologically similar organisms, such as *Fusarium dimerum*, remain in the genus. Teleomorphs of this group are not seen in the clinical laboratory and are not described here.

This loosely related group of Hypocrealean anamorphs was regarded as rarely pathogenic, except in ocular and nail infections, until the number of neutropenic patients increased dramatically in connection with new anticancer chemotherapies in the 1980s (302, 412, 413). The widespread use of amphotericin B therapy or prophylaxis has also contributed to this emergence, since the isolates of these species are relatively likely to manifest typical Hypocrealean resistance to this drug (susceptible strains do occur) (412–414). Since this emergence, several species have been recognized as particularly significant (412). As in *Aspergillus*, in *Fusarium* there are a small number of major opportunistic species and a much greater number of marginal opportunists. The major agents are *Fusarium oxysporum*, *F. solani*, *F. verticillioides* (=*F. moniliforme*), and *F. proliferatum*.

1. General Pattern of Involvement by Major Opportunistic *Fusarium* Species

The major opportunistic fusaria show a similar pattern of involvement in disseminated infection of the neutropenic patient. Generally patients with acute leukemia are involved, although chronic leukemia, aplastic anemia, and lymphoma-related cases have also been observed repeatedly (412, 415). Many other strongly immunocompromising conditions yield occasional cases. Cases frequently feature rapid blood-borne dissemination and the development of local lesions where inoculum lodges in capillary beds. This is dramatically seen as the appearance of ecthyma gangrenosa-like, necrotic skin lesions, but also may be seen in organ and bone lesions, (e.g., in magnetic resonance images) (416). The mediation of this pattern by copious microconidial production in the host tissue and circulatory system was proposed by Richardson et al. (302) based on the extensive conidial production seen in these species in liquid culture, and was definitively confirmed in a series of studies by W. Schell and collaborators at Duke University Medical Center. In one of these studies, Liu et al. (276) showed microconidia and phialides in numerous histopathological preparations. Even multicelled *Fusarium* conidia could rarely be seen in skin lesions and peritoneal fluid. The proliferation of conidia in situ facilitates rapid diagnosis in that positive blood cultures are readily obtained (302, 413, 417).

Common portals of entry for these disseminated infections appear to be the sinuses, the lungs, and the skin. Emergence from onychomycosis has been documented several times (415, 418). The common environmental association between generalist decomposer *Fusarium* species (a group overlapping strongly with the opportunistic mammalian pathogens) and water systems (74) appears to have been responsible for some nosocomial infections (419); patients may acquire the organism, for example, by inhaling aerosols while showering.

Although localized pulmonary infection is rare in immunocompromised patients, presumably because infections disseminate too readily, localized skin and soft tissue infections are uncommonly but repeatedly seen. Several related cases are contrasted by Guarro and Gené (412).

As with most fungi causing disseminated infections in immunosuppressed patients, opportunistic *Fusarium* species may also be isolated from highly vulnerable sites in immunocompetent patients with major barrier breaks, giving rise to cases of catheter-related peritonitis, endocarditis, keratitis, exogenous endophthalmitis, and trauma-related osteomyelitis and septic arthritis (412, 415). With regard to ocular infections, there is a noteworthy tendency for *F. solani* (see below) to be the predominant organism involved as well as the most aggressive, but the other major opportunistic fusaria may also cause these infections (229, 393, 420).

Even though disseminated *Fusarium* infection may not infrequently begin as sinusitis, members of this genus appear to have little involvement in allergic fungal sinusitis of otherwise healthy individuals (421). Likewise there appear to be few or no well-substantiated records from etiologically similar colonizations of surfaces exposed to air within immunocompetent patients, such as otomycosis, allergic bronchopulmonary mycosis, and pulmonary fungus ball.

Fusarium species have long been known to cause onychomycosis, particularly of the superficial white variety (49). Here *F. oxysporum* appears to be the main agent involved, but other major opportunistic fusaria are also seen in this role.

One underpublicized but frequently encountered aspect of opportunistic fusarial biology is the tendency of these fungi to grow saprobically on wound and ulcer surfaces, as well as on suppurating intertriginous fissures (422, 423). Such growths are heavily positive for fungal filaments in direct microscopy, but detailed analysis shows that generally no tissue penetration occurs in immunocompetent hosts. Exceptions are discussed under *F. oxysporum* and *F. solani* below. Burn patients also may experience superficial *Fusarium* growth, which may then progress to life-threatening systemic infection (424, 425).

In mammals other than humans, *Fusarium* species may cause keratitis (426), especially in horses (395). Infections of reptiles and amphibians are occasionally encountered (see individual *Fusarium* species discussions below), and occasional skin infections, often not well confirmed, have been reported from a variety of mostly aquatic mammals (426). Infections of arthropods and of bird and reptile eggs are not discussed in this chapter. The complex topic of mycotoxicoses caused in humans and animals ingesting food colonized by *Fusarium* species is also beyond the scope of this chapter.

2.   Unidentified *Fusarium* in Opportunistic Pathogenesis

Fusarium species require some skill to identify morphologically, and molecular identification is at least temporarily entangled in the remarkable complexity of sibling speciation that has been disclosed within this genus (e.g., the existence of at least 26 entities within the morphospecies *F. solani* that would be considered distinct species using a DNA-based phylogenetic species concept) (427), with more such entities undoubtedly still uncharacterized. Generic-level identification is therefore still common in case reports. Since the major opportunistic fusaria so far have not been shown to have strong clinical differentiation in terms of their involvement in serious systemic infection or in their responses to therapy, generic identification is not as problematical as it is in the more heterogeneous *Paecilomyces* and *Acremonium*. The information in the reports that lack species identification is essentially summarized in Sec. 1 above. Species identification is still recommended in this genus, however, so that emerging patterns of species-specific behavior can be revealed, particularly in connection with novel infections or therapies. *F. solani* in particular is not especially closely related to the other major opportunistic fusaria, and is easily distinguished by its elongate microconidial phialides.

3. Identification

For practical purposes in the clinical laboratory, correct genus-level identification of *Fusarium* species is the most critically important action to enable correct treatment and infection control. Although antifungal susceptibility testing is indicated for each *Fusarium*-like isolate causing a confirmed systemic infection, genus-level identification as *Fusarium* suggests that itraconazole and other azole drugs more directed at *Aspergillus* or *Candida* infections may need to be replaced or supplemented with amphotericin B.

When *Fusarium* forms characteristic elongate, canoe-shaped macroconidia rapidly, genus-level identification can be extremely straightforward. The laboratorian need only be aware that *Cylindrocarpon* species may have somewhat similar macroconidia. The macroconidia of this genus, however, differ by being either straight or sausage-shaped, with rounded apices. The distinctive bootlike *pedicel* that is found at the base of most *Fusarium* macroconidia is not present. The toelike part of this structure serves in most *Fusarium* species to stabilize forming macroconidia against the side of the phialide apex so that they do not break off due to shear stress while still immature. Only a few species, such as *F. incarnatum*, are adapted to form macroconidia in ways that do not involve such stabilization structures. The *Fusarium solani* complex has perhaps the most *Cylindrocarpon*-like macroconidia, yet these conidia do have a small but perceptible pedicellate structure. Also, in any comparisons with most similar-looking organisms, the elongate phialides of *F. solani* are distinctive.

In heavily microconidial *Fusarium* species such as *F. proliferatum*, in which macroconidia may form belatedly or only after special stimulation, the isolates may need to be distinguished from *Acremonium* species. With experience this is normally done at a glance. The underlying characters permitting this are the much faster growth rate of *Fusarium* species, and in most species, the much greater amount of floccose aerial mycelium produced. As mentioned previously, the colonial growth rate of *Acremonium* species is generally under 35 mm in 7 days at 25°C, often under 20 mm, whereas *Fusarium* species generally grow over 50 mm in 7 days, and in most cases exceed 60 mm. Many *Fusarium* species, including members of subgenera *Liseola* (e.g., *F. proliferatum*, *F. verticillioides*) most likely to be purely microconidial in early growth, are notably floccose. Among the medically important fusaria in general, only *F. dimerum*, which normally forms copious pointed two-celled macroconidia, is relatively flat, slimy, and *Acremonium*-like. (It grows significantly more quickly than an *Acremonium* in any case.) Note, however, that heavily bacterially contaminated *Fusarium* cultures of all kinds may be flat, slimy, and intensely pigmented, probably in part because mycolytic bacteria have digested most of the aerial hyphae. These bacteria may be antibiotic-resistant pseudomonads, not eliminated simply by culturing to antibiotic media. Microscopically, *Fusarium* has hyphae that are generally relatively broad with diameters mostly over 2.5 μm. *Acremonium* species have thin filaments mostly below 2.5 μm in diameter.

Means of rapid molecular identification of *Fusarium* species have recently appeared. Hue et al. (428) identified a ribosomal primer pair that selectively amplified DNA from *Fusarium* species with affinities to the teleomorph genera *Gibberella* and *Nectria*, including *F. oxysporum*, *F. solani*, and *Fusarium* subgenus *Liseola* (e.g., *F. proliferatum*, *F. verticillioides*), in addition to related nectriaceous species such as *Fusarium dimerum* and *Neocosmospora vasinfecta*. A *Monographella nivalis* (referred to by former name *Fusarium nivale*) isolate giving a positive response appears to have been misidentified, since the real *M. nivalis* (see section on this fungus below) is much more distantly related to *Fusarium* species than are many of the isolates that gave negative reactions with the primers of Hue et al. The sole *Acremonium* species tested, *A. strictum*, was not amplified. Given the taxonomic heterogeneity of *Acremonium* species and their dispersion throughout the Hypocreales and several related orders (296), it seems likely that at least some species of this genus would probably cross-react with any probes reacting to all *Fusarium* species. Whether medically important species would be included is difficult to predict. *Fusarium*, like *Acremonium*, is by no means monophyletic as currently defined. It may be argued that phylogenetic clade identification is more predictive than current anamorph genus identification in any case.

Hue et al. (428) also identified a primer pair apparently specific to the intersection of the phylogenetically related *Fusarium* subgenus *Elegans* and *F.* subgenus *Liseola* clades, as delimited by O'Donnell et al. (429), as well as to the more distant *F.* subgenus *Martiella* (e.g., *F. solani*). Not enough species or isolates were tested to advance this result beyond the preliminary stage, however.

A more specific but clearly more specialized rapid 28S ribosomal sequence identification technique for six medically important *Fusarium* species was described by Hennequin et al. (430). Since most of the "species" tested are in fact species complexes that were by no means fully explored in this study, it must be considered very preliminary, although more widely applicable in principle.

A number of papers have recently appeared on detection and identification of *Fusarium* from infected eyes using polymerase chain reaction in combination with panfungal 18S rDNA (431) or cutinase (432) primers. The studies are again in a preliminary stage, neither dealing with taxonomic diversity in the species tested nor with the problem of false positives derived from airborne contamination. They nonetheless show considerable promise.

Presumptive genus-level recognition of *Fusarium* elements in histopathology, and possible differentiation from *Aspergillus* elements in some cases by means of detecting in vivo phialide and conidium formation, is admirably summarized by Liu et al. (276). Still clearer genus-specific histopathology using cross-adsorbed fluorescent antibody reagents was developed by Kaufman et al. (433) to distinguish *Fusarium* from *Aspergillus* and *Pseudallescheria* in fixed tissue specimens. Due to close antigenic similarity, *Fusarium* filaments could not be distinguished from those of another hypocrealean fungus tested, *Paecilomyces lilacinus*, and even distinction from the Microascalean *Pseudallescheria boydii* required the use of two reagents. *Microascales* are relatively closely related to *Hypocreales* phylogenetically (296).

Morphological species identification in *Fusarium* may be relatively easy or very difficult, depending on the species involved and on the attention paid to newly recognized phylogenetic sibling species. The most important character by far in most medically important species is the shape of the phialides producing microconidia. It is important to distinguish *monophialides*, which have only a single apical opening, from *polyphialides*, which begin as a monophialide but rapidly grow extra toothlike extensions that give rise to additional clumps or chains of microconidia. In some species, such as *F. chlamydosporum*, that rarely cause human or animal disease, such proliferating extensions may each give rise to only a single conidium. The phialides showing this pattern are called "polyblastic phialides," and the resulting conidia, whether micro- or macro- in morphology, are referred to by some authors as "blastoconidia" or "mesoconidia." (See text below for *F. chlamydosporum* and *F. incarnatum*.)

The next most important character in species identification is to observe whether any microconidia that form do so in sticky heads or chains. Chain formation is not reliable on most common medical mycology laboratory media, which are too rich (434). *F. verticillioides* may form chains on PDA, but in at least some isolates of this species as well as many isolates of *F. proliferatum*, the use of special *Fusarium* media such as synthetic nutrient agar (435) or carnation leaf agar (436) may be required. As mentioned below, the present author has had good preliminary results using modified Leonian's agar (396). The best way to detect chains is to examine plates dry under the 10× compound microscope lens, or to use a dissecting microscope at its highest available magnification. The long and robust chains of *F. verticillioides* are especially conspicuous. In the absence of actual chain formation, the species potentially forming these structures all form radially symmetrical microconidia with a flattened base, suitable for balancing in chains. The conidia may be club-shaped, pyriform (pear-shaped), or napiform (turnip-shaped) in all cases with the truncate base just mentioned.

An intuitively appealing but much more difficult character to use in identification is the shape of macroconidia. This is relatively straightforward in ususual species such as *F. dimerum*, but otherwise can be very difficult to apply without extensive experience. The common medically important species with the most distinctive macroconidia is *F. solani* (i.e., from the phylogenetic point of view, the *F. solani* species complex). In many isolates, these conidia may be distinctively thick and blunt-ended, ruling out all such species as *F. oxysporum* with more sharply pointed macroconidia. There are, however, *F. solani* isolates with relatively pointed macroconidia. [See the range of shapes drawn by Gerlach and Nirenberg (437)]. Phialide shape is far superior for making the distinction among the less obviously distinguished members of these two species.

Also useful but relatively difficult to use is colony pigmentation. Many *Fusarium* species have characteristic colors, at least in uncontaminated isolates growing on PDA. Many begin as pale colonies, and *F. oxysporum*, *F. solani*, *F. verticillioides*, and *F. proliferatum* may remain quite pale in some isolates. As a multimembered species complex, *F. solani* is especially variable. The chestnut red-brown colors that form in some of its subtypes need to be distinguished from the vinaceous (red-wine-colored) and violaceous (purplish) colors that are characteristic of *F. oxysporum* and members of *F.* subgenus *Liseola*, as well as the carmine (vivid lipstick red) pigments that form in *F. chlamydosporum* and a variety of *Fusarium* species not known from human infection. It cannot be overstressed that these characters are highly variable when different growth media are used and as is frequently true in the identification of non-Onygenalean molds, Sabouraud agar is particularly unsuitable. The media and conditions specified in descriptive literature must be used to obtain interpretable results and make correct identifications. Note that many *Fusarium* species undergo a dermatophyte-like cultural degeneration, and isolates should if possible be identified before such processes can begin. Persons organizing quality control surveys must ensure that *Fusarium* isolates sent out to participants are still in a representative natural condition.

Distinction of most common *Fusarium* species at the biomedically useful level in which some closely related species complexes are lumped as single species (e.g., *F. solani*) may be accomplished using the monograph of Nelson et al. (438). Distinction of common medically important species at this level may be aided by exoantigen testing (439). The 20+-year-old synoptic key to common *Fusarium* species in *Compendium of Soil Fungi* by Domsch et al. (74) remains remarkably effective, although the treatment of the "*F. moniliforme*" complex in that manual must be compared with more modern concepts, and a few other names (e.g., that of the former *Fusarium semitectum*, now *F. incarnatum*) have changed. Higher-level morphological identification may be carried out using the pictorial atlas published by Gerlach and Nirenberg (437). A useful key for many species is Nirenberg's (440) key to fusaria on European crop plants. It includes all the repeatedly seen medically important fungi in temperate areas (the seldom-seen *F. chlamydosporum* is absent), as well as most potential clinical contaminant species in these areas. It also includes a key to the temperate and cosmopolitan subgenera of *Fusarium*. The recent molecularly informed revision of morphological distinctions in subgenus *Liseola* by Nirenberg and O'Donnell (441) must be considered. In general, accurate *Fusarium* identification at the reference level requires constant vigilance with regard to new literature.

The current state of the art in molecular identification of most medically important *Fusarium* species and species complexes is summarized by O'Donnell et al. (429) and O'Donnell (427). Considerably more work must be done, however, to place all the medically important *Fusarium* types into phylogenetically defined clusters and species concepts. There is currently a tendency for molecular studies less ambitious than those of O'Donnell and collaborators to be based on a naively small representation of the relevant biodiversity in this genus, and caution must be used in interpretation of such studies.

4. *Fusarium {aff. Gibberella} anthophilum* (A. Braun) Wollenw.

A disseminated infection in a Japanese acute lymphocytic leukemia patient gave rise to numerous skin lesions that were separately investigated via one biopsy and nine surface scrapings (442). Biopsy culture was negative, but cultures from the scrapings of six lesions gave rise to *Fusarium solani*, while two lesions grew *F. anthophilum*. The latter fungus was depicted and described, and its identity was stated to have been determined by *Fusarium* authority P. Nelson (after mis-identification as *F. oxysporum* by two less specialized authorities). The authors, however, explicitly stated that "septate hyphae *with chlamydospores*" (emphasis added) were seen in direct microscopy of all scraping samples, while their *F. anthophilum* isolate in culture was noted not to produce these structures. Indeed, the species is not known to do so (437). *F. solani* may produce abundant chlamydospores. The balance of probabilities in this case suggests that *F. solani* was the sole etiologic agent in the skin lesions and *F. anthophilum* was insignificant. (See the discussion under *F. proliferatum* for the distinction of *F. anthophilum* from similar-looking, known medically important members of *Fusarium* subgenus *Liseola*.)

5. *Fusarium {aff. Cosmospora} aquaeductuum* (Radl. &
    Rabenh.) Lagerh. Anamorph of *Nectria episphaeria*
    (Tode: Fr.) Fr. (*Fusarium* Subgenus: Eupionnotes)

This species is indexed as being of medical significance by Guarro and Gené
(443) and de Hoog et al. (69) on the basis of a record of *"Fusarium episphaeria"*
keratitis in a review by Pflugfelder et al. (342). The original authors of the record
[Mandelbaum, Forster et al. (444)], however, probably used this name in the
broad sense of Snyder and Hansen [see Booth (445)] to indicate *F. dimerum*, as
has commonly been done by several ophthalmological authors, particularly R. K.
Forster, the second author of the Pflugfelder et al. report. (See Ref. 393.) *F.
aquaeductuum* is not known to date from human or animal disease.

6. *Fusarium {aff. Gibberella} chlamydosporum* Wollenw. &
    Reink. (*Fusarium* Subgenus: *Sporotrichiella*)

A non-neutropenic American woman with lymphocytic lymphoma presented
with fever and chills that transpired to be at least partially caused by a persistent
F. chlamydosporum fungemia associated with colonization of the tip of her Brov-
iac catheter (446). Catheter removal and amphotericin B cured the infection. A
neutropenic woman was found to have a limited lesion of the nasal turbinate
caused by this fungus (447). Chlamydospores consistent with the species were
found intermixed with fungal filaments in the lesion and initially raised suspicions
of phaeohyphomycosis because of their dark color and positive Masson–Fontana
staining reaction. The infection was eliminated with surgery and amphotericin
B-lipid complex.

   This fungus is a soil fungus, associated with rhizospheres of many plants
as well as with various decaying plant materials, mostly in the tropics and subtropics. Both reported medical cases, however, were in the New York–Washington area. *F. chlamydosporum* produces the mycotoxin moniliformin (448).

Description.   Colonies grow on PDA 65.5 to 70 mm in 7 days. They are
floccose, and sometimes powdery with conidia. They are whitish at first, then
ochraceous to brownish to bright pink or rose. The reverse begins pinkish and
becomes carmine, wine-red to reddish-brown (437). *Conidiophores* begin as phialides arising singly on aerial hyphae, soon forming branched clusters, and less
commonly, closely appressed sporodochial clumps. *Phialides* appear monophialidic at first, but soon often proliferate, especially near the apex, to form polyphia-
lide-like "polyblastic phialides." These structures at maturity bear multiple,
short, toothlike fertile protuberances on which they give rise to so-called mesoconidia (412, 449) [terminology not supported by some *Fusarium* authorities
(415)] or blastoconidia (69, 450). In this species, these conidia are shaped for
the most part like typical *Fusarium* microconidia and are referred to as such by

Segal et al. (447). They differ from the conidia ordinarily formed on fusarial polyphialides (as seen in *F. proliferatum*, e.g.) not just in their size but also in their ontogeny; only a single conidium is formed per fertile denticle, not a multiconidial cluster or chain. Mature *polyblastic phialides*, then, are mostly irregularly elongate, straight, or curved, with up to 10 protuberant conidiogenous openings, 8–18 × 2–3 μm, giving an overall spiny appearance. *Sporodochial phialides* producing macroconidia are monophialidic and cylindrical with narrow apex, 10–16 × 2–3 μm. *Microconidia* are strongly predominant, mostly spindle-shaped and smooth walled. They are commonly one- or two-celled, but some conidia with up to five cells may be formed on polyblastic phialides. Single-celled forms are mostly (5-) 9–10 (-14) × 2–3 (-4) μm, two-celled forms up to 17 μm long. *Macroconidia* are formed sparsely or not at all, generally only in sporodochia. They are hyaline and gently crescent-shaped, with a hooked and sharply pointed apical cell and a small basal pedicel. They are mostly three-septate but commonly up to five-septate. When three-septate they are (21-) 30–34 (-41) × (2.5-) 3–4 (-4.5) μm; when five-septate commonly to 38 μm and uncommonly to 47 μm. *Chlamydospores* are abundant, intercalary or terminal, and often in chains or clusters. They are smooth or strongly roughened and globose or nearly so, becoming brown and measuring 7 to 17 μm when formed singly.

For distinction from other medically important fusaria, see the discussion under *F. incarnatum* below, as well as Guarro and Gené (443). Morphological distinction from species traditionally clustered with *F. chlamydosporum* in *Fusarium* subgenus *Sporotrichiella* is briefly summarized by Kiehn et al. (446); molecular distinction has also been possible since the 25S ribosomal sequence study of Logrieco et al. (451).

7. *Fusarium {aff. Nectria} coeruleum* (Libert) Sacc. (Common Synonym: *Fusarium solani* var. *coeruleum*; *Fusarium* subgenus: *Martiella*)

This strongly pigmented fungus related to *F. solani* was indicated as causing black-grain mycetoma in the ankle of a Thai farmer (452). The identification was confirmed (as *F. solani* var. *coeruleum*) by *Fusarium* authority C. Booth of CMI (CABI). Recent molecular studies have cast doubt on this then-credible identification, as molecularly confirmed *F. coeruleum* isolates derive exclusively from infected potatoes (452a).

This species is primarily known as a pathogen of stored potatoes, but has been also reported from other plant species and from nematode eggs. The other reports from substrates other than potatoes are currently considered dubious until similar reports are molecularly substantiated.

Colonies on PDA grow 47.5 to 52.5 mm in 7 days. They are usually thinly or nearly appressed-felty. They are locally low-floccose, and beige, yellow-brown, dull purple to bluish, with cream or purple, slimy sporodochial masses. The reverse is typically strong ink blue-purple, sometimes yellow-brown to dull violet. *Conidiophores* consist of single phialides arising laterally on aerial hyphae, or small groups of phialides formed on rudimentary branching structures, or closely appressed sporodochial clumps (Fig. 77). *Phialides* produce microconidia in aerial mycelium and are strictly monophialidic, long-cylindrical, and 15–25 × 2.5–3.5 µm. *Microconidia* are not formed as a separate conidial morph, but a few microconidium-like, single-celled, ''subdeveloped'' (437) conidia may be present, always in sticky heads, and never in chains. They are hyaline, smooth-walled, ellipsoidal to obovoidal, or sometimes slightly curved, and (8-) 10–17 (-22) × 4–5 µm. *Macroconidia* are overwhelmingly predominant, sometimes in sporodochia, hyaline, ''only slightly curved'' (437), mostly three-septate, less commonly 4 to 5-septate, relatively broad in relation to length, when three-septate (21-) 28–45 (-50) × 4–6 µm, with little or no pedicellate character in the basal cell, and blunt apical cells. *Chlamydospores* are usually abundant, smooth-walled, terminal or intercalary, single, in chains or clustered, pale or brownish, sometimes in macroconidia, and 7 to 10 µm.

This species is distinguished from most *F. solani* isolates by its heavy ink-blue colony reverse pigmentation. Occasional *F. solani* isolates will produce bluish or dark reverse pigments, but can be distinguished by their abundant and

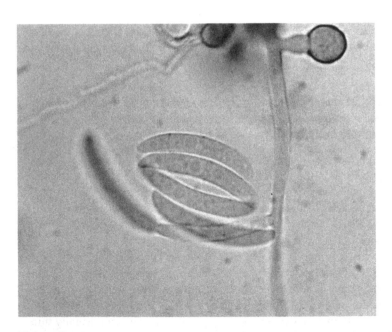

Figure 77 *Fusarium coeruleum* (CBS 133.73 macroconidia, phialide, and chlamydospore).

well-differentiated production of microconidia on PDA or specialized *Fusarium* media. Although *F. coeruleum* has relatively long and thin phialides for a *Fusarium* species, *F. solani* isolates may have phialides that are considerably longer and thinner. *F. solani* phialides commonly approach 40 μm, while *F. coeruleum* phialides are seldom longer than 25 μm.

8. *Fusarium {aff.* unknown} *dimerum* Penzig (in Sacc.)
    [Common Synonym: *Fusarium episphaeria* (Tode: Fr.)
    Emend Snyder & Hansen *Pro Parte* (i.e., Lumped with
    Other Taxa Under This Name)

Fusarium *Subgenus*: Eupionnotes.   This fungus is uncommonly but regularly isolated from keratitis; cases have been documented by Zapater and collaborators (453, 454) as well as by more recent investigators (455) in Argentina and by Liesegang and Forster (393) in Florida. (See also the discussion of *F. aquaeductuum* above.) Identifications in the Zapater studies were confirmed at CBS. Like most fusaria in eye infections, *F. dimerum* generally responds to pimaricin (456). *F. dimerum* has also been confirmed as an agent of endocarditis in France (457).

Fatal *F. dimerum* fungemia has been reported in a catheterized Slovak leukemia patient who had experienced prolonged neutropenia (417). Poirot et al. (458) reported an acute myeloblastic leukemia patient who grew *F. dimerum* in several urine cultures just prior to bone marrow regeneration, but became culturenegative with regeneration and amphotericin B treatment.

In the present author's experience in Ontario, Canada, this is a common contaminant producing, saprobic, noninvasive growth on wounds and ulcers. Unpublished cases of CAPD-related peritonitis are under investigation. *F. dimerum* is a soil fungus also found on various plant substrates.

Description.   Colonies grow on PDA 27.3 to 31.5 mm in 7 days. They are completely slimy with conidia or produce limited amounts of aerial mycelium. They are orange on both the surface and reverse.

Conidiophores consisting of phialides arise singly on both aerial and submerged hyphae, or in small or moderately complex branched tufts (Fig. 78). *Phialides* are monophialidic, mostly short and inflated to wine-bottle-shaped with a narrowed apical neck. They are occasionally longer and cylindrical, and 6–32 × 2.5–5 μm. *Microconidia* are not considered to be formed; there are, however, a small proportion of single-celled conidia mixed with and intergrading in size and shape with the macroconidia, mostly 7–11 × 2.0–2.8 μm. *Macroconidia* are abundant, hyaline, and crescent-shaped. They are unicellular at first, but mostly become one-septate at maturity, less commonly two- or three-septate; when one-septate 7–30 × 2–4 μm, with pointed apical ends and minutely blunted or pedicellate basal ends. *Chlamydospores* are uncommon to common, mostly intercalary, usually smooth, globose or nearly so, and 6 to 12 μm.

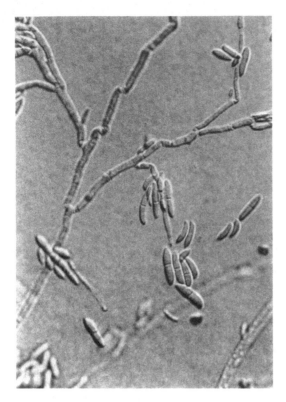

Figure 78   *Fusarium dimerum.*

     Isolates from clinical settings often appear to correspond to the subtaxon *F. dimerum* var. *pusillum* (Wollenw.) Wollenw., with a preponderance of unicellular or belatedly one-septate macroconidia mostly in the 7 to 9 μm length range. This little-recognized (437) subtaxon requires molecular investigation.

9.   *Fusarium equiseti* (Cda.) Sacc. Anamorph of *Gibberella intricans* Wollenw.

Expressed blisters from a pustular dermatitis in a dog in 1928 grew *Mucor racemosus* as well as a fungus that was investigated biomedically by Leinati (459) and described taxonomically by Curzi (460) as *Fusarium moronei* Curzi. Wollenweber and Reinking (461) later synonymized this taxon with *Fusarium scirpi* var. *caudatum*, which in turn was synonymized with *F. equiseti* by Booth (445). The animal disease record was indexed by Austwick (426) as "one of the earliest records of *Fusarium* infection." The case report, however, did not include direct microscopy of the pustule contents, and the concurrently isolated *M. racemosus*, assuming it was correctly identified, is probably nonpathogenic (69). Moreover, *F. equiseti* isolates studied so far have a maximum growth temperature of 28°C (74), and there have been no subsequent valid records of zoopathogenesis by this fungus. This, then, appears to be a spurious record based on contamination.

10. *Fusarium fujikoroi* Nirenberg Anamorph of *Gibberella fujikoroi* (Sawada) Wollenw. (*Fusarium* Subgenus: *Liseola*)

This organism was reported without mycological substantiation under the teleomorph name *Gibberella fujikoroi* from an etiologically well-documented North Carolina keratitis case by Anderson et al. (92). It seems unlikely that a teleomorph was actually observed for this heterothallic species, in which laborious mating with tester isolates must be done to make this observation in culture. Possibly a member of *Fusarium* subgenus *Liseola* was isolated and linked to *G. fujikoroi* (used in the broad sense for the related complex of then mostly unnamed species) via literature descriptions or by an individual with phytopathological experience. (See the discussion of *Aspergillus versicolor* and *Penicillium spinulosum* case records above for further analysis of the eccentric but by no means wholly discreditable mycology in the Anderson et al. publication.)

F. fujikoroi is a rice pathogen known only from Asia and Australia (437). In culture it is scarcely distinguishable from *F. proliferatum* except that very few polyphialides are formed, and conidial chains are short. Its production of the mycotoxin and plant hormone gibberellic acid is distinctively high compared to that of other members of subgenus *Liseola* (437).

Distinction of *F. fujikoroi* from numerous newly described and minutely morphologically distinguished species in the "*Gibberella fujikoroi* species complex," most not known from mammalian pathogenesis, is provided by Nirenberg and O'Donnell (441). Molecular characterization is coordinately given by O'Donnell et al. (429).

11. *Fusarium {aff. Gibberella} incarnatum* (Rob.) Sacc. (Common Synonyms: *F. Semitectum*, *F. pallidoroseum*; *Fusarium* Subgenus: *Arthrosporiella*)

A Brazilian man suffered fatal endocarditis after a newly implanted metallic prosthetic heart valve became heavily colonized with *F. incarnatum* (reported as *F. pallidoroseum*) (462). The identity of the organism was well documented. An earlier report of the same organism (as *F. semitectum*) from an electroshock-related burn wound in Thailand (463) described "dark-brown" filaments in tissue stained with hematoxylin and eosin. The discord between this finding and the general finding that this fungus produces only hyaline hyphae was queried by McGinnis et al. (462). *F. incarnatum* may produce short chains of pale to moderately brownish chlamydospores, and these may possibly be seen in tissue (see notes on *F. chlamydosporum* above), but the repeated and exclusive use of dark brown to describe the structures seen makes it appear more likely that the burn was infected by more than one species or that *F. incarnatum* occurred as a surface contaminant or colonizer and overgrew an invasive melanized fungus in culture. Another case reported in the same paper featured electrical burns colonized by *Curvularia lunata*, a melanized fungus. Rush-Munro et al. (464) depicted filaments attributed to *F. incarnatum* in intertriginous toe skin, but gave no supporting etiologic or mycological details.

    F. incarnatum is a mostly tropical and subtropical soil fungus causing rots in stored crops with high water content, particularly potatoes and fruits. It produces moniliformin as well as mycotoxins in the zearalenone class (448).

Description. Colonies grow on PDA 65.5 to 70 mm in 7 days. They are floccose and sometimes powdery with conidia. They are dirty-white to cinnamon-brown above, very rarely with small (<1 mm) and pale brown sclerotial bodies. The colony reverse begins pale to pinkish or peach, soon becoming ochraceous to cinnamon-brown. [Gerlach and Nirenberg (437) stress "red, violet or bluish pigment lacking".] *Conidiophores* beginning as phialides arise singly on aerial hyphae, soon forming branched clusters (Fig. 79). *Phialides* are mostly monophialidic at first, but soon often proliferate sympodially, especially near the apex, to give rise to new phialidic extensions. In addition, polyphialidelike "polyblastic phialides" with multiple, short, toothlike fertile protuberances are formed, giving rise to so-called mesoconidia (412, 449) or blastoconidia (69). These conidia in *F. incarnatum* (unlike those in *F. chlamydosporum*; see above) are shaped for the most part like typical *Fusarium* macroconidia and will be referred to here as such. They differ from normal macroconidia by lacking the foot cell typically seen in *Fusarium*. In comparison with the microconidia ordinarily formed on fusarial polyphialides (as seen in *F. proliferatum*, e.g.), they differ not just in their size but also in their ontogeny; only a single conidium is formed per fertile denticle, not a multiconidial cluster or chain. Mature polyblastic phialides, then, are mostly irregular, with two to four toothlike conidiogenous openings, 8–25 × 2.5–5 μm. *Microconidia* are relatively uncommon, intergrading with macroconidia. They are mostly spindle-shaped or slightly curved, and smooth-walled. They are usually one-celled, but some two-celled conidia also present. The single-celled forms are mostly (4-) 7–12 (-17) × 2–3.5 (-4) μm. *Macroconidia* are predominant, mostly scattered in the aerial mycelium. They are hyaline and nearly straight in the midsection, especially in the inner wall of curvature, with a hooked and sharply pointed apical cell and a thinly truncate-conical, apedicellate basal cell. They are mostly three-septate, with an admixture of commonly four to five- (rarely to seven-) septate conidia; when three-septate (13-) 20–40 (-50) × (2.5-) 3.5–5 (-6.5) μm. *Chlamydospores* are sparse, mostly intercalary, and smooth. They are globose or nearly so, become brown, and are 6 to 12 μm.

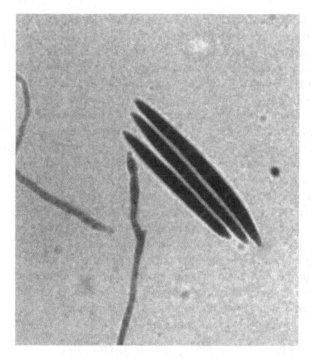

Figure 79 *Fusarium incarnatum* CBS 132.73 polyphialide.

This species can be distinguished from most other *Fusarium* species involved in human disease by its macroconidia that arise as solitary blastoconidia
on individual denticles of polyblastic phialides. It also completely lacks the purple
colors of *Fusarium* subgenera *Liseola* and *Elegans* (e.g., *F. proliferatum*, *F. oxysporum*) and the red colors of another rarely etiologic species with polyblastic
phialides, *F. chlamydosporum*. At least in fresh isolates, it forms macroconidia
commonly on PDA, unlike *F. chlamydosporum*, in which macroconidia are
scarce and strictly associated with sporodochia.

12. *Fusarium* {aff. unknown} *lacertarum* Subrahmanyam
     (Published in Orthographic Error as *Fusarium laceratum*;
     Fusarium Subgenus: Compared in Its Original Description
     with Members of *F.* subgenus *Discolor*; Unlikely to Be
     a true *Fusarium*)

This species, which needs further study to determine its correct taxonomic position, has only been recorded from parasitized lizards. It has *Fusarium*-like conidia, but the existing descriptions of its conidiogenesis suggest it may belong
to another genus. The species description by Subrahmanyam (465) was based on

an isolate from an infected "house lizard" from Poona, India. The animal was profusely colonized by white mycelium, which bore copious conidia. This finding appeared belatedly to confirm the detailed description of a similar fungus by Blanchard (466) from equally profusely colonized skin cankers on a green lizard (*Lacerta viridis*) of Italian provenance. Blanchard compared the fungus to a description of *Selenosporium urticearum* Corda, now considered a synonym of *Fusarium lateritium* [*F. lateritium* var. *mori fide* Booth (445)]. Austwick (426) ascribed Blanchard's record to this species. He was evidently unaware that Blanchard's organism had subsequently been named *Fusarium cuticola* (Blanchard) Guégen (467, 468) and had also been recorded in additional lizards, including a chameleon. In any case, the now long-lost isolate described by Blanchard has no resemblance to the tree-parasitizing *F. lateritium*, but is similar to *F. lacertarum* in several distinctive respects; namely, yellow colony pigmentation, a rapid growth rate on agar medium, sharply pointed macroconidia mostly under 30 μm with one to three indistinct (465) or extremely thin (466) septa, and most notably the production of most conidia on short, lateral protuberances rather than well-defined phialides, as well as the apparent production of at least some conidia by direct blastoconidiation from the sides of hyphae (possibly from adelophialides or short *Microdochium*-like sympodial polyblastic initials). The description below is after Subrahmanyam (465), with direct quotes indicated.

Description. Colonies on potato sucrose agar at 28°C grow 70 mm in 3 days, with thinly floccose, white mycelium. At first they are pale on the surface and reverse, but after 30 days become buff above and pale lemon-yellow below or near the colony center. *Conidiophores* are "short, lateral, branched or unbranched" up to 7 μm long, and disposed laterally along the hyphae. Discrete "phialides" are rarely present, 2.5–4 × 1–1.5 μm. *Conidia* are found mostly with a nearly straight or gently curved midregion and curved and sharply pointed apical and basal regions. They are hyaline, smooth-walled, zero- to four-septate, and 6.6–30.8 × 2.2–3.3 μm, with notably thin or obscure septa and with basal cell lacking a pedicel. A small proportion of small, clavate to broadly pyriform no to one-septate conidia are also produced. *Chlamydospores* are common, terminal or intercalary, solitary, catenulate, or "in clumps." They are hyaline, smooth or rarely roughened, and 6.6–18.5 × 5.5–18.7 μm.

Blanchard (466) depicts mostly two to three-septate conidia but his drawings also include one five-septate conidium. He gives 25 μm as the maximal conidial length seen.

13. *Fusarium {aff. Gibberella} moniliforme* Sheldon
    (See *F. verticillioides* Below)

Fusarium *Subgenus:* Liseola. Discussion of why the earlier epithet, combined as *F. verticillioides*, is preferred to *F. moniliforme* is given by Gams (469).

14.   *Fusarium {aff. Gibberella} napiforme* Marasas, Nelson &
      Rabie (*Fusarium* Subgenus: *Liseola* (Referred to by Some
      Authors as *F.* Subgenus *Dlaminia*, an Unacceptably
      Artificial, Paraphyletic Group Created to Encompass
      Organisms with Affinities to *F.* Subgenus *Liseola* but
      Possessing Chlamydospores)

This species has been verified from a disseminated infection in a leukemia patient
(470). *F. napiforme* is mainly known from millet and sorghum from Africa, but
has also been found in warm grassland soils elsewhere (471) (Fig. 80).

Colonies grow on PDA 46 to 82 mm in 7 days. They are floccose and
whitish above, with pale to deep purple reverse. *Conidiophores* consisting of
single phialides arise laterally on aerial hyphae, or small groups of phialides
formed on rudimentary branching structures, or closely appressed sporodochial
clumps. *Phialides* are monophialidic, subulate (tapered, awl shaped), and 14–31
× 1.5–3 μm. *Microconidia* are formed in short chains or sticky heads. They
are hyaline, smooth-walled, and always with truncate bases but otherwise very
variable, with a proportion of conidia obovoid to fusoid, no to one-(less commonly to three-) septate and 6–23 × 1.5–4.5 μm. Others are lemon-shaped or
the characteristic napiform (turnip or rutabaga-shaped), no to one-septate, and
9–15 × 6.5–10.5 μm.

Macroconidia are produced from scattered monophialides or from sporo-

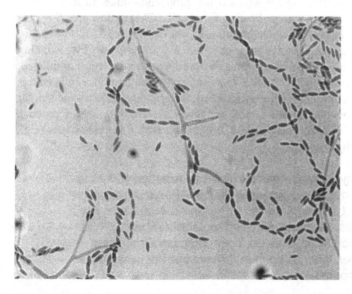

Figure 80   *Fusarium napiforme* CBS 673.94, phialides and microconidia.

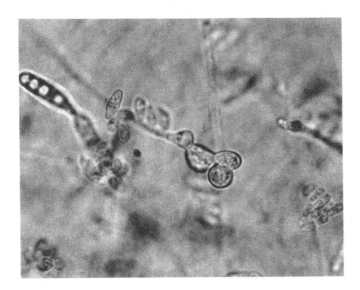

Figure 81 *Fusarium napiforme* CBS 673.94, chlamydospores.

dochia (in the latter case forming orange conidial masses). They are hyaline, falcate, typically with the inner wall nearly straight and the outer wall more distinctly curved, mostly five-septate with a lower proportion three- or four-septate and 29–90 × 3.5–5.0 μm, with pedicellate basal cells. Chlamydospores (Fig. 81) are common. They are smooth-walled, hyaline to pale brown, and 5–12 × 4–8 μm.

This fungus differs from the more common *F. verticillioides* by producing well-developed chlamydospores, as well as by producing a proportion of inflated, napiform, or limoniform microconidia. *F. proliferatum* may produce inflated microconidia but never produces chlamydospores. Several other related and similar chlamydospore-producing species, most notably *F. oxysporum*, never produce microconidia in chains, and may have microconidia with different shapes (e.g., allantoid in *F. oxysporum*).

15. *Fusarium nygamai* Burgess & Trimboli Anamorph of
     Gibberella nygamai Klaasen & Nelson [*Fusarium*
     Subgenus: *Liseola* (Referred to by Some Authors as
     F. Subgenus *Dlaminia*, an Unacceptably Artificial,
     Paraphyletic Group Created to Encompass Organisms
     with Affinities to *F*. Subgenus *Liseola* but Possessing
     Chlamydospores)]

This species caused disseminated infection in a granulocytopenic lymphoma patient who had recently traveled to Egypt (472). The fungus in nature appears to be pantropical in distribution, and is associated with plant roots, particularly of sorghum, maize, and beans, as well as with sorghum grain heads (473).

Description. Colonies grow on PDA 58 to 82 mm in 7 days. They are floccose to powdery, whitish above at first, later dull violet with a greyish-orange or dark violet central conidial mass with pale to deep purple reverse. *Conidiophores* consisting of single phialides arise laterally on aerial hyphae, or small groups of phialides are formed on rudimentary branching structures or closely appressed sporodochial clumps (Fig. 82). *Phialides* are usually monophialidic and subulate (tapered, awl shaped) and 13–40 × 1–3 μm, with polyphialides also uncommonly present. *Microconidia* form in sticky heads or—especially on specialized media such as carnation leaf agar or synthetic nutrient agar—in a mixture of heads and short (usually 10 or fewer conidia) chains. They are hyaline, smooth-walled, obovoid to fusoid, no- to one-septate, and 5–27 × 2–3 μm, with truncate bases. *Macroconidia* produced mainly from sporodochia, forming greyish-orange, slimy masses, appear under the microscope as hyaline, nearly straight to falcate, mostly three-septate with a lower proportion one-, two-, or four-septate. They are 25–54 × 2–5 μm, with pedicellate basal cells. Chlamydospores are usually common but uncommon in some isolates. They are often in chains or clumps, smooth-walled, and hyaline to pale yellow-brown (Fig. 83).

This fungus is distinguished from the superficially similar *F. oxysporum* by the sometimes difficult to detect production of conidia in chains, as well as

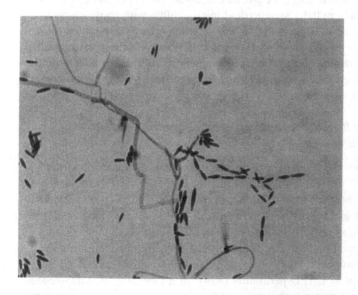

Figure 82 *Fusarium nygamai* CBS 675.94, phialides and microconidia in heads and chains.

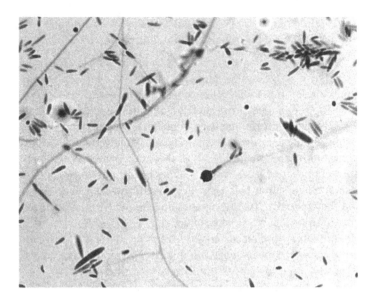

Figure 83   *Fusarium nygamai* CBS 675.94, chlamydospore.

the production of predominantly catenulate (in chains) or clumped chlamydospores, as opposed to the single or paired chlamydospores generally seen in *F.
oxysporum*. Detecting radially symmetrical microconidia with conspicuous truncate bases is often an indication that a fungus other than *F. oxysporum* should
be considered. *F. verticillioides* and *F. proliferatum* are normally distinguished
by their lack of chlamydospores (but see comment about thickened cells in description of *F. verticillioides* below). *F. napiforme* differs by its high proportion
of swollen, napiform microconidia.

16.   *Fusarium {aff. Gibberella} oxysporum* Schlecht. (*Fusarium* subgenus: *Elegans*)

F. oxysporum shows the "general pattern of involvement by major opportunistic
Fusarium species" described in the introductory section for *Fusarium* above.
Case literature for this well-established opportunist has been reviewed previously
(see same section) and will not be reviewed in detail here. De Hoog et al. (69)
have recently given a species-specific synopsis. In general, this fungus is a major
agent of disseminated and cutaneous infection in the immunocompromised—
especially the neutropenic—patient. It also causes infections related to major
barrier breaks, such as CAPD peritonitis (474). In a rare case of AIDS-related
Fusarium infection, a port-a-cath implanted for cytomegalovirus therapy was the

locus of entry of a disseminated *F. oxysporum* infection in an HIV patient (475). The infection responded to liposomal amphotericin B. An apparently nonimmunocompromised patient who had adult respiratory distress syndrome and was receiving extracorporal membrane oxygenation died of a disseminated *F. oxysporum* infection that was first detected as a positive bronchial lavage specimen (476).

There are some distinctive features about the pathogenic record of *F. oxysporum*. In keratitis, it is frequently seen in some surveys (420), but is generally significantly less common than *F. solani* (393, 456). It is correspondingly less common than *F. solani* in exogenous endophthalmitis (412). It may be less virulent and therefore less likely to spread spontaneously from keratitis to the posterior chamber and the vitreous cavity (342). Rosa et al. (30) reported that *F. oxysporum* became more common than *F. solani* in Florida keratitis and endophthalmitis in the period from 1982 to 1992, and also caused most cases associated with the use of soft contact lenses. The large numbers of cases reported for *F. oxysporum* statistically overwhelm the etiologic confirmation shortcomings of this paper as discussed above under *Aspergillus glaucus* (e.g., potentially as few as one-third of the cases confirmed by direct microscopy). No identification criteria were given for the species, but the distinction among the major *Fusarium* species may have been considered routine after many years of investigation in Florida. Chodosh et al. (477) recently reported a keratitis ascribed to *F. oxysporum*, apparently contracted during routine ophthalmic examination, responding to tobramycin in vivo while the fungal isolate also showed high susceptibility in vitro. Natamycin therapy was begun as soon as a fungus was grown, but the eye had already substantially improved with antibacterials. Corneal scrapings showed no fungal elements, making it quite possible that the fungal isolation was purely fortuitous; however, no bacteria were obtained in culture. The identification was again undocumented. Ocular infection by *F. oxysporum* has also been recorded in snakes (426).

F. oxysporum is by far the most commonly reported *Fusarium* from onychomycosis, and shares a distinctive niche with a few other fungi, most notably *Trichophyton mentagrophytes* and members of *Acremonium* subgenus *Acremonium*, in the causation of superficial white onychomycosis (49, 50, 464). Another distinctive regularity of *F. oxysporum* onychomycosis is that the nails of the hand are often affected, as in the first well-substantiated case reported by Ritchie and Pinkerton (478). Regular occurrence of fingernail infections, including superficial, distal–subungual, and paronychial onychomycoses, was documented by Gianni et al. (479). An unusual type of nail infection, proximal subungual onychomycosis, was well documented from both finger- and toenails in cases in which paronychia was also seen (480, 481). Four pedal and three digital onychomycoses, as well as two cases of pedal intertrigo, were studied by Romano et al. (482). All cases were meticulously confirmed as *Fusarium* infections. Species identification criteria were given in a later paper (483). In the intertrigo cases, in response to earlier demonstrations that *Fusarium* species could grow saprobi-cally on wounds, ulcers, and other moist lesions of the foot, including tinea pedis lesions, without penetrating tissue (422, 423), the authors made biopsy slides that unequivocally demonstrated fungal tissue penetration. Also, they repeated culture and direct microscopy four times at 20-day intervals to probabilistically exclude cryptic dermatophytosis.

Very rarely, *F. oxysporum* may cause an extensive subcutaneous infection in an immunocompetent patient (e.g., a case in which it invaded an arterial leg ulcer with invasion of nearby blood vessels). In this case it failed to respond to amphotericin B and local debridement, and the leg was ultimately amputated (484). Some tissue penetration in less severe ulcer cases was seen by English (485). Landau et al. (486) described a severe foot ulcer that had extended over a 1-year period in a 69-year-old man and that had a heavy superficial growth of *F. oxysporum* with copious filaments in direct smears but no detectable tissue penetration in biopsies. Despite the lack of tissue penetration, treatment with ketoconazole appeared to cause rapid resolution of the lesion. Interestingly, *F. oxysporum* is almost completely resistant to this drug in vitro (414). Extensive, purely cutaneous infections also occur in immunocompromised patients whose immunodeficiencies differ from those seen in the leukemia and aplastic anemia patients most likely to experience *Fusarium* infection. A 16-year-old Sri Lankan girl, thought likely to have an uncharacterized immunodeficiency, suffered a life-long syndrome characterized by superficially chromoblastomycosis-like areas of subcutaneous *Fusarium* infection on her arms (487). The tissue was heavily in-vested with hyphae and showed microabscess formation and a chronic granulo-matous reaction. The identification of *F. oxysporum* was confirmed by L. Ajello of CDC.

F. oxysporum has a widespread distribution in soils and has many geneti-cally differentiated forms causing wilt, damping off, and other diseases in various plants. Thrane (448) states that it produces moniliform and a variety of other secondary metabolites, although many isolates are not discernibly toxigenic.

Description. Colonies on PDA grow 65.5 to 70 mm in 7 days. They are felty to floccose, not infrequently bald and wet with bacterial contamination in primary isolates. They are whitish above or with a purple tinge, deep vinaceous to violet in heavily bacterially affected isolates, sometimes spotted with orange sporodochia, and very rarely (in isolates seen clinically) with small (<1 mm), cauli-flowerlike, pale brown to bluish sclerotial bodies. The colony reverse is usually pale vinaceous to deep violet. *Conidiophores* arise singly on both aerial and submerged hyphae, in small branched tufts, or uncommonly, in sporodochial clumps (Fig. 84). *Phialides* are monophialidic, essentially flask-shaped, cylindrical or wine-bottle-shaped with a narrowed apical neck, but often somewhat inflated near the base and sometimes gently curved. They are rather short, 8–14 × 2.5–3.0 μm (10 to 25 μm long in sporodochia), and sometimes reduced to short, adelophialidic fertile

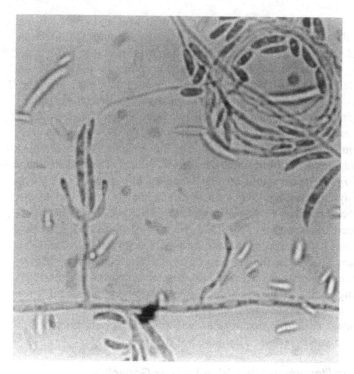

Figure 84 *Fusarium oxysporum.* From Laboratory Handbook of Dermatophytes, by J. Kane et al., 1997, Star Publishing Company, Belmont, CA (see Ref. 89). Used with permission.

openings on the sides of hyphae in submerged mycelium. *Microconidia* are in sticky heads, never chains. They vary from ellipsoidal to cylindrical, to slightly curved to allantoid (sausage-shaped) to reniform (kidney-shaped). They are hyaline and smooth-walled. They are usually one-celled but some two-celled conidia are also present. Single-celled forms are mostly 5–9 (-13) × 2.4–3.5 μm. *Macroconidia* are commonly formed, in sporodochia or in ordinary aerial mycelium. They are hyaline, nearly straight in the midsection or more often moderately curved, mostly three-septate, less commonly 4- to 5 (-7)-septate; when three-septate (18-) 27–42 (-54) × 3–5 μm. They have pedicellate basal cells and pointed, often somewhat hooked apical cells. *Chlamydospores* are commonly formed, often delayed until after 7 days, and terminal or intercalary. They are usually smooth, globose or nearly so, and 7 to 11 μm.

This variable species can easily be confused with opportunistic members of *Fusarium* subgenus *Liseola*, especially *F. nygamai*, *F. napiforme*, and immature *F. proliferatum*, as well as a host of species not known from human or animal infection. The distinction should not be attempted with Sabouraud agar.

Specialized *Fusarium* media such as synthetic nutrient agar and carnation leaf agar are optimal. PDA is less so but serviceable in cases of confirmed infection. Modified Leonian's agar seems to work well for the present author but has not been subjected to formal comparison. The chlamydospores of *F. oxysporum*, when suitably promptly formed, can distinguish it from immature *F. prolifera-tum*. Examination of aerial mycelium under the dissecting microscope usually reveals at least some conidial chains in *F. proliferatum*, at least on specialized *Fusarium* media and Leonian's. *F. proliferatum* isolates may sometimes form polyphialides rapidly and be easily distinguishable. Some isolates, however, have only sparse polyphialides until after approximately 10 days. The clavate shape of most *F. proliferatum* microconidia strongly suggests this species even in iso-lates in which polyphialides and chains are not immediately seen, and further investigation is necessary in all *F. oxysporum*-like strains in which clavate micro-conidia with truncate bases predominate. This symmetrical shape with a flat base suggests adaptation to formation of the carefully balanced conidial chains that are typical in many species of the subgenus *Liseola*. The kidney- to sausage-shaped microconidia commonly seen in *F. oxysporum* would not be physically capable of forming chains; however, some *Liseola* members may form a few curving conidia from submerged mycelium. (See the additional discussion under *F. nygamai* for distinction from that species, which so far is only known from tropical areas. For distinction from *F. solani*, see that species.)

17. *Fusarium proliferatum* (Mats.) Nirenberg ex Gerlach & Nirenberg, Anamorph of an Unnamed *Gibberella* Species [Mating Population D of the *G. fujikoroi* complex (488), *Fusarium* Subgenus: *Liseola*]

The involvement of this fungus in human disease is primarily known from three disseminated infections in leukemia patients (489–491), two of which were suc-cessfully cured (490, 491). Based on several unpublished cases seen by the pres-ent author in Ontario, where this fungus was a predominant cause of opportunistic mycosis in connection with neutropenia, it probably shares the "general pattern of involvement by major opportunistic *Fusarium* species" described above in the introductory section for *Fusarium*. Isolates with sparse or delayed formation of polyphialides could easily be misidentified as *F. oxysporum*, and some of the medical history of *F. proliferatum* may be concealed under this name.

In nature, *F. proliferatum* has been frequently isolated from corn (*Zea mays*), sorghum, wheat, figs, diverse other plant substrates, various insects, and soil (429). It produces moniliformin as a secondary metabolite (448), as well as fumonisins (492).

Description.   Colonies on PDA grow 65.5 to 70 mm in 7 days. They are thinly floccose to raggedly tufted. They are whitish above, sometimes becoming dull violet, with pale to deep vinaceous reverse. *Conidiophores* consisting of

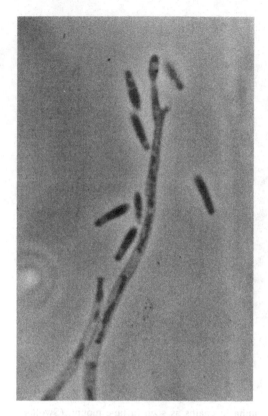

Figure 85 *Fusarium proliferatum* polyphialide and monophialide.

phialides arise singly on aerial hyphae, in small branched tufts, or in more elaborate antlerlike branching structures (Fig. 85). Uncommonly, sporodochial clumps
may appear, especially in cultures over 14 days old. *Phialides* are monophialidic
and candle-shaped at first but after some days producing spinelike lateral proliferations, usually from just beneath the apex, giving rise to asymmetrically forking
phialides, occasionally producing further lateral proliferations, giving an overall
antlerlike appearance, 11–32 × 2.3–3.5 (-4.2) μm. *Microconidia* are in sticky
heads or long chains, with the latter most abundantly formed on specialized media
such as carnation leaf agar or synthetic nutrient agar (Fig. 86). They are hyaline,
smooth-walled, variable in shape, mostly clavate with a flattened base but intermixed with inflated, pyriform to globose-apiculate forms, usually aseptate, and
mostly 7–9 (-11 when inflated) × 2.2–3.2 (-7.7 or occasionally as wide as 12.5
when inflated) μm. *Macroconidia* are often formed only after extended culture
or not at all, sometimes in sporodochia, hyaline, nearly straight to falcate, mostly

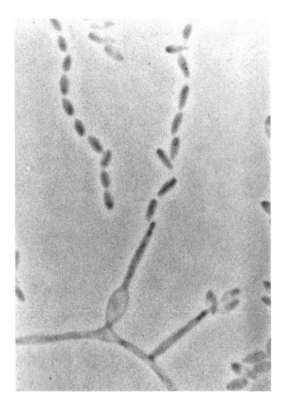

Figure 86 *Fusarium proliferatum* conidia in chains as seen in tape mount. (Swollen conidiophore base seen is not typical.)

three- or five-septate, (19-) 30–58 (-79) × 2.6–5 μm, with pedicellate basal cells. *Chlamydospores* are absent.

Distinction of *F. proliferatum* from *F. oxysporum* can be achieved by show-ing polyphialides, at least some conidial chains, mainly clavate microconidia with a truncate base (i.e., blackjack- or cosh-shaped), often mixed with some pyriform conidia, and an absence of chlamydospores. (For additional details, see the de-scription of *F. oxysporum.*)

F. verticillioides is readily distinguished by its lack of polyphialides. *F. anthophilum* produces polyphialides but differs from *F. proliferatum* by produc-ing only sticky heads, never chains, of mostly broadly pyriform microconidia, especially in early growth on media with relatively low nutrient content. Later mi-croconidia may be almost cylindrical to allantoid (sausage-shaped) (437). There are a number of *Fusarium* species specific to common household plant substrates [e.g., *F. lactis* Pirotta & Riboni from figs and *F. phyllophilum* Nirenb. & O'Donn. (=*F. proliferatum* var. *minus* Nirenb.) from potted *Sansevieria* ("good luck plant") and *Dracaena*] that strongly resemble the more ecologically catholic *F. proliferatum*. Causation of opportunistic infection by such fungi is not recorded. With patience and special media, these species can be distinguished morphologically using the key provided by Nirenberg and O'Donnell (441); otherwise, they may be identified by β-tubulin gene sequencing as noted by O'Donnell et al. (429). As an example of the types of morphological distinctions made, here are Nirenberg and O'Donnell's summaries of salient characters distinguishing *F. proliferatum* and *F. lactis*: *F. proliferatum* "produces clavate conidia mainly in long, linear, crowded chains from mono- and polyphialides in the dark" on synthetic nutrient agar, while *F. lactis* "produces obovoid conidia in zigzaglike, short (<15 conidia) to medium (15–30 conidia), crowded chains, mainly on polyphialides." *F. phyllophilum*, formerly called *F. proliferatum* var. *minus*, differs from *F. proliferatum* by producing "short (<15 conidia) linear conidial chains formed abundantly only in the dark." The conidial chains of *F. proliferatum* in these conditions on synthetic nutrient agar comprise more than 30 conidia.

18. *Fusarium sacchari* (Butler) W. Gams Anamorph of an Unnamed *Gibberella* Species [Mating Population B of the *G. fujikoroi* Complex (488); *Fusarium* Subgenus: *Liseola*]

A keratitis was ascribed to this fungus without etiologic or mycological substantiation by Zapater (456). It is unlikely that this experienced investigator of mycotic keratitis was incorrect about the etiology, but since *F. sacchari* is in a complex of species that are particularly difficult to identify correctly, mycological substantiation is necessary for the record to be accepted.

Guarro et al. (493) reported a case of fungemia caused by *F. sacchari* in a nonneutropenic Brazilian renal transplant patient. Both etiology and identification were well documented. A small dose of amphotericin B was sufficient to cure the infection, which had no obvious portal of entry in the uncatheterized patient.

F. sacchari in nature is commonly isolated from a variety of tropical plants, usually sugar cane and banana, but also such other diverse plants as *Cattleya* orchids and sorghum (429). It produces moniliformin as a secondary metabolite.

Description. Colonies on PDA grow 65.5 to 70 mm in 7 days. They are floccose with abundant hyphal strands. They are whitish above, sometimes becoming dark vinaceous, with yellowish or ochre to deep vinaceous reverse. *Conidiophores* consisting initially of phialides arise singly or in small, branched groups along aerial hyphal strands, soon forming more elaborate antlerlike branching structures (Fig. 87). *Phialides* are monophialidic and candle-shaped at first, but rapidly produce spinelike lateral proliferations, usually from just beneath the apex or just above the base, giving rise to asymmetrically bifurcate or trifur-

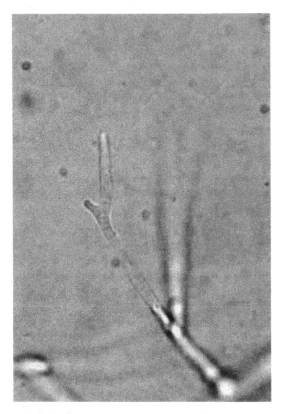

Figure 87   *Fusarium sacchari* CBS 223.76 polyphialide.

cate phialides and often giving rise to additional proliferations giving an overall antlerlike appearance, 17–30 × 2.5–3.8 μm. *Microconidia* are in sticky heads, never in chains. They are hyaline, smooth-walled, ellipsoidal, ovoid, or allantoid (curved, sausage-shaped), mostly aseptate but occasionally two-celled, and mostly (4-) 7–11 (-18) × 2–3 μm. *Macroconidia* are often formed only after near-ultraviolet light stimulation or not at all. They are hyaline, nearly straight to falcate, mostly three-septate, and (24-) 33–43 (-52) × 3.3–3.5 μm, with slightly beaked, sharply pointed apical cells and pedicellate basal cells. *Chlamydospores* are absent. (For distinction from other taxa, see discussion under *F. subglutinans.*)

19.   *Fusarium solani* (Mart.) Sacc. Anamorph of *Nectria haematococca* Berk. & Br. (*Fusarium* Subgenus: *Martiella*)

F. solani shows the "general pattern of involvement by major opportunistic *Fusarium* species" described in the introductory section for *Fusarium* above. Case literature for this well-established opportunist has been reviewed previously (see same section) and will not be reviewed in detail here. De Hoog et al. (69) have recently given a species-specific synopsis.

Along with *F. oxysporum*, this fungus is one of the most common fusarial agents of disseminated and cutaneous infection in the immunocompromised— especially the neutropenic—patient.

One very distinctive feature of *F. solani*'s biology is its high importance worldwide as an agent of keratitis and as a major cause of human blindness in tropical areas. Most surveys show the incidence of *F. solani* keratitis to be far more frequent than keratitis caused by other fusaria; for example, in Miami, Florida [11 identified *F. solani*: one *F. oxysporum* (229); 76 *F. solani*: six other identified *Fusarium* isolates (393)], in Enugu, Nigeria [12 *F. solani*: no other fusaria (230)], and in Buenos Aires, Argentina [20 *F. solani*: nine other identified fusaria (456)]. Affected persons often live in rural areas (456). Trauma involving vegetative matter is the normal predisposing factor. Use of soft contact lenses may also predispose significantly, but can be controlled by good hygiene and use of overnight storage fluids containing polyhexamide biguanide, which inhibits *F. solani* (494). A dramatic contact lens-related case recently reported from an immunocompetent patient featured keratitis that initially required penetrating keratoplasty, then progressed to endophthalmitis successfully treated with amphotericin B lipid complex (495). Filaments were not seen in corneal scrapings but were seen in in vivo confocal microscopy of the cornea and confirmed later in an anterior chamber tap. Recently an epidemiological study by Mselle (496) in Dar es Salaam, Tanzania, showed that HIV infection is now a powerful predisposing factor in Africa, with 81.2% of studied fungal keratitis patients and only 33% of nonfungal keratitis patients being HIV+. *F. solani* is attested (without identification characters) to have caused 75% of the total fungal keratitis seen. In leukemia patients, endogenous endophthalmitis caused by *F. solani* may develop as a sole disease manifestation or as part of a dissemination (497). There are a small number of cases in which persons with no or slight predisposing factors acquired *F. solani* endophthalmitis. The origin and portal of entry of inoculum in such cases is uncertain (497). It appears likely that the *F. solani* isolates involved in oculomycosis will ultimately be ascribed to only a limited number of the biological entities in this species complex. (See taxonomic discussion below description.) Jones et al. (229) showed that 16 isolates of this species from infected eyes in Miami, San Francisco, and Singapore were morphologically similar, grew well at 37°C, and survived at 40°C. Four comparison isolates of plant-pathogenic *F. solani* grew poorly at 37°C. Their survival success at 40°C is not recorded except in a general comment that most plant-pathogenic fusaria of the various species tested did not survive. English (485) similarly found an *F. solani* isolate from keratitis significantly more thermotolerant than those from leg ulcers or plants.

Some additional connections between *F. solani* and HIV-related syndromes have been noted. A female AIDS patient with late-stage non-Hodgkins malignant lymphoma was well demonstrated to have a soft palatal ulceration caused by *F. solani* (498). The identity of the isolate was determined by C. de Bièvre and C. Hennequin of Institut Pasteur.

F. solani is sufficiently virulent to rarely cause mycetoma and other subcutaneous infection (possibly prodromal mycetoma) subsequent to dermal trauma in otherwise healthy patients. A mycetoma from South America was reported by Luque et al. (499), while an erythematous, indurated lesion with copious fungal filaments developed 14 days after surgical removal of a stingray barb from the hand of a diver (500). According to Austwick (426), the U.K. National Collection of Pathogenic Fungi held two *F. solani* isolates from confirmed mycetoma as of 1984, as well as one *F. oxysporum* isolate. Three other relatively recent case reports involving *F. solani* mycetoma are reviewed by Guarro and Gené (412).

A case of intertrigo in an otherwise healthy male, a very recent immigrant from Senegal to Italy, was well linked by Romano et al. (483) to *F. solani* by repeated antifungal culture and biopsy examination. Similar cases of intertrigo linked by the same authors to *F. oxysporum*, discussed above, were also in African patients. A nearly identical, well-confirmed case reported from France (411) concerned a patient whose travel history was not discussed, but who was described as a "practicing Muslim . . . who washes his feet 5 times per day (prior to) his prayers." (See further discussion of this matter above under *Cylindrocarpon lichenicola*.) Both the Italian and the French cases failed to respond to terbinafine therapy; the isolate in the latter case showed resistance in vitro to all other topical and systemic drugs considered.

In veterinary medicine, *F. solani* is particularly noted for causing cutaneous infection in turtles, particularly in young turtles and their eggshells (501). The genetic type causing one such infection was markedly different from a selection of culture collection isolates from various sources, as gauged by random amplified polymorphic DNA (RAPD) analysis (502). Given the diversity of *F. solani* as outlined below, however, this is perhaps to be expected. Outbreaks of infection caused by *F. solani* have been observed in Australian crocodile farms (503, 504). A dramatic macular skin infection, authoritatively linked to *F. solani*, was described in sea lions (*Zalophus californianus*) and grey seals (*Halichoerus grypus*) held in chlorinated freshwater pools (505). The infections responded poorly to various antifungals (all retrospectively well known to be ineffective against *F. solani* in vitro) and were best controlled by manipulating environmental conditions. The identifications in this study were credited to *Fusarium* authority C. Booth of CMI (CABI). A report by Jacobson (506) of necrotizing skin lesions caused by *F. solani* in a caged Burmese python is etiologically well demonstrated but does not substantiate the fungal identification.

In nature, *F. solani* is well known from various soils and plant associations. Like *F. oxysporum*, it has numerous forms associated with root or other diseases of various plants. It is also well known from ponds, rivers, sewage facilities, and water pipes (74). Its mycotoxins include fusaric acid and naphthoquinones (448).

Description. Colonies on PDA grow 64 to 70 mm in 7 days. They are usually thinly or nearly appressed-felty. They are locally low-floccose, whitish-cream to buff, pale brownish, pale red-brown, or in some isolates pale blue-green above, sometimes spotted with creamy or dusky blue-green, slimy sporodochial masses, with pale, tea- or tea-with-milk-brown, red-brown, or very uncommonly (in clinical isolates) blue-green to ink-blue reverse. *Conidiophores* consisting of single phialides arise laterally on aerial hyphae, or small groups of phialides formed on rudimentary branching structures, or closely appressed sporodochial clumps. *Phialides* when producing microconidia in aerial mycelium are strictly monophialidic, very characteristically filiform (elongated and slender), and 15–40 × 2–3 μm (Fig. 88). Those producing macroconidia are shorter and more inflated, subcylindric to flask-shaped, and 10–25 × 3–4.5 μm. *Microconidia* are al-

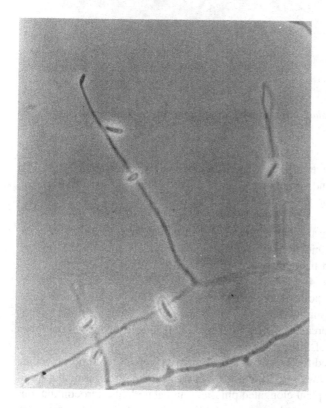

Figure 88 *Fusarium solani* microconidial phialide.

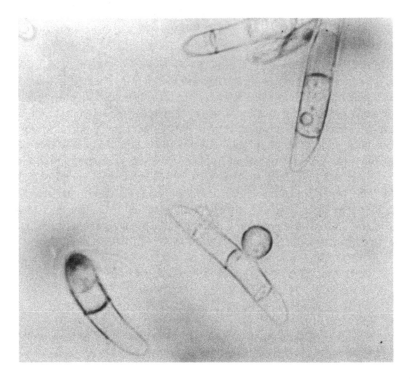

Figure 89   *Fusarium solani* macroconidia, one with a chlamydospore attached.

ways in sticky heads, never in chains. They are hyaline, smooth-walled, ellipsoidal to nearly cylindrical, and sometimes slightly curved. They are usually aseptate, less commonly one- or two-septate. When aseptate they are (5-) 8–13 (-17) × 3–4 (-5) µm.

Macroconidia are usually common, sometimes in sporodochia. They are hyaline, gently curved, mostly three-septate, less commonly four-to-five-septate, characteristically broad in relation to length (Fig. 89). When three-septate they are (22-) 27–50 (-58) × 4–6 (-7) µm, with scarcely pedicellate, thick basal cells and relatively broadly pointed to tapered but nearly round-ended apical cells. *Chlamydospores* are usually abundant. They are smooth or more often rough-walled, terminal or intercalary, single or clustered, pale or brownish, sometimes protruding from sides of macroconidia, and 6 to 11 µm.

Pale to tea-colored forms within this complex taxon are common from internal infection. A form with vivid red-brown reverse, sometimes overshadowed with bluish (but with typical elongated phialides and abundant microconidia, unlike the deeply blue-pigmented potato-rot and soil fungus *Fusarium coeruleum*, discussed above) may also be seen from such infections, but is particularly common from onychomycosis. Such a form producing red to brown colony colors and blue-greenish sporodochia on oatmeal agar, as well as yellow to greenish colonies on more osmotically active (2–5% glucose-supplemented) media, was described in detail (in Chinese) from a keratitis isolate by Ming and Yu (507). The isolate grew at 37°C. Ming and Yu's proposal of a new taxonomic forma (a rank below variety) based on this study was nomenclaturally invalid; a Latin diagnosis was lacking, and no type material was designated.

Recent molecular studies of plant-derived and saprobic *F. solani* isolates have disclosed that this complex consists of at least 26 entities that would be considered distinct species using a phylogenetic species concept (427). This confirmed numerous earlier mating studies, reviewed by O'Donnell (427), that showed numerous intersterile heterothallic mating groups within *F. solani*, all of which would be considered separate species under the older biological species concept. In addition, there are several homothallic entities, and some possibly clonal entities. The medically important isolates have not yet been placed in terms of their relation to elements of this complex. The variation seen in culture suggests that several of these sibling species are involved in pathogenesis.

The *F. solani* complex as a whole is readily distinguished from the sometimes similar *F. oxysporum* by the stark contrast between its long, thin phialides and the stubby, inflated phialides of the latter. Although macroconidia in *F. solani* are generally distinctly thicker and blunter than those of *F. oxysporum*, there is a degree of overlap in certain isolates, so this character can be a screening character revealing obvious *F. solani* isolates as such, but it cannot be used for consistent mutual exclusion of the two species. *F. solani* seldom has the violaceous purplish colors frequently seen in *F. oxysporum*; the red-brown and dusky blue-green to ink-blue colors occasionally seen in the former species are all readily distinguished with experience.

20. *Fusarium subglutinans* (Wollenw. & Reink.) Nelson et al. Anamorph of *Gibberella fujikoroi* (Sawada) Wollenw. var. *subglutinans* Edwards [Synonyms: *Fusarium sacchari* var. *Subglutinans*, *F. moniliforme*, var. *subglutinans* Pro Parte (i.e., the Latter Name Was Applied to a Broad Concept Including This and Other Species) *Fusarium* Subgenus: *Liseola*]

An isolate from well-verified mycotic keratitis in northern Italy (508) was identified prior to March 1975 by medical mycology authority R. Vanbreuseghem as *Fusarium moniliforme* Sheldon var. *subglutinans* Wollenw. & Reink., consistent with contemporary taxonomic concepts. On the basis of this record, *F. subglutinans* has been indexed in reviews of medically important fungi (69, 443). The complex group of isolates referred to as *F. moniliforme* var. *subglutinans*, however, was split into two separate varietal concepts and placed into *F. sacchari* along with the type variety by Nirenberg in 1976 (435). These varieties have since been elevated to species status as *F. subglutinans*, *F. bulbicola*, and *F. sacchari ss. str.* (= in the strict sense), based on molecular investigations (429). Moreover, another closely related medically important species, *Fusarium proliferatum*, may appear very morphologically similar to *F. subglutinans* unless characters related to microconidial chain formation are emphasized. Only in the mid-1970s, however, did these characters begin to be more broadly publicized and made more accessible by the introduction of better media such as carnation leaf agar (434). *F. proliferatum* was only distinguished as a separate species from the long-standing Wollenweber and Reinking concept of *F. moniliforme* by Nirenberg in 1976 (435) [apart from being described in 1971 as a *Cephalosporium* species by Matsushima (509), who did not see macroconidia in his material], and was therefore not well conceptualized in 1975. An example of early 1970s confusion about these fungi was given by Gerlach and Nirenberg (437), who pointed that the microphoto alleged to depict *F. moniliforme* var. *subglutinans* in the authoritative *Fusarium* monograph by Booth (445) showed a species that "produces its microconidia in chains, because the microconidia in the picture are clavate with a flattened base." In other words, the concept of *F. moniliforme* var. *subglutinans* was mistakenly illustrated in the most modern monograph available in 1975 by a species that was similar to *F. proliferatum* in its conidiogenesis. There is thus considerable doubt about which currently recognized species the isolate examined by Vanbreuseghem belonged to, and this isolate, RV 32739, should be re-examined if it remains available.

   F. subglutinans is very similar to *F. sacchari* (see above) in colony characters and conidial morphology. It differs in that a high proportion of its microconidia are septate, showing an intergradation between microconidia and macroconidia, and in that macroconidia tend to form readily in cultures on PDA. *F. sacchari*, on the other hand, has mainly single-celled microconidia and forms macroconidia only rarely after induction with near-ultraviolet light. The conidiophores in *F. subglutinans* are often erect, arising directly from the substrate as small treelike formations. In *F. sacchari*, however, the conidiophores mostly arise from the aerial mycelium and are therefore referred to as "prostrate" by Nirenberg and O'Donnell (441). *F. subglutinans* is a cosmopolitan decomposer, whereas *F. sacchari* is a tropical organism usually associated with sugar cane and banana plants. Both species have slightly to distinctly curved microconidia with more or less rounded bases, and are unable to form microconidia in chains. The club-shaped microconidia of *F. proliferatum*, however, with their flattened bases, form chains readily whenever suitable nutrient and moisture conditions permit. *F. verticillioides* (=*F. moniliforme*) differs strongly from all the species mentioned so far by not forming polyphialides. Molecular distinction of all these species is summarized by O'Donnell et al. (429), and a morphological key is provided by Nirenberg and O'Donnell (441).

21. *Fusarium verticillioides* (Sacc.) Nirenberg Anamorph of *Gibberella moniliformis* Wineland (Common Synonym: *Fusarium moniliforme* Sheldon; *Fusarium* Subgenus: *Liseola*)

F. verticillioides, frequently referred to as *F. moniliforme* in the literature, shows the "general pattern of involvement by major opportunistic *Fusarium* species" described in the introductory section for *Fusarium* above. Case literature for this well-established opportunist has been reviewed previously (see the same section) and will not be reviewed in detail here. De Hoog et al. (69) have recently given a species-specific synopsis. Although not as frequently reported as *F. solani*, this fungus vies with *F. oxysporum* as the second most common fusarial agents of disseminated and cutaneous infection in neutropenic and other severely immuno-compromised patients (412). A prototypical case is that published by Young et al. (510) regarding disseminated infection in a granulocytopenic lymphoma patient. Skin disrupted by varicella zoster infection was thought to be the portal of entry. In some areas, *F. verticillioides* can become the predominant *Fusarium* species infecting the compromised patient (511).

Like *F. solani*, *F. verticillioides* may occasionally cause subcutaneous infection in apparently immunocompetent individuals. Collins and Rinaldi (512) described a case of a man with an inflamed hand pustule that grew this species. Although no fungal filaments were seen in the expressed pus on Gram staining, destaining and restaining with periodic-acid-Schiff (PAS) stain allowed confirmation of fungal filaments. The patient was an avid gardener but recalled no injury. A well-developed *F. verticillioides* mycetoma in an Italian miner was described by Ajello et al. (513). The case had originally been reported as being caused by *Acremonium* sp., since only microconidia were then seen (514). The case isolate is now in most major fungal culture collections. On the other hand, an infected hyperkeratotic heel lesion attributed by Pereiro et al. (515) to *F. verticillioides* in a prostatic adenocarcinoma patient appears to be better ascribed to an *Acremonium* sp. The authors state that they made the identification based on formation of small phialoconidia in chains from a floccose, off-white colony. No purple coloration was noted. A photograph depicts solitary, thin, sometimes flexuose phialides rather than the rigid-looking primary phialides of *F. verticillioides*, which are not infrequently borne on branched conidiophores. The authors' description of the conidia as "8–9 × 2–3 μm"; that is, as elongated conidia with a length–width ratio of not less than 2.5:1 [the average for *F. vert-*

icillioides is 2.7:1, based on the measurements of Gerlach and Nirenberg (437)] appears to be contradicted by their unscaled but otherwise excellent photo showing short, obovate conidia with a length–width ratio of not more than 1.5:1.

F. verticillioides is occasionally involved in ocular infections, particularly keratitis, although the incidence of such infection is much less than the incidence of *F. solani* infections (393, 456, 516). The strong ecological affinity of this organism for maize corn may be significant in etiology; for example, Zapater (456) recounts a keratitis that began after a moth flying from a fungally colonized bin of chicken-feed corn struck a woman in the eye without causing immediate discernible injury. One case of apparent *F. verticillioides* keratitis (direct microscopy was not performed on ocular material) was successfully treated with cyclopiroxolamine after disk susceptibility testing suggested it was resistant to the other agents available for testing (pimaricin was not tried, but resistance was shown to two other polyenes) (516).

In veterinary mycology, *F. verticillioides* has been reported to cause fatal pneumonia in a captive American alligator (517). The isolate was atypical, but its identity was later confirmed by sequencing (429).

In nature, *F. verticillioides* is frequently associated with corn (*Zea mays*), but also with various other plants (grains, pines, cabbage family) and also insects (429). It is the best-known producer of fumonisins, mycotoxins implicated in the etiology of the fatal brain degeneration known as equine leukoencephalomalacia in horses eating contaminated feed. A variety of other secondary metabolites such as fusaric acid and moniliformin are also produced (448).

Description. Colonies on PDA grow 88 to 93 mm in 7 days. They are floccose to powdery. They are whitish above, sometimes becoming dull violet with pale to deep vinaceous reverse. *Conidiophores* consisting of single phialides arise laterally on aerial hyphae, or small groups of phialides formed on rudimentary branching structures, or closely appressed sporodochial clumps (Fig. 90). *Phialides* are strictly monophialidic and subulate (tapered, awl shaped), and 11–32 × 2–3.5 (-4.5) μm. *Microconidia* have a strong tendency to form in long chains, especially on specialized media such as carnation leaf agar or synthetic nutrient agar. They are sometimes also in sticky heads, especially on rich media. They are hyaline, smooth-walled, clavate with a flattened base, usually aseptate, rarely one- or two-septate, and 7–10 (-19) × 2.5–3.2 (-4.2) μm. *Macroconidia* are often formed only after extended culture or not at all, sometimes in sporodochia. They are hyaline, nearly straight to falcate, mostly three- or five-septate, and (18-) 30–58 (-73) × 2–4.3 μm, with pedicellate basal cells. Chlamydospores are conventionally said to be absent; however, according to Gerlach and Nirenberg (437) "inflated cells with thickened walls occur."

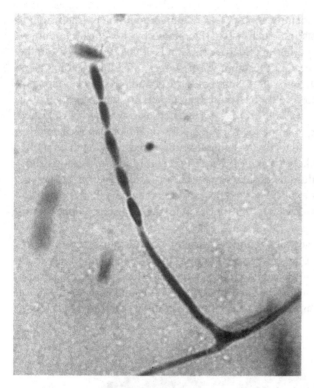

Figure 90   *Fusarium verticillioides*, monophialide with conidia in a chain.

E. *Microdochium*

1. *Microdochium nivalis* (Fries) Samuels & Hallett Anamorph
    of *Monographella nivalis* (Schaffnit) E. Müller (Synonym:
    Fusarium nivale Ces. ex Sacc.)

This is one of numerous problematic fungi that have several records from human
disease despite lacking the ability to grow at body temperature.

An allegedly significant isolation of this fungus, recorded as *F. nivale*, was
included without specific etiologic or mycological documentation in an omnibus
report on fungi causing keratitis in south Florida (393). In general, this study
accepted only fungi that grew from specimens on multiple media or those that
grew on a single medium and also correlated with positive direct microscopy.
Mycology was performed at the Mayo Clinic. Since this fungus has a maximum
growth temperature of 28°C (74, 518), this record must be questioned. It may

possibly reflect either misidentification or a contamination problem. Inclusion of some contamination events into such lists is facilitated by the "positive culture on >1 media" acceptance criterion. Such a criterion is ethically necessary in the clinic, since it is conservatively presumptive of fungal infection in ambivalent cases requiring rapid treatment, but it is mycologically dubious, since nonpathogenic airborne conidia often settle on surfaces in coherent clumps or chains or in conidia-laden aqueous aerosols from sources such as tap water or splashed outdoor leaf surfaces. Dispersal of such aggregations on the ocular surface prior to sampling would lead to clinically insignificant, single-species growth on multiple media. Another less likely source of low-temperature fungi in keratitis samples is retention of inoculum as dormant material on residual debris at the site of trauma. Finally, serous exudates on bodily surfaces may be colonized by saprobes, especially fusaria (422, 423, 485); a possible ocular instance appears immediately below.

A seemingly very carefully confirmed case of keratitis attributed to *M. nivalis* var. *major*, a variant of *M. nivalis* with mostly three-septate, relatively thick macroconidia, was presented by Perz et al. (519) (Fig. 91). A patient under unspecified treatment for kidney stones but otherwise healthy and with no history of ocular trauma, spontaneously developed a fungal keratitis, confirmed by biopsy.

Figure 91   *Microdochium nivalis* var. *major* CBS 105.90.

Scrapings grew *M. nivalis*, and a compatible description was published, along with nonspecific but nondiscrepant photographs. K. Mańka, a forest phytopathologist experienced with hypocrealean soil fungi, was a coinvestigator. According to Gams and Müller (518), *M. nivalis* var. *major* is ecologically indistinguishable from the type variety; it would therefore not be predicted to grow at body temperature, but appears not to have been tested. A typical isolate, CBS 106.90, was tested ad hoc by me for purposes of this chapter, and was found to have a maximal growth temperature of approximately 27°C. Perz et al. noted that conjunctival sacs and ocular secretions of their patient grew *M. nivalis* in both the affected and unaffected eyes, as well as *Geotrichum candidum*. It is conceivable that this surface colonization overgrew inoculum of an unelucidated true agent of the keratitis; otherwise, an atypical heat-tolerant variant of *M. nivalis* var. *major* or a fungus very similar to it must cause occasional keratitis. The culture described had a brown reverse on PDA, and formed mostly (1-) three-septate apedicellate macroconidia with very few aseptate conidia and no chlamydospores, a combination of characters ruling out most common *Fusarium* and *Cylindrocarpon* species involved in keratitis.

Microdochium oryzae (Hashioka & Yokogi) Samuels & Hallett (anamorph of *Monographella albescens* [von Thümen] Parkinson et al.), the agent of rice leaf scald disease, is morphologically similar to *M. nivalis* but grows up to 36°C (518). This fungus, only isolated from infected rice plants so far, may be considered if *M. nivalis*-like fungi are seen in areas of suitable climate. It has conidia with a somewhat swollen basal cell.

Description. Colonies on PDA grow 63 to 70 mm in 7 days. They are sparsely cobwebby or floccose to felty. They are whitish above, becoming pale pinkish to peach, orange, or amber, with these colors often visible on the surface but most prominent on the reverse. *Conidiophores* consisting of annellides arise laterally on aerial hyphae or in small groups on rudimentary branching structures, and also commonly on closely appressed sporodochial clumps. *Annellides* have a nearly cylindrical to bulbous, barrel- to pear-shaped base and an elongating, thin percurrently proliferating apical tubule, overall 6–15 × 2.2–4.0 (basal measurement) μm. *Microconidia* are not considered to be formed; there is, however, a small proportion of single-celled conidia mixed with and intergrading in size and shape with the macroconidia, mostly 8–12 × 2.0–2.8 μm. *Macroconidia* are crescent-shaped, sometimes with a relatively straight basal half, hyaline, smooth-walled, mostly one-septate, fairly commonly two- or three-septate, and exceptionally up to seven-septate. When one-septate they are (9-) 13–18 (-23) × 2.2–3.0 (-4.5) μm; when three-septate they are up to 36 μm long, with a wedge-shaped, obtuse base and a sharply pointed apex. In *M. nivalis* var. *major* conidia are mostly three-septate, ranging from one- to 7-septate, and are 19–30 (-37) × 3.5–4.5 (-6) μm. Chlamydospores are absent.

The production of annellides by this fungus and its connection to the teleomorph genus *Monographella* in the family *Amphisphaeriaceae* show that any resemblance between the former *F. nivale* and the genus *Fusarium* is coincidental. Nonetheless, the resemblance of the conidia to small *Fusarium* conidia is high. The maximum growth temperature, 28°C, can easily be used to distinguish typical isolates of this fungus from mammalian opportunists.

F. *Neocosmospora*

1. *Neocosmospora vasinfecta* E. F. Smith [Common Synonym: *Fusarium vasinfectum* (an Inappropriate Combination of an Anamorph Genus Name with a Species Concept Including the Teleomorph)]

A case of allergic bronchopulmonary mycosis in a patient with asthmalike symptoms and hemoptysis was attributed to *N. vasinfecta* (as *Fusarium vasinfectum*) by Backman et al. (520). The fungus was never cultured. A panel of mold allergy tests indicated this fungus as the only one to which the patient reacted; further tests showed positive precipitins against *N. vasinfecta* antigens. *N. vasinfecta*, however, happened to be the only Hypocrealean fungus included in the panel; thus the reaction seen could have been a cross-reaction against an indefinite number of fungi from this group or its relatives. [See Kaufman et al. (433).] The possibility of a coincidentally cross-reacting antigen from an entirely different group of organisms also cannot be ruled out. Diagnoses of allergic bronchopulmonary mycosis caused by fungi other than *Aspergillus fumigatus* clearly need to be established by repeated isolation of positive cultures, correlated with specific immunological responses. The criteria designed to make the troublesome culturing step unnecessary for classic allergic bronchopulmonary aspergillosis do not apply to other molds. The case record is rejected.

N. vasinfecta was, however, authoritatively associated with a localized mycotic cyst of soft tissue in the leg of a renal transplant and dialysis patient (521, 522). No trauma had occurred at the infection site, suggesting a circulatory source for the inoculum. Surgery and a small dose of ketoconazole (truncated prematurely because of a cyclosporin toxicity side effect) eliminated the infection. A French parachutist who suffered an open ankle dislocation in Senegal contracted an *N. vasinfecta* osteoarthritis, ultimately requiring amputation (523). The fungus was not seen in several histopathologic studies but grew repeatedly from the wound and from the progressing osteitis despite extensive cleaning measures and treatment with amphotericin B.

The fungus in nature is common in tropical and subtropical soils. Molecular phylogeny studies show that it is closely related to the species in the *Fusarium solani* complex (427).

Description. Colonies are fast-growing—40 mm in diameter after 7 days at 25°C. They are thinly floccose, white to buff or pinkish, soon becoming stippled with production of white (immature) to orange-brown or red-brown (mature) ascomata. *Conidiophores* are acremonium-like, consisting of short to long lateral branches, sometimes with one or rarely more branch points (Fig. 92). *Phialides* are subulate (awl-shaped, long, and tapered) to cylindrical, 30–100 × 1–2 µm, and bearing conidia in sticky heads. *Conidia* are hyaline, smooth, aseptate or less often one-septate, long-ellipsoidal, cylindrical, or fusoid (Fig. 93). They are sometimes slightly curved, and are 5–13 × 2–3.5 µm. *Ascomata* are bright orange to orange-brown or red-brown overall. They are nearly spherical, 200 to 500 µm in diameter, with a short, protruding apical neck bearing an ostiole (opening), and with thick walls that become yellow at maturity and are composed of *textura angularis* (composed of polygonal cells as seen in face view) tissue. *Asci* are cylindrical and 80–100 × 11–15 µm, containing eight linearly arranged ascospores. *Ascospores* are ellipsoidal to subglobose and orange-brown, with thick, heavily roughened walls. They are 10–16 × 7.5–12 µm, and lack a germ pore. *Chlamydospores* are present. Growth occurs at 37°C in vitro (Fig. 94).

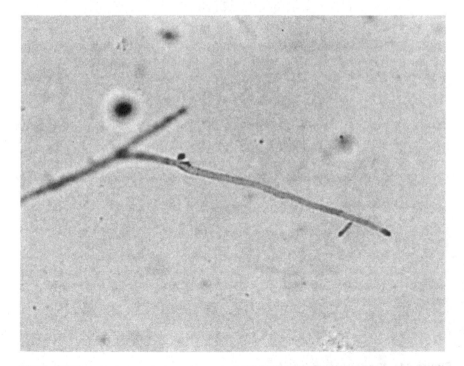

Figure 92 *Neocosmospora vasinfecta* conidiophores CBS 554.94 conidiophore.

Figure 93 *Neocosmospora vasinfecta* CBS 554.94 zero- and one-septate conidia.

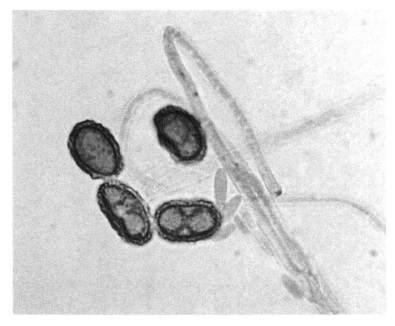

Figure 94 *Neocosmospora vasinfecta* ascospores (large, rough structures) and conidia.

G. *Trichoderma*

Trichoderma species, like *Chaetomium* species, form a large group of difficult to distinguish, often thermotolerant common saprobes, which, for reasons that remain unclear, are much less frequently involved in opportunistic pathogenesis than other common thermotolerant saprobes. In recent years, however, a small number of legitimate cases of *Trichoderma* infection have been recorded in humans, and at least one species, *T. longibrachiatum*, may reasonably be called an emerging opportunist of the severely immunocompromised patient.

Trichoderma species as classically defined by Rifai (524) and Bissett (525) were known aggregates of several closely related, all but indistinguishable sibling species, some of which were associated with distinguishable *Hypocrea* teleomorphs on natural substrata. With the advent of molecular methodologies, the species aggregates that had been used in identification schemes were seen as inadequately precise to specify ecologically distinct biological entities. A strong impetus for this judgment came from the appearance of a genetically distinct organism generally fitting the aggregate description of *Trichoderma harzianum*, but causing an economically important disease of cultivated mushroom beds that other genetically distinct *T. harzianum* groups did not cause (526, 527). This mushroom pathogen is in the process of being described as a distinct species subsequent to delineation by sequencing. (Morphological characters have also been found to distinguish it.) The "gold standard" for species identification in *Trichoderma* has therefore been removed to the molecular level, which is most accurately predictive of natural associations and interactions. Molecular confirmation of species identification has been suggested as necessary for precision in case reports wherever possible (528). At the same time, however, molecular clarifications of species relationships have allowed more precise morphological characters to be delineated, so the most up-to-date works such as the keys and descriptions published by Gams and Bissett (529) may for careful pheneticists facilitate purely morphological identification at a level comparable to that made possible by molecular methodologies.

1. Identification

The procedures for morphological identification of *Trichoderma* species are unique. The organism is best grown on malt extract, oatmeal, or cornmeal agar (529). Within approximately 5 to 7 days in most isolates, well-developed sporodochial pustules will be seen. These pustules may be found in four stages: (1) immature, with branching incomplete and phialides sparse or nil, white to yellow in species with green conidia; (2) submature, with branching complete and newly formed conidial heads present, yellow to pale green in species with green conidia; (3) mature, with phialides mostly intact and large numbers of mature conidia present, green, dark green, intense yellow-green, or hoary blue-green in species with green conidia; and (4) overmature, with phialides collapsed and masses of conidia present, deep yellow-green to deep green in green-conidial species. *Trichoderma* species can only be analyzed in stage 2. Stage 3, which for any given sporodochium occurs approximately 24 hr after stage 2, tends to feature large numbers of conidia that make discernment of structure in the still well-formed conidiophores impossible. Stage 4, the typical 7-day stage that most persons unfamiliar with *Trichoderma* would attempt to analyze, is almost completely devoid of information useful for species identification, except in submature sporodochia that may still remain near the colony margin. *Trichoderma* colonies for identification should therefore be inoculated so that they may be observed between their fourth and seventh days of growth. Structures form well at a wide range of room temperatures in light, as well as at 25°C. Recent species identification keys by Gams and Bissett (529), however, establish a standard of 20°C for growth rate measurements—a standard that unfortunately will seldom be convenient in the clinical laboratory. A recalibration based on 25°C growth is not currently available.

Before sporodochia are mounted for microscopy they are best examined under the dissecting microscope or under the 10× compound microscope lens to discern any sterile setae that are produced, as well as the general branching structure. Sporodochia of some species bear a characteristic down composed of long, sinuous setae that arise from the ends of the branches. A few species [e.g., *Trichoderma* (formerly *Gliocladium*) *virens*] have convergent, *Penicillium*-like conidial structures, and their conidia may coalesce into large, sticky masses that will be very conspicuous at low magnification. Heavily bacterially affected colonies of other *Trichoderma* species may mimic this gloeoid appearance, so it is advisable to ascertain that one is working with a pure culture if such a manifestation is seen in gross overview.

For making microscopic slide mounts, analysts must be aware that *Trichoderma* sporodochia are highly water-repellent, and simply inserting them into ordinary mounting media generally results only in the trapping of an intractable air bubble that prevents meaningful microscopy. Any tearing or heavy squashing of the material destroys the branching structure that must be observed. The best technique to use is to carefully pick submature (see above) sporodochia off the growth medium, being careful to cut from under the base so that the entire structure is intact. Place onto a slide and add a drop of 95% ethanol to defat the hydrophobic components, making sure that the sporodochium appears to be wetted. Before the ethanol can dry, add the mounting medium (lactic acid is optimal; cotton blue may be added) and gently allow a coverslip to angle over the top of the sporodochium, pressing it as little as possible to make a useable slide. If a key using conidial measurements is being employed [e.g., that of Gams and Bissett (529)], the same growth medium should be used as was used by the authors of the key.

The main character that careful mounting should allow detection of is the branching pattern of structures within the sporodochia. Some species, such as *Trichoderma koningii*, have verticillate structures, in which branch points tend to give rise to approximately three branches arranged radiately, like spokes in a wheel. Where only two branches diverge from a branch point in such structures, they tend to diverge either as opposite pairs, or as two "spokes" diverging at approximately 120° from each other with the third radially symmetrical spoke missing. As this verticillate structure iterates through major branches, smaller branches, and finally the clusters of phialides themselves, a very symmetrical bushy structure, sometimes referred to as "pyramidal," is generated. This type of structure, typical of *Trichoderma* species in sections *Pachybasium* and *Trichoderma*, is distinguished from structures featuring long main branches that do not rebranch extensively, as seen in the section *Longibrachiatum*. The secondary branches that do arise from the main branches in this section often arise singly, and in many species the finest tips of the conidiophores often feature a zone in which a number of phialides are attached singly and spaced irregularly apart, along an otherwise nearly unbranched elongated section of the conidiophore.

Another morphological character that has been more emphasized since molecular methods clarified species distinctions in this group is the shape of phialides. In section *Pachybasium*, they are typically short and stout, whereas in section *Longibrachiatum*, they are elongate, lageniform (wine-bottle-shaped) or nearly cylindrical. An excellent key to the species, along with descriptions and illustrations, is provided by Gams and Bissett (529). Many *Trichoderma* species form similar chlamydospores, and this character is seldom used in species descriptions.

A large number of molecular identification techniques have been investigated for *Trichoderma* species (530). Since relatively few of the species in this genus are known or suspected to have clinical significance so far, some of these methodologies deal primarily with species and groups not known to be biomedically relevant. Recently Kuhls et al. (530) have shown that the widely used M13 and (GACA)$_4$ primers for moderately repetitive loci can readily be used in PCR to generate fingerprints that allow species identification of clinical *Trichoderma* isolates. Ribosomal internal transcribed spacer (ITS) sequencing may then be used to confirm identifications, especially for isolates giving atypical banding patterns. *Trichoderma longibrachiatum* isolates from confirmed human disease, however, were shown to have very similar or identical banding patterns, while saprobic isolates of the same species were more heterogeneous. This would tend to increase the ease of confirming the identity of clinically significant isolates by PCR fingerprinting alone.

The majority of credible case reports involving *Trichoderma* spp. have proposed a species identification for the causal agent. An exception is an etiologically credible report of infection attributed to *Trichoderma* sp. in two ball pythons

(*Python regius*) kept as cagemates (506). The generic identification itself is un-substantiated, and a long-outdated generic name (*Oospora*) is used for one of the other fungi identified in the same paper, suggesting use of a questionably credible laboratory. Nonetheless, an incorrect generic identification of *Trichoderma* seems unlikely.

2. *Trichoderma citrinoviride* Bissett Anamorph of *Hypocrea schweinitzii* (Fr.:Fr.) Sacc. (*Trichoderma* Section: *Longibrachiatum*)

An isolate of *T. citrinoviride* was among the human-pathogenic isolates included in a recent study on identification of clinical *Trichoderma* isolates with molecular methods (530). The isolate was said to have been from ''hemocultures'' of a neutropenic lymphoma patient. No other case information was given, and assurance that the isolate was not a contaminant lies primarily in the *s* at the end of the word hemocultures. *T. citrinoviride* is a common organism on temperate soil and wood, so far mostly known from northern North America and Europe (531).

Description. Colonies grow 45 to 75 mm in 7 days at 20°C. They are whitish at first, with effuse or loosely tufted bright green to yellow-green conidiation developing mostly at the colony margin. The reverse is usually yellowish-green. *Conidiophores* have long primary branches and with secondary branches that are short, often rebranched once or twice (Fig. 95). *Phialides* are variously arranged, in verticils of three or in pairs (or solitary), with a strong tendency to

Figure 95 *Trichoderma citrinoviride* conidiophore structure.

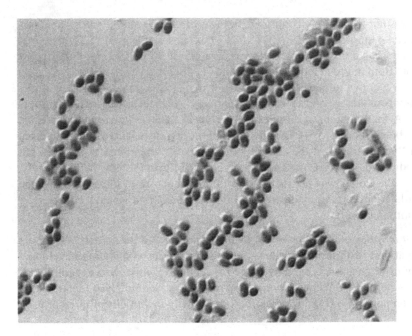

Figure 96 *Trichoderma citrinoviride* conidia.

appear singly at irregular intervals along the main branch near the branch apex. They are mostly distinctly constricted at the base, especially when laterally disposed, lageniform (wine-bottle-shaped), or when somewhat shorter, ampulliform (ampoule-shaped, with a slightly extended apex and a bulge just above the basal constriction). They are 3.5–6.6 × 2.0–3.2 μm, with terminal phialides (at major branch ends) longer to 12 μm. *Conidia* are green, smooth, ellipsoidal, and 2.2–3.7 × 1.5–2.1 μm. *T. citrinoviride* grows at 40°C (Fig. 96).

This species is morphologically distinguished from *T. longibrachiatum* by its conidiophores bearing side branches that are often rebranched once or twice. Its phialides are mostly distinctly constricted at the base, and its conidia are smaller than 4.0 × 2.5 μm. It can be readily distinguished in PCR using M13 and (GACA)$_4$ primers (530). (For morphological distinction from other *Trichoderma* species, see the discussion under *T. harzianum*.)

3. *Trichoderma {aff. Hypocrea} harzianum* Rifai (*Trichoderma* Section: *Trichoderma*)

A renal transplant patient in Spain died of autopsy-confirmed, disseminated *T. harzianum* infection, particularly affecting the brain and lungs (532). The case isolate was not molecularly confirmed, but was carefully morphologically characterized and deposited in CBS (CBS 102174). Gams and Bissett (529) state that this species has a maximum growth temperature of 36°C. Clearly, culturing of an isolate from multiple human brain lesions is discordant with this and suggests the matter requires further study. As Richter et al. (528) stated, "Growth at elevated temperatures is one of the well known virulence factors of neurotropic fungi." A fatal peritonitis caused by a member of the *T. harzianum* aggregate has also been reported (533). Identification of the organism, anomalously referred to as a "yeast strain" in the case report, was not substantiated in the text, but was credited to an unspecified staff member of Institut Pasteur. Unfortunately, however, the isolate was not included in the molecular re-evaluations published 3 years later concerning medically important *Trichoderma* isolates collected by that institution (530).

Description. Colonies grow 45 to 75 mm in 7 days at 20°C. They are whitish at first, with profuse yellow-green to dark green conidiation developing effusely or in pustules fringed by white mycelium. *Conidiophores* regularly rebranch in verticils or pairs, forming a "pyramidal structure" (see Identification section on page 437) (Fig. 97). *Phialides* are usually in verticils of three or four,

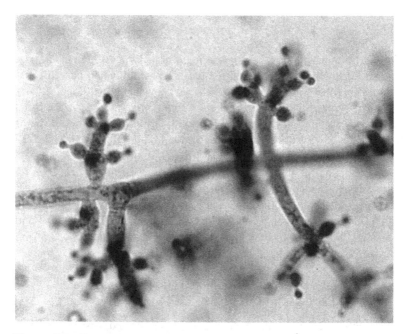

Figure 97 *Trichoderma harzianum* conidiophores.

less commonly in pairs and seldom solitary. They are mostly distinctly constricted at the base, especially when laterally disposed, ampulliform (ampoule-shaped, with a slightly extended apex and a bulge just above the basal constriction), or when somewhat longer and especially when terminal, lageniform (wine-bottle-shaped). Terminal phialides are occasionally curved or flexuous (e.g., with a gentle S-curve). Most phialides are 3.5–7.5 × 2.5–3.8 μm, with terminal phialides (at major branch ends) longer, to 10 μm. *Conidia* are nearly hyaline to pale green in transmitted light. They are smooth, obovoid (egg-shaped with broad end produced first) to subglobose, and (2.5-) 2.7–3.5 × 2.1–2.6 (-3) μm. Maximum growth temperature is 36°C (Fig. 98).

Trichoderma viride and similar species also producing broad conidia are distinguished from *T. harzianum* by the rough ornamentation of their conidial walls. *T. koningii* and similar species share the highly verticillate "pyramidal" conidiophore structure of *T. harzianum* but are distinguished by their narrower, ellipsoidal, not broadly ovoid or subglobose conidia. *T. longibrachiatum*, *T. citrinoviride*, and other members of section *Longibrachiatum* are distinguished by their mostly ellipsoidal conidia, and most importantly by their seldom rebranched, largely nonverticillate conidiophore structure, featuring conidiophore branch

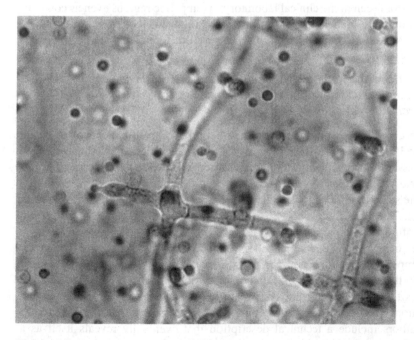

Figure 98   *Trichoderma harzianum* conidia.

ends with irregular rows of singly or asymmetrically paired phialides and few or no verticils. Many members of section *Longibrachiatum* are thermotolerant, growing well at 40°C (531). *T. atroviride* and *T. inhamatum*, two species not recorded from human and animal disease, are more similar to *T. harzianum*. The former has dark green conidia in transmitted light, in contrast to the paler conidia of *T. harzianum*, and also has slightly larger conidia, 2.6–3.8 (-4.2) × 2.2–3.4 (-3.8) μm. The latter has relatively strictly pustular conidiation with little effuse development of conidia, and has shorter, broadly flask-shaped phialides that are 4–5 (-7) × 2.3–3 μm. *T. inhamatum* is also similar to *T. harzianum* molecularly, but can be distinguished by a two-nucleotide difference in its ribosomal ITS1 sequence (534). The *T. harzianum* subgroups causing mushroom crop loss, soon to be described as separate from *T. harzianum*, can be distinguished from *T. harzianum ss. str.* by failure to grow at 36°C (North American Th2 group) or by conidia that decolorize ("bleach out") when placed in plain lactophenol (British Th4 group) (527). The distinction can be made more definitively by molecular means (534).

 Trichoderma species with distinct long, undulate sterile spines extending from the ends of the conidiophore branches and unusually thick main conidiophore branches over 10 μm belong to *Trichoderma* section *Pachybasium*, a group not reported from human and animal disease and, in the present author's experience, seldom seen in the clinical laboratory in temperate regions even as contaminants. Immature branch ends in underdeveloped pustules of other *Trichoderma* sections should not be confused with these well-differentiated sterile spines.

4. *Trichoderma koningii* Oudemans Anamorph of *Hypocrea koningii* Lieckfeldt et al. (*Trichoderma* Section: *Trichoderma*)

This species has been reported to cause infection in dialysis patients by Ragnaud et al. (535) and Campos-Herrero et al. (536). Molecular methods have allowed the disassembly of the complex of species once included in the *T. koningii* aggregate, and a *T. koningii sensu stricto* has been delimited based on re-collecting from the same locality and materials as Oudemans originally collected from prior to 1902. The "real" *T. koningii* is thus a species that has a maximum growth temperature of approximately 33°C (537) and is unlikely to play any role in diseases of warm-blooded animals. Its morphological distinction from the medically important species is noted in the discussion under *T. harzianum* above.

 In the Ragnaud et al. (535) dialysis case, identification is credited to C. de Bièvre of Institut Pasteur, but the isolate appears not to have survived at that institution to be included in the later molecular reassessment by Kuhls et al. (530). The authors include a technical description that eventually reveals itself as a textbook-derived generic-level description of *Trichoderma* grown in pure culture (e.g., it states "conidia subglobose to oblong, smooth or echinulate, with color hyaline or most often green"). No clue remains as to the specific characters of the case isolate.

In the Campos-Herrera et al. case (536), an excellent microphotograph shows three long terminal conidiophore branches with phialides borne singly or in offset pairs, completely lacking verticillate structure. This, along with the ellipsoidal-oblong conidia shown, allows the causal agent to be definitively recognized as a member of *Trichoderma* section *Longibrachiatum*, although the species is not determinable. As such a fungus could not be mistaken for *T. koningii* by anyone following Rifai's 1969 monograph (524) of *Trichoderma* or works based on it [e.g., Domsch et al. (74)], this shows that the concept of *T. koningii* used in this 1996 paper followed pre-1969 concepts dividing all green *Trichoderma* spp. into *T. viride* (or *T. lignorum*) for round-conidial species and *T. koningii* for ellipsoidal-conidial species.

An isolate from a liver transplant patient reported in a meeting abstract (538) as *T. koningii* was later reidentified as *T. longibrachiatum* (539). The very thorough compilation by de Hoog et al. (69) indexes both reports separately under the names given.

5. *Trichoderma* {aff. *Hypocrea*} *longibrachiatum* Rifai
   (*Trichoderma* Section: *Longibrachiatum*)

This fungus has now repeatedly been confirmed as an opportunist of the severely immunocompromised patient. Fatal disseminated infections in bone marrow transplant patients have been well confirmed on two occasions (528, 540). On one of those occasions the organism was originally identified as *T. pseudokoningii* (540), but this judgment was later amended after molecular study (530). Brain lesions occurred in both cases. (Compare the similar case mentioned under *T. harzianum* above for a possible emerging pattern in *Trichoderma* infections.) An isolated brain abscess in a neutropenic leukemia patient was cured by resection and prolonged therapy with ketoconazole and itraconazole (541). Susceptibility testing had shown the isolate to be resistant to the amphotericin B that had been used in initial therapy. An invasive sinusitis caused by *T. longibrachiatum* in a liver and small bowel transplant recipient was successfully treated with debridement, amphotericin B, and subsequent oral itraconazole (539). A neutropenic child with severe aplastic anemia had an infection of skin around an intravenous line site; *T. longibrachiatum* from the lesion was molecularly confirmed by *Trichoderma* authority G. Samuels and was successfully treated with amphotericin B lipid complex prior to bone marrow transplantation (542). In keeping with another emerging pattern in *Trichoderma* infections, a fatal peritonitis was caused by *T. longibrachiatum* in a CAPD patient (543). The isolate in this case was identified by hypocrealean authority W. Gams of CBS.

Description. Colonies grow 60 to 70 mm in 7 days at 20°C. They are whitish at first, with conidiation first occurring in strands, then coalescing into crusts that are dark green. The reverse is pale to greenish-yellow.

Conidiophores consist of long primary branches and relatively few secondary branches with little rebranching (Fig. 99). *Phialides* are often solitary, sometimes paired or in verticils of three, with a strong tendency to appear singly or in offset (not precisely opposite) pairs at irregular intervals along the main branch near the branch apex. They are lageniform (wine-bottle-shaped), usually with little or no basal constriction, and are 5.3–11.6 × 2.0–3.2 μm, with terminal phialides (at major branch ends) longer to 14 μm. *Conidia* are pale to medium green. They are smooth, obovoid, ellipsoidal or narrowly ellipsoidal, and 3.4–6.6 × 2.3–3.5 μm. *T. longibrachiatum* grows well at 42°C.

For distinction from *T. citrinoviride*, see that species. For distinction from other *Trichoderma* groups, see discussion under *T. harzianum*. Richter et al. (528) reported that potato flake agar elicited strongly yellow diffusing pigment from their isolate of *T. longibrachiatum*, but other species were not compared. *T. longibrachiatum* and *T. citrinoviride* may both produce such pigments; Gams and Bissett (529) note the production of "bright greenish-yellow pigments" as a character of the section *Longibrachiatum* as a whole and also note about other *Trichoderma* groups that "dull yellowish pigments are common in many species

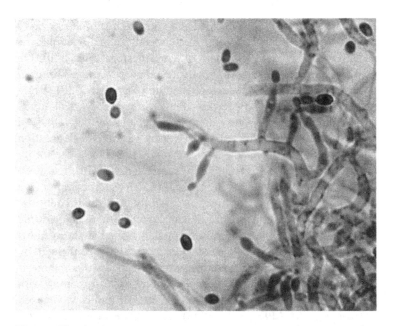

Figure 99 *Trichoderma longibrachiatum* conidiophore structure showing several solitary phialides and conidia.

but are not very distinctive'' (i.e., not very useful in distinguishing species). *T. harzianum*, in *Trichoderma* section *Trichoderma*, is one species in which such "dull yellow" pigments are often present. Another species in that group, *T. aureoviride* Rifai, is named for its conspicuous production of brownish-yellow reverse pigments forming yellow crystals in the agar. The ability of *T. longibrachiatum* to grow above 40°C may assist in separating it from many other *Trichoderma* types, but temperature ranges are by no means well defined so far for many species and variants. (See discussion of *T. harzianum* above.)

6. *Trichoderma pseudokoningii* Rifai Anamorph of an Unnamed *Hypocrea* Species (*Trichoderma* Section: *Longibrachiatum*)

This species was reported by Gautheret et al. (540) as an agent of a fatal infection in a bone marrow transplant recipient. The isolate, however, was later reidentified as *T. longibrachiatum* using molecular methods (530). Studies by Turner et al. (544) and Kuhls et al. (545) have suggested that the genuine *T. pseudokoningii* is very rare outside Australia and New Zealand. The name was long used as an aggregate designator, as suggested by Rifai (524), and was mainly applied to isolates that molecular studies now reveal to be *T. longibrachiatum* or *T. citrinoviride*. Another case listed by de Hoog et al. (69) for this species based on a meeting abstract by Degeilh et al. (546) concerns an isolate later identified as *T. longibrachiatum* (541).

7. *Trichoderma viride* Pers. Anamorph of *Hypocrea rufa* (Pers.: Fr.) Fr. (*Trichoderma* Section: *Trichoderma*)

As with *T. koningii*, until recently the name *T. viride* was applied to a highly phylogenetically diverse group of similar-looking *Trichoderma* isolates, essentially all those with rounded, roughened conidia. Prior to 1969 the name was applied to an even greater range of *Trichoderma* species with rounded conidia. It was in the context of these former aggregate concepts that *T. viride* was noted by Escudero et al. (547) as forming an aspergilloma-like pulmonary fungus ball, by Loeppky et al. (548) as causing CAPD-related peritonitis in an amyloidosis patient, and by Jacobs et al. (549) as causing perihepatic hematoma in a liver transplant patient. More recently, a *T. viride sensu stricto* was defined based on isolates definitely derived from the *H. rufa* teleomorph (550), and this species has a maximal growth temperature below 35°C. Species from this complex with higher growth maxima are beginning to be described [e.g., *T. asperellum* (551)], but no isolates have been preserved from the *T. viride* disease reports to allow corrected identification of the pathogenic types involved.

The Escudero et al. (547) report on *Trichoderma* pulmonary colonization describes conidia of the causal agent only as "small and rounded . . . forming heads typical of the genus *Trichoderma*," without mentioning the characteristic roughening of *T. viride* aggr. *sensu* Rifai. The discussion reveals that species identification was arrived at without reference to Rifai's (524) monograph published 6 years earlier, and the significance of the distinction between smooth- and rough-walled, rounded *Trichoderma* conidia was unknown to the authors. The isolate may have been *T. harzianum* or another species. The identification given in the peritonitis report of Loeppky et al. (548) is completely unsubstantiated. The report by Jacobs et al. (549) of *T. viride* infection in a liver transplant recipient includes a capsule description that in retrospect rules out identification as *T. viride*. The report involved an organism with a conidia reported as "slightly rough." Based on the measurements given, their l/w ratio was approximately 1.75, corresponding well with the "ovoid" shape described by Jacobs et al. The isolate produced chlamydospores and grew at 45°C. A photograph showed a long branch bearing conspicuous single phialides. *T. viride* has distinctly rough, round conidia with a median l/w ratio of 1.1 (551). Its temperature tolerance is mentioned above. Only a minority of isolates produce chlamydospores within usual incubation periods. The newly described *T. asperellum* does have finely roughened conidia, of which some may be ovoid, but most are subglobose to globose, and the median l/w is 1.2. Its branching has symmetrical, verticillate "pyramidal" structure; structures with series of single phialides would not be common. Its maximum growth temperature is not recorded, but photos published by Samuels et al. (551) clearly show that growth is reduced by over 50% between 30°C and 35°C, a finding that seldom if ever corresponds with thermotolerance. Jacobs et al.'s isolate, therefore, does not fit either described *Trichoderma* species with uniformly rough conidia or taxa such as *Trichoderma saturnisporum* or *T. ghanense* with irregular conidial roughening. On the other hand, only the mention of fine roughening prevents this isolate from fitting a description of *T. longibrachiatum*. If the authors mistook cytoplasmic granulation or attached debris on the conidial wall for fine conidial roughening, as persons inexperienced with *Trichoderma* commonly do, *T. longibrachiatum* would be the most likely identification of their isolate, with a small chance of *T. harzianum*.

In general, the *T. viride* aggregate is similar in most respects to *T. harzianum* (see discussion of that species), but differs by having conidia with distinctly roughened walls. Isolates from medical and veterinary cases should be molecularly characterized or deposited in major collections, at least until the taxonomy of this group has become stabilized in molecularly informed species concepts.

H. *Tubercularia*

1. *Tubercularia vulgaris* Tode: Fr. Anamorph of *Nectria cinnabarina* (Tode: Fr.) Fr.

This distinctive species forms large, stout, heavily conidial synnemata on the natural substrate (newly killed and weakened branches of various trees and shrubs), but only occasionally does so in culture. Indeed, it is seldom identified or described from cultures. The use of this name to indicate a putative etiologic agent in a tabulation of Florida endophthalmitis cases (407) is therefore unexplained. Cases were evidenced only by multiple culture on one occasion. (See discussion under *Microdochium nivalis* above.) The record may well be correct, but cannot be accepted.

I. *Verticillium*

1. *Verticillium {aff.? Glomerella} nigrescens* Pethybr.
   (Synonym: *Cephalosporium serrae* Maffei)

The ex-type isolate of the now-synonymized name *C. serrae* was obtained from well-demonstrated keratitis by G. M. Serra, and was described by Maffei (552), who also recapitulated relevant clinical details. Direct microscopy was positive for fungal filaments. The isolate is in CBS and other collections, but is degenerated, no longer producing the characteristic dark chlamydospores figured by Maffei.

A severe case of "*C. serrae*" keratitis reported by Gingrich (420) had neither etiologic nor mycological substantiation. The same *C. serrae* case was summarized earlier by Gingrich in recorded discussion notes following a seemingly unrelated paper on the experimental destruction of rabbit corneas by enzymes from an unidentified "*Cephalosporium*" (553). Explaining the identification of his isolate, Gingrich stated: "*Cephalosporium serrae* [is] distinct from all other *Cephalosporium* species in that [its] conidiophores are branched." This statement bore no relation to scientific reality at that (1960) or any earlier time, making this record uninterpretable. Connection with a microconidial *Fusarium* may be speculated, since the case isolate was highly resistant to amphotericin B in vitro. Gingrich (420) later attributed the findings of the rabbit cornea study to *C. serrae*. Apart from its identification as a "*Cephalosporium*" isolate from keratitis, however, the fungus used in the experiments was not identified as to species or provenance. Gingrich's (420) statement that "*C. serrae*" is strongly enzymatically destructive toward the eyes thus appears to be unfounded, or if his own case isolate was the unidentified strain used in the experiments, at least not to apply correctly to *V. nigrescens.*

An isolate from mycetoma in Venezuela identified as *C. serrae* by de Albornoz (554) was more recently reidentified by W. Gams as representative of a newly described species, *Phaeoacremonium inflatipes* W. Gams, Crous et Wingf. (555). The isolate is in various culture collections: CBS 651.85, ATCC 32628, and the University of Alberta Microfungus Collection and Herbarium (UAMH) 4034. Contrary to a statement by Guarro et al. (327) in their review of infections caused by *Acremonium* and species formerly considered *Cephalosporium* spp., the authentic *V. nigrescens* is not known from mycetoma. *V. nigrescens* is a soil fungus of areas of temperate climate. It is especially associated with roots of the family Solanaceae (potato, tomato, pepper, eggplant) (74).

Description.   Colonies grow 23 to 28 cm in 7 days on PDA at 25°C. They are white and velvety, soon becoming greyish. The reverse is pale at first, then grey to black as chlamydospores form. It is sometimes also orange-brown at the center. *Conidiophores* are mostly erect, forming one, two, or uncommonly more successively elevated branch points giving rise to 1–2 (-3) phialides. Sporulation is also seen within the agar on short phialides (Fig. 100). Aerial *phialides* are subulate (awl-shaped) to candle-shaped, 20–35 (-50) × 1.5–3 µm. *Conidia* are ellipsoidal or short cylindrical, usually aseptate, rarely one-septate, and 4–8.5 × 1.5–2.5 µm. Chlamydospores are produced after 10 days, dark, singly or in chains, terminal or intercalary, usually distinctly swollen to near globose, and 5 to 10 µm.

The normal maximal temperature for growth of this fungus is 34°C. The isolate ex-type of *C. serrae* appears not to have been tested. *V. cinerescens* is one of the *Verticillium* species shown by molecular phylogeny to have affinities with the order *Phyllachorales*, and is remotely related to the teleomorph genera *Glomerella* and *Plectosphaerella*. It can be distinguished from *Acremonium* species by the combination of its erect conidiophores and dark chlamydospores.

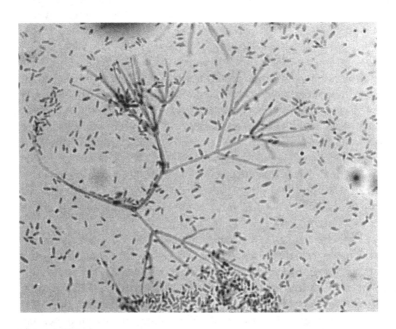

Figure 100   *Verticillium nigrescens* CBS 109724 conidiophores.

J. *Volutella*

1. *Volutella {aff. ?Pseudonectria} cinerescens* (Cesati) Sacc.

This fungus was isolated in New York in 1958 by Foster et al. (556) from three cases of postsurgical endophthalmitis, one of which was thoroughly documented. Theodore (557) later illuminated the source of this outbreak by stating that the same organism was isolated from "the cocaine used during the surgery," possibly explaining the paucity of similar outbreaks in recent times. Identification was credited to G. A. de Vries of CBS, and consistent photos and partial descriptions were given. Curiously, no culture has been retained at CBS. A fourth confirmed case was reported under the generic name only by Theodore et al. (557). A keratitis case from Florida was ascribed without details to a member of the genus *Volutella* by Liesegang and Forster (393). The CBS collection contains an isolate of *V. cinerescens*, CBS 832.71, listed as isolated in 1971 from "keratomycosis" by R. C. Zapater in Argentina. No case report definitely documenting this isolation has so far been traced by the present author. Zapater (558) does, however, refer in a summary table to a postsurgical ocular infection caused by *V. cinerescens*.

Description. Colonies grow moderately rapidly. They are pale above and below, with profuse formation of cushionlike sporodochial tufts bearing conidia in slimy masses. *Conidiophores* are sometimes produced as short single side branches from aerial or submerged hyphae, but are generally also produced in tangled, bushlike sporodochial complexes over 50 μm wide (Fig. 101). They are loosely interwoven, multiply rebranched in a monopodial (maple- or fig-tree-like) fashion, bundling together more tightly at the base of the sporodochium to form a compacted stipelike region. *Phialides* are candle-shaped to elongate-cylindrical, often with a slightly swollen basal region, especially in nonsporodochial phialides, and 8.5–20 × 1–2 μm. Short adelophialides (phialidelike, fertile short side branches not delimited from the subtending hypha by a septum) are also present. The *conidia* are ellipsoidal, cylindrical, or somewhat curved, smooth, hyaline, and 2–4 × 0.8–1.2 μm. Long, thick-walled, erect *setae* (spines) with rounded or pointed tips are also commonly formed from substrate mycelium.

The nonsporodochial phialides and conidia of this species are somewhat reminiscent of those seen in the otherwise quite different *Acremonium kiliense*. Growth on PDA or another non-Sabouraud fungal sporulation medium should be performed to elicit sporodochia in potential *Volutella* isolates.

IV. *OPHIOSTOMATALES*

This order is generally well known for plant pathogenicity; for example, *Ophiostoma ulmi* and its offshoot *O. novo-ulmi* are the agents of Dutch elm disease.

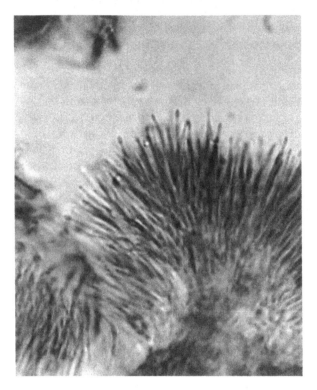

Figure 101   *Volutella cinerescens* CBS 832.71 conidioma.

It also contains one notorious animal-pathogenic anamorph species, *Sporothrix schenckii*, the dimorphic fungus that is the agent of sporotrichosis. Many anamorphs in this group, whether or not they are associated with a teleomorph, are currently placed in the genus *Sporothrix*, and some species with no pathogenic record (e.g., *Sporothrix inflata*) are commonly seen (although not commonly identified) as contaminants from skin and nails. This therefore is a group of fungi that are reasonably frequently seen in the clinical laboratory.

A.   Identification

Most Ophiostomatalean anamorphs can be identified at least to the level of species complex by morphological studies conducted primarily with slide cultures. Many species produce a compact sympodially proliferating conidiogenous cell bearing an apical "rosette" of conidia, as *S. schenckii* does. Most of these species differ from *S. schenckii* by producing secondary conidia on the still attached conidia of the primary rosette, significantly longer conidia, well-organized synnemata, or sympodial conidiogenous cells with an inflated apex. [See de Hoog (559) for very useful keys and descriptions, but note that the synonymy of *S. schenckii* and *Ophiostoma stenoceras* mooted in that work was later revealed to be incorrect.] *S. schenckii* is also tested for conversion to a budding yeast phase at 35° and 37°C. (Different isolates tend to convert better at one of these temperatures than the other, so using both temperatures speeds the finalization of the test.) It is not the only species in the group that produces a yeast phase at these temperatures, but it converts more thoroughly than other species and forms a high proportion of distinctive, elongated "cigar-shaped" buds, often in pairs (forming so-called rabbit ears). *S. schenckii* is also unique among described Ophiostomatalean anamorphs in producing copious dark "secondary conidia" that in fresh isolates arise densely as single dark subglobose or rarely triangular cells on short stalks from the substrate mycelium. Their density is often so great that they are said to form "sleeves" along the hyphae (560). The augmentation in numbers of these dark conidia rapidly turns colonies dark brown to black. These conidia are important in helping to distinguish *S. schenckii* from poorly known, possibly undescribed *Sporothrix* spp. sharing its environmental habitats (e.g., peat moss) as well as the very similar early growth stages of the *Sporothrix* state of *O. stenoceras*. The mimics lack secondary conidia. Molecular methods can also make the distinction between pathogenic and nonpathogenic types (561).

According to Dixon et al. (560), some mimics of *S. schenckii* occur that are nonpathogenic in animal models and possess dark secondary conidia, differing phenotypically from pathogenic *S. schenckii* in vitro only in their ability to grow at 37°C. These authors therefore recommended a test for 37° growth as a reference test for all *S. schenckii*-like isolates converting to a yeast phase and forming dark secondary conidia. Definitive studies, however, have shown that some proven etiologically significant *S. schenckii* isolates, especially from fixed cutaneous sporotrichosis, grow minimally or apparently not at all at 37°C, despite growing well at 35° [see review by Kwon-Chung and Bennett (68)]; this is also the present author's experience. The small number of apparently nonpathogenic isolates forming dark secondary conidia isolated from peat moss by Dixon et al. (560) are the only such isolates known; therefore, pending clarification of the nature of these isolates, all isolates obtained from suspected sporotrichosis and matching *S. schenckii* in all characters except the ability to grow at 37°C in vitro (while still growing at 35°C) should be reported as *S. schenckii*. Molecular confirmation (152, 561) may be done if there is significant uncertainty about the identification. A recommended molecular analysis is mitochondrial DNA *Hae*III restriction fragment analysis, which allows recognition of geographic variants of *S. schenckii* (562) based on studies of a large number of isolates from around the world.

B. *Sporothrix*

1. *Sporothrix {aff. Ophiostoma} schenckii* Hektoen & Perkins
    (Common Synonym in Older Literature: *Sporotrichum
    schenckii*)

This species, the agent of sporotrichosis, is mainly known as an agent of subcutaneous infection subsequent to traumatic inoculation. Most primary infections develop into acute lymphocutaneous infections by the spread of the organism into regional lymph nodes (68). The usual source of inoculum is plant material, particularly of a grassy or strawlike nature (563); a large outbreak in 1988 in North America was associated with poorly stored peat moss (560). In Central and South America, a nonprogressive form of the disease, fixed cutaneous sporotrichosis, is seen especially in persons who are repeatedly exposed to inoculum of *S. schenckii*, and who thus manifest an immune response that tends to prevent lymphocutaneous spread. Inoculation of *S. schenckii* into sites other than skin can cause other local infections (e.g., keratitis, endophthalmitis, invasive otitis, and mycotic arthritis) (69). In immunocompromised patients, more severe infections may be seen; chronic alcoholic debilitation has been noted as a relatively frequent predisposing factor for rare pulmonary cases of sporotrichosis (68). England and Hochholzer (564) report that the combination of alcoholism and chronic obstructive pulmonary disease is seen in most patients with pulmonary sporotrichosis; the majority also are male and middle-aged. Sporotrichosis in general is much more common in males than females (68), as is often the case with thermodimorphic mycoses. Rare idiopathic cases of pulmonary and other extracutaneous sporotrichosis may occur in individuals with no known predisposing factors (68). Disseminated sporotrichosis is occasionally seen in patients with AIDS (565–567), but on the other hand, very few cases of sporotrichosis have been seen in connection with neoplasms (568), and a Medline search of "sporot* AND neutropeni*" elicits no records. The disease may affect a wide range of animal species and is particularly common in cats.

   Description.   Colonies are flat and slow-growing—4 to 15 mm in 7 days on Sabouraud peptone glucose agar at 25°C. They are whitish then partly deep brown to black, sometimes near black from the start. The reverse is concolorous with the surface. *Conidiogenous cells* arise more or less at right angles to surface hyphae. They are sometimes terminal, cylindrical, and 10 to 40 × 1 to 2 μm at maturity, with a slightly swollen, compact sympodial rhachis of toothlike denticles at the apex. There are also often sparsely scattered denticles along the stalk. *Primary conidia* are formed in sympodial rosettes on conidiophore apices. They are obovoidal (egg-shaped, attached at the thin end), and 3–6 × 1.5–3 μm (Fig. 102).
   The secondary conidia are deep brown, often subglobose, and approximately 3 μm in diameter (Fig. 103). They are sometimes triangular, are attached

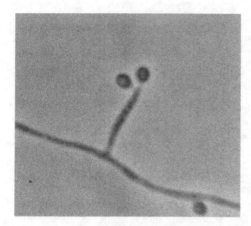

Figure 102 *Sporothrix schenckii* primary sympodial conidia.

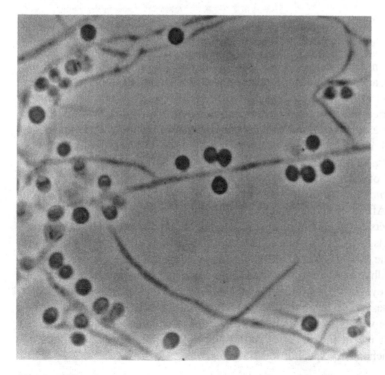

Figure 103 *Sporothrix schenckii* secondary sessile conidia.

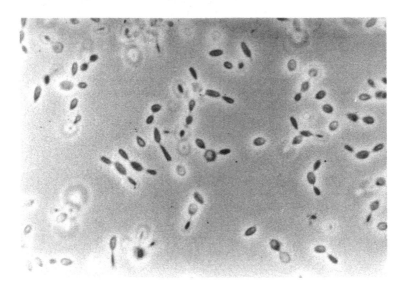

Figure 104 *Sporothrix schenckii* yeast phase at 37°C in vitro.

individually on denticles on the sides of the hyphae, especially substrate (in the agar) hyphae, and are often densely packed and surrounding the hyphae in sleeve-like formation in fresh isolates.

Converted phase on brain–heart infusion agar overlaid with a few drops of 10% yeast extract solution and incubated in parallel cultures at 35° and 37°C (as explained above) gives rise to white, glossy masses of yeast cells, 3–10 × 1–3 μm, with long obclavate (‘‘cigar-shaped’’) buds (Fig. 104). Sympodially formed pairs of such buds (rabbit ears) are typically seen.

S. schenckii has a distinct variant, *S. schenckii* var. *luriei*, generally seen in southern Africa or India, which is very similar to the type variety in most cultural aspects but differs in its host phase, producing unusually large, irregular yeast-like cells as well as septate cells resembling fission cells, rather like hyaline versions of the fission cells of chromoblastomycosis fungi. A recent case investigation by Padhye et al. (569) reported large numbers of incompletely separated cells in an ‘‘eyeglass’’ configuration as characteristic of *S. schenckii* var. *luriei* infection. The variety has a distinct mitochondrial DNA type (570) and differs from the type variety in not assimilating creatine, creatinine, or guanidinoacetic acid (571). A recent Indian case that did not yield a culture was confirmed by immunohistochemistry specific for *S. schenckii.*

C. *Ophiostoma*

1. *Ophiostoma stenoceras* (Robak) Melin & Nannf.

This fungus, closely related to *S. schenckii*, was reported by Summerbell et al. (572) as associated with long-term paronychia in a 65-year-old woman who had acquired the infection while laboring in forestry and agricultural work. The lesion was direct-microscopic-positive for filaments but not for yeast elements. Unfortunately for the case report (but not for the patient), the infection was rapidly cured with nonspecific iodine therapy while the laboratory report was pending, and no repeat isolation for confirmation of etiology could be performed. Ophiostomatalean fungi are among the few pathogenic fungi highly susceptible to iodine, and the cure tended to correlate the proposed etiology; however, additional and better confirmed cases are needed before *O. stenoceras* can be formally listed as a pathogen.

 O. stenoceras forms colonies and conidiogenous structures that are similar to *S. schenckii* colonies within their first 10 to 14 days of growth, except for the following features: (1) no dark secondary conidia are formed; (2) conidia may appear more elongate and clavate than typical *S. schenckii* primary conidia [according to de Hoog (559) the difference between the two species is not statistically significant, but restudy of this matter using fresh isolates would be ideal]; (3) dark brownish to black spots not associated with secondary conidia may de-

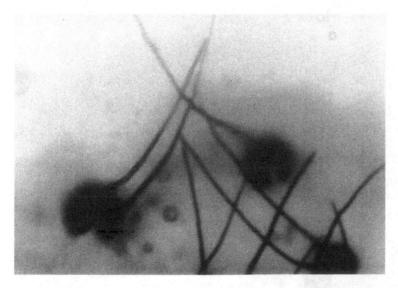

Figure 105 *Ophiostoma stenoceras* ascomata (immersed in cloud of spores).

velop on the colony surface; and (4) attempts to convert the organism to the yeast phase give a partial conversion, with most elements still recognizably hyphae or mycelial conidia in microscopic examination even though the colony has a pasty appearance. After 14 days on modified Leonian's agar, ascomata form (Fig. 105). They are black, with a bulbous base 80 to 180 μm, and with an elongate (up to 2 mm), thin neck bearing a crownlike ring of short, sharp, spinelike setae 18 to 35 (-55) μm, at the ostiole, reflexed outwardly to support an extruded, slimy mass of pale, smooth-walled, bean- or orange-segment-shaped ascospores (2.0-) 2.5–4.5 × 1.0–1.5 μm (Figs. 106 and 107). Unlike *S. schenckii*, *O. stenoceras* degrades starch in vitro (572).

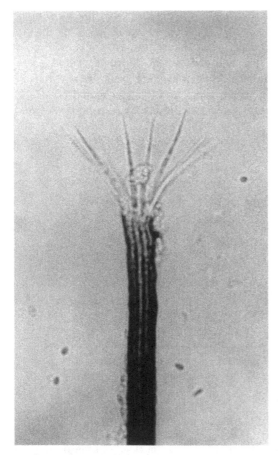

Figure 106 *Ophiostoma stenoceras* ascoma apex.

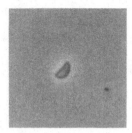

Figure 107 *Ophiostoma stenoceras* ascospore.

V. *DOTHIDEALES, PRO PARTE* (OTHER MEMBERS OF THIS GROUP WILL APPEAR IN OTHER CHAPTERS)

A. *Neotestudina*

1. *i. Neotestudina rosatii* Ségr. & Destombes (Synonyms: *Zopfia rosatii, Pseudophaeotrichum sudanense, Pseudodelitschia coriandri*)

This ascomycetous fungus is a well-characterized agent of mycetoma (573). Like many such fungi, it is currently identified by the characters of its ascomata rather than from general cultural characters (Fig. 108 and 109).

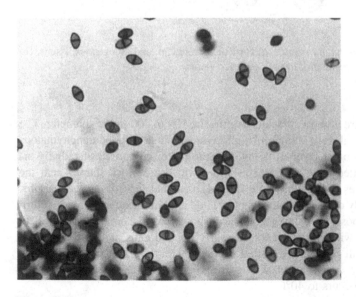

Figure 108 *Neotestudina rosatii* ascospores.

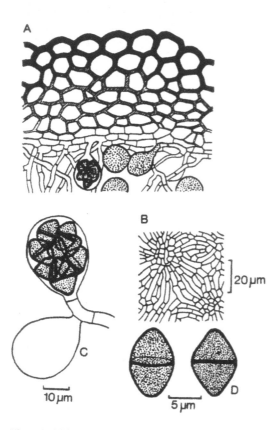

Figure 109 *Neotestudina rosatii*: (a) cortical, subcortical, and ascogenous tissues; (b) surface view of the peridium; (c) asci; (d) ascospores.

Ascomata are globose, black cleistothecia, 100 to 200 μm in diameter. The peridium (ascomatal wall) is smooth, multilayered, with a pseudoparenchymatous (cellular-looking) outer layer. It is divided into plates composed of radially arranged, thick-walled cells 4 to 10 μm wide, which are block-shaped near the centers of the plates and become elongate and somewhat curved near the margins. Asci form laterally on ascogenous hyphae. They are bitunicate and broadly ellipsoidal to subglobose, containing eight ascospores. Ascospores are two-celled, rhomboidal (i.e., with abruptly curved, nearly pointed ends and with an abrupt change of cell wall angle at the central septum), with thick, smooth walls. They are brown, 9–12.5 × 4.5–8 μm, with germ pores at both ends. No anamorph is present. Growth occurs to 40°C.

B. *Piedraia*

1. *Piedraia hortae* (Brumpt) da Fonseca & Arêa Leão

This is one of the few truly contagious fungal parasites of humans. It also occurs on other primates. Black concretions are attached to the shaft of scalp hairs, and ascomata form within these concretions. Identification is therefore by direct microscopic analysis of the concretions; the fungus may be grown in pure culture, but is hard to isolate, slow-growing, and generally nonsporulating. The disease, piedra, may be acquired by contact with soil and is cultivated as a sign of beauty in some cultures (11). A synopsis of recent references on the biology of *P. hortae* is provided by de Hoog et al. (69). Extensive overviews are given by Rippon (11) and Kwon-Chung and Bennett (68).

   Ascomata are ascostromata, and are compact, black, very variable in size, up to 1 mm long by 0.33 mm wide, and irregularly lumpy with formation of fertile locules (Fig. 110). Stromatal material is composed of vertical rows of thick-walled, dark, polygonal cells that are 2–10 × 2.5–6.5 μm. Fertile *locules* are 30 to 60 μm in inner diameter, opening to the surface by means of an inconspicuous ostiolar pore. *Asci* are bitunicate, ellipsoidal, and evanescent, with two to eight hyaline. They are elongate, sinuously curved, and spindle shaped (i.e., with acutely angled, rounded ends). The *ascospores* are 30–45 × 5.5–10 μm, and bear whip-like extensions at both ends. A similar species, *P. quintanilhae* van Uden et al., differs by lacking the whip-like ascospore appendages, and occurs in wild populations of chimpanzees (574).

VI. *PLEOSPORALES, PRO PARTE* (OTHER MEMBERS OF THIS GROUP WILL APPEAR IN OTHER CHAPTERS)

A. *Leptosphaeria*

1. *Leptosphaeria senegalensis* Ségretain et al.

This species was described based on isolations from mycetoma in Senegal and Mauritania (575) (Figs. 111, 112, and 113).

   Description. Colonies grow slowly, according to de Hoog et al. (69), but show "great variability in growth rate" and grow more quickly at 37° than at 30°C, according to De Vries (3). They are velvety to wooly, pale at first, then becoming dark brownish-grey with a nearly black reverse. Brownish soluble pigment may be formed. *Ascomata* form after 30 to 60 days on oatmeal agar. They are black, 100 to 300 μm in diameter, and covered with short, flexuous, dark hyphae. They are subglobose, lacking an ostiole, with a thick peridium composed of several layers of isodiametric cells, and the inner ones are less dark than the

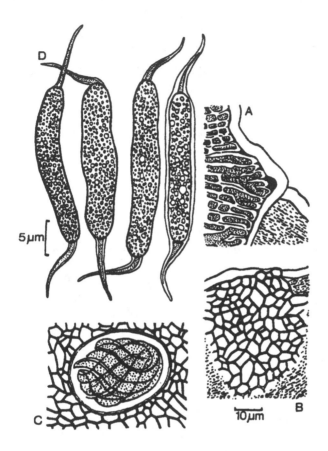

Figure 110 *Piedraia hortae*: (a) border of subcuticular zone of a mature ascostroma; (b) young ascotroma; (c) mature ascus; and (d) ascospores.

outer, cortical layer. *Asci* are bitunicate, clavate, and 80–100 × 17–22 µm. They are eight-spored, arising from a stroma, and intermingled with thin, occasionally branched paraphyses. The ascospores are hyaline to pale brown, ellipsoidal, and 23–30 × 8–10 µm, usually with five linearly arranged cells, less commonly six, distinctly constricted at the septa. Each ascospore is embedded in a thick, hyaline to slightly colored slimy sheath.

This species is distinguished from *L. tompkinsii* below by producing mostly five-celled ascospores with rounded ends rather than mostly seven-celled ascospores with pointed ends. The ascospores of *L. senegalensis* have a broad gel sheath described by El-Ani (576) as "turbinate [top-shaped] greatly enlarged near the spore apex," while sheaths of *L. tompkinsii* are "fusoid [spindle-shaped]

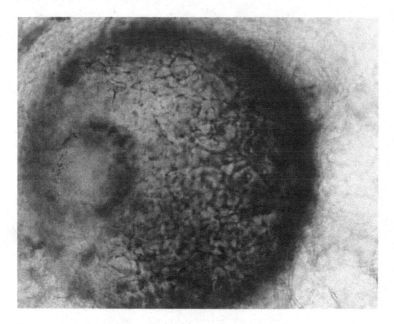

Figure 111 *Leptosphaeria senegalensis* ascoma peridium with ostiole.

like the spore." Further details of the distinction are given by El-Ani (576). Ségretain et al. (575) and de Vries (3) note rare ascospores with up to eight cells in *L. senegalensis*, but this appears to be questioned by El-Ani (576), who studied 823 spores of four *L. senegalensis* isolates and saw 810 four-celled spores, 13 five-celled, and none with more cells. Because there may have been some early confusion between the two species, I have cited El-Ani's measurements for *L. senegalensis* microscopic structures above in preference to those from other sources.

2. *Leptosphaeria tompkinsii* El-Ani

This species was described based on an isolate from mycetoma in Mauritania (576) (Fig. 114).

Description. Colonies grow slowly, according to de Hoog et al. (69), but show "great variability in growth rate" and grow more quickly at 37°C than at 30°C, according to de Vries (3). They are velvety to woolly, pale at first, then becoming dark brownish-grey with a nearly black reverse. Brownish soluble pigment may be formed. *Ascomata* form after 30 to 60 days on oatmeal agar. They are black, 214 to 535 μm in diameter, and covered with short, flexuous, dark

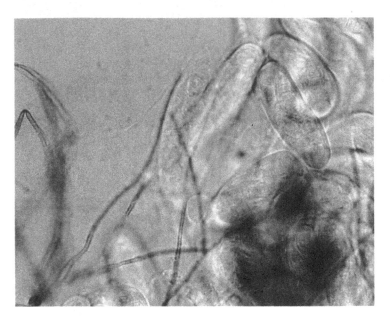

Figure 112   *Leptosphaeria senegalensis* asci containing ascospores.

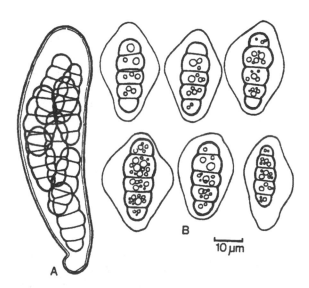

Figure 113   *Leptosphaeria senegalensis*, Segretain et al. (IP 614): (a) mature ascus; (b) ascospores.

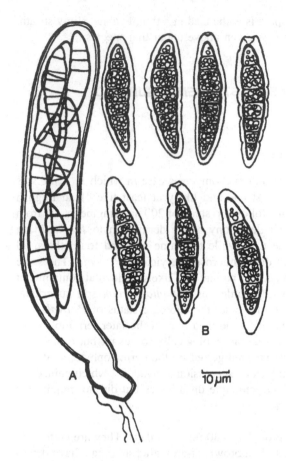

Figure 114 *Leptosphaeria tompkinsii El-Ani* (IP 1151.76): (a) mature asci; (b) ascospores.

hyphae. They are subglobose to compressed, lacking an ostiole, with a thick peridium composed of several layers of isodiametric cells, the inner ones less dark than the outer, cortical layer. The upper portion of the peridium is often thinner than the lower portion. *Asci* are bitunicate, clavate, and 80–115 × 20–25 μm. They are eight-spored, arising from a stroma, intermingled with thin, rarely branched paraphyses. The ascospores are hyaline, fusoid (spindle-shaped; i.e., with ends tapering to a relatively acute conical point), and 32–45 × 8.8–11 μm, with (5-) 7 to 8 (-9) linearly arranged cells, slightly constricted at the septa, with the second or third cell from the apical end often somewhat inflated compared to the others. Each ascospore is embedded in a thin, hyaline, slimy sheath. Distinction from *L. senegalensis* is given above under that species.

VII. *PSEUDEUROTIACEAE SS. STR. SENSU* SUH AND BLACKWELL (1)

A. *Pseudeurotium*

1. *Pseudeurotium ovale* Stolk

English et al. (577) investigated an onychomycosis case in which *P. ovale* was shown to be established in at least one and likely at least five of an English patient's 10 dystrophic nails. In a follow-up study of 200 nail fragments, 20 from each of the patient's nails, a single colony of the dermatophyte *Trichophyton rubrum* was obtained from one nail, while the same nail and four others, all positive for fungal elements in direct microscopy, yielded *P. ovale*. The same species was also isolated from another patient as a likely incidental contaminant from a typical *Acremonium* sp. (recorded as *Cephalosporium* sp.) superficial white onychomycosis. In the former case, the etiologic status of *P. ovale* was not entirely clarified, as English et al. pointed out. In the later terminology of Summerbell (578), it may have been an etiologically irrelevant but established "secondary colonizer" of material predigested by the dermatophyte, or it may have been an etiologically significant "successional invader," which gained access to the nail as a result of dermatophyte disturbance but did not require the continuing presence of *T. rubrum* to persist.

Description. Colonies grow 25 to 30 mm in 7 days. They are cottony to somewhat granular, pale at first, then brownish-grey after ascomata have developed. The reverse is white, greenish, or brownish. *Ascomata* are cleistothecia (i.e., rounded structures lacking an ostiole or differentiated opening), mostly 90 to 180 μm in diameter (Fig. 115). They are dark, bald, with a rather tough peridium composed of polygonal cells. *Asci* are ellipsoidal to spherical, and 7–9 × 6.5–8 μm. *Ascospores* are ellipsoidal to ovoidal, 4–6 × 2.5–4 μm, smooth-walled, and hyaline to pale olive-brown. *Conidiophores* are hyaline, vaguely *Sporothrix*-like, but lacking toothlike denticles at conidial attachment points. They consist of minimal stub-like side branches not delimited by a crosswall or more developed structures, with the most differentiated being delimited, slightly swollen, smooth-walled, single-celled lateral branches up to 20 μm long with tapering tips bearing at maturity an apical rosettelike cluster of several sympodially produced conidia attached by flat bases. *Conidia* are ellipsoidal to slightly irregular, smooth, pale, and mostly 4–6 × 2.5–3.5 μm. The related *Pseudeurotium zonatum* has spherical, not ellipsoidal ascospores.

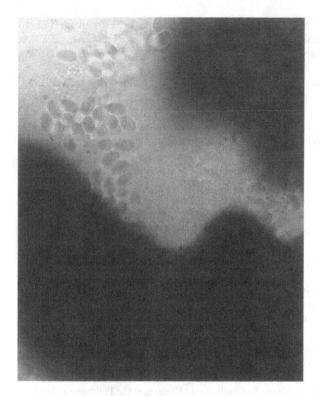

Figure 115 *Pseudeurotium ovale* CBS 247.89 ascoma breaking open to reveal ascospores.

ACKNOWLEDGMENTS

This chapter is dedicated to Dr. K-J. (June) Kwon-Chung, who was associated with an impressive number of the best-quality case reports seen in its preparation.

I would like to thank Katsuhiko Ando for Japanese translation, Walter Gams for German translation and Latin advice, Marjan Erich-Vermaas for photography, Rob Samson for managerial intervention, and Arien van Iperen for preparation of cultures. Gerard de Vries is thanked for drawings, and the Instructional Media Centre, Ontario Ministry of Health, for inking them.

REFERENCES

1. S-O Suh, M Blackwell. Molecular phylogeny of the cleistothecial fungi placed in the *Cephalothecaceae* and *Pseudeurotiaceae*. Mycologia 91:836–848, 1999.

2. M Goodman, CA Porter, J Czelusniak, SL Page, H Schneider, J Shoshani, G Gunnell, CP Groves. Toward a phylogenetic classification of primates based on DNA evidence complemented by fossil evidence. *Molec Phylogen Evol* 9:585–598, 1998.
3. GA de Vries. *Ascomycetes: Eurotiales, Sphaeriales*, and *Dothideales*. In: DH Howard, ed. Fungi Pathogenic for Humans and Animals. Part A, Biology. New York: Marcel Dekker, 1983, pp. 81–111.
4. DW Malloch. Moulds: Their Isolation, Cultivation and Identification. Toronto: University of Toronto Press, 1981 (online version: http://www.botany.utoronto.ca/ ResearchLabs/MallochLab/Malloch/Moulds/Moulds.html).
5. RA Samson. Taxonomy—Current concepts of *Aspergillus* systematics. In: JE Smith, ed. *Aspergillus*. New York: Plenum, 1994, pp. 1–22.
6. SW Peterson. Phylogenetic relationships in *Aspergillus* based on rDNA sequence analysis. In: RA Samson, JI Pitt, eds. Integration of Modern Taxonomic Methods for *Penicillium* and *Aspergillus*. Amsterdam: Harwood Academic, 2000, pp. 323–355.
7. KB Raper, DI Fennell. The Genus *Aspergillus*. Baltimore: Williams and Wilkins, 1965.
8. JC Frisvad, U Thrane. Mycotoxin production by food-borne fungi. In: RA Samson, ES Hoekstra, JC Frisvad, O Filtenborg, eds. Introduction to Food-Borne Fungi. 5th ed. Baarn, Netherlands: Centraalbureau voor Schimmelcultures, 1996, pp. 251–260.
9. RC Young, A Jennings, JE Bennett. Species identification of invasive aspergillosis in man. *Amer J Clin Path* 58:554–557, 1972.
10. MG Rinaldi. Invasive aspergillosis. *Rev Infect Dis* 5:1061–1077, 1983.
11. JW Rippon. Medical Mycology: The Pathogenic Fungi and the Pathogenic Actinomycetes. 3rd ed. Philadelphia: Saunders, 1988.
12. M David, A Charlin, J Morice, [initial unknown] Naudascher. Infiltration mycosique à *Aspergillus amstelodami* du lobe temporal simulant un abscès encapsulé. Ablation en masse. Guérison opératoire. *Rev Neurol* 85:121–124, 1951.
13. LK Georg, WM Williamson, EB Tilden, RE Getty. Mycotic pulmonary disease of captive giant tortoises due to *Beauveria bassiana* and *Paecilomyces fumosoroseus*. Sabouraudia 2:80–86, 1962.
14. P Negroni, JA Tey. Estudio mycologico del primer caso argentino de micetoma maduromicósico de granos negros. *Rev Inst Bacteriol* 9:176–189, 1939.
15. PK Shukla, M Jain, B Lal, PK Agrawal, OP Srivastava. A study on keratomycoses caused by the species of *Aspergillus*. *Biol Mem* 11:161–166, 1985.
16. RV Talice, JE Mackinnon. *Aspergillus (Eurotion) montevidensis n. sp.* aislado de una otomicosis del hombre. *C R Séanc Soc Biol Argent* 108:1007–1009, 1931.
17. K Wadhwani, AK Srivastava. Fungi from otitis media of agricultural field workers. *Mycopathologia* 88:155–159, 1984.
18. D Grigoriu, A Grigoriu. Les onychomycoses. *Rev. Méd. Suisse Romande* 95:839–849, 1975.
19. D Janke. Kasuistik seltener Mykosen. *Dtsch Dermatol Ges* 4:387–390, 1953.
20. JR Person, MJ Ossi. A case of possible *Penicillium* tinea capitis. *Arch Derm* 119:4, 1983.

21. UW Leavell Jr, EB Tucker, R Muelling. Blue dot infection of the scalp of two brothers. J Ky Med Assoc 64:1107–1110, 1966.

22. O da Fonseca Jr. Mycetoma por *Aspergillus amstelodami*. Rev Medico-Cirurg Bras 38:415–423, 1930.

23. C Thom, MB Church. The Aspergilli. Baltimore: Williams & Wilkins, 1926.

24. JI Pitt. Nomenclatorial and taxonomic problems in the genus *Eurotium*. In: RA Samson, JI Pitt, eds. Advances in *Penicillium* and *Aspergillus* Systematics. New York: Plenum, 1985, pp. 383–396.

25. J Naidu, SM Singh. *Aspergillus chevalieri* (Mangin) Thom and Church: A new opportunistic pathogen of human cutaneous aspergillosis. Mycoses 37:271–274, 1994.

26. J Naidu. Growing incidence of cutaneous and ungual infections by non-dermatophyte fungi at Jabalpur (M.P.). Indian J Path Microbiol 36:113–118, 1993.

27. AEW Gregson, CJ La Touche. Otomycosis: A neglected disease. J Laryn Otol 75: 45–69, 1961.

28. SD Deshpande, GV Koppikar. A study of mycotic keratitis in Mumbai. Indian J Path Microbiol 42:81–87, 1999.

29. J Chander, A Sharma. Prevalence of fungal corneal ulcers in northern India. Infection 22:207–209, 1994.

30. RH Rosa, D Miller, EC Alfonso. The changing spectrum of fungal keratitis in south Florida. Ophthalmol 101:1005–1013, 1994.

31. S Dreizen, GP Bodey, KB McCredie, MJ Keating. Orofacial aspergillosis in acute leukemia. Oral Surg Oral Med Oral Path 59:499–504, 1985.

32. D Shetty, N Giri, CE Gonzalez, PA Pizzo, TJ Walsh. Invasive aspergillosis in human immunodeficiency virus-infected children. Pediat Infec Dis J 16:216–221, 1997.

33. S Abbasi, JL Shenep, WT Hughes, PM Flynn. Aspergillosis in children with cancer: A 34-year experience. Clin Infec Dis 29:1210–1219, 1999.

34. JS Weingarten, DM Crockett, RP Lusk. Fulminant aspergillosis: Early cutaneous manifestations and the disease process in the immunocompromised host. Otolaryn Head Neck Surg 97:495–499, 1987.

35. KR Gupta, B Udhayakumar, PB Rao, M Madhavan, LR Das Gupta. Aspergilloma of the frontal bone. J Laryn Otol 87:1007–1011, 1973.

36. PV Venugopal, TV Venugopal, A Gomathi, ES Ramakrishna, S Ilavarasi. Mycotic keratitis in Madras. Indian J Path Microbiol 32:190–197, 1989.

37. AA Gage, DC Dean, G Schimert, N Minsley. *Aspergillus* infection after cardiac surgery. Arch Surg 101:384–387, 1970.

38. J Bille, L Stockman, GD Roberts. Detection of yeasts and filamentous fungi in blood cultures during a 10-year period (1972 to 1981). J Clin Microbiol 16:968–970, 1982.

39. KW Sell, T Folks, KJ Kwon-Chung, J Coligan, WL Maloy. Cyclosporin immunosuppression as the possible cause of AIDS. N Eng J Med 309:1065, 1983.

40. HR Schelbert, OF Müller. Detection of fungal vegetations involving a Starr–Edwards mitral prosthesis by means of ultrasound. Vasc Surg 6:20–25, 1972.

41. G Bambule, M Savary, D Grigoriu, J Delacretaz. Les otomycoses. Ann Oto-Laryng (Paris) 99:537–540, 1982.

42. K Wulf. Aspergillose der Paukenhöhle. In: H Grimmer, H Rieth, eds. Krankheiten durch Schimmelpilze bei Mensch und Tier. Berlin: Springer-Verlag, 1965, pp. 89–92.

43. I Vennewald, M Henker, E Klemm, C Seebacher. Fungal colonization of the paranasal sinuses. Mycoses 42, suppl. 2:33–36, 1999.

44. JU Ponikau, DA Sherris, EB Kern, HA Homburger, E Frigas, TA Gaffey, GD Roberts. The diagnosis and incidence of allergic fungal sinusitis. Mayo Clin Proc 74:877–884, 1999.

45. ES Bereston, WS Waring. *Aspergillus* infection of the nails. Arch Derm Syphilol 54:552–557, 1946.

46. MM Walshe, MP English. Fungi in nails. Brit J Derm 78:198–207, 1966.

47. RC Summerbell, J Kane, S Krajden. Onychomycosis, tinea pedis and tinea manuum caused by non-dermatophytic fungi. Mycoses 32:609–619, 1989.

48. N Contet-Audonneau, O Salvini, AM Basile, G Percebois. Onychomycosis provoked by moulds. Importance of the biopsy for diagnosis. Frequency of the pathogenic species. Sensitivity to antifungal products. Nouv Dermatol 14:330–340, 1995.

49. N Zaias. Superficial white onychomycosis. Sabouraudia 5:99–103, 1966.

50. N Zaias. Onychomycosis. Arch Derm 105:263–274, 1972.

51. JI Pitt, AD Hocking. Fungi and Food Spoilage. 2nd ed. London: Blackie Academic & Professional, 1997.

52. C Aznar, C de Bièvre, C Guiguen. Maxillary sinusitis from *Microascus cinereus* and *Aspergillus repens*. Mycopathologia 105:93–97, 1989.

53. K Mencl, M Otčenášek, J Spacek, E Rehulova. *Aspergillus restrictus* and *Candida parapsilosis*—Agents of endocarditis following heart valve replacement (in German). Mykosen 28:127–133, 1985.

54. K Pospisil, V Straka, M Otčenášek, K Mencl, M Pospisil, J Spacek, A Hamet, V Pidrman. Mycotic ulcerative aortitis after replacement of the aortic valve caused by the fungus *Aspergillus restrictus* (in Czech). Vnitr Lek 30:292–297, 1984.

55. M Resl, M Otčenášek, I Steiner. Mycoses of the heart (in Czech). Cesk Patol 29:32–35, 1993.

56. AW Chen, DM Griffin. Soil physical factors and the ecology of fungi. 6. Interaction between temperature and soil moisture. Trans Brit Mycol Soc 49:551–561, 1966.

57. SL Smith, ST Hill. Influence of temperature and water activity on germination and growth of *Aspergillus restrictus* and *A. versicolor*. Trans Brit Mycol Soc 79:558–560, 1982.

58. F Estrader, H Longefait, G Lalavée, C Coury, P Constans. *Aspergillus restrictus*, forme rare d'aspergillome intra-cavitaire associée à une tuberculose active. J Fr Méd Chirurg Thorac 26:241–249, 1972.

59. S Tamaki, T Danbara, H Natori, H Kanamori, M Koike, N Motoyama, Y Tamano, S Kira. A resected case of endobronchial aspergilloma due to *Aspergillus restrictus*. Japan J Thorac Dis 18:464–469, 1980.

60. C Schönborn, H Schmoranzer. Untersuchungen über Schimmelpilzinfektionen der Zehennägel. Mykosen 13:253–272, 1970.

61. R Yang, WR Kenealy. Regulation of restrictocin production in *Aspergillus restrictus*. J Gen Microbiol 138:1421–1427, 1992.

62. R Kao, J Davies. Fungal ribotoxins: A family of naturally engineered targeted toxins? Biochem Cell Bio 73:1151–1159, 1995.

63. JI Pitt, RA Samson. Taxonomy of *Aspergillus* section *Restricta*. In: RA Samson, JI Pitt, eds. Modern Concepts in *Penicillium* and *Aspergillus* Classification. New York: Plenum, 1990, pp. 249–257.

64. M Otčenášek, V Janečková, R Kaupa, B Medek, D Nevludová. On the etiology of pulmonary aspergillomas (in Czech). Cesk Epidemiol Mikrobiol Imunol 25: 263–268, 1976.

65. DB Jones. Therapy of postsurgical fungal endophthalmitis. Ophthalmol 85:357–373, 1978.

66. E Maršálek, Z Žižka, V Říha, J Dušek, Č Dvořaček. Plicní aspergilóza s generalizací vyvolaná druhem *Aspergillus restrictus*. Čas Lék Česk 99:1285–1292, 1960.

67. P Fragner, J Vítovec, P Vladík, Z Záhoř. *Aspergillus penicillioides* v. solitárním plicním aspergilomu u srny. Česká Mykol 27:151–155, 1973.

68. KJ Kwon-Chung, JE Bennett. Medical Mycology. Philadelphia: Lea & Febiger, 1992.

69. GS de Hoog, J Guarro, J Gené, MJ Figueras. Atlas of Clinical Fungi. 2nd ed. Utrecht, Netherlands: Centraalbureau voor Schimmelcultures, 2000.

70. BH Segal, ES DeCarlo, KJ Kwon-Chung, HL Malech, JI Gallin, SM Holland. *Aspergillus nidulans* infection in chronic granulomatous disease. Medicine (Baltimore) 77:345–354, 1998.

71. el-S Mahgoub. Maduromycetoma caused by *Aspergillus nidulans*. J Trop Med Hyg 74:60–61, 1971.

72. J Guillot, C Collobert, E Guého, M Mialot, E Lagarde. *Emericella nidulans* as an agent of guttural pouch mycosis in a horse. J Med Vet Mycol 35:433–435, 1997.

73. RG Mitchell, AJ Chaplin, DW Mackenzie. *Emericella nidulans* in a maxillary sinus. J Med Vet Mycol 25:339–341, 1987.

74. KH Domsch, W Gams, T-H Anderson. Compendium of Soil Fungi. Eching, Germany: IHW Verlag, 1993.

75. DM Geiser, ML Arnold, WE Timberlake. Sexual origins of British *Aspergillus nidulans* isolates. Proc Natl Acad Sci USA 91:2349–2352, 1994.

76. A Coenen, F Debets, R Hoekstra. Additive action of partial heterokaryon incompatibility (partial-het) genes in *Aspergillus nidulans*. Curr Genet 26:233–237, 1994.

77. DK Sandhu, RS Sandhu. A new variety of *Aspergillus nidulans*. Mycologia 55: 297–299, 1963.

78. CJ White, KJ Kwon-Chung, JI Gallin. Chronic granulomatous disease of childhood: An unusual case of infection with *Aspergillus nidulans* var. *echinulatus*. Amer J Clin Path 90:312–316, 1988.

79. I Polachek, A Nagler, E Okon, P Drakos, J Plaskowitz, KJ Kwon-Chung. *Aspergillus quadrilineatus*, a new causative agent of fungal sinusitis. J Clin Microbiol 30: 3290–3293, 1992.

80. MP Singh, CM Singh. Fungi associated with superficial mycoses of cattle and sheep in India. Ind J Animal Hlth 9:75–77, 1970.

81. WU Knudtson, CA Kirkbride. Fungi associated with bovine abortion in the northern plains states (USA). J Vet Diag Invest 4:181–185, 1992.

82. P Émile-Weil, L Gaudin. Contribution à l'étude des onychomycoses. Onycho-

mycoses à *Penicillium*, à *Scopulariopsis*, à *Sterigmatocystis*, à *Spicaria*. Arch Méd Expér Anat 28:452–467, 1919.

83. MP English, R Atkinson. Onychomycosis in elderly chiropody patients. Brit J Derm 91:67–72, 1974.

84. JM Torres-Rodríguez, N Madrenys-Brunet, M Siddat, O López-Jodra, T Jimenez. *Aspergillus versicolor* as cause of onychomycosis: Report of 12 cases and susceptibility testing to antifungal drugs. J Eur Acad Derm Venereol 11:25–31, 1998.

85. JM Torres-Rodríguez, O López-Jodra. Epidemiology of nail infection due to keratinophilic fungi. In: KS Kushwaha, J Guarro, eds. Biology of Dermatophytes and Other Keratinophilic Fungi. Bilbao, Spain: Revista Iberoamericana de Micología, 2000, pp. 122–135.

86. A Kornerup, JH Wanscher. Methuen Handbook of Colour. 3rd ed. London: Eyre Methuen, 1978.

87. JI Pitt. A Laboratory Guide to Common *Penicillium* Species. 2nd ed. North Ryde, NSW, Australia: Commonwealth Scientific and Industrial Research Organization, 1988.

88. G St. Germain, RC Summerbell. Identifying Filamentous Fungi. Belmont, CA: Star, 1997.

89. RC Summerbell. Non-dermatophytic fungi causing onychomycosis and tinea. In: J Kane, RC Summerbell, L Sigler, S Krajden, G Land, eds. Laboratory Handbook of Dermatophytes. Belmont, CA: Star, 1997, pp. 213–259.

90. PV Venugopal, TV Venugopal, K Thiruneelakantan, S Subramanian, BM Shetty. Cerebral aspergillosis: Reports of two cases. Sabouraudia 15:225–230, 1977.

91. Z Liu, T Hou, Q Shen, W Liao, H Xu. Osteomyelitis of sacral spine caused by *Aspergillus versicolor* with neurological deficits. Chinese Med J 108:472–475, 1995.

92. B Anderson, SS Roberts, C Gonzalez, EW Chick. Mycotic ulcerative keratitis. Arch Ophth 62:169–179, 1959.

93. R Galimberti, A Kowalczuk, IH Parra, M Gonzalez Ramos, V Flores. Cutaneous aspergillosis: A report of six cases. Brit J Derm 139:522–526, 1998.

94. KG Keegan, CL Dillavou, SE Turnquist, WH Fales. Subcutaneous mycetoma-like granuloma in a horse caused by *Aspergillus versicolor*. J Vet Diag Invest 7:564–567, 1995.

95. A Yassin, A Maher, MK Moawad. Otomycosis: A survey in the eastern province of Saudi Arabia. J Laryn Otol 92:869–876, 1978.

96. MG Fakih, GE Barden, CA Oakes, CS Berenson. First reported case of *Aspergillus granulosus* infection in a cardiac transplant patient. J Clin Microbio 33:471–473, 1995.

97. Th Neuhann. Clotrimazol in der Behandlung von Keratomykosen. Klin Monatsbl Augenheilk 169:459–462, 1976.

98. IC da Alecrim, AF Vital. O *Aspergillus sydowii* (Bain. & Sart.) Thom e Church numa lesão ungueal. An Fac Med Univ Recife 15:229–240, 1955.

99. S Prasad, HV Nema. Mycotic infections of cornea. Ind J Ophth 30:81–85, 1982.

100. LE Zimmerman. Fatal fungus infections complicating other diseases. Amer J Clin Path 25:16–65, 1955.

101. JM Torres-Rodríguez, J Balaguer-Melez, AT Martinez, J Antonell-Reixach. Onychomycosis due to a member of the *Aspergillus versicolor* group. Mycoses 31: 579–583, 1988.
102. LM Weiss, WA Thiemke. Disseminated *Aspergillus ustus* infection following cardiac surgery. Amer J Clin Path 80:408–411, 1983.
103. MJ Stiller, L Teperman, SA Rosenthal, A Riordan, J Potter, JL Shupack, MA Gordon. Primary cutaneous infection by *Aspergillus ustus* in a 62-year-old liver transplant recipient. J Amer Acad Derm 31:344–347, 1994.
104. S Bretagne, A Marmorat-Khuong, M Kuentz, J-P Latgé, E Bart-Delabesse, C Cordonnier. Serum *Aspergillus* galactomannan antigen testing by sandwich ELISA: Practical use in neutropenic patients. J Infec 35:7–15, 1997.
105. PC Iwen, ME Rupp, MR Bishop, MG Rinaldi, DA Sutton, S Tarantolo, SH Heinrichs. Disseminated aspergillosis caused by *Aspergillus ustus* in a patient following allogeneic peripheral stem cell transplantation. J Clin Microbiol 36:3711–3717, 1998.
106. RM Ricci, JC Evans, JJ Meffert, L Kaufman, LC Sadkowski. Primary cutaneous *Aspergillus ustus* infection: Second reported case. J Amer Acad Derm 3:797–798, 1998.
107. PE Verweij, MF van den Burgh, PM Rath, BE de Pauw, A Voss, JF Meis. Invasive aspergillosis caused by *Aspergillus ustus*: A case report and review. J Clin Microbiol 37:1606–1609, 1999.
108. K Sandner, C Schönborn. Schimmelpilzinfektion der Haut bei ausgedehnter Verbrennung. Dtsch Gesundheitswesen 28:125–128, 1973.
109. J Carrizosa, ME Levison, T Lawrence, D Kaye. Cure of *Aspergillus ustus* endocarditis on a prosthetic valve. Arch Int Med 133:486–490, 1974.
110. AK Gupta, EA Cooper, P McDonald, RC Summerbell. Inoculum counting (Walshe/English criteria) in the clinical diagnosis of onychomycosis caused by nondermatophytic filamentous fungi. J Clin Microbiol 39:2115–2121, 2000.
111. RA Samson, ES Hoekstra, JC Frisvad, O Filtenborg. Introduction to Food-Borne Fungi. 5th ed. Baarn, Netherlands: Centraalbureau voor Schimmelcultures, 1996.
112. SS Jang, TE Dorr, EL Biberstein, A Wong. *Aspergillus deflectus* infection in four dogs. J Med Vet Mycol 24:95–104, 1986.
113. JS Kahler, MW Leach, S Jang, A Wong. Disseminated aspergillosis attributable to *Aspergillus deflectus* in a springer spaniel. J Amer Vet Med Assoc 197:871–874, 1990.
114. WJ Barson, FB Ruymann. Palmar aspergillosis in immunocompromised children. Pediat Infec Dis 5:264–268, 1986.
115. GA Roselle, IM Baird. *Aspergillus flavipes* group osteomyelitis. Arch Intern Med 139:590–592, 1979.
116. S Yokoyama, H Taniguchi, Y Kondo, K Matsumoto, A Okada. A case of bronchopulmonary aspergillosis recurred in a residual tuberculosis cavity. Kekkaku 64: 579–584, 1989.
117. R Staib, M Seibold, G Grosse. *Aspergillus* findings in AIDS patients suffering from cryptococcosis. Mycoses 32:516–523, 1989.
118. JB Wiley, EG Simmons. New species and new genus of Plectomycetes with *Aspergillus* states. Mycologia 65:934–938, 1973.

119. JH Seabury, M Samuels. The pathogenic spectrum of aspergillosis. Amer J Clin Path 40:21–33, 1963.

120. JB Wiley, DI Fennell. Ascocarps of *Aspergillus stromatoides*, *A. niveus* and *A. flavipes*. Mycologia 65:752–760, 1973.

121. R Morquer, L Enjalbert. Étude morphologique et physiologique d'un *Aspergillus* nouvellement isolé au cours d'une affection pulmonaire de l'homme. C R Acad Sci 244:1405–1408, 1957.

122. MA Klich, JI Pitt. A laboratory guide to common *Aspergillus* species and their teleomorphs. North Ryde, NSW, Australia: Commonwealth Scientific and Industrial Research Organization, 1988.

123. DM Tritz, GL Woods. Fatal disseminated infection with *Aspergillus terreus* in immunocompromised hosts. Clin Infec Dis 16:118–122, 1993.

124. RC Summerbell. Taxonomy and ecology of *Aspergillus* species associated with colonizing infections of the respiratory tract. In: VP Kurup, AJ Apter, Allergic Bronchopulmonary Aspergillosis. Immunol Allerg Clin N Amer 18(3):549–573, 1998.

125. S Tiwari, SM Singh, S Jain. Chronic bilateral suppurative otitis media caused by *Aspergillus terreus*. Mycoses 38:297–300, 1995.

126. P Onsberg, D Stahl, NK Veien. Onychomycosis caused by *Aspergillus terreus*. Sabouraudia 16:39–46, 1978.

127. PC Iwen, ME Rupp, AN Langnas, EC Reed, SH Hinrichs. Invasive pulmonary aspergillosis due to *Aspergillus terreus*: 12-year experience and review of the literature. Clin Infec Dis 26:1092–1097, 1998.

128. SL Tracy, MR McGinnis, JE Peacock Jr, MS Cohen, DH Walker. Disseminated infection by *Aspergillus terreus*. Amer J Clin Path 80:728–733, 1983.

129. MJ Kabay, WF Robinson, CRR Huxtable, R MacAleer. The pathology of disseminated *Aspergillus terreus* infection in the dog. Vet Path 22:540–547, 1985.

130. MJ Day, WJ Penhale, CE Eger, SE Shaw, MJ Kabay, WF Robinson, CRR Huxtable, JN Mills, RS Wyburn. Disseminated aspergillosis in dogs. Aust Vet J 16:55–59, 1986.

131. ZU Khan, M Kortom, R Marouf, R Chandy, MG Rinaldi, DA Sutton. Bilateral pulmonary aspergilloma caused by an atypical isolate of *Aspergillus terreus*. J Clin Microbiol 38:2010–2014, 2000.

132. G Linares, PA McGarry, RD Baker. Solid solitary aspergillotic granuloma of the brain: Report of a case due to *Aspergillus candidus* and review of the literature. Neurology 21:177–184, 1971.

133. F Avanzini, A Bigoni, G Nicoletti. Su un raro caso di aspergilloma isolato del seno sfenoidale. Isolated sphenoid sinus aspergillosis. Acta Otorhinol Ital 11:483–489, 1991.

134. N Falser. Pilzbefall des Ohres—Harmloser Saprophyt oder pathognomonischer Risikofaktor? Laryng Rhinol Otol 62:140–146, 1983.

135. P Fragner, V Kubíčková. Onychomykóza vyvolaná *Aspergillus candidus*. Cesk Dermatol 49:322–324, 1974.

136. L Zaror, MI Moreno. Onicomicosis por *Aspergillus candidus* Link. Rev Arg Micologia 3:13–15, 1980.

137. BM Cornere, M Eastman. Onychomycosis due to *Aspergillus candidus*: case report. NZ Med J 82:13–15, 1975.

138. K Iwasaki, T Tategami, Y Sakamoto, T Yasutake, S Otsubo. An operated case report of pulmonary aspergillosis by saprophytic infection of *Aspergillus candidus* in congenital bronchial cyst of right lower right lower lobe. Kyobu Geka (Jpn J Thor Surg) 44:429–432, 1991.

139. L Rahbaek, JC Frisvad, C Christophersen. An amendment of *Aspergillus* section *Candidi* based on chemotaxonomical evidence. Phytochemistry 53:581–586, 2000.

140. S-I Udagawa, S Uchiyama, S Kamiya. *Petromyces muricatus*, a new species with an *Aspergillus* anamorph. Mycotaxon 52:207–214, 1994.

141. JC Frisvad, RA Samson. *Neoptromyces* gen. nov. and an overview of teleomorphs of *Aspergillus* subgenus *Circumdati*. Stud Mycol 45:201–207, 2000.

142. HS Novey, ID Wells. Allergic bronchopulmonary aspergillosis caused by *Aspergillus ochraceus*. Amer J Clin Path 70:840–843, 1978.

142a. L Bellucci. Di alcuni importanti reperti micotici in oto-rino-laringologia. (Acremoniosi tonsillare—ozena nasale—otite media da *Sterigmatocystis ochracea*). Atti Reale Accad Fisiocritici Siena 27:17–43, 1925.

143. G Williams, F Billson, R Husain, SA Howlader, N Islam, K McClellan. Microbiological diagnosis of suppurative keratitis in Bangladesh. Brit J Ophth 71:315–321, 1987.

144. M Wierzbicka, B Podsiadło, M Janczarski. Inwazyjna aspergiloza płuc wywołana przez *Aspergillus ochraceus*. Pneumonol Alergol Polska 65:254–260, 1997.

145. CM Muñoz, M González, E Alvarez. Placentitis micótica y aborto por *Aspergillus ochraceus* en una vaca. Rev Salud Anim 11:190–195, 1989.

146. A Bassiouny, A Maher, TJ Bucci, MK Moawad, DS Hendawy. Non-invasive antromycosis (diagnosis and treatment). J Laryn Otol 96:215–228, 1982.

147. M Feuilhade de Chauvin, C de Bièvre. Onyxis et perionyxis à *Aspergillus sclerotiorum*. Bull Soc Fr Mycol Méd 14:77–79, 1985.

148. H Koenig, C de Bièvre, J Waller, C Conraux. *Aspergillus alliaceus*, agent d'otorrhée chronique. Bull Soc Fr Mycol Méd 14:84–87, 1984.

149. DW Malloch, RF Cain. The Trichocomataceae: Ascomycetes with *Aspergillus*, *Paecilomyces* and *Penicillium* imperfect states. Can J Bot 50:2613–2628, 1972.

150. R Patterson. Allergic bronchopulmonary aspergillosis: A historical perspective. Immunol Allerg Clin N Amer 18:471–478, 1998.

151. R Shapira, N Paster, O Eyal, M Menasherov, A Mett, R Salomon. Detection of aflatoxigenic molds in grains by PCR. Appl Environ Microbiol 62:3270–3273, 1996.

152. GS Sandhu, BC Kline, L Stockman, GD Roberts. Molecular probes for diagnosis of fungal infections. J Clin Microbiol 33:2913–2919, 1995.

153. CM Tang, DW Holden, A Aufavre-Brown, J Cohen. The detection of *Aspergillus* spp. by the polymerase chain reaction and its evaluation in bronchoalveolar lavage fluid. Amer Rev Resp Dis 148:1313–1317, 1993.

154. TJ Walsh, A Francesconi, M Kasai, SJ Chanock. PCR and single-strand conformational polymorphism for recognition of medically important opportunistic fungi. J Clin Microbiol 33:3216–3220, 1995.

155. DM Geiser, JI Pitt, JW Taylor. Cryptic speciation and recombination in the afla-toxin-producing fungus *Aspergillus flavus*. Proc Natl Acad Sci USA 95:388–393, 1998.

156. K Akiyama, H Takizawa, M Suzuki, S Miyachi, M Ichinohe, Y Yanagihara. Aller-gic bronchopulmonary aspergillosis due to *Aspergillus oryzae*. Chest 91:285–286, 1987.

157. O Kobayashi, M Narita, H Kawamoto, R Asano, J Toyama, A Nagai, S Kioi, M Arakawa. A case of allergic broncho-pulmonary aspergillosis due to *Aspergillus oryzae* and its subtype. Nihon Kyobu Shikkan Gakkai Zasshi (Jpn J Thor Dis) 22: 925–931, 1984.

158. MA Gordon, RS Holzman, H Senter, EW Lapa, MJ Kupersmith. *Aspergillus oryzae* meningitis. JAMA 235:2122–2123, 1976.

159. RW Byard, RA Bonin, AU Haq. Invasion of paranasal sinuses by *Aspergillus ory-zae*. Mycopathologia 96:41–34, 1986.

160. RG Washburn, DW Kennedy, MG Begley, DK Henderson, JE Bennett. Chronic fungal sinusitis in apparently normal hosts. Medicine 67:231–247, 1988.

161. R Degos, G Ségretain, G Badillet, A Maisler. Aspergillose de la paupière. Bull Soc Fr Derm Syph 77:732–734, 1970.

162. M Bayó, M Agut, M Àngels Calvo. Otitis externas infecciosas: Etiología en el área de Terrassa, métodos de cultivo y consideraciones sobre la otomicosis. Micro-biología 10:279–284, 1994.

163. TN Sharma, PR Gupta, AK Mehrota, SD Purohit, HN Mangal. Aspergilloma with ABPA due to *Aspergillus niger*. J Assoc Physicians India 33:748, 1985.

164. MB Kurrein, GH Green, SL Rowles. Localized deposition of calcium oxalate around a pulmonary *Aspergillus niger* fungus ball. Amer J Clin Path 64:556–563, 1975.

165. F Staib, J Steffen, D Krumhaar, G Kapetanakis, C Minck, G Grosse. Lokalisierte Aspergillose und Oxalose der Lunge durch *Aspergillus niger*. Dtsch Med Wschr 104:1176–1179, 1979.

166. MM Landry, CW Parkins. Calcium oxalate crystal deposition in necrotizing oto-mycosis caused by *Aspergillus niger*. Modern Path 6:493–496, 1993.

167. KM Cahill, AM El Mofty, TP Kawaguchi. Primary cutaneous aspergillosis. Arch Derm 96:545–547, 1967.

168. H Paldrok. Report on a case of subcutaneous dissemination of *Aspergillus niger*, type *awamori*. Acta Dermato-Venereol 45:275–282, 1965.

169. AK Gupta, N Konnikov, P MacDonald, P Rich, NW Rodger, MW Edmonds, R McManus, RC Summerbell. Prevalence and epidemiology of toenail onycho-mycosis in diabetic subjects: A multicentre study. Brit J Derm 139:665–671, 1998.

170. A Tosti, BM Piraccini. Proximal subungual onychomycosis due to *Aspergillus ni-ger*: Report of two cases. Brit J Derm 139:152–169, 1998.

171. M Birch, MJ Anderson, DW Denning. Molecular typing of *Aspergillus* species. J Hosp Infec 30 (suppl):339–351, 1995.

172. B Williams, B Popoola, SK Ogundana. A possible new pathogenic *Aspergillus* isolation and general mycological properties of the fungus. Afr J Med Medical Sci 13:111–115, 1984.

173. AA Padhye, JH Godfrey, FW Chandler, SW Peterson. Osteomyelitis caused by *Neosartorya pseudofischeri*. J Clin Microbiol 32:2832–2836, 1994.
174. RA Samson, PV Nielsen, JC Frisvad. The genus *Neosartorya*: Differentiation by scanning electron microscopy and mycotoxin profiles. In: RA Samson, JI Pitt, eds. Modern Concepts in *Penicillium* and *Aspergillus* Classification. New York: Plenum, 1990, pp. 455–467.
175. J Varga, Z Vida, B Tóth, F Debets, Y Horie. Phylogenetic analysis of newly described *Neosartorya* species. Antonie van Leeuwenhoek 77:235–239, 2000.
176. S Lonial, L Williams, G Carrum, M Ostrowski, P McCarthy Jr. *Neosartorya fischeri*: An invasive fungal pathogen in an allogeneic bone marrow transplant patient. Bone Marr Transpl 19:753–755, 1997.
177. RC Summerbell, L de Repentigny, C Chartrand, G St-Germain. Graft-related endocarditis caused by *Neosartorya fischeri* var. *spinosa*. J Clin Microbio 30:1580–1582, 1992.
178. CS Chim, PL Ho, KY Yuen. Simultaneous *Aspergillus fischeri* and *Herpes simplex* pneumonia in a patient with multiple myeloma. Scand J Infec Dis 30:190–191, 1998.
179. JAH van Burik, R Colven, DH Sprach. Itraconazole therapy for primary cutaneous aspergillosis in patients with AIDS. Clin Infec Dis 27:643–644, 1998.
180. PT Redig. Avian aspergillosis. In: ME Fowler, ed. Zoo and Wild Animals Medicine. Philadelphia: Saunders, 1993, pp. 178–181.
181. AS Sekhon, PG Standard, L Kaufman, AK Garg, P Cifuentes. Grouping of *Aspergillus* species with exoantigens. Diag Immunol 4:112–116, 1986.
182. K Makimura, SY Muruyama, H Yamaguchi. Specific detection of *Aspergillus* and *Penicillium* species from respiratory specimens by polymerase chain reaction (PCR). Jpn J Med Sci Bio 47:141–156, 1994.
183. WJ Melchers, PE Verweij, P van den Hurk, A van Belkum, BE De Pauw, JA Hoogkamp-Korstanje, JF Meis. General primer-mediated PCR for detection of *Aspergillus* species. J Clin Microbio 32:1710–1717, 1994.
184. R Kappe, CN Okeke, C Fauser, M Maiwald, HG Sonntag. Molecular probes in the detection of pathogenic fungi in the presence of human tissue. J Med Microbio 47:811–820, 1998.
185. Y Yamakami, A Hashimoto, I Tokimatsu, M Nasu. PCR detection of DNA specific for *Aspergillus* species in serum of patients with invasive aspergillosis. J Clin Microbio 34:2464–2468, 1996.
186. EE Jaeger, NM Carroll, S Choudhury, AA Dunlop, HM Towler, MM Matheson, P Adamson, N Okhravi, S Lightman. Rapid detection and identification of *Candida*, *Aspergillus* and *Fusarium* species in ocular samples using nested PCR. J Clin Microbio 38:2902–2908, 2000.
187. KA Haynes, TJ Westerneng, JW Fell, W Moens. Rapid detection and identification of pathogenic fungi by polymerase chain reaction amplification of large subunit ribosomal RNA. J Med Vet Mycol 33:319–325, 1995.
188. T Henry, PC Iwen, SH Hinrichs. Identification of *Aspergillus* species using internal transcribed spacer regions 1 and 2. J Clin Microbio 38:1510–1515, 2000.
189. K Ohgoe, S Miyanishi, M Aihara, S Matsuo. Significance of *Aspergillus fumigatus*

rDNA detected by polymerase chain reaction in diagnosis of pulmonary aspergillosis. Kansenshogaku Zasshi 71:507–512, 1997.

190.   C Spreadbury, D Holden, A Aufavre-Brown, B Bainbridge, J Cohen. Detection of
       Aspergillus fumigatus by polymerase chain reaction. J Clin Microbio 31:615–621,
       1993.

191.   Z Erjavec, M Brinker, HZ Apperloo-Renkema, JP Arends, HG de Vries-Hospers,
       MH Ruiters. Applicability of random primer R143 for determination of *Aspergillus
       fumigatus* DNA. J Med Vet Mycol 35:399–403, 1997.

192.   ME Brandt, AA Padhye, LW Mayer, BP Holloway. Utility of random amplified
       polymorphic DNA PCR and TaqMan automated detection in molecular identifica-
       tion of *Aspergillus fumigatus*. J Clin Microbio 36:2057–2062, 1998.

193.   S Bretagne, JM Costa, A Mormorat-Khuong, F Poron, C Cordonnier, M Vidaud,
       J Fleury-Feith. Detection of *Aspergillus* species DNA in bronchoalveolar lavage
       samples by competitive PCR. J Clin Microbio 33:1164–1168, 1995.

194.   ME Kambouris, U Reichard, NJ Legakis, A Velegraki. Sequences from the asper-
       gillopepsin PEP gene of *Aspergillus fumigatus*: Evidence on their use in selective
       PCR identification of *Aspergillus* species in infected clinical samples. FEMS Im-
       munol Med Microbiol 25:255–264, 1999.

195.   LV Reddy, A Kumar, VP Kurup. Specific amplification of *Aspergillus fumigatus*
       DNA by polymerase chain reaction. Molec Cell Probes 7:121–126, 1993.

196.   HA Fletcher, RC Barton, PE Verweij, EG Evans. Detection of *Aspergillus fumiga-
       tus* PCR products by a microtitre plate based DNA hybridisation assay. J Clin Path
       51:617–620, 1998.

197.   J Löffler, N Henke, H Hebart, D Schmidt, L Hagmeyer, U Schumacher, H Einsele.
       Quantification of fungal DNA by using fluorescence resonance energy transfer and
       the light cycler system. J Clin Microbiol 38:586–590, 2000.

198.   E Rodriguez, T De Meeus, M Mallie, F Renaud, F Symoens, P Mondon, MA Piens,
       B Lebeau, MA Viviani, R Grillot, N Nolard, F Chapuis, AM Tortorano, J-M
       Bastide. Multicentric epidemiological study of *Aspergillus fumigatus* isolates
       by multilocus enzyme electrophoresis. J Clin Microbiol 34:2559–2568, 1996.

199.   SE Gochenaur. Fungi of a Long Island oak-birch forest. II. Population dynamics
       and hydrolase patterns for the soil Penicillia. Mycologia 76:218–231, 1984.

200.   SM Opal, LB Reller, G Harrington, P Cannady Jr. *Aspergillus clavatus* endocarditis
       involving a normal aortic valve following coronary artery surgery. Rev Infec Dis
       8:781–785, 1986.

201.   A Batista, H da Silva Maia, IC Alecrim. Onicomicose produzida por *Aspergillus
       clavatonanica n. sp.* An Fac Med Univ Recife 15:197–203, 1955.

202.   JI Pitt. The genus *Penicillium* and its teleomorphic states *Eupenicillium* and *Talaro-
       myces*. London: Academic, 1979.

203.   H Veléz, F Díaz. Onychomycosis due to saprophytic fungi. Mycopathologia 91:
       87–92, 1985.

204.   R Ramani, A Ramani, PG Shivananda. *Penicillium* species causing onycho-
       mycosis. J Postgrad Med 40:87–88, 1994.

205.   GA Liebler, GJ Magovern, P Sadighi, SB Park, WJ Cushing. *Penicillium* granu-
       loma of the lung presenting as a solitary pulmonary nodule. JAMA 237:671,
       1977.

206. MS Gelfland, FH Cole Jr, RC Baskin. Invasive pulmonary penicilliosis: Successful therapy with amphotericin B. South Med J 83:701–704, 1990.

207. L Lombardo, A Pera, L Genovesio, G Verme. Duodenal mycosis during carbenoxolone and cimetidine treatment. Lancet i:607–608, 1979.

208. J Fahhoum, MS Gelfland. Peritonitis due to *Penicillium* sp. in a patient receiving continuous ambulatory peritoneal dialysis. South Med J 89:87–88, 1996.

209. O Equils, JG Deville, A Shapiro, CP Sanchez. *Penicillium* peritonitis in an adolescent receiving chronic peritoneal dialysis. Pediat Neph 13:771–772, 1999.

210. WJ Hall III. *Penicillium* endocarditis following open heart surgery and prosthetic valve insertion. Amer Heart J 87:501–506, 1974.

211. AJ DelRossi, D Morse, PM Spagna, GM Lemold. Successful management of *Penicillium* endocarditis. J Thor Cardiovasc Surg 80:945–947, 1980.

212. CE Harvey, JA O'Brien, PJ Felsburg, HL Izenberg, MH Goldschmidt. Nasal penicilliosis in six dogs. J Amer Vet Med Assoc 178:1084–1087, 1981.

213. RL Peiffer Jr, PV Belkin, BH Janke. Orbital cellulitis, sinusitis, and pneumonitis caused by *Penicillium* species in a cat. J Amer Vet Med Assoc 176:449–451, 1980.

214. AL Chute. An infection of the bladder with *Penicillium glaucum*. Boston Med Surg J 164:420–422, 1911.

215. SK Swan, RA Wagner, JP Myers, AB Cinelli. Mycotic endophthalmitis caused by *Penicillium* sp. after parenteral drug abuse. Amer J Ophth 100:408–410, 1985.

216. RG Berger. Chronic olecranon bursitis caused by *Penicillium*. Arth Rheum 32: 239–240, 1989.

217. G Leigheb, A Mossini, P Boggio, M Gattoni, G Bornacina, P Griffanti. Sporotrichosis-like lesions caused by a *Paecilomyces* genus fungus. Internat J Derm 33: 275–276, 1994.

218. JI Pitt. The current role of *Aspergillus* and *Penicillium* in human and animal health. J Med Vet Mycol 32, suppl 1:17–32, 1994.

219. AC Stolk, RA Samson. The ascomycete genus *Eupenicillium* and related *Penicillium* anamorphs. Stud Mycol 23:1–149, 1983.

220. RV Talice, JE Mackinnon. *Penicillium bertai n. sp.* agent d'une mycose bronchopulmonaire de l'homme. Ann Parasitol 7:97–106, 1929.

221. K Yoshida, T Hiraoka, M Ando, U Katsuhisa, V Mohsenin. *Penicillium decumbens*: A new cause of fungus ball. Chest 101:1152–1153, 1992.

222. S Alvarez. Systemic infection caused by *Penicillium decumbens* in a patient with acquired immunodeficiency syndrome. J Infec Dis 162:283, 1990.

223. BH Senturia, FT Wolf. Treatment of external otitis. II. Action of sulfonamide compounds on fungi isolated from cases of otomycosis. Arch Otolaryngol 41:56–63, 1945.

224. P Delore, J Coudert, R Lambert, J Fayolle. Un cas de mycose bronchique avec localisations musculaires septicémiques. Presse Méd 63:1580–1582, 1955.

225. T Mori, M Matsumura, T Kohara, Y Watanabe, T Ishiyama, Y Wakabayashi, H Ikemoto, A Watanabe, M Tanno, T Shirai, M Ichinoe. A fatal case of pulmonary penicilliosis. Jpn J Med Mycol 28:341–348, 1987.

226. T Mori, T Ebe, M Takahashi, T Kohara, H Isonuma, M Matsumura. Clinical aspects of penicilliosis, a rare infection. Jpn J Med Mycol 34:145–153, 1993 (in Japanese).

227. T Mok, AP Koehler, MY Yu, DH Ellis, PJ Johnson, NWR Wickham. Fatal *Penicillium citrinum* pneumonia with pericarditis in a patient with acute leukemia. J Clin Microbio 2654–2656, 1997.
228. JS Gilliam, SA Vest. *Penicillium* infection of the urinary tract. J Urol 65:484–489, 1951.
229. DB Jones, R Sexton, G Rebell. Mycotic keratitis in south Florida: A review of thirty-nine cases. Trans Ophth Soc 89:781–797, 1969.
230. HC Gugnani, RS Talwar, ANU Njuko-obi, HC Kodilinye. Mycotic keratitis in Nigeria. Brit J Ophth 60:607–613, 1976.
231. HC Gugnani, S Gupta, RS Talwar. Role of opportunistic fungi in ocular infections in Nigeria. Mycopathologia 65:155–166, 1978.
232. MGM Hove, J Badalamenti, GL Woods. *Penicillium* peritonitis in a patient receiving continuous ambulatory peritoneal dialysis. Diag Microbio Infec Dis 25:97–99, 1996.
233. C Bates, RC Read, AH Morice. A malicious mould. Lancet 349:1598, 1997.
234. N Rodríquez de Kopp, G Vidal. Micosis ocular postraumática por *Penicillium oxalicum*. Rev Iberoam Micol 15:103–106, 1998.
235. R de la Cámara, I Pinilla, E Muñoz, B Buendía, JL Steegman, JM Fernández-Rañada. *Penicillium brevicompactum* as the cause of a necrotic lung ball in an allogeneic bone marrow transplant recipient. Bone Marr Transpl 18:1189–1193, 1996.
236. S-N Huang, LS Harris. Acute disseminated penicilliosis: Report of a case and review of pertinent literature. Amer J Clin Path 39:167–174, 1963.
237. TD Bui, D Dabdub, SC George. Modeling bronchial circulation with application to soluble gas exchange: Description and sensitivity analysis. J Appl Physiol 84:2070–2088, 1998.
238. J Solway, AR Leff, I Dreshaj, NM Munoz, EP Ingenito, D Michaels, RH Ingram Jr, JM Drazen. Circulatory heat sources for canine respiratory heat exchange. J Clin Invest 78:1015–1019, 1986.
239. M Hoffman, E Bash, SA Berger, M Burke, I Yust. Fatal necrotizing esophagitis due to *Penicillium chrysogenum* in a patient with acquired immunodeficiency syndrome. Eur J Clin Microbio Infec Dis 11:1158–1160, 1992.
240. D D'Antonio, B Violante, C Farina, R Sacco, D Angelucci, M Masciulli, A Iacone, F Romano. Necrotizing pneumonia caused by *Penicillium chrysogenum*. J Clin Microbio 35:3335–3337, 1997.
241. CB Upshaw. *Penicillium* endocarditis of aortic valve prosthesis. J Thor Cardiovasc Surg 68:428–431, 1974.
242. Y-K Keung, R Kimbrough III, K-Y Yuen, W-C Wong, E Cobos. *Penicillium chrysogenum* infection in a cotton farmer with acute myeloid leukemia. Infec Dis Clin Prac 6:482–483, 1997.
243. K-Y Yuen, SS Wong, DN Tsang, P-Y Chau. Serodiagnosis of *Penicillium marneffei* infection. Lancet 344:444–445, 1994.
244. ML Eschete, JW King, BC West, A Oberle. *Penicillium chrysogenum* endophthalmitis: First reported case. Mycopathologia 74:125–127, 1981.
245. E Nouri. Penicillinose der Nasennebenhöhlen. Laryng Rhinol Otol 65:420–422, 1986.
246. A Bengoa, V Briones, MB López, MJ Payá. Beak infection by *Penicillium cyclopium* in a macaw (*Ara ararauna*). Avian Dis 38:922–927, 1994.

247. KB Raper, C Thom. A manual of the Penicillia. Baltimore: Williams and Wilkins, 1949.

248. JC Frisvad, O Filtenborg, F Lund, RA Samson. The homogeneous species and series in subgenus *Penicillium* are related to mammal nutrition and excretion. In: RA Samson, JI Pitt, eds. Integration of Modern Taxonomic Methods for *Penicillium* and *Aspergillus*. Amsterdam: Harwood Academic, 2000, pp. 265–283.

249. KA Seifert, G Louis-Seize. Phylogeny and species concepts in the *Penicillium aurantiogriseum* complex as inferred from partial β-tubulin gene DNA sequences. In: RA Samson, JI Pitt, eds. Integration of Modern Taxonomic Methods for *Penicillium* and *Aspergillus*. Amsterdam: Harwood Academic, 2000, pp. 189–198.

250. R Aho, B Westerling, L Ajello, AA Padhye, RA Samson. Avian penicilliosis caused by *Penicillium griseofulvum* in a captive toucanet. J Med Vet Mycol 28: 349–354, 1990.

251. J Orós, AS Ramírez, JB Poveda, JL Rodríguez, A Fernández. Systemic mycosis caused by *Penicillium griseofulvum* in a Seychelles giant tortoise (*Megalochelys gigantea*). Vet Rec 139:295–296, 1996.

252. B Cimon, J Carrere, JP Chazalette, JF Vinatier, D Chabasse, JP Bouchara. Chronic airway colonization by *Penicillium emersonii* in a patient with cystic fibrosis. Med Mycol 37:291–293, 1999.

253. JI Pitt. *Geosmithia* gen. nov. for *Penicillium lavendulum* and related species. Can J Bot 57:2021–2030, 1979.

254. AC Stolk, RA Samson. The genus *Talaromyces*: Studies in *Talaromyces* and related genera II. Stud Mycol 2:1–65, 1972.

255. L Kaufman. Penicilliosis marneffei and pythiosis: Emerging tropical diseases. Mycopathologia 143:3–7, 1998.

256. TA Duong. Infection due to *Penicillium marneffei*, an emerging pathogen: Review of 155 reported cases. Clin Infec Dis 23:125–130, 1996.

257. S Chariyalertsak, T Sirisanthana, K Supparatpinyo, KE Nelson. Seasonal variation of disseminated *Penicillium marneffei* infections in northern Thailand: A clue to the reservoir? J Infec Dis 6:1490–1493, 1996.

258. W Nittayananta. Penicilliosis marneffei: Another AIDS-defining illness in Southeast Asia. Oral Dis 5:286–293, 1999.

259. N Vanittanakom, CR Cooper Jr, S Chariyalertsak, S Youngchim, KE Nelson, T Sirisanthana. Restriction endonuclease analysis of *Penicillium marneffei*. J Clin Microbiol 34:1834–1836, 1996.

260. PKC Austwick, JAJ Venn. Mycotic abortion in England and Wales 1954–1960. Proceedings IVth International Congress on Animal Reproduction, the Hague, 1961, pp. 562–568.

261. R Horré, S Gilges, P Breig, K Kupfer, GS de Hoog, E Hoekstra, N Poonwan, KP Schaal. Case report. Fungemia due to *Penicillium piceum*, a member of the *Penicillium marneffei* complex. Mycoses 44:502–504, 2001.

262. O Morin, P Germaud, M Miegeville, N Milpied. Mycose pulmonaire opportuniste à *Penicillium purpurogenum*: À propos d'une observation chez un malade immunodeprimé. Bull Soc Mycol Méd 15:441–448, 1986.

263. P Breton, P Germaud, O Morin, AF Audouin, N Milpied, JL Harousseau. Mycoses pulmonaires rares chez le patient d'hématologie. Rev Pneumol Clin 54:253–257, 1998.

264. I Świetliczkowa, E Szusterowska-Martinowa, W Braciak. Ocena kliniczna 1% maści Clotrimazol w leczeniu grzybic rogówki. Klin Oczna 86:221–223, 1984.

265. DI Wigney, GS Allan, LE Hay, AD Hocking. Osteomyelitis associated with *Penicillium verruculosum* in a German Shepherd dog. J Small Animal Prac 31:449–452, 1990.

266. K Yamashita. Fungus problems in otolaryngology. J Otolaryngol Jpn 59:129–149, 1956 (in Japanese, English abs.).

267. K Hikita. Mycological, clinical and experimental studies of otomycosis. Practica Otolaryngol (Kyoto) 50:432–479, 1957 (in Japanese).

268. G Smith. *Polypaecilum* gen. nov. Trans Brit Mycol Soc 44:437–440, 1961.

269. K Yamashita, T Yamashita. *Polypaecilum insolitum* (=*Scopulariopsis divaricata*) isolated from cases of otomycosis. Sabouraudia 10:128–131, 1972.

270. E Piontelli, MA Toro, J Testar. Un raro caso de hialohifomicosis en uñas por *Polypaecilum insolitum* G. Smith. Bol Micol 4:155–159, 1989.

271. F Coutelen, J Biguet, G Cochet, S Mullet, M Doby-Dubois. Etude d'un champignon nouveau isolé d'une tumeur mycosique pulmonaire. Ann Parasitol Hum Comp 30: 395–419, 1955.

272. GY Cole, RA Samson. Pattern of Development in Conidial Fungi. London: Pittman, 1979.

273. J Sugiyama. Relatedness, phylogeny, and evolution of the fungi. Mycoscience 39: 487–511, 1998.

274. LGM Castro, A Salebian, MN Sotto. Hyalohyphomycosis by *Paecilomyces lilacinus* in a renal transplant patient and a review of human *Paecilomyces* species infections. J Med Vet Mycol 28:15–26, 1990.

275. C Aguilar, I Pujol, J Sala, J Guarro. Antifungal susceptibilities of *Paecilomyces* species. Antimicrob Agents Chemother 42:1601–1604, 1998.

276. K Liu, DN Howell, JR Perfect, WA Schell. Morphologic criteria for the preliminary identification of *Fusarium*, *Paecilomyces*, and *Acremonium* species by histopathology. Amer J Clin Pathol 109:45–54, 1998.

277. JH Sillevis Smitt, JHW Leusen, HG Stas, AH Teeuw, RS Weening. Chronic bullous disease of childhood and a paecilomyces [sic] lung infection in chronic granulomatous disease. Arch Dis Child 77:150–152, 1997.

278. PA March, K Knowles, CL Dillavou, R Jakowski, G Freden. Diagnosis, treatment and temporary remission of disseminated paecilomycosis in a vizsla. J Amer Animal Hosp Assoc 32:509–514, 1996.

279. CH Crompton, JW Balfe, RC Summerbell, MM Silver. Peritonitis with *Paecilomyces* complicating peritoneal dialysis. Pediat Infec Dis J 10:869–871, 1991.

280. TH Chan, A Koehler, PK Li. *Paecilomyces variotii* peritonitis in patients on continuous ambulatory peritoneal dialysis. Amer J Kidney Dis 27:138–142, 1996.

281. MMK Shing, M Ip, CK Li, KW Chik, PMP Yuen. *Paecilomyces variotii* fungemia in a bone marrow transplant patient. Bone Marr Transpl 17:281–283, 1996.

282. PR Williamson, KJ Kwon-Chung, JI Gallin. Successful treatment of *Paecilomyces variotii* infection in a patient with chronic granulomatous disease and a review of *Paecilomyces* species infections. Clin Infec Dis 14:1023–1026, 1992.

283. A Cohen-Abbo, KM Edwards. Multifocal osteomyelitis caused by *Paecilomyces variotii* in a patient with chronic granulomatous disease. Infection 23:55–57, 1995.

284. DS Lam, AP Koehler, DS Fan, W Cheuk, AT Leung, JS Ng. Endogenous fungal endophthalmitis caused by *Paecilomyces variotii*. Eye 37:57–60, 1999.

285. SB Kalish, R Goldschmidt, C Li, R Knop, FV Cook, G Wilner, TA Victor. Infective endocarditis caused by *Paecilomyces variotii*. Amer J Clin Path 78:249–252, 1982.

286. W Lawson, A Blitzer. Fungal infections of the nose and paranasal sinuses. Part II. Otolaryn Clin N Amer 26:1037–1068, 1993.

287. MP Littman, MH Goldschmidt. Systemic paecilomycosis in a dog. J Amer Vet Med Assoc 191:445–447, 1987.

288. JP Patterson, S Rosendal, J Humphrey, WG Teeter. A case of disseminated paecilomycosis in the dog. J Amer Animal Hosp Assoc 19:569–574, 1983.

289. AK Patnaik, S-K Liu, RJ Wilkins, GF Johnson, PE Seitz. Paecilomycosis in a dog. J Amer Vet Med Assoc 161:806–813, 1972.

290. SS Jang, EL Biberstein, DO Slauson, PF Suter. Paecilomycosis in a dog. J Amer Vet Med Assoc 159:1775–1779, 1971.

291. R Arenas, M Arce, H Muñoz, J Ruiz-Esmenjaud. Onychomycosis due to *Paecilomyces variotii*: Case report and review. J Mycol Méd 8:32–33, 1998.

292. RA Samson. *Paecilomyces* and some allied Hyphomycetes. Stud Mycol 6:1–119, 1974.

293. G Ségretain, J Fromentain, P Destombes, É-P Brygoo, A Dodin. *Paecilomyces viridis, n. sp.*, champignon dimorphique, agent d'une mycose géneralisée de *Chamaeleon unilateralis* Gray. C R Acad Sci 251:258–261, 1964.

294. M Rodrigues, D MacLeod. Exogenous fungal endophthalmitis caused by *Paecilomyces*. Amer J Ophth 79:687–690, 1975.

295. FW Chandler, W Kaplan, L Ajello. A Colour Atlas and Textbook of the Histopathology of Mycotic Diseases. London: Wolf Medical Publishing, 1980.

296. AE Glenn, CW Bacon, R Price, RT Hanlin. Molecular phylogeny of *Acremonium* and its taxonomic implications. Mycologia 88:369–383, 1996.

297. W Hennig. Phylogenetic Systematics. Urbana, IL: University of Illinois Press, 1979. (English transl. by D. D. Davis and R. Zangerl).

298. E Mayr. What is a species and what is not? Philos Sci 63:262–277, 1996.

299. KA Seifert, W Gams, PW Crous, GJ Samuels. Molecules, morphology and classification: Towards monophyletic genera in the *Ascomycetes*. Afterword. Stud Mycol 45:223–224, 2000.

300. M Del Poeta, WA Schell, JR Perfect. In vitro antifungal activity of pneumocandin L-743,872 against a variety of clinically important molds. Antimicrob Agents Chemo 41:1835–1836, 1997.

301. A Espinel-Ingroff. Comparison of in vitro activities of the new triazole SCH56592 and the echinocandins MK-0991 (L-743,872) and LY303366 against opportunistic filamentous and dimorphic fungi and yeasts. J Clin Microbiol 36:2950–2956, 1998.

302. SE Richardson, RM Bannatyne, RC Summerbell, J Milliken, R Gold, SS Weitzman. Disseminated fusarial infection in the immunocompromised host. Rev Infec Dis 10:1171–1181, 1988.

303. WA Schell, JR Perfect. Fatal, disseminated *Acremonium strictum* infection in a neutropenic host. J Clin Microbiol 34:1333–1336, 1996.

304. T Heinz, J Perfect, W Schell, E Ritter, G Ruff, D Serafin. Soft-tissue fungal infections: Surgical management of 12 immunocompromised patients. Plast Reconstr Surg 97:1391–1399, 1996.

305. AHS Brown, G Smith. The genus *Paecilomyces* Bainier and its perfect stage *Byssochlamys* Westling. Trans Brit Mycol Soc 40:17–89, 1957.

306. MS Whitney, WM Reed, JF Tuite. Antemortem diagnosis of paecilomycosis in a cat. J Amer Vet Med Assoc 184:93–94, 1984.

307. PA Allevato, JM Ohorodnik, E Mezger, JF Eisses. *Paecilomyces javanicus* endocarditis of native and prosthetic aortic valve. Amer J Clin Pathol 82:247–252, 1984.

308. K-L Ho, PA Allevato. Hirano body in an inflammatory cell of leptomeningeal vessel infected by the fungus *Paecilomyces*. Acta Neuropathol (Berl) 71:159–162, 1986.

309. K-L Ho, PA Allevato, P King, JL Chason. Cerebral *Paecilomyces javanicus* infection: An ultrastructural study. Acta Neuropathol (Berl) 72:134–141, 1986.

310. B Orth, R Frei, PH Itin, MG Rinaldi, B Speck, A Gratwohl, AF Widmer. Outbreak of invasive mycoses caused by *Paecilomyces lilacinus* from a contaminated skin lotion. Ann Intern Med 125:799–806, 1996.

311. F Westenfeld, WK Alston, WC Winn. Complicated soft tissue infection with prepatellar bursitis caused by *Paecilomyces lilacinus* in an immunocompetent host: Case report and review. J Clin Microbio 34:559–562, 1996.

312. KM Chan-Tack, CL Thio, NS Miller, CL Karp, C Ho, WG Merz. *Paecilomyces lilacinus* fungemia in an adult bone marrow transplant recipient. Med Mycol 37: 57–60, 1999.

313. TQ Tan, AK Ogden, J Tillman, GJ Demmler, MG Rinaldi. *Paecilomyces lilacinus* catheter-related fungemia in an immunocompromised pediatric patient. J Clin Microbiol 30:2479–2483, 1992.

314. R Gucalp, P Carlisle, P Gialanella, S Mitsudo, J McKitrick, J Dutcher. *Paecilomyces* sinusitis in an immunocompromised adult patient: Case report and review. Clin Infec Dis 23:391–393, 1996.

315. AM Kozarsky, D Stulting, GO Waring III, FM Cornell, LA Wilson, HD Cavanagh. Penetrating keratoplasty for exogenous *Paecilomyces* keratitis followed by postoperative endophthalmitis. Amer J Ophth 98:552–557, 1984.

316. N Ono, K Sato, H Yokomise, K Tamura. Lung abscess caused by *Paecilomyces lilacinus*. Respiration 66:85–87, 1999.

317. MB Starr. *Paecilomyces lilacinus* keratitis: Two case reports in extended wear contact lens wearers. CLAO J 13:95–101, 1987.

318. FH Theodore. Etiology and diagnosis of fungal postoperative endophthalmitis. Ophth 85:327–340, 1978.

319. MA Mosier, B Lusk, TH Pettit, DH Howard, J Rhodes. Fungal endophthalmitis following intraocular lens implantation. Amer J Ophth 83:1–8, 1977.

320. RC Rockhill, MD Klein. *Paecilomyces lilacinus* as the cause of chronic maxillary sinusitis. J Clin Microbiol 11:737–739, 1980.

321. CL Fletcher, RJ Hay, G Midgley, M Moore. Onychomycosis caused by infection with *Paecilomyces lilacinus*. Brit J Derm 139:1111–1137, 1998.

322. IF Keymer. Diseases of chelonians: (2) necropsy survey of terrapins and turtles. Vet Rec 103:577–582, 1978.

323. DJ Heard, GH Cantor, ER Jacobson, B Purich, L Ajello, AA Padhye. Hyalohyphomycosis caused by *Paecilomyces lilacinus* in an Aldabra tortoise. J Amer Vet Med Assoc 189:1143–1145, 1986.

324. M Maslen, J Whitehead, WM Forsyth, H McCracken, AD Hocking. Systemic mycotic disease of captive crocodile hatchling (*Crocodylus porosus*) caused by *Paecilomyces lilacinus*. J Med Vet Mycol 26:219–225, 1988.

325. MA Gordon. *Paecilomyces lilacinus* (Thom) Samson from systemic infection in an armadillo (*Dasypus novemcinctus*). Sabouraudia 22:109–116, 1984.

326. LF Harris, BM Dan, LB Lefkowitz Jr, RH Alfonso. *Paecilomyces* cellulitis in a renal transplant patient: Successful treatment with intravenous miconazole. South Med J 72:897–898, 1979.

327. J Guarro, W Gams, I Pujol, J Gené. *Acremonium* species: New emerging fungal opportunists—in vitro antifungal susceptibilities and review. Clin Infec Dis 25: 1222–1229, 1997.

328. R-ME Fincher, JF Fisher, RD Lovell, CL Newman, A Espinel-Ingroff, HJ Shadomy. Infection due to the fungus *Acremonium* (*Cephalosporium*). Medicine 70: 398–409, 1991.

329. SM Kennedy, GS Shankland, WR Lee. Keratitis due to the fungus *Acremonium* (*Cephalosporium*). Eye 8:692–716, 1994.

330. D Dunphy, D Andrews, C Seamone, M Ramsay. Fungal keratitis following laser excimer photorefractive keratectomy. Can J Ophth 34:286–289, 1999.

331. MR Mascarenhas, KL McGowan, E Ruchelli, B Athreya, SM Altschuler. *Acremonium* infection of the esophagus. J Pediat Gastroent Nutr 24:356–358, 1997.

332. MH Grunwald, M Cagnano, M Mosovich, S Halevy. Cutaneous infection due to *Acremonium*. J Eur Acad Derm Venereol 10:58–61, 1998.

333. KW Simpson, KNM Khan, M Podell, SE Johnson, DA Wilkie. Systemic mycosis caused by *Acremonium* sp. in a dog. J Amer Med Vet Assoc 203:1296–1299, 1993.

334. W Gams. *Cephalosporium*-artige Schimmelpilze. Stuttgart, Germany: Gustav Fischer, 1971.

335. W Gams. Connected and disconnected chains of phialoconidia and *Sagenomella* gen. nov. segregated from *Acremonium*. Persoonia 10:97–112, 1978.

336. CV Wetli, SD Weiss, TJ Cleary, E Gyori. Fungal cerebritis from intravenous drug use. J Foren Sci 29:260–268, 1984.

337. G Morgan-Jones. Notes on Hyphomycetes. V. A new thermophilic species of *Acremonium*. Can J Bot 52:429–431, 1974.

338. RW Read, RSH Chuck, NA Rao, RE Smith. Traumatic *Acremonium atrogriseum* keratitis following laser-assisted in situ keratomileusis. Arch Ophth 118:418–421, 2000.

339. A Chatterjee, K Mitra, GR Saha. Isolation of *Acremonium blochii* (Matr.) W. Gams from cutaneous lesions in Asian buffaloes. Ind J Animal Hlth 26:171–173, 1987.

340. B Bloch, A Vischer. Die Kladiose, eine durch einen bisher nicht bekannten Pilz (*Mastigocladium*) hervorgerufene Dermatomykose. Arch Derm Syphil 108:477–512, plates 9–12, 1911.

341. G Bolognesi, GA Chiurco. Cladiosi. In G Pollacci, ed. Micosi Chirurgiche, vol. 2. Trattato di Micopatologia Umana. Siena, Italy: Libreria Editrice Siena, 1927, pp. 688–692.

342. SC Pflugfelder, HW Flynn, TA Zwicksey, RK Forster, A Tsiligianni, WW Culbertson, S Mandelbaum. Exogenous fungal endophthalmitis. Ophthalmology 95:19–30, 1988.

343. JG McCormack, PB McIntyre, MH Tilse, DH Ellis. Mycetoma associated with *Acremonium falciforme* infection. Med J Aust 147:187–188, 1987.

344. C Halde, AA Padhye, LD Haley, MG Rinaldi, D Kay, R Leeper. *Acremonium falciforme* as a cause of mycetoma in California. Sabouraudia 14:319–326, 1976.
345. PV Venugopal, TV Venugopal. Pale grain eumycetomas in Madras. Australas J Derm 36:149–151, 1995.
346. MW Lee, JC Kim, JS Choi, KH Kim, DL Greer. Mycetoma caused by *Acremonium falciforme*: Successful treatment with itraconazole. J Amer Acad Derm 32:897–900, 1995.
347. LL Van Etta, LR Peterson, DN Gerding. *Acremonium falciforme* (*Cephalosporium falciforme*) mycetoma in a renal transplant patient. Arch Derm 119:707–708, 1983.
348. O Miró, J Ferrando, V Lecha, JM Campistol. Abscesos subcutáneos por *Acremonium falciforme* en un trasplantado renal. Med Clin (Barcelona) 102:316, 1994.
349. RC Noble, J Salgado, SW Newell, NL Goodman. Endophthalmitis and lumbar diskitis due to *Acremonium falciforme* in a splenectomized patient. Clin Infec Dis 24:277–278, 1997.
350. JA Cameron, EM Badawi, PA Hoffman, KF Tabbara. Chronic endophthalmitis caused by *Acremonium falciforme*. Can J Ophth 31:367–368, 1996.
351. YL Lau, KY Yuen, CW Lee, CF Chan. Invasive *Acremonium falciforme* infection in a patient with severe combined immunodeficiency. Clin Infec Dis 20:197–198, 1995.
352. CEM Hay, RK Loveday, BMT Spencer, de B Scott. Bilateral mycotic myositis, osteomyelitis and nephritis in a dog caused by a *Cephalosporium*-like hyphomycete. J S Afr Vet Assoc 49:359–361, 1978.
353. C Lacroix, JL Jacquemin, F Guilhot, MH Rabot, C Burucoa, C de Bièvre. Septicémie à *Acremonium kiliense* avec dissémination secondaire chez une patiente atteinte d'un myélome à forte masse tumorale. Bull Soc Fr Mycol Méd 17:93–98, 1988.
354. G Simon, G Rákóczy, J Galgóczy, T Verebély, J Bókay. *Acremonium kiliense* in oesophagus stenosis. Mycoses 34:257–260, 1991.
355. JO Lopes, SH Alves, AC Rosa, CB Silva, JC Sarturi, CAR Souza. *Acremonium kiliense* peritonitis complicating continuous ambulatory peritoneal dialysis: Report of two cases. Mycopathologia 131:83–85, 1995.
356. W Brabender, J Ketcherside, GR Hodges, S Rengachary, WG Barnes. *Acremonium kiliense* osteomyelitis of the calvarium. Neurosurgery 16:554–556, 1985.
357. C da S Lacaz, E Porto, JJ Carneiro, IO Pazianni, WP Pimenta. Endocardite em prótese de dura-mater provocada pelo *Acremonium kiliense*. Rev Inst Med Trop São Paulo 23:274–279, 1981.
358. SK Fridkin, FB Kremer, LA Bland, A Padhye, MM McNeil, WR Jarvis. *Acremonium kiliense* endophthalmitis that occurred after cataract extraction in an ambulatory surgical center and was traced to an environmental reservoir. Clin Infec Dis 22:222–227, 1996.
359. DJ Weissgold, AM Maguire, AJ Brucker. Management of post-operative *Acremonium* endophthalmitis. Ophthalmology 103:749–756, 1996.
360. DJ Weissgold, SE Orlin, ME Sulewski, WC Frayer, RC Eagle Jr. Delayed-onset fungal keratitis after endophthalmitis. Ophthalmology 105:258–262, 1998.
361. C Paiva, A Chaves Batista, A Gomes. Endoftalmite micótica pós-operatória por *Hyalopus bogolepofii*. Rev Bras Oftalmol 19:193–202, 1960.

362. MX Wang, DJ Shen, JC Liu, SC Pflugfelder, EC Alfonso, RK Forster. Recurrent fungal keratitis and endophthalmitis. Cornea 19:558–560, 2000.

363. L Mendoza, A Donato, AA Padhye. Canine mycotic keratoconjunctivitis caused by *Acremonium kiliense*. Sabouraudia 23:447–450, 1985.

364. O-E Lund, HM Miño de Kaspar, V Klauss. Strategie der Untersuchung und Therapie bei mykotischer Keratitis. Klin Monatsbl Augenheilkd 202:188–194, 1993.

365. JO Lopes, LC Killing, W Neumaier. Kerion-like lesion of the scalp due to *Acremonium kiliense* in a noncompromised boy. Rev Inst Med Trop São Paulo 37:365–368, 1995.

366. H Bargman, J Kane, M-L Baxter, RC Summerbell. *Trichophyton rubrum* tinea capitis in adult women. Mycoses 38:231–234, 1995.

367. W-Q Liao, Y-S Xue, P-M Chen, D-Q Xu, J-Z Zhang, Q-T Chen. *Cephalosporium acremonium*, a new strain of fungus causing white piedra. Chinese Med J 104:425–427, 1991.

368. M Lahourcade, L Texier. À propos d'un cas original de céphalosporiose cutanée superficielle provoquée par *Cephalosporium acremonium* Corda 1839. Bull Soc Fr Mycol Méd 5:127–131, 1976.

369. L Texier, R Lahourcade, R Despinis, Y Gauthier, O Gauthier, G Ducombs, J-M Tamisier, P Bioulac. Granulome fungique dû à un *Cephalosporium*. Bull Soc Fr Derm Syphil 79:504–507, 1972.

370. WM Dion, TW Dukes. Bovine mycotic abortion caused by *Acremonium kiliense* Grutz. Sabouraudia 17:355–361, 1979.

371. RK Forster, G Rebell, W Stiles. Recurrent keratitis due to *Acremonium potronii*. Amer J Ophth 79:126–128, 1975.

372. S van den Akker. Een schimmelinfectie (*Cephalosporium potronii*) in de mondholte van een kat. Tijdschr. Dierengeneesk. 77:514–516, 1952.

373. T Rodriguez-Ares, VDR Silva, MP Ferreiros, EP Becerra, CC Tome, M Sanchez-Salorio. *Acremonium* keratitis in a patient with herpetic neurotrophic corneal disease. Acta Ophthalmol Scand 78:107–109, 2000.

374. M Potron, G Noisette. Un cas de mycose. Rev Méd Est 43:132–139, 1911.

375. P Vuillemin. Les conidiosporées. Bull Séances Soc Sci Nancy (2 juin): 19, 1910.

376. G Pollacci, A Nannizzi. I miceti patogeni dell'uomo e degli animali, fasc. IV, no. 32. Siena, Italy: Libreria Editrice Senese, 1925.

376a. G Zanni. Micosi primitiva tonsillare e faringea da *Acremonium potronii*. Il Valsava 6:258–264, 1926.

377. J Montpellier, A Catanei. Résultats de l'étude d'un nouveau mycétome du pied observé à Alger. Bull Soc Path Exot 27:209–214, 1934.

378. JS Kinnas. Ophthalmic disease caused by a mycete of the giant cane. Brit J Ophth 49:327–329, 1965.

379. AE Arêa Leão, J Lobo. Mycétome du pied à *Cephalosporium recifei* n. sp. Mycétome à grains blancs. C R Soc Biol (Paris) 107:303–305, 1934.

380. G Koshi, AA Padhye, L Ajello, FW Chandler. *Acremonium recifei* as an agent of mycetoma in India. Amer J Trop Med Hyg 28:692–696, 1979.

381. C Zaitz, E Porto, EM Heins-Vaccari, A Sadahiro, LRB Ruiz, CS Lacaz. Subcutaneous hyalohyphomycosis caused by *Acremonium recifei*: case report. Rev Inst Med Trop São Paulo 37:267–270, 1995.

382. HJ Simonsz. Keratomycosis caused by *Acremonium recifei*, treated with keratoplasty, miconazole and ketoconazole. Doc Ophth 56:131–135, 1983.

383. S Moulias, E Hazouard, M Delain, A Barrabes, M Therizol-Ferly, A Legras. Fongémie prolongée à *Acremonium recifei* après autogreffe de moelle. J Mycol Méd 8:26–29, 1998.

384. HP Ward, WJ Martin, JC Ivins, LA Weed. *Cephalosporium* arthritis. Proc Staff Mtgs Mayo Clinic 36:337–343, 1961.

385. P Negroni. Onicomicosis por *Cephalosporium spinosus n. sp.* Negroni, 1933. Rev Soc Arg Bio 9:16–22, 1933.

386. P Negroni. Onycomicose par *Cephalosporium spinosus n. sp.* Negroni 1933. CR Séanc Soc Biol Buenos Aires 113:478–480, 1933.

387. O Morin, N Milpied, AF Audouin, M Maillot. Mycose opportuniste invasive à *Acremonium strictum* chez une malade atteint de myelome. Bull Soc Fr Mycol Méd 17:357–362, 1988.

388. J Trupl, M Májek, J Mardiak, Z Jesenská, V Krcméry Jr. *Acremonium* infection in two compromised patients. J Hosp Infec 25:299–301, 1993.

389. H Boltansky, KJ Kwon-Chung, AM Macher, JI Gallin. *Acremonium strictum*-related pulmonary infection in a patient with chronic granulomatous disease. J Infec Dis 149:653, 1984.

390. S Medek, A Nemes, A Khoor, A Széll, C Dobolyi, E Novák. Tartós steroidkezelés alatt kialakult *Acremonium strictum* okozta meningitis. Orv Hetil 128:2529–2532, 1987.

391. AN Koç, C Utaş, O Oymak, E Sehmen. Peritonitis due to *Acremonium strictum* in a patient on continuous ambulatory peritoneal dialysis. Nephron 79:357–358, 1998.

392. A Warris, F Wesenberg, P Gaustad, PE Verweij, TG Abrahamsen. *Acremonium strictum* fungemia in a pediatric patient with acute leukemia. Scand J Infec Dis 32:442–444, 2000.

393. TJ Liesegang, RK Forster. Spectrum of microbial keratitis in south Florida. Amer J Ophth 90:38–47, 1980.

394. DE Brooks, SE Andrew, CL Dillavou, G Ellis, PS Kubilis. Antimicrobial susceptibility patterns of fungi isolated from horses with ulcerative keratitis. Amer J Vet Res 59:138–142, 1998.

395. SE Andrew, DE Brooks, PJ Smith, KN Gelatt, NT Chmielewski, CJ Whittaker. Equine ulcerative keratomycosis: Visual outcome and ocular survival in 39 cases (1987–1996). Equine Vet J 30:109–116, 1998.

396. J Kane, RC Summerbell, L Sigler, S Krajden, G Land, eds. Laboratory Handbook of Dermatophytes. Belmont, CA: Star, 1997, pp. 213–259.

397. HP de Bruyn, JM Broekman, GA de Vries, AH Klokke, JM Greep. Een patiënt met eumycetoma in Nederland. Ned Tijdschr Geneeskd 129:1099–1101, 1985.

398. GA de Vries, GS de Hoog, HP de Bruyn. *Phialophora cyanescens* sp. nov. with *Phaeosclera*-like synanamorph, causing white-grain mycetoma in man. Antonie van Leeuwenhoek 50:149–153, 1984.

399. DE Zoutman, L Sigler. Mycetoma of the foot caused by *Cylindrocarpon destructans*. J Clin Microbiol 29:1855–1859, 1991.

400. DVH Hendrix, NT Chmielewski, PJ Smith, DE Brooks, KN Gelatt, C Whittaker.

Keratomycosis in four horses caused by *Cylindrocarpon destructans*. Vet Compar Ophth 6:252–257, 1996.

401. C Booth, YM Clayton, M Usherwood. *Cylindrocarpon* species associated with mycotic keratitis. Proc Indian Acad Sci (Plant Sci) 94:433–436, 1985.
402. S Laverde, LH Moncada, A Restrepo, CL Vera. Mycotic keratitis: 5 cases caused by unusual fungi. Sabouraudia 11:119–123, 1973.
403. PC Iwen, SR Tarantolo, DA Sutton, MG Rinaldi, SH Hinrichs. Cutaneous infection caused by *Cylindrocarpon lichenicola* in a patient with acute myelogenous leukemia. J Clin Microbio 38:3375–3378, 2000.
404. C Booth. The genus *Cylindrocarpon*. Mycol Papers 104:1–58, 1966.
405. T Matsumoto, J Masaki, T Okabe. *Cylindrocarpon tonkinense*: As a cause of keratomycosis. Trans Brit Mycol Soc 72:503–504, 1979.
406. M Mangiaterra, G Giusiano, G Smilasky, L Zamar, G Amado, C Vicentín. Keratomycosis caused by *Cylindrocarpon lichenicola*. Med Mycol 39:143–145, 2001.
407. JC Affeldt, HW Flynn, RK Forster, S Mandelbaum, JG Clarkson, GD Jarus. Microbial endophthalmitis resulting from ocular trauma. Ophthalmol 94:407–413, 1987.
408. R Sharma, CKT Farmer, WR Grandsen, CS Ogg. Peritonitis in continuous ambulatory peritoneal dialysis due to *Cylindrocarpon lichenicola* infection. Nephrol Dial Transplant 13:2662–2664, 1998.
409. EA James, K Orchard, PH McWhinney, DW Warnock, EM Johnson, AB Mehta, CC Kibbler. Disseminated infection due to *Cylindrocarpon lichenicola* in a patient with acute myeloid leukaemia. J Infec 65–67, 1997.
410. B Lamey, Ch Blanc, J Lapalu. Le *Cylindrocarpon*: Nouvel agent d'intertrigo. Bull Soc Fr Mycol Méd 14:73–76, 1985.
411. S Comparot, G Reboux, H van Landuyt, L Guetarni, Th Barale. *Fusarium solani*: Un case rebelle d'intertrigo. J Mycol Méd 5:119–121, 1995.
412. J Guarro, J Gené. Opportunistic fusarial infection in humans. Eur J Clin Microbiol Infec Dis 14:741–754, 1995.
413. E Anaissie, H Kantarjian, J Ro, R Hopfer, K Rolston, V Fainstein, G Bodey. The emerging role of *Fusarium* infections in patients with cancer. Medicine 67:77–83, 1988.
414. I Pujol, P Guarro, J Gené, J Sala. In-vitro antifungal susceptibility of clinical and environmental *Fusarium* spp. strains. J Antimicrob Chemo 39:163–167, 1997.
415. PJ Nelson, MC Dignani, EJ Anaissie. Taxonomy, biology and clinical aspects of *Fusarium* species. Clin Microbiol Rev 7:479–504, 1994.
416. JL Sunshine, A Gentili. Imaging of disseminated infection by a rare fungal pathogen, *Fusarium*. Clin Nucl Med 19:435–437, 1994.
417. V Krcméry Jr, Z Jesenská, S Spanik, J Gyarfas, J Nogova, R Botek, J Mardiak, J Sufliarsky, J Sisolakova, M Vanickova, A Kunova, M Studena, J Trupl. Fungemia due to *Fusarium* spp. in cancer patients. J Hosp Infec 36:223–228, 1997.
418. C Girmenia, W Arcese, A Micozzi, P Martino, P Bianco, G Morace. Onychomycosis as a possible origin of disseminated *Fusarium solani* infection in a patient with severe aplastic anemia. Clin Infec Dis 14:1167, 1992.
419. E Anaissie, R Kuchar, J Rex, R Summerbell, T Walsh. The hospital water system as a reservoir of *Fusarium* spp. 37th Interscience Conference on Antimicrobial Agents and Chemotherapy, Toronto, Sept. 28–Oct. 1, 1997.

420.   WD Gingrich. Keratomycosis. JAMA 179:602–608, 1962.
421.   WA Schell. Unusual fungal pathogens in fungal rhinosinusitis. Otolaryn Clin N
       Amer 33:367–373, 2000.
422.   MP English. Invasion of skin by filamentous non-dermatophyte fungi. Brit J Derm
       80:282–286, 1968.
423.   MP English, RJ Smith, RR Harman. The fungal flora of ulcerated legs. Brit J Derm
       84:567–581, 1971.
424.   K Holzegel, HJ Kempf. *Fusarium* mycosis of the skin of a burned patient. Derm
       Monatsschr 150:651, 1964.
425.   MS Wheeler, MR McGinnis, WA Schell, DH Walker. *Fusarium* infection in
       burned patients. Amer J Clin Pathol 75:304–311, 1981.
426.   PKC Austwick. *Fusarium* infection in man and animals. In: MO Moss, JE Smith,
       eds. The Applied Mycology of *Fusarium*. Cambridge: Cambridge University Press,
       1984, pp. 129–140.
427.   K O'Donnell. Molecular phylogeny of the *Nectria haematococca-Fusarium solani*
       species complex. Mycologia 92:919–938, 2000.
428.   F-X Hue, M Huerre, MA Rouffault, C de Bièvre. Specific identification of *Fu-
       sarium* species in blood and tissue by a PCR technique. J Clin Microbiol 37:2434–
       2438, 1999.
429.   K O'Donnell, E Cigelnik, HI Nirenberg. Molecular systematics and phylogeo-
       graphy of the *Gibberella fujikoroi* species complex. Mycologia 90:465–493,
       1998.
430.   C Hennequin, E Abachin, F Symoens, V Lavarde, G Reboux, N Nolard, P Berche.
       Identification of *Fusarium* species involved in human infections by 28S rRNA gene
       sequencing. J Clin Microbiol 37:3586–3589, 1999.
431.   EE Jaeger, NM Carroll, S Choudhury, AA Dunlop, HM Towler, MM Matheson,
       P Adamson, N Okhravi, S Lightman. Rapid detection and identification of *Candida*,
       Aspergillus, and *Fusarium* species in ocular samples using nested PCR. J Clin
       Microbiol 38:2902–2908, 2000.
432.   G Alexandrakis, S Jalali, P Gloor. Diagnosis of *Fusarium* keratitis in an animal
       model using the polymerase chain reaction. Brit J Ophth 82:306–311, 1998.
433.   L Kaufman, PG Standard, M Jalbert, DE Kraft. Immunohistologic identification
       of *Aspergillus* spp. and other hyaline fungi by using polyclonal fluorescent antibod-
       ies. J Clin Microbiol 35:2206–2209, 1997.
434.   NL Fisher, WFO Marasas, TA Toussoun. Taxonomic importance of microconidial
       chains in *Fusarium* section *Liseola* and effects of water potential on their formation.
       Mycologia 75:693–698, 1983.
435.   H Nirenberg. Untersuchungen über die morphologische und biologische Dif-
       ferenzierung in der *Fusarium*-Sektion *Liseola*. Mitt. Biol. Bundesanst. Land-
       Forstwirtsch. 119:1–117, 1976.
436.   NL Fisher, LW Burgess, TA Toussoun, PE Nelson. Carnation leaves as a substrate
       and for preserving *Fusarium* species. Phytopathology 72:151–153, 1982.
437.   W Gerlach, H Nirenberg. The Genus *Fusarium*—A pictorial atlas. Mitt. Biol. Bun-
       desanst. Land-Forstwirtsch. 209:1–406, 1982.
438.   PE Nelson, TA Toussoun, WFO Marasas. *Fusarium* species: An illustrated manual
       for identification. University Park, PA: Pennsylvania State University Press, 1983.

439. AS Sekhon, L Kaufman, N Moledina, RC Summerbell, AA Padhye. An exoantigen test for the rapid identification of medically significant *Fusarium* species. J Med Vet Mycol 33:287–289, 1995.

440. HI Nirenberg. Identification of fusaria occurring in Europe on cereals and potatoes. In: J Chełkowski, ed. *Fusarium*, Mycotoxins, Taxonomy and Pathogenicity. Amsterdam: Elsevier, 1989, pp. 179–193.

441. HI Nirenberg, K O'Donnell. New *Fusarium* species and combinations within the *Gibberella fujikoroi* species complex. Mycologia 90:434–458, 1998.

442. C Okuda, M Ito, Y Sato, K Oka, M Hotchi. Disseminated cutaneous *Fusarium* infection with vascular invasion in a leukemic patient. J Med Vet Mycol 25:177–186, 1987.

443. J Guarro, J Gené. *Fusarium* infections: Criteria for the identification of the responsible species. Mycoses 35:109–114, 1992.

444. S Mandelbaum, RK Forster, H Gelender, W Culbertson. Late onset endophthalmitis associated with filtering blebs. Ophthalmol 92:964–972, 1985.

445. C Booth. The genus *Fusarium*. Kew, Surrey, UK: Commonwealth Mycological Institute, 1971.

446. TE Kiehn, PE Nelson, EM Bernard, FF Edwards, B Koziner, D Armstrong. Catheter-associated fungemia caused by *Fusarium chlamydosporum* in a patient with lymphocytic lymphoma. J Clin Microbiol 21:501–509, 1985.

447. BH Segal, TJ Walsh, JM Liu, JD Wilson, KJ Kwon-Chung. Invasive infection with *Fusarium chlamydosporum* in a patient with aplastic anemia. J Clin Microbiol 36: 1772–1776, 1998.

448. U Thrane. *Fusarium* species and their specific profiles of secondary metabolites. In: J Chełkowski, ed. *Fusarium*, Mycotoxins, Taxonomy and Pathogenicity. Amsterdam: Elsevier, 1989, pp. 199–225.

449. IG Pascoe. *Fusarium* morphology I: Identification and characterization of a third conidial type, the mesoconidia. Mycotaxon 32:121–160, 1990.

450. HI Nirenberg. Recent advances in the taxonomy of *Fusarium*. Stud Mycol 32:91–101, 1990.

451. A Logrieco, S Peterson, A Bottalico. Phylogenetic affinities of the species in *Fusarium* section *Sporotrichiella*. Exp Mycol 15:174–179, 1991.

452. M Thianprasit, A Sivayathorn. Black dot mycetoma. Mykosen 27:219–226, 1983.

453. RC Zapater, A Arrechea, VH Guevara. Queratomicosis por *Fusarium dimerum*. Sabouraudia 10:274–275, 1972.

454. RC Zapater, A Arrechea. Mycotic keratitis by *Fusarium*: A review and report of two cases. Ophthalmologia 170:1–12, 1975.

455. S Sallaber, G Lori, I Galeppi. Queratomicosis por *Fusarium dimerum*. Enferm Infec Microbio Clin 17:146–147, 1999.

456. RC Zapater. "Opportunistic fungus infections"—*Fusarium* infections—(Keratomycosis by *Fusarium*). Jpn J Med Mycol 27:68–69, 1986.

457. A-M Camin, C Michelet, T Langanay, C de Place, S Chevrier, E Guého, C Guiguen. Endocarditis due to *Fusarium dimerum* four years after coronary artery bypass grafting. Clin Infec Dis 28:150, 1999.

458. JL Poirot, JP Laporte, E Guého, A Verny, NC Gorin, A Najman, M Marteau, P Roux. Mycose profonde à *Fusarium*. Presse Méd 14:2300–2301, 1985.

459. F Leinati. Sull'azione patogena di una specie nuova di *Fusarium*. Riv Biol 10: 141–154, 1928.
460. M Curzi. Intorno alla posizione sistematica di un *Fusarium* isolato dalla pelle del cane. Atti Ist. Bot Univ Pavia, Ser IV, 1:95–105, 1929.
461. HW Wollenweber, OA Reinking. Die Fusarien, ihre Beschreibung, Schadwirkung und Bekämpfung. Berlin: Verl. Paul Parey, 1935.
462. MR McGinnis, LC Severo, R Kalil, PT Falleiro. Endocarditis caused by *Fusarium pallidoroseum*. J Mycol Méd 4:45–47, 1994.
463. S Imwidthaya, C Chuntrasakul, N Chantarakul. Opportunistic fungal infection of the burn wound. J Med Assoc Thai 67:242–248, 1984.
464. FM Rush-Munro, H Black, JM Dingley. Onychomycosis caused by *Fusarium oxysporum*. Aust J Derm 12:18–29, 1971.
465. A Subrahmanyam. *Fusarium laceratum*. Mykosen 26:478–480, 1983.
466. R Blanchard. Sur une remarquable dermatose causée chez le lézard vert par un champignon du genre *Selenosporium*. Mém Soc Zool Fr 3:241–255, 1890.
467. R Blanchard. Parasites végétaux à l'exclusion des bactéries. In: C Bouchard, ed. Traité de Pathologie Générale. vol 2. Paris, G. Masson, 1896, pp. 811–926.
468. F Guégen. Les Champignons Parasites de l'Homme et des Animaux. Paris: Maison d'éditions, 1904, p. 262.
469. W Gams. Generic names for synanamorphs? Mycotaxon 15:459–454, 1982.
470. GP Melcher, DA McGough, AW Fothergill, C Norris, MG Rinaldi. Disseminated hyalohyphomycosis caused by a novel human pathogen, *Fusarium napiforme*. J Clin Microbio 16:528–530, 1993.
471. WFO Marasas, CJ Rabie, A Lübben, PE Nelson, TA Toussoun, PS van Wyk. *Fusarium napiforme*, a new species from millet and sorghum in southern Africa. Mycologia 79:910–914, 1987.
472. JWM Krulder, RW Brimicombe, PW Wijermans, W Gams. Systemic *Fusarium nygamai* infection in a patient with lymphoblastic non-Hodgkins lymphoma. Mycoses 39:121–123, 1996.
473. LW Burgess, D Trimboli. Characterization and distribution of *Fusarium nygamai*, sp. nov. Mycologia 78:223–229, 1986.
474. D Farrell, L Abbey, C Payne. *Fusarium oxysporum* peritonitis as a complication of continuous peritoneal dialysis (CAPD): A case report and review. Abstr. ISHAM Congress, Adelaide, Australia, 1994.
475. J Eljaschewitsch, J Sandfort, K Tintelnot, I Horbach, B Ruf. Port-a-cath-related *Fusarium oxysporum* infection in an HIV-infected patient: treatment with liposomal amphotericin B. Mycoses 39:115–119, 1996.
476. A Sander, U Beyer, R Amberg. Systemic *Fusarium oxysporum* infection in an immunocompetent patient with an adult respiratory distress syndrome (ARDS) and extracorporal membrane oxygenation (ECMO). Mycoses 41:109–111, 1998.
477. J Chodosh, D Miller, EY Tu, WW Culbertson. Tobramycin-responsive *Fusarium oxysporum* keratitis. Can J Ophth 35:29–30, 2000.
478. EB Ritchie, ME Pinkerton. *Fusarium oxysporum* infection of the nail. Arch Derm 79:705–708, 1959.
479. C Gianni, A Cerri, C Crosti. Unusual clinical features of fingernail infection by *Fusarium oxysporum*. Mycoses 40:455–459, 1997.

480. R Baran, A Tosti, BM Piraccini. Uncommon clinical patterns of *Fusarium* nail infection. Brit J Derm 136:424–427, 1997.
481. ML Dordain-Bigot, R Baran, MT Baixench, J Bazex. Onychomycose à *Fusarium*. Ann Derm Venereol 123:191–193, 1996.
482. C Romano, C Miracco, EM Difonzo. Skin and nail infections due to *Fusarium oxysporum* in Tuscany, Italy. Mycoses 41:433–437, 1998.
483. C Romano, L Presenti, L Massai. Interdigital intertrigo of the feet due to therapy-resistant *Fusarium solani*. Dermatology 199:177–179, 1999.
484. MJ Willemsen, AL de Coninck, JE Coremans-Pelseneer, MA Marichal-Pipeleers, DI Roseeuw. Parasitic invasion of *Fusarium oxysporum* in an arterial ulcer in an otherwise healthy patient. Mykosen 29:248–252, 1986.
485. MP English. Observations on strains of *Fusarium solani*, *F. oxysporum* and *Candida parapsilosis* from ulcerated legs. Sabouraudia 10:35–42, 1972.
486. M Landau, A Srebrnik, R Wolf, E Bashi, S Brenner. Systemic ketoconazole for *Fusarium* leg ulcers. Internat J Derm 31:511–512, 1992.
487. MC Attapattu, C Anandakrishnan. Extensive subcutaneous hyphomycosis caused by *Fusarium oxysporum*. J Med Vet Mycol 24:105–111, 1986.
488. JF Leslie. *Gibberella fujikoroi*: Available populations and variable traits. Can J Bot 73:S282–S291, 1995.
489. RC Summerbell, SE Richardson, J Kane. *Fusarium proliferatum* as an agent of disseminated infection in an immunocompromised patient. J Clin Microbio 26:82–87, 1988.
490. TN Helm, DL Longworth, GS Hall, BJ Bolwell, B Fernández, KJ Tomecki. Case report and review of resolved fusariosis. J Amer Acad Derm 23:393–398, 1990.
491. NK Barrios, DV Kirkpatrick, A Murciano, K Stine, RB Van Dyke, JR Humbert. Successful treatment of disseminated *Fusarium* infection in an immunocompromised child. Amer J Pediat Hematol/Oncol 12:319–324, 1990.
492. WP Norred, CW Bacon, RT Riley, KA Voss, FI Meredith. Screening of fungal species for fumonisin production and fumonisin-like disruption of sphingolipid biosynthesis. Mycopathologia 146:91–98, 1999.
493. J Guarro, M Nucci, T Akiti, J Gené, MDGC Barreiro, RT Gonçalves. Fungemia due to *Fusarium sacchari* in an immunosuppressed patient. J Clin Microbio 38: 419–421, 2000.
494. R Foroozan, RC Eagle Jr, EJ Cohen. Fungal keratitis in a soft contact lens wearer. CLAO J 26:166–168, 2000.
495. D Goldblum, BE Frueh, S Zimmerli, M Böhnke. Treatment of postkeratitis *Fusarium* endophthalmitis with amphotericin B lipid complex. Cornea 19:853–856, 2000.
496. J Mselle. Fungal keratitis as an indicator of HIV infection in Africa. Trop Doct 29:133–135, 1999.
497. T Louie, F el Baba, M Shulman, V Jimenez-Lucho. Endogenous endophthalmitis due to *Fusarium*: Case report and review. Clin Infec Dis 18:585–588, 1994.
498. A Paugam, M-T Baixench, N Frank, P Bossi, G de Pinieux, C Tourte-Schaefer, J Dupouy-Camet. Localized oral *Fusarium* infection in an AIDS patient with malignant lymphoma. J Infec 39:153–162, 1999.
499. AG Luque, MT Mugica, ML D'Anna, DP Alvárez. Micetoma podal por *Fusarium solani* (Mart.) Appel & Wollenweber. Bol Micol 6:55–57, 1991.

500. JW Hiemenz, B Kennedy, KJ Kwon-Chung. Invasive fusariosis associated with an injury by a stingray barb. J Med Vet Mycol 28:209–213, 1990.

501. G Rebell. *Fusarium* infections in human and veterinary medicine. In: PE Nelson, TA Toussoun, RJ Cook, eds. *Fusarium*: Diseases, Biology, Taxonomy. University Park, PA: Pennsylvania State University Press, 1981, pp. 212–220.

502. G Castello, J Cano, J Guarro, FJ Cabanes. DNA fingerprinting of *Fusarium solani* isolates related to a cutaneous infection in a sea turtle. Med Mycol 37:223–226, 1999.

503. Crocodile Specialist Group Newsletter. WWW edition. 17 (3):2, 1998; *http:// www.flmnh.ufl.edu/natsci/herpetology/newsletter/news173b.htm.*

504. EMA Hibberd, KM Harrower. Mycoses in crocodiles. Mycologist 7:32–37, 1993.

505. RJ Montali, M Bush, JD Strandberg, DL Janssen, DJ Boness, JC Whitla. Cyclic dermatitis associated with *Fusarium* sp. infection in pinnipeds. J Amer Vet Med Assoc 179:1198–1202, 1981.

506. ER Jacobson. Necrotizing mycotic dermatitis in snakes: Clinical and pathologic features. J Amer Vet Med Assoc 177:838–841, 1980.

507. Y-N Ming, T-F Yu. Identification of a *Fusarium* species isolated from corneal ulcer. Acta Microbiol Sinica 12:180–186, 1966 (in Chinese).

508. F Polenghi, A Lasagni. Observations on a case of mycokeratitis and its treatment with BAY b 5097 (Canesten). Mykosen 19:223–226, 1976.

509. T Matsushima. Microfungi of the Solomon Islands and Papua-New Guinea. Kobe, Japan: Nippon Printing Co., 1971, p. 11.

510. NA Young, KJ Kwon-Chung, TT Kubota, AE Jennings, RI Fisher. Disseminated infection by *Fusarium moniliforme* during treatment for malignant lymphoma. J Clin Microbio 7:589–594, 1978.

511. C Farina, F Vailati, A Manisco, A Goglio. Fungemia survey: A 10-year experience in Bergamo, Italy. Mycoses 42:543–548, 1999.

512. MS Collins, MG Rinaldi. Cutaneous infection in man caused by *Fusarium moniliforme*. Sabouraudia 15:151–160, 1977.

513. L Ajello, AA Padhye, FW Chandler, MR McGinnis, L Morganti, F Alberici. *Fusarium moniliforme*, a new mycetoma agent: Restudy of a European case. Eur J Epidem 1:5–10, 1985.

514. F Alberici, L Morganti, F Suter, A Dei Cas. Sul primo caso di micetoma del piede da *Acremonium* sp. osservato in Italia. Giorn Mal Infett Parassit 30:34–37, 1978.

515. M Pereiro Jr, J Labandeira, J Toribio. Plantar hyperkeratosis due to *Fusarium verticillioides* in a patient with malignancy. Clin Exp Derm 24:175–178, 1999.

516. JA Durán, A Malvar, M Pereiro, M Pereiro. *Fusarium moniliforme* keratitis. Acta Ophthalmol 67:710–713, 1989.

517. PF Frelier, L Sigler, PE Nelson. Mycotic pneumonia caused by *Fusarium moniliforme* in an alligator. Sabouraudia 23:399–402, 1985.

518. W Gams, E Müller. Conidiogenesis of *Fusarium nivale* and *Rhynchosporium oryzae* and its taxonomic implications. Neth J Pl Pathol 86:45–53, 1980.

519. M Perz, C Majewski, K Mańka. *Fusarium nivale* jako przyczyna grzybicy rogówki. Klin Oczna 36:609–612, 1966.

520. KS Backman, M Roberts, R Patterson. Allergic bronchopulmonary mycosis caused by *Fusarium vasinfectum*. Amer J Resp Crit Care Med 152:1379–1381, 1995.

521. J Chandenier, MP Hayette, C de Bièvre, PF Westeal, J Petit, JM Achard, N Bove, B Carme. Tuméfaction de la jambe à *Neocosmospora vasinfecta* chez un transplanté rénal. J Mycol Méd 3:165–168, 1993.

522. F Ben Hamida, JM Achard, PF Westeel, J Chandenier, M Bouzernidj, J Petit, B Carme, A Fournier. Leg granuloma due to *Neocosmospora vasinfecta* in a renal graft recipient. Transpl Proc 25:2292, 1993.

523. G Kac, P Piriou, E Guého, P Roux, J Trémoulet, M Denis, T Judet. Osteoarthritis caused by *Neocosmospora vasinfecta*. Med Mycol 37:213–217, 1999.

524. MA Rifai. A revision of the genus *Trichoderma*. Mycol Papers 116:1–56, 1969.

525. J Bissett. A revision of the genus *Trichoderma*. IV. Additional notes on section *Longibrachiatum*. Can J Bot 69:2418–2420, 1991.

526. S Muthumeenakshi, PR Mills, AE Brown, DA Seaby. Intraspecific molecular variation among *Trichoderma harzianum* isolates colonizing mushroom compost in the British Isles. Microbiology 140:769–777, 1994.

527. DA Seaby. Differentiation of *Trichoderma* taxa associated with mushroom production. Plant Pathol 45:905–912, 1996.

528. S Richter, MG Cormican, MA Pfaller, CK Lee, R Gingrich, MG Rinaldi, DA Sutton. Fatal disseminated *Trichoderma longibrachiatum* infection in an adult bone marrow transplant patient: Species identification and review of the literature. J Clin Microbio 37:1154–1160, 1999.

529. W Gams, J Bissett. Morphology and identification of *Trichoderma*. In: CP Kubicek, GE Harman, eds. *Trichoderma* and *Gliocladium*. vol. 1. Basic Biology, Taxonomy and Genetics. London: Taylor & Francis, 1998, pp. 3–34.

530. K Kuhls, E Lieckfeldt, T Börner, E Guého. Molecular reidentification of human pathogenic *Trichoderma* isolates as *Trichoderma longibrachiatum* and *Trichoderma citrinoviride*. Med Mycol 37:25–33, 1999.

531. GJ Samuels, O Petrini, K Kuhls, E Lieckfeldt, CP Kubicek. The *Hypocrea schweinitzii* complex and *Trichoderma* section *Longibrachiatum*. Stud Mycol 41:1–54, 1998.

532. J Guarro, MI Antolín-Ayala, J Gené, J Gutiérrez-Calzada, C Nieves-Díez, M Ortoneda. Fatal case of *Trichoderma harzianum* infection in a renal transplant patient. J Clin Microbiol 37:3751–3755, 1999.

533. J Guiserix, M Ramdane, P Finielz, A Michault, P Rajaonarivelo. *Trichoderma harzianum* peritonitis in peritoneal dialysis. Nephron 74:473–474, 1996.

534. W Gams, W Meyer. What exactly is *Trichoderma harzianum* Rifai? Mycologia 90:904–915, 1998.

535. JM Ragnaud, C Marceau, MC Roche-Bezian, C Wone. Infection péritonéale à *Trichoderma koningii* sur dialyse péritonéale continué ambulatoire. Méd Malad Infect 7:402–405, 1984.

536. MI Campos-Herrero, A Bordes, A Perera, MC Ruiz, A Fernandez. *Trichoderma koningii* peritonitis in a patient undergoing peritoneal dialysis. Clin Microbiol Newsl 18:150–152, 1996.

537. E Lieckfeldt, GJ Samuels, W Gams. Neotypification of *Trichoderma koningii* and its *Hypocrea koningii* teleomorph. Can J Bot 76:1519–1522, 1998.

538. DA McGough, AW Fothergill, S Kusne, J Furukawa, MG Rinaldi. *Trichoderma koningii*: Yet another new agent of contemporary mycoses. Washington, DC, Abstract General Meeting, American Society of Microbiology, 1994, p. 602.

539. H Furukawa, S Kusne, DA Sutton, R Manez, R Carrau, L Nichols, K Abu-Elmagd, D Skedros, S Todo, MG Rinaldi. Acute invasive sinusitis due to *Trichoderma longibrachiatum* in a liver and small bowel transplant recipient. Clin Infec Dis 26: 487–489, 1998.

540. A Gautheret, F Dromer, JH Bourhis, A Andremont. *Trichoderma pseudokoningii* as a cause of fatal infection in a bone marrow transplant recipient. Clin Infec Dis 20:1063–1064, 1995.

541. P Seguin, B Degeilh, I Grulois, A Gacouin, S Maugendre, T Dufour, B Dupont, C Camus. Successful treatment of a brain abscess due to *Trichoderma longibrachiatum* after surgical resection. Eur J Clin Microbiol Infec Dis 14:445–448, 1995.

542. FM Muñoz, G Demmler, WR Travis, AK Ogden, SN Rossmann, MG Rinaldi. *Trichoderma longibrachiatum* infection in a pediatric patient with aplastic anemia. J Clin Microbiol 35:499–503, 1997.

543. BC Tanis, H van der Pijl, ML van Ogtrop, RC Kibbelaar, PC Chang. Fatal fungal peritonitis by *Trichoderma longibrachiatum* complicating peritoneal dialysis. Nephrol Dial Transpl 10:114–116, 1995.

544. D Turner, W Kovacs, K Kuhls, E Lieckfeldt, B Peter, I Arisan-Atac, J Strauss, GJ Samuels, T Börner, CP Kubicek. Biogeography and phenotype variation in *Trichoderma* sect. *Longibrachiatum* and associated *Hypocrea* species. Mycol Res 101:449–459, 1997.

545. K Kuhls, E Lieckfeldt, GJ Samuels, W Meyer, CP Kubicek, T Börner. Revision of *Trichoderma* section *Longibrachiatum* including related teleomorphs based on analysis of ribosomal DNA internal transcribed spacer sequences. Mycologia 89: 442–460, 1995.

546. B Degeilh, P Seguin, P Brasy, S Maugendre, C Guiguen. Abscès cérebral à *Trichoderma pseudokoningii* chez une jeune leucemique. Abstr. 1st Congress Eur Congr Med Mycol, Paris, 1993, p. 121.

547. MR Escudero Gil, E Pino Corral, R Muñoz Muñoz. Pulmonary mycoma caused by *Trichoderma viride*. Actas Dermosifiliogr 67:673–680, 1976.

548. CB Loeppky, RF Sprouse, JV Carlson, ED Everett. *Trichoderma viride* peritonitis. South Med J 76:798–799, 1983.

549. R Jacobs, B Byl, N Bourgeois, J Coremans-Pelseneer, S Florquin, G Depré, J Van de Stadt, M Adler, M Gelin, JP Thys. *Trichoderma viride* infection in a liver transplant recipient. Mycoses 35:301–303, 1992.

550. E Lieckfeldt, GJ Samuels, HI Nirenberg, O Petrini. A morphological and molecular perspective of *Trichoderma viride*: is it one or two species? Appl Environ Microbiol 65:2418–2428, 1999.

551. GJ Samuels, E Lieckfeldt, H Nirenberg. *Trichoderma asperellum*, a new species with warted conidia, and a redescription of *T. viride*. Sydowia 51:71–88, 1999.

552. L Maffei. Nuova specie di *Cephalosporium* causa di una cheratomicosi dell'uomo. Atti Ist Bot Univ Pavia, ser IV 1:183–198, 1929.

553. CD Burda, E Fisher Jr. Corneal destruction by extracts of *Cephalosporium* mycelium. Amer J Ophth 50:926–937, 1960.

554. MB de Albornoz. *Cephalosporium serrae*, agente etiologico de micetomas. Mycopath Mycol Appl 54:485–498, 1974.

555. PW Crous, W Gams, MJ Wingfield, PS van Wyk. *Phaeoacremonium* gen. nov.

associated with wilt and decline diseases of woody hosts and human infections. Mycologia 88:786–796, 1996.

556. JBT Foster, E Almeda, ML Littman, ME Wilson. Some intraocular and conjunctival effects of amphotericin B in man and in the rabbit. Arch Ophth 60:555–564, 1958.

557. FH Theodore, ML Littman, E Almeda. The diagnosis and management of fungus endophthalmitis following cataract extraction. Arch Ophth 66:163–175, 1961.

558. RC Zapater. Las micoses oculares. Prensa Med Arg 65:203–206, 1978.

559. GS De Hoog. The genera *Blastobotrys*, *Sporothrix*, *Calcarisporium* and *Calcarisporiella gen. nov.* Stud Mycol 7:1–84, 1974.

560. D Dixon, IF Salkin, RA Duncan, NJ Hurd, JH Haines, ME Kemna, FB Coles. Isolation and characterization of *Sporothrix schenckii* from clinical and environmental sources associated with the largest U.S. epidemic of sporotrichosis. J Clin Microbiol 29:1106–1113, 1991.

561. CR Cooper Jr, BJ Breslin, DM Dixon, IF Salkin. DNA typing of isolates associated with the 1988 sporotrichosis epidemic. J Clin Microbiol 30:1631–1635, 1992.

562. H Ishizaki, M Kawasaki, M Aoki, H Vismer, D Muir. Mitochondrial DNA analysis of *Sporothrix schenckii* in South Africa and Australia. Med Mycol 38:433–436, 2000.

563. JE Mackinnon, IA Conti-Diaz, E Gezuele, E Civila, S da Luz. Isolation of *Sporothrix schenckii* from nature and considerations on its pathogenicity and ecology. Sabouraudia 7:38–45, 1969.

564. DM England, L Hochholzer. *Sporothrix* infection of the lung without cutaneous disease. Primary pulmonary sporotrichosis. Arch Path Lab Med 111:298–300, 1987.

565. AJ Ware, CJ Cockerell, DJ Skiest, HM Kussman. Disseminated sporotrichosis with extensive cutaneous involvement in a patient with AIDS. J Amer Acad Derm 40:350–355, 1999.

566. JA al-Tawfiq, KK Wools. Disseminated sporotrichosis and *Sporothrix schenckii* fungemia as the initial presentation of human immunodeficiency virus infection. Clin Infec Dis 26:1403–1406, 1998.

567. HM Heller, J Fuhrer. Disseminated sporotrichosis in patients with AIDS: Case report and review of the literature. AIDS 5:1243–1246, 1991.

568. S Kumar, D Kumar, WK Gourley, JB Alperin. Sporotrichosis as a presenting manifestation of hairy cell leukemia. Amer J Hematol 46:134–137, 1994.

569. AA Padhye, L Kaufman, E Durry, CK Banerjee, SK Jindal, P Talwar, AA Chakrabarti. Fatal pulmonary sporotrichosis caused by *Sporothrix schenckii* var. *luriei* in India. J Clin Microbiol 30:2492–2494, 1992.

570. K Suzuki, M Kawasaki, H Ishizaki. Analysis of restriction profiles of mitochondrial DNA from *Sporothrix schenckii* and related fungi. Mycopathologia 103:147–151, 1988.

571. F Staib, A Blisse. Stellungnahme zu *Sporothrix schenckii* var. *luriei*. Ein Beitrag zum diagnostischen Wert der Assimilation von Kreatinin, Kreatin und Guanidinoessigsäure durch *Sporothrix schenckii*. Zbl Bakt, Parasitenkde, Infektionskr Hyg, I Abt, Orig, Reihe A 229:261–263, 1974.

572. RC Summerbell, J Kane, S Krajden, EE Duke. Medically important *Sporothrix*

species and related ophiostomatoid fungi. In: MJ Wingfield, KA Seifert, JF Webber, eds. *Ceratocystis* and *Ophiostoma*. Taxonomy, Ecology and Pathogenicity. St. Paul, MN: APS, 1993, pp. 185–192.

573. G Ségretain, P Destombes. Description d'un nouvel agent de maduromycose, *Neotestudina rosatii* n. gen., n. sp., isolé en Afrique. CR Séanc. Acad Sci (Paris) 253: 2577–2579, 1961.

574. M Takashio, C de Vroey. Piedra noire chez les chimpanzes du Zaire. Sabouraudia 13:58–62, 1975.

575. G Ségretain, J Baylet, H Darasse, R Camain. *Leptosphaeria senegalensis*, n. sp., agent de mycétome à grains noirs. CR Acad Sci (Paris) 248:3730–3772, 1959.

576. AS El-Ani. A new species of *Leptosphaeria*, an etiologic agent of mycetoma. Mycologia 58:406–411, 1966.

577. MP English, RRM Harman, JWJ Turvey. *Pseudeurotium ovalis* in toenails. Brit J Derm 79:553–556, 1967.

578. RC Summerbell. Epidemiology and ecology of onychomycosis. Dermatology 194 (suppl. 1):32–36, 1997.

8

Yeasts

Blastomycetes and Endomycetes

Kevin C. Hazen
University of Virginia Health System, Charlottesville, Virginia, U.S.A.

Susan A. Howell
St. John's Institute of Dermatology, King's College of London, London, England

I. INTRODUCTION

The Blastomycetes and Endomycetes contain a large number of medically important yeast species, most notably species within the genus *Candida*, which is referred to as a form genus because the genus itself is a repository for yeast species having certain characteristics and lacking sexual reproduction. The lack of sexuality limits taxonomists in their phylogenetic attempts to relate the various species when using traditional identification criteria. Molecular methods, however, have provided a new and exciting means to establish the phylogenetic relationships of not only the blastomycetous yeasts but also the endomycetous yeasts. In this chapter, we attempt to review the results of traditional and molecular studies of yeast taxonomy and relate those results to the clinical mycology laboratory. It is evident that yeast taxonomy will continue to evolve as more refined techniques are developed.

II. TAXONOMY AND CLASSIFICATION

The class Blastomycetes belongs in the division Deuteromycota, and the class Endomycetes are yeasts within the division Ascomycota (Table 1). Each order has two families of medically important organisms (Table 1).

499

Table 1 Taxonomy and Classification of Yeasts Within the Blastomycetes and Endomycetes

Division	Class	Order	Family	Description
Ascomycona	Endomycetes			Produce asci.
				Asci not contained in ascocarp (naked asci).
		Saccharomycetales		Form asci directly from zygote or zygote's diploid progeny.
			Dipodascaceae	Anamorphs with arthroconidia.
			Saccharomycetaceae	Anamorphs with budding cells.
Deuteromycota	Blastomycetes			Lack meiotic spores.
				Vegetative cells are yeasts. May produce pseudo-mycelium. When present, true mycelium is poorly developed.
		Sporobolomycetales		Produce ballistospores.
			Sporobolomycetaceae	Reproduction by fission or budding. Produce ballis-tospores on sterigmata arising from vegetative cells.
		Cryptococcales		Nonsexual yeasts that do not produce sterigmata.
			Cryptococcaceae	Budding cells always present. May produce pseudo-mycelia, mycelium, and arthroconidia. Cells may be hyaline or pigmented but rarely brown or black.

Yeast taxonomy is under constant revision. Recent advances in molecular methods for phylogenetic analysis have particularly caused significant revisions of taxonomic assignments. These advances have helped to demonstrate ascomycetous and basidiomycetous affinities for certain imperfect (nonsexual) yeasts, such as *Candida albicans*. In the absence of a sexual structure, however, these organisms remain in the fungi imperfecti (Deuteromycota)—a pseudotaxonomic repository for organisms with no known sexual state.

This entire division, Deuteromycota, is sometimes referred to as a "form division" to indicate that the term division is used arbitrarily, and unlike the true divisions, implies no taxonomic (or evolutionary) significance. When sexual reproduction is observed along with the attendant sexual reproductive structures (e.g., asci), the form-genus species is reassigned to the appropriate taxonomically legitimate division and renamed with an epithet consistent with the division. Such renaming has in practice led to confusion in scientific and clinical discussions about yeast infections, but adherence to this classification principle is important because it provides insights into the relationship of the organism to similar organisms. The demonstration that some species of *Candida* belong to the ascomycetous yeasts and others to the basidiomycetous yeasts illustrates the pseudotaxonomic value of the form-order term Cryptococcales. Knowing the correct taxonomic assignment provides information useful for identifying these organisms isolated from patient specimens.

As noted above, a number of traditional mating and recent molecular analyses have led to the reassignment of various blastomycetous yeasts to either the endomycetes or one of the basidiomycetous yeast genera. This chapter will not describe the basidiomycetous yeasts but will focus only on the strictly asexual yeasts and the endomycetous yeasts. For blastomycetes reassigned to endomycetous genera (Table 2) we will primarily use the more common (albeit taxonomically invalid) blastomycetous species binomial because these organisms are typically seen only in the anamorphic form. Blastomycetous yeasts with basidiomycetous affinities (Table 2) will be described only relative to their blastomycetous features.

III. TRADITIONAL METHODS OF IDENTIFICATION AND CLASSIFICATION

Several characteristics of fungi can be used to separate different classes. The definitive characteristics for studying ascomycetous fungi are sexual propagules and fruiting structures. Other characteristics used to differentiate the ascomycetous yeasts include vegetative cell morphology, cell wall polysaccharides, coenzyme Q families (e.g., coenzyme Q_{10} is present only in basidiomycetous fungi),

Table 2 Classification of the Medically Important Yeast Genera Within the Blastomycetes and Endomycetes

Class	Order	Teleomorph genus-species	Anamorph
Endomycetes	Saccharomycetales	*Arxiozyma telluris*	*Candida pintolopesii*
		Citeromyces matritensis	*Candida globosa*
		Clavispora lusitaniae	*Candida lusitaniae*
		Clavispora capitatus	*Blastoschizomyces capitatus*
		Debaryomyces hansenii	*Candida famata*
		Galactomyces geotrichum	*Geotrichum candidum*
		Hansenula anomala	*Candida pelliculosa*
		Issatchenkia orientalis	*Candida krusei*
		Kluyveromyces lactis	*Candida sphaerica*
		Kluyveromyces marxianus	*Candida kefyr*
		Metschnikowia pulcherrima	*Candida pulcherrima*
		Pichia guilliermondii	*Candida guilliermondii*
		Pichia fermentans	*Candida lambica*
		Pichia jadinii	*Candida utilis*
		Pichia membranaefaciens	*Candida valida*
		Pichia norvegensis	*Candida norvegensis*
		Saccharomyces cerevisiae	
		Saccharomyces exiguus	*Candida holmii*
		Stephanoascus ciferrii	*Candida ciferrii*
		Yarrowia lipolytica	*Candida lipolytica*
Blastomycetes	Sporobolomycetales		*Sporobolomyces*
	Cryptococcales		*Blastoschizomyces*
			Candida
			Cryptococcus
			Malassezia
			Rhodotorula
			Trichosporon

morphology of septa (e.g., the dolipore-parenthesome septum of most basidiomycetes versus the ascomycetous simple pore septum with Woronin body), sensitivity to killer toxins, membrane fatty acid components, ascus and ascospore features, temperature tolerance, urease production, fermentation, nitrate assimilation, and intranuclear mitosis. In contrast, the imperfect yeasts have features that are either ascomycetous or basidiomycetous but lack sexual structures (asci and basidia), therefore key characteristics to help identify the asexual blastomycetes include most of those listed above for ascomycetous fungi plus diazonium B blue reaction, ballistospore formation, and production of carotenoids (1–3). Molecular methods have significantly enhanced the ability to classify imperfect yeasts. The recent development of proteomics may also provide useful classification criteria to separate the yeasts.

The definitive methods of determining classification of an unknown yeast are not readily available to the clinical laboratory. Clinical laboratories rely on a limited set of characteristics that provide a most likely approximation of an organism's taxonomic position. Morphology can provide some information (e.g., in the case of the genus *Kloeckera* or species *Metschnikowia lunata*), but other tests, such as assimilation and fermentation characteristics and exoenzyme production, may be used. When possible, the clinical laboratory may be able to induce mating reactions and obtain the arrangement and morphology of the resultant meiotic progeny. Mating reactions are relatively easy to induce if the yeast isolate is homothallic. When the organism is heterothallic, it may be necessary to try several different species along with both mating types before the right mating cognate is obtained to allow identification. Ascospore formation of most hemiascomycetous yeasts can be induced by growing them on dilute V-8 juice agar, but some medically important genera may require specific media, such as ascospore agar. The incubation temperature is typically 25°C, but *Metschnikowia* species may require temperatures between 12–15°C, and the optimum temperature for sporulation of *Debaryomyces* species is below 20°C.

The essential characteristics for separating ascomycetous yeasts are the sexual spores and accompanying structures. (See the Appendix.) The color, number, size, morphology, ornamentation, and arrangement of the ascospores within the ascus and the characteristics of the asci provide definitive information to ascertain the classification of an unknown yeast. In the absence of sexual reproduction, clinical laboratories will use morphology and assimilation/fermentation patterns as the key criteria to determine an organism's taxonomic position. Particularly useful tests for clinical laboratories include urease production (generally ascomycetous yeasts are negative and basidiomycetous yeasts are positive), inositol assimilation (generally ascomycetous yeasts are negative and basidiomycetous yeasts are positive), and nitrate assimilation. Commercial test systems (e.g., the API 20C and the IDS RapID) provide good accuracy for common yeasts but may not be particularly useful for unusual yeasts. In addition, these tests can sometimes provide species identifications that are at best questionable. Similarly, *C. inconspicua* may be misidentified as *C. krusei*. Karyotyping, in this case, determined the correct identity (4).

The most useful assimilation/fermentation tests are the auxanographic methods modeled after the work of Wickerham and Burton (5). Taxonomic treatises provide tables of the reactions with the various substrates based on the Wickerham method. Commercial assimilation systems do not show 100% agreement with the Wickerham method, particularly with the more unusual clinical isolates. A clinical laboratory using assimilation results from commercial kits must be aware that some results will not correlate with the Wickerham and Burton method.

The implications from such studies are significant, given the growing pharmaceutical interest in developing new agents for treatment of mycoses. The agents generally show narrower spectra of species efficacy than older agents, such as amphotericin B. Antibiograms of *Candida* species have shown that species identification is important before the choice of appropriate antifungal agent for a given clinical situation can be made.

One of the common problems faced by clinical laboratories—which rely on phenotypic tables of biochemical/physiological characteristics for yeast identification—is the consistency between the methods employed to generate the characteristics defined on the table and the methods used in the clinical laboratory. For example, urease production is essentially a characteristic of the basidiomycetous yeasts. When the presence of urease in taxonomically different yeasts is evaluated, however, positive tests are obtained for many yeast species if the test medium is not sufficiently buffered. This factor of buffering may account for the variable urease results reported among isolates of *C. krusei.* The conditions for urease production thus must be identical to those used to generate the table of phenotypic characteristics.

A similar situation occurs with cycloheximide (CHX) sensitivity testing. Cycloheximide sensitivity is a useful characteristic for separating closely related species. Clinical laboratories typically use media containing CHX at concentrations of 400 to 500 µg/ml, and may presume that such media are suitable for testing sensitivity to CHX. At least two major compendia of yeast species (1, 3), however, noted that CHX concentrations in media may vary from 100 to 1000 µg/ml and that species can fall into groups based on their level of sensitivity. This point was also determined by Whiffen (6), who originally reported the use of CHX for species discrimination. Some isolates may develop tolerance to lower concentrations, making the reliability of species discrimination based on sensitivity to low concentrations of CHX unreliable. Which specific concentration is used to assess CHX sensitivity for inclusion in a table of yeast phenotypic characteristics is often not stated, but clinical laboratories should be aware that it is likely the concentration is not the one found in standard clinical fungal growth media. The incubation temperature at which the CHX sensitivity test is performed may also influence the final result.

Nitrate assimilation is also a useful characteristic that can help to identify the species of an organism and help demonstrate taxonomic associations. The presence of nitrate reductase (as detected by a rapid swab test) does not imply that an organism can assimilate nitrate. The test thus must be consistent with the methods used to generate the phenotypic table.

"Definitive" taxonomic and classification evaluation of an isolate with subsequent placement into a species is not always possible with assimilation and fermentation tests. Additional tests are always necessary if the organism is under consideration as the type species of a new taxonomic epitaph. In this situation, molecular methods provide precise information.

IV. NEWER METHODS OF IDENTIFICATION AND CLASSIFICATION

In recent years there has been an explosion in the variety of molecular methods developed to examine taxonomic relationships. Two of the original DNA-based methodologies used the mol % G+C ratio and DNA:DNA hybridization to determine the species identity. Isolates of the same species would be expected to have similar mol % G+C and DNA:DNA reassociation values. The values indicate if the organisms belong to the same species or genus or are unrelated, but do not provide information of taxonomic relationships. Phylogenetic relationships between organisms are measured by using nucleotide sequence divergence. Sequences of conserved genes have been studied, but the most popular targets for these analyses have been the rRNA genes.

Alternative methods that do not require sequencing of nucleic acids have been developed for identification purposes rather than strictly for taxonomy, and are more suited to the clinical and research environments. The comparison of the sizes and numbers of chromosomes by karyotyping with pulsed-field gel electrophoresis has been used for both epidemiology and for examining the genetic organization of similar species. DNA probes have been produced that were species-specific or generated species-specific profiles. Restriction enzyme digests of DNA have been used to produce profiles characteristic of species or strains. Fingerprinting techniques that use the polymerase chain reaction have been developed to demonstrate intraspecies differences for epidemiology and interspecies differences for identification of yeasts.

V. EFFECT OF MOLECULAR ANALYSIS ON CLASSIFICATION AND IDENTIFICATION

The morphological and physiological tests for traditional species identification of yeasts can be unreliable or difficult to interpret, especially if the organism is metabolically unreactive. Furthermore, the characteristics obtained by these tests provide little evidence of the evolution of species. In such cases, various cell components may be examined for taxonomic significance and usefulness in assessing the relatedness of species and genera. Only the methods using nucleic acids will be discussed here, although many of the cited papers contain references to alternative phenotypic, physiological, or chemotaxonomic analyses, applicable to the study of blastomycetous and endomycetous yeasts.

A. Phylogeny Deduced from rRNA Sequence Data

The impact of molecular methods on the phenotypically defined classification of yeasts was reviewed by Kurtzman (7). Based on the partial sequences of small and large subunit (SSU and LSU, respectively) rRNA the ascomycetous yeasts were found to be monophyletic, and the fission yeasts (*Schizosaccharomyces*) were separated from budding and filamentous yeasts. Phylogenetic trees constructed from the rRNA sequences of the D2 region of the LSU demonstrated the basidiomycetous yeasts to be divided into two clades broadly reflecting the type of hyphal septum, the presence or absence of teliospores in the sexual state, and the occurrence of cellular xylose. Clade A contained species of *Rhodotorula* and *Sporobolomyces*, among others, which have simple septal pores and lack xylose. Clade B contained species, including *Malassezia*, *Trichosporon*, and *Cryptococcus*, which have dolipore septa and xylose. Kurtzman (8) also used nucleotide divergence of the LSU and SSU rRNAs to demonstrate that genera as diverse as *Saccharomyces*, *Debaryomyces*, *Metschnikowia*, and *Galactomyces* belong to a single order, the Endomycetales (now designated Saccharomycetales).

The evolutionary relationships of species within genera have been examined by sequencing data. *Metschnikowia* species possess a unique characteristic among the ascomycetous yeasts, a large deletion in the LSU rRNA 25S-635-initiated region (9). Analysis of the sequence divergence within this genus suggests that the aquatic species *M. australis*, *M. bicuspidata*, *M. krissii*, and *M. zobellii* form a separate group from the terrestrial species *M. hawaiiensis*, *M. lunata*, *M. pulcherrima*, and *M. reukaufii*. Two of the terrestrial species—*M. hawaiiensis* and *M. lunata*—contained even larger deletions in the LSU region and were well separated from the other members of the group. The authors comment on the surprisingly large sequence divergence for a group of organisms with relative phenotypic homogeneity, and suggest that either the genus is very old or underwent rapid evolution, given the parasitic associations of this genus and the need to adapt to specific niches.

In contrast, isolates identified as *Galactomyces geotrichum* were heterogeneous, yielding inconclusive data from mating reactions with test strains belonging to *G. geotrichum* sensu stricto and *G. citri-aurantii*, and exhibited variable levels of DNA similarity (10). Examination of 57 isolates representing *G. geotrichum*, *G. citri-aurantii*, and *G. reessii* identified six groups on the basis of mol % G+C and DNA:DNA reassociation values; isolates of *G. geotrichum* formed four of these groups. The only isolates recovered from human sources were all placed in *G. geotrichum* group A along with many other environmental isolates. The authors suggest that the reasons for such heterogeneity in this species could be that the groups are at an early stage of species differentiation or that sexual reproduction was induced by the organisms themselves.

Phylogenetic studies of some groups of organisms demonstrate relationships not apparent from traditional classification. Ando et al. (11) used partial sequences of 18S and 26S rRNA to examine phylogenetic relationships of species of the genus *Kluyveromyces*, and compared the data with a strain of *Saccharomyces cerevisiae*. The genus was heterogeneous, with several species closely related to *S. cerevisiae*. Phenotypically *Saccharomyces* and *Kluyveromyces* are similar in some aspects (e.g., the major ubiquinone is Q-6, and nitrate is not used as the only source of nitrogen), but differ in both the number of ascospores per ascus and the morphology of the ascopores. James et al. (12) examined the 18S rRNA gene sequence of species of *Saccharomyces*, *Kluyveromyces*, and *Zygosaccharomyces*, and demonstrated *Saccharomyces* to be heterogeneous with some species of different genera more closely associated with each other than with members of their own genus. Some close associations were confirmed by sequence analysis, such as for the four species of the *Saccharomyces* sensu stricto complex (*S. bayanus*, *S. cerevisiae*, *S. paradoxus*, and *S. pastorianus*) and *S. exiguus* with its anamorph *Candida holmii*, and these groupings also shared close phenotype description. Phylogeny could therefore be useful in determining the phenotype characteristics that have greatest importance in taxonomic classification.

Another example where the comparison of phylogenetic and phenotypic classification needed to be carefully considered concerned the differentiation of *Issatchenkia* and *Pichia* species.

Comparison of the partial sequences of 18S and 26S rRNA of 10 *Issatchenkia* species indicated *Issatchenkia orientalis* to be more closely related to *Pichia membranaefaciens* than to *P. anomola* or the other *Issatchenkia* species, and the genus was described as phylogenetically divergent (13). *Pichia* species were demonstrated to be heterogeneous when a phylogenetic tree of 204 ascomycetous yeasts had 20 species of *Pichia* dispersed throughout (14). An important criterion of yeast classification is morphology of ascospores. *Pichia* species produce smooth hat-shaped ascospores, whereas *Issatchenkia* produce roughened, round ascospores. The conflict between morphology and sequencing analysis indicates that the criteria defining these two genera should be re-examined.

The evolutionary relationships in SSU rRNA sequences between pathogens of the genus *Candida* and related species were examined, and organisms were found to cluster according to the existence of a known teleomorph and disease-causing capability (15). *C. albicans*, *C. tropicalis*, *C. parapsilosis*, and *C. viswanathii* form a closely related subgroup, all of which can cause disease in humans and have no determined teleomorph species. The next branches on the phylogenetic tree contained *C. guilliermondii* and then *C. lusitaniae*, both disease-causing agents but with known teleomorphs, *Pichia guilliermondii* and *Clavispora lusi-*

taniae, respectively. The evolutionary distance then increased in order with *Candida glabrata, Hansenula polymorpha, Candida kefyr, Kluyveromyces marxianus* var. *lactis, Saccharomyces cerevisiae, Candida krusei*, and distantly *Yarrowia lipolytica*. Of these latter species only *C. glabrata* does not have an identified teleomorph and is closely associated with *S. cerevisiae*. Similar groupings of these species were produced in the survey of nucleotide divergence in the LSU rDNA gene of 204 ascomycetous yeasts (14). One clade contained *Lodderomyces elongisporus, Candida parapsilosis, C. tropicalis, C. maltosa, C. viswanathii* (and its synonym *C. lodderae), C. albicans*, and *C. dubliniensis. C. glabrata* was again more closely associated with *S. cerevisiae*. Pathogenic species of yeasts were not confined to a particular clade, however, as *C. guilliermondii, C. zeylanoides*, and *Clavispora lusitaniae* were placed in other distinct branches of the phylogenetic tree.

B. Species Identification and Relationships Demonstrated by Other Molecular Methods

Yeasts that are germ-tube- and chlamydoconidium-positive have been studied using a wide spectrum of molecular methods to determine if the yeasts should be considered. *C. albicans, C. stellatoidea*, or *C. dubliniensis*. The karyotypes and genetic organization of *C. albicans* and *C. stellatoidea* type I were shown to be different, although *C. stellatoidea* type II was indistinguishable from *C. albicans*. The results agreed with the description type II as a sucrose-negative variant of *C. albicans* (16, 17). The literature in the early 1990s, however, contained many references to atypical *C. albicans*, yeasts that differed in their assimilation profiles and produced distinctive patterns with the *C. albicans*-specific probe 27A (18–20).

In 1995 Sullivan et al. (21) proposed these atypical *C. albicans* yeasts as a separate species, *C. dubliniensis*, on the basis of genetic comparisons. Isolates of *C. albicans* and *C. stellatoidea* type II were indistinguishable from each other by fingerprinting with oligonucleotide probes, RAPD patterns, karyotype, and sequence of the V3 LSU rRNA region; *C. stellatoidea* type I could be distinguished only by karyotype. Isolates of *C. dubliniensis*, however, were distinguishable by these methods, but most importantly by the sequencing data. The *C. dubliniensis* strains differed from *C. albicans* by 14 nucleotides and from *C. stellatoidea* by 13 positions. *C. albicans* and *C. stellatoidea* differed only at one position, however. This information suggests that *C. albicans* and *C. stellatoidea* should not be in separate species and that *C. dubliniensis* should be considered separate with a sequence divergence of more than 2%. Investigations examining different loci have reached similar conclusions. Sequencing the *ACT1* gene demonstrated *C. dubliniensis* to be a unique taxon in *Candida* (22), while amplification of ITS regions, or the V3 region of the 23S rRNA, followed by restriction enzyme analysis (REA) showed *C. albicans* and *C. stellatoidea* to be indistinguishable and *C. dubliniensis* to be different (23).

Similar techniques have been used to assess relationships between other yeast species. Karyotyping by pulsed-field gel electrophoresis and hybridization to restriction enzyme digests with species-specific probes supported the finding that *C. tropicalis* was conspecific with *C. paratropicalis* and that *C. krusei* was conspecific with *Issatchenkia orientalis* (24).

The results of nucleotide sequences do not, however, always agree with the results from alternative methods. DNA base composition and sequence similarity of strains of *Candida utilis* and *Hansenula jadinii* showed these two species to be indistinguishable (25). The electrophoretic karyotypes of 13 *C. utilis* isolates and one *H. jadinii* isolate, however, demonstrated a large degree of chromosome length polymorphism, although the mtDNA restriction digest patterns were all similar (26). The electrophoretic karyotype is known to be variable for some species, such as *C. albicans* and *C. glabrata*, where it has been used as an epidemiological tool (27, 28). For other organisms the karyotype appears to be remarkably consistent within species. Different karyotypes were shown to exist among the *Malassezia* species, and some corresponded to the different cell types of *Malassezia furfur* (28, 29). Currently the genus *Malassezia* is divided into seven species, based on differences in phenotype and genotype. Each species has a characteristic karyotype, and is molecularly defined by differences in the DNA base composition, reassociation values, and LSU rRNA sequence data (30–32). Clearly taxonomic deductions based on karyotyping are highly dependent on the organism and the frequency of genetic reorganization characteristic for isolates of that species.

A more rapid method of species identification involves PCR amplification of sections of the rRNA genes followed by restriction enzyme digestion of the PCR products to give characteristic electrophoretic patterns. The yeasts *C. albicans*, *C. tropicalis*, *C. krusei*, *C. kefyr*, *C. lusitaniae*, *C. guilliermondii*, *C. glabrata*, and *S. cerevisiae* were identified by digestion of the 18S rDNA gene with six restriction enzymes. The method failed to distinguish between *C. parapsilosis* and *C. viswanathii* and between *Cryptococcus neoformans* and *Trichosporon beigelii* (33), however. Isolates of *C. krusei*, *C. inconspicua*, and *C. norvegensis*, all species that are fluconazole-resistant and occasionally fail to be distinguished by assimilation tests incorporated in the API 32C strips, were distinguished by amplification of the internal transcribed spacer (ITS) region and digestion with the enzyme *Hha*I (34). Species of the *Saccharomyces* sensu stricto complex analyzed by PCR–REA differentiated *S. cerevisiae* and *S. paradoxus* from each other and from *S. bayanus* and *S. pastorianus*, which were indistinguishable (35). Messner and Prillinger (36) used digests of the 18S rDNA with internal transcribed spacer regions and the 25S rDNA to identify 10 type strains of recognized *Saccharomyces* species. The results were reproducible and found to be in accordance with sequencing analyses of the same locus. The reproducibility of the method makes it suitable for isolate identification once a database of restriction patterns of reference stains is obtained. This contrasts with the results generated by random amplification of polymorphic DNA (RAPD), where the DNA fingerprint can be affected by many factors.

RAPD is increasingly being applied to species identification. *Candida paratropicalis* was shown to be a sucrose-negative variant of *C. tropicalis*, as isolates of both species produced nearly identical patterns with a panel of 10-mer primers (37). *Candida guilliermondii* and *C. fermentati* are very similar yeasts, but can be distinguished phenotypically by the ability of the latter to ferment galactose at 30°C after 21 days. Rapid identification of these species by RAPD demonstrated that *C. fermentati* was not unusual among clinical specimens (38). Isolates from culture collections representing nine species were compared by combining the patterns generated by using five 10-mer primers (39). The results demonstrated *C. intermedia* and *Issatchenkia orientalis* to be homogeneous species, while *C. catenulata, Debaryomyces hansenii, C. sake, C. rugosa,* and *Arxiozyma telluris* were heterogeneous. RAPD patterns also distinguished between the three subgroups of *C. parapsilosis. Arxiozyma telluris* was the assigned teleomorph species of *C. pintolopesii*; however, the RAPD pattern did not resemble either of the anamorph varieties of *C. pintolopesii*. The authors caution that micro-organisms assigned to the same species may not necessarily be closely related.

Doubt was cast on the validity of the holomorph pairing of *C. pintolopesii* and *A. telluris* by Meyer et al. (40), who found 16 other holomorph pairs to produce highly similar patterns following amplification with the minisatellite M13 sequence. Biochemical and physiological tests revealed that there were marked differences between *C. (Torulopsis) pintolopesii* and *A. (Saccharomyces) telluris.* Although they shared similar mol % G+C ratios of 31.8–34.9% *C. pintolopesii* was respiration-deficient while *A. telluris* was respiration-competent, and they differed in their fatty acid and cytochrome composition (41). It is therefore possible that these yeasts are members of the same taxon but are not holomorphic pairs.

Others, examining a wider range of species, have found RAPD to be a suitable tool for recognizing yeast species (42, 43), and that isolates from anamorph–teleomorph pairs gave almost identical patterns (42). Molecular methods are constantly demonstrating the existence of subgroups within species, however. *Candida parapsilosis* has been shown to contain three distinct subgroups by mol % G+C content, DNA:DNA reassociation, and RFLP patterns, although group II isolates were not encountered among a panel of clinical strains (44). *Candida haemulonii* has been shown to contain two distinct groups when examined by isoenzyme analysis and DNA:DNA reassociation, which is in some agreement with physiological tests (45). For RAPD to be used as a tool for species identification of medically important yeasts, a suitable panel of reference strains would therefore be required along with standardized primers and PCR conditions.

C. Molecular Techniques for Epidemiology

The epidemiology of infection requires the gathering of large amounts of circumstantial evidence in order to demonstrate a most likely scenario. For example, following increased recovery of *C. glabrata* from patients in an intensive care unit over a 10-week period, the information on the arrival date, location of the patients in the unit, and dates of isolation of the yeast all suggested that person-to-person transmission was likely. Molecular typing of the isolates, however, demonstrated that five of the seven patients affected had distinguishable strains and that the rise was unlikely to have been due to cross-infection (46). Only the direct comparison of the yeasts in question could show that infection control procedures had not broken down.

It is important to use the most appropriate method for each organism under investigation (Table 3). Karyotyping by pulsed field gel electrophoresis (PFGE) has been successful when used for *C. glabrata*, but was less discriminatory for *C. albicans*, *C. parapsilosis*, and *C. lusitaniae*. Restriction enzyme analysis has been moderately successful in distinguishing among isolates of *C. albicans*, *C. tropicalis*, and *S. cerevisiae*, but its sensitivity and interpretation is subject to the enzyme used. Another contributing factor is the frequency of occurrence of types; certain types may be more commonly found than others. Clemons et al. (27) compared large numbers of *C. albicans* by PFGE and REA. Restriction enzyme analysis detected 71 types from 112 geographically distinct isolates, of which 32 isolates were assigned to one group IA2, while PFGE detected 18 types and placed 57 isolates into a single type. Similarly, a study using PFGE (CHEF) to examine the epidemiology of oral candidiasis in HIV-infected patients found one type of *C. albicans* was recovered at some stage from 49 of the 66 patients who were studied (47). Ideally, therefore, investigations using molecular typing for any yeast species should include a suitable panel of epidemiologically distinct isolates and reference strains to establish the discriminatory ability of the method for that organism. This is particularly important in comparisons of small numbers of isolates. For example, an investigation of an outbreak of sternal wound infections due to *C. tropicalis* involved six patients and one nurse. Restriction enzyme analysis showed them all to be of the same type but distinct from nine epidemiologically unrelated isolates that were each distinguishable (48). For analyses by REA appropriate restriction enzymes should be selected to maximize the discrimination between isolates. The restriction enzyme *Hin*fI has been demonstrated to provide reproducible and discriminatory results for species identification and to be suitable for strain typing of some species but not all (49). Eleven isolates of *C. dubliniensis* and 49 isolates of *C. parapsilosis* produced only two types,

Table 3 Examples of Epidemiologic Studies and Results in Recent Nosocomial
Outbreaks of Yeast Infections

Organism	Reference	Number of isolates/number of subjects	Methods	Number types
C. albicans	27	112 isolates/112 subjects	REA	71
			RAPD	58
			PFGE	18
	47	66 patients	PFGE	34
	52	32 isolates/32 patients	RAPD	22
	136	36 isolates/9 patients	PCR	7/36, 8/8
		8 isolates/8 reference cultures	27A probe	6/36, 8/8
	137	14 isolates/14 patients	PFGE	8
			PFGE-REA	8
C. glabrata	138	23 isolates/20 patients	PFGE	17
			RAPD	8
	51	22 isolates/21 patients	PFGE	22
			mtDNA-REA	3
			RAPD	6 or 9
C. krusei	56	7 isolates/7 patients	REA	2
		3 isolates/3 reference cultures		3
	139	131 isolates/95 subjects	PCR	95
C. parapsilosis	140	60 isolates/54 subjects	PFGE-REA	8
			PFGE	26
	137	15 isolate/15 patients	PFGE	10
			PFGE-REA	5 or 8
	141	21 isolates/21 patients	PFGE	11
C. tropicalis	48	8 isolates/6 patients + 1 nurse	REA	1
		9 isolates/9 control subjects		9
C. lusitaniae	142	47 isolates/33 patients	PFGE-REA	25
			PFGE	28
	55	29 isolates/7 patients + 5 controls	REA	8
C. inconspicua	143	5 isolates/3 patients + 1 reference	REA	2
			RAPD	2
S. cerevisiae	144	60 isolates/49 clinical + 11 other	REA	41
	145	15 isolates	RAPD	6
			PCR-REA	4

whereas 48 types were detected among 111 isolates of *C. albicans.* When REA
with one restriction enzyme fails to give satisfactory discrimination of isolates,
using a second enzyme can improve results. PFGE-REA with *Bss*HII demonstrated isolates of *C. albicans* to be identical, but they could be distinguished by
digestion with *Sfi*I (50).

The ideal typing technique would be both fast and sensitive; however, REA
requires at least 2 days and PFGE can take at least 4 days if time for extraction
and electrophoresis is included. RAPD is a rapid and sensitive method, but suc-

cess depends heavily on the choice of primer and the organism being tested. Greater discrimination was achieved for *C. glabrata* by PFGE than by RAPD, but the reverse appeared to apply to studies of *C. albicans* (Table 3). Twenty-two isolates of *C. glabrata* were divided into either six or nine types, depending on the primer used (51). More usually several primers are screened for their discriminatory ability, and the primer yielding the largest number of types is selected for use. Indeed, Robert et al. (52) screened 12 primers before choosing one to compare 32 isolates of *C. albicans*.

Once an appropriate technique has been selected and the conditions optimized many different areas of microbial ecology and disease can be examined. The possibility of a patient being infected at different sites with different strains of the same yeast species was demonstrated in an AIDS patient with meningitis and oral candidiasis caused by distinct strains of fluconazole-resistant *C. albicans* (53). The dynamics of antifungal resistance within populations of *C. albicans* causing recurrent oropharyngeal candidiasis has been examined and the development of several different mechanisms of resistance identified (54). Simultaneous oral carriage in HIV-positive and -negative patients of more than one type of *C. albicans* and with more than one species has been shown (18). Nosocomial transmission of yeasts such as *C. lusitaniae* (55) and *C. krusei* (56) within intensive care settings has been demonstrated. Typing of *Cryptococcus neoformans* revealed geographical groupings and also that patients acquired their infecting strain from different locations and that the infection had remained dormant for many months or years before diagnosis (57).

D. Probes that Demonstrate Microevolution

One of the first species-specific DNA probes used for epidemiological investigations was the *C. albicans*-specific probe 27A (58). This probe distinguished between isolates from different patients and between isolates from different anatomical locations of individual patients. Minor differences, however, in band pattern were observed in a proportion of colonies from four strains following unselected laboratory passage on laboratory media and in the spontaneously laboratory-produced 5-fluorocytosine mutants of one strain. These results demonstrated that genetic reorganization occurred and was detectable. Another *C. albicans*-specific probe, Ca3, also used to assess strain relatedness, was found to provide stable patterns for three strains tested through 400 generations (59). Lockhart et al. (60), however, used the Ca3 probe to compare multiple isolates from individual patients and found that the patterns within each patient population could vary by up to three bands. In addition, another probe derived from Ca3, the C1 fragment, demonstrated single band differences in populations shown to be identical with Ca3. Such results suggest the process of microevolution within the ecology of an individual patient's flora. Recently, similar studies have been conducted on other *Candida* species. Two probes, Cg6 and Cg12, have been used to demonstrate minor genetic changes in isolates of *C. glabrata* recovered from four vaginitis patients over a 50-month period (61). Similarly, isolates of *C. tropicalis* were compared by means of the Ct3 probe (62). This probe produced stable patterns for three strains over 600 generations, but could distinguish minor differences in the hybridization patterns of sequential isolates of some patients and not others. Neither of these studies involved multiple isolates recovered at each sampling time to reveal the ecology at the site of isolation or the proportion of genotypes present. Sequential minor changes in highly related hybridization patterns, however, were demonstrated for isolates recovered from some patients but not all, indicating that microevolution also occurs with *C. glabrata* and *C. tropicalis*.

E. Molecular Methods in the Clinical Laboratory

Routine phenotypic and physiological identification of yeasts works well for the majority of clinical specimens. As the numbers of immune-compromised and immune-suppressed patients and the use of antifungals increase, however, so does the array of yeast species with which patients can become colonized or infected. Commercial yeast identification kits are commonplace in clinical laboratories, and the efficiency of these systems is periodically evaluated. In 1994 the Vitek Yeast Biochemical Card was reported to have failed to identify 42% (10/24) of *C. krusei* isolates, four of five *C. lambica* isolates, and seven of eight *Trichosporon beigelii* isolates (63). A new API Candida system was evaluated in 1996 and failed to identify 23% (4/17) *C. famata*, or any isolates of *C. sphaerica*, *C. sake*, *C. rugosa*, *C. pelliculosa*, *C. lipolytica*, and *C. intermedia* that were not included in the database (64). More recently the identification of 19 species of yeasts by seven commercial systems varied from 59.6–80.8%, and all failed to identify *C. norvegensis*, *C. catenula*, *C. haemulonii*, and *C. dubliniensis* (65). *Candida sake* was a common misidentification by the ID 32C system of the recently described species *C. dubliniensis* (20); however the accuracy of identification of this species varies between commercial systems (66). Other species that may be misidentified with biochemical methods are *C. inconspicua*, *C. krusei*, and *C. norvegensis*. Molecular methods have been developed to aid in the identification of some of these species. Karyotyping by PFGE (67) and RAPD (4) differentiated isolates of *C. krusei* and *C. inconspicua*, and PCR-REA differentiated *C. krusei*, *C. inconspicua*, and *C. norvegensis* (34).

There is a variety of molecular techniques that can be used in the clinical laboratory to aid in species identification. PCR-REA (68, 69) and RAPD (42, 70) provide characteristic fingerprints of organisms, while PCR reactions with fungus-specific primers have been used for yeast identification (71) and to detect yeasts in clinical fluids (72). Requiring more time and dedicated staff is the use of species-specific probes for identification (73), although PCR combined with an enzyme immunoassay using species-specific probes can reduce the overall identification time (74). The application of probes to clinical specimens could decrease the identification time of yeast species isolated from candidemias (75) and has potential for use with in situ hybridization to histology sections (76). The development of automated sequencing and array-based hybridization schemes could lead to rapid identification of many yeast species as sequence databases are constructed and made available (77).

VI. OCCURRENCE OF THE SPECIES AS PATHOGENS

The endomycetous and blastomycetous yeasts represent a diverse collection of organisms that cause an equally diverse range of disease manifestations. The most prominent genus is *Candida*, with at least 19 species capable of causing infection in humans. As described in earlier sections of this chapter, however, some of these species have teleomorphs that are in the endomycetous yeasts. In this section, the teleomorphs of the holomorph will be described individually, but the anamorphs, depending on the genus, will be considered as a group.

A. *Arxiozyma*

Arxiozyma telluris (anamorph, *Candida pintolopesii*, previously called *Torulopsis pintolopesii*) is a rare agent of fungal infection. It has been isolated from the pleura and lungs of a single patient and has been associated with oral leukoplakia from one individual (78, 79).

B. *Blastoschizomyces*

B. capitatus (formerly *Trichosporon capitatum*) has been reported to cause systemic infection, including endocarditis (80), onychomycosis (81), osteomyelitis and discitis (82–84), urinary tract infections (85), possibly hepatitis (86), and septicemia (86). The organism, which is widely distributed in nature, may be considered an emerging cause of invasive disease in leukemia patients (87). It has also been implicated in a fatal case of fungemia following infusion of organism-laden fluids (88).

C. *Candida*

Candida species infect a wide range of tissues and are, with the possible exception of dermatophytes, the most common cause of fungal infections in humans. The predominant diseases include vaginitis, oral thrush, and fungemia. Onychomycosis caused by *Candida* species has been reported, but definitive evidence that the species are the primary agents of the disease rather than secondary agents or colonizers following dermatophytosis is lacking. *Candida* species are a common cause of infection in immunocompromised individuals, including patients with hematogenous disorders and transplant recipients. Some species (e.g., *C. albicans*, *C. tropicalis*, *C. glabrata*) are members of the normal microbiota of the oral cavity, gastrointestinal tract, and vagina. Other species are normally found in the environment.

   Candida species represent the fourth leading cause of nosocomial bloodstream infection (89, 90). Recent reports indicate that long-term azole therapy (primarily fluconazole) may lead to selection in patients of fluconazole-resistant pathogenic yeasts, such as *C. glabrata*, *C. krusei*, *C. inconspicua*, and *C. dubliniensis* (91, 92).

D. *Clavispora*

The two medically important species are *Cl. capitatus* and *Cl. lusitaniae*. The former is the teleomorph of *Blastoschizomyces capitatus* and is discussed in Sec. VI.B. *Cl. lusitaniae*, the teleomorph of *Candida lusitaniae*, is commonly identified as the anamorph in the clinical laboratory. *Candida lusitaniae* has been isolated from cases of fungemia, meningitis, systemic disease, and urinary tract infections (55, 93–95). A distinctive feature of the organism is its resistance to the polyene amphotericin B. (For example, see Ref. 94.) Such resistance may not be evident on initial testing in vitro, but may develop when the isolate is exposed to the drug for extended periods.

E. *Citeromyces*

Citeromyces matritensis (anamorph, *Candida globosa*) has not been shown to cause human disease, but it may be occasionally isolated from clinical specimens.

F. *Cryptococcus*

Combined with *Candida* species, *Cryptococcus* species represent >90% of all yeast infections in humans. The primary disease caused by *Cr. neoformans* is meningitis, which was a relatively infrequent disease prior to the AIDS epidemic. Cryptococcal meningitis occurs in individuals who are immunocompromised. Four serotypes, A, B, C, and D, have been reported as pathogens, although additional serotypes may be involved (combinations of the serotypes). Serotype A (*Cr. neoformans* var. grubii) and D (*Cr. neoformans* var. *neoformans*) are the most common serotypes recovered from patients. Along with meningitis, *Cr. neoformans* can cause a variety of disease entities, most notably pulmonary and cutaneous infections (96).

G. *Debaryomyces*

This genus is represented by *D. hansenii*, the teleomorph of *Candida famata* (previously known as *Torulopsis candida*). *D. hansenii* has been isolated from fresh droppings and cloacal samples of feral pigeons (97). In the clinical laboratory, the organism is typically reported as *C. famata*. This species has been associated with fungemia and endophthalmitis (98, 99). It is an infrequent cause of both diseases. A recent study has revealed that several strains of *Torulaspora delbrueckii*, a species with industrial applications, are actually *Debaryomyces* spp., including one which may be *D. hansenii* (100).

H. *Galactomyces*

Galactomyces geotrichum is the teleomorph of *Geotrichum candidum*, which belongs to the order moniliales in the hyphomycetes. The anamorph does not produce blastoconidia and therefore does not belong in the blastomycetes. *Geotrichum candidum* has been reported to cause oral disease (101) and disseminated disease (102). It is possible that true disease caused by this organism is rare; it may have been reported as geotrichosis caused by *Blastoschizomyces capitatus*, the recently revised genus-species of *Geotrichum capitatum*.

I. *Hansenula*

Hansenula anomala (anamorph *Candida pelliculosa*) is an emerging pathogen. It has been associated with cases of fungemia, urinary tract infection, and endocarditis (103–106). Invasive disease caused by this organism other than those already listed has also been suggested (107).

J. *Issatchenkia*

The medically important species *I. orientalis* (anamorph *Candida krusei*) has been associated with cases of fungemia (108) and possibly sinusitis and pneumonia (109). This organism is typically obtained from patients who are undergoing treatment with fluconazole and from whom a fluconazole-resistant yeast is isolated.

K. *Kluyveromyces*

Kluyveromyces marxianus (anamorph *Candida kefyr*; formerly *Candida pseudo-tropicalis*) is a rare agent of fungemia and invasive disease in immunocompromised patients (110).

L. *Malassezia*

The most common clinically important species in this genus are *M. furfur*, *M. pachydermatis*, and *M. sympodialis*. Another four species have been isolated from human specimens (32), however. Some of the species were previously considered members of the genus *Pityrosporum*. The disease pityriasis versicolor is a common manifestation of superficial disease caused by *Malassezia* species. In addition, the species have been reported to cause folliculitis, fungemia (associated with hyperalimentation fluids), dermatitis, and lung invasion (111–113).

M. *Metschnikowia*

Metschnikowia pulcherrima (anamorph, *Candida pulcherrima*) has been implicated in onychomycosis and has been recovered in other clinical specimens, such as sputum and cutaneous tissues (114). The overall significance of this organism in disease is uncertain.

N. *Pichia*

This genus contains several medically important yeasts with anamorphs in the genus *Candida*. Diseases caused by these species are manifold. The most frequent species in clinical material are *P. guilliermondii* (anamorph, *Candida guilliermondii*) and *P. norvegensis* (anamorph, *Candida norvegensis*). Both species have been reported to cause fungemia and invasive disease (115, 116).

O. *Rhodotorula*

The red yeasts *Rhodotorula* have basidiomycetous affinities. These organisms have been reported as agents of fungemia, primarily associated with central venous catheters (117, 118). They have also been implicated in at least one case each of peritonitis, meningitis, and extrinsic allergic alveolitis (95, 119, 120). These organisms are found in the environment, particularly in water sources. The primary species involved in human disease are *R. rubra* (synonymous with *R. mucilaginosa*) and *R. glutinis*.

P. Saccharomyces

Saccharomyces cerevisiae is the primary medically important agent in this genus. This organism has been associated with cases of vaginitis, oral disease, fungemia, and empyema (121–123). The organism's habitat is the environment, but it is found in a number of basic foods (bread). When ingested, it can transiently colonize the gastrointestinal tract, hence its isolation from GI-related sites may not imply clinical significance.

Q. Sporobolomyces

This form genus, which has a teleomorph in the basidiomycetes, is an uncommon agent of human disease. Several infectious manifestations have been reported, including lymphadenitis, nasal polyps, fungemia, and dermatitis (124–126). Sporobolomycosis is an opportunistic disease of immunocompromised individuals. At least three species have been isolated from clinical materials and include *S. salmonicolor, S. holsaticus,* and *S. roseus.*

R. Stephanoascus

The anamorph of *S. ciferrii, Candida ciferrii,* has been reported as a possible agent of onychomycosis in elderly patients (127, 128) and possibly otomycosis (128). The clinical significance of the organism is doubtful (128), however.

S. Trichosporon

Trichosporon beigelii was once considered the only pathogenic species in this genus, but recent revisions to the genus suggest that six species may be pathogens. The epithet *T. beigelii* is considered by some authorities as invalid, as this species actually contained multiple variants that have been reassigned to new species (129, 130). The five species have been implicated as agents of superficial disease (e.g., white piedra) and two species, *T. mucoides* and *T. asahii,* have been suggested to cause disseminated infections in leukemia patients. The organisms are resistant to amphotericin B, making therapeutic management difficult (131, 132). "*T. beigelii*" has been isolated from cases of fungemia (131–133). *Trichosporon cutaneum* has been shown to colonize and subsequently cause fungemia in neonates (134).

T. Yarrowia

Yarrowia lipolytica (anamorph, *Candida lipolytica*) has been isolated from a case of fungemia and a case of sinusitis (135). Due to the difficulty of identifying this organism using standard clinical tests, misidentifications may have occurred and this organism may be more common than suspected.

APPENDIX: DIFFERENTIAL CHARACTERISTICS OF THE GENERA

Arxiozyma

Sexual spores and morphologic features: Asci with 1–2, spheroidal to ellipsoidal, verrucose to tuberculate, ascospores. Yeasts diploid, globose to ellipsoidal.

Biochemical characteristics: Urease −, CHX S, nitrate −, inositol assim. −, DBB −, fermentation +.

Medically significant species: A. telluris.

Blastoschizomyces

Sexual spores and morphologic features: Asexual genus. Produces arthroconidia and annelloconidia on tip of percurrently proliferating conidiogenous cell; blastoconidia present in young cultures.

Biochemical characteristics: Urease −, CHX R, nitrate −, inositol assim. −, DBB −, fermentation −.

Medically significant species: B. capitatus.

Candida

Sexual spores and morphologic features: Asexual genus. Budding cells, pseudo- and true-septate mycelium possible.

Biochemical characteristics: Urease −, CHX R/S, nitrate ±, inositol assim. −, DBB −, fermentation +.

Predominant medically significant species: C. albicans (including C. dubliniensis), C. ciferrii, C. famata, C. guilliermondii, C. glabrata, C. haemulonii, C. kefyr, C. krusei, C. lipolytica, C. lusitaniae, C. norvegensis, C. parapsilosis, C. tropicalis, C. utilis, C. viswanathii, C. zeylanoides.

Clavispora

Sexual spores and morphologic features: Asci with 1–4 conical or clavate ascospores; budding cells, pseudomycelium possible.

Source: Refs. 1, 3, 129, 146–148.

Biochemical characteristics: Urease −, CHX ±, nitrate −, inositol assim.
−, DBB −, fermentation +.

Medically significant species: C. lusitaniae, C. capitatus.

Citeromyces

Sexual spores and morphologic features: Asci with one (rarely two) round, warty ascospore; spheroidal to ellipsoidal budding cells, no hyphae.

Biochemical characteristics: Urease −, CHX?, nitrate +, inositol assim.
−, DBB −, fermentation +.

Medically significant species (isolated as contaminant, not pathogen): C. matritensis.

Cryptococcus

Sexual spores and morphologic features: Asexual genus. Budding cells. Encapsulated. Occasional pseudo- and true hyphae.

Biochemical characteristics: Urease +, CHX S (C. laurentii occasionally R), nitrate ±, inositol + (occasionally−), DBB +, fermentation −.

Medically significant species: C. neoformans.

Debaryomyces

Sexual spores and morphologic features: Asci with 1–4 round or oval, warty ascospores. Conjugation occurs between cell and bud.

Biochemical characteristics: Urease −, CHX S/R, nitrate −, inositol −, DBB −, fermentation ±.

Medically significant species: D. hansenii.

Galactomyces

Sexual spores and morphologic features: Spherical asci with 1 to 2 (rare) ellipsoidal, echinate ascospores; ascospores may contain equatorial furrow. Produces arthoconidia.

Biochemical characteristics: Urease −, CHX?, nitrate −, inositol −, DBB −, fermentation ±.

Medically significant species: G. geotrichum.

Hansenula

Sexual spores and morphologic features: Asci with 1 to 4 hat-shaped ascospores; budding cells, pseudo- or true hyphae possible.

Biochemical characteristics: Urease −, CHX S/R, nitrate +, inositol −, DBB −, fermentation ±.

Medically significant species: *H. anomala.*

Issatchenkia

Sexual spores and morphologic features: Asci with 1 to 4 roughened, round ascospores; budding cells, pseudohyphae possible.

Biochemical characteristics: Urease −, CHX S/R, nitrate −, inositol −, DBB −, fermentation +.

Medically significant species: *I. orientalis.*

Kluyveromyces

Sexual spores and morphologic features: Evanescent asci with 1 to typically 4, 8, or 16 ascospores (one species produces up to 60 ascospores/ascus); ascospores are oval, crescent shaped, or reniform; budding cells spheroidal to elongate.

Biochemical characteristics: Urease −, CHX S/R, nitrate −, inositol −, DBB −, fermentation +.

Medically significant species: *K. marxianus.*

Malassezia

Sexual spores and morphologic features: Asexual genus; monopolar budding on a broad base (percurrent); hyphae may be produced.

Biochemical characteristics: Urease +, CHX R, nitrate +?, inositol ?, DBB +, fermentation −.

Medically significant species: *M. furfur, M. globosa, M. obtusa, M. pachydermatis, M. restricta, M. slooffiae, M. sympodialis.*

Metschnikowia

Sexual spores and morphologic features: Club-shaped asci with one or two needle-shaped ascospores/spheroidal to ellipsoidal budding cells; rudimentary pseudomycelium usually present.

Biochemical characteristics: Urease −, CHX ?, nitrate −, inositol −, DBB −, fermentation +.

Medically significant species: *M. pulcherrima.*

Pichia

Sexual spores and morphologic features: Generally dehiscent asci with 1 to 4, smooth, hat-shaped, hemispherical, saturnine, or spheroidal ascospores.

Biochemical characteristics: Urease −, CHX S/R, nitrate −, inositol −, DBB −, fermentation ±.

Medically significant species: P. guilliermondii, P. fermentans, P. jadinii, P. membranaefaciens, P. norvegensis.

Rhodotorula

Sexual spores and morphologic features: Asexual genus; budding cells, red or yellow carotenoids produced.

Biochemical characteristics: Urease +, CHX S/R, nitrate ±, inositol −, DBB +, fermentation −.

Medically significant species: R. rubra, R. glutinis.

Saccharomyces

Sexual spores and morphologic features: Persistent asci with 1 to 4, smooth-walled, globose to short ellipsoidal ascospores; budding cells, globose to ellipsoidal; pseudohyphae may be formed.

Biochemical characteristics: Urease −, CHX S/R, nitrate −, inositol −, DBB −, fermentation +.

Medically significant species: S. cerevisiae, S. exiguus.

Sporobolomyces

Sexual spores and morphologic features: Asexual genus. Budding cells generally ellipsoidal but variable, percurrent budding. Pseudohyphae present. Ballistoconidia on sterigmata. Produces salmon-pink carotenoids.

Biochemical characteristics: Urease +, nitrate ±, inositol ±, DBB +, fermentation −.

Medically significant species: S. salmonicolor, S. holsaticus, and S. roseu.

Stephanoascus

Sexual spores and morphologic features: Asci formed after hyphal conjugation, contain 1 to 4 hat-shaped ascospores when immature, hemispherical when mature.

Biochemical characteristics: Urease −, nitrate −, inositol +, DBB −, fermentation −.
Medically significant species: *Stephanoascus ciferrii.*

Trichosporon

Sexual spores and morphologic features: Asexual genus. Budding may be absent or present. Arthroconidia produced.
Biochemical characteristics: Urease +, CHX S/R, nitrate −, inositol ±, DBB +, fermentation −.
Medically significant species: *T. asahii, T. asteroides, T. cutaneum, T. inkin, T. mucoides, T. ovoides.*

Yarrowia

Sexual spores and morphologic features: Asci with 1 to 4 round, oval, walnut-shaped, hat-shaped, or saturnoid ascospores; budding cells; pseudohyphae and hyphae usually present.
Biochemical characteristics: Urease +, nitrate −, inositol −, DBB −, fermentation −.
Medically significant species: *Y. lipolytica.*

REFERENCES

1. JA Barnett, RW Payne, D Yarrow. Yeasts: Characteristics and Identification. Cambridge: Cambridge University Press, 1983.
2. AN Hagler, DG Ahearn. Rapid diazonium blue B test to detect basidiomycetous yeasts. Internat J Syst Bact 31:204–208, 1981.
3. NJW Kreger-van Rij. The Yeasts: A Taxonomic Study. New York: Elsevier Science, 1984.
4. GG Baily, CB Moore, SM Essayag, S De Wit, JP Burnie, DW Denning. *Candida inconspicua*, a fluconazole-resistant pathogen in patients infected with human immunodeficiency virus. Clin Infec Dis 25:161–163, 1997.
5. LJ Wickerham, KA Burton. Carbon assimilation tests for the classification of yeasts. J Bacteriol 56:363–371, 1948.
6. AJ Whiffen. The production, assay, and antibiotic activity of actidione, an antibiotic from *Streptomyces griseus*. J Bacteriol 56:283–291, 1948.
7. CP Kurtzman. Molecular taxonomy of the yeasts. Yeast 10:1727–1740, 1994.
8. CP Kurtzman. Systematics of the ascomycetous yeasts assessed from ribosomal RNA sequence divergence. Antonie van Leeuwenhoek 63:165–174, 1993.

9. LC Mendonça-Hagler, AN Hagler, CP Kurtzman. Phylogeny of *Metschnikowia* species estimated from partial rRNA sequences. Internat J Syst Bact 43:368–373, 1993.

10. MT Smith, AWAM De Cock, GA Poot, HY Steensma. Genome comparisons in the yeastlike fungal genus *Galactomyces* Redhead et Malloch. Internat J Syst Bact 45:826–831, 1995.

11. S Ando, K Mikata, Y Tahara, Y Yamada. Phylogenetic relationships of species of the genus *Kluyveromyces* Van der Walt (Saccharomycetaceae) deduced from partial base sequences of 18S and 26S ribosomal RNAs. Biosci Biotech Biochem 60: 1063–1069, 1996.

12. SA James, J Cai, IN Roberts, MD Collins. A phylogenetic analysis of the genus *Saccharomyces* based on 18S rRNA gene sequences: Description of *Saccharomyces kunashiresis* sp. nov. and *Saccharomyces martiniae* sp. nov. Internat J Syst Bact 47:453–460, 1997.

13. Y Yamada, J-I Yano, T Suzuki, K Mikata. The phylogeny of species of the genus *Issatchenkia* Kudriavzev (Saccharomycetaceae) based on the partial sequences of 18S and 26S ribosomal RNAs. Biosci Biotech Biochem 61:577–582, 1997.

14. CP Kurtzman, CJ Robnett. Identification of clinically important ascomycetous yeasts based on nucleotide divergence in the 5′ end of the large-subunit (26S) ribosomal DNA gene. J Clin Microbio 35:1216–1223, 1997.

15. SM Barns, DJ Lane, ML Sogin, C Bibeau, WG Weisburg. Evolutionary relationships among pathogenic *Candida* species and relatives. J Bacteriol 173:2250–2255, 1991.

16. KJ Kwon-Chung, WS Riggsby, RA Uphoff, JB Hicks, WL Whelan, E Reiss, BB Magee, BL Wickes. Genetic differences between type I and type II *Candida stellatoidea*. Infec Immun 57:527–532, 1989.

17. EHA Rikkerink, BB Magee, PT Magee. Genomic structure of *Candida stellatoidea*: Extra chromosomes and gene duplication. Infec Immun 58:949–954, 1990.

18. RM Anthony, J Midgley, SP Sweet, SA Howell. Multiple strains of *Candida albicans* in the oral cavity of HIV positive and HIV negative patients. Microbial Ecol Hlth Dis 8:23–30, 1995.

19. MJ McCullough, BC Ross, BD Dwyer, PC Reade. Genotype and phenotype of oral *Candida albicans* from patients infected with the human immunodeficiency virus. Microbiology 140:1195–1202, 1994.

20. D Sullivan, D Bennett, M Henman, P Harwood, S Flint, F Mulcahy, D Shanley, D Coleman. Oligonucleotide fingerprinting of isolates of *Candida* species other than *C. albicans* and of atypical *Candida* species from human immunodeficiency virus-positive and AIDS patients. J Clin Microbio 31:2124–2133, 1993.

21. DJ Sullivan, TJ Westerneng, KA Haynes, DE Bennett, DC Coleman. *Candida dubliniensis* sp. nov.: Phenotypic and molecular characterization of a novel species associated with oral candidosis in HIV-infected individuals. Microbiology 141: 1507–1521, 1995.

22. SM Donnelly, DJ Sullivan, DB Shanley, DC Coleman. Phylogenetic analysis and rapid identification of *Candida dubliniensis* based on analysis of *ACT1* intron and exon sequences. Microbiology 145:1871–1882, 1999.

23. MJ McCullough, KV Clemons, DA Stevens. Molecular and phenotypic character-

ization of genotypic *Candida albicans* subgroups and comparison with *Candida dubliniensis* and *Candida stellatoidea*. J Clin Microbio 37:417–421, 1999.

24. BL Wickes, JB Hicks, WG Merz, KJ Kwon-Chung. The molecular analysis of synonymy among medically important yeasts within the genus *Candida*. J Gen Microbio 138:901–907, 1992.

25. PL Manachini. DNA sequence similarity, cell wall mannans and physiological characteristics in some strains of *Candida utilis*, *Hansenula jadinii* and *Hansenula petersonii*. Antonie van Leeuwenhoek 45:451–463, 1979.

26. R Stoltenburg, U Klinner, P Ritzerfeld, M Zimmermann, CC Emeiss. Genetic diversity of the yeast *Candida utilis*. Curr Genet 22:441–446, 1992.

27. KV Clemons, F Feroze, K Holmberg, DA Stevens. Comparative analysis of genetic variability among *Candida albicans* isolates from different geographic locales by three genotypic methods. J Clin Microbio 35:1332–1336, 1997.

28. SA Howell, C Quin, G Midgley. Karyotypes of oval cell forms of *Malassezia furfur*. Mycoses 36:263–266, 1993.

29. T Boekhout, RW Bosboom. Karyotyping of *Malassezia* yeasts: Taxonomic and epidemiological implications. Syst Appl Microbio 17:146–153, 1994.

30. RM Anthony. Molecular typing methods applied to yeasts from the genus *Malassezia*. Ph.D. thesis, University of London, London, 1996.

31. E Gueho, G Midgley, J Guillot. The genus *Malassezia* with description of four new species. Antonie van Leeuwenhoek 69:337–355, 1996.

32. J Guillot, E Guého. The diversity of *Malassezia* yeasts confirmed by rRNA sequence and nuclear DNA comparisons. Antonie van Leeuwenhoek 67:297–314, 1995.

33. M Maiwald, R Kappe, H-G Sonntag. Rapid presumptive identification of medically relevant yeasts to the species level by polymerase chain reaction and restriction enzyme analysis. J Med Vet Mycol 32:115–122, 1994.

34. S Nho, MJ Anderson, CB Moore, DW Denning. Species differentiation by internally transcribed spacer PCR and *Hha*I digestion of fluconazole-resistant *Candida krusei*, *Candida inconspicua*, and *Candida norvegensis* strains. J Clin Microbio 35:1036–1039, 1997.

35. MJ McCullough, KV Clemons, JH McCuster, DA Stevens. Intergenic transcribed spacer PCR ribotyping for differentiation of *Saccharomyces* species and interspecific hybrids. J Clin Microbio 36:1035–1038, 1998.

36. R Messner, H Prillinger. *Saccharomyces* species assignment by long range ribotyping. Antonie van Leeuwenhoek 67:363–370, 1995.

37. D Lin, PF Lehmann. Random amplified polymorphic DNA for strain delineation within *Candida tropicalis*. J Med Vet Mycol 33:241–246, 1995.

38. RM San Millan, L-C Wu, IF Salkin, PF Lehmann. Clinical isolates of *Candida guilliermondii* include *Candida fermentati*. Internat J Syst Bact 47:385–393, 1997.

39. S Zeng, L-C Wu, PF Lehmann. Random amplified polymorphic DNA analysis of culture collection strains of *Candida* species. J Med Vet Mycol 34:293–297, 1996.

40. W Meyer, GN Latouche, H-M Daniel, M Thanos, TG Mitchell, D Yarrow, G Schönian, TC Sorrell. Identification of pathogenic yeasts of the imperfect genus *Candida* by polymerase chain reaction fingerprinting. Electrophoresis 18:1548–1559, 1997.

41. K Watson, H Arthur, M Blakey. Biochemical correlations among the thermophilic enteric yeasts *Torulopsis bovina, Torulopsis pintolopesii, Saccharomyces telluris*, and *Candida slooffii*. J Bacteriol 143:693–702, 1980.

42. GN Latouche, H-M Daniel, OC Lee, TG Mitchell, TC Sorrell, W Meyer. Comparison of use of phenotypic and genotypic characteristics for identification of species of the anamorph genus *Candida* and related teleomorph yeast species. J Clin Microbiol 35:3171–3180, 1997.

43. M Thanos, G Schonian, W Meyer, C Schweynoch, Y Graser, TG Mitchell, W Presber, H-J Tietz. Rapid identification of *Candida* species by DNA fingerprinting with PCR. J Clin Microbio 34:615–621, 1996.

44. B Roy, SA Meyer. Confirmation of the distinct genotype groups within the form species *Candida parapsilosis*. J Clin Microbio 36:216–218, 1998.

45. PF Lehmann, L-C Wu, WR Pruitt, SA Meyer, DG Ahearn. Unrelatedness of groups of yeasts within the *Candida haemulonii* complex. J Clin Microbio 31:1683–1687, 1993.

46. S Arif, T Barkham, EG Power, SA Howell. Techniques for investigation of an apparent outbreak of infections with *Candida glabrata*. J Clin Microbio 34:2205–2209, 1996.

47. JA Sangeorzan, SF Bradley, X He, LT Zarins, GL Ridenour, RN Tiballi, CA Kauffman. Epidemiology of oral candidiasis in HIV-infected patients: Colonization, infection, treatment, and emergence of fluconazole resistance. Amer J Med 97:339–346, 1994.

48. BN Doebbeling, RJ Hollis, HD Isenberg, RP Wenzel, MA Pfaller. Restriction fragment analysis of a *Candida tropicalis* outbreak of sternal wound infections. J Clin Microbio 29:1268–1270, 1991.

49. S-I Fujita, T Hashimoto. DNA fingerprinting patterns of *Candida* species using *Hin*fI endonuclease. Internat J Syst Evol Microbio 50:1381–1389, 2000.

50. A Voss, MA Pfaller, RJ Hollis, J Rhine-Chalberg, BN Doebbeling. Investigation of *Candida albicans* transmission in a surgical intensive care unit cluster by using genomic DNA typing methods. J Clin Microbio 33:576–580, 1995.

51. A Defontaine, M Coarer, JP Bouchara. Contribution of various techniques of molecular analysis to strain identification of *Candida glabrata*. Microbial Ecol Hlth Dis 9:27–33, 1996.

52. F Robert, F Lebreton, ME Bougnoux, A Paugam, D Wassermann, M Schlotterer, C Tourte-Schaefer, J Dupouy-Camet. Use of random amplified polymorphic DNA as a typing method for *Candida albicans* in epidemiological surveillance of a burn unit. J Clin Microbio 33:2366–2371, 1995.

53. J Berenguer, TM Diaz-Guerra, B Ruiz-Diez, JCLB De Quiros, JL Rodriguez-Tudela, JV Martinez-Suarez. Genetic dissimilarity of two fluconazole-resistant *Candida albicans* strains causing meningitis and oral candidiasis in the same AIDS patient. J Clin Microbio 34:1542–1545, 1996.

54. JL Lopez-Ribot, RK McAtee, S Perea, WR Kirkpatrick, MG Rinaldi, TF Patterson. Multiple resistant phenotypes of *Candida albicans* coexist during episodes of oropharyngeal candidiasis in human immunodeficiency virus-infected patients. Antimicrob Agents Chemo 43:1621–1630, 1999.

55. V Sanchez, JA Vazquez, D Barth-Jones, L Dembry, JD Sobel, MJ Zervos. Epidemiology of nosocomial acquisition of *Candida lusitaniae*. J Clin Microbio 30:3005–3008, 1992.

56. GA Noskin, J Lee, DM Hacek, M Postelnick, BE Reisberg, V Stosor, SA Weitzman, LR Peterson. Molecular typing for investigating an outbreak of *Candida krusei*. Diag Microbio Infec Dis 26:117–123, 1996.

57. D Garcia-Hermoso, G Janbon, F Dromer. Epidemiological evidence for dormant *Cryptococcus neoformans* infection. J Clin Microbio 37:3204–3209, 1999.

58. S Scherer, DA Stevens. A *Candida albicans* dispersed, repeated gene family and its epidemiologic applications. Proc Natl Acad Sci 85:1452–1456, 1998.

59. J Schmid, E Voss, DR Soll. Computer-assisted methods for assessing strain relatedness in *Candida albicans* by fingerprinting with the moderately repetitive sequence Ca3. J Clin Microbio 28:1236–1243, 1990.

60. SR Lockhart, JJ Fritch, AS Meier, K Schröppel, T Srikantha, R Galask, DR Soll. Colonizing populations of *Candida albicans* are clonal in origin but undergo microevolution through C1 fragment reorganization as demonstrated by DNA fingerprinting and C1 sequencing. J Clin Microbio 33:1501–1509, 1995.

61. SR Lockhart, S Joly, C Pujol, JD Sobel, MA Pfaller, DR Soll. Development and verification of fingerprinting probes for *Candida glabrata*. Microbiology 143:3733–3746, 1997.

62. S Joly, C Pujol, K Schröppel, DR Soll. Development of two species-specific fingerprinting probes for broad computer-assisted epidemiological studies of *Candida tropicalis*. J Clin Microbio 34:3063–3071, 1996.

63. DP Dooley, ML Beckius, BS Jeffrey. Misidentification of clinical yeast isolates by using the updated Vitek Yeast Biochemical Card. J Clin Microbio 32:2889–2892, 1994.

64. H Fricker-Hidalgo, O Vandapel, M-A Duchesne, M-A Mazoyer, D Monget, B Lardy, B Lebeau, J Freney, P Ambroise-Thomas, R Grillot. Comparison of the new API Candida system to the ID 32C system for identification of clinically important yeast species. J Clin Microbio 34:1846–1848, 1996.

65. PE Verweij, IM Breuker, AJMM Rijs, JFGM Meis. Comparative study of seven commercial yeast identification systems. J Clin Path 52:271–273, 1999.

66. DH Pincus, DC Coleman, WR Pruitt, AA Padhye, IF Salkin, M Geimer, A Bassel, DJ Sullivan, M Clarke, V Hearn. Rapid identification of *Candida dubliniensis* with commercial yeast identification systems. J Clin Microbio 37:3533–3539, 1999.

67. SM Essayag, GG Baily, DW Denning, JP Burnie. Karyotyping of fluconazole-resistant yeasts with phenotype reported as *Candida krusei* or *Candida inconspicua*. Internat J Syst Bact 46:35–40, 1996.

68. RL Hopfer, P Walden, S Setterquist, WE Highsmith. Detection and differentiation of fungi in clinical specimens using polymerase chain reaction (PCR) amplification and restriction enzyme analysis. J Med Vet Mycol 31:65–75, 1993.

69. DW Williams, MJ Wilson, MAO Lewis, AJC Potts. Identification of *Candida* species by PCR and restriction fragment length polymorphism analysis of intergenic spacer regions of ribosomal DNA. J Clin Microbio 33:2476–2479, 1995.

70. P Steffan, JA Vazquez, D Boikov, C Xu, JD Sobel, RA Akins. Identification of *Candida* species by randomly amplified polymorphic DNA fingerprinting of colony lysates. J Clin Microbio 35:2031–2039, 1997.

71. BM Mannarelli, CP Kurtzman. Rapid identification of *Candida albicans* and other

human pathogenic yeasts by using short oligonucleotides in a PCR. J Clin Microbio 36:1634–1641, 1998.

72. P Burgener-Kairuz, J-P Zuber, P Jaunin, TG Buchman, J Bille, M Rossier. Rapid detection and identification of *Candida albicans* and *Torulopsis (Candida) glabrata* in clinical specimens by species-specific nested PCR amplification of a cytochrome P-450 lanosterol-α-demethylase (L1A1) gene fragment. J Clin Microbio 32:1902–1907, 1994.

73. GS Sandhu, BC Kline, L Stockman, GD Roberts. Molecular probes for diagnosis of fungal infections. J Clin Microbio 33:2913–2919, 1995.

74. CM Elie, TJ Lott, E Reiss, CJ Morrison. Rapid identification of *Candida* species with species-specific DNA probes. J Clin Microbio 36:3260–3265, 1998.

75. JH Shin, FS Nolte, CJ Morrison. Rapid identification of *Candida* species in blood cultures by a clinically useful PCR method. J Clin Microbio 35:1454–1459, 1997.

76. A Lischewski, RI Amann, D Harmsen, H Merkert, J Hacker, J Morschhäuser. Specific detection of *Candida albicans* and *Candida tropicalis* by fluorescent in situ hybridisation with an 18S rRNA-targeted oligonucleotide probe. Microbiology 142:2731–2740, 1996.

77. YC Chen, JD Eisner, MM Kattar, SL Rassoulian-Barrett, K LaFe, SL Yarfitz, AP Limaye, BT Cookson. Identification of medically important yeasts using PCR-based detection of DNA sequence polymorphisms in the internal transcribed spacer 2 region of the rRNA genes. J Clin Microbio 38:2302–2310, 2000.

78. E Anaissie, GP Bodey, H Kantarjian, J Ro, SE Vartivarian, R Hopfer, J Hoy, K Rolston. New spectrum of fungal infections in patients with cancer. Rev Infec Dis 11:369–378, 1989.

79. P Krogh, P Holmstrup, JJ Thorn, P Vedtofte, JJ Pindborg. Yeast species and biotypes associated with oral leukoplakia and lichen planus. Oral Surg Oral Med Oral Path 63:48–54, 1987.

80. I Polacheck, IF Salkin, R Kitzes-Cohen, R Raz. Endocarditis caused by *Blastoschizomyces capitatus* and taxonomic review of the genus. J Clin Microbio 30:2318–2322, 1992.

81. D D'Antonio, F Romano, A Iacone, B Violante, P Fazii, E Pontieri, T Staniscia, C Caracciolo, S Bianchini, R Sferra, A Vetuschi, E Gaudio, G Carruba. Onychomycosis caused by *Blastoschizomyces capitatus*. J Clin Microbio 37:2927–2930, 1999.

82. D D'Antonio, R Piccolomini, G Fioritoni, A Iacone, S Betti, P Fazii, A Mazzoni. Osteomyelitis and intervertebral discitis caused by *Blastoschizomyces capitatus* in a patient with acute leukemia. J Clin Microbio 32:224–227, 1994.

83. AM Ortiz, C Sanz-Rodriguez, J Culebras, B Buendía, I González-Álvaro, E Ocón, R De la Cámara. Multiple spondylodiscitis caused by *Blastoschizomyces capitatus* in an allogenic bone marrow transplantation recipient. J Rheum 25:2276–2278, 1998.

84. MY Cheung, NC Chiu, SH Chen, HC Liu, CT Ou, DC Liang. Mandibular osteomyelitis caused by *Blastoschizomyces capitatus* in a child with acute myelogenous leukemia. J Formos Med Assoc 98:787–789, 1999.

85. S Krcmery, M Dubrava, V Krcmery Jr. Fungal urinary tract infections in patients at risk. Internat J Antimicrob Ag 11:289–891, 1999.

86. P Martino, M Venditti, A Micozzi, G Morace, L Polonelli, MP Mantovani, MC Petti, VL Burgio, C Santini, P Serra, F Mandelli. *Blastoschizomyces capitatus*: An emerging cause of invasive fungal disease in leukemia patients. Rev Infec Dis 12: 570–582, 1990.
87. KC Hazen. New and emerging yeast pathogens. Clin Microbio Rev 8:462–478, 1995.
88. MS Matthews, S Sen. *Blastoschizomyces capitatus* infection after contamination of fluids for intravenous application. Mycoses 41:427–428, 1998.
89. CM Beck-Sagué, WR Jarvis, TNNIS System. Secular trends in the epidemiology of nosocomial fungal infections in the United States, 1980–1990. J Infec Dis 167: 1247–1251, 1993.
90. WR Jarvis. Epidemiology of nosocomial fungal infections, with emphasis on *Candida* species. Clin Infec Dis 20:1526–1530, 1995.
91. S Nho, MJ Anderson, CB Moore, DW Denning. Species differentiation by internally transcribed spacer PCR and *Hha*I digestion of fluconazole-resistant *Candida krusei*, *Candida inconspicua*, and *Candida norvegensis* strains. J Clin Microbio 35: 1036–1039, 1997.
92. DJ Sullivan, MC Henman, GP Moran, LC O'Neill, DE Bennett, DB Shanley, DC Coleman. Molecular genetic approaches to identification, epidemiology and taxonomy of non-*albicans Candida* species. J Microbio Meth 44:399–408, 1996.
93. K Bartizal, G Abruzzo, C Trainor, D Krupa, K Nollstadt, D Schmatz, R Schwartz, M Hammond, J Balkovec, F Vanmiddlesworth. In vitro antifungal activities and in vivo efficacies of 1,3-β-D-glucan synthesis inhibitors L-671,329, L-646,991, tetrahydroechinocandin B, and L-687,781, a papulacandin. Antimicrob Agents Chemo 36:1648–1657, 1992.
94. V Krcmery Jr, F Mateicka, S Grausova, A Kunova, J Hanzen. Invasive infections due to *Clavispora lusitaniae*. FEMS Immun Med Microbio 23:75–78, 1999.
95. M Huttova, K Kralinsky, J Horn, I Marinova, K Iligova, J Fric, S Spanik, J Filka, Uher, J Kurak, V Krcmery Jr. Prospective study of nosocomial fungal meningitis in children—Report of 10 cases. Scand J Infec Dis 30:485–487, 1998.
96. A Casadevall, JR Perfect. Cryptococcus neoformans. Washington, DC: ASM Press, 1998, pp. 541.
97. R Mattsson, PD Haemig, B Olsen. Feral pigeons as carriers of *Cryptococcus laurentii*, *Cryptococcus unguttulatus* and *Debaryomyces hansenii*. Med Mycol 37:367–369, 1999.
98. NA Rao, AV Nerenberg, DJ Forster. *Torulopsis candida (Candida famata)* endophthalmitis simulating *Propionibacterium acnes* syndrome. Arch Ophth 109:1718–1721, 1991.
99. G St.-Germain, M Laverdière. *Torulopsis candida*, a new opportunistic pathogen. J Clin Microbio 24:884–885, 1986.
100. Y Oda, M Yabuki, K Tonomura, M Fukunaga. Reexamination of yeast strains classified as *Torulaspora delbrueckii* (Lindner). Internat J Syst Bact 47:1102–1106, 1997.
101. GS Heinic, D Greenspan, LA MacPhail, JS Greenspan. Oral *Geotrichum candidum* infection associated with HIV infection. Oral Surg Oral Med Oral Path 73:726–728, 1992.

102. H Kassamali, E Anaissie, J Ro, K Rolston, H Kantarjian, V Fainstein, GP Bodey. Disseminated *Geotrichum candidum* infection. J Clin Microbio 25:1782–1783, 1987.

103. E Haron, E Anaissie, F Dumphy, K McCredie, V Fainstein. *Hansenula anomala* fungemia. Rev Infec Dis 10:1182–1186, 1988.

104. P Muñoz, M-EG Leoni, J Berenguer, JCL Bernaldo de Quiros, E Bouza. Catheter-related fungemia by *Hansenula anomala*. Arch Intern Med 149:709, 712, 1989.

105. B Nohinek, C-S Zee-Cheng, WG Barnes, L Dall, HR Gibbs. Infective endocarditis of a bicuspid aortic valve caused by *Hansenula anomala*. Amer J Med 82:165–168, 1987.

106. LCS Thuler, S Faivichenco, E Valasco, CA Martins, CRG Nascimento, IAMA Castilho. Fungaemia caused by *Hansenula anomala*—An outbreak in a cancer hospital. Mycoses 40:193–196, 1997.

107. KJ Kwon-Chung, JE Bennet. Medical Mycology. Philadelphia: Lea and Febiger, 1992.

108. M Goldman, JC Pottage, DC Weaver. *Candida krusei* fungemia. Medicine 72:143–150, 1993.

109. JR Wingard, WG Merz, MG Rinaldi, TR Johnson, JE Karp, R Saral. Increase in *Candida krusei* infection among patients with bone marrow transplantation and neutropenia treated prophylactically with fluconazole. New Eng J Med 325:1274–1277, 1991.

110. MA Morgan, CJ Wilkowske, GD Roberts. *Candida pseudotropicalis* fungemia and invasive disease in an immunocompromised patient. J Clin Microbio 20:1006–1007, 1984.

111. E Guého, RB Simmons, WR Pruitt, SA Meyer, DG Ahearn. Association of *Malassezia pachydermatis* with systemic infections of humans. J Clin Microbio 25:1789–1790, 1987.

112. MJ Marcon, DA Powell. Human infections due to *Malassezia* spp. Clin Microbio Rev 5:101–119, 1992.

113. O Teglia, PE Schoch, BA Cunha. *Malassezia furfur* infections. Infec Con Hosp Epid 12:676–681, 1991.

114. L Pospisil. The significance of *Candida pulcherrima* findings in human clinical specimens. Mycoses 32:581–583, 1989.

115. H Nielsen, J Stenderup, B Bruun, J Ladefoged. *Candida norvegensis* peritonitis and invasive disease in a patient on continuous ambulatory peritoneal dialysis. J Clin Microbio 28:1664–1665, 1990.

116. G Samonis, D Bafaloukos. Fungal infections in cancer patients: An escalating problem. In Vivo 6:183–194, 1992.

117. DK Braun, CA Kauffman. *Rhodotorula* fungaemia: A life-threatening complication of indwelling central venous catheters. Mycoses 305:308, 1992.

118. TE Kiehn, E Gorey, AE Brown, FF Edwards, D Armstrong. Sepsis due to *Rhodotorula* related to use of indwelling central venous catheters. Clin Infec Dis 14:841–846, 1992.

119. ES Eisenberg, BE Alpert, RA Weiss, N Mittman, R Soeiro. *Rhodotorula rubra* peritonitis in patients undergoing continuous ambulatory peritoneal dialysis. Amer J Med 75:349–352, 1983.

120. HC Siersted, S Gravesen. Extrinsic allergic alveolitis after exposure to the yeast *Rhodotorula rubra*. Allergy 48:298–299, 1993.

121. GM Chertow, ER Marcantonio, RG Wells. *Saccharomyces cerevisiae* empyema in a patient with esophago-pleural fistula complicating variceal sclerotherapy. Chest 99:1518–1519, 1991.

122. RHK Eng, R Drehmel, SM Smith, EJC Goldstein. *Saccharomyces cerevisiae* infections in man. Sabouraudia: J Med Vet Mycol 22:403–407, 1984.

123. A Oriol, J-M Ribera, J Arnal, F Milla, M Batlle, E Filiu. *Saccharomyces cerevisiae* septicemia in a patient with myelodysplastic syndrome. Amer J Hematol 43:325–326, 1993.

124. AG Bergman, CA Kauffman. Dermatitis due to *Sporobolomyces* infection. Arch Derm 120:1059–1060, 1984.

125. JT Morris, M Beckius, CK McAllister. *Sporobolomyces* infection in an AIDS patient. J Infec Dis 164:623–624, 1991.

126. J Plazas, J Portilla, V Boix, M Pérez-Mateo. *Sporobolomyces salmonicolor* lymphadenitis in an AIDS patient: Pathogen or passenger? AIDS 8:387–398, 1994.

127. L De Gentile, JP Bouchara, B Cimon, D Chabasse. *Candida ciferrii*: Clinical and microbiological features of an emerging pathogen. Mycoses 34:125–128, 1991.

128. RM Furman, DG Ahearn. *Candida ciferrii* and *Candida chiropterum* isolated from clinical specimens. J Clin Microbio 18:1252–1255, 1983.

129. GS de Hoog, J Guarro. Atlas of clinical fungi. In: GS De Hoog, J Guarro, eds. Delft, the Netherlands: Centraalbureau voor Schimmelcultures, 1995.

130. E Guého, L Improvisi, GS de Hoog, B Dupont. *Trichosporon* on humans: A practical account. Mycoses 37:3–10, 1994.

131. E Anaissie, A Gokaslan, R Hachem, R Rubin, G Griffin, R Robinson, J Sobel, G Bodey. Azole therapy for trichosporonosis: Clinical evaluation of eight patients, experimental therapy for murine infection, and review. Clin Infec Dis 15:781–787, 1992.

132. TJ Walsh. Trichosporonosis. Infec Dis Clin N Amer 3:43–52, 1989.

133. V Krcmery Jr, F Mateicka, A Kunova, S Spanik, J Gyarfas, Z Sycova, J Trupl. Hematogenous trichosporonosis in cancer patients: Report of 12 cases including 5 during prophylaxis with itraconazole. Supp Care Canc 7:39–43, 1999.

134. K Singh, A Chakrabarti, A Narang, S Gopalan. Yeast colonisation and fungaemia in preterm neonates in a tertiary care centre. Indian J Med Res 110:169–173, 1999.

135. P Wherspann, U Fullbrandt. *Yarrowia lipolytica* (Wickerham et al.) van der Walt and von Arx isolated from a blood culture. Mykosen 28:217–222, 1985.

136. B Ruiz-Diez, V Martinez, M Alvarez, JL Rodriguez-Tudela, JV Martinez-Suarez. Molecular tracking of *Candida albicans* in a neonatal intensive care unit: Long-term colonizations versus catheter-related infections. J Clin Microbio 35:3032–3036, 1997.

137. K Reiderer, P Fozo, R Khatib. Typing of *Candida albicans* and *Candida parapsilosis*: Species-related limitations of electrophoretic karyotyping and restriction endonuclease analysis of genomic DNA. Mycoses 41:397–402, 1998.

138. U Schwab, F Chernomas, L Larcom, J Weems. Molecular typing and fluconazole susceptibility of urinary *Candida glabrata* isolates from hospitalized patients. Diag Microbio Infec Dis 28:11–17, 1997.

139. A Carlotti, F Chaib, A Couble, N Bourgeois, V Blanchard, J Villard. Rapid identification and fingerprinting of *Candida krusei* by PCR-based amplification of the species-specific repetitive polymorphic sequence CKRS-1. J Clin Microbio 35: 1337–1343, 1997.

140. MA Pfaller, SA Messer, RJ Hollis. Variations in DNA subtype, antifungal susceptibility, and slime production among clinical isolates of *Candida parapsilosis*. Diag Microbio Infec Dis 21:9–14, 1995.

141. F De Bernardis, F Mondello, R San Millan, J Ponton, A Cassone. Biotyping and virulence properties of skin isolates of *Candida parapsilosis*. J Clin Microbio 37: 3481–3486, 1999.

142. MA Pfaller, SA Messer, RJ Hollis. Strain delineation and antifungal susceptibilities of epidemiologically related and unrelated isolates of *Candida lusitaniae*. Diag Microbio Infec Dis 20:127–133, 1994.

143. D D'Antonio, B Violante, A Mazzoni, T Bonfini, MA Capuani, F D'Aloia, A Iacone, F Schioppa, F Romano. A nosocomial cluster of *Candida inconspicua* infections in patients with hematological malignancies. J Clin Microbio 36:792–795, 1998.

144. KV Clemons, PS Park, JH McCusker, MJ McCullough, RW Davis, DA Stevens. Application of DNA typing methods and genetic analysis to epidemiology and taxonomy of *Saccharomyces* isolates. J Clin Microbio 35:1822–1828, 1997.

145. MM Balieras Couto, BEH Hofstra, JHJ Huis in't Veld, JMBM Van der Vossen. Evaluation of molecular typing techniques to assign genetic diversity among *Saccharomyces cerevisiae* strains. Appl Environ Microbio 62:41–46, 1996.

146. DL Larone. Medically important fungi: A guide to identification. Washington, DC: ASM Press, 1995.

147. JP Van der Walt, D Yarrow. The genus *Arxiozyme* gen. nov. (Saccharomycetaceae). S Afr J Bot 3:340–342, 1984.

148. JA von Arx, JP Van der Walt. Ophiostomatales and endomycetales. Stud Mycol 30:167–176, 1987.

9
Basidiomycetous Yeasts

Teun Boekhout
*Centraalbureau voor Schimmelcultures, Institute of the Royal
Netherlands Academy of Arts and Sciences, Utrecht, The Netherlands*

Eveline Guého
Mauves sur Huisne, France

I. INTRODUCTION

Yeasts are generally defined as unicellar fungi. Numerous yeasts are able to form hyphae and/or pseudohyphae, however. Yeasts are polyphyletic in origin and belong to the Ascomycetes and the Basidiomycetes (1). Within the Basidiomycetes, they occur in all three main phylogenetic lines, namely the Hymenomycetes [Cystofilobasidiales, Trichosporonales, Tremellales (jelly fungi), and Filobasidiales], the Urediniomycetes (Sporidiales, and including the obligate plant parasitic rust fungi), and the Ustilaginomycetes (Malasseziales, and including the plant parasitic smut fungi; Fig. 1) (2). Medically important basidiomycetous yeasts belong to the genera *Cryptococcus* Vuillemin, *Trichosporon* Behrend (both Hymenomycetes), and *Malassezia* Baillon (Ustilaginomycetes). Other basidiomycetous yeasts that have been reported from patients occur in the genera *Rhodotorula* FC Harrison and *Sporobolomyces* Kluyver & van Niel (both Urediniomycetes). Yeasts, including the basidiomycetous ones, are usually identified by using medical physiological characteristics. Summarized physiological data of the most important basidiomycetous yeasts are presented in Tables 1 and 2.

II. *CRYPTOCOCCUS* VUILLEMIN

Cryptococcus is an anamorphic basidiomycetous yeast belonging to the Hymenomycetes (2), comprising in its current circumscription 34 species. The genus con-

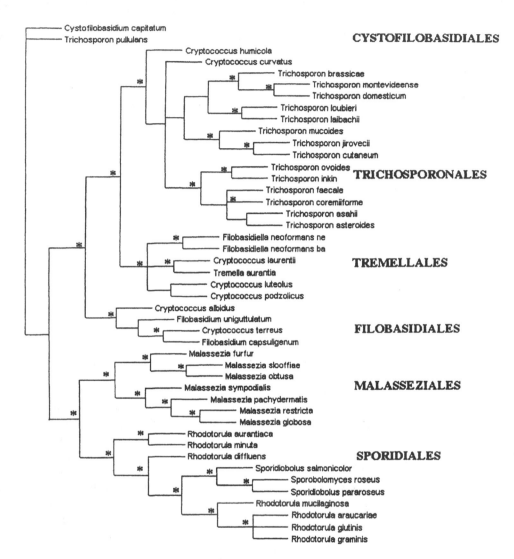

Figure 1 Phylogenetic tree of selected basidiomycetous yeasts based on the D1/D2 domain of the large subunit ribosomal DNA. Cystofilobasidiales are used as an outgroup. Asterisks indicate bootstrap values of 70% and higher.

Table 1 Specific Characters of *C. albidus*, *C. curvatus*, *C. laurentii*, and *C. neoformans*

Tests	*C. albidus* CBS 142	*C. curvatus*[a] CBS 570	*C. laurentii* CBS 139	*C. neoformans* NIH 271
ID 32C pattern[b]				
Galactose	−	+	+	+
0.01% cycloheximide	−	D	−	−
DL-lactate	−	+	−	−
N acetyl-glucosamine	−	+	+	+
L-arabinose	+	−	+	+
Cellobiose	+	+	+	−
Raffinose	+	+	+	−
Ribose	−	+	−	+
Glycerol	−	+	−	−
Erythritol	−	−	+	+
Melibiose	−	−	+	−
Gluconate	−	+	+	+
Lactose	−	+	+	−
Sorbose	−	−	+	+
Urea hydrolosis[c]	+	+	D	+
Phenoloxydase	−	−	−	+
Growth at 37°C	+	D	D	+/D

[a] Characteristics were the same for five clinical isolates of *C. curvatus*, except that inositol was negative for the type culture CBS 570.
[b] For the four species, growth was negative (−) on levulinate and positive (+) on the other substrates of the ID 32C strips.
[c] D = delayed.

tains *C. neoformans*, the most important basidiomycetous yeast from a clinical perspective (3), as well as a number of other species encountered in the medical literature. The name of the genus was conserved as *Cryptococcus* Vuillemin to improve nomenclatural stability (4).

A. *Cryptococcus neoformans*

Cryptococcus neoformans is known in both its asexual (anamorph) and sexual (teleomorph) states, for which the respective names *Cryptococcus neoformans* and *Filobasidiella neoformans* Kwon-Chung are used. Estimates on the incidence rate in AIDS patients range from 5–30%, with the highest numbers occurring in sub-Saharan Africa (5). The fungus can cause serious infections, especially in immunocompromised patients. The main sites of infection are the lungs and the

Table 2 Physiological Patterns of Clinical Basidiomycetous Yeasts Analyzed with ID 32C (bioMérieux, Marcy l'Étoile, France)

	T. asahii	T. asteroides	T. ovoides	T. inkin	T. cutaneum	T. montevideense	T. mucoides	C. neoformans	C. albidus	C. curvatus	C. humicola	C. laurentii	Filobasidium uniguttulatum	R. glutinis	R. mucilaginosa	Pseudozyma species
Esculine	v	+	-	-	+	+	+	v	+	+	+	+	-	-	-	+
Glucosamine	+	v	+	v	v	+	+	+	-	+	+	+	+	-	-	+
Sorbose	-	-	-	-	-	-	v	v	-	-	+	-	+	+	+	+
Glucose	+	+	+	+	+	+	+	+	+	+	+	+	+	+	+	+
Inositol	-	+	-	+	+	+	+	+	v	+	+	+	+	-	-	+
Lactose	+	+	+	+	+	+	+	-	-	+	+	+	+	-	-	-
Mannitol	-	-	v	+	-	+	-	+	+	-	+	+	-	v	+	+
Laevulinate	-	+	-	-	-	-	-	-	-	+	+	-	-	-	-	-
Gluconate	+	+	+	+	+	+	+	+	+	+	+	+	+	+	+	+
Melezitose	v	v	+	+	v	v	v	+	+	-	+	+	+	+	+	+
Glucuronate	+	+	+	+	+	+	+	+	+	+	+	+	-	-	-	-
Melibiose	-	-	-	-	-	-	-	-	-	-	+	-	-	-	-	+
Erythritol	+	+	-	+	+	+	+	v	-	-	+	+	-	-	-	+
Palatinose	+	+	+	+	+	+	+	+	+	+	+	+	+	+	+	+
Rhamnose	v	v	v	-	v	v	v	+	+	-	+	+	v	+	+	+
Glycerol	v	v	v	+	+	+	+	-	+	-	+	-	-	+	+	+
Ribose	+	+	+	+	+	+	+	+	+	+	+	+	+	+	+	+
D-Xylose	+	+	+	+	+	+	+	+	+	+	+	+	+	+	+	+
Sorbitol	-	-	-	-	-	-	+	+	-	-	+	+	-	v	v	+
α-methyl-D-glucoside	+	+	+	+	+	+	+	+	v	+	+	+	+	-	-	+
2-Keto-D-gluconate	+	+	+	+	+	+	+	+	+	+	+	+	+	+	+	+
Trehalose	v	+	+	+	v	v	+	+	+	+	+	+	+	+	+	+
Maltose	+	+	+	+	+	+	+	+	+	+	+	+	+	+	+	+
Raffinose	-	-	-	-	-	-	+	v	+	-	-	+	+	+	+	+
Cellobiose	+	+	+	+	v	+	+	v	+	+	-	+	+	v	-	+
L-Arabinose	+	v	+	-	+	+	+	+	+	+	+	+	+	+	+	+
DL-Lactate	+	+	+	+	+	+	+	-	-	+	+	-	-	-	-	+
N-Acetyl-glucosamine	+	+	+	+	-	+	+	+	+	+	+	+	+	+	+	+
Sucrose	+	v	+	+	+	+	+	+	+	+	+	+	+	+	+	+
0.01% Cycloheximide	+	v	+	v	-	+	+	-	-	-	+	-	-	-	-	-
Galactose	+	v	+	+	+	+	+	+	+	+	+	+	+	+	+	-

T. = *Trichosporon*; *C.* = *Cryptococcus*; *R.* = *Rhodotorula*.

central nervous system, including cerebrospinal fluid (CSF). In the brain it causes meningitis, meningoencephalitis, and granuloma formation. Primary cutaneous infections are rare, and most skin infections are due to disseminated systemic infections (6, 7). *Cryptococcus neoformans* var. *neoformans* is encountered in nearly all of the AIDS-related infections (7, 8).

According to the current classification, the species consists of three varieties: *C. neoformans* var. *neoformans* (serotype D), *C. neoformans* var. *grubii* (serotype A), and *C. neoformans* var. *gattii* (serotypes B and C). In contrast, the teleomorph *Filobasidiella neoformans* var. *neoformans* corresponds to both serotypes A and D, and *F. neoformans* var. *bacillispora* to serotypes B and C (9–18).

The occurrence of recombinants between strains of *C. neoformans* var. *neoformans* and *C. neoformans* var. *gattii*, and the presence of genetic recombination in the first filial generation (F1) suggested their varietal status (15, 19), but no genetic analysis of the second generation (F2) has been made. In contrast, rather low DNA-DNA reassociation values ranging between 55–63% (20) were observed between isolates of the two varieties, suggesting a considerable genetic divergence between the taxa. Both varieties differ in their karyotypes (21, 22), DNA fingerprints (23, 24), and a number of physiological characteristics; for example, assimilation of D-proline, D-tryptophane, and L-malic acid, regulation of creatinine deaminase by ammonia production (25–28), and susceptibility to killer toxins of *C. laurentii* CBS 139 (29). The varieties also differ in geographic distribution and habitat. *C. neoformans* var. *grubii* and *C. neoformans* var. *neoformans* seem to occur worldwide in clinical and veterinary sources, bird droppings, and occasionally in such substrates as fermenting fruit juice, drinking water, wood, and air (5, 7, 30–31). Both varieties occur in AIDS patients, and serotype D isolates seem to be relatively common in Europe (32). Recently, serotype A isolates were also isolated from trees in Brazil (33). *C. neoformans* var. *gattii* seems to occur mainly in the tropics, the southern hemisphere, southern Europe, and the southern United States. Clinical isolates are mainly known from non-AIDS patients (e.g., CSF, skin, tumor) and animals (e.g., cats and goat, as well as cows with mastitis). Saprobic isolates are usually associated with *Eucalyptus* species (3, 16, 34–36), but have been isolated from bat guano (33), and more recently from almond trees (37–38) as well.

Selective isolation and presumptive identification of the species is usually based on its ability to form melanin on plates containing niger seeds (*Guizotia abyssinica*) (39) or dopamines such as L-DOPA or norepiniphrine (40), and the presence of an extracellular capsule visible in india ink preparations. Other identification tests are based on its ability to hydrolyze urea (41, 42), but this is a characteristic for all basidiomycetous yeasts, and urease negative strains of the species have rarely been reported (43). Distinction between the varieties is usually performed on 1-canavanine-glycine-bromothymol blue medium (CGB-medium;

14, 44), or by testing D-proline or D-tryptophan assimilation (26, 27). The four serotypes can be identified by agglutination with polyclonal antibodies (45) or by immunofluorescence with monoclonal antibodies (46). The four serotypes have the same physiological growth characteristics as are usually evaluated in clinical laboratories (4) with a commercial testing system (Table 1).

The observed genetic, biochemical, ecological, pathological, and geographical differences raise strong questions about the conspecific status of the varieties. If they interbreed, it is hard to understand how all the observed differences are being maintained, as one would expect homogenization to occur within the population. Recent observations using amplified fragment length polymorphism (AFLP) (47) and nucleotide sequences of the intergenic spacer (IGS) of the ribosomal DNA (48) seem to support the existence of two species, therefore the presence of two separate teleomorphic species is proposed: *F. neoformans* Kwon-Chung and *F. bacillispora* Kwon-Chung with their respective anamorphs *C. neoformans* (Sanfelice) Vuillemin and *C. bacillisporus* Kwon-Chung & Bennett (= *C. neoformans* var. *gattii* Vanbreuseghem & Takashio).

Using molecular data, Franzot et al. (49) recently described a new variety for serotype A isolates of *C. neoformans* as *C. neoformans* var. *grubii*. Our AFLP analysis and IGS sequence data largely support the existence of this variety, but also showed that the resulting genotypic clusters do not entirely correspond with the serotype boundaries (47). The same can be concluded from the *URA5* sequence and CNRE-1 results presented by Franzot et al. (50), as a serotype D isolate was incidentally found to cluster among the serotype A isolates they studied. These observations suggest that serotyping alone cannot be used to distinguish *C. neoformans* var. *neoformans* from *C. neoformans* var. *grubii*. We explain these nonconcordant results by proposing the presence of sexual or parasexual hybridization between serotype A and D isolates of *C. neoformans* and between serotype B and C isolates in *C. bacillisporus*. So far we have only evidence for the occurrence of hybridization within—and not between—the species. Presently we favor the scenario in which the fungus uses both sexual and asexual reproduction mechanisms. The role of the sexual *Filobasidiella* states in the hybridization events is not yet clear. After recombination, strains with the hybrid genotypes may disseminate clonally. Hybrid strains may have a selective (dis)advantage toward changes that occur in the environment. Such environmental changes may have happened several times in the history of the species (51). The hybrids may also differ in virulence or resistance to antibiotics. Our hypothesis about the reproductive biology of the species differs from the previous proposed strictly clonal reproduction of *C. neoformans*, as was deduced from linkage disequilibrium studies in which multilocus electrophoretic enzyme patterns were used (52, 53). Additional support for the previous proposal derived from the concordance between different molecular parameters in geographically separated populations, as well as the dominance of a few genotypes in the population (54).

We think, however, that the molecular markers used in these studies probably yield lower levels of resolution than the AFLP markers we employed.

B. Other *Cryptococcus* Species

Several cryptococci other than *C. neoformans* have been reported from clinical sources. In a critical discussion of the literature, Krajden et al. (55) noticed the problems related with a correct interpretation of earlier clinical reports on cryptococcal non-*C. neoformans* infections. This uncertainty is due to the possible contamination with saprobically growing yeasts, unreliable identifications, and the fact that most isolates were not deposited in public culture collections. Summarized physiological data of some of the clinically reported *Cryptococcus* species are given in Tables 1 and 2.

Cryptococcus luteolus was reported from pulmonary cryptococcosis (56). The limited physiological data given, however, suggested a proper identification of this isolate as *C. neoformans*, thus supporting the suggestion of Krajden et al. (55, 57).

Clinical isolates of *C. albidus* have mainly been reported to occur in immunocompromised patients. Isolates have been reported from AIDS patients and leukemic patients with pemphigus foliaceus, pulmonary cryptococcosis, and meningitis, as well as a nosocomial infection (58–67). Wieser (68) reported the presence of *C. albidus* in cerebrospinal fluid of five patients with meningitis. The identity of an isolate described by Krumholz (59) was strongly questioned by Gordon (69). In most of these cases the identity of the isolates should be interpreted with care. The limited data presented in the literature usually hamper a reliable reidentification, and in most cases no isolates have been preserved in a public culture collection. To complicate the situation, recent molecular taxonomic investigations of *C. albidus* suggest that many species occur in this complex (70).

Cryptococcus curvatus was reported to occur in the cerebrospinal fluid (CSF) of a 30-year-old male AIDS patient (71). Clinical isolates of this species present in the CBS yeast collection (72) were obtained from sputum, urine, feces, and the uterus of a cow.

Another cryptococcal species, *C. laurentii*, has been reported from pulmonary lesions, peritoneal fluid, fungemia occurring in a bone marrow transplant patient, a cutaneous infection and granulomas, endophalmitis, CSF of an AIDS patient, and from a catheter-associated fungaemia (67, 73–80). The isolates from two cases of pulmonary lesions showed a cross-reaction with fluorescent antibodies against *C. neoformans*, and maintained their fluorescence after absorption with a *C. laurentii* antigen. It was therefore concluded that the disease very likely was not caused by *C. laurentii* but by *C. neoformans* (55).

Cryptococcus ater has been isolated from an ulcer of a leg of an 18-year-old man in Portugal. No proof of its pathogenicity was presented (81), however.

Cryptococcus humicola has been isolated from skin and bronchial excretion, as well as soil and mushrooms. Isolates of *Filobasidium uniguttulatum*, the perfect state of *Cryptococcus uniguttulatus* (82), have been obtained from an infected fingernail, bronchus, sputum, the sole of a foot, and white patches in the mouth (72). No proof for pathogenicity exists for the latter two species, however. *Filobasidium uniguttulatum* does not grow at human body temperature in vitro.

III. *TRICHOSPORON* BEHREND

The genus *Trichosporon* consists of anamorphs for which no sexual state (teleomorph) has been described. In 1992, the genus was limited to yeastlike fungi combining a layered cell wall ultrastructure, the presence of a dolipore, positive urease and Diazonium Blue B (DBB) reactions, and the ability to both produce true filaments (hyphae) and multiply asexually by arthroconidia and under certain conditions blastoconidia (83). Behrend had created the genus in 1890 with the species *T. ovoides* to accommodate a fungus isolated from mustache white piedra nodules (84). *T. ovoides* was reintroduced as the type species of the genus to replace the name *T. beigelii* (83). The epithet *beigelii* first appeared in 1867 when Küchenmeister and Rabenhorst (85) observed round cells composing white piedra nodules on "postiche chignon" hair. They named the micro-organism *Pleurococcus beigelii*, because they thought that these nodules were composed of algae. They did not get a culture or provide a description, however. In contrast, the description given by Behrend (84) for *T. ovoides* was macroscopically and microscopically well detailed. Unfortunately, his original culture was not maintained, but a neotype culture of the species with the same characteristics as the original material was selected and deposited in culture collections (86, 87). In 1992 all isolates preserved in culture collections and differing from this type of culture were shown to represent other known species or undescribed species [e.g., see *T. mucoides* described in 1992 (83)]. The genus was enlarged to 19 species. *Trichosporon pullulans*, which resembles the other species only morphologically, needs to be recombined in another genus, as it belongs to a distinct phylogenetic lineage. This species belongs to the order Cystofilobasidiales, which contains psychrophilic genera such as *Mrakia* and *Cystofilobasidium*. The genus *Trichosporon sensu stricto* is phylogenetically related to the Tremellales and the Filobasidiales, and has recently been classified in the Trichosporonales (2, 88–90). The species of the genus are serologically similar to *Cryptococcus neoformans* (91). The psychrophily of *T. pullulans* (no growth above 20°C) precludes any pathogenicity of this species in humans or other warm-blooded animals. The rare papers describing *T. pullulans* as a pathogen may be based on misidentified isolates (92). Contrary to the other *Trichosporon* species, *T. pullulans* is able to grow with nitrates as a nitrogen source.

The results of the revision of 1992 were confirmed by subsequent authors (93), which transferred three taxa recognized as species to the varietal level. *T. faecale* and *T. coremiiforme* were made varieties of *T. asahii*, and *T. laibachii* a variety of *T. loubieri* (93). Recently, several species, either from humans or the environment, have been described (94–97). Consequently, the genus presently again includes 19 species, and more species are expected to be discovered in the near future. The genus is not as large as *Candida* or *Cryptococcus*, but the lack of well-defined key characteristics makes recognition of its species difficult. Any identification of a medical *Trichosporon* isolate should involve the following characteristics: macromorphology (colony texture and rate of growth) and micromorphology with specific features, such as appressoria, fusoid or meristematic cells, and globose or ovoid blastoconidia, obtained after growth of the organism in slide cultures on 2% malt extract agar or cornmeal agar; growth at 37°C or higher temperatures; and cycloheximide sensitivity at 0.01 and 0.1%. The assimilation tests must include a sufficient number of substrates (Table 2). Also, the origin of isolates is important since each species occupies its preferential ecological niche (83, 98, 99).

The second most important source of *Trichosporon* yeast isolates has generally been the human skin. Since the description of *T. beigelii*, and its synonym *T. cutaneum* (85), the two names were used for most *Trichosporon* isolates irrespective their origin (83, 86). In 1970, Watson and Kallichurum (100) published a case of deep-seated infection (brain abscess) caused by a species of *Trichosporon* (in this case named *T. cutaneum*). Unfortunately, no isolate was maintained in a culture collection. During the following three decades many cases of trichosporonosis were published, most of them listing *T. beigelii* as the causative agent (101, 102). In the 1992 taxonomic revision six species adapted to humans were included (83, 98, 102). Among these species, only three were recognized as occurring in deep-seated trichosporonosis, and five were found in nodules of white piedra (103, 104). In order of decreasing frequency and virulence, the six species infecting humans are *T. asahii*, *T inkin*, *T. mucoides*, *T. ovoides*, *T. asteroides*, and *T. cutaneum*. It is recommended that clinical laboratories should identify at least the three agents of deep-seated trichosporonosis, *T. asahii*, *T. inkin*, and *T. mucoides*. Because it is difficult to distinguish the cutaneous species, a PCR detection assay for *T. asahii* was developed by using sequences of the internal transcribed spacer (ITS) regions of the ribosomal DNA (105). Later, the same methodology was extended to the five other clinically relevant species, and a database was constructed (106, 107). According to recent findings *T. montevideense*, isolated from nails of a patient with AIDS (E. Guého, unpublished data), and the new species *T. domesticum* (94) should be considered of clinical importance as well. All these species show irregular hyphae and conidia which appear as pseudohyphae and blastoconidia in primary cultures, and all are able to hydrolyze urea in liquid urea-indol medium at 37°C if heavily inoculated. None of them uses nitrates as a nitrogen source, but all are able to use most of the carbon substrates commonly used in yeast identification (83).

A. *Trichosporon asahii*

Although originally isolated as a contaminant from psoriatic nails, this species is becoming recognized as the major causative agent of deep-seated trichosporonosis (102, 108), in particular in immunocompromised patients. It is also isolated from cutaneous lesions, nails, and white piedra nodules, although not from capital white piedra (103, 104). It is able to grow at 37°C, but the typical colonies, measuring 16 to 24 mm in diameter in 10 days and with a wrinkled white farinose surface, are better developed at 25–28°C on Sabouraud glucose agar (SGA) without antibiotics. Hyphae disarticulate into regular arthroconidia (Fig. 2a), which become barrel-shaped with age on slide cultures. Appressoria and blastoconidia are missing (83). Its physiological pattern is characterized by lack of assimilation of raffinose and sorbitol, which are substrates present in most commercial kits for yeast identification (98, 104). It should be noted, however, that since the physiological data presented in the fourth edition of *The Yeasts: A Taxonomic Study* (99) are based on relatively long incubation periods, the results may differ from those obtained with commercial kits. This discrepancy in assimilation patterns may cause confusion to medical microbiologists. Consequently, only a few clinicians such as Mahal and coworkers (109) have risked publishing any of the correct, newer names instead of the obsolete *T. beigelii*. *T. asahii* was proven to correspond to the serotype II standard culture, one of the causative agents of summer-type hypersensitivity pneumonitis (110). The species as currently defined includes previously recognized variants, such as *T. infestans* (83, 99), now

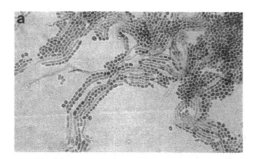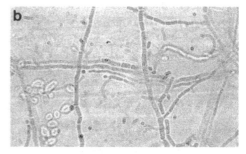

Figure 2   Microscopic morphology of *Trichosporon mucoides* and *T asahii*. (a) Arthroconidia of *T. asahii*; (b) Hyphae, arthroconidia, and lateral refringent blastoconidia of *T. mucoides*.

considered a synonym, and retains the former *T. faecale* and *T. coremiiforme* as varieties (93).

B. *Trichosporon inkin*

The fungus was first isolated from scrotal skin lesions. Most isolates from the genital area isolates, including those from white piedra on pubic hairs, were shown to belong to the same genetic entity. The fungus may be particularly well adapted to the genital microniche (83, 87, 103, 104). The species is also found in urine, however, and occasionally causes deep-seated trichosporonosis (102). Reports with the right identification of *T. inkin* (111) are rare, and therefore its clinical importance is not yet known.

Because of the presence of sarcinae, round multiseptate cells that occur in the cutaneous tissues of scrotum, *T. inkin* has been classified as *Sarcinomyces inkin* or *Sarcinosporon inkin*. These monotypic genera are presently considered to be synonyms of *T. inkin* (E. Guého, unpublished data), however. Consequently, the report of a *Sarcinosporon inkin* invasive infection (112) must be attributed to *T. inkin*. The species was also thought to be related to algae of the genus *Prototheca*, because the sarcinae are reminiscent of the morphology of this genus, and to *Fissuricella filamenta* multiplying by meristematic cleavage. (See *T. asteriodes*.) Sarcinae are restricted to the parasitic phase of *T. inkin*, however, and are formed in vitro on culture media with high sugar contents (Fig. 3a).

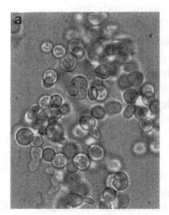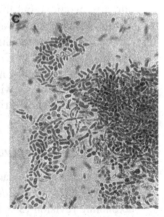

Figure 3 Micromorphology of *Trichosporon inkin*: (a) sarcina-like cells formed on high sugar containing media; (b) arthroconidia and appressoria formed on solid substrates (e.g., glass slides); (c) primary culture from urine with uninformative arthroconidia and blasto-conidia-like cells, which can be observed in other species as well.

Hyphae and arthroconidia are similar to those of other *Trichosporon* species, and appressoria similar to those of *T. ovoides* (98, 113) (Figs 3b, c). Colonies on SGA measure only 9 to 12 mm in diameter. They are dry and finely cerebriform with a white farinose covering, do not have a marginal zone, and often crack the agar medium. Appressoria, mostly borne laterally on long arthroconidia are present in slide cultures (Fig. 3b). The species grows at 37°C, and L-arabinose and raffinose are not assimilated (83, 98, 104).

C. *Trichosporon mucoides*

Several strains were found to be identical with the agent of a cerebral trichosporonosis (114), but did not agree with any known species. Therefore a new species, *T. mucoides*, was described from the characters of a strain isolated from a brain as its type culture (83). It is unknown if the first reported agent of a deep-seated trichosporonosis from a brain abscess and a case of chronic meningitis (83, 100, 114) corresponded to this species. Presently it is clear that *T. mucoides* is medically very important (102), and it seems to be the only *Trichosporon* species able to cross the cerebral barrier. Besides, it has been isolated from lesions of skin (115), nails, and from pubic white piedra (103, 104). In Japan, it has been found to occur in the environment of a patient with summer-type hypersensitivity pneumonitis and was identified as serotype I standard culture (110). The species is able to grow at 37°C. Colonies at 25–28°C reach 13 to 17 mm in diameter, and are mucoid, cerebriform at the center, and have deep furrows toward the margin. Morphologically, *T. mucoides* is characterized by hyphae that can disarticulate into arthroconidia, but the species also produces lateral blastoconidia that at the mature state are thick-walled and refringent (Fig. 2b). The species utilizes most carbon substrates, including L-arabinose and raffinose (98). Its physiological profile is similar to that of *Cryptococcus humicola*. This latter species, however, was never found to be implicated in any pathology, nor does it produces arthroconidia. Moreover, in contrast with *T. mucoides* it assimilates levulinate present in the ID 32C identification strips (bioMérieux, Marcy l'Etoile, France; Table 2).

D. *Trichosporon ovoides*

White piedra on the head and beard hair shaft is nowadays rare, and so is its causative agent, *T. ovoides* (83, 104). Only a few isolates were included in the taxonomic revision (83), but Sugita and co-workers recently isolated *T. ovoides* from the urine of a patient with trichosporonosis (108). The species grows very slowly at 37°C. Colonies at 25–28°C resemble those of *T. asahii*, while arthroconidia and appressoria are similar to those of *T. inkin*. Physiologically, *T. ovoides* is also intermediate between these two species. Like *T. asahii*, it does not grow with raffinose, and unlike *T. inkin* it grows weakly with L-arabinose.

Phylogenetically, *T. ovoides* clusters together with *T. inkin* and the new species *T. japonicum* (95) at a short distance from *T. asahii*.

E. *Trichosporon asteroides*

This species is mainly known from skin, but incidentally was also isolated from blood (83). The distinction from the phylogenetically closely related *T. asahii* is difficult. Growth at 37°C is weak, but colonies at 25–28°C are very similar to those of *T. asahii*. Appressoria are not present, and arthroconidia are often thick-walled and septate. Assimilation reactions, although much slower, are the same as in *T. asahii*. *Fissuricella filamenta*, which morphologically differs from *T. asahii*, was found to be conspecific because of high DNA/DNA reassociation values and rRNA sequence similarity (83). Colonies of *F. filamenta* are dry, finely cerebriform, and brownish, and measure only 5 to 6 mm in diameter after 10 days. Growth of the filamentous phase is very brief and typical meristematic packs of cells quickly formed (98).

F. *Trichosporon cutaneum*

This species, long synonymized with *T. beigelii* (83), occurs on skin and hairs (98, 104, 108). It does not grow at 37°C or in the presence of cycloheximide. Its colonies are similar to those of *T. mucoides*, and produce hyphae, arthroconidia, and lateral blastoconidia. The primary cultures may be strictly yeastlike, however, and may be confused with *C. neoformans*. Like the latter species, *T. cutaneum* is also urease-positive, but it lacks a visible capsule when stained with india ink.

G. *Trichosporon montevideense* and *T. domesticum*

Trichosporon montevideense has recently been isolated from human nails (E. Guého, unpublished data), and *T. domesticum* from skin (94). They are close relatives (94, 107), but *T. domesticum* grows at 37°C and *T. montevideense* does not. Both species produce only hyphae and arthroconidia. *Trichosporon montevideense* was represented by only two strains in the taxonomic revision (83), but it does not seem to be a rare species since it also corresponds to the standard culture of serotype III, another causative agent of the summer-type hypersensitivity pneumonitis (110).

IV. *MALASSEZIA* BAILLON

The genus *Malassezia* belongs to the basidiomycetous yeasts because of its multi-layered cell wall, enteroblastic budding, urease activity, and a positive staining reaction with DBB. Phylogenetically it belongs to the Ustilaginomycetes, where it forms a well-defined cluster, classified as the Malasseziales R.T. Moore emend. (Fig. 1), which is closely related to the Exobasidiales and Ustilaginales (2, 116). The recognition of species in *Malassezia* has caused considerable confusion (117). The genus was created in 1889 by Baillon (118) with the species *M. furfur* to accommodate the filamentous fungus observed in scales of the human skin disease pityriasis versicolor (PV). In fact, filaments and round, yeastlike cells had been observed to occur in PV scales much before, and earlier names were proposed for the yeastlike cells (117). *Pityrosporum* (119) was proposed as an alternative generic name, but unfortunately no living culture was preserved. Because the name *Malassezia* was published before *Pityrosporum*, the former has nomenclatural priority. For a long time the genus remained limited to *M. furfur* and *M. pachydermatis*. The first species concept included *Malassezia* isolates that were lipophilic and lipid-dependent and that occur only in humans. *M. pachydermatis*, still regarded as a valid single species, is lipophilic but not lipid-dependent, and usually occurs on animals (120). In keeping with this taxonomy, for many years all pathologies caused by *M. furfur sensu lato*, particularly disorders of the skin (PV, dandruff, seborrhoeic dermatitis, and folliculitis), were ascribed to a single species (121). Only the recent recognition of new clinically important species changed this approach (122).

Ribosomal RNA sequences published in 1995 revealed seven genetic entities (123), which later were described as species with distinct morphological and physiological key characteristics (117). These characteristics were established in liquid medium (117), but a practical identification scheme based on tests performable on agar media was established for the laboratory routine (124). The recent addition of esculin and cremophor EL (castor oil) tests (125) further improved this identification scheme. Primary tests for the identification of *Malassezia* species are catalase and β-glucosidase activities (esculin tubes), followed by evaluation of growth with Tweens 20, 40, 60, 80, and cremophor EL by using the diffusion method on Sabouraud glucose agar. In this method each compound is deposited in a well cut into an agar plate or on a paper disk placed on the plate (Figs. 4a, b). Because three species have their maximum temperatures for growth at 37°C, all tests are performed at 32–34°C. The ability to grow at 37°C is evaluated as a useful but not indispensable characteristic for identification (126). Fortunately, easy-to-apply molecular techniques, such as PCR-restriction fragment length polymorphism (PCR-RFLP) (126a,b) and amplified fragment length polymorphism (AFLP) (126c) analysis, have shortened and facilitated considerably the identification of *Malassezia* yeasts, in particular that of the strongly lipid-dependent species.

A. *Malassezia furfur*

This species is represented by two neotype cultures, namely CBS 1878 from a scalp and CBS 7019 from lesions of PV. The two cultures proved to be conspe-

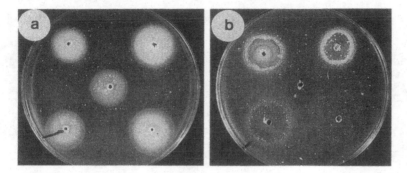

Figure 4 Plates showing growth of *M. furfur* (a), and *Malassezia globosa* (b) on media with Tween 20, Tween 40, Tween 60, and Tween 80 (start at black bar and clockwise) and cremophor EL (castor oil) (center). *M. furfur* shows distinct growth with all substrates, whereas *M. globosa* shows weak or no growth with these substrates. (Notice the white precipitate around Tween 40 and 60.)

cific, and *M. furfur* is maintained as the generic type species (117). Several questions remain about this species, however. All strains retained as *M. furfur* showed high percentages of DNA/DNA reassociation and high ribosomal RNA similarity (123), but two karyotypes were observed (127). Moreover, the species is morphologically heterogeneous with globose, oval, or cylindrical yeast cells, even though it is physiologically homogeneous. *Malassezia furfur* can be routinely identified by the combined characteristics of its ability to grow at 37°C, a strong catalase reaction, absence or a very weak β-glucosidase activity, and equal growth in the presence of Tweens 20, 40, 60, 80 and cremophor EL as sole lipid sources (126) (Fig. 4a). Some strains are able to produce filaments, either spontaneously or under particular culture conditions (128) (Fig. 5a). These filaments should be interpreted as pseudohyphae, not as true hyphae, because they lack septa with a central pore. Strains of the species originate from various hosts, sites, and diseases. *M. furfur* was not observed in recent epidemiological surveys of healthy persons and patients with PV or seborrhoeic dermatitis (SD) (129) and healthy volunteers (130). This absence may perhaps be caused by the isolation protocols used, or may arise from competition between different skin-inhabiting species of *Malassezia* (131). When the same protocol is used, however, the species has been isolated from systemic and mucosal sites, such as urine, vagina, and blood (127), or exposed sites such as nails (E. Guého, unpublished data). As a preliminary conclusion, we can state that the species is a pathogen, but its role in disease remains to be elucidated. Because *M. furfur* is mildly lipid-dependent, it survives—particularly in collections—better than the other species. It has also been isolated from animals (132–134).

Malassezia slooffiae, one of the four new species, may be misidentified as

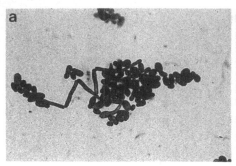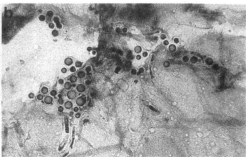

Figure 5 Micromorphology of *M. furfur* (a), and *Malassezia globosa* (b): (a) gram-stained oval yeast cells and filaments of *M. furfur* CBS 7019 originally isolated from pityriasis versicolor; (b) scales of pityriasis versicolor with parker ink-stained globose yeast cells and short filaments of *M. globosa*.

M. furfur. It can be differentiated from the latter species, however, by its unique absence of growth with cremophor EL (126). The species is regularly isolated from human skin and is mostly found in association with *M. sympodialis* or *M. globosa*. Its role as human pathogen is likely very weak, but it may be better adapted to animals, especially pigs (132, 133).

  Malassezia obtusa, another new species, resembles *M. furfur* morphologically but differs physiologically. It does not grow at 37°C, nor with any of the five lipids used in the tests. It darkens esculin medium (126), however. It is a very rare species that so far is known to occur only in the healthy skin of humans.

B. *Malassezia pachydermatis*

Malassezia pachydermatis is the only lipophilic *Malassezia* species able to grow without supplementation of long-chain fatty acids or their esters. So far, the non-lipid-dependent *Malassezia* yeasts are considered to represent a single species, for which the epithet *pachydermatis* has priority (135). The species may be in the process of speciation (123), however, likely in relation to host specificity (136). In contrast to *M. furfur*, it has a constant karyotype (127), and minor phenotypic differences seen among isolates correlate with differences observed in rDNA genotypes. All isolates grow well at 37°C, and some primary cultures show a certain lipid dependence (137, 138), therefore epidemiological surveys of *Malassezia* yeasts from any animal or human source should utilize lipid-supplemented media. The weakly lipid-dependent isolates have smaller colonies than those that lack a response to lipid supplements. Differences in catalase and β-glucosidase expression, in Tweens 20, 40, 60, 80, and cremophor EL reactivity occur in all rDNA genotypes (126). These different compounds, particularly Tween 20 and cremophor EL, may be more or less inhibitory, as growth may occur only at some distance from the compounds (139). Even if these phenotypic differences are found to be reproducible, however, they may not be sufficient to discriminate consistent separate species among the non-lipid-dependent isolates. *M. pachydermatis* is rare in humans, although it has been found to cause septic epidemics, usually in infants as a complication of prematurity (140, 141). In one case, the contamination was linked to a nurse's dog (142). *M. pachydermatis* is well known as a normal cutaneous inhabitant of numerous warm-blooded animals. Seborrhoeic dermatitis and otitis associated with this lipophilic yeast are now commonly recognized, especially in dogs (143). So far, however, it has not been possible to relate differences in pathogenicity to phenotypic or genotypic traits (136, 144).

C. *Malassezia sympodialis*

This lipophilic and lipid-dependent *Malassezia* yeast was described in 1990 (145), but only the comparison of a large number of strains resulted in a more accurate circumscription of the species (117). It correlates to the former *M. furfur* serovar A (146) and can be characterized in routine identifications by a strong β-glucosidase activity, which deeply darkens the esculin medium in 24 hr. The species grows at 37°C. Cremophor EL as a unique lipid supplement does not allow good growth (as also seen in *M. slooffiae*, but this species does not split esculin). Primary isolates may develop a ring of tiny colonies at some distance from the cremophor EL source. Morphologically, the yeast cells are small and ovoid, and they are not able to form filaments in culture. The species is commonly isolated from healthy as well as diseased skin (129). Its role as a pathogen is not yet elucidated. Indeed, in skin lesions *M. sympodialis* is often present, but usually associated with the more abundantly occurring *M. globosa* (130). The species has also been isolated from healthy feline skin (147).

D. *Malassezia globosa*

This species has a stable micromorphology in that its yeast cells remain spherical even after several transfers (117). Buds are also spherical and emerge from the mother yeast through a narrow site, contrary to the patterns seen in other *Malassezia* species. The species corresponds to the original description of *Pityrosporum orbiculare* obtained from a PV case (148). It seems likely, however, that many data published in the past as *P. orbiculare* correspond in large part to other species. Like *M. furfur*, *M. globosa* is able to produce filaments. Particularly in primary cultures, a certain number of yeasts produce germination tubes of various lengths (Fig. 5b), but this ability disappears after transfer. *M. globosa* correlates to former *M. furfur* serovar B (146). The yeast does not grow at 37°C or does so poorly, does not grow on the five lipid substrates (Fig. 4b), and does not split esculin. *Malassezia globosa* is the most important species in PV (148a), either alone or associated with other species, particularly *M. sympodialis* (131). The species is present in other types of cutaneous lesions as well, but more studies are necessary to examine its role in different pathologies, alone or in synergy with other species. *Malassezia globosa* has also been isolated from a cat (147), but it mainly occurs on humans (122, 131, 148a).

E. *Malassezia restricta*

This is the only lipid-dependent species lacking catalase activity (117). Like *M. globosa*, *M. restricta* lacks β-glucosidase activity, does not grow at 37°C, and is strongly lipid-dependent. Growth of the colonies is very restricted. The species is isolated almost exclusively from the head, including scalp, neck, and face (130), and it corresponds to serovar C (146). Because of its localization on the human head and its small oval to round yeast cells, *M. restricta* resembles the former concept of *Pityrosporum ovale* (149). Unfortunately, no culture has been preserved to represent the latter name. *Malassezia restricta* does not produce any filaments. Although *M. restricta* is very fastidious, more studies are needed to understand its implication in *Malassezia*-associated diseases, in particular dandruff and seborrhoeic dermatitis. This species is not known to occur in animals (132, 133).

V. *RHODOTORULA* HC HARRISON

The genus *Rhodotorula* is an anamorphic genus with 34 species of mainly red-pigmented basidiomycetous yeasts, belonging to the Urediniomycetes (2, 150). All species are considered to be nonpathogenic or of low pathogenicity, but may cause severe infections after gaining access to normally sterile regions of the human body, mainly through indwelling catheters (151–155). During a 29-month survey *Rhodotorula* yeasts comprised 0.71% of the total of 8062 fungal specimens cultured at a clinical center (156). In one year, sepsis due to *Rhodotorula* species occurred in 7 out of 47 patients at another clinical center after implantation of central venous catheters. From 1985 onwards an increase of the incidence of catheter-related infections with *Rhodotorula* yeasts was observed (157). In 5 years 36 patients were found to be infected, and all had underlying diseases such as tumors, leukemia, and AIDS, and almost all had catheters. Of the 23 clinically relevant cases 22 were caused by *Rhodotorula mucilaginosa* (= *Rh. rubra*), and one was caused by *Rh. minuta* (155).

　　Rhodotorula mucilaginosa (= *Rh. rubra*) is the only species considered to be clinically relevant (151). Isolates of this species present in the CBS yeast collection have been obtained from a wide variety of sources, including lung, nail, lymph nodes, skin, and feces (72), although by no means were all of these isolations etiologically significant. The species is mainly reported from patients receiving total parenteral nutrition, but also from patients with AIDS, carcinoma, endocarditis, meningitis, sepsis, and keratitis (158–164).

Another species, *Rhodotorula glutinis* var. *glutinis*, has been reported from patients receiving nutrition or antibiotics through indwelling catheters, and also from keratitis (165–167). *Rhodotorula minuta* has been reported from an AIDS patient (168). Unidentified *Rhodotorula* yeasts were isolated from patients suffering from endocarditis, carcinoma, leukemia, corneal infection, or diabetes, or receiving neurosurgery (151–154, 169–171).

VI. *SPOROBOLOMYCES* KLUYVER & CB NIEL

The genus *Sporobolomyces* is an anamorphic yeast genus containing 21 species and belonging to the Urediniomycetes (2, 172). All species are considered to be nonpathogenic, but may be involved in AIDS-related infections (173, 174). *Sporobolomyces holsaticus* was reported to cause dermatitis, and *Sporobolomyces roseus* was isolated from a case of madura foot and from mycotic lesions of mucous membranes (72, 175, 176). The ballistoconidia of *Sporobolomyces* species may cause respiratory allergy (177). Isolates of *S. salmonicolor* (including *S. johnsonii* and *S. holsaticus*) are known from infected skin, but also from cerebrospinal fluid, and *S. roseus* has been isolated from a mycotic lesion on skin and a case of madura foot (72). As these yeasts widely occur in nature, however, and produce air-borne conidia, these isolates may represent contaminants.

VII. *TILLETIOPSIS* DERX EX DERX AND *PSEUDOZYMA* BANDONI EMEND. BOEKHOUT

These yeastlike fungi represent anamorphs of the plant pathogenic smut fungi (Ustilaginomycetes) (2, 116, 178, 179). *Tilletiopsis* species from ballistoconidia, whereas acropetally branched chains of fusiform to cylindrical blastoconidia occur on sterigmalike outgrowths in *Pseudozyma* (180, 181). *Tilletiopsis albescens* is known from a lesion of colostomy, and *T. minor* has been isolated from the human cervix and urethra and also from a puncture from an eye of a 70-year-old human (178). *Pseudozyma* strains have been isolated from diverse substrates, including skin and blood (182). Pathogenicity has not been proven in any of these cases, however, and the isolates may represent contaminants.

REFERENCES

1. CP Kurtzman, JW Fell. The Yeasts: A Taxonomic Study. 4th ed. Amsterdam: Elsevier, 1998.

2. JW Fell, T Boekhout, A Fonseca, G Scorzetti, A Statzell-Tallman. Biodiversity and systematics of basidiomycetous yeasts as determined by large subunit rDNA D1/D2 domain sequence analysis. Internat J Syst Evol Microbio 50:1351–1371, 2000.

3. DH Howard, KJ Kwon-Chung. Zoopathogenic basidiomycetous yeasts. Stud Mycol 38:59–66, 1995.

4. JW Fell, A Statzell-Tallman. *Cryptococcus* Vuillemin. In: CP Kurtzman, JW Fell, eds. The Yeasts: A Taxonomic study. 4th ed. Amsterdam: Elsevier, 1998, pp. 742–767.

5. TG Mitchell, JR Perfect. Cryptococcosis in the era of AIDS—100 years after the discovery of *Cryptococcus neoformans*. Clin Microbio Rev 8:515–548, 1995.

6. CW Schupbach, CE Wheeler, RA Briggaman, NA Warner, EP Kanof. Cutaneous manifestations of disseminated cryptococcosis. Arch Derm 112:1734–1740, 1976.

7. A Casadevall, JR Perfect. *Cryptococcus neoformans*. Washington, DC: ASM Press, 1998.

8. JA Kovacs, AA Kovacs, M Polis, WG Wright, VJ Gill, CU Tuazon, EP Gellmann, HC Lane, R Longfield, G Overturf, AM Macher, AS Fauci, JE Parrillo, JE Bennett, H Masur. Cryptococcosis in the acquired immunodeficiency syndrome. Ann Intern Med 103:533–538, 1985.

9. R Ikeda, R Shinoda, Y Fukuzawa, L Kaufman. Antigenic characterization of *Cryptococcus neoformans* serotypes and its application to serotyping of clinical isolates. J Clin Microbio 6:22–29, 1982.

10. KJ Kwon-Chung. A new genus, *Filobasidiella*, the perfect state of *Cryptococcus neoformans*. Mycologia 67:1197–1200, 1975.

11. KJ Kwon-Chung. A new species of *Filobasidiella*, the sexual state of *Cryptococcus neoformans* B and C serotypes. Mycologia 68:942–946, 1976.

12. KJ Kwon-Chung, JE Bennett. Epidemiologic differences between the two varieties of *Cryptococcus neoformans*. Amer J Epidem 120:123–130, 1984.

13. KJ Kwon-Chung, JE Bennett, TS Theodore. *Cryptococcus bacillisporus* sp. nov.: Serotype B-C of *Cryptococcus neoformans*. Internat J Syst Bact 28:616–620, 1978.

14. KJ Kwon-Chung, I Polacheck, JE Bennett. Improved diagnostic medium for separation of *Cryptococcus neoformans* var. *neoformans* (serotypes A and D) and *Cryptococcus neoformans* var. *gattii* (serotypes B and C). J Clin Microbio 5:535–537, 1982.

15. KJ Kwon-Chung, JE Bennett, JC Rhodes. Taxonomic studies on *Filobasidiella* species and their anamorphs. Antonie van Leeuwenhoek 48:25–38, 1982.

16. TJ Pfeiffer, DH Ellis. Serotypes of Australian environmental and clinical isolates of *Cryptococcus neoformans*. J Med Vet Mycol 31:401–404, 1993.

17. DE Wilson, JE Bennett, JW Bailey. Serologic grouping of *Cryptococcus neoformans*. Proc Soc Exp Bio 27:820–823, 1968.

18. SP Franzot, IF Salkin, A Casadevall. *Cryptococcus neoformans* var. *grubii*: Separate varietal status for *Cryptococcus neoformans* serotype A isolates. J Clin Microbio 7:838–840, 1999.

19. KA Schmeding, SO Jong, R Hugh. Sexual compatibility between serotypes of *Filobasidiella neoformans (Cryptococcus neoformans)*. Curr Microbio 5:133–138, 1981.

20. HS Aulakh, SE Straus, KJ Kwon-Chung. Genetic relatedness of *Filobasidiella neoformans (Cryptococcus neoformans)* and *Filobasidiella bacillispora (Cryptococcus bacillisporus)* as determined by deoxyribonucleic acid base composition and sequence homology studies. Internat J Syst Bact 31:97–103, 1981.

21. BL Wickes, TDE Moore, KJ Kwon-Chung. Comparison of the electrophoretic karyotypes and chromosomal location of ten genes in the two varieties of *Cryptococcus neoformans*. Microbiology 40:543–550, 1994.

22. T Boekhout, A van Belkum, ACAP Leenders, HA Verbrugh, P Mukamurangwa, D Swinne, WA Scheffers. Molecular typing of *Cryptococcus neoformans*: Taxonomic and epidemiological aspects. Internat J Syst Bact 47:432–442, 1997.

23. A Varma, D Swinne, F Staib, JE Bennett, KJ Kwon-Chung. Diversity of DNA fingerprints in *Cryptococcus neoformans*. J Clin Microbio 33:1807–1814, 1995.

24. W Meyer, K Marszewska, S Kidd, J Holland, T Sorrell. Molecular epidemiology of *Cryptococcus neoformans*—Standardization of techniques for a detailed global genotypic analysis. 4th International Conference *Cryptococcus* and Cryptococcosis, London, Sept. 12–16, 1999.

25. JE Bennett, KJ Kwon-Chung, TS Theodore. Biochemical differences between serotypes of *Cryptococcus neoformans*. Sabouraudia 16:167–174, 1978.

26. R Dufait, R Velho, C De Vroey. Rapid identification of the two varieties of *Cryptococcus neoformans* by D-proline assimilation. Mykosen 30:483, 1987.

27. P Mukaramangwa, C Raes Wuytack, C De Vroey. *Cryptococcus neoformans* var. *gattii* can be separated from var. *neoformans* by its ability to assimilate D-tryptophan. J Med Vet Mycol 33:419–420, 1995.

28. I Polacheck, KJ Kwon-Chung. Creatinine metabolism in *Cryptococcus neoformans* and *Cryptococcus bacillisporus*. J Bact 42:15–20, 1980.

29. T Boekhout, G Scorzetti. Differential killer toxin sensitivity patterns of varieties of *Cryptococcus neoformans*. J Med Vet Mycol 35:147–149, 1987.

30. SM Levitz. The ecology of *Cryptococcus neoformans* and the epidemiology of cryptococcosis. Rev Infec Dis 3:1163–1169, 1991.

31. D Swinne-Desgain. *Cryptococcus neoformans* of saprophytic origin. Sabouraudia 13:303–308, 1975.

32. F Dromer, A Varma, O Ronin, S Mathoulin, B Dupont. Molecular typing of *Cryptococcus neoformans* serotype D clinical isolates. J Clin Microbio 32:2364–2371, 1994.

33. MS Lazera, B Wanke, NM Nishikawa. Isolation of both varieties of *Cryptococcus neoformans* from saprophytic sources in the city of Rio de Janeiro, Brazil. J Med Vet Mycol 31:449–454, 1993.

34. DH Ellis, TJ Pfeiffer. Ecology, life cycle, and infectious propagule of *Cryptococcus neoformans*. 1990, 336:923–925, Lancet.

35. D Ellis, T Pfeiffer. The ecology of *Cryptococcus neoformans*. Eur J Epidem 8: 321–325, 1992.

36. TC Sorrell, AG Brownlee, P Ruma, R Malik, TJ Pfeiffer, DH Ellis. Natural environmental sources of *Cryptococcus neoformans* var. *gattii*. J Clin Microbio 34:1261–1263, 1996.

37. E Castañeda, N Ordoñez, A Callejas, MC Rodríguiez, A Castañeda, S Huérfano. In search of the habitat of *Cryptococcus neoformans* var. *gattii* in Colombia. 4th

International Conference *Cryptococcus* and Cryptococcosis, London, Sept. 12–16, 1999.

38. A Callejas, N Ordoñez, MC Rodríguez, E Castañeda. First isolation of *Cryptococcus neoformans* var. *gattii* serotype C, from the environment in Colombia. Med Mycol 36:341–344, 1998.

39. F Staib. *Cryptococcus neoformans* und *Guizotia abyssinica* (syn. *G. oleiferea* D.C.) (Farbereaktion für *C. neoformans*). Z Hyg 148:466–475, 1962.

40. KJ Kwon-Chung. *Filobasidiella* Kwon-Chung. In: CP Kurtzman, JW Fell, eds. The Yeasts: A Taxonomic Study. 4th ed. Amsterdam: Elsevier, 1998, pp. 656–662.

41. BL Zimmer, GD Roberts. Rapid selective urease test for presumptive identification of *Cryptococcus neoformans*. J Clin Microbio 10:380–381, 1979.

42. CE Canteros, L Rodero, MO Rivas, G Davel. A rapid urease test for presumptive identification of *Cryptococcus neoformans*. Mycopathologia 136:21–23, 1996.

43. AJ Bava, R Negroni, M Bianchi. Cryptococcosis produced by a urease negative strain of *Cryptococcus neoformans*. J Med Vet Mycol 31:87–89, 1993.

44. D Swinne. Study of *Cryptococcus neoformans* varieties. Mykosen 27:137–141, 1984.

45. R Ikeda, T Shinoda, Y Fukazawa, L Kaufman. Antigenic characterization of *Cryptococcus neoformans* serotypes and its application to serotyping of clinical isolates. J Clin Microbio 16:22–29, 1982.

46. F Dromer, E Guého, O Ronin, B Dupont. Serotyping of *Cryptococcus neoformans* by using a monoclonal antibody specific for capsular polysaccharide. J Clin Microbio 31:359–363, 1993.

47. T Boekhout, B Theelen, M Diaz, JW Fell, WC Hop, E Abeln, F Dromer, W Meyer. Hybrid genotypes in the pathogenic yeast *Cryptococcus neoformans*. Microbiology 147:891–907, 2001.

48. M Diaz, T Boekhout, B Theelen, JW Fell. Molecular sequence analysis of the intergenic spacer (IGS) associated with rDNA of the two varieties of the pathogenic yeast *Cryptococcus neoformans*. Syst Appl Microbio 23:535–545, 2000.

49. SP Franzot, IF Salkin, A Casadevall. *Cryptococcus neoformans* var. *grubii*: separate varietal status for *Cryptococcus neoformans* serotype A isolates. J Clin Microbio 37:838–840, 1999.

50. SP Franzot, BC Fries, W Cleare, A Casadevall. Genetic relationship between *Cryptococcus neoformans* var. *neoformans* strains of serotypes A and D. J Clin Microbio 36:2200–2204, 1998.

51. T Boekhout, B Theelen, A Abeln, M Diaz, JW Fell. Population genetics based epidemiology of *Cryptococcus neoformans*. 4th International Conference *Cryptococcus* and Cryptococcosis, London, Sept. 12–16, 1999.

52. ME Brandt, LC Hutwagner, RW Pinner and the Cryptococcal Disease Active Surveillance Group. Comparison of multilocus enzyme electrophoresis and random amplification of polymorphic DNA analysis for molecular subtyping of *Cryptococcus neoformans*. J Clin Microbio 33:1890–1895, 1995.

53. ME Brandt, LC Hutwagner, LA Klug, WS Baughman, D Rimland, EA Graviss, RJ Hamill, C Thomas, PG Pappas, AL Reingold, RW Pinner, and the Cryptococcal Disease Active Surveillance Group. Molecular subtype distribution of *Cryptococ-*

cus neoformans in four areas of the United States. J Clin Microbio 34:912–917, 1996.

54. SP Franzot, JS Hamdan, BP Currie, A Casadevall. Molecular epidemiology of *Cryptococcus neoformans* in Brazil and the United States: Evidence for both local genetic differences and a global clonal population structure. J Clin Microbio 35: 2243–2251, 1997.

55. S Krajden, RC Summerbell, J Kane, IF Salkin, ME Kemna, MG Rinaldi, M Fuksa, E Spratt, C Rodrigues, J Choe. Normally saprobic cryptococci isolated from *Cryptococcus neoformans* infections. J Clin Microbio 29:1883–1887, 1991.

56. L Binder, A Csillag, G Tóth. Diffuse infiltration of the lungs associated with *Cryptococcus luteolus*. Lancet 260:1043–1045, 1956.

57. J Barnett, R Payne, D Yarrow. The Yeasts, Characteristics and Identification. 3rd ed. Cambridge: Cambridge University Press, 2000.

58. S-R Lin, C-F Peng, S-A Yang, H-S Yu. Isolation of *Cryptococcus albidus* var. *albidus* in patient with pemphigus foliaceus. Kaohsiung J Med Sci 4:126–128, 1988.

59. RA Krumholz. Pulmonary cryptococcosis. Amer Rev Resp Dis 105:421–424, 1972.

60. ID Horowitz, EA Blumberg, L Krevolin. *Cryptococcus albidus* and mucormycosis empyema in a patient receiving hemodialysis. South Med J 86:1070–1072, 1993.

61. J Loison, JP Bouchara, E Guého, L de Gentile, B Cimon, JM Chennebault, D Chabasse. First report of *Cryptococcus albidus* septicaemia in an HIV patient. J Infec 33:139–140, 1996.

62. JL Gluck, JP Myers, LM Pass. Cryptococcemia due to *Cryptococcus albidus*. South Med J 80:511–513, 1987.

63. JC Melo, S Srinivasan, ML Scott, MJ Raff. *Cryptococcus albidus* meningitis. J Infec 2:79–82, 1980.

64. T DaCunha, J Lusins. *Cryptococcus albidus* meningitis. South Med J 66:1230–1243, 1973.

65. GD Taylor, M Buchanan-Chell, T Kirkland, M McKenzie, R Wiens. Trends and sources of nosocomial fungaemia. Mycoses 37:187–190, 1994.

66. GM Wells, A Gajjar, TA Pearson, KL Hale, JL Shenep. Pulmonary cryptosporidiosis and *Cryptococcus albidus* fungemia in a child with acute lymphicytic leukemia. Med Ped Oncol 31:544–546, 1998.

67. T Kordossis, A Avlami, A Velegraki, I Stefanou, G Georgakopoulos, C Papalambrou, NJ Legakis. First report *of Cryptococcus laurentii* meningitis and a fatal case of *Cryptococcus albidus* cryptococcaemia in AIDS patients. Med Mycol 36:335–339, 1998.

68. HG Wieser. Zur Frage der Pathogenität des *Cryptococcus albidus*. Schweiz Med Wschr 103:475–481, 1973.

69. MA Gordon. Pulmonary cryptococcosis: A case due to *Cryptococcus albidus*. Amer Rev Resp Dis 106:786–787, 1972.

70. A Fonseca, G Scorzetti, JW Fell. Diversity in the yeast *Cryptococcus albidus* and related species as revealed by ribosomal DNA sequence analysis. Can J Microbio 46:7–27, 2000.

71. F Dromer, A Moulignier, B Dupont, E Guého, M Baudrimont, L Improvisi, F Provost, G Gonzalez-Canali. Myeloradiculitis due to *Cryptococcus curvatus* in AIDS. AIDS 9:395–408, 1995.

72. http://www.cbs.knaw.nl.

73. JP Lynch III, DR Schaberg, DG Kissner, CA Kauffman. *Cryptococcus laurentii* lung abscess. Amer Rev Resp Dis 123:135–138, 1981.

74. JT Sinnott, J Rodnite, PJ Emmanuel, A Campos. *Cryptococcus laurentii* infection complicating peritoneal dialysis. Pediat Inf Dis J 8:803–805, 1989.

75. H Mocan, AV Murphy, TJ Beattie, TA McAllister. Fungal peritonitis in children on continuous ambulatory peritoneal dialysis. Scott Med J 34:494–496, 1989.

76. V Krcméry, A Kunova, J Mardiak. Nosocomial *Cryptococcus laurentii* fungemia in a bone marrow transplant patient after prophylaxis with ketoconazole successfully treated with oral fluconazole. Infection 2:130, 1997.

77. A Kamalan, AS Thambiah. A study of 3891 cases of mycoses in the tropics. Sabouraudia 14:129–148, 1976.

78. A Kamalan, P Yesudian, AS Thambiah. Cutaneous infection by *Cryptococcus laurentii*. Brit J Derm 97:221–223, 1977.

79. PH Custis, JA Haller, E de Juan. An unusual case of cryptococcal endophthalmitis. Retina 15:300–304, 1995.

80. LB Johnson, SF Bradley, CA Kauffman. Fungaemia due to *Cryptococcus laurentii* and a review of non-*neoformans* cryptococcaemia. Mycoses 41:277–280, 1998.

81. A Castellani. A capsulated yeast producing black pigment: *Cryptococcus ater* n.sp. J Trop Med Hyg 63:27–30, 1960.

82. KJ Kwon-Chung. Perfect state of *Cryptococcus uniguttulatus*. Internat J Sys Bac 27:293–299, 1977.

83. E Guého, MTh Smith, GS de Hoog, G Billon-Grand, R Christen, WH Batenburg-van der Vegte. Contribution to a revision of the genus *Trichosporon*. Antonie van Leeuwenhoek 61:289–316, 1992.

84. G Behrend. Über Trichomycosis nodosa (Juhel-Renoy): Piedra (Osario). Berlin Klin Wochenschr 27:464–467, 1890.

85. L Rabenhorst. Zwei Parasiten an den todten Haaren der Chignons. Hedwigia 4:1, 1867.

86. E Guého, GS de Hoog, MTh Smith. Neotypification of the genus *Trichosporon*. Antonie van Leeuwenhoek 61:285–288, 1992.

87. GS de Hoog, E Guého, MTh Smith. Nomenclatural notes on some arthroconidial yeasts. Mycotaxon 63:345–347, 1997.

88. E Guého, L Improvisi, R Christen, GS de Hoog. Phylogenetic relationships of *Cryptococcus neoformans* and some related basidiomycetous yeasts determined from partial large subunit rRNA sequences. Antonie van Leeuwenhoek 63:175–189, 1993.

89. JW Fell, A Statzell-Tallman, MJ Lutz, CP. Kurtzman. Partial sequences in marine yeasts: A model for identification of marine eukaryotes. Molec Marine Bio Biotech 1:175–186, 1992.

90. JW Fell, H Roeijmans, T Boekhout. Cystofilobasidiales; a new order of basidiomycetous yeasts. Internat Syst Bacteriol 49:907–913, 1999.

91. EJ McManus, JM Jones. Detection of a *Trichosporon beigelii* capsular polysaccha-

ride in serum from a patient with disseminated *Trichosporon* infection. J Clin Microbio 21:681–685, 1985.

92. CE Hughes, D Serstock, BD Wilson, W Payne. Infection with *Trichosporon pullulans*. Ann L Intern Med 108:772–773, 1988.
93. T Sugita, A Nishikawa, T Shinoda. Reclassification of *Trichosporon cutaneum* by DNA relatedness by the spectrophotometric method and the chemiluminometric method. J Gen Appl Microbio 40:397–408, 1994.
94. T Sugita, A Nishikawa, T Shinoda, K Yoshima, M Ando. A new species, *Trichosporon domesticum*, isolated from the house of a summer-type hypersensitivity pneumonitis patient in Japan. J Gen Appl Microbio 41:429–436, 1995.
95. T Sugita, T Nakase. *Trichosporon japonicum* sp. nov. isolated from the air. Internat J Syst Bacteriol 48:1425–1429, 1998.
96. WJ Middelhoven, G Scorzetti, JW Fell. *Trichosporon guehoae* sp. nov., an anamorphic basidiomycetous yeast. Can J Bot 45:686–690, 1999.
97. WJ Middelhoven, G Scorzetti, JW Fell. *Trichosporon veenhuisii* sp. nov., an alkane-assimilating anamorphic basidiomycetous yeast. Internat J Syst Bacteriol 50:381–387, 2000.
98. E Guého, L Improvisi, GS de Hoog, B Dupont. *Trichosporon* on humans: A practical account. Mycoses 37:3–10, 1994.
99. E Guého, MTh Smith, GS de Hoog. *Trichosporon* Behrend. In: CP Kurtzman, JW Fell, eds. The Veasts: A Taxonomic Study. 4th ed. Amsterdam: Elsevier, 1998; pp. 854–872.
100. KC Watson, S Kallichurum. Brain abscess due to *T. cutaneum* J Med Microbio 3: 191–193, 1970.
101. GM Cox, JR Perfect. *Cryptococcus neoformans* var. *neoformans* and *gattii* and *Trichosporon* species. In: L Ajello, RJ Hay, eds. Topley and Wilson's Microbiology and Microbial Infections. Medical Mycology. 9th ed., vol 4. London: Arnold, 1998, pp. 461–484.
102. R Herbrecht, H Koenig, J Waller, L Liu, E Guého. *Trichosporon* infections: Clinical manifestations and treatment. J Mycol Méd 3:129–136, 1993.
103. M Therizol-Ferly, M Kombila, M Gomez de Diaz, TH Duong, D Richard-Lenoble. White piedra and *Trichosporon* species in equatorial Africa. I. History and clinical aspects: An analysis of 449 superficial inguinal specimens. Mycoses 37:279–253, 1994.
104. GS de Hoog, E Guého. Agents of white piedra, black piedra and tinea nigra. In: L Ajello, & RJ Hay, eds. Topley and Wilson's Microbiology and Microbial Infections. Medical Mycology. 9th ed, vol. 4. London: Arnold, 1998, pp. 191–197.
105. T Sugita, A Nishikawa, T Shinoda. Identification of *Trichosporon asahii* by PCR based on sequences of the internal transcribed spacer regions. J Clin Microbio 36: 2742–2744, 1998.
106. T Sugita, A Nishikawa, T Shinoda. Rapid detection of species of the opportunistic yeast *Trichosporon* by PCR. J Clin Microbio 36:1458–1460, 1998.
107. T Sugita, A Nishikawa, R Ikeda, T Shinoda. Identification of medically relevant *Trichosporon* species based on sequences of internal transcribed spacer regions and construction of a database for *Trichosporon* identification. J Clin Microbio 37: 1985–1993, 1999.

108. T Sugita, A Nishikawa, T Shinoda, H Kume. Taxonomic position of deep-seated, mucosa-associated, and superficial isolates of *Trichosporon cutaneum* from trichosporonosis patient. J Clin Microbio 33:1368–1370, 1995.

109. M Mahal, L Saiman, L Bitman, L Weitzman, M Grossman, F Dembitzer, P Della-Latta, J Garvin. Review of trichosporonosis with a report of a case of disseminated *Trichosporon asahii* infection. Infec Dis Clin Prac 7:175–179, 1998.

110. Y Nishiura, K Nakagawa-Yoshida, M Suga, T Shinoda, E Guého, M Ando. Assignment and serotyping of *Trichosporon* species: The causative agents of summer-type hypersensitivity pneumonitis. J Med Vet Mycol 35:45–52, 1997.

111. JO Lopes, SH Alves, C Klock, LTO Oliveira, NRF Dal Forno. *Trichosporon inkin* peritonitis during continuous ambulatory peritoneal dialysis with bibliography review. Mycopathologia 139:15–18, 1997.

112. RT Kenney, KJ Kwon-Chung, FG Witebski, DA Melnick, HL Malech, JI Gallin. Invasive infection with *Sarcinosporon inkin* in a patient with chronic granulomatous disease. Amer J Clin Path 94:344–350, 1990.

113. Thérizol-Ferly, M Kombila, M Gomez de Diaz, C Douchet, Y Salaun, A Barrabes, TH Duong, D Richard-Lenoble. White piedra and *Trichosporon* species in equatorial Africa. II. Clinical and mycological associations: An analysis of 449 superficial inguinal specimens. Mycoses 37:255–260, 1994.

114. I Surmont, B Vergauwen, L Marcelis, L Verbist, G Verhoef, M Boogaerts. First report of chronic meningitis caused by *Trichosporon beigelii*. Eur J Clin Microbio Infec Dis 9:226–229, 1990.

115. T Sugita, A Nishikawa, T Shinoda, T Kusunoki. Taxonomic studies on clinical isolates from superficial trichosporonosis patients by DNA relatedness. Jpn J Med Mycol 37:107–110, 1996.

116. D Begerow, R Bauer, T Boekhout. Phylogenetic placement of ustilaginomycetes anamorphs as deduced from nuclear LSU rDNA sequences. Mycol Res 104:53–60, 2000.

117. E Guého, G Midgley, J Guillot. The genus *Malassezia* with description of four new species. Antonie van Leeuwenhoek 69:337–355, 1996.

118. H Baillon. Traité de botanique médicale cryptogamique. Paris, Octave Douin 234–239, 1889.

119. R Sabouraud. Maladies du cuir chevelu. II—Les maladies desquamatives. Paris, Masson 296, 1904.

120. D Ahearn, D Yarrow. *Malassezia* Baillon. In: NJW Kreger-van Rij, ed. The Yeasts: A Taxonomic Study. 3rd ed. Amsterdam: Elsevier, 1984, pp. 882–885.

121. MJ Marcon, DA Powell. Human infections due to *Malassezia* spp. Clin Microbio Rev 5:101–119, 1992.

122. G Midgley, E Guého, J Guillot. Diseases caused by *Malassezia*. In: L Ajello, RJ Hay, eds. Topley and Wilson's Microbiology and Microbial Infections. Medical Mycology. 9th ed., vol 4. London: Arnold, 1998, pp. 201–211.

123. J Guillot, E Guého. The diversity of *Malassezia* yeasts confirmed by rRNA sequence and nuclear DNA comparisons. Antonie van Leeuwenhoek 67:297–314, 1995.

124. J Guillot, E Guého, M Lesourd, G Midgley, B Dupont. Identification of *Malassezia* species: A practical approach. J Mycol Méd 6:103–110, 1996.

125. P Mayser, P Haze, C Papavassilis, M Pickel, M Gründer, E Guého. Differentiation of *Malassezia* spp. Selectivity of cremophor EL, castor oil and ricinoleic acid for *M. furfur*. Brit J Derm 137:208–213, 1997.

126. E Guého, T Boekhout, HR Ashbee J Guillot, A Van Belkum, J Faergemann. The role of *Malassezia* species in the ecology of human skin and as pathogens. Med Mycol 36:220–229, 1998.

126a. J Guillot, M Deville, M Berthelemy, F Provost, E Guého. A single PCR-restriction endonuclease analysis for rapid identification of *Malassezia* species. Lett Appl Microbio 31:400–403, 2000.

126b. JG Gaitanis, A Velegraki, E Frangoulis, A Mitroussia, A Tsigonia, A Tzimogianni, A Katsambas, NJ Velegraki. Identification of *Malassezia* species from patient skin scales by PCR-RFLP. Clin Microbio Infec 8:162–173, 2002.

126c. B Theelen, M Silvestri, E Guého, A van Belkum, T Boekhout. Identification and typing of *Malassezia* yeasts using amplified fragment length polymorphism (AFLP™), random amplified polymorphic DNA (RAPD) and denaturing gradient gel electrophoresis (DGGE). FEMS Yeast Res 1:79–86, 2001.

127. T Boekhout, M Kamp, E Guého. Molecular typing of *Malassezia* species with PFGE and RAPD. Med Mycol 36:365–372, 1998.

128. J Guillot, E Guého, M-C Prévost. Ultrastructural features of the dimorphic yeast *Malassezia furfur*. J Mycol Méd 5:86–91, 1995.

129. HR Ashbee, E Ingham, KT Holland, WJ Cunliffe. The carriage of *Malassezia furfur* serovars A, B and C in patients with pityriasis versicolor, seborrhoeic dermatitis and controls. Brit J Derm 29:533–540, 1993.

130. C Aspiroz, L-M Moreno, A Rezusta, C Rubio. Differentiation of three biotypes of *Malassezia* species on human normal skin: Correspondence with *M. globosa*, *M. sympodialis* and *M. restricta*. Mycopathologia 145:69–74, 1999.

131. V Crespo Erchiga, A Ojeda Martos, A Vera Casano, A Crespo Erchiga, F Sanchez Fajardo, E Guého. Mycology of pityriasis versicolor. J Mycol Méd 9:143–148, 1999.

132. J Guillot, E Guého, M Mialot, R Chermette. Importance des levures du genre *Malassezia* en pratique vétérinaire. Point Vétér 29:21–31, 1998.

133. J Guillot, E Guého, R Chermette. Infections animales à *Malassezia*. Rev Praticien 49:1840–1843, 1999.

134. MJ Crespo, ML Abarca, FJ Cabanes. Isolation of *Malassezia furfur* from a cat. J Clin Microbio 37:1573–1574, 1999.

135. J Guillot, E Guého, R Chermette. Confirmation of the nomenclatural status of *Malassezia pachydermatis*. Antonie van Leeuwenhoek 67:173–176, 1995.

136. F Midreuil, J Guillot, E Guého, F Renaud, M Mallié, J-M Bastide. Genetic diversity in the yeast species *Malassezia pachydermatis* analysed by multilocus enzyme electrophoresis. Internat J Syst Bacteriol 49:1287–1294, 1999.

137. R Bond, RM Anthony. Characterization of markedly lipid-dependent *Malassezia pachydermatis* isolates from healthy dogs. J Appl Bacteriol 78:537–542, 1995.

138. D Senczek, U Siesenop, KH Böhm. Characterization of *Malassezia* species by means of phenotypic characteristics and detection of electrophoretic karyotypes by pulsed-field gel electrophoresis. Mycoses 42:409–414, 1999.

139. E Guého, J Guillot. Comments on *Malassezia* species from dogs and cats. Mycoses 42:673–674, 1999.

140. PA Mickelsen, MC Viano-Paulson, DA Stevens, P Diaz. Clinical and microbiological features of infection with *Malassezia pachydermatis* in high-risk infants. J Infec Dis 157:1163–1168, 1988.

141. A Van Belkum, T Boekhout, R Bosboom. Monitoring spread of *Malassezia* infections in a neonatal intensive care unit by PCR-mediating genetic typing. J Clin Microbio 32:2528–2532, 1994.

142. HJ Chang, HL Miller, N Watkins, M Arduino, DA Ashford, G Midgley, SM Aguero, R Pinto-Powell, CF von Reyn, W Edwards, R Pruitt, M McNeil, WR Jarvis. An epidemic of *Malassezia pachydermatis* in an intensive care nursery associated with colonization of health care workers' pet dogs. N Eng J Med 338:706–711, 1998.

143. J Guillot, R Bond. *Malassezia pachydermatis*: A review. Med Mycol 37:295–306, 1999.

144. J Guillot, E Guého, G Chévrier, R Chermette. Epidemiological analysis of *Malassezia pachydermatis* isolates by partial sequencing of the large subunit ribosomal RNA. Res Vet Sci 62:22–25, 1997.

145. RB Simmons, E Guého. A new species of *Malassezia*. Mycol Res 94:1146–1149, 1990.

146. AC Cunningham, JP Leeming, E Ingham, G Gowland. Differentiation of three serovars of *Malassezia furfur*. J Appl Bacteriol 68:439–446, 1990.

147. R Bond, SA Howell, PJ Haywood, DH Lloyd. Isolation of *Malassezia sympodialis* and *Malassezia globosa* from healthy pet cats. Vet Rec 141:200–201, 1997.

148. MA Gordon. The lipophilic mycoflora of the skin I: In vitro culture of *Pityrosporum orbiculare* n. sp. Mycologia 43:524–535, 1951.

148a. V Crespo Erchiga, A Ojeda, A Vera Casaño, A Crespo Erchiga, F Sanchez Fajardo. *Malassezia globosa* as the causative agent of pityriasis versicolor. Br J Derm 143:799–803, 2000.

149. A Castellani, AJ Chalmers. Manual of tropical medicine. London: Tindall Baillière, 1923.

150. JW Fell, A Statzell-Tallman. *Rhodotorula* FC Harrison. In: CP Kurtzman, JW Fell, eds. The Yeasts: A Taxonomic Study. 4th ed. Amsterdam: Elsevier, 800–827, pp. 1998.

151. DB Louria, SM Greenberg, DW Molander. Fungemia caused by certain nonpathogenic strains of the family Cryptococcaceae, New Eng J Med 263:1281–1284, 1960.

152. DB Louria, A Blevins, D Armstrong, R Burdick, P Lieberman. Fungemia caused by "nonpathogenic" yeasts. Arch Intern Med 19:247–252, 1967.

153. PA Leeber, I Scheer. *Rhodotorula* fungemia presenting as "endotoxic" shock. Arch Int Med 123:78–81, 1969.

154. I Marinová, V Szabadosová, O Brandeburová, V Krcméry. *Rhodotorula* spp. Fungemia in an immunocompromised boy after neurosurgery successfully treated with miconazole and 5-flucytosine: case report and review of the literature. Chemotherapy 40:287–289, 1994.

155. TE Kiehn, E Gorey, AE Brown, FF Edwards, D Armstrong. Sepsis due to *Rhodotorula* related to use of indwelling central venous catheters. Clin Infec Dis 4:841–846, 1992.

156. AE Jennings, JE Bennett. The isolation of red yeast-like fungi in a diagnostic laboratory. J Med Microbio 5:391–394, 1972.
157. KJ Kwon-Chung, JW Bennett. Medical Mycology. Philadelphia: Lea & Febiger, 1992.
158. DK Braun, CA Kauffman. *Rhodotorula* fungaemia: A life-threatening complication of indwelling central venous catheters. Mycoses 35:305–308, 1992.
159. Y Naveh, A Friedman, D Merzbach, N Hashman. Endocarditis caused by *Rhodotorula* successfully treated with 5-fluorocytosine. Brit Heart J 37:101–104, 1975.
160. RS Pore, J Chen. Meningitis caused by *Rhodotorula*. Sabouraudia 14:331–335, 1976.
161. E Segal, A Romano, E Eylan, R Stein, T Ben-Tovim. *Rhodotorula rubra*—Cause of eye infection. Mykosen 18:107–111, 1972.
162. H Papadogeorgakis, E Frangoulis, C Papaefstathiou, A Katsambas. *Rhodotorula rubra* fungaemia in an immunosuppressed patient. J Eur Acad Derm Venereol 12: 169–170, 1999.
163. AY Lui, GS Turett, DL Karter, PC Bellman, JW Kislak. Amphotericin B lipid complex therapy in an AIDS patient with *Rhodotorula rubra* fungemia. Clin Infec Dis 27:892–893, 1998.
164. OH Gyaurgieva, TS Bogomolova, GI Gorshkova. Menigitis caused by *Rhodotorula rubra* in an HIV-infected patient. J Med Vet Mycol 34:357–359, 1996.
165. FD Pien, RL Thompson, D Deye, GD Roberts. *Rhodotorula* septicemia. Mayo Clin Proc 55:258–260, 1980.
166. G Bertoli, F Rivasi, U Fabio. *Rhodotorula glutinis* keratitis. Int Ophthal 16:187–190, 1992.
167. C Casolari, A Nanetti, CM Cavallini, F Rivasi, U Fabio, A Mazoni. Keratomycosis with an unusual etiology (*Rhodotorula glutinis*): A case report. Microbiologica 15: 83–88, 1992.
168. LZ Goldani, DE Craven, AM Sugar. Central venous catheter infection with *Rhodotorula minuta* in a patient with AIDS taking suppressive doses of fluconazole. J Med Vet Mycol 33:267–270, 1995.
169. PF Shelburne, RJ Carey. *Rhodotorula* fungemia complicating staphylococcal endocarditis. JAMA 180:38–42, 1962.
170. A Panda, N Pushker, S Nainiwal, G Satpathy, N Nayak. *Rhodotorula* sp. infection in corneal interface following lamellar keratoplasty—A case report. Act Ophthal Scand 77:227–228, 1999.
171. JJ Rusthoven, R Feld, PG Tuffnell. Systemic infection by *Rhodotorula* spp. in the immunocompromised host. J Infec 8:241–246, 1984.
172. T Boekhout, T Nakase. *Sporobolomyces* Kluyver & van Niel. In: CP Kurtzman, JW Fell, eds. The Yeasts: A Taxonomic study. 4th ed. Amsterdam: Elsevier, 1998, pp. 828–843.
173. JT Morris, M Beckius, CK McAllister. *Sporobolomyces* infection in an AIDS patient. J Infec Dis 164:623–624, 1991.
174. J Plazas, J Portilla, V Boix, M Perez-Mateo. *Sporobolomyces salmonicolor* lymphadenitis in an AIDS patient. Pathogen or passenger? AIDS 8:387–388, 1994.
175. AG Bergman, CA Kauffman. Dermatitis due to *Sporobolomyces* infection. Arch Derm 20:1059–1060, 1984.

176. A Janke. *Sporobolomyces roseus* var. *madurae* var. nov. und die beziehungen zwischen den genera *Sporobolomyces* und *Rhodotorula*. Zentrlbl Bakteriol Parasitenk 161:514–520, 1954.

177. RG Evans. *Sporobolomyces* as a cause of respiratory allergy. Acta Allergol. 20: 197–205, 1965.

178. T Boekhout. A revision of ballistoconidia-forming yeast and fungi. Stud Mycol 33: 1–194, 1991.

179. T Boekhout, RJ Bandoni, JW Fell, KJ Kwon-Chung. Discussion of teleomorphic and anamorphic genera of heterobasidiomycetous yeasts. In: CP Kurtzman, JW Fell, eds. The Yeasts: A Taxonomic study. 4th ed. Amsterdam: Elsevier, 1998; pp. 609–625.

180. T Boekhout, JW Fell. *Pseudozyma* Bandoni emend: Boekhout and a comparison with the yeast stage of *Ustilago maydis* (De Candolle) Corda. In: CP Kurtzman, JW Fell, eds. The Yeasts: A Taxonomic Study. 4th ed. Amsterdam: Elsevier, 1998; pp. 790–797.

181. T Boekhout. *Tilletiopsis* Derx ex Derx. In: CP Kurtzman, JW Fell, eds. The Yeasts: A Taxonomic Study. 4th ed. Amsterdam: Elsevier, 1998; pp. 848–853.

182. T Boekhout. Systematics of anamorphs of Ustilaginales (smut fungi)—a preliminary survey. Stud Mycol 30:137–149, 1987.

10

Dematiaceous Hyphomycetes

Wiley A. Schell
Duke University Medical Center, Durham, North Carolina, U.S.A.

I. DESCRIPTION AND NATURAL HABITATS

Dematiaceous fungi comprise a heterogeneous group that is characterized by the presence of a melanin compound within the cell walls of the hyphae or spores (or both). Cell coloration varies from hyaline (colorless) to pale or mid-brown, depending on the concentration of melanin within individual cells. This causes colony coloration to range from gray, olive, light to dark brown, to black. The presence of melanin traditionally has been used to group these fungi as a practical basis for their identification. Dematiaceous fungi are represented in the Hyphomycetes, Ascomycetes, Coelomycetes, and Zygomycetes. Most are pathogens of plants or saprobes of senescent and decaying matter, and can be found in soil. They are widely disseminated in the environment, primarily via dispersal of their spores, and most can be encountered in the laboratory setting as insignificant isolates from nonsterile clinical specimens or as outright laboratory contaminants. In addition, they can cause opportunistic infections in humans and animals (1–5).

II. CLINICAL OVERVIEW

Dematiaceous fungi can cause chromoblastomycosis, mycetoma, phaeohyphomycosis, and sporotrichosis in normal and compromised hosts. In most cases, infection occurs following implantation of the fungus during abrasion or penetrating injury to the host. In contrast, cases of fungal sinusitis and at least some cases of pulmonary or disseminated infection are presumed to begin after conidia are inhaled. Chromoblastomycosis is a chronic, localized infection of skin and subcutaneous tissue that is caused mainly by three species of molds. When in host tissue, these molds undergo a morphologic conversion into subglobose, often multicellular (muriform) bodies. These bodies propagate by splitting along their septa. The finding of muriform bodies in cutaneous or subcutaneous tissue is pathognomonic of chromoblastomycosis, although very rarely they can be seen in other mycotic infections as well (1). In some cases, dematiaceous hyphae are present. Because all agents of chromoblastomycosis have the same morphology in host tissue, the identity of the fungus in a particular case can be determined only by culture.

Phaeohyphomycosis similarly begins with traumatic implantation or inhalation of cells, but the fungus appears in tissue as dematiaceous yeastlike cells, pseudohyphae, hyphae, or any combination of these forms. Most agents of phaeohyphomycosis form at least some dark cells in infected tissue, but some, such as species of *Alternaria*, *Bipolaris*, and *Curvularia*, often are hyaline in tissue because of scant melanin production. Still, these species are considered agents of phaeohyphomycoses because they clearly are dematiaceous when grown in culture. More than 100 fungi are documented causes of phaeohyphomycoses (2), and because the morphology of these agents is similar in infected tissue, they must be recovered in culture media before identification is possible. Historical usage and distinctions for the terms phaeohyphomycosis, chromoblastomycosis, and chromomycosis have been reviewed (1).

Sporotrichosis is caused only by *Sporothrix schenckii*. In host tissue, this mold converts to budding yeast cells. Most cases begin with implantation of the fungus, resulting in chronic lesions of skin and subcutaneous tissue that tend to spread via the lymphatic system. Musculoskeletal involvement and disseminated infection sometimes occur, and nasal and pulmonary infections following inhalation of conidia also have been documented (6–9). Sporotrichosis can be a zoonotic disease as well, as shown by numerous infections acquired from cats or other animals (1,10,11).

Mycetoma is a chronic, localized granulomatous infection of subcutaneous and cutaneous tissue that can spread to and destroy adjacent bone tissue. Infection follows traumatic implantation of the fungus. At least 26 species of fungi cause mycetoma, and several of them are dematiaceous molds (12,13).

III. IDENTIFICATION

Taxonomy, nomenclature, and identification of some dematiaceous fungi are problematic because of their phenotypic plasticity, and because of complexities and uncertainties in their life cycles. Application of molecular sequencing techniques coupled with powerful analytic algorithms has done much to establish apparent phylogenetic relationships among described taxa, as well as determining the homogeneity of isolates within various taxa. Still, in most settings, especially clinical laboratories, the traditional morphocentric approach remains the primary if not sole means by which dematiaceous fungi are identified.

Some species of dematiaceous fungi readily form a teleomorph during culture, and the identification accordingly is based on the teleomorph whether or not an anamorph is present (14). Other species are able to form only an anamorph, and the identification necessarily is based on this asexual morph. Still other species lack a teleomorph but are able to form multiple asexual forms, termed synanamorphs. Each of these synanamorphs may be referred to by a separate name as needed. One approach to identifying these synanamorphic fungi bases the name on the anamorph that is judged to be the predominant or most distinctive. An accompanying synanamorph then can be referred to, if need be, by using a genus-level name. For example, in the case of the *Fonsecaea pedrosoi*, the name is based on the synanamorph that forms a compactly sympodial, branched conidiophore. A *Rhinocladiella* synanamorph and a *Phialophora* synanamorph also may be seen, but these are not essential to the identification of an isolate as *F. pedrosoi* (15).

IV. DESCRIPTIONS

A. *Acrophialophora* Edward 1959

Acrophialophora fusispora (Saksena) M.B. Ellis 1971, has been identified as causing keratitis (16). Colonies are dense, matted, and feltlike at the base, with floccose aerial hyphae, initially dull white, becoming grayish-brown. Colony reverse is black. Conidiophores are erect, septate, smooth or roughened, thick-walled and brown at base, becoming pale toward apex (Fig. 1). Phialides are flask-shaped and swollen near base, with tapered neck, borne singly, in pairs or verticils on conidiophores or sometimes on vegetative hyphae. Conidia are one-celled, colorless to pale brown, broadly ellipsoidal to limoniform, with fine echinulations formed in spiral bands. Conidia are borne in long basipetal chains. Compare with *Paecilomyces* (17, 18).

B. *Alternaria* Nees 1816

Alternaria species have been reported in infections of bone, cutaneous tissue, ears, eyes, paranasal sinuses, and the urinary tract (19,20). The great majority of infections are caused by *A. alternata* (synonym *A. tenuis*) (21). *A. chlamydospora* has infected hyponychium (22) and skin (23), both *A. infectoria* and *A. longipes* have caused cutaneous phaeohyphomycosis (24,25), and *A. chartarum* has infected the nasal septum (26). Cutaneous infections have been attributed to *A.*

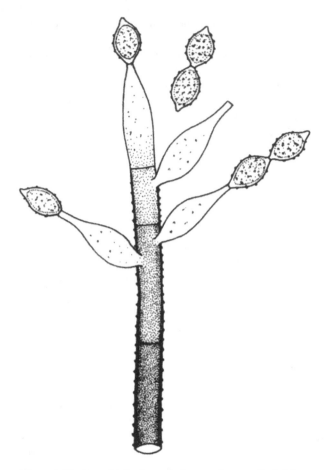

Figure 1 *Acrophialophora fusispora.* Source: Ref. 26a.

tenuissima (27–29). Colonies are rapidly growing, floccose, white to gray, becoming brown, reverse brown to black. Conidiophores are erect, dark, septate, simple, or branched. Conidia are muriform, obclavate, with beak (tapering apex), darkly pigmented, smooth or rough, in simple or branched acropetal chains. Compare with *Ulocladium, Stemphilium, Embeliasia* (4,5,17,30).

1.   *Alternaria alternata* (Fr.) Keissler 1912

A teleomorph is not known. Conidiophores are pale yellow-brown to midbrown, usually unbranched, mostly 30 to 50 μm long and one-to-three septate, bearing a single apical scar, sometimes with one to two subterminal scars (Fig. 2).

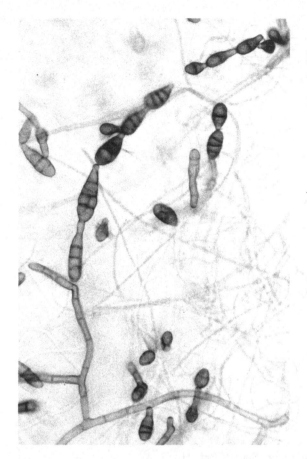

Figure 2 *Alternaria alternata.*

Conidia are smooth to minutely roughened, obpyriform or obclavate, sometimes ovoid, 20–63 (37) μm long, 9–18 (13) μm wide, three to eight transverse septa, one or more longitudinal septa, apical beak when present not exceeding one-fourth to one-third total length of conidium (Fig. 2). Conidia sometimes are solitary, but usually form in acropetal chains that sometimes branch (17).

2. *Alternaria chlamydospora* Mouchacca 1973

A teleomorph is not known. Conidia mostly are 20–50 μm × 7–20 μm, obclavate to obpyriform, and lack a defined beak (Fig. 3). Cells of most conidia enlarge and bulge in a manner reminiscent of chlamydospore development, resulting in distorted conidia. Conidia sometimes arise directly from hyphae (31).

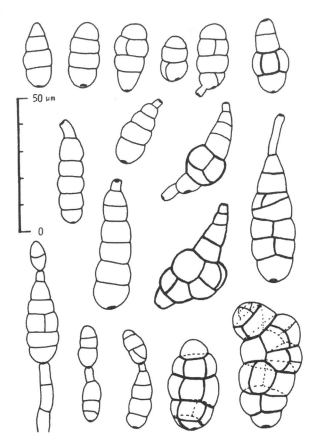

Figure 3 *Alternaria chlamydospora*. Source: Ref. 31.

3. *Alternaria longipes* (Ellis & Everhart) Mason 1928

This species is one of the small-spored species, and can be confused with *A. alternata*. The most prominent difference is the tendency of the conidia to form a beak that immediately continues its development into a conidiophore (Fig. 4). This transitional region of the conidium often is marked by a slight constriction (31). A teleomorph is not known.

4. *Alternaria tenuissima* (Kunze ex Persoon) Wiltshire 1933

This species is characterized by conidia that are straight or curved, obclavate or tapering to a beak, and solitary or borne in short chains (Fig. 5). The beak usually

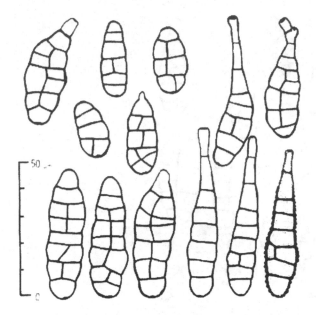

Figure 4 *Alternaria longipes.* Source: Ref. 31.

is swollen at the tip, but can be pointed. The beak can be as long as one-half the length of the conidium (17,31).

C. *Aureobasidium* Viala & Boyer 1891

Aerobasidium pullulans reportedly has caused infections of the peritoneum, nail, skin, and subcutaneous and deeper tissues (32–39). *Hormonema* species often are mistaken for *Aureobasidium* species, in part because they have been illustrated in the past under the name *Aureobasidium.* It is likely that some infections attributed to *A. pullulans* instead were caused by misidentified isolates of *Hormonema* species. *A. pullulans* differs from *Hormonema* spp. by having conidia that arise synchronously, and differs from *Phaeococcomyces* spp. by the lack of dematiaceous yeast cells. Compare with *Hormonema, Scytalidium* (5,17,30,40).

1. *Aureobasidium pullulans* (de Bary) Arnaud 1910

A. pullulans is the most common of the several described species. Colonies are smooth and moist, off-white, cream, light pink, or light brown, finally becoming dark brown from formation of dematiaceous arthroconidia. Conidiogenous cells are undifferentiated from hyphae; intercalary or terminal (Fig. 6). Conidia are

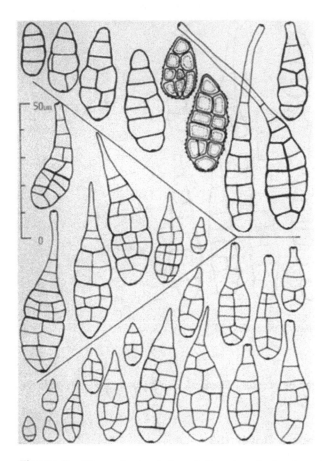

Figure 5 *Alternaria tenuissima*, left center; *A. longipes*, top; *A. alternata*, bottom. Source: Ref. 31.

hyaline, smooth, straight, ellipsoidal, one-celled, ranging in size from (7.5–) 9–11 (−16) × (3.5–) 4–5.5 (−7) μm (40). Secondary blastoconidia often are produced. Large, dark one-to-two-celled, thick-walled arthroconidia (i.e., a *Scytalidium* synanamorph) usually are present. In addition, conidia sometimes can be formed endogenously in clusters within hyphal cells. The teleomorph of *A. pullulans* might be *Discosphaerina fulvida* (41).

D. *Bipolaris* Shoemaker 1959

Several species, including *B. australiensis*, *B. hawaiiensis*, and *B. spicifera*, have caused meningitis; paranasal sinusitis; and subcutaneous, eye, pulmonary, and

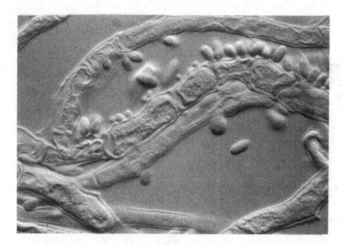

Figure 6   *Aureobasidium pullulans.*

disseminated infections (42–46). Species of *Drechslera, Exserohilum,* and *Helminthosporium* were confused with *Bipolaris* species in the past, but subsequent taxonomic studies established useful criteria for separating species of the genera. *Exserohilium* is distinguished by a protuberant hilum on each conidium. *Drechslera* differs from *Bipolaris* by its ability to germinate from any cell of its conidia rather than just the two end cells, and further by its conidial germ tube arising at a strong angle to the main axis of the conidium (43, 47). Colonies are rapidly growing, woolly, gray to black. Conidiophores are dark, erect, simple or branched, septate, and geniculate. Conidia are multidistoseptate, cylindrical to oblong, dark, with hila protruding only slightly. Conidia germinate only from end cells (with little exception), and the germination hyphae initially grow parallel to the long axis of the conidium. Compare with *Curvularia, Drechslera, Exserohilum, Helminthosporium, Nakateae* (4,5,43,47,48)

1.   *Bipolaris australiensis* (Ellis) Tsuda & Ueyama 1981

The teleomorph of this species is *Cochliobolus australiensis* and is heterothallic. Conidia are similar to *B. spicifera* except they (rarely) can have as many as four to five distosepta, are mostly 10 or fewer μm wide (most range from 18–33 μm × 8–10 μm), and do not exhibit a pale region just above the hila (Fig. 7).

2.   *Bipolaris hawaiiensis* (Ellis) Uchida & Aragaki 1979

The teleomorph is *Cochliobolus hawaiiensis* and is heterothallic. Conidia mostly have five distosepta but can have as few as two and as many as seven distosepta (Fig. 8). Conidial size usually ranges from 12–37 μm × 5–11 μm.

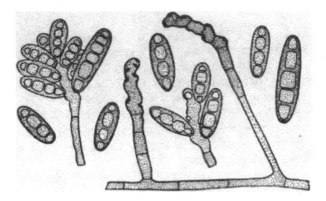

Figure 7   *Bipolaris australiensis*. Source: Ref. 17.

3.   *Bipolaris spicifera* (Banier) Subramanian 1971

The teleomorph of this species is *Cochliobolus spicifera* and is heterothallic. Conidia consistently have three (rarely four) distosepta (Fig. 9). Distosepta are evenly colored, none being darker than others. Most conidia are less than 40 μm long and more than 10 μm wide (typical size range is 30–36 μ × 11–13 μm). Mature conidia are straight and pale brown in color except for a small region just above the hila.

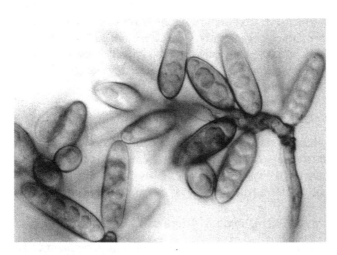

Figure 8   *Bipolaris hawaiiensis*.

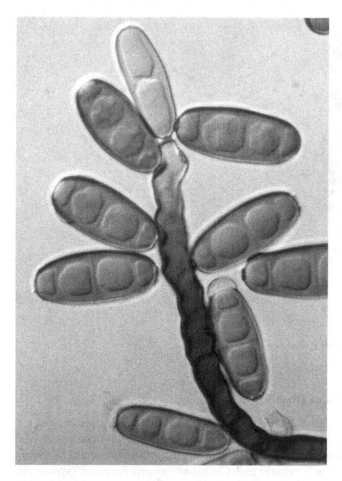

Figure 9 *Bipolaris spicifera.*

E. *Botryomyces* de Hoog & Rubio 1982

1. *Botryomyces caespitosus* de Hoog & Rubio 1982

The teleomorph is unknown. Skin infections have been reported, and its tissue forms are consistent both with phaeohyphomycosis and chromoblastomycosis (49,50). Colonies are black, restricted in growth, raised, and dry in texture. Hyphae are absent. Cells are unicellular to muriform, subglobose, hyaline when young, brown in age, 7.5 to 13 µm. Budding is absent, and cells instead adhere in aggregations and become septate, with portions of the aggregates eventually separating (Fig. 10). Compare with *Sarcinomyces.*

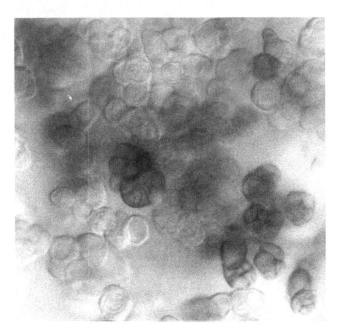

Figure 10   *Botryomyces caespitosus.*

F.   *Cladophialophora* Borelli 1980

The genus *Cladophialophora* Borelli (1980) now accommodates certain species previously classified in other genera, including *Cladosporium* and *Xylohypha* (51). More notable species in *Cladophialophora* are *Cladophialophora carrionii* (formerly *Cladosporium carrionii*) and *Cladophialophora bantiana* (synonyms *Cladosporium bantianum, Xylohypha bantiana, Cladosporium trichoides*). This taxonomic revision reflects phylogeny inferred from molecular sequencing data. (52,53). *Xylohypha emmonsii* in 1995 was regarded as conspecific with *Clado-phialophora bantiana* based on nuclear DNA homology analysis (52). New data, however, have led to a reversal of this disposition, and the binomial *Xylohypha emmonsii* has been reintroduced (54). *Cladophialophora bantiana* has caused dozens of cerebral infections (53) and also has been involved occasionally in cutaneous and subcutaneous infections (55–57). Because most cases are cerebral infection with no evidence of cutaneous lesions, it is assumed that the organism can gain entrance via the lungs. For this reason it has been recommended that isolates be manipulated only within a biological safety cabinet (58). Recently an isolate from a cerebral infection was reidentified and described as a new species, *Cladophialophora modesta* (59). *Cladophialophora carrionii* is a leading agent of chromoblastomycosis in Africa, Australia, and Madagascar (1,60). *Cladophialophora emmonsii* is known from cutaneous and subcutaneous tissue and from spleen, but not from brain (54,56). Compare with *Cladosporium, Taeniolella.*

1. *Cladophialophora bantiana* (Saccardo) de Hoog et al. 1995

C. bantiana conidiophores are hyphalike, poorly differentiated, pale brown. Conidia occur in very long sparsely branched chains that in most isolates are poorly differentiated from conidiophores and vegetative hyphae (Fig. 11). Conidia of

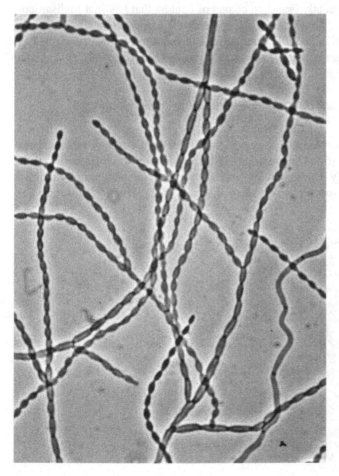

Figure 11 *Cladophialophora bantiana.*

C. bantiana measure 2–2.5 × 4–7 µm (sometimes larger). All isolates tested thus far possess a 558 base pair intron at position 1768 of the small subunit rDNA gene (54). Isolates of *C. emmonsii* tested thus far lack this distinctive intron. Morphologically, *C. emmonsii* differs by having conidia that are asymmetric to sigmoid, sometimes two-celled, occurring in shorter chains, and failing to grow at 42–43°C. Compare with *Cladosporium* (4, 56, 61).

2. *Cladophialophora carrionii* (Trejos) de Hoog et al. 1995

Colonies are velvety, olivaceous gray to black, slow to moderate in growth rate. Gelatin hydrolysis is negative (rarely weak), there is growth at 37°C, and isolates are not susceptible to cycloheximide. Conidiophores are short, lateral or terminal, and bear long, infrequently branched chains of conidia that tend not to disarticulate completely (Figs. 12, 13). Conidia are one-celled, approximately 2.2–2.6 × 4.5–6.0 µm, elliptical, bilaterally symmetrical, mainly uniform in size and shape. A *Phialophora* anamorph can be seen when grown on nutritionally poor media (Fig. 13) (4).

3. *Cladophialophora devriesii* (Padhye & Ajello) de Hoog et al. 1995

This species was described as a new species in 1984 from a case of subcutaneous phaeohyphomycosis (62). Diagnosed in 1981, the infection was refractory to therapy and subsequently disseminated to the liver, killing the patient in 1988 (63). Morphology of the species is similar to that of *Cladosporium cladosporioides*

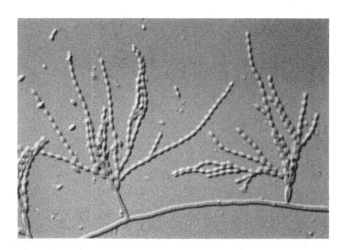

Figure 12 *Cladophialophora carrionii.*

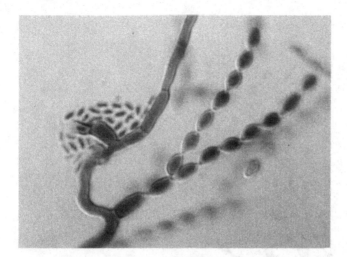

Figure 13 *Phialophora* synanamorph (left) of *Cladophialophora carrionii*.

and *C. sphaerospermum*, but the conidia taper more sharply toward both ends in such a way that the spores are lemon-shaped to fusiform and their connectives often are narrow and elongated (Fig. 14). Spore walls remain smooth. Further in contrast to *C. cladosporioides* and *C. sphaerospermum*, colonies will grow at 37°C. There is no growth at or above 40°C. Gelatin hydrolysis is negative and urease is formed. Only one isolate has been described. Subsequently described was a species, *Cladophialophora arxii* (64), that seems to have no significant morphologic difference from *C. devriesii*, but that grows at 40°C and exhibited differences in partial 26S rRNA sequence. Recently an isolate similar in morphology to *C. arxii* and *C. devriesii*, but differing in molecular makeup was described (65).

4. *Cladophialophora emmonsii* (Padhye et al.) de Hoog & Padhye 1999

This species closely resembles *C. bantiana*. It differs by conidia that are asymmetric to sigmoid, one- (rarely two-) celled, occurring in shorter chains than seen with *C. bantiana*, and failing to grow at 42–43°C (Fig. 15). Also, molecular analysis shows that it lacks a distinctive intron found in *C. bantiana* (54).

G. *Cladosporium* Link 1816

Cladosporium species are widespread in the environment, and there is little evidence that they can cause infection in humans. Reported cases consist of cutane-

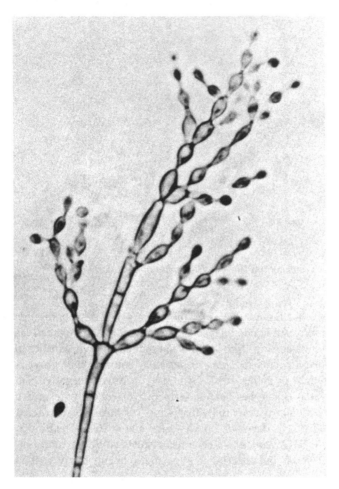

Figure 14   *Cladophialophora devriesii*. Source: Ref. 62.

ous, subcutaneous, and eye infections (5,66–71), and colonization of a pre-existing lung cavity (72). Species of *Cladosporium*, especially *C. cladosporioides* and *C. sphaerospermum*, are among the most common dematiaceous mold contaminants recovered by clinical laboratories. Colonies are rapidly growing, velvety or cottony, olive-gray to olive-brown or black. Conidiophores are dark, erect, long, often septate and branching. Conidia are one-celled (several-celled in some species), smooth or rough, with dark prominent hila, occurring in long, fragile, profusely branched acropetal chains. Ramoconidia are one-to-three-celled, usu-

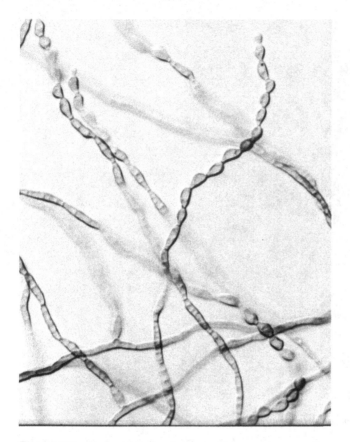

Figure 15 *Cladophialophora emmonsii.*

ally shield-shaped. Usually no growth at 37°C; gelatin hydrolysis usually positive. Compare with *Fonsecaea, Cladophialophora* (4,5,17,30,62,73).

1. *Cladosporium cladosporioides* (Fresen.) de Vries 1952

Conidia are smooth in most strains, ellipsoidal to lemon-shaped, one- (rarely two-) celled, mostly 3-7 × 2-4 μm with one or more hila (Fig. 16). No teleomorph is known.

2. *Cladosporium herbarum* (Pers.) Link ex S.F. Gray 1821

Conidia are pale brown to brown, more or less verrucose, mostly one-to-two-celled with some up to four-celled. One-celled conidia mostly are 4–5 × 4.5–

Figure 16 *Cladosporium cladosporioides.*

11 μm. Conidiophores are up to 225 μm long, often nodose, and somewhat geniculate (Fig. 17). No teleomorph is known.

3. *Cladosporium sphaerospermum* Penz. 1882

Conidia rough-walled, mostly globose to subglobose and one-celled, 3–4.5 μm in diameter. Ramoconidia smooth or rough-walled, one-to-three-celled, with one or more hila (Fig. 18). No teleomorph is known.

H. *Curvularia* Boedijn 1933

Collectively, *Curvularia* species are leading agents of fungal sinusitis and keratitis and may cause endocarditis, mycetoma, pulmonary infection, cerebral infection, and subcutaneous phaeohyphomycosis as well (42,74–78). *Curvularia*

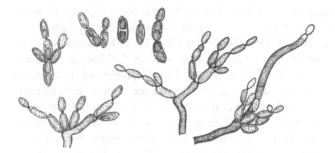

Figure 17 *Cladosporium herbarum.* Source: Ref. 17.

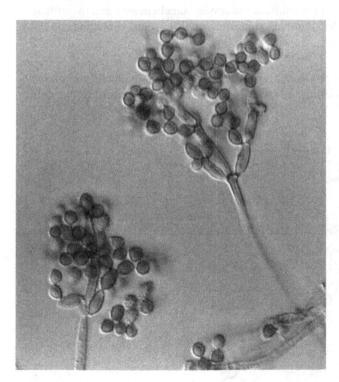

Figure 18 *Cladosporium sphaerospermum.*

brachyspora, C. clavata, C. geniculata, C. lunata, C. pallescens, C. senegalensis, and C. verruculosa specifically have been identified from infections, with C. lunata being most common by far. Descriptions by Ellis (17,30,79) should be consulted if a Curvularia isolate must be identified to species. Colonies are rapidly growing, woolly, gray to grayish-black or brown. Conidiophores are dark, erect, and geniculate due to sympodial development. Conidia are multiseptate, usually curved, with central cell larger and darker than end cells, thickness of septa and outer cell wall approximately the same; hilum dark. Some species will form large, erect, cylindrical, sometimes branched, black stromata when grown on rice grains and sometimes on potato dextrose agar (PDA). Compare with Alternaria, Bipolaris, Drechslera, Exserohilum, Nakataea.

1. Curvularia brachyspora Boedijn 1933

Conidiophores often are nodose (Fig. 19). Conidia are three-septate; middle septum is median, occurs at widest part of conidium, and is the thickest and darkest septum. Conidia mostly are somewhat curved, broadly fusiform, end cells are subhyaline to very pale brown, and intermediate cells are brown, smooth-walled, 19–26 µm × 10–14 µm. Conidia are borne in apical clusters and in verticils at the conidiophore nodes. Stromata are formed in rice grains, sometimes on PDA.

2. Curvularia geniculata Nelson 1964

The teleomorph is Cochliobolus geniculatus and is heterothallic. Conidia usually are curved, rather fusiform in shape, 18–37 × 8–14 µm, uniformly five-celled,

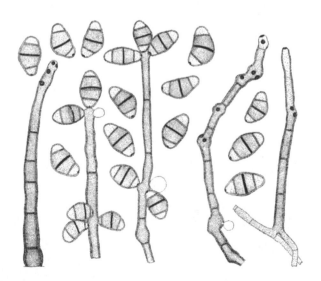

Figure 19 Curvularia brachyspora. Source: Ref. 79.

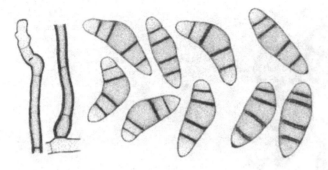

Figure 20   *Curvularia geniculata.*

end cells subhyaline or only pale brown, remaining cells plainly brown (Fig. 20). Stromata are not formed on rice or PDA.

3.   *Curvularia lunata* (Wakker) Boedijn 1933

The teleomorph is *Cochliobolus lunatus* Nelson & Haasis 1964, and is heterothallic. Stromata can be present on rice or PDA and are large, black, and sometimes branched. Conidiophores often are geniculate (Fig. 21). Conidia are straight to curved, smooth, 18–30 × 8–15 µm, mostly four-celled; some septa may be thick and dark, and the third cell from the base often is darker and larger. End cells of the conidia usually are paler; hila not protuberant.

4.   *Curvularia pallescens* Boedijn 1933

The teleomorph is *Cochliobolus pallescens* and is heterothallic. Conidia are straight to slightly curved, 17–32 × 7–12 µm, smooth, all cells pale, end cells sometimes more so, four-celled, with third cell from bottom often being the widest; hila not protuberant (Fig. 22). Stromata are absent.

5.   *Curvularia senegalensis* (Spegazzini) C.V. Subramanian 1956

The teleomorph is not known. Conidia mostly are curved, 19–30 × 10–14 µm, mostly five-celled, smooth, end cells colorless to pale brown, other cells plainly brown, hila not protuberant (Fig. 23). Stromata are not formed.

6.   *Curvularia verruculosa* Tandon & Bilgrami 1962

Conidia are straight or curved, ellipsoidal to clavate, 20–35 × 12–17 µm, four-celled, basal cell or both end cells smooth and colorless to pale brown; central

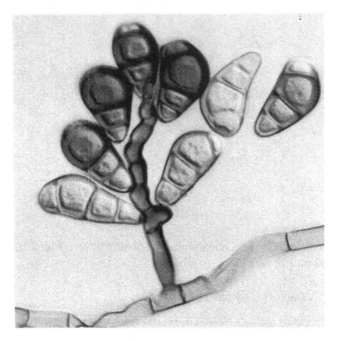

Figure 21 *Curvularia lunata.*

cells are brown and verrucose (Fig. 24). Stromata are formed on rice and sometimes PDA. No teleomorph is known.

I. *Dactylaria* Saccardo 1880

Dactylaria is a genus noted in medical mycology because of the binomial *D.
gallopava*, which was proposed for a species described originally as *Diplorhinotrichum gallopavum* (80,81). In addition to causing epizootic encephalitis in poultry flocks, this species is a well-documented cause of disseminated mycoses in
immunocompromised humans (82–84). Two varieties of this binomial were proposed (85): *D. constricta* var. *gallopava* and *D. constricta* var. *constricta*. Because the genus *Dactylaria* is characterized in part by conidia that are released
cleanly from their conidiophore through an enzymatic (rhexolytic) process, however, and because the conidia of *gallopava* separate from their conidiophores by
cell wall breakage (schizolytic), it was suggested that *Dactylaria* cannot satisfac-
torily accommodate this species (Fig. 59). As a result, the new combination *Ochroconis gallopava* was proposed (86). The distinction between the genera *Ochroconis* and *Scolecobasidium* is not clear, however, and so an alternative binomial,

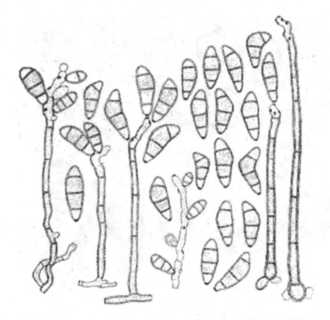

Figure 22   *Curvularia pallescens*. Source: Ref. 79.

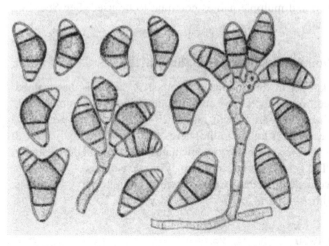

Figure 23   *Curvularia senegalensis*. Source: Ref. 79.

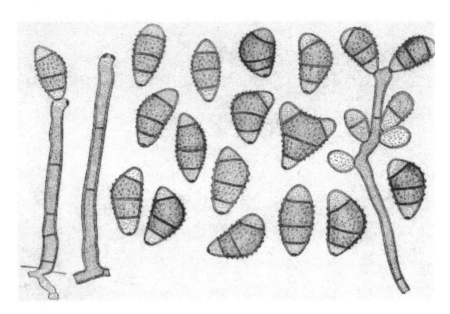

Figure 24 *Curvularia verruculosa.* Source: Ref. 79.

Scolecobasidium gallopavum (87), probably would be more satisfactory. Recent data from molecular sequencing also would support this disposition (88). The matter remains unsettled and multiple names continue to be used. See Sec. IV. A' for further discussion.

J.   *Dissitimurus* Simmons, McGinnis & Rinaldi 1987

A teleomorph is not known in this monotypic genus. A single isolate named *D. exedrus* was cultured from nasopharynx and turbinate lesions in a human (89). The colony is blackish-brown, and growth rate at 20–25°C is 1 cm per week. Growth is subsurface except for long conidiophores that develop slowly over 2 to 4 weeks. Conidiophores are sympodial to geniculate, with conidia borne singly or in chains (Fig. 25). The genus is quite similar to dysgonic isolates of *Alternaria* except for a distinctive disjunctor that occurs between conidia borne in chains (Fig. 26). Compare with *Alternaria.*

K.   *Drechslera* Ito 1930

Only one species of *Drechslera* is known to have caused infection in a human host (90). Several *Bipolaris* species that can cause infection in humans were

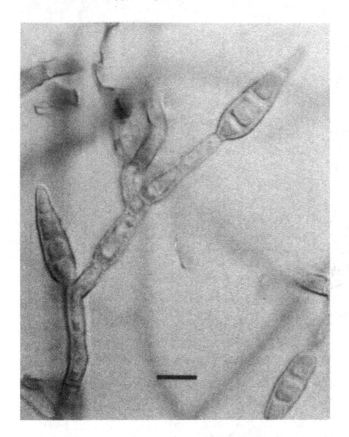

Figure 25 *Dissitimurus exedrus.* Source: Ref. 89.

accommodated briefly in the genus *Drechslera* before being returned to *Bipolaris*. Compare with *Bipolaris, Curvularia, Exserohilum, Helminthosporium, Nakateae*.

1. *Drechslera biseptata* (Saccardo & Roumeguère) Richardson & Fraser 1968

Colonies are rapidly growing, velvety to lanose, gray becoming brown to blackish. Conidia are straight, usually obovoid to slightly clavate, without a protuberant hilum, mostly less than 40 µm long and more than 10 µm wide, with two to three distosepta, pale to mid-brown (Fig. 27). Conidia can germinate from middle as well as end cells. Germ tubes arise at a large angle to the main axis of the conidium.

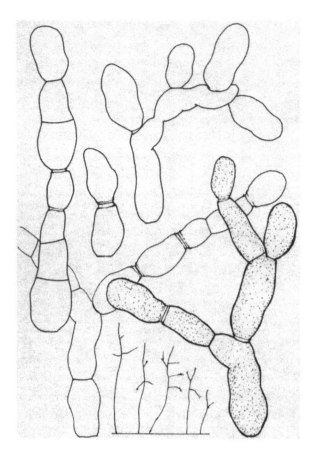

Figure 26 *Dissitimurus exedrus*. Source: Ref. 89.

L.   *Exophiala* Carmichael 1966

Exophiala jeanselmei is a leading etiologic agent of the subcutaneous phaeohy-phomycoses and has caused mycetoma and peritonitis (91–93). *Exophiala moni-liae*, *E. pisciphila*, and *E. spinifera* also have been reported as agents of phaeohy-phomycoses (94–96). *Exophiala werneckii*, the cause of "black palm," has been renamed *Phaeoannellomyces werneckii* (pro parte *Hortae werneckii*). The species known most recently as *Wangiella heteromorpha* and as *E. jeanselmei* var. *heter-omorpha* is reported to have been the cause of human infections, but at least some of these isolates were reidentified as *Wangiella dermatitidis* (97). This fun-gus appears by molecular analysis to be different enough to warrant species sta-tus, but opinion is divided as to whether the appropriate genus is *Wangiella* or

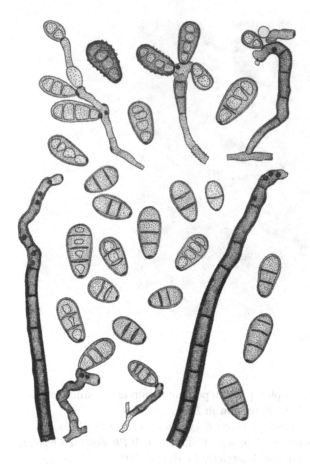

Figure 27 *Drechslera biseptata.* Source: Ref. 17.

Exophiala (98,99). Reidentification of isolates to which pathogenicity has been attributed is warranted.

The colony morphology of *Exophiala* species is varied. Most isolates initially grow in the form of a brown yeast (*Phaeoannellomyces* synanamorph) that is succeeded by the hyphal *Exophiala* synanamorph. As a result, colonies are moist and yeastlike at first, becoming velvety to woolly with age, and pale brown to black in color. Some isolates that consist predominantly of the *Phaeoannellomyces* synanamorph of *E. jeanselmei* may remain yeastlike in colony texture. Conidiogenous cells are annellides. Conidia are one-celled (one-to-three-celled in *E. salmonis*), hyaline to pale brown, accummulating in balls at apices of the annellides. For all species described below, the teleomorph is not known, a

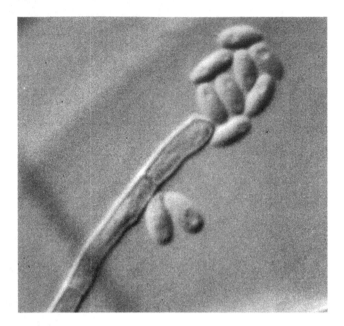

Figure 28   *Exophiala castellanii.*

Phaeoannellomyces synanamorph is present, potassium nitrate is assimilated, and conidia are one-celled (with the exception in *E. pisciphila*).

   Phaeoannellomyces werneckii (syn. *Exophiala werneckii*) (100) is discussed under the genus *Phaeoannellomyces*. Compare with *Phaeoannellomyces, Phaeococcomyces, Rhinocladiella, Wangiella* (4,40,101–107)

1.   *Exophiala castellanii* Iwatsu, Nishimura & Miyaji 1984

Exophiala castellanii is characterized by poorly differentiated conidiogenous cells having inconspicuous annellations (Fig. 28). There is no growth at 40°C. It has been suggested that *E. mansonii* is conspecific with *E. castellanii* (108).

2.   *Exophiala jeanselmei* (Langeron) McGinnis & Padhye 1977

E. jeanselmei is the most common species in the genus. Two varieties are recognized. *Exophiala jeanselmei* var. *lecanii-corni* bears annellations arising directly from micronematous annellides (Fig. 29). *Exophiala jeanselmei* var. *jeanselmei* bears well-developed, erect, lageniform to cylindrical annellides (Fig. 30), and in addition may exhibit some micronematous annellides. There is no growth at

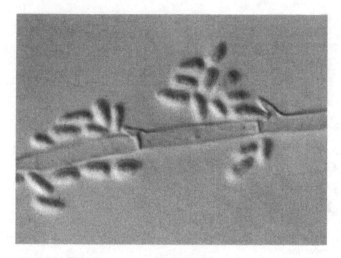

Figure 29   *Exophiala jeanselmei* var. *lecanii-corni.*

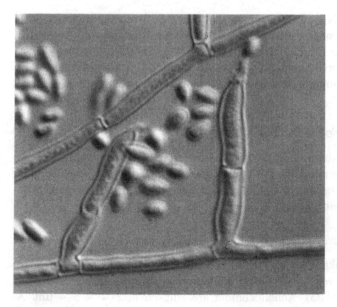

Figure 30   *Exophiala jeanselmei* var. *jeanselmei.*

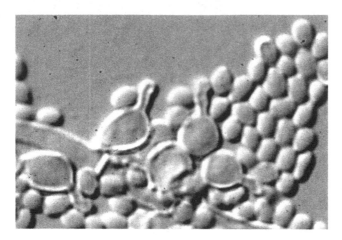

Figure 31   *Exophiala moniliae.*

40°C. It has been suggested that these two organisms might be regarded as separate species (109).

3.   *Exophiala moniliae* de Hoog 1977

The species is morphologically distinguished from *E. jeanselmei* by annellides that are swollen or bulging in shape, and that exhibit a prominently elongated annellated apical zone (Fig. 31). Conidia are 2.5–4 × 1.5–2.5 µm, broadly ellipsoidal, sometimes curved cylindrical. Growth at 40°C is variable. Analysis of mitochondrial DNA suggests that *E. moniliae* may be conspecific with *E. jeanselmei* (110).

4.   *Exophiala pisciphila* McGinnis & Ajello 1974

This species is very similar to *E. jeanselmei*, differing by its inability to grow at 37°C and by its larger conidia 3–8 µm × 2–4 µm, which rarely may be one-septate (Fig. 32). It was noted without further details that the single isolate from man grew at 35°C but not at 41°C.

5.   *Exophiala spinifera* (Nelson & Conant) McGinnis 1977

The teleomorph is unknown. Conidiophores are distinctly spinelike, multicellular, often are darker than vegetative hyphae, and terminate in a prominent, annellated, sporogenous tip (Fig. 33). Annelloconidia are ellipsoidal, 2–4 × 3–4 µm. A *Phialophora* synanamorph bearing small (1.5 µm) globose conidia may be

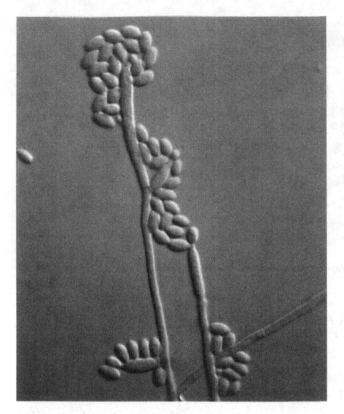

Figure 32 *Exophiala pisciphila.*

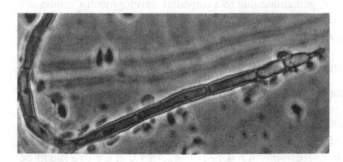

Figure 33 *Exophiala spinifera.*

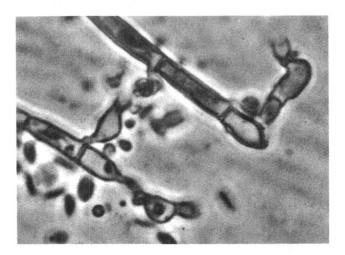

Figure 34   *Phialophora* synanamorph of *Exophiala spinifera.*

present (Fig. 34). A *Phaeoannellomyces* synanamorph always is present (111). There is growth at 37°C and a few isolates have shown weak growth at 40°C (112).

M.   *Exserohilum* Leonard & Suggs 1974

Phaeohyphomycosis of skin, subcutaneous tissue, cornea, and nasal sinuses has been documented (42,43,113–116). *Exserohilum* contains three recognized opportunistic pathogens: *E. longirostratum*, *E. mcginnisii*, and *E. rostratum*. Members of the genus previously had been confused with species of *Bipolaris*, *Drechslera*, and *Helminthosporium*, but ensuing taxonomic studies have clarified the distinctions (43,47). Colonies are rapidly growing, woolly, gray to black. Conidiophores are dark, erect, geniculate due to sympodial development. Conidia are multiseptate, cylindrical to oblong, dark, each with a strongly protruding hilum. Compare with *Bipolaris*, *Drechslera*, *Helminthosporium*, Nakateae.

1.   *Exserohilum longirostratum* (Subramanian) Sivanesan 1987

This species is characterized by a dichotomy in the size of the conidia. The smaller conidia are very much like those of *E. rostratum*. The larger conidia range from 100 to 400 μm in length, exhibit as many as 21 distosepta, and are rostrate (beaked) in shape (Fig. 35). Because of morphologic overlap between *E. longirostratum* and *E. rostratum* it has been suggested that the two might be conspecific (5,43,48).

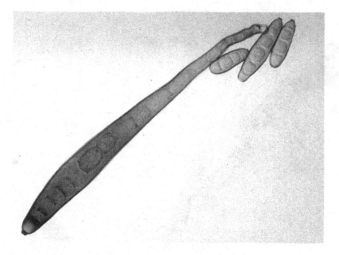

Figure 35   *Exserohilum longirostratum.*

2.   *Exserohilum mcginnisii* Padhye & Ajello 1986

This species is similar to *E. rostratum* but differs by the lack of thick-walled distosepta and by the presence of unevenly roughened cell walls in age (Fig. 36) (117).

3.   *Exserohilum rostratum* (Drechslera) Leonard & Suggs 1974

The teleomorph of this fungus is *Setosphaeria rostrata* Leonard 1976 and is heterothallic. Conidia are rostrate, straight or slightly curved, cylindrical or sometimes ellipsoidal, usually having 6–8 (−16) distosepta (Fig. 37). The distoseptum closest to each end of the conidium is dark and thick-walled, and each end cell is pale compared to the other cells. Conidia commonly measure 60–90 × 11–20 μm and are smooth-walled.

N.   *Fonsecaea* Negroni 1936

Fonsecaea pedrosoi (Brumpt) Negroni 1936 and *F. compacta* Carrión 1940 are the only species in the genus, and both can cause chromoblastomycosis. *Fonsecaea pedrosoi* is the leading agent worldwide, while *F. compacta* is extremely rare (1,118,119). *Fonsecaea pedrosoi* also has caused phaeohyphomycosis in a few cases (120–123). Isolates are pleomorphic and no teleomorph is known. They are characterized by the formation of one-celled conidia on erect, dark, compactly sympodial conidiophores. These conidia directly give rise to a second level of

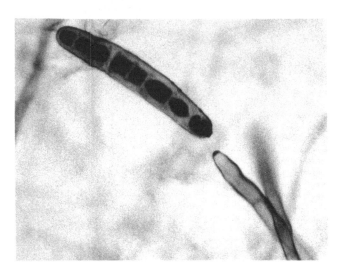

Figure 36 *Exserohilum mcginnisii.*

conidia, and usually a third level is formed similarly. This limitation to two or
three levels of conidia contrasts with *Cladosporium* spp. and constitutes the distinctive morphology for the genus. In addition, *Rhinocladiella* and *Phialophora*
synanamorphs can be present. Colonies of both species are slow-growing, velvety, olivaceous black. Conidiophores are pale to mid-brown, usually erect, with
slight apical swelling of main axis. Conidia are one-celled, pale to mid-brown.
Fonsecaea compacta (Fig. 38) differs from *F. pedrosoi* (Fig. 39) by conidia that
are subglobose, broadly attached, and formed in compact conidial heads. Both
species grow at 37°C. Molecular analysis has shown substantial ITS sequence
diversity among isolates in this genus (99, 124). Compare with *Cladosporium*,
Rhinocladiella (4,15)

O. *Hormonema* Lagerberg & Mellin 1927

Hormonema species often have been identified incorrectly as *Aureobasidium* species even though the mechanism of sporogenesis differs between the two genera.
In *Hormonema* species, conidia arise by percurrent succession from micronematous loci on the surface of hyaline to dematiaceous hyphae (Fig. 40). In contrast,
Aureobasidium species produce conidia synchronously. Brown arthroconidia develop in most isolates of both *Hormonema dematioides* and *Aureobasidium pullulans*. Cutaneous phaeohyphomycosis and peritonitis have been reported from
Hormonema dematioides (125,126). Colonies are yeastlike, white or slightly
pinkish in color, becoming dark brown centrally, at the margin, in sectors, or

Figure 37 *Exserohilum rostratum.*

entirely. True hyphae are present with age, and are most easily found submerged within the agar. Hyphal formation and observation of sporulation can be facilitated by inoculating an agar plate and then placing a cover glass over the inoculum. The teleomorph of *Hormonema dematioides* is *Sydowia polyspora*. Compare with *Aureobasidium*, *Scytalidium* (40).

P. *Madurella* Brumpt 1905

Madurella mycetomatis is the world's leading cause of mycotic mycetoma with more than 1,000 cases having been reported, while *Madurella grisea* has been

Figure 38 *Fonsecaea compacta.*

reported fewer than 100 times (127). *Madurella grisea* colonies are slow-growing, velvety, folded and heaped, dark brown with grayish tints, reverse dark brown without diffusing pigment. Isolates of *M. grisea* are sterile, though some reportedly have formed pycnidia. Conidia are not present. Growth is better at 30°C than at 37°C. Sucrose is assimilated; lactose assimilation is variable. Colonies of *M. mycetomatis* are slow-growing, cream-colored, glabrous, folded, tough; with age becoming velvety, dark brown, staining the agar with a diffusing brown pigment. Growth is enhanced at 37°C; lactose is assimilated but sucrose

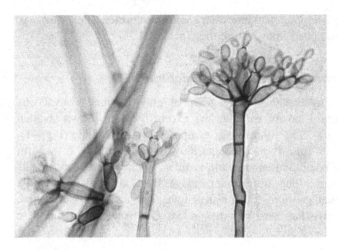

Figure 39 *Fonsecaea pedrosoi.*

is not. Colonies are sterile on routine media; sporulation is present in about 50% of *M. mycetomatis* isolates grown on nutrient-poor media. Conidia are subglobose to pyriform, 3 to 5 μm, with truncate basal scar, occurring in balls, rarely in fragile chains. Phialides are variable, usually 9 to 11 μm long, tapering, often with collarette; but range from long (15 μm) and tubular to short (3 μm) and integrated. Occasionally two or three phialides may be borne upon a single

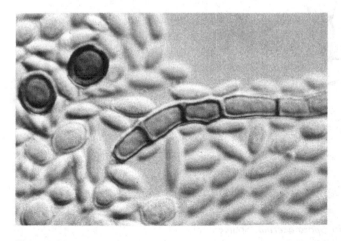

Figure 40 *Hormonema dematioides.*

branch. Large vesicles, terminal and intercalary, often are present. Sclerotia may be present. No teleomorph is known for either species (12).

Q. *Mycocentrospora acerina* (Hartig) Deighton 1972

Colonies initially are hyaline, becoming green or grayish, finally olivaceous black. Hyphae in culture mostly are colorless except for dark brown swollen cells. Conidiophores are 25 to 50 μm long, sympodial and geniculate (Fig. 41). Conidia have four to 24 (mostly 8–11) septa, 60-250-150 μm long, 5 to 7 μm wide. Middle cells of conidia often are dark, swollen, thick-walled. Apical cells of conidia taper to form a long, filiform appendage. Basal cell of some conidia forms a filiform appendage up to 150 μm in length. One case of disseminated, fatal phaeohyphomycosis has been reported (128). The isolate from this case

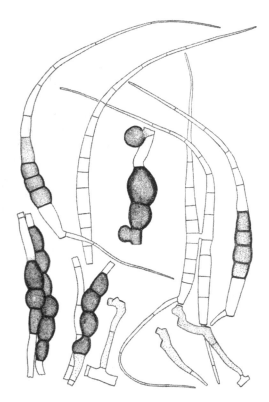

Figure 41   *Mycocentrospora acerina.* Source: Ref. 17.

grew better at 30°C than at 25°C, but did not grow at 37°C. No teleomorph is known. Compare with *Cercospora*, *Vermispora* (17).

R. *Mycoleptodiscus* Ostazeski 1968

1. *Mycoleptodiscus indicus* (Sahni) Sutton 1973

Colonies initially are spreading with long, thin aerial hyphae, yellow-brown to tan in color, becoming grayish-black. Conidial formation is delayed. Conidia are approximately 15 × 7 μm, hyaline, one-celled, reniform, and sharply tapered to both ends, with a hairlike appendage at either end (Fig. 42). Conidia arise from brown ampulliform phialides with collarettes. Phialides may or may not be organized as part of sporodochia. Dark brown ellipsoidal appressoria may be present. A teleomorph (*Omnidemptus*) has been reported in the genus. At least two cases

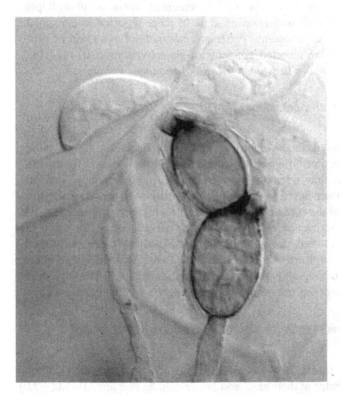

Figure 42 *Mycoleptodiscus indicus.*

of phaeohyphomycosis are known, one of which is published (129). Compare with *Ciliphora* (30,130).

S.   *Phaeoacremonium* W. Gams, Crous & M.J. Wingf. 1996

The genus *Phaeoacremonium* is morphologically intermediate between *Acremonium* and *Phialophora*. Its morphology is quite similar to that of *Acremonium* and differs by hyphae that become dematiaceous in part. Phialides range from micronematous to macronematous and their morphology is quite variable, but the most distinctive shapes are elongate and taper steadily from base to tip in an awllike manner. The collarette usually is fairly well developed but is thin-walled and somewhat inconspicuous; collarette walls are parallel to slightly divergent. Teleomorphs are not known. Six species have been proposed for the genus, three of which are known to cause opportunistic infections in humans. Prior to the 1996 revision all three were considered to be isolates of *Phialophora parasitica*. *Phaeoacremonium parasiticum* is a well-documented agent of phaeohyphomycosis (131–133). The identification of some of these reported isolates might be changed under current taxonomy. *Phaeoacremonium inflatipes* and *P. rubrigenum* recently have been reported as agents of phaeohyphomycosis (134,135). Compare with *Acremonium, Lecythophora, Phialemonium, Phialophora*.

1.   *Phaeoacremonium parasiticum* (Ajello et al.) W. Gams et al. 1996

Colonies are buff-colored, almost glabrous when young, becoming velvety during maturity. Colonies become brown either in sectors or entirely except for the margin. Phialides are variable in length, some isolates forming extremely long phialides swollen near their bases, with prominent wartlike cell wall encrustations on the lower section of the conidiophore and the adjoinining hyphal sections (Fig. 43). Conidia are elliptical to cylindrical, often curved, range in size from 2–6 × 1–2 μm, and accumulate in a mass at the phialide tip (136). An inconspicuous yeastlike synanamorph arises from conidia that do not immediately germinate but instead enlarge and begin to sporulate in a phialidic manner. These yeastlike forms are one-to-two-celled and bear a collarette at the sporogenous orifice.

2.   *Phaeoacremonium inflatipes* W. Gams, Crous & M.J. Wingf. 1996

This species is quite similar to *P. parasiticum*. It differs by phialides that have a more pronounced bulge at their base, and larger conidia that are 3.0–5.0(−7.0) × (1.0) 1.5–2.0(−2.5) μm (136).

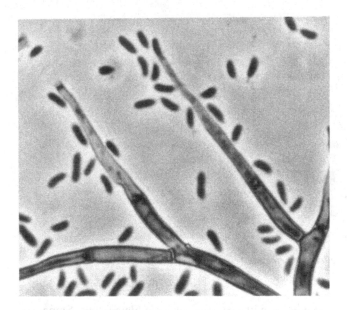

Figure 43 *Phaeoacremonium parasiticum.*

3. *Phaeoacremonium rubrigenum* W. Gams et al. 1996

This species is highly similar to *P. inflatipes* but differs by its vinaceous red reverse on malt extract agar. Conidia are the same size range as with *P. inflatipes*, although rarely may be very slightly larger (136).

T. *Phaeoannellomyces* McGinnis & Schell 1985

The genus *Phaeoannellomyces* was established for dematiaceous yeast morphs that are characterized by annellidic sporulation (100). A teleomorph is not known. The two clinically important species are *P. werneckii* and *P. elegans*. Both are agents of phaeohyphomycosis (121,122,137,138). The otherwise identical genus *Phaeococcomyces* differs by forming conidia in a holoblastic manner. Compare with *Phaeococcomyces, Exophiala, Wangiella.*

1. *Phaeoannellomyces elegans* McGinnis & Schell 1985

This species usually is found as a yeast synanamorph accompanying isolates of various species of *Exophiala*. In such circumstances, colonies at first are mucoid, slow-growing, smooth, yeastlike, pale brown to black. As the mold synanamorph

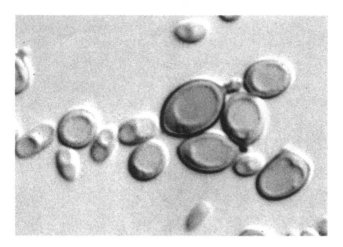

Figure 44   *Phaeoannellomyces elegans.*

develops and becomes predominant, the colonies become filamentous. Some isolates of *P. elegans*, however, have little or no development of a hyphal synanamorph (139). *Phaeoannellomyces elegans* exhibits one-celled annellated yeast
cells that are subhyaline to pale brown (Fig. 44). Pseudohyphae may be formed.
Compare with *Exophiala, Phaeococcomyces, Wangiella* (100,106).

2.   *Phaeoannellomyces werneckii* (Horta) McGinnis & Schell 1985 (syn. *Cladosporium werneckii, Exophiala werneckii, Hortaea werneckii* pro parte)

This species is the cause of a superficial phaeohyphomycosis known as black
palm (6). The species exhibits a distinctive one-to-two-celled, very broadly annellated yeast morph that is the predominant and most stable synanamorph seen in
this pleomorphic species and that provides the basis of the name (Fig. 45). The
binomial *Hortaea werneckii* has been proposed for the mold synanamorph that
sometimes is present (Fig. 46). This hyphal synanamorph, however, often is
poorly developed and thus is not of primary importance to a morphologically
based identification of isolates.

U.   *Phaeotrichonis* Subramanian 1956

1.   *Phaeotrichonis crotalariae* (Salam & Rao) Subramanian 1956

This species was reported from a case of keratitis (140, 141). Colonies are lanose,
dark gray to grayish-brown, sometimes with sclerotia forming. Conidiophores

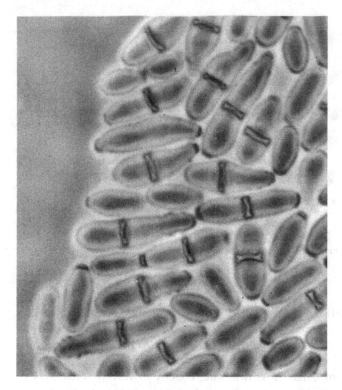

Figure 45 *Phaeoannellomyces werneckii.*

are macronematous, multicellular, thick-walled, brown, sympodial, at times geniculate (Fig. 47). Conidia are borne singly, three to eight (mostly 5–6) septate, obclavate, with prominent filiform beak 30 μm or more in length (17). A teleomorph is not known. Compare with *Bipolaris, Drechslera, Exserohilum*.

V. *Phialophora* Medlar 1915

Phialophora verrucosa long has been known as a leading cause of chromoblastomycosis, and has caused a few cases of keratitis and phaeohyphomycosis (1, 142,143). *Phialophora americana* has been regarded by many mycologists as a synonym of *P. verrucosa*, but now molecular sequence data seem to correlate well with the proposed morphologic distinction (144). *Phialophora americana* has caused chromoblastomycosis. *Phialophora bubakii*, *P. repens*, and *P. richardsiae* are agents of phaeohyphomycosis (5,145–147). Other reported infections include endocarditis, keratitis, osteomyelitis, atypical eumycetoma, and opportu-

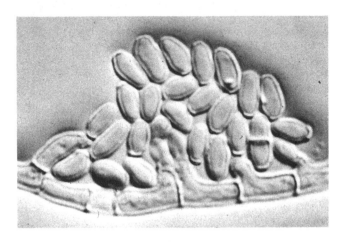

Figure 46 *Phaeoannellomyces werneckii*, showing the filamentous synanamorph (genus *Hortaea*).

nistic infections in AIDS patients (148–151). Two additional agents of phaeohyphomycosis previously classified in *Phialophora* have been transferred to the genus *Lecythophora* as *L. hoffmannii* and *L. mutabilis* (152). (See Chap. 11.) *Phialophora parasitica* has been reclassified as *Phaeoacremonium parasiticum* (136). Colonies are moderate in growth rate, cottony to velvety, olive-gray to black. Conidiophores (if present) usually are short and pale brown. Conidiogenous cells are phialides with distinct collarettes. Conidia are one-celled, hyaline to pale brown, accumulating in balls at the phialide apices. *Capronia semiimmersa* has been shown as the teleomorph of *Phialophora americana* (144); otherwise, teleomorphs are not proved. Compare *Phialophora* with *Acremonium*, *Lecythophora*, *Phaeoacremonium*, *Phialemonium* (4,5,30,73,131,136,145,152).

1. *Phialophora americana* (Nannfeldt) S. Hughes 1958

This species produces prominent flask-shaped phialides with dark collarettes that are deeper than wide, resulting in a vaselike shape (Fig. 48). Conidia are elliptical. *Phialophora americana* is extremely similar in morphology to *P. verrucosa*, differing only in the shape of the collarette. Colonies are olivaceous black.

2. *Phialophora repens* (Davidson) Conant 1937

This species shows a wide variation in the size and shape of its phialides, similar to that seen with *Phaeoacremonium parasiticum*. Phialides can be adelophialides, or can be up to 20 μm long and subtended by a supporting cell. Phialides mostly are cylindrical to slightly lageniform with a delicate collarette (Fig. 49). Very

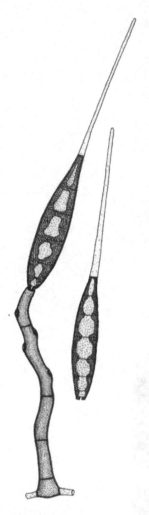

Figure 47 *Phaeotrichonis crotalariae*. Source: Ref. 26a.

rarely they can occur in branched clusters (Fig. 50). Conidia are cylindrical and often curved.

3. *Phialophora richardsiae* (Nannfeldt) Conant 1937

This species produces phialides of variable size and shape. Some phialides are long with flattened, saucer-shaped collarettes, and these give rise to conidia that are almost spherical and 2.5 to 3.5 μm in diameter (Fig. 51). The remaining

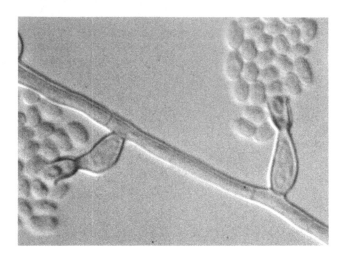

Figure 48   *Phialophora americana.*

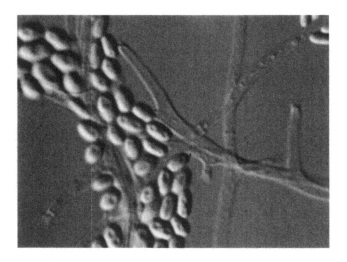

Figure 49   *Phialophora repens.*

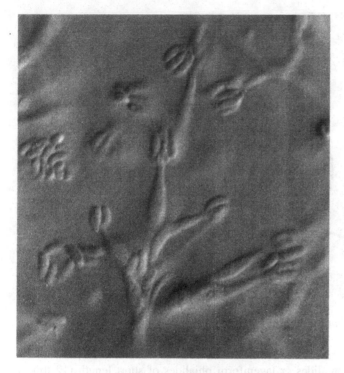

Figure 50 *Phialophora repens* clustered phialides on branched conidiophores.

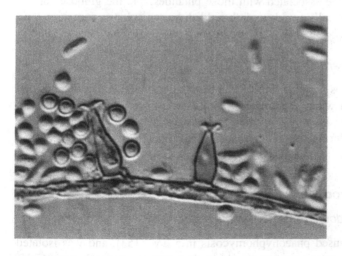

Figure 51 *Phialophora richardsiae*.

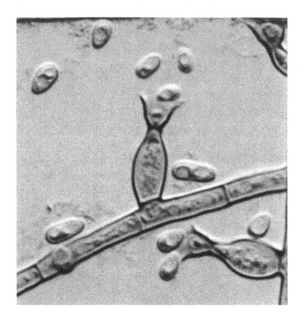

Figure 52 *Phialophora verrucosa.*

phialides are adelophialides or lageniform phialides of short length (12 μm or less), bearing an inconspicuous collarette, that form ellipsoidal to cylindrical, often curved conidia. Melanin concentration in those phialides having prominent collarettes, the hyphae associated with those phialides, and the globose conidia is high, giving these cells a mid- to dark reddish-brown color as compared with the remaining cells.

4. *Phialophora verrucosa* Medlar 1915

This species produces prominent flask-shaped phialides with dark collarettes that are not deeper than wide, resulting in a cuplike or funnellike shape (Fig. 52). Conidia are elliptical. *Phialophora americana* is extremely similar in morphology to *P. verrucosa*, differing only in the shape of the collarette, which is deeper than wide, resulting in a vaselike shape.

W. *Pseudomicrodochium* B. Sutton 1975

1. *Pseudomicrodochium suttonii* Ajello et al. 1980

This species has caused phaeohyphomycosis in a dog (153), and was isolated from subcutaneous lesions in a human (154). A teleomorph is not known. Colo-

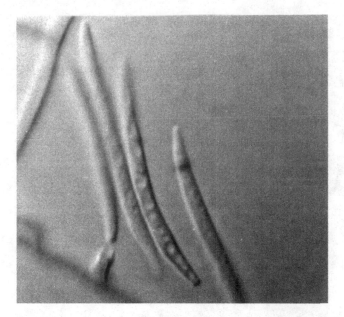

Figure 53 *Pseudomicrodochium suttonii.*

nies are velvety, olivaceous black with olivaceous gray aerial hyphae. Conidiogenous cells are monophialidic, typically micronematous and integrated, often
with collarettes, not separated from the hyphae by a septum (Fig. 53). Erect lateral
or terminal phialides are rarely present. Conidia are holoblastic, schizolytic, subhyaline to pale brown, zero to three septate, acerose to falcate when mature, 14–
30 × 1–1.8 μm. Anastomosis between adjacent conidia occurs readily, and conidia sometimes function as phialides by directly producing phialoconidia from
a lateral focus.

X. *Ramichloridium* Staehl ex de Hoog 1977

A teleomorph is not known. Colonies are velvety, pale brown, brown, or olivaceous brown, sometimes with a yellow to orange hue on the reverse. *Ramichloridium* has no distinct morphologic difference compared with *Rhinocladiella*. If
these genera were to be regarded as congeneric, *Rhinocladiella* would retain
priority. The most medically significant species, *Ramichloridium obovoideum*,
was described originally as a species of *Rhinocladiella* (i.e., *Rhinocladiella obo-
voidea* Matsushima 1975). Compare with *Exophiala, Fonsecaea, Rhinocladiella,
Veronae* (15,17,30,155).

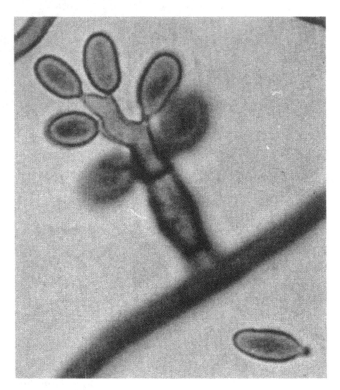

Figure 54   *Ramichloridium obovoideum.* Source: Ref. 156.

1. *Ramichloridium obovoideum* (Matsushima) de Hoog 1977

Several brain infections have been caused by *R. obovoideum* (156). Most of these were attributed to a newly described species, *R. mackenziei*. It is not clear, however, that this species is different from the previously described *Ramichloridium obovoideum*. Further study is needed, and for the purposes here the two will be treated as synonyms. Conidiogenous cells are macronematous, sympodial, almost geniculate at times, and somewhat thick-walled in the lower portion (Fig. 54). Conidia are borne on prominent denticles, are obovoid in shape, 3–5 × 5–12 μm, pale brown, with basal scar sometimes protuberant. Growth is very slow at 25°C and much faster at 30–35°C (156,157).

Y. *Rhinocladiella* Nannfeldt 1934

Colonies are rapidly growing, velvety, blackish-brown to olive-black on obverse and reverse. Conidia from sympodial cells are one-celled, ellipsoidal to obovoid to fusiform. A similar genus, *Veroneae*, differs by conidia that often are two-celled, and one report of skin infection has been noted for *V. botryosa* (5). Compare with *Exophiala*, *Fonsecaea*, *Rhamichloridium*, *Veronae* (15,17,30).

1. *Rhinocladiella aquaspersa* (Borelli) Schell et al. 1983

The teleomorph is not known. *Rhinocladiella aquaspersa* has been reported only rarely as causing chromoblastomycosis, these cases being from Brazil, Mexico, Colombia (158), and Venezuela (155,159,160). Colonies are velvety, elevated, olivaceous black, hyphae are mid- to pale brown. Conidiophores usually are darker than vegetative hyphae, well developed, 21–82 × 1.3–3.5 µm, zero to four septate, macronematous, erect, unbranched, often tapering slightly (Fig. 55). Sporulation is sympodial; fertile region is long and scarred with partially to prom-

Figure 55 *Rhinocladiella aquaspersa.*

inently raised denticles. Conidia are fusiform, ellipsoidal, or obovate, one-celled, smooth, mid-brown, 4–9 × 1.3–2.6 µm. An *Exophiala* synanamorph and a *Wangiella* synanamorph can be present.

2. *Rhinocladiella atrovirens* Nannfeldt 1934

Rhinocladiella atrovirens was suspected of causing cerebral infection in an AIDS patient (161) and mycetoma of the foot (162). Isolates sometimes are confused with *Fonsecaea pedrosoi*. There often is an annellated yeast (*Phaeoannellomyces elegans*) synanamorph inconspicuously present, however, and this is not the case with *F. pedrosoi*. Also, *R. atrovirens* differs by its sympodial conidiophore that tends to resume growing after sporulating and then sporulates again at the new apex (Fig. 56). A *Phialophora* synanamorph can occur rarely (Fig. 57) and an annellidic anamorph (as seen in *Exophiala* spp.) may be present. Sympodial conidiophores are pale to mid-brown, erect, and usually bear distinct crowded scars. Conidia are one-celled, ellipsoidal to obovate to fusiform, pale brown, with a flat basal scar. Conidia occur along the elongating conidiophore. Occasionally a conidium will give rise to a second conidium, allowing potential confusion with *Fonsecaea pedrosoi*. A teleomorph is not known.

Z.   Sarcinomyces Lindner 1898

1.   *Sarcinomyces phaeomuriformis* Matsumoto et al. 1986

A teleomorph is unknown. Colonies are black, restricted, raised, dry, friable, mulberrylike in texture. Hyphae are absent. Cells are unicellular to muriform,

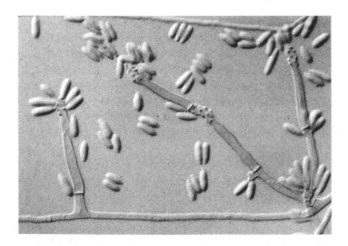

Figure 56   *Rhinocladiella atrovirens.*

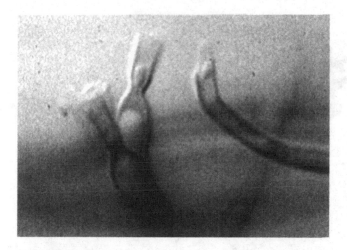

Figure 57   *Phialophora* synanamorph of *Rhinocladiella atrovirens*, as seen in paratypus specimen No. 59.36 (160:7).

subglobose, hyaline when young, brown with age, forming multilateral holoblastic buds, often adhering in large masses (Fig. 58). Skin infections have been documented (163,164). Compare with *Botryomyces*, *Phaeococcomyces*, *Phaeoannellomyces*, *Wangiella*.

A'.   *Scolecobasidium* Abbot 1927

A teleomorph is not known. Distinction between *Scolecobasidium* and *Ochroconis* de Hoog & von Arx 1973 has been uncertain, and the two genera will be considered here as synonyms (87,165). This disposition also is supported by recent molecular sequencing data (88). The species *gallopava* can cause disseminated (often cerebral) phaeohyphomycosis in immunocompromised persons (53, 82–84), and localized infection has been reported (166). Colonies are flat, olivaceous-gray to brown, usually with brown to reddish-brown diffusing pigment evident. Conidiophores are hyaline, erect, sympodial, occasionally geniculate (Fig. 59). Conidia are two-celled, dark, cylindrical to oblong, with basal frill resulting from rhexolytic dehiscence (breakage of the supporting cell wall). One-celled, globose phialoconidia may be present in young colonies. The species grows at 45°C, but is inhibited on media containing 5% NaCl when incubated at 30°C (88). *Scolecobasidium constrictum* does not grow at or above 37°C, but does grow on media containing 5% NaCl when incubated at 30°C. No naturally occurring infections are known. A teleomorph is not known. Compare with *Dactylaria* (4,85,167).

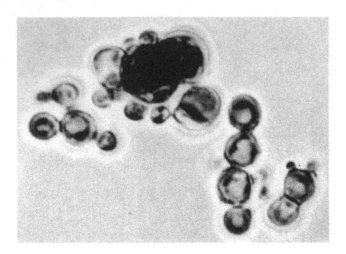

Figure 58   *Sarcinomyces phaeomuriformis.* Source: Ref. 163.

B′.   *Sporothrix* Hektoen & Perkins 1900

Colony color varies from white to shades of brown, and texture varies from glabrous to granular or floccose. Conidiogenous cells are sympodial, and the fertile
zone in some of the species is restricted to the apex of the cell. Conidia are one-
celled, hyaline, elliptical, ovoid, clavate, or curved cylindrical. No teleomorph
has been proved.

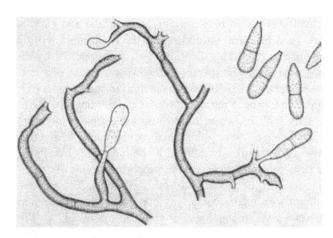

Figure 59   Species *gallopava.*

Sporothrix has accommodated a heterogeneous group of morphologically similar species. One of these, known in the medical literature as *S. cyanescens*, was removed from the genus because its septal ultrastructure suggests affinity to basidiomycetes (168). This species, distinctive for its whitish colony and a purple-blue diffusing pigment, has been recovered in culture of clinical specimens, but virulence studies in animals have shown no pathogenicity (169–171). More than one dozen species of *Sporothrix* are described, but only *S. schenckii* (in two varieties, *S. schenckii* var. *schenckii* and *S. schenckii* var. *luriei*) is a known pathogen. Subcutaneous–cutaneous infection is most common, but any part of the body (including lung) can be infected (172–174).

1. *Sporothrix schenckii* Hektoen & Perkins 1900

Colonies of most isolates initially are cream-colored, smooth, moist, yeastlike; gradually turning brown in irregular patches; becoming velvety to lanose as aerial hyphae develop. Conidiophores are elongate, compactly sympodial, with swollen apex; bearing one-celled, hyaline, elliptical to obovate conidia in a radial "rosette" appearance (Fig. 60). Conidia of a second kind arise directly from the vegetative hyphae, are one-celled, dematiaceous, thick-walled, and oval (rarely triangular) (Fig. 61). These second conidia are chiefly responsible for the dark color of the colony. *Sporothrix schenckii* can be converted to a budding yeast form at 35–37°C by using enriched media such as brain–heart infusion agar, chocolate agar, or brain–heart infusion both containing 0.1% agar (Fig. 62). Morphologic identification of an isolate as *S. schenckii* requires this demonstration of dimorphism. Some isolates are difficult to convert and might require multiple subcultures and extended incubation time. Conversion of all cells is not necessary, and a mixture of hyphal and budding cells is commonly observed. Several *Sporothrix* sp. isolated from environmental samples during investigation of a spo-

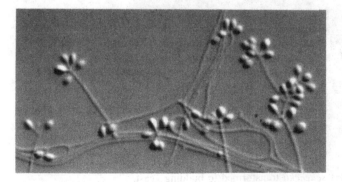

Figure 60 *Sporothrix schenckii* sympodial conidiophores.

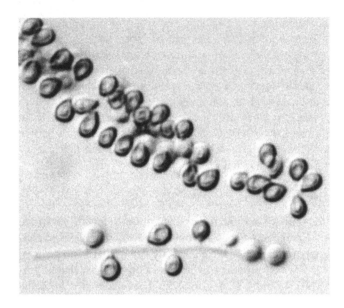

Figure 61 *Sporothrix schenckii* thick-walled dematiaceous conidia attached directly to vegetative hyphae.

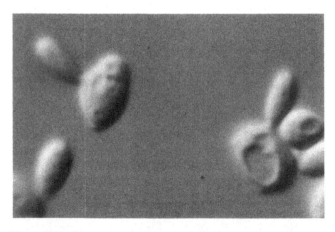

Figure 62 *Sporothrix schenckii* transformed to budding yeast.

rotrichosis outbreak appeared to have the morphology of *Sporothrix schenckii*, except for their lack of thick-walled, dematiaceous conidia borne on vegetative hyphae. These isolates did convert to the yeast form at 35–37°C, but were avirulent in mice. It was thus suggested that the presence of these conidia might constitute a significant identification criterion for *S. schenckii* (175).

2. *Sporothrix schenckii* var. *luriei*

Three isolates of *Sporothrix schenckii* var. *luriei* have been reported (176–178). Its appearance differs from *S. schenckii* var. *schenckii* only by its morphology in host tissue. In culture, whether as mold or yeast, the two varieties are indistinguishable. Assessment of mitochondrial DNA seems to confirm that *S. schenckii* var. *luriei* is markedly different from *S. schenckii* (179). The fungus exhibits large (10–30 μm), subglobose to elongated, thick-walled (up to 2 μm) cells. Each of these forms a single septum that precedes its separation into two cells. During separation, a fragment of the cell wall previously shared by the two cells briefly remains attached in a bridgelike manner, giving a distinctive appearance of eyeglasses. Sometimes a second septum forms, giving the fungus a muriform appearance. It is not known whether there might be serologic, therapeutic, or prognostic differences associated with this variety. *Sporothrix schenckii* and *P. parasiticum* can be confused because of highly similar colony coloration and the presence of a (phialidic) yeast synanamorph in *P. parasiticum*. Compare with *Rhinocladiella* and *Phaeoacremonium parasiticum*.

C′. *Taeniolella* Hughes 1958

Both *T. stilbaspora* and *T. exilis* reportedly have caused cutaneous and subcutaneous lesions (5). A teleomorph is not known. Colonies are filamentous, immersed or superficial, brown to olivaceous black. Conidiophores are little differentiated from hyphae (Fig. 63). Conidia sometimes are solitary, but more often are borne in acropetal moniliform, unbranched, or sparsely branched chains. Conidia are brown, thick-walled, cylindrical, truncate, not readily released from chains (17). *Taeniolella stilbaspora* Hughes 1958 is characterizd by conidia that are 7 to 11 μm wide and have 3 to 24 septa. *Taeniolella exilis* Hughes 1958 is distinguished by conidia that are 12 to 15 μm thick and composed of two to four cells. Compare with *Cladophialophora*, *Xylohypha*, *Septonema*.

D′. *Tetraploa* Berkeley & Broome 1850

1. *Tetraploa aristata* Berkeley & Broome 1850

The teleomorph is not known. Colonies are velvety, dark grayish-brown. Conidiophores are micronematous, inconspicuous with conidia borne singly (Fig. 64).

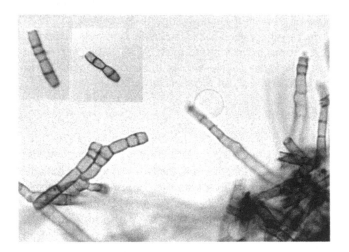

Figure 63   *Taeniolella exilis.*

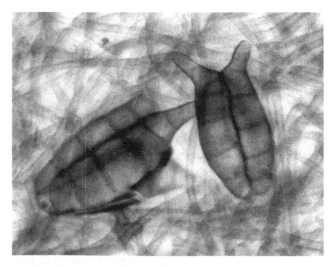

Figure 64   *Tetraploa aristata.*

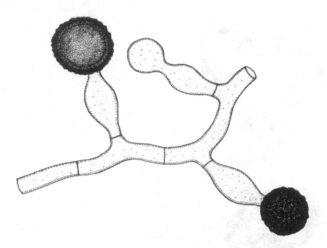

Figure 65 *Thermomyces lanuginosus*. Source: Ref. 26a.

Conidia usually are brown, roughened, six- to 16-celled, consisting of four bundled rows of two or four cells per bundle, with a slight furrow between rows. Conidia measure 14–29 × 25–39 μm and are further distinguished by a long (12–80 μm), filiform, multicellular appendage at the apex of each bundle (17). Keratitis and subcutaneous infection have been reported (180,181). Compare to *Hughesinia*.

E'. *Thermomyces* Tsiklinsky 1899

1. *Thermomyces lanuginosus* Tsiklinsky 1899

A teleomorph not known. Colonies are velvety to lanose, grayish-white, becoming greenish, then brown to black, often with a pink or vinaceous diffusing pigment. Conidia are borne singly on short stalks and are one-celled, dark brown, subglobose, with a roughened surface, 6–12 μm in diameter (17) (Fig. 65). Endocarditis has been reported (182). Compare to *Harzia*, *Acremoniula*.

F'. *Ulocladium* Preuss 1851

1. *Ulocladium chartarum* (Preuss) Simmons 1967

A teleomorph is not known. Colonies are velvety to floccose, rapidly growing, olivaceous brown to black. Conidia are obovoid to short ellipsoidal, golden brown to blackish-brown, roughened, with one to five oblique or longitudinal septa and one to five lateral septa, borne singly or in short chains from sympodial conidio-

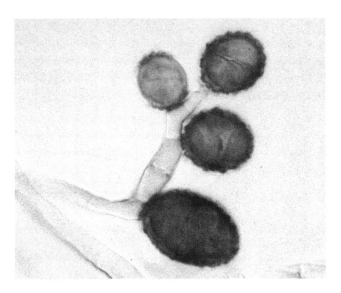

Figure 66 *Ulocladium chartarum.*

phores (21) (Fig. 66). Subcutaneous infection has been attributed (5). Compare to *Alternaria, Stemphilium, Embeliasia.*

G'. *Wangiella* McGinnis 1977

1. *Wangiella dermatitidis* (Kano) McGinnis 1977

This species caused phaeohyphomycoses of cutaneous and subcutaneous tissue, and infections of the eye, brain, joint, and catheter-related fungemia also have been documented (97,164,183–185). Most isolates initially grow in a yeast form (*Phaeococcomyces* synanamorph) with development of toruloid hyphae that is succeeded by the hyphal *W. dermatitidis* synanamorph. Colonies are slow-growing, initially yeastlike, smooth, viscous, pale brown to black, becoming filamentous to velvety. Conidia are one-celled, subglobose to elliptical or obovoid, subhyaline to pale brown, not catenate. Growth occurs at 40°C; potassium nitrate is not assimilated. A teleomorph has not been proved. Conidiogenous cells are phialides that lack collarettes (Fig. 67) (97,186). Some conidiogenous cells exhibit a group of slightly raised, truncate protrusions at their apices that occur as a result of a polyphialidic development. In addition, annellides of the kind seen in *Exophiala* may be formed rarely, and this has led some mycologists to propose that the fungus be reclassified as *Exophiala dermatitidis*. Results from molecular

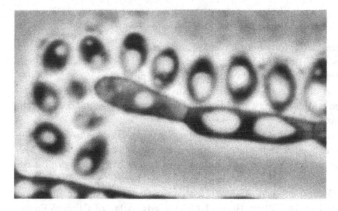

Figure 67 *Wangiella dermatitidis.*

sequencing analyses have not yet resolved this question (52,144). A *Phialophora* synanamorph may be present (Fig. 68). Another fungus known lately as *Exophiala jeanselmei* var. *heteromorpha* has been reevaluated by morphologic criteria, and a new combination for it, *W. heteromorpha*, has been proposed (98). (See *Exophiala.*) Compare with *Phaeoannellomyces, Phaeococcomyces, Phialophora* (3,104,187).

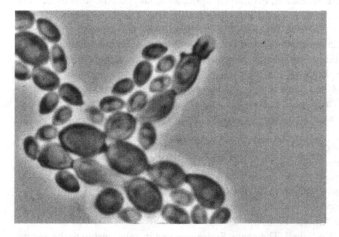

Figure 68 *Phialophora* synanamorph (upper right) of *Wangiella dermatitidis.*

REFERENCES

1. WA Schell. Agents of chromoblastomycosis and sporotrichosis. In: L Ajello, RJ Hay, eds. Topley & Wilson's Microbiology and Microbial Infections: Vol. 4, Medical Mycology, 9th ed. London: Edward Arnold 1998; pp. 315–336.
2. T Matsumoto, L Ajello. Agents of Phaeohyphomycosis. In: L Ajello, RJ Hay, eds. Topley & Wilson's Microbiology and Microbial Infections: Vol. 4, 9th ed. London: Arnold, 1998; pp. 503–524.
3. WA Schell, L Pasarell, IF Salkin, MR McGinnis. *Bipolaris, Exophiala, Scedosporium, Sporothrix* and other dematiaceous fungi. In: PR Murray, EJ Baron, MA Pfaller, FC Tenover, RH Yolken, eds. Manual of Clinical Microbiology. 7th ed. Washington, DC: American Society for Microbiology, 1999; pp. 1295–1375.
4. GS de Hoog, J Guarro, CS Tan, RGF Wintermans, J Gene. Pathogenic fungi and common opportunists. In: GS de Hoog, J Guarro, eds. Atlas of Clinical Fungi. Baarn: Centraalbureau voor Schimmelcultures, 1995; pp. 1–243.
5. J Guarro, GS de Hoog, MJ Figueras, J Gene. Rare opportunistic fungi. In: J Guarro, GS de Hoog, MJ Figueras, J Gene, eds. Atlas of Clinical Fungi. Baarn: Centraalbureau voor Schimmelcultures, 1995; pp. 243–668.
6. KJ Kwon-Chung, JE Bennett. Medical Mycology. Philadelphia: Lea & Febiger, 1992.
7. BM Clay, VK Anand. Sporotrichosis: A nasal obstruction in an infant. Amer J Otolaryngol 17:75–77, 1996.
8. S Gori, A Lupetti, M Moscato, M Parenti, A Lofaro. Pulmonary sporotrichosis with hyphae in a human immunodeficiency virus-infected patient—A case report. Acta Cytol 41:519–521, 1997.
9. OV Castrejon, M Robles, OE Zubieta Arroyo. Fatal fungaemia due to *Sporothrix schenckii*. Mycoses 38:373–376, 1995.
10. PS Saravanakumar, P Eslami, FA Zar. Lymphocutaneous sporotrichosis associated with a squirrel bite: Case report and review. Clin Infec Dis 23:647–648, 1996.
11. A Amaya Tapia, E Uribe Jimenez, R Diaz Perez, MA Covarrubias Velasco, D Diaz Santa Bustamante, G Aguirre Avalos, A Rodriguez Toledo. Esporotricosis cutanea transmitida por mordedura de tejon. Medicina Cutanea Ibero-Latino-Americana 24:87–89, 1996.
12. AA Padhye, MR McGinnis. Fungi causing eumycotic mycetoma. In: PR Murray, EJ Baron, MA Pfaller, FC Tenover, RH Yolken, eds. Manual of Clinical Microbiology. 7th ed. Washington, DC: American Society for Microbiology, 1999; pp. 1318–1326.
13. RJ Hay. Agents of eumycotic mycetomas. In: L Ajello, RJ Hay, eds. Topley & Wilson's Microbiology and Microbial Infections: Vol. 4, Medical Mycology. 9th ed. London: Arnold, 1998; pp. 487–496.
14. DL Hawksworth. Kingdom fungi: Fungal phylogeny and systematics. In: L Ajello, RJ Hay, eds. Topley & Wilson's Microbiology and Microbial Infections. Vol. 4, 9th ed. London: Arnold, 1998; pp. 43–55.
15. MR McGinnis, WA Schell. The genus *Fonsecaea* and its relationship to the genera *Cladosporium, Phialophora, Ramichloridium,* and *Rhinocladiella.* Proceedings of

the 5th International Conference on the Mycoses: Superficial, Cutaneous, and Subcutaneous Infections. Pan American Health Organization, 1980; pp. 215–224.

16. PK Shukla, ZA Khan, B Lal, PK Agrawal, OP Srivastava. Clinical and experimental keratitis caused by the *Colletotrichum* state of *Glomerella cingulata* and *Acrophialophora fusispora*. Sabouraudia 21:137–147, 1983.

17. MB Ellis. Dematiaceous Hyphomycetes. Kew, England: Commonwealth Mycological Institute, 1971.

18. JC Edward. A new genus of the moniliaceae. Mycologia 51:781–786, 1959.

19. MA Viviani, AM Tortorano, G Laria, A Giannetti, G Bignotti. Two new cases of cutaneous alternariosis with a review of the literature. Mycopathologia 96:3–12, 1986.

20. PM Wiest, K Wiese, MR Jacobs, AB Morrissey, TI Abelson, W Witt, MM Ledermann. *Alternaria* infection in a patient with acquired immune deficiency syndrome: Case report and review of invasive *Alternaria* infections. Rev Infec Dis 9:799–803, 1987.

21. EG Simmons. Typification of *Alternaria*, *Stemphylium*, and *Ulocladium*. Mycologia 59:67–92, 1967.

22. SM Singh, J Naidu, M Pouranik. Ungual and cutaneous phaeohyphomycosis caused by *Alternaria alternata* and *Alternaria chlamydospora*. J Med Vet Mycol 28:275–282, 1990.

23. B Bartolome, R Valks, J Fraga, V Buendia, J Fernandez-Herrera, A Garcia-Diez. Cutaneous alternariosis due to *Alternaria chlamydospora* after bone marrow transplantation. Acta Dermato-Venereologica 79:244, 1999.

24. J Gene, A Azon-Masoliver, J Guarro, F Ballester, I Pujol, M Llovera, C Ferrer. Cutaneous phaeohyphomycosis caused by *Alternaria longipes* in an immunosuppressed patient. J Clin Microbiol 33:2774–2776, 1995.

25. C Laumaille, FI Gall, B Degeilh, E Guého, M Huerre. Infection cutanee a *Alternaria infectoria* apres greffe hepatique. Annalles de Pathologie 18:192–194, 1998.

26. A Magina, C Lisboa, P Santos, G Oliveira, J Lopes, M Rocha, J Mesquita-Guimaraes. Cutaneous alternariosis by *Alternaria chartarum* in a renal transplanted patient. Br J Derm 142:1261–1262, 2000.

26a. WB Kendrick, JW Carmichael. Hyphomycetes, In: The Fungi: An Advanced Treatise, vol. IVA. New York: Academic Press, 1973, pp. 323–509.

27. C Romano, L Valenti, C Miracco, C Alessandrini, E Paccagnini, E Faggi, EM Difonzo. Two cases of cutaneous phaeohyphomycosis by *Alternaria alternata* and *Alternaria tenuissima*. Mycopathologia 137:65–74, 1997.

28. C Romano, M Fimiani, M Pellegrino, L Valenti, L Casini, C Miracco, E Faggi. Cutaneous phaeohyphomycosis due to *Alternaria tenuissima*. Mycoses 39:211–215, 1996.

29. J Castanet, JP Lacour, M Toussaint-Gary, C Perrin, S Rodot, JP Ortonne. Infection cutanee pluri-focale a *Alternaria tenuissima*. Annales de Dermatologie et de Venereologie 122:115–118, 1995.

30. MB Ellis. More Dematiaceous Hyphomycetes. Kew, England: Commonwealth Mycological Institute, 1976.

31. EG Simmons. *Alternaria* themes and variations. Mycotaxon 13:16–34, 1981.

32. IF Salkin, JA Martinez, ME Kemna. Opportunistic infection of the spleen caused by *Aureobasidium pullulans*. J Clin Microbiol 23:828–831, 1986.
33. NE Caporale, L Calegari, D Perez, E Gezuele. Peritoneal catheter colonization and peritonitis with *Aureobasidium pullulans*. Perit Dial Inter 16:97–98, 1996.
34. BE Hirsch, BF Farber, JF Shapiro, S Kennelly. Successful treatment of *Aureobasidium pullulans* prosthetic hip infection. Infec Dis Clin Prac 5:205–207, 1996.
35. EC Clark, SM Silver, GE Hollick, MG Rinaldi. Continuous ambulatory peritoneal dialysis complicated by *Aureobasidium pullulans* peritonitis. Amer Nephr 15:353–355, 1995.
36. A Franco, I Aranda, MJ Fernandez, MA Arroyo, F Navas, D Albero, J Olivares. Chromomycosis in a European renal transplant recipient. Nephrol Dial Transplant 11:715–716, 1996.
37. H Fletcher, NP Williams, A Nicholson, L Rainford, H Phillip, A East-Innis. Systemic phaeohyphomycosis in pregnancy and the puerperium. West Ind Med J 49:79–82, 2000.
38. R Ibanez Perez, J Chacon, A Fidalgo, J Martin, V Paraiso, JL Munoz-Bellido. Peritonitis by *Aureobasidium pullulans* in continuous ambulatory peritoneal dialysis. Nephrol Dial Transplant 12:1544–1545, 1997.
39. P Redondo-Bellon, M Idoate, M Rubio, HJ Ignacio. Chromoblastomycosis produced by *Aureobasidium pullulans* in an immunosuppressed patient. Arch Derm 133:663–664, 1997.
40. GS de Hoog, EJ Hermanides-Nijhof. Studies in Mycology 15: The Black Yeasts and Allied Hyphomycetes. Baarn, Netherlands: Centraalbureau voor Schimmelcultures, 1977.
41. NA Yurlova, GS de Hoog, AH Gerrits van den Ende. Taxonomy of *Aureobasidium* and allied genera. In: GS de Hoog, ed. Studies in Mycology 43: Ecology and Evolution of Black Yeasts and Their Relatives. 1999; pp. 63–69.
42. SB Kupferberg, JP Bent III, FA Kuhn. Prognosis for allergic fungal sinusitis. Otolaryn Head Neck Surg 117:35–41, 1997.
43. MR McGinnis, MG Rinaldi, RE Winn. Emerging agents of phaeohyphomycosis: Pathogenic species of *Bipolaris* and *Exserohilum*. J Clin Microbiol 24:250–259, 1986.
44. RH Latham. *Bipolaris spicifera* meningitis complicating a neurosurgical procedure. Scand J Infec Dis 32:102–103, 2000.
45. R Pauzner, A Goldschmied-Reouven, I Hay, Z Vared, Z Ziskind, N Hassin, Z Farfel. Phaeohyphomycosis following cardiac surgery: Case report and review of serious infection due to *Bipolaris* and *Exserohilum* species. Clin Infec Dis 25:921–923, 1997.
46. KL Flanagan, AD Bryceson. Disseminated infection due to *Bipolaris australiensis* in a young immunocompetent man: Case report and review. Clin Infec Dis 25:311–313, 1997.
47. JL Alcorn. Generic concepts in *Drechslera*, *Bipolaris* and *Exserohilum*. Mycotaxon 17:1–86, 1983.
48. L Pasarell, MR McGinnis, PG Standard. Differentiation of medically important isolates of *Bipolaris* and *Exserohilum* with exoantigens. J Clin Microbiol 28:1655–1657, 1990.

49. GS de Hoog, C Rubio. A new dematiaceous fungus from human skin. Sabouraudia 20:15–20, 1982.

50. D Benoldi, A Alinovi, L Polonelli, S Conti, M Gerloni, L Ajello, A Padhye, GS de Hoog. *Botryomyces caespitosus* as an agent of cutaneous phaeohyphomycosis. J Med Vet Mycol 29:9–13, 1991.

51. GS de Hoog, E Guého, F Masclaux, AH Gerrits van den Ende, KJ Kwon-Chung, MR McGinnis. Nutritional physiology and taxonomy of human-pathogenic *Cladosporium-Xylohypha* species. J Med Vet Mycol 33:339–347, 1995.

52. F Masclaux, E Guého, GS de Hoog, R Christen. Phylogenetic relationships of human-pathogenic *Cladosporium* (*Xylohypha*) species inferred from partial LS rRNA sequences. J Med Vet Mycol 33:327–338, 1995.

53. R Horre, GS de Hoog. Primary cerebral infections by melanized fungi: A review. In: GS de Hoog, ed. Studies in Mycology 43: Ecology and Evolution of Black Yeasts and Their Relatives. 1999; pp. 176–193.

54. AH Gerrits van den Ende, GS de Hoog. Variability and molecular diagnostics of the neurotropic species *Cladophialophora bantiana*. In: GS de Hoog, ed. Studies in Mycology 43: Ecology and Evolution of Black Yeasts and Their Relatives. 1999; pp. 151–162.

55. WK Jacyk, JH Du Bruyn, N Holm, H Gryffenberg, VO Karusseit. Cutaneous infection due to *Cladophialophora bantiana* in a patient receiving immunosuppressive therapy. Brit J Derm 136:428–430, 1997.

56. AA Padhye, MR McGinnis, L Ajello, FW Chandler. *Xylohypha emmonsii* sp. nov., a new agent of phaeohyphomycosis. J Clin Microbio 26:702–708, 1988.

57. JW Patterson, NG Warren, LW Kelly. Cutaneous phaeohyphomycosis due to *Cladophialophora bantiana*. J Amer Acad Derm 40:364–366, 1999.

58. MR McGinnis. Dematiaceous fungi. In: EH Lennette, A Balows, WJ Hausler, Jr, HJ Shadomy, eds. Manual of Clinical Microbiology. 4th ed. Washington, DC: 1985; pp. 561–574.

59. MR McGinnis, SM Lemon, DH Walker, GS de Hoog, G Haase. Fatal cerebritis caused by a new species of *Cladophialophora*. Studies in Mycology 43:166–171, 1999.

60. P Esterre, A Andriantsimahavandy, ER Ramarcel, J-L Pecarrere. Forty years of chromoblastomycosis in Madagascar: A review. Amer J Trop Med Hyg 55:45–47, 1996.

61. MR McGinnis, D Borelli, AA Padhye, L Ajello. Reclassification of *Cladosporium bantianum* in the genus *Xylohypha*. J Clin Microbiol 23:1148–1151, 1986.

62. MS Gonzalez, B Alfonso, D Seckinger, AA Padhye, L Ajello. Subcutaneous phaeohyphomycosis caused by *Cladosporium devriesii*. Sabouraudia 22:427–432, 1984.

63. DM Mitchell, M Fitz-Henley, J Horner-Bryce. A case of disseminated phaeohyphomycosis caused by *Cladosporium devriesii*. West Ind Med J 39:118–123, 1990.

64. K Tintelnot, P von Hunnius, GS de Hoog, A Polak-Wyss, E Guého, F Masclaux. Systemic mycosis caused by a new *Cladophialophora* species. J Med Vet Mycol 33:349–354, 1995.

65. AA Padhye, JD Dunkel, RM Winn, S Weber, EP Ewing, GS de Hoog, Jr. Subcutaneous phaeohyphomycosis caused by an undescribed *Cladophialophora* species.

In: GS de Hoog, ed. Studies in Mycology 43: Ecology and Evolution of Black Yeasts and Their Relatives. 1999; pp. 172–175.

66. C Lopez, L Ramos, G Weisburd, S Margasin, R Ramirez. Feohifomicosis causada por *Cladosporium cladosporioides*. J Med Microbiol 46:699–703, 1997.

67. WA Schell. Oculomycosis caused by dematiaceous fungi. Proceedings of the VI International Conference on the Mycoses. Vol. 879. Pan American Health Organization, 1986; pp. 105–109.

68. HC Gugnani, S Neelam, B Singh, R Makkar. Subcutaneous phaeohyphomycosis due to *Cladosporium cladosporioides*. Mycoses 43:85–87, 2000.

69. C Romano, R Bilenchi, C Alessandrini, C Miracco. Cutaneous phaeohyphomycosis caused by *Cladosporium oxysporum*. Mycoses 42:111–115, 1999.

70. M Pereiro Jr, J Jo-Chu, J Toribio. Phaeohyphomycotic cyst due to *Cladosporium cladosporioides*. Dermatology 197:90–92, 1998.

71. MR Vieira, A Milheiro, FA Pacheco. Phaeohyphomycosis due to *Cladosporium cladosporioides*. Med Mycol 39:135–137, 2001.

72. KJ Kwon-Chung, IS Schwarts, BJ Rybak. A pulmonary fungus ball produced by *Cladosporium cladosporioides*. Amer J Clin Pathol 64:564–568, 1975.

73. CJK Wang. Microfungi. In: CJK Wang, RA Zabel, eds. Identification manual for fungi from utility poles in the eastern United States. Rockville, MD: American Type Culture Collection, 1990.

74. MG Rinaldi, P Phillips, JG Schwartz, RE Winn, GR Holt, FW Shagets, J Elrod, G Nishioka, TB Aufdemorte. Human *Curvularia* infections: Report of five cases and review of the literature. Diag Microbiol Infect Dis 6:27–39, 1987.

75. YC Yau, J de Nanassy, RC Summerbell, AG Matlow, SE Richardson. Fungal sternal wound infection due to *Curvularia lunata* in a neonate with congenital heart disease: Case report and review. Clin Infect Dis 19:735–740, 1994.

76. JR Ebright, PH Chandrasekar, S Marks, MR Fairfax, A Aneziokori, MR McGinnis. Invasive sinusitis and cerebritis due to *Curvularia clavata* in an immunocompetent adult. Clin Infec Dis 28:687–689, 1999.

77. C Janaki, G Sentamilselvi, VR Janaki, S Devesh, K Ajithados. Eumycetoma due to *Curvularia lunata*. Mycoses 42:345–346, 1999.

78. M Fernandez, DE Noyola, SN Rosemann, MS Edwards. Cutaneous phaeohyphomycosis caused by *Curvularia lunata* and a review of *Curvularia* infections in pediatrics. Pediat Infec Dis J 18:727–731, 1999.

79. MB Ellis. Dematiaceous hyphomycetes. VII: *Curvularia, Brachysporium*, etc. Mycological Papers no. 106, 1966.

80. LK Georg, BW Bierer, WB Cooke. Encephalitis in turkey poults due to a new fungus species. Sabouraudia 3:239–244, 1964.

81. GC Bhatt, WB Kendrick. *Diplorhinotrichum* and *Dactylaria* and a description of a new species. Can J Bot 46:1253–1257, 1968.

82. MC Mancini, MR McGinnis. *Dactylaria* infection of a human being: Pulmonary disease in a heart transplant recipient. J Heart Lung Transpl 11:827–830, 1992.

83. EH Sides, JD Benson, AA Padhye. Phaeohyphomycotic brain abscess due to *Ochroconis gallopavum* in a patient with malignant lymphoma of a large cell type. J Med Vet Mycol 29:317–322, 1991.

84. SM Kralovic, JC Rhodes. Phaeohyphomycosis caused by *Dactylaria* (human dac-

tylariosis): Report of a case with review of the literature. J Infect 31:107–113, 1995.

85. DM Dixon, IF Salkin. Morphologic and physiologic studies of three dematiaceous pathogens. J Clin Microbiol 24:12–15, 1986.

86. GS de Hoog. On the potentially pathogenic dematiaceous hyphomycetes. In: DH Howard, ed. The Fungi Pathogenic to Humans and Animals. Part A. 1983; pp. 149–216.

87. KH Domsch, W Gams, TH Anderson. Compendium of Soil Fungi. vols I and II. New York: Academic, 1980.

88. R Horre, GS de Hoog, C Kluczny, G Marklein, KP Schaal. rDNA diversity and physiology of *Ochroconis* and *Scolecobasidium* species reported from humans and other vertebrates. In: GS de Hoog, ed. Studies in Mycology 43: Ecology and Evolution of Black Yeasts and Their Relatives. 1999; pp. 194–205.

89. EG Simmons, MR McGinnis, MG Rinaldi. *Dissitimuris*, a new dematiaceous genus of hyphomycetes. Mycotaxon 30:247–252, 1987.

90. RG Washburn, DW Kennedy, MG Begley, DK Henderson, JE Bennett. Chronic fungal sinusitis in apparently normal hosts. Medicine 67:231–247, 1988.

91. EJ Sudduth, AJ Crumbley, WE Farrar. Phaeohyphomycosis due to *Exophiala* species: Clinical spectrum of disease in humans. Clin Infec Dis 15:639–644, 1992.

92. LC Severo, FM Oliveira, G Vettorato, AT Londero. Mycetoma caused by *Exophiala jeanselmei*: Report of a case successfully treated with itraconazole and review of the literature. Revista Iberica de Micologia 16:57–59, 1999.

93. S Kinkead, V Jancic, T Stasko, AS Boyd. Chromoblastomycosis in a patient with a cardiac transplant. Cutis 58:367–370, 1996.

94. JF Barba-Gomez, J Mayorga, MR McGinnis, A Gonzalez-Mendoza. Chromoblastomycosis caused by *Exophiala spinifera*. J Amer Acad Derm 26:367–370, 1992.

95. T Matsumoto, K Nishimoto, K Kimura, AA Padhye, L Ajello, MR McGinnis. Phaeohyphomycosis caused by *Exophiala moniliae*. Sabouraudia 22:17–26, 1984.

96. M Sughayer, PC DeGirolami, U Khettry, D Korzeniowski, A Grumney, L Pasarell. MR McGinnis. Human infection caused by *Exophiala pisciphila*: Case report and review. Rev Infect Dis 13:379–382, 1991.

97. T Matsumoto, AA Padhye, L Ajello, PG Standard. Critical review of human isolates of *Wangiella dermatitidis*. Mycologia 76:232–249, 1984.

98. JM McKemy, SO Rogers, CJK Wang. Emendation of the genus *Wangiella* and a new combination, *W. heteromorpha*. Mycologia 91:200–205, 1999.

99. G Haase, L Sonntag, B Melzer-Krick, GS de Hoog. Phylogenetic inference by SSU-gene analysis of members of the *Herpotrichiellaceae* with special reference to human pathogenic species. In: GS de Hoog, ed. Studies in Mycology 43: Ecology and Evolution of Black Yeasts and Their Relatives. 1999; pp. 80–97.

100. MR McGinnis, WA Schell, J Carson. *Phaeoannellomyces* and the Phaeoannellomycetaceae, new dematiaceous Blastomycete taxa. Sabouraudia 23:179–188, 1985.

101. DM Dixon, A Polak-Wyss. The medically important dematiaceous fungi and their identification. Mycoses 34:1–18, 1991.

102. A Espinel-Ingroff, PR Goldson, MR McGinnis, TM Kerkering. Evaluation of pro-

teolytic activity to differentiate some dematiaceous fungi. J Clin Microbiol 26:
301–307, 1988.

103. T Iwatsu, K Sishimura, M Makoto. *Exophiala castellanii* sp. nov. Mycotaxon 20:
307–314, 1984.

104. MR McGinnis. Human pathogenic species of *Exophiala*, *Phialophora*, and *Wangiella*. Pan Amer Hlth Org Sci Pub 356:35–59. 1978.

105. MR McGinnis. Taxonomy of *Exophiala jeanselmei* (Langeron) McGinnis and Padhye. Mycopathologia 65:79–87, 1978.

106. MR McGinnis. Taxonomy of *Exophiala werneckii* and its relationship to *Microsporum mansonii*. Sabouraudia 17:145–154, 1979.

107. AA Padhye. Comparative study of *Phialophora jeanselmei* and *P. gougerotii* by morphological, biochemical, and immunological methods. Pan Amer Hlth Org Sci Pub 356:60–65, 1978.

108. G Haase. *Exophiala jeanselmei* variety *castellanii* and *Exophiala mansonii* are synonyms. Clin Infec Dis 23:852–853, 1996.

109. M Kawasaki, H Ishizaki, T Matsumoto, T Matsuda, K Nishimura, M Miyaji. Mitochondrial DNA analysis of *Exophiala jeanselmei* var. *lecanii-corni* and *Exophiala castellanii*. Mycopathologia 146:75–77, 1999.

110. M Kawasaki, H Ishizaki, K Nishimura, M Miyaji. Mitochondrial DNA analysis of *Exophiala moniliae*. Mycopathologia 121:7–10, 1993.

111. HS Nielsen Jr, NF Conant. A new human pathogenic *Phialophora*. Sabouraudia 6:228–231, 1968.

112. GS de Hoog, N Poonwan, AH Gerrits van den Ende. Taxonomy of *Exophiala spinifera* and its relationship to *E. jeanselmei*. In: GS de Hoog, ed. Studies in Mycology 43: Ecology and Evolution of Black Yeasts and Their Relatives. 1999, pp. 133–142.

113. MM Hsu, Y-Y Lee. Cutaneous and subcutaneous phaeohyphomycosis caused by *Exserohilum rostratum*. J Amer Acad Derm 28:340–344, 1993.

114. AA Padhye, L Ajello, MA Wieden, KK Steinbronn. Phaeohyphomycosis of the nasal sinuses caused by a new species of *Exserohilum*. J Clin Microbiol 24:245–249, 1986.

115. VM Aquino, JM Norvell, K Krisher, MM Mustafa. Fatal disseminated infection due to *Exserohilum rostratum* in a patient with aplastic anemia: Case report and review. Clin Infec Dis 20:176–178, 1995.

116. MS Mathews, SV Maharajan. *Exserohilum rostratum* causing keratitis in India. Med Mycol 37:131–132, 1999.

117. AA Padhye, L Ajello, MA Wieden, KK Steinbronn. Phaeohyphomycosis of the nasal sinuses caused by a new species of *Exserohilum*. J Clin Microbiol 24:245–249, 1986.

118. JP Silva, W Souza, S Rosenthal. Chromoblastomycosis: A retrospective study of 325 cases on Amazonic region (Brazil). Mycopathologia 143:171–175, 1998.

119. MC Attapattu. Chromoblastomycosis—A clinical and mycological study of 71 cases from Sri Lanka. Mycopathologia 137:145–151, 1997.

120. A Morris, WA Schell, D McDonagh, S Chafee, JR Perfect. *Fonsecaea pedrosoi* pneumonia and *Emericella nidulans* cerebral abscesses in a bone marrow transplant patient. Clin Infec Dis 21:1346–1348, 1995.

121. MR McGinnis. Chromoblastomycosis and phaeohyphomycosis: New concepts, diagnosis and mycology. Am Acad Derm 8:1–16, 1983.

122. RC Fader, MR McGinnis. Infections caused by dematiaceous fungi: Chromoblastomycosis and phaeohyphomycosis. Infect Dis Clin N Amer 2:925–938, 1988.

123. K Barton, D Miller, SC Pflungfelder. Corneal chromoblastomycosis. Cornea 16: 235–239, 1997.

124. DS Attili, GS de Hoog, AA Pizzirani-Kleiner. rDNA-RFLP and ITS1 sequencing of species of the genus *Fonsecaea*, agents of chromoblastomycosis. Med Mycol 36:219–225, 1998.

125. BM Coldiron, EL Wiley, MG Rinaldi. Cutaneous phaeohyphomycosis caused by a rare fungal pathogen, *Hormonema dematioides*: Successful treatment with ketoconazole. J Amer Acad Derm 23:363–367, 1990.

126. J Shin, L SangKu, S Suh, D Ryang, N Kim, MG Rinaldi, DA Sutton. Fatal *Hormonema dematioides* peritonitis in a patient on continuous ambulatory peritoneal dialysis: Criteria for organism identification and review of other known fungal etiologic agents. J Clin Microbiol 36:2157–2163, 1998.

127. WA Schell, MR McGinnis. Molds involved in subcutaneous infections: In: BB Wentworth, ed. Diagnostic Procedures for Bacterial, Mycotic, and Parasitic Infections, 7th ed. Washington, DC: American Public Health Association, 1988; pp. 99–171.

128. TE Lie-Kian, S Njo-Injo. A new verrucous mycosis caused by *Cercospora apii*. Arch Derm 75:864–870, 1957.

129. AA Padhye, MS Davis, A Reddick, MF Bell. ED Gearhart, L Von Moll. *Mycoleptodiscus indicus*: A new etiologic agent of phaeohyphomycosis. J Clin Microbiol 33:2796–2797, 1995.

130. TR Nag Raj. Coelomycetous Anamorphs with Appendage-bearing Conidia. Waterloo: Mycologue Publications, 2001.

131. RME Fincher, JF Fisher, AA Padhye, L Ajello, JCH Steele Jr. Subcutaneous phaeohyphomycotic abscess caused by *Phialophora parasitica* in a renal allograft recipient. J Med Vet Mycol 26:311–314, 1988.

132. SV Hood, CB Moore, JS Cheesbrough, A Mene, DW Denning. Atypical eumycetoma caused by *Phialophora parasitica* successfully treated with itraconazole and flucytosine. Brit J Derm 136:953–956, 1997.

133. CH Heath, JL Lendrum, BL Wetherall, SL Wesselingh, DL Gordon. *Phaeoacremonium parasiticum* infective endocarditis following liver transplantation. Clin Infect Dis 25:1251–1252, 1997.

134. AA Padhye, D Davis, PN Baer, A Reddick, KK Sinha, J Ott. Phaeohyphomycosis caused by *Phaeoacremonium inflatipes*. J Clin Microbiol 36:2763–2765, 1998.

135. T Matsu, K Nishimoto, S Udagawa, H Ishihara, T Ono. Subcutaneous phaeohyphomycosis caused by *Phaeoacremonium rubrigenum* in an immunosuppressed patient. Japn J Med Mycol 40:99–102, 1999.

136. PW Crous, W Gams, MJ Wingfield, PSV Wyk. *Phaeoacremonium* gen. nov. associated with wilt and decline disease of woody hosts and human infections. Mycologia 88:786–796, 1996.

137. L Ajello. Phaeohyphomycosis: definition and etiology. Pan Amer Hlth Org 304: 126–133, 1975.

138. CE Huber, T LaBarge, T Schwiesow, K Carroll, PS Bernstein, N Mamalis. *Exophiala werneckii* endophthalmitis following cataract surgery in an immunocompetent individual. Ophth Surg Lasers 31:417–422, 2000.

139. NC Engleberg, J Johnson, J Bluestein, K Madden, MG Rinaldi. Phaeohyphomycotic cyst caused by a recently described species, *Phaeoannellomyces elegans.* J Clin Microbiol 25:605–608, 1987.

140. PK Shukla, ZA Kahn, B Lal, PK Agrawal, OP Srivatava. A study of the association of fungi in human corneal ulcers and their therapy. Mykosen 27:385–390, 1984.

141. PK Shukla, M Jain, B Lal, PK Agrawal, OP Srivastauna. Mycotic keratitis caused by *Phaeotrichoconis crotalariae.* Mycoses 32:230–232, 1986.

142. TS Lundstrom, MR Fairfax, MC Dugan, JA Vazquez, PH Chandrasekar, E Abella, C Kasten-Sportes. *Phialophora verrucosa* infection in a BMT patient. Bone Marr Transpl 20:789–791, 1997.

143. UM Tendolkar, P Kerkar, H Jerajani, A Gogate, AA Padhye. Phaeohyphomycotic ulcer caused by *Phialophora verrucosa*: Successful treatment with itraconazole. J Infect 36:122–125, 1998.

144. WA Untereiner, FA Naveau. Molecular systematics of the Herpotrichiellaceae with an assessment of the phylogenetic positions of *Exophiala dermatitidis* and *Phialophora americana.* Mycologia 91:67–83, 1999.

145. M Hironaga, K Nakano, I Yokoyama, J Kitajima. *Phialophora repens*, an emerging agent of subcutaneous phaeohyphomycosis in humans. J Clin Microbiol 27:394–399, 1989.

146. DL Pitrak, EW Koneman, RC Estupinan, J Jackson. *Phialophora richardsiae* in humans. Rev Infec Dis 10:1195–1203, 1988.

147. E Guého, A Bonnefoy, J Luboinski, J-C Petit, GS de Hoog. Subcutaneous granuloma caused by *Phialophora richardsiae*: Case report and review of the literature. Mycoses 32:219–223, 1989.

148. JM Duggan, MD Wolf, CA Kauffman. *Phialophora verrucosa* infection in an AIDS patient. Mycoses 38:215–218, 1995.

149. C Uberti-Foppa, L Fumagalli, N Gianotti, AM Viviani, R Vaiani, E Guého. First case of osteomyelitis due to *Phialophora richardsiae* in a patient with HIV infection. AIDS 9:975–976, 1995.

150. GW Turiansky, PM Benson, LC Sperling, P Sau, IF Salkin, MR McGinnis, WD James. *Phialophora verrucosa*: A new cause of mycetoma. J Amer Acad Derm 32:311–315, 1995.

151. A Juma. *Phialophora richardsiae* endocarditis of aortic and mitral valves in a diabetic man with a porcine mitral valve. J Infect 27:173–175, 1993.

152. W Gams, MR McGinnis. *Phialemonium*, a new anamorph genus intermediate between *Phialophora* and *Acremonium.* Mycologia 75:977–987, 1983.

153. L Ajello, AA Padhye. Phaeohyphomycosis in a dog caused by *Pseudomicrodochium suttonii* sp. nov. Mycotaxon 12:131–136, 1980.

154. WA Schell, JR Perfect. *Pseudomicrodochium suttonii* isolated from subcutaneous lesion in a sarcoid patient. Abstract 127. Washington, D.C.: 95th American Society for Microbiology, 1995.

155. WA Schell, MR McGinnis, D Borelli. *Rhinocladiella aquaspersa*, a new combination for *Acrotheca aquaspersa.* Mycotaxon 17:341–348, 1983.

156. CK Campbell, SSA Al-Hedaithy. Phaeohyphomycosis of the brain caused by *Ramichloridium mackenziei* sp. nov. in Middle Eastern countries. J Med Vet Mycol 31:325–332, 1993.

157. DA Sutton, M Slifkin, R Yakulis, MG Rinaldi. U.S. case report of cerebral phaeohyphomycosis caused by *Ramichloridium obovoideum* (*R. mackenziei*): Criteria for identification, therapy, and review of other known dematiaceous neurotropic taxa. J Clin Microbiol 36:708–715, 1998.

158. M Arango, C Jaramillo, A Cortes, A Restrepo. Auricular chromoblastomycosis caused by *Rhinocladiella aquaspersa*. Med Mycol 36:43–45, 1998.

159. M Perez-Blanco, G Fernandez-Zeppenfeldt, VR Hernandez, F Yegres, D Borelli. Cromomicosis por *Rhinocladiella aquaspersa*: Description del primer caso en Venezuela. Revista Iberica de Micologia 15:51–54, 1998.

160. JJC Sidrim, RHO Menezes, GC Paixao, MFG Rocha, RSN Brilhante, AMA Oliveria, MJN Diogenes. *Rhinocladiella aquaspersa*: Limite imprecise entre chromoblastomycose et phaeohyphomycose? Journal de Mycologie Medicale 9:114–118, 1999.

161. A del Palacio-Hernanz, MK Moore, CK Campbell, A del Palacio-Perez-Mendel, R del Castillo-Cantero. Infection of the central nervous system by *Rhinocladiella atrovirens* in a patient with acquired immunodeficiency syndrome. J Med Vet Mycol 27:127–130, 1989.

162. B Ndiaye, M Develoux, MT Dieng, A Kane, O Ndir, G Raphenon, M Huerre. Current report of mycetoma in Senegal: Report of 109 cases. Journal de Mycologie Medicale 10:140–144, 2000.

163. T Matsumoto, AA Padhye, L Ajello, MR McGinnis. *Sarcinomyces phaeomuriformis*: A new dematiaceous hyphomycete. J Med Vet Mycol 24:395–400, 1986.

164. T Matsumoto, T Matsuda, MR McGinnis, L Ajello. Clinical and mycological spectra of *Wangiella dermatitidis* infections. Mycoses 36:145–155, 1992.

165. JW Carmichael, B Kendrick, IL Conners, L Sigler. Genera of Hyphomycetes. Edmonton, Alberta, Canada, University of Alberta Press, 1980.

166. KEA Burns, NP Ohori, AT Iacono. *Dactylaria gallopava* infection presenting as a pulmonary nodule in a single-lung transplant recipient. J Heart Lung Transpl 19:900–902, 2000.

167. IF Salkin, DM Dixon. *Dactylaria constricta*: Description of two varieties. Mycotaxon 29:377–381, 1987.

168. MTh Smith, WH Batenburg-Van der Vegte. Ultrastructure of septa in *Blastobotrys* and *Sporothrix*. Antonie Van Leeuwenhoek 51:121–128, 1985.

169. GS de Hoog, GA de Vries. Two new species of *Sporothrix* and their relations to *Blastobotrys nivea*. Antonie van Leeuwenhoek J Microbio Serol 39:515–520, 1973.

170. R Tambini, C Farina, R Fiocchi, B Dupont, E Guého, G Delvecchio, F Mamprin, G Gavazzeni. Possible pathogenic role of *Sporothrix cyanescens* isolated from a lung lesion in a heart transplant patient. J Med Vet Mycol 34:195–198, 1996.

171. L Sigler, JL Harris, DM Dixon, AL Flis, IF Salkin, M Kemna, RA Duncan. Microbiology and potential virulence of *Sporothrix cyanescens*, a fungus rarely isolated from blood and skin. J Clin Microbiol 28:1009–1015, 1990.

172. JP Wang, KF Granlund, SA Bozzette, MJ Botte, J Fierer. Bursal sporotrichosis: Case report and review. Clin Infect Dis 31:615–616, 2000.

173. LZ Goldani, VR Aquino, AA Dargel. Disseminated cutaneous sporotrichosis in an AIDS patient receiving maintenance therapy with fluconazole for previous crypto-coccal meningitis. Clin Infec Dis 28:1337–1338, 1999.

174. CA Kauffman. Sporotrichosis. Clin Infec Dis 29:231–237, 1999.

175. DM Dixon, IF Salkin, RA Duncan, NJ Hurd, JH Haines, ME Kemna, FB Coles. Isolation and characterization of *Sporothrix schenckii* from clinical and environ-mental sources associated with the largest U.S. epidemic of sporotrichosis. J Clin Microbiol 29:1106–1113, 1991.

176. F Alberici, CT Patsies, G Lombardi, L Ajello, L Kaufman, F Chandler. *Sporothrix schenckii* var *luriei* as the cause of sporotrichosis in Italy. Eur J Epidem 5:173–177, 1989.

177. AA Padhye, L Kaufman, E Durry, CK Banerjee, SK Jindal, P Talwar, A Chakra-barti. Fatal pulmonary sporotrichosis caused by *Sporothrix schenckii* var. *luriei* in India. J Clin Microbiol 30:2492–2494, 1992.

178. L Ajello, W Kaplan. A new variant of *Sporothrix schenckii*. Mykosen 12:633–644, 1969.

179. K Suzuki, M Kawasaki, H Ishizaki. Analysis of restriction profiles of mitochondrial DNA from *Sporothrix schenckii* and related fungi. Mycopathologia 103:147–151, 1988.

180. E Newmark, FM Polack. *Tetraploa* keratomycosis: Amer J Ophth 70:1013–1015, 1970.

181. WD Markham, RD Key, AA Padhye, L Ajello. Phaeohyphomycotic cyst caused by *Tetraploa aristata*. J Med Vet Mycol 28:147–150, 1990.

182. M Leeso-Bornet, E Gúeho, G Barbier-Boehm, G Berthelot, M Gaildrat, D Tara-vella. Prosthetic valve endocarditis due to *Thermomyces lanuginosus* Tsiklinsky—First case report. J Med Vet Mycol 29:205–209, 1991.

183. N Ajanee, M Alam, K Holmberg, J Khan. Brain abscess caused by *Wangiella dermatitidis*: Case report. Clin Infect Dis 23:197–198, 1996.

184. A Woollons, CR Darley, S Pandian, P Arnstein, J Blackee, J Paul. Phaeohypho-mycosis caused by *Exophiala dermatitidis* following intra-articular steroid injec-tion. Brit J Derm 135:475–477, 1996.

185. S Nachman, O Alpan, R Malowitz, ED Spitzer. Catheter-associated fungemia due to *Wangiella* (*Exophiala*) *dermatitidis*. J Clin Microbiol 34:1011–1013, 1996.

186. MR McGinnis. *Wangiella*, a new genus to accommodate *Hormiscium dermatitidis*. Mycotaxon 5:353–363, 1977.

187. G St-Germain, R Summerbell. Identifying Filamentous Fungi. Belmont: Star, 1996.

11

Miscellaneous Opportunistic Fungi

Microascaceae and Other Ascomycetes, Hyphomycetes, Coelomycetes, and Basidiomycetes

Lynne Sigler
University of Alberta, Edmonton, Alberta, Canada

I. INTRODUCTION

The fungi treated in this chapter have different phylogenetic affinities. Pathogenic members of two ascomycete families—Microascaceae and Chaetomiaceae—are treated first. The family Microascaceae includes *Scedosporium* anamorphs and teleomorphs in *Pseudallescheria* and *Petriella* in addition to *Scopulariopsis* anamorphs and teleomorphs in *Microascus*. The remaining fungi represent diverse ascomycetes and basidiomycetes that are recognized in culture usually by their conspicuous conidial stages or by vegetative mycelial features. The conidia are produced from conidiogenous cells borne diffusely on the mycelium, on mononematous or synnematous conidiophores, or in sporodochial or pycnidial conidiomata. These species traditionally have been classified in the Hyphomycetes and the Coelomycetes. Although connections to teleomorphs have been established for many of the species, they are described here under their better-known anamorphic names. The hyphomycetous fungi included in this chapter are predominantly moniliaceous; that is, they produce lightly colored conidia. Fungi producing darkly pigmented conidia are treated in Chap. 10.

Isolates that are initially sterile in culture may be ones that form fruiting bodies representative of the ascomycetes, the basidiomycetes, or the coelomycetes. Sexually producing fungi may be heterothallic, and the teleomorph will not be obtained unless compatible mating strains are crossed. Isolates of homothallic species will form fruiting bodies (ascomata or basidiomata) when grown on a suitable medium or under appropriate conditions of light or temperature. Media that stimulate sporulation include oatmeal-salts agar, cornmeal agar, and Takashio agar or low-glucose media amended with natural substrates (e.g., carnation leaves, wood shavings, hair, or other keratins) (1). It is often necessary to try more than one medium and to hold cultures for extended time periods (4–8 weeks or longer) to obtain mature structures. Inspection of plates with a dissecting microscope (with illumination from above and below) is of great value in locating pycnidia or ascomata that may be embedded in the agar or hidden under mats of mycelium and/or exudate droplets, and for observing the presence of ostioles, appendages, or setae and the manner of spore release (e.g., ascospores in cirrhi, conidia in slime).

II. FAMILY MICROASCACEAE (ORDER MICROASCALES)

A. Taxonomic Concepts

The Microascaceae were formerly classified with the Ophiostomataceae in the Microascales, but recent molecular phylogenetic analyses have placed these families in separate orders (reviewed in Ref. 2). These are the Microascales and Ophiostomatales (3–6). This separation is supported by physiological characters, as the Ophiostomataceae demonstrate a tolerance to cycloheximide (7), while the Microascaceae are tolerant of benomyl (8, 9). Members of the Microascaceae produce ostiolate or nonostiolate membranaceous ascocarps (perithecia and cleistothecia) that are small (usually less than 300 μm in diameter), brown to black, smooth or hairy, and contain asci irregularly arranged within the centrum. Asci are eight-spored, ovoid, and prototunicate; that is, they have thin walls that lyse, allowing for passive discharge of the ascospores. Ascospores are one-celled, smooth, yellowish to reddish-brown (straw to copper-colored), dextrinoid when immature, and commonly have one or two germ pores. Teleomorphic genera included in the Microascaceae are *Microascus, Pseudallescheria, Petriella, Pithoascus, Kernia,* and *Lophotrichus* (5). Anamorphs are present in many species and are placed in the genera *Scopulariopsis, Scedosporium, Graphium* (in part), *Cephalotrichum* (=*Doratomyces*), *Trichurus, Wardomyces, Wardomycopsis,* and *Echinobotryum.* Conidia are produced from annellidic conidiogenous cells borne on mononematous or synnematous conidiophores or are solitary. The main pathogenic genera are *Microascus* and *Pseudallescheria*, and these are distinguished by ascomatal type and ascospore and conidial features. Species of *Microascus* produce perithecia, yellowish to reddish-brown asymmetrical, reniform, heart-shaped or triangular ascospores bearing a single pore (10–12). As perithecia age, ascospores are commonly extruded in a long column of spores (a cirrhus). Conidia are produced in dry chains and the anamorphs are placed in the genus *Scopulariopsis*. Members of the genus *Pseudallescheria* produce cleistothecia, yellowish-brown ellipsoid ascospores having two germ pores (12, 13). Conidia are slimy and anamorphs are placed in *Scedosporium* and *Graphium*.

B. Species of Medical Relevance

1. *Pseudallescheria boydii* (Shear) McGinnis, Padhye & Ajello, Mycotaxon 14:97, 1982 (Figs. 1a–c, 2, 3)

> Synonyms: *Petriellidium boydii* (Shear) Malloch, Mycologia 62:739, 1970; *Allescheria boydii* Shear, Mycologia 14:242, 1922
> Synanamorphs: *Scedosporium apiospermum* (Saccardo) Castellani & Chalmers, Manual of Tropical Medicine Ed. 3, p 1122, 1919; *Graphium eumorphum* Saccardo, Michelia 2:560, 1881

Occurrence as a Pathogen. Clinical problems mimic those presented by *Aspergillus fumigatus* (14–18). Patients have similar risk factors and symptomology. It can be difficult to distinguish between colonization and invasive infection, and infections are often refractory to antifungal therapy. As well, hyphae in tissue resemble those of *Aspergillus*. Disease occurs in the healthy or immunocompromised host, and the spectrum of clinical syndromes includes otomycosis, saprophytic pulmonary colonization, allergic sinusitis and bronchopulmonary mycosis, fungus ball, keratitis, endocarditis, osteomyelitis, keratitis, endophthalmitis, invasive sinusitis, and pulmonary or disseminated disease. *Pseudallescheria boydii* may affect the brain, causing cerebral abscesses or meningitis. It also causes white grain mycetoma. Persistent respiratory tract colonization is a problem for patients with cystic fibrosis and chronic suppurative lung disease who may be excluded for consideration of a lung transplant due to therapeutic difficulties (19).

Description. Colonies grow moderately rapidly, and are pale to dark brownish-gray ("mouse-gray"), with light to olivaceous brown reverse. Yellow to brown diffusible pigment is often produced. Isolates grow well at 37°C, and many isolates grow at 40°C. Growth is equivalent or somewhat inhibited on medium containing cycloheximide. In culture, this homothallic ascomycete may produce three states. The diffuse *Scedosporium apiospermum* anamorph is produced by all isolates, but the *Graphium* synanamorph and the *Pseudallescheria* teleomorph are produced only by some. Ascomata are nonostiolate (cleistothecia), discrete, solitary, globose, brownish-black, wall of textura epidermoidea, 55 to 180 µm in diameter, and often immersed in the agar. Asci are globose to ovoid, evanescent, and eight-spored. Ascospores are ellipsoidal, symmetrical, yellowish-brown, smooth with two germ pores, and measure 6 to 8.5 µm long by 3.5 to

5.5 μm wide. Ascospores characteristically have a de Bary bubble. Conidia of *Scedosporium apiospermum* are oval or ellipsoidal, 5 to 11 μm long by 3 to 6.5 μm wide, dilute yellow-brown, and are solitary on the side of the hypha or formed in slimy masses from short, usually unswollen conidiogenous cells (annellides). *Graphium* conidia are cylindrical or clavate, 3 to 11.5 (22.5) μm long by 1.5 to 3.5 (6.5) μm wide, dilute yellow-brown, and formed in slimy heads from annellides borne at the tip of a synnema.

Comments. This ubiquitous, soil-borne thermotolerant fungus and its relative, *Scedosporium prolificans*, are the most important pathogenic species of the Microascaceae. *Scedosporium prolificans* is distinguished by annellidic conidiogenous cells that are basally swollen, and by the absence of a teleomorph and a *Graphium* stage. Development of ascomata and synnemata in *P. boydii* isolates is enhanced on oatmeal-salts agar or other medium. Examination with a dissecting microscope and basal illumination may reveal ascomata that are often immersed in the agar.

Comparison of ribosomal internal transcribed spacer (ITS) region sequences revealed considerable genetic variability within *Pseudallescheria boydii* and confirmed its distinction from *S. prolificans* (20, 21). *Pseudallescheria angusta*, *P. ellipsoidea*, and *P. fusoidea*, known thus far only from soil (12), are considered synonyms based on their grouping within a clade of clinical and environmental isolates of *P. boydii* (21), as had been suggested by morphological similarity (13). Molecular typing (19, 21) has been found useful to characterize individual strains but has not yet elucidated the source of infection and means of transmission of this species (19). Patients with chronic suppurative lung disease demonstrated persistent carriage of the same molecular type over long periods (19), but isolates of different types could be obtained from individual patients (19) and from identical environmental sites (21).

2. *Scedosporium prolificans* (Hennebert & Desai) Guého & de Hoog, J Mycol Med 118:9, 1991 (Fig. 1a)

   Synonyms: *Lomentospora prolificans* Hennebert & Desai, Mycotaxon 1: 47, 1974; *Scedosporium inflatum* Malloch & Salkin, Mycotaxon 21:249, 1984

Occurrence as a Pathogen. Although the spectrum of disease is similar to that of *S. apiospermum*, *S. inflatum* is more commonly associated with osteomyelitis and arthritis in humans and animals (16, 22, 23). Idigoras et al. (24) reviewed 18 Spanish cases that occurred over a 10-year period and grouped clinical conditions into three types: (1) pulmonary colonization, particularly in patients with cystic fibrosis, AIDS, or transplant recipients; (2) localized infection in immunocompetent and immunocompromised patients involving the bone, skin,

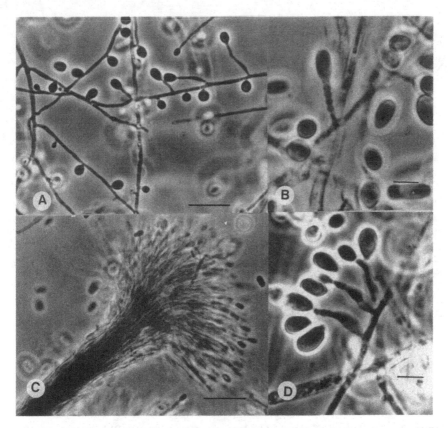

Figure 1 *Pseudallescheria boydii* and *Scedosporium prolificans*. (A) *Scedosporium apiospermum* conidiogenous cells (annellides). (B) *Scedosporium apiospermum* annellides. (C) *Graphium* synnema. (D) *Scedosporium prolificans* annellides. Note: A, C—bar = 20 μm. B, D—bar = 5 μm.

eye, lung, heart valve, and peritoneum; and (3) disseminated infection in immunocompromised patients. They reported the frequent isolation from blood cultures. The species is notorious for its antifungal drug resistance in vitro, and this is correlated with clinical therapeutic failure, especially if there is no reversal of the underlying deficiency.

Description. Colonies grow slightly slower than *S. apiospermum* and are denser and darker gray-brown to dark brown. Isolates grow well at 37°C. Growth is inhibited on medium containing cycloheximide. Conidiogenous cells (annellides) are borne singly on the hypha or in a cluster of two to six at the apex of

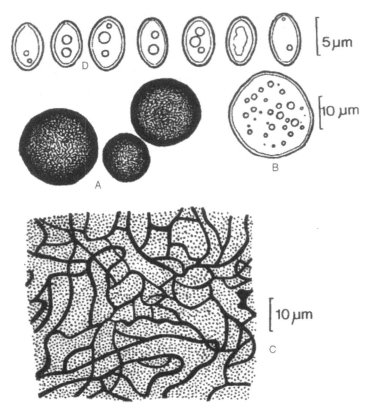

Figure 2 *Pseudallescheria boydii.* (A) Ascomata (cleistothecia). (B) Immature asci. (C) Wall of ascoma. (D) Ascospores. Source: Reprinted from Ref. 119, p. 96, by courtesy of Marcel Dekker, Inc.

a short conidiophore, and are distinctly basally swollen and tapered at the tip. The tip elongates as successive conidia are formed, and conidia often remain attached laterally. Conidia are formed in slimy masses and are subhyaline to light brown, ovoid, and 5.5 to 8 μm long by 3.5 to 5.5 μm wide.

   Comments. *S. prolificans* is distinguished from *S. apiospermum* by annellides with inflated bases commonly borne in a cluster of two to four at the ends of short conidiophores, an absence of synnematal and sexual stages, and inhibition by cycloheximide. Although there was initial disagreement, it is now accepted that *Lomentospora prolificans* is an earlier name for *Scedosporium inflatum* (25). No teleomorph has been found, but molecular data show a relationship with species of *Petriella* (26), a genus traditionally separated from *Pseudalles-*

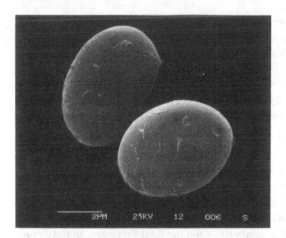

Figure 3 *Pseudallescheria boydii* ascospore. Note: bar = 2 μm.

cheria on the basis of ostiolate perithecia but sharing similar anamorphs. Isolates of *S. prolificans* demonstrate less genetic variability than those of *S. apiospermum* (20, 21). Clinical isolates have been typed using randomly amplified polymorphic DNA analysis (27,28).

3. *Scopulariopsis brevicaulis* (Saccardo) Bainier, Bulletin Societé Mycologique de France 23:99 (Fig. 4)

> Synonym: *Scopulariopsis koningii* (Oudemans) Vuillemin, Bulletin Societé Mycologique de France 27:143, 1911
> Teleomorph: *Microascus brevicaulis* S.P. Abbott, Mycologia 90:298, 1998

Occurrence as a Pathogen. *S. brevicaulis* mainly causes onychomycosis, usually of the toenail, sometimes in mixed infections with dermatophytes (29, 30), and exceptionally involves plantar skin (30, 31). Sporadic reports include endocarditis (32, 33), keratitis (34), and deep tissue invasion in immunocompetent and immunocompromised individuals (18, 35–39).

Description. Colonies are sandy-tan to avellaneous, rapid growing, and coarsely powdery. Growth is lightly to moderately inhibited on medium with cycloheximide and at 37°C. Conidiogenous cells are annellides borne on penicillately branched conidiophores, and are cylindrical to slightly ampulliform, 10 to 25 μm long by 3 to 5 μm wide. Conidia are borne in dry chains and are globose to subglobose or lemon-shaped with broadly truncate base, coarsely roughened, occasionally smooth or finely roughened, and measure 6 to 9 μm long by 5.5 to 9 μm wide. Heterothallic with perithecia forming in mated strains and uncommonly in wild type isolates having both mating types (40, 41). Perithecia are subglobose to globose, with a papilla or short neck, have a wall of textura angularis, and are black and not hairy. Asci are 8 to 10 µm in diameter, subglobose, and evanescent. Ascospores are broadly reniform (plano-convex to concavo-convex), 5 to 6 µm long by 3.5 to 4 µm wide in face view, and flattened 2.5 to 3 µm in end view. They are subhyaline to orange in mass and smooth. De Bary bubbles, guttules, and germ pores are not evident (40, 41).

Comments. Morton and Smith (10) described six species within the "*Scopulariopsis brevicaulis* series" as having annellides of similar shape; that is, cylindrical to slightly ampulliform and broad at the apex. These differed, however, in conidial ornamentation and colony color (white avellaneous, or fuscous). Mating studies have confirmed that *S. brevicaulis*, *S. candida*, and *S. asperula* are anamorphs of *M. brevicaulis*, *M. manginii*, and *M. niger*, respectively (41). There are occasional reports of *S. koningii* in the medical literature, mainly concerning onychomycosis. This species has been shown to be a synonym of *S. brevicaulis*, however, and such reports should be considered as probably referring to *S. brevicaulis* variants that have conidia that roughen tardily (10, 41).

4. *Scopulariopsis brumptii* Salvanet-Duval, Thèse Fac Pharm Paris 23:58, 1935

Occurrence. Reports of onychomycosis by this species appear to be validated both by positive direct microscopy showing nondermatophytic filaments and by repeat isolation (30, 42). Other reports are more difficult to evaluate. Few details of the fungus are given in a report concerning brain abscess in a liver transplant (43). *Microascus* species may be confused with *S. brumptii* if not grown on a medium to induce the sexual stage.

Description. Colonies are slow-growing, reaching a diameter of 2 to 3 cm in 21 days, and are charcoal gray to grayish-black, usually with a narrow white margin. Growth occurs at 37°C, but isolates vary in their ability to grow on medium with cycloheximide (22). Annellides are borne in clusters on branched conidiophores, are basally swollen, and are 5 to 11 µm long by 2 to 2.5 µm wide at the base. Conidia are ovoid to subglobose, smooth but often surrounded by mucilage or a thin membrane, and 3 to 5 µm long by 2 to 4 µm wide.

Comment. Colonies of *S. chartarum* are similar in color but faster-growing, and annellides are mostly solitary. See also comments under *Microascus cinereus.*

5. *Scopulariopsis candida* Vuillemin, Bulletin Societé Mycologique de France 27:143, 1911 (Figs. 5, 6)

Teleomorph: *Microascus manginii* (Loubière) Curzi, Bollettino Stazione di Patologia Vegetale di Roma (NS) 11:60, 1931

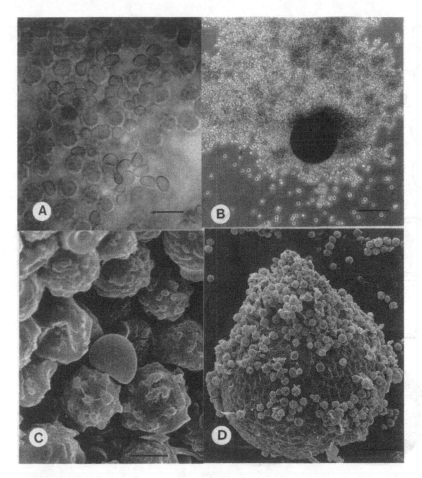

Figure 4 *Microascus brevicaulis.* Anamorph *Scopulariopsis brevicaulis.* (A) Conidia and ascospores (bar = 10 μm). (B) Ascoma (bar = 75 μm). (C) SEM of conidia and ascospores (bar = 3 μm). (D) Ascoma (bar = 25 μm).

Occurrence as a Pathogen. *Scopulanopsis candida* is an uncommon agent of onychomycosis (29, 30). It was reported to cause invasive sinusitis in a girl with non-Hodgkin's lymphoma (44).

Description. Colonies are rapid-growing, coarsely powdery, white to yellowish-white, sometimes darkening centrally with the development of infertile black perithecia ("sclerotia"). Growth is tolerant of cycloheximide and restricted at 37°C. Conidiogenous cells are annellides that are solitary or borne on clusters on short conidiophores. Annellides are cylindrical to slightly ampulliform, and

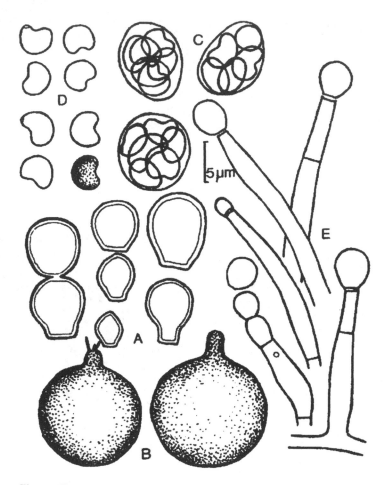

Figure 5 *Microascus manginii.* Anamorph *Scopulariopsis candida.* (A) Conidia. (B) Ascomata (perithecia). (C) Asci. (D) Ascospores. (E) Conidiogenous cells (annellides). Source: Reprinted from Ref. 119, p. 97, by courtesy of Marcel Dekker, Inc.

10 to 25 μm long by 2.5 to 4 μm wide. Conidia are borne in chains and are subglobose or broadly ovate, usually rounded, sometimes tapered at the apex, broadly truncate at the base. They are smooth and measure 5 to 8 μm long by 4 to 7 μm wide. *Scopulariopsis candida* is heterothallic with perithecia forming in mated strains and uncommonly in wild type isolates having both mating types; infertile ascomata are sometimes formed (10, 41). Perithecia are black, globose, and papillate, with a wall of textura angularis, and are not hairy. Asci are 8 to

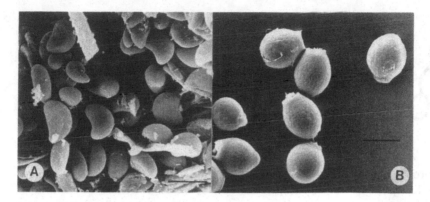

Figure 6 *Microascus manginii.* (A) Ascospores. (B) *Scopulariopsis candida* conidia (bar = 4 μm).

12 μm in diameter, subglobose to ovoid, and evanescent. Ascospores are pale reddish-brown, smooth, broadly reniform to heart-shaped (markedly concavo-convex), 5 to 6 μm long by 4 to 6 μm wide, and extruded in a cirrhus in age (10, 11, 41).

Comments.   *S. candida* is distinguished from *S. brevicaulis* by its smooth conidia and white to yellowish-white colonies.

6. *Microascus cinereus* (Émile-Weil & Gaudin) Curzi, Bollettino Stazione de Patologia Vegetale di Roma 11:60, 1931 (Figs. 7, 8)

Anamorph: *Scopulariopsis*

Occurrence as a Pathogen.   Émile-Weil and Guidin (45) first described this fungus as the cause of onychomycosis, and it is considered a confirmed but rare agent of nail infection (29). Infections also include maxillary sinusitis in which biopsy specimens demonstrated characteristic perithecia (46), endocarditis in a prosthetic valve recipient (47), skin lesions in a patient with chronic granulomatous disease (48), and a brain abscess in a bone marrow transplant recipient (49). Marques et al. (48) and Baddley et al. (49) reviewed other reports concerning this species.

Description.   Colonies are moderately fast-growing (approximately 2.5 to 3.5 cm in 14 days). They are velvety becoming granular, initially pale becoming mid- to dark gray or brownish-black, sometimes appearing reddish-brown, especially with development of ascospores. Tested isolates are tolerant of cyclohexi-

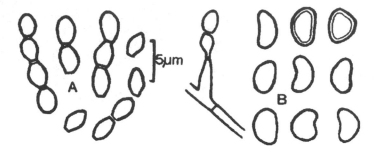

Figure 7 *Microascus cinereus.* Anamorph *Scopulariopsis.* (A) Conidia. (B) Ascospores. Source: Reprinted from Ref. 119, p. 99, by courtesy of Marcel Dekker, Inc.

mide (9) and grow at 40°C (49). Conidiogenous cells are annellides borne solitary on the hyphae or on short conidiophores in verticils or in a penicillately branched arrangement. Annellides are slightly swollen near the middle and measure 4 to 9 μm long, narrowing to 1.5 to 2 μm wide at the tip. Conidia are formed in chains, and are brown, smooth, oval, slightly pointed at the tip and truncate at the base, and measure 3.5 to 4 μm long by 2.5 to 4 μm wide. Perithecia are smooth to finely hairy, black, globose, and usually with a small nipplelike protuberance, sometimes extending to form a neck up to 100 μm long. The ascospores are plano-convex to slightly concavo-convex (almost flattened on one side and hemispherical on

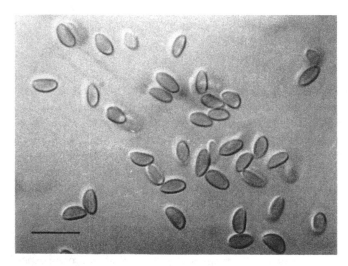

Figure 8 *Microascus cinereus.* Ascospores (bar = 10 μm).

other side), with two germ pores. They are pale reddish-brown, measuring 5 to 7 μm long by 3 to 4 μm wide, usually extruded in a cirrhus in age (10–12, 29).

Comments. There are many *Scopulariopsis* species with darkly pigmented conidia, and it can be difficult to distinguish among them on the basis of anamorph alone. Isolates of *M. cinereus* and *M. cirrosus* will fruit in culture if grown on a suitable medium such as oatmeal-salts or other suitable agar (1). This approach is also beneficial in uncovering isolates that form synnemata. Darkly pigmented synnematous isolates are classified in the genus *Cephalotrichum* (synonym *Doratomyces*).

M. cirrosus differs from *M. cinereus* in having fruiting bodies with short necks and ascospores that are broadly reniform to almost heart-shaped. *Pseudallescheria boydii* forms cleisothecia (globose ascomata without an opening), and conidia are produced in slimy masses.

7. *Microascus cirrosus* Curzi. 1930. Bollettino. Stazione de Patologia Vegetale di Roma 10:308 (Figs. 9, 10)

Anamorph: *Scopulariopsis*

Occurrence as a Pathogen. *Microascus cirrosus*, originally reported under the name *M. desmosporus* is a confirmed but rare agent of onychomycosis (29). An invasive cutaneous infection in a bone marrow transplant recipient has also been documented (50).

Description. Colonies and *Scopulariopsis* anamorph are as described for *M. cinereus*. Perithecia are black, globose, with short, well-developed necks up to 80 μm long. Ascospores are concavo-convex (curved on one side and hemispherical on the other), irregular to almost heart-shaped, 4.5 to 6 μm long and 3 to 4.5 μm wide (10–12, 29).

Comments. Isolates described as *M. desmosporus* by Morton and Smith (10) represent *M. cirrosus.* (See also discussion under *M. cinereus.*)

8. Additional Comments

The species listed above are reliably documented as agents of onychomycosis, but reports concerning other species (51, 52) have not been validated by stringent criteria, including observation of nondermatophytic filaments by direct microscopy and isolation of the same species from repeat specimens (29). *Scopulariopsis chartarum* was reported as a cause of disseminated disease involving the brain in a dog (53), but the case isolate was reidentified as *Acrophialophora fusispora* (54). (See Chap. 10 of this volume.) *Scopulariopsis acremonium* was cited as the cause of invasive sinusitis in a patient with leukemia, but no details of the fungus were given (55). Colonies of *S. acremonium* are white to cream,

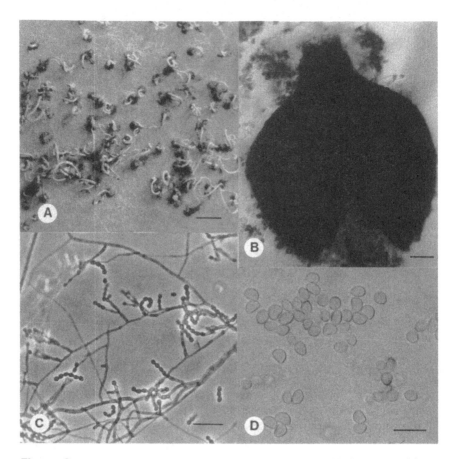

Figure 9 *Microascus cirrosus.* Anamorph *Scopulariopsis.* (A) Ascomata with asco-spores in cirrhi viewed under dissecting microscope. (B) Ascoma (perithecium) with short neck and ruptured at the base (bar = 20 μm). (C) Conidia (bar = 20 μm). (D) Ascospores (bar = 10 μm).

like those of *S. candida*, which has also been reported from sinus (44), but conidia of the former are elongate and tapered to pointed at the tip. Neglia et al. (56) reported *Scopulariopsis* species as the cause of persistent ear infection and dis-seminated disease involving the brain in two leukemic patients. One of the iso-lates was later identified as *S. candida*, but without explanation (38). Pneumonia in patients with leukemia (57), brain abscess following bone marrow transplanta-tion (58), and nasal disease in a normal host (59) have been attributed to unidenti-

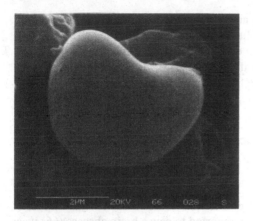

Figure 10   *Microascus cirrosus.* Ascospore (bar = 2 μm).

fied species. Guarro (60) has pointed out that a new genus *Ascosubramania* was
described for a misidentified *Microascus* isolate.

III. FAMILY CHAETOMIACEAE (ORDER SORDARIALES)

A. Taxonomic Concepts

Approximately 300 species of *Chaetomium* have been described, but only about
100 are currently recognized and species identification is often difficult (61). The
genus *Chaetomium* is recognized by the distinctive lateral and terminal setae or
hairs covering the perithecia and by the single-celled, brown, typically lemon-
shaped (lemoniform) ascospores with one (sometimes two) germ pores (61). The
shape and branching pattern of the setae and differences in ascospore size, shape,
and symmetry have been considered important characters in distinguishing species (61). Perithecia are usually subglobose to ovoid and lack a neck. Similar
fungi lacking hairs on the perithecia have been assigned to the genus *Achaetomium*, but the relationship between these genera and to other ascomycetes has not
been clear (2). Cannon (62) accepted *Achaetomium*, but transferred *A. strumarium*
to *Chaetomium*. Small subunit rDNA sequence analysis did not support this transfer (63).

Colonies are fast-growing, commonly filling a petri dish within 10 days at
25°C, and are yellowish-green to yellowish-gray. Isolates may remain sterile on
rich media. Most species are homothallic and will usually fruit in culture if grown
on oatmeal salts, cornmeal agar, or other appropriate medium (1). These fungi

are vigorous cellulose decomposers and may fruit on sterilized filter paper or cellophane membrane. Phialidic or solitary conidia are produced in a few species. About 20 cases of infection have been described, and these mainly concern three species (17, 64, 65). It is significant that two species, *C. atrobrunneum* and *A. strumarium*, which demonstrate a higher optimum temperature for growth, are neurotropic (64).

B. Species of Medical Relevance

1. *Chaetomium globosum* Kunze ex Fries, Syst Mycol 3:226, 1829

Occurrence as a Pathogen.   *C. globosum* is a rare cause of onychomycosis (29, 64, 65). An isolate of *C. globosum* reported to cause brain abscess (66) was reidentified as *C. atrobrunneum* (64). An isolate of *C. perpulchrum* from nail (67) was determined to be *C. globosum* (64, 65). The pathogenic role of *C. globosum* in a bone marrow transplant recipient with sepsis was unclear due to the recovery of a single colony from sterile fluids and lack of evidence of tissue invasion (68).

Description.   Colonies on cornmeal agar are yellowish or grayish-green and measure 4 to 5 cm in diameter after 5 days at 25°C. Growth is similar at 35°C, and no growth occurs at 42°C (64). Perithecia are 175 to 280 μm in diameter and have olivaceous green, unbranched, undulate or wavy, thick-walled hairs (setae or appendages). Ascospores are lemon-shaped, bilaterally flattened, with a single germ pore. They are brown and 8.5 to 11 μm long by 7 to 8.5 μm wide.

Comments.   Although *C. globosum* is the more commonly reported species in the medical literature, it is difficult to evaluate its role in deep infections because some literature reports are based on misidentified isolates. (See the discussion above.) *C. globosum* is known to produce the mycotoxins chaetoglobosins and chaetomin (Fig. 11).

2. *Chaetomium atrobrunneum* Ames, Mycologia 41:641, 1949 (Fig. 11A–B)

Occurrence as a Pathogen.   Three cases of fatal cerebral infection have been reported: a leukemic patient (69), a diabetic male who received a kidney transplant (66), and a bone marrow recipient with multiple myeloma (70, 71). The latter patient had concomitant pulmonary abscess.

Description.   Colonies on cornmeal agar are dark gray to black and slower-growing than *C. globosum*, measuring 1.5 to 2 cm in diameter at 25°C. Growth is faster at 35°C and 42°C (colonies 3 to 3.5 cm and 4 to 4.5 cm diameter)

Figure 11 A–B: *Chaetomium atrobrunneum*. (A) Ascoma (perithecium) with straight hairs. (B) Ascospores. C–D: *Achaetomium strumarium*. (C) Ascoma covered with fine hairs. (D) Ascospores. Note: A, C—bar = 50 μm. B, D—bar = 10 μm.

(64). Perithecia are 70 to 150 μm in diameter and have few hairs, which are dark brown, straight, and occasionally branched in age. Ascospores are brown, narrowly fusoidal, with a single slightly subapical germ pore. They are 9 to 11 μm long by 4.5 to 6 μm wide.

Comments. Both *C. atrobrunneum* and *A. strumarium* cause brain abscess with high mortality. These species differ from *C. globosum* in the smaller size of the ascocarps, the nature of the ascomatal hairs, and in their ability to grow at 42°C (64).

3. *Achaetomium strumarium* Rai, Tewari & Mukerji, Can J Bot
    42:693, 1964 (Fig. 11c–d)

Synonyms: *Chaetomium strumarium* (Rai et al.) Cannon, Trans Brit Mycol
    Soc 87:65, 1986; *Achaetomium cristalliferum* Faurel & Locquin-Linard,
    in Locquin-Linard, Cryptogamie Mycologie 1:235, 1980

Occurrence as a Pathogen. Three cases of fatal cerebral infection in
males with prior histories of intravenous drug use have been reported under the
name *Chaetomium strumarium* (64).

Description. Colonies on cornmeal agar are white with sparse aerial mycelium, developing patches of pale yellowish-green to yellow hyphae in which
ascomata form, and typically developing an intense pink to pinkish-brown diffusible pigment. The colony diameter is 4.5 to 5.5 cm at 25°C and >9 cm at 35°C
after 5 days (64). Growth at 42°C is similar to that at 35°C, but ascomata are
produced more abundantly at 35°C. Perithecia are 100 to 250 μm in diameter,
subglobose, pale brown, and have few thin-walled straight or flexuous (curved),
finely roughened hyphalike setae (hairs). Ascospores are smooth, brown, fusoidal,
and rarely inequilateral, with single apical germ pore. They are 13 to 17.5 (21)
μm long by 8.5 to 11 μm wide. Crystals are produced on hyphae or associated
with ascomata. An anamorph is usually present on sporulation medium. Small
conidia (up to 6 μm long by 3 μm wide) are produced from short, solitary phialides.

Comments. Cannon (62) retained the genus *Achaetomium* for the type
species but transferred *A. strumarium* and *A. luteum* to *Chaetomium* based on
similarities of asci and ascospores. Although these species share a common ancestor with *Chaetomium* species, *A. strumarium* and *A. macrosporum* grouped together suggest their separate placement was warranted (64). The rapid growth at
37° and 42°C, development of pink diffusing pigment on most media, and the
perithecia with less distinct hairs distinguish this neurotropic species from *C.
atrobrunneum.*

4. Additional Comments

Yeghen et al. (72) identified *C. globosum* as the cause of fatal pulmonary mycosis
in a leukemic patient, but their illustration suggests a different species, possibly
C. atrobrunneum (18, 60). Schulze et al. (73) reported a similar case, except
that *Aspergillus fumigatus* was also isolated. Their isolate was identified as *C.
homopilatum*, but the length of the hairs and shape of the ascospores appear
dissimilar to published descriptions for this species (61, 62) and more similar to
C. atrobrunneum.

IV. MISCELLANEOUS ANAMORPHIC FUNGI

A. Moniliaceous Hyphomycetes

1. *Arthrographis kalrae* (Tewari & MacPherson) Sigler & JW Carmich, Mycotaxon 4:360, 1976 (Figs. 12–13)

> Teleomorph: *Eremomyces langeronii* (Arx) Malloch & Sigler, Can J Bot 66:1931, 1987
>
> Synonyms: *Pithoascus langeronii* Arx, Persoonia 18:24, 1978; *Pithoascina langeronii* (Arx) Valmaseda, Martinez, Barrasa, Can J Bot 65:102, 1987

Occurrence as a Pathogen. This thermotolerant fungus was first described from chronic skin infection and has been implicated in pulmonary lesions and keratitis (18, 74). Other reports include keratitis with severe photophobia in a healthy woman (75) and sinusitis and meningitis in an AIDS patient (76).

Description. Colonies are slow-growing and variable in texture and color. Initial growth is yellowish-white and pasty (yeastlike), but colonies become powdery or fasciculate with development of aerial mycelium and are yellowish-white, buff, or tan. Isolates are cycloheximide-tolerant and most will grow at 45°C. Initial growth consists of budding cells, but these may not be formed by all isolates and may be influenced by the medium. Small cylindrical arthroconidia are formed by schizolytic dehiscence of fertile branches borne in a dendritic (treelike) arrangement at the ends of short conidiophores. Vegetative hyphae also fragment to form longer arthroconidia. Solitary conidia are usually produced in the submerged mycelium and may be observed in slide culture preparations. The teleomorph is rarely produced. Cleistothecia are globose, brown, 75 to 160 µm in diameter, and are usually submerged in the agar. Ascospores are phaseoliform (shaped like orange sections), hyaline to yellowish-brown, smooth, lacking germ pores, and 2.7 to 5 µm long by 1.8 to 2.6 µm wide.

Comments. Although the yeast phase is sometimes transitory and may not be present in cultures that have been maintained for long periods, it is regularly present in primary isolates from soil and human sources (Sigler, unpublished data). Isolates may be mistaken for a yeast and subjected to methods normally used for yeast identification (18, 76). Subculture onto cereal or oatmeal-salts (or other) agar (1), however, yields the typical hyphal growth and arthroconidial development. *A. kalrae* is a common fungus in north temperate soil, and soil contamination of a contact lens was thought to contribute to infection in one instance (75). The teleomorph is classified in the Eremomycetaceae which appears to belong in the Eurotiales. The connection between anamorph and teleomorph has been questioned, but sequence analysis is required to help resolve the uncertainty (74, 77).

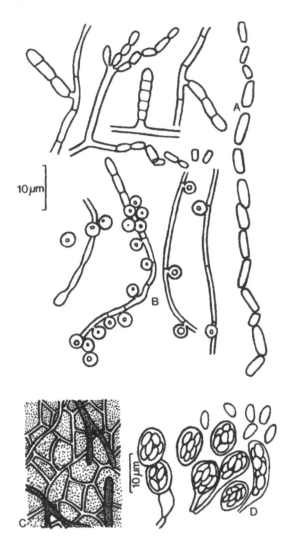

Figure 12 *Eremomyces langeronii*. Anamorph *Arthrographis kalrae*. (A) Arthroconidia. (B) Solitary conidia. (C) Wall of ascoma. (D) Asci and ascospores. Source: Reprinted from Ref. 119, p. 101, by courtesy of Marcel Dekker, Inc.

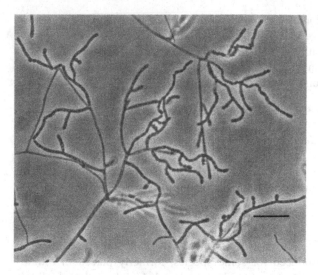

Figure 13 *Arthrographis kalrae* arthroconidia (bar = 20 μm).

2. *Beauveria bassiana* (Bals.) Vuillemin, Bull Soc Bot Fr 59: 40, 1912

Occurrence. Mycotic keratitis in humans (78, 79) and pulmonary infection in alligator and tortoise (80, 81) have been reported, but human pulmonary infection has not been substantiated (18).

Description. Colonies are moderately fast-growing, yellowish-white, dense, and cottony to somewhat powdery. Conidiogenous cells are borne in dense clusters (sporodochia) and are basally swollen, tapering at the apex and proliferating sympodially to form a zigzag (geniculate) rachis. Conidia are single-celled, subglobose, and 2 to 3 μm long by 1.5 to 2.5 μm wide.

Comment. *Beauveria bassiana* is an insect pathogen with a wide distribution. *Engyodontium album* is closely related (17) and was formerly placed in the genus *Beauveria*. (See comments under *E. album*.)

3. *Engyodontium album* (Limber) de Hoog 1978 Persoonia 10:53, 1978

Synonyms: *Beauveria alba* (Limber) Saccas, Revue Mycol 13:64, 1948; *Tritirachium album* Limber, Mycologia 32:27, 1940

Occurrence. *E. album* is a rare agent of eye infection (82) and native valve endocarditis (83).

Description. Colonies are slow-growing, white, and velvety. Conidiogenous cells are solitary or in whorls of one to three, and are cylindrical, tapering at the apex and proliferating sympodially to form a narrow zigzag rachis from 7 to 15 μm long and 1 μm wide. Conidia are single-celled, oval with an apiculate base, and 1.8 to 2 μm long by 1 to 1.5 μm wide.

Comment. *E. album* has had a complicated nomenclatural history, with various authors disagreeing about its correct generic placement (83). Although it was transferred from *Beauveria* to the new genus *Engyodontium* by de Hoog in 1978, it was not selected as type species. On the basis of 18S rDNA sequences, *E. album* has been shown to be a close relative of *Beauveria* and *Tolypocladium* species, but its relationship with other members of the genus *Engyodontium* was not elucidated (17).

4. *Lecythophora* Species

Lecythophora was reintroduced for some species, formerly placed in *Phialophora*, that form slimy conidia from peglike or reduced phialides (adelophialides) lacking a basal septum. Adelophialides are often conical in shape, and the collarette is usually distinct (84). Phialides may also be longer, slender, and resemble those of *Acremonium*. Teleomorphs are in the genus *Coniochaeta* (84).

   a. Lecythophora hoffmannii (Beyma) Gams & McGinnis, Mycologia 75: 985, 1983

> Synonym: *Phialophora hoffmannii* (Beyma) Schol-Schwarz, Persoonia 6:79, 1970

Occurrence. Reports concerning *Lecythophora hoffmannii* may refer to *Phaeoacremonium* species. (See Chap. 10 of this volume.) *Phaeoacremonium* was erected in 1996 for *Phialophora parasitica*, a more commonly reported pathogen, and some other species occasionally involved in infection. Rinaldi et al. (85) described a patient with subcutaneous abscess of the buttock, but the features of the case isolate, including cinnamon-colored colonies on Sabouraud dextrose agar, initial yeastlike growth, and phialides darkening to pale brown, are suggestive of a *Phaeoacremonium* species. An isolate from the sinus of a male patient with AIDS (86) was reidentified as *P. parasiticum* (18).

Description. Colonies are pale salmon to orange, sometimes creamy white in degenerate strains, moist but becoming hairy with development of hyphae. Conidia are oval or ellipsoidal to slightly curved and measure 3.5 to 7 μm long by 1 to 2.5 μm wide.

   b. Lecythophora mutabilis Nannf. in Melin & Nannf. in Svenska Skogsvardsforen Tidskr 32:432, 1934

Occurrence. Endocarditis, eye infections, and peritonitis in a patient on chronic ambulatory peritoneal dialysis have been reported (16, 17, 87, 88).

Description. Colonies are initially white to yellowish-white, but darken to gray with the development of brown chlamydospores. Conidia are oval to ellipsoidal and 2.5 to 4.5 μm long by 1 to 1.5 μm wide. Chlamydospores are brown, thick-walled, solitary, terminal or lateral, 0–1 septate, 7 to 9 μm long by 4 to 5 μm wide, or intercalary and in short chains.

c. Additional Comments. Ramani et al. (89) reported *Tilletiopsis minor*, a member of the Ustilaginales, as the cause of subcutaneous cyst in an immuno-compromised man, but the case isolate appears to be a *Lecythophora* species. Colonies become dark and have yellowish-brown mycelium but lack the pigmented chlamydospores of *L. mutabilis*. Conidia are formed from phialides as illustrated in the case report (89) and are not ballistic.

5. *Metarhizium anisopliae* (Metschnikov) Sorokin, Plant Parasites of Man & Animals, p 267, 1883 (orthographic variant '*Metarrhizium*') (Fig. 14)

Occurrence. Cases include keratitis (90) and sinusitis (91) in immuno-competent hosts and invasive infection manifesting with skin lesions in a boy with leukemia (92). An otherwise healthy cat developed invasive rhinitis with extension through the nasal bones (93).

Description. Colonies are moderately fast-growing, initially floccose and white, turning olivaceous green or buff with the development of conidia, usually sporulating heavily. The optimum temperature is 25–28°C, with a maximum near 37°C (92). Conidiogenous cells are cylindrical phialides borne on verticillately or irregularly branched conidiophores formed in sporodochia. Conidia are formed in adherent columns and are cylindrical, smooth, yellowish-green, and measure 6.5 to 8 μm long by 2 to 3 μm wide. Irregularly shaped appresoria are formed and may be observed in slide culture preparations.

Comments. Known as the green muscardine fungus, *M. anisopliae* is an important insect pathogen worldwide and has commercial application as a biological control agent (92).

6. *Myceliophthora thermophila* (Apinis) Oorschot, Persoonia 9: 403, 1977 (Fig. 15)

Synonyms: *Sporotrichum thermophile* Apinis, Nova Hedwigia 5:74, 1962; *Chrysosporium thermophilum* (Apinis) Klopotek, Arch Microbiol 98: 366, 1974

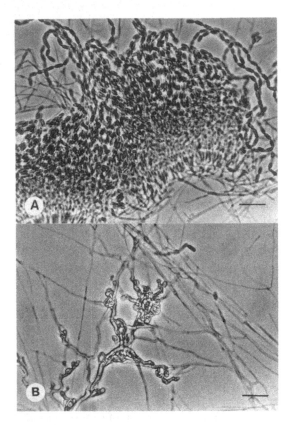

Figure 14 *Metarhizium anisopliae.* (A) Conidiogenous cells (phialides) in sporodochium. (B) Appressoria (bar = 20 μm).

Teleomorph: *Corynascus heterothallicus* (Klopotek) von Arx, Sydowia 34: 25, 1981

Occurrence. Aortic vasculitis with fatal outcome occurred in two patients, secondary to invasive pulmonary infection in a leukemic patient (94), and to idiopathic cystic medial necrosis (95). Secondary infection of a bacterial cerebral abscess developed after trauma (96).

Description. *M. thermophila* is thermophilic. Colonies are fast-growing at optimum temperature of 37–45°C, and are pink to buff, then cinnamon-brown. Pink to cinnamon-brown diffusible pigment is usually present. One to three conidia are formed on small denticles borne on the sides or the ends of short stalks,

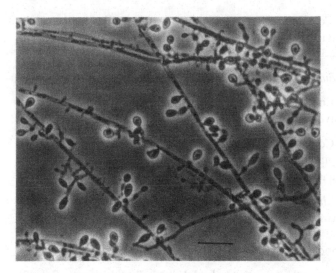

Figure 15 *Myceliophthora thermophila* conidia (bar = 20 μm).

which are usually slightly swollen. Conidia are yellowish to reddish-brown oval, and rough-walled. Ascomata (cleistothecia) are produced only in mated cultures.

Comments. Found in molding silage, pulp, and soil, this species was described originally in *Sporotrichum*, now used for anamorphs of some basidiomycetes. It was transferred to *Chrysosporium* because of its formation of solitary aleurioconidia. (See Chap. 6 of this volume for a discussion on *Chrysosporium*.) Placement in *Myceliophthora* is appropriate because this genus includes anamorphs of *Corynascus* (Sordariaceae) that are cellulolytic and form conidia on swollen stalks and excludes keratinolytic dermatophytoids (97).

7. *Phialemonium* Species

Species intermediate between *Acremonium* and *Phialophora* are placed in *Phialemonium*, but distinction from *Lecythophora* can be difficult. Slimy conidia are formed from short cylindrical lateral phialides (adelophialides lacking a basal septum) or from longer *Acremonium*-like phialides. Collarettes are lacking or inconspicuous (84).

a.   Phialemonium obovatum Gams & McGinnis, Mycologia 76:978, 1983

Occurrence. Disseminated infection in a burned child, peritonitis in a renal transplant recipient, and mycetoma, subcutaneous, and disseminated infections in dogs have been reported (16, 17, 98).

Description. Colonies are slow-growing, creamy white, and initially moist, becoming grayish, velvety to hairy. Conidia are obovoid (egg-shaped), hyaline, smooth, and 3.5 to 5 μm long by 1.2 to 1.7 μm wide.

    b. *Phialemonium curvatum* Gams & McGinnis, Mycologia 76:980, 1983

Occurrence. Endocarditis, subcutaneous cyst in a renal transplant recipient, and fungemia have been reported (16, 17, 99).

Description. Colonies are white, sometimes developing light brown to grayish patches, and glabrous (smooth) to hairy. Conidia are cylindrical to curved (allantoid), hyaline, and 2.5 to 4.5 (10) μm long by 0.8 to 1.5 μm wide.

Comment. *P. dimorphosporum* is considered a synonym based on PCR-RFLP banding patterns (99).

8. *Scytalidium* Species

Scytalidium dimidiatum is the synanamorph of *Nattrassia mangiferae*. *S. hyalinum* is a white variant. (See description under Sec. B.3.)

B. Coelomycetes

Plant-associated anamorphic fungi forming conidia within a conidioma have traditionally been assigned to the Coelomycetes. Sutton (100) re-evaluated 370 genera and recognized six suborders based on conidial ontogeny and structure of conidiomata—that is, whether pycnothyrial, pycnidial, sporodochial, or acervular conidiomata. Nag Raj (101) used a similar approach to redescribe 142 genera forming conidia with appendages. Coelomycetes are recognized as occasional agents of infection, but there are complexities in identifying clinical isolates since conidiomatal structures produced in culture may be less elaborate than those formed on the plant host, and descriptions of many species lack cultural details. Also, important information concerning the plant host is unavailable. Synopses of the identifying features of medially important species are found in several sources (17, 102–104); therefore only some species are treated here. Several species are reported from single cases of infection. In order to authenticate some of these reports, it would be prudent to verify the identity of case isolates; however, in many instances these have not been retained in culture collections.

1. *Colletotrichum* Species (Fig. 16)

    Teleomorph: *Glomerella*

Occurrence. *C. dematium, C. gloeosporioides, C. coccodes,* and *C. graminicola* have been implicated in eye infection (103), while *C. gloeosporioides* and *C. crassipes* were the cause of subcutaneous infections (105, 106).

Figure 16 *Colletotrichum gloeosporioides.* (A) Sporodochium and setae. (B) Conidia and appressoria (bar = 20 μm).

Description. Conidiogenous cells are phialides formed in sporodochial conidiomata (acervular on plant host) or occasionally solitary. Thick-walled, brown septate setae are formed in conidiomata by many species. Sclerotia are sometimes present and may bear setae. Appressoria are usually present and are brown, smooth, mostly solitary, and oval, clavate, or irregularly shaped and lobate to crenate. Conidia are hyaline, straight, and cylindrical or fusiform to falcate (curved and tapered at the ends).

Comments. Species are distinguished by the size and shape of conidia (straight or falcate), shape and size of appressoria, and presence of setae (17, 100, 105). The most common species, *C. gloeosporioides*, has conidia that are straight with tapered ends, 9 to 24 μm long by 3 to 4.5 μm wide. Appressoria are clavate to irregular and 6 to 20 μm long by 4 to 12 μm wide (100).

2. *Lasiodiplodia theobromae* (Pat.) Griffon & Maublanc, Bull Trimest Soc Mycol Fr 25:57, 1909

> Synonym: *Botryodiplodia theobromae* Pat., Bull Trimest Soc Mycol Fr 8:136, 1892
>
> Teleomorph: *Botryosphaeria rhodina* (Berk. & Curt.) Arx, The Genera of Fungi Sporulating in Pure Culture p 143, 1970

Occurrence.   Mainly causing eye and nail infection, this fungus is rarely seen in temperate areas in patients who have immigrated from tropical and subtropical areas (16, 17, 29, 102, 103). Subcutaneous infection followed intramuscular injection in an otherwise healthy patient (107).

Description.   Colonies are very fast-growing, gray to black, dense, with aerial tufts of ropy mycelium. Conidia are produced from cylindrical phialides within uni- or multiloculate, often hairy, black stromatic pycnidia. Conidiogenous cells are intermixed with paraphyses (sterile hyphae) within the cavity. Conidia are broadly ellipsoidal, initially hyaline, and single-celled, but at maturity are two-celled, brown, and have longitudinal striations.

Comments.   *L. theobromae* resembles *Nattrassia mangiferae* in geographic distribution, clinical presentation, and colonial morphology. Isolates do not produce arthroconidia.

3. *Nattrassia mangiferae* (H. Sydow & P. Sydow) Sutton & Dyko, Mycol Res 93:484, 1989 (Fig. 17)

> Synanamorph: *Scytalidium dimidiatum* (Penz.) Sutton & Dyko, Mycol Res 93:484, 1989

Occurrence.   *Nattrassia mangiferae* causes dermatomycoses of the nail, toe webs, and glabrous skin, especially in immigrants from parts of Africa, India, and the Caribbean (16–18, 29, 108, 109). Deep infections involve subcutaneous or verrucose lesions on arm, foot, face, and finger, mycetoma, maxillary sinusitis, endophthalmitis, fungemia, lesions of abdomen, and inguinal lymph node in patients with underlying immunosuppression, diabetes, chronic obstructive lung disease, tuberculosis, arthritis, AIDS, or prior injury from handling wood (16–18, 86, 108, 110). Some individuals had concomitant onychomycosis. This species is recognized and often identified under the name of the arthroconidial synanamorph *Scytalidium dimidiatum*, which is the state found in primary subculture.

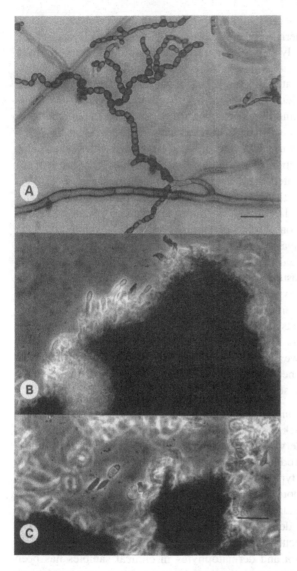

Figure 17 *Nattrassia mangiferae.* Synanamorph *Scytalidium dimidiatum.* (A) Arthro-conidia. (B, C) Pycnidial conidia (bar = 20 μm).

Description. Colonies are rapid-growing (filling petri dish in 7 days), black, and ropy with deep aerial mycelium, producing diffusible black pigment, sensitive to cycloheximide. Hyphae vary in width, ranging from 2.5 to 10 µm in diameter, and are hyaline to subhyaline or dark brown, and often surrounded by brown slime. Hyphae aggregate into strands and form loops. Very broad hyphae do not fragment, but narrower hyphae form arthroconidia by schizolytic dehiscence. Arthroconidia are cylindrical, barrel-shaped to subglobose, subhyaline to dark brown, nonseptate or one-septate, and measure 4 to 15 µm long by 2.5 to 7 µm wide. Morphological variants are recognized by growth rates, abundance of arthroconidia, and production of pycnidia (108, 110). Fast-growing isolates (form 1 or type A) are associated with individuals from subtropical areas of West Africa, the Caribbean, or South America, and from infected plant hosts. Form 1 isolates can usually be induced to form flask-shaped to irregularly shaped pycnidia or pycniostromata if grown on appropriate media (108, 110). Pycnidial conidia are ellipsoidal and initially hyaline and may remain so for prolonged periods, but eventually develop one—usually two—septa. The central cell is brown with the end cells paler. Conidia measure 10 to 15 µm long by 4 to 5.5 µm wide. Slow-growing isolates (form 3 or type B) are associated only with human hosts from East Africa, the Indian subcontinent, and Southeast Asia, and it has been suggested that this form represents an adaptation to anthropophilic transmission. Colonies are slower growing, gray rather than black, and aerial mycelium is scanty. Hyphae are generally narrower, often form loops, and produce few or no arthroconidia. Pycnidia have not been obtained in isolates of this type. Intermediate forms have also been noted, some of which will produce pycnidia (108).

Comments. *Scytalidum hyalinum* Campbell & Mulder was described for fast-growing isolates that lack pigmentation. Colonies are yellowish-white, less dense, and somewhat powdery. Arthroconidia are abundant, hyaline to subhyaline, but no pycnidia are produced. This fungus is found exclusively from the human host causing similar types of infections as those described for *S. dimidiatum* (17, 18, 108, 110). Moore's (108) hypothesis that *S. hyalinum* could be a white variant based on similarities in clinical presentation, immunological and biochemical properties, and demonstration of mixed infections with *S. dimidiatum*, was confirmed by molecular typing of ribosomal genes (109). A PCR-RFLP method to detect *Scytalidium* and dermatophytes in clinical samples has been described (111). *Scytalidium lignicola* has not been reliably documented as an agent of infection; cases referring to this species are based on misidentified *S. dimidiatum* isolates (108, 110).

4. *Phoma* Species

Occurrence. Several *Phoma* species, including *P. glomerata*, *P. hibernica*, *P. minutella*, *P. eupyrena*, *P. minutispora*, *P. sorghina*, and *Phoma* species, have been reported mainly from subcutaneous infection (17, 103, 104), but reports concerning *P. minutispora* and *P. sorghina* as well as *P. cruris-hominis* and *P. oculo-hominis* are doubtful (104).

    a.   *P. glomerata* (Corda) Wollenw. & Hochapf. Z Parasitkde 8:592, 1936

Description.   Colonies often appear slimy and pale to reddish-orange, becoming greenish-gray to dark brown to black in hyphal areas. Pycnidia are immersed or superficial, globose or subglobose, and more pigmented around the ostiole. Conidia are borne from minute phialides that line the inner cavity and are unicellular, occasionally two-celled, ellipsoidal, hyaline, and measure 5 to 9 μm long by 2.5 to 3 μm wide. Chains of brown dictyochlamydospores (brown, swollen structures having longitudinal and transverse septa) are common in the hyphae.

Comment.   There are a large number of species of *Phoma*, and a number have been well characterized in culture. The dictyochlamydospores make *P. glomerata* one of the easier species to identify. These may appear superficially similar to the dictyoconidia of *Alternaria* species.

5. *Pleurophomopsis lignicola* Petrak, Ann Mycol 22:165, 1924

Occurrence.   Subcutaneous infection and maxillary sinusitis have been described (17, 103, 104).

Description.   Colonies are light yellowish to grayish-brown and floccose. Pycnidia are immersed or superficial, reddish-brown, subglobose to flask-shaped, mostly uniloculate, and ostiolate. Conidiophores have one to two septa. Conidia are hyaline, cylindrical, phialidic, and acropleurogenous (i.e., formed at the ends and on the side with a lateral branch forming immediately below the terminal septum) and measure 2.5 to 3 μm long by 1.5 μm wide (104).

6. Additional Comments

Microsphaeropsis olivacea, implicated in dermatomycosis (17), is similar to *Phoma* species in forming ostiolate pycnidia lined with minute phialides, but conidia are brown rather than hyaline. *Pleurophoma* species, rare agents of subcutaneous and systemic infection, differ in having septate, filiform conidiophores from which conidia develop at the tip or from small lateral openings below a septum (17, 100, 103). Species of *Pyrenochaeta* have a similar conidiogenesis to *Pleurophoma* species, but pycnidia are setose. *Pyrenochaeta mackinnonii* and *P. romeroi* are often listed as agents of mycetoma, but the former species is known only from the original description and the ex-type culture is sterile. Isolates identified as *P. romeroi* are suspected to represent a *Phoma* species, possibly *P. leveillei* (17). *Phomopsis* species are distinguished by forming conidia of two types; alpha conidia are shorter and broader, ellipsoid to fusiform, while beta conidia are longer and narrow, filiform, straight or slightly bent or hooked. An undetermined *Phomopsis* species was found to cause osteomyelitis (103, 112).

V. BASIDIOMYCETES

A. Species of Medical Relevance

1. *Schizophyllum commune* Fr., Syst Mycol 1:330, 1821
    (Schizophyllaceae, Aphyllophorales)

Occurrence. *S. commune* has increasingly been reported as the cause of chronic or allergic sinusitis, allergic bronchopulmonary mycosis, and other allergy-related pulmonary diseases. Invasive infections of the brain, lung, and buccal mucosa in immunosuppressed and immunocompetent individuals have also been described (17, 18, 113–115). Sigler and Kennedy (18) suggested that a case of sinusitis attributed to *Myriodontium keratinophilum* may refer to *S. commune*.

Description. Colonies are white, moderately fast-growing, dense, tough, and cottony to woolly. They are thermotolerant with good growth at 37°C, tolerant of benomyl (2 µg/ml), and sensitive to cycloheximide. No anamorph is formed. Isolates having clamp connections on the hyphae will normally produce fruiting bodies (fan-shaped basidiomata with split gills) on appropriate media in conditions of light. Many hyphae also bear small pegs (or spicules). Monokaryotic isolates are flat and thinly cottony, have a pronounced unpleasant odor, and lack clamps. Some isolates also lack spicules (115–116).

Comments. Clamped isolates with spicules can be identified as *S. commune*, but clampless isolates lacking these structures may be difficult to identify (116). Vegetative compatibility tests and comparison of ITS1 region sequences have been used to identify atypical and aberrant isolates (113, 115). A clampless monokaryon, if grown in paired culture with a compatible single basidiospore isolate, will develop clamp connections at the interface of the paired mycelia.

2. *Coprinus cinereus* (Schaeffer: Fries) S.F. Gray, Nat Arr Brit
    Plants p 634, 1821 (Coprinaceae, Agaricales)

Anamorph: *Hormographiella aspergillata*, Guarro, Gené & de Vroey, Mycotaxon 45:182, 1992

Occurrence. Prosthetic valve endocarditis, fatal pulmonary infection in a leukemic patient, and skin lesions have been attributed to *C. cinereus* or its anamorph, *Hormographiella aspergillata* (17, 18).

Description. Colonies are white to amber or tan, cottony, and dense. They are thermotolerant, growing well at 37°C, and tolerant of benomyl. Clinical isolates are normally encountered in the anamorphic state. Hyphae are simple septate, lacking clamps. Conidiophores bear several branches on the sides or at the tip, and these branches divide schizolytically to form arthroconidia that remain adherent at the ends of the conidiopores. Arthroconidia are hyaline, cylindrical, thin-walled, and 3.5 to 6 μm long by 2 to 3 μm wide. Sclerotia may be formed by some isolates.

Comments. The anamorph genus *Hormographiella* was described for some clinical and environmental isolates that were thought to have basidiomycetous affinity. Study of additional clinical isolates yielded some that formed *Coprinus*-type basidiocarps in culture. Compatibility tests and molecular analyses verified relationship between *H. aspergillata* and *Coprinus cinereus* and confirmed that *H. verticillata* in a distinct species with an undermined *Coprinus* teleomorph (117, 118).

REFERENCES

1. J Kane, RC Summerbell, L Sigler, S Krajden, GA Land. Laboratory Handbook of Dermatophytes. Belmont, CA: Star, 1997.
2. CJ Alexopoulos, CW Mims, M Blackwell. Introductory Mycology. New York: Wiley, 1996.
3. G Hausner, J Reid, GR Klassen. On the phylogeny of *Ophiostoma, Ceratocystis* s.s., and *Microascus*, and relationships within *Ophiostoma* based on partial ribosomal DNA sequences. Can J Bot 71:1249–1265, 1993.
4. JW Spatafora, M Blackwell. The polyphyletic origins of ophiostomatoid fungi. Mycol Res 98:1–9, 1994.
5. OE Eriksson, DL Hawksworth. Outline of the Ascomycetes—1998. Systema Ascomycetum 16:83–301, 1998.
6. OE Eriksson, H-O Baral, RS Currah, K Hansen, CP Kurtzman, G Rambold, T Laessoe (Eds.). Outline of Ascomata. Myconet 7:1–88, 2001.
7. MJ Wingfield, KA Seifert, JF Webber, eds. *Ceratocystis* and *Ophiostoma*, Taxonomy, Ecology and Pathogenicity. St. Paul, MN: APS, 1993.
8. RC Summerbell. The benomyl test as a fundamental diagnostic method for medical mycology. J Clin Microbio 31:572–577, 1993.
9. SP Abbott. Holomorph studies of the Microascaceae. PhD dissertation, University of Edmonton, Alberta, Canada, 2000.
10. FJ Morton, G Smith. The genera Scopulariopsis Bainier, Microascus Zukal, and Doratomyces Corda. Mycol Pap 86:1–96, 1963.
11. GL Barron, RF Cain, JC Gilman. The genus *Microascus*. Can J Bot 39:1609–1631, 1961.
12. JA von Arx, MJ Figueras, J Guarro. Sordariaceous ascomycetes without ascospore ejaculation. Nova Hedwigia 94:1–104, 1988.
13. MR McGinnis, AA Padhye, L. Ajello. Pseudallescheria Negroni et Fischer, 1943 and its later synonym Petriellidium Malloch, 1970. Mycotaxon 14:94–102, 1982.

14. JW Rippon. Medical Mycology: The Pathogenic Fungi and the Pathogenic Actinomycetes. Philadelphia, PA: Saunders, 1988.

15. KJ Kwon-Chung, JW Bennett. Medical Mycology. Malvern, PA: Lea & Febiger, 1992.

16. WA Schell, IF Salkin, L Pasarell, MR McGinnis. Bipolaris, Exophiala, Scedosporium, Sporothrix and other dematiaceous fungi. In PR Murray, EJ Baron, MA Pfaller, FC Tenover, RH Yolken, eds. Manual of Clinical Microbiology. Washington, DC: American Society for Microbiology, 1999, pp. 1295–1317.

17. GS De Hoog, J Guarro, J Gene, MJ Figueras. Atlas of Clinical Fungi. Utrecht, Netherlands: Centraalbureau voor Schimmelcultures, 2000.

18. L Sigler, MJ Kennedy. Aspergillus, Fusarium, and other opportunistic moniliaceous fungi. In PR Murray, EJ Baron, MA Pfaller, FC Tenover, RH Yolken, eds. Manual of Clinical Microbiology. Washington, DC: American Society for Microbiology, 1999, pp. 1212–1241.

19. ECM Williamson, D Speers, IH Arthur, G Harnett, G Ryan, TJJ Inglis. Molecular epidemiology of Scedosporium apiospermum infection determined by PCR application of ribosomal intergenic spacer sequences in patients with chronic lung disease. J Clin Microbio 39:47–50, 2001.

20. M Wedde, D Muller, K Tintelnot, GS De Hoog, U Stahl. PCR-based identification of clinically relevant Pseudallescheria/Scedosporium strains. Med Mycol 36:61–67, 1998.

21. J Rainier, GS De Hoog, M Wedde, Y Graser, S Gilges. Molecular variability of Pseudallescheria boydii, a neurotropic opportunist. J Clin Microbio 38:3267–3273, 2000.

22. IF Salkin, MR McGinnis, MJ Dykstra, MG Rinaldi. Scedosporium inflatum, an emerging pathogen. J Clin Microbio 26:498–503, 1988.

23. TW Swerczek, JM Donahue, RJ Hunt. Scedosporium prolificans infection associated with arthritis and osteomyelitis in a horse. J Amer Vet Med Assoc 218:1800–1802, 2001.

24. P Idigoras, E Perez-Trallero, L Pineiro, J Larruskain, MC Lopez-Lopategui, N Rodriguez, J Marin Gonzalez. Disseminated infection and colonization by Scedosporium prolificans: A review of 18 cases, 1990–1999. Clin Infec Dis 32:e158–el65, 2001.

25. PA Lennon, CR Cooper Jr, IF Salkin, SB Lee. Ribosomal DNA internal transcribed spacer analysis supports synonymy of Scedosporium inflatum and Lomentospora prolificans. J Clin Microbio 32:2413–2416, 1994.

26. J Issakainen, J Jalava, E Eerola, CK Campbell. Relatedness of Pseudallescheria, Scedosporium and Graphium pro parte based on SSU rDNA sequences. J Med Vet Mycol 35:389–398, 197.

27. R San Millan, G Quindos, J Garaizar, R Salesa, J Guarro, J Ponton. Characterization of Scedosporium prolificans clinical isolates by randomly amplified polymorphic DNA analysis. J Clin Microbio 35:2270–2274, 1997.

28. B Ruiz-Diez, F Martin-Diez, JL Rodriguez-Tudela, M Alvarez, JV Martinez-Suarez. Use of random amplification of polymorphic DNA (RAPD) and PCR-fingerprinting for genotyping a Scedosporium prolificans (inflatum) outbreak in four leukemic patients. Curr Microbio 35:186–190, 1997.

29. RC Summerbell. Non dermatophytic molds causing dermatophytosis-like nail and

skin infection. In: J Kane, RC Summerbell, L Sigler, S Krajden, GA Land. Laboratory Handbook of Dermatophytes. Belmont, CA: Star, 1997, pp. 213–259.

30. E Piontelli, LMA Toro. Biomorphological and clinical commentaries on the genus Scopulariopsis Bainier. Hyalohyphomycosis in nails and skin. II. Boletin Micrológico 3:259–273, 1988.

31. M Ginarte, M Pereiro Jr, V Fernandez-Redondo, J Toribio. Plantar infection by Scopulariopsis brevicaulis. Dermatology 193:149–151, 1996.

32. RQ Migrino, GS Hall, DL Longworth. Deep tissue infections caused by Scopulariopsis brevicaulis: Report of a case of prosthetic valve endocarditis and review. Clin Infec Dis 21:672–674, 1995.

33. LD Gentry, MM Nasser, M Kleihofner. Scopulariopsis endocarditis associated with duran ring valvuloplasty. Tex Heart Inst J 22:81–84, 1995.

34. AJ Lotery, JR Kerr, BA Page. Fungal keratitis caused by Scopulariopsis brevicaulis: Successful treatment with topical amphotericin B and chloramphenicol without the need for surgical debridement. Brit J Ophthalmol 78:730, 1994.

35. AW Sekhon, DJ Willans, JH Harvey JH. Deep scopulariopsosis: A case report and sensitivity studies. J Clin Path 27:837–843, 1974.

36. L Creus, P Umbert, JM Torres-Rodriguez, F Lopez-Gill. Ulcerous granulomatous cheilitis with lymphatic invasion caused by Scopulariopsis brevicaulis infection. J Amer Acad Derm 31:881–883, 1994.

37. P Phillips, WS Wood, G Phillips, MG Rinaldi. Invasive hyalohyphomycosis caused by Scopulariopsis brevicaulis in a patient undergoing allogenic bone marrow transplant. Diag Microbio Infect Dis 12:429–432, 1989.

38. EJ Anaissie, GP Bodey, MG Rinaldi. Emerging fungal pathogens. Eur J Clin Microbio Infect Dis 8:323–330, 1989.

39. P Sellier, JJ Monsuez, C Lacroix, C Feray, J Evans, C Minozzi, F. Vayre, P Del Giudice, M Feuilhade, C Pinel, D Vittecoq, J Passeron. Recurrent subcutaneous infection due to Scopulariopsis brevicaulis in a liver transpalnt recipient. Clin Infect Dis 30:820–823, 2000.

40. SP Abbott, L Sigler, RS Currah. Microascus brevicaulis sp. nov., the teleomorph of Scopulariopsis brevicaulis, supports placement of Scopulariopsis with the Microascaceae. Mycologia 90:297–302, 1998.

41. SP Abbott, L Sigler. Heterothallism in the Microascaceae demonstrated by three species in the Scopulariopsis brevicaulis series. Mycologia 93:1211–1220, 2001.

42. J Naidu, SM Singh, M Pouranik. Onychomycosis caused by Scopulariopsis brumptii: A case report and sensitivity studies. Mycopathologia 113:159–164, 1991.

43. R Patel, CA Gustaferro, RAF Krom, RH Wiesner, GD Roberts, CV Paya. Phaeohyphomycosis due to Scopulariopsis brumptii in a liver transplant patient. Clin Infec Dis 19:198–200, 1994.

44. JD Kriesel, EE Adderson, WM Gooch III, AT Pavia. Invasive sinonasal disease due to Scopulariopsis candida: Case report and review of Scopulariopsis. Clin Infec Dis 19:317–319, 1994.

45. P Emile-Weil, L Gaudin. Contribution to a study of onychomycoses. Arch Méd Expér Anat 28:452–467, 1919.

46. C Azner, C de Bievre, C Guiguen. Maxillary sinusitis from Microascus cinereus and Aspergillus repens. Mycopathologia 105:93–97, 1989.

47. M Celard, E Dannaoui, MA Piens, E Guého, G Kirkorian, T Greenland, F Vandenesch, S Picot. Early Microascus endocarditis of a prosthetic valve implanted after Staphylococcus aureus endocarditis of the native valve. Clin Infec Dis 29: 691–692, 1999.
48. AR Marques, KJ Kwon-Chung, SM Holland, ML Turner, JI Gallin. Suppurative cutaneous granulomata caused by Microascus cinereus in a patient with chronic granulomatous disease. Clin Infec Dis 20:110–114, 1995.
49. JW Baddley, SA Moser, DA Sutton, PG Pappas. Microascus cinereus (anamorph Scopulariopsis) brain abscess in a bone marrow transplant recipient. J Clin Microbio 38:395–397, 2000.
50. KK Krisher, NB Holdridge, MM Mustafa, MG Rinaldi, DA McGough. Disseminated Microascus cirrosus infection in pediatric bone marrow transplant recipient. J Clin Microbio 33:735–737, 1995.
51. L Krempl-Lamprecht. Scopulariopsisarten bei onychomykosen. Proceedings 2nd International Symposium Med Mycol, Poznan, Poland, 1967, pp. 45–48.
52. C Schönborn, H. Schmoranzer. Untersuchungen über schimmelpilzinfektionen der zehennägel. Mykosen 13:253–272, 1970.
53. RD Welsh, RW Ely. Scopulariopsis chartarum systemic mycosis in a dog. J Clin Microbio 37:2102–2103, 1999.
54. IZ Al-Mohsen, DA Sutton, L Sigler, E Almodovar, N Mahboub, H Frayha, S Al-Hajjar, MG Rinaldi, T Walsh. *Acrophialophora fusispora* brain abscess in a child with acute lymphoblastic leukemia: Review of cases and taxonomy. J Clin Microbio 38:4569–4576, 2000.
55. MD Ellison, RT Hung, K Harris, BH Campbell. Report of the first case of invasive fungal sinusitis caused by Scopulariopsis acremonium: Review of Scopulariopsis infections. Archive Otolaryngol Head Neck Surg 124:1014–1016, 1998.
56. JP Neglia, DD Hurd, P Ferrieri, DC Snover. Invasive Scopulariopsis in the immunocompromised host. Amer J Med 83:1163–1166, 1987.
57. LJ Wheat, M Bartlett, M Ciccarelli, JW Smith. Opportunistic Scopulariopsis pneumonia in an immunocompromised host. South Med J 77:1608–1609, 1984.
58. ME Hagensee, JE Bauwens, B Kjos, RA Bowden. Brain abscess following marrow transplantation: Experience at the Fred Hutchinson Cancer Research Center, 1984–1992. Clin Infec Dis 19:402–408, 1994.
59. MA Jabor, DL Greer, RG Amedee. Scopulariopsis: An invasive nasal infection. Amer J Rhinol 12:367–371, 1998.
60. J Guarro. Comments on recent human infections caused by ascomycetes. Med Mycol 36:349, 1998.
61. JA von Arx, J Guarro, MJ Figueras. The ascomycete genus Chaetomium. Beih Nova Hedwigia 84:1–162, 1986.
62. PF Cannon. A revision of Achaetomium, Achaetomiella and Subramaniula, and some similar species of Chaetomium. Trans Brit Mycol Soc 87:45–76, 1986.
63. S Lee, RT Hanlin. Phylogenetic relationships of *Chaetomium* and similar genera based on ribosomal DNA sequences. Mycologia 91:434–442, 1999.
64. SP Abbott, L Sigler, R McAleer, DA McGough, MG Rinaldi, G Mizell. Fatal cerebral mycoses caused by the ascomycete Chaetomium strumarium. J Clin Microbio 33:2692–2698, 1995.

65. J Guarro, L Soler, MG Rinaldi. Pathogenicity and antifungal susceptibility of Chaetomium species. Eur J Clin Microbio Infec Dis 14:613–618, 1995.
66. V Anandi, TJ John, A Walter, JCM Shastry, MK Lalitha, AA Padhye, L Ajello, FW Chandler. Cerebral phaeohyphomycosis caused by Chaetomium globosum in a renal transplant recipient. J Clin Microbio 27:2226–2229, 1989.
67. AR Costa, E Porto, CS Lacaz, NT Melo, MJF Calux, NTS Valente. Cutaneous and ungual phaeohyphomycosis caused by species of *Chaetomium* Kunze (1817) ex Fresenius, 1829. J Med Vet Mycol 26:261–268, 1988.
68. V Lesire, E Hazouard, P-F Dequin, M Delain, M Therizol-Fery, A Legras. Possible role of Chaetomium globosum in infection after autologous bone marrow transplantation. Intensive Care Medicine 25:124–125, 1999.
69. MG Rinaldi, CB Inderlied, V Mohnovski, H Monforte, GL Lam, AW Fothergill, DA McGough. Fatal Chaetomium atrobrunneum Ames 1949, systemic mycosis in a patient with acute lymphoblastic leukemia. ISHAM abstr PS2.69, 1991.
70. KH Guppy, C Thomas, K Thomas, D Anderson. Cerebral fungal infections in the immunocompromised host: A literature review and a new pathogen—Chaetomium atrobrunneum: case report. Neurosurgery 43:1463–1469, 1998.
71. C Thomas, D Mileusnic, RB Carey, M Kampert, D Anderson. Fatal Chaetomium cerebritis in a bone marrow transplant patient. Human Path 30:874–879, 1999.
72. T Yeghen, L Fenelon, CK Campbell, DW Warnock, AV Hoffbrand, HG Prentice, CC Kibbler. Chaetomium pneumonia in patient with acute myeloid leukemia. J Clin Path 49:184–186, 1996.
73. H Schulze, A Aptroot, A Grote-Metke, L Balleisen. Aspergillus fumigatus and Chaetomium homopilatum in a leukemic patient: Pathogenic significance of Chaetomium species. Mycoses 40 (suppl 1): 104–109, 1997.
74. L Sigler, JW Carmichael. Redisposition of some fungi referred to Oidium microspermum and a review of Arthrographis. Mycotaxon 18:495–507, 1983.
75. EM Perlman, L Binns. Intense photophobia caused by Arthrographis kalrae in a contact lens wearing patient. Amer J Ophthalmol 123:547–549, 1997.
76. PV Chin-Hong, DA Sutton, M Roemer, MA Jacobson, JA Aberg. Invasive fungal sinusitis and meningitis due to Arthrographis kalrae in a patient with AIDS. J Clin Microbio 39:804–807, 2001.
77. J Gené, JM Guillamon, K Ulfig, J Guarro. Studies on keratinophilic fungi. X. Arthrographis alba sp. nov. Can J Microbio 42:1185–1189, 1996.
78. SW Sachs, J Baum, C Mies. Beauveria bassiana keratitis. Brit J Ophthalmol 69: 548–550, 1985.
79. TA Kisla, A Cu-Unjieng, L Sigler, J Sugar. Medical management of *Beauveria bassiana* keratitis. Cornea 19:405–406, 2000.
80. RA Fromtling, SD Kosanke, JM Jensen, GS Bulmer. Fatal Beauveria bassiana infection in a captive American alligator. Amer Vet Med Assoc 175:934–936, 1979.
81. JF Gonzalez Cabo, J Espejo Serrano, MC Barcena Asensio. Mycotic pulmonary disease by Beauveria bassiana in a captive tortoise. Mycoses 38:67–169, 1995.
82. PJ McDonnell, TP Werblin, L Sigler, WR Green. Mycotic keratitis due to Beauveria alba. Cornea 3:213–216, 1985.
83. J Augustinsky, P Kammeyer, A Husain, GS de Hoog, CR Libertin. Engyodontium album endocarditis. J Clin Microbio 28:1479–1481, 1990.

84. W Gams, MR McGinnis. Phialemonium, a new anamorphic genus intermediate between Phialemonium and Acremonium. Mycologia 75:977–987, 1983.

85. MG Rinaldi, EL McCoy, DF Winn. Gluteal abscess caused by Phialophora hoffmannii and review of the role of this organism in human mycoses. J Clin Microbio 16:181–185, 1982.

86. DJ Marriott, KH Wong, E Azner, JL Harkness, DA Cooper, D Muir. Scytalidium dimidiatum and Lecythophora hoffmannii: Unusual causes of fungal infections in patients with AIDS. J Clin Microbio 35:2949–2952, 1997.

87. DM Marcus, DS Hull, RM Rubin, CL Newman. Lecythophora mutabilis endophthalmitis after long-term corneal cyanoacrylate. Retina 19:351–353, 1999.

88. S Ahmad, RJ Johnson, S Hillier, WR Shelton, MG Rinaldi. Fungal peritonitis caused by Lecythophora mutabilis. J Clin Microbio 22;182–186, 1985.

89. R Ramani, BT Kahn, V Chaturvedi. Tilletiopsis minor: A new etiologic agent of human subcutaneous mycosis in an immunocompromised host. J Clin Microbio 35:2992–2995, 1997.

90. MC Cepero de Garcia, ML Arboleda, F Barraquer, E Grose. Fungal keratitis caused by Metarhizium anisopliae. J Med Vet Mycol 35:361–363, 1997.

91. SG Revankar, DA Sutton, SE Sanche, J Rao, M Zervos, F Dashti, MG Rinaldi. Metarhizium anisopliae as a cause of sinusitis in immunocompetent hosts. J Clin Microbio 37:195–198, 1999.

92. D Bugner, G Eagles, M Burgess, P Procopis, M Rogers, D Muir, R Pritchard, A Hocking, M Priest. Disseminated invasive infection due to Metarhizium anisopliae in an immunocompromised child. J Clin Microbio 36:1146–1150, 1998.

93. D Muir, P Martin, K Kendall, R Malik. Invasive hyphomycotic rhinitis in a cat due to Metarhizium anisopliae. Med Mycol 36:51–54, 1998.

94. P Bourbeau, DA McGough, H Fraser, N Shah, MG Rinaldi. Fatal disseminated infection caused by Myceliophthora thermophila, a new agent of mycosis: Case history and laboratory characteristics. J Clin Microbio 30:3019–3023, 1992.

95. C Farina, A Gamba, R Tambini, H Beguin, JL Trouillet. Fatal aortic Myceliophthora thermophila infection in a patient affected by cystic medial necrosis. Med Mycol 36:113–118, 1998.

96. IH Tekkok, MJ Higgins, EC Ventureyra. Posttraumatic gas-containing brain abscess caused by Clostridium perfringens with unique simultaneous fungal suppuration by Myceliopthora thermophilia: case report. Neurosurgery 39:1247–1251.

97. L Sigler. Chrysosporium and molds resembling dermatophytes. In: J Kane, PC Summerbell, L Sigler, S Krajden, GA Land. Laboratory Handbook of Dermatophytes. Belmont, CA: Star, 1997, pp. 261–311.

98. AN Smith, JA Spencer, JS Stringfellow, KR Vygantas, JA Welch. Disseminated infection with Phialemonium obovatum in a German shepherd dog. J Amer Vet Med Assoc 216:708–712, 2000.

99. J Guarro, M Nucci, T Akiti, J Gené, J Cano, MD Barreiro, C Aguilar. Phialemonium fungemia: Two documented nosocomial cases. J Clin Microbio 37:2493–2497, 1999.

100. BC Sutton. The Coelomycetes. Kew, UK: Commonwealth Mycological Institute, 1980.

101. TR Nag Raj. Coelomycetous anamorphs with appendage-bearing conidia. Waterloo, Canada: Mycologue, 1993.

102. E Punithalingam. Sphaeropsidales in culture from humans. Nova Hedwigia 31: 119–158, 1979.
103. DA Sutton. Coelomycetous fungi in human disease. A review: clinical entities, pathogenesis, identification and therapy. Rev Iberoam Micol 16:171–179, 1999.
104. AA Padhye, RW Gutekunst, DJ Smith, E Punithalingam. Maxillary sinusitis caused by Pleurophomopsis lignicola. J Clin Microbio 35:2136–2141, 1997.
105. J Guarro, TE Svidzinski, L Zaror, MH Forjaz, J Gené, O Fischman. Subcutaneous hyalohyphomycosis caused by Colletotrichum gloeosporioides. J Clin Microbio 36: 3060–3065, 1998.
106. LGM Castro, C da Silva Lacaz, J Guarro, J Gené, EM Heins-Vaccari, RS de Freitas Leite, GLH Arriagada, MMO Reguera, EM Ito, NYS Valente, RS Nunes. Phaeohyphomycotic cyst caused by Colletotrichum crassipes. J Clin Microbio 39:2321–2324, 2001.
107. MM Maslen, T Collis, R Stuart. Lasiodiplodia theobromae isolated from a subcutaneous abscess in a Cambodian immigrant to Australia. J Med Vet Mycol 34:279–283, 1996.
108. MK Moore. The infection of human skin and nail by Scytalidium species. In: M Borgers et al., eds. Current Topics in Medical Mycology. New York: Springer-Verlag, 1992, pp. 1–42.
109. HJ Roeijmans, GS De Hoog, CS Tan, MJ Figge. Molecular taxonomy and GC/MS of metabolites of Scytalidium hyalinum and Nattrassia mangiferae (Hendersonula toruloidea). J Med Vet Mycol 35:181–188, 1997.
110. L Sigler, RC Summerbell, L Poole, M Wieden, DA Sutton, MG Rinaldi, M Aguirre, GW Estes, JN Galgiani. Invasive Nattrassia mangiferae infections: Case report, literature review, therapeutic and taxonomic appraisal. J Clin Microbio 35:433–440, 1997.
111. M Machouart-Dubach, C Lacroix, M Feuilhade de Chauvin, I Le Gall, C Giudicelli, F Lorenzo, F Derouin. Rapid discrimination among dermatophytes, Scytalidum spp., and other fungi with a PCR-RFLP ribotyping method. J Clin Microbio 39: 685–690, 2001.
112. DA Sutton, WD Timm, G Morgan-Jones, MG Rinaldi. Human phaeohyphomycotic osteomyelitis caused by the coelomycete Phomopsis Saccardo 1905: Criteria for identification, case history, and therapy. J Clin Microbio 37:807–811, 1999.
113. L Sigler, S Estrada, NA Montealegre, E Jaramillo, M Arango, C De Bedout, A Restrepo. Maxillary sinusitis caused by Schizophyllum commune and experience with treatment. J Med Vet Mycol 35:365–370, 1997.
114. JD Rihs, AA Padhye, CB Good. Brain abscess caused by Schizophyllum commune: An emerging basidiomycete pathogen. J Clin Microbio 34:628–632, 1996.
115. L Sigler, L de la Maza, G Tan, KN Egger, RK Sherburne. Diagnostic difficulties caused by a nonclamped Schizophyllum commune isolate in a case of fungus ball of the lung. J Clin Microbio 33:1979–1983, 1985.
116. L Sigler, SP Abbott. Characterizing and conserving diversity of filamentous basidiomycetes from human sources. Microbio Cult Coll 13:21–27, 1997.
117. J Gené, JM Guillamon, J Guarro, J Pujol, K Ulfig. Hormographiella aspergillata Anamorph of Coprinus cinereus, a human opportunistic fungus: Molecular characterization and antifungal susceptibility. Ant v Leeuw 70:49–57, 1996.

118. GS De Hoog, AHG van den Ende. Molecular diagnostics of clinical strains of filamentous basidiomycetes. Mycoses 41:183–189, 1997.

119. GA De Vries. Ascomycetes: Eurotiales, Sphaeriales, and Dothideales. In: DH Howard, ed. Fungi Pathogenic for Humans and Animals. Part A. Biology. New York: Marcel Dekker, 1983, 81–111.

12

Molecular Methods to Identify Pathogenic Fungi

Thomas G. Mitchell
Duke University Medical Center, Durham, North Carolina, U.S.A.

Jianping Xu
McMaster University, Hamilton, Ontario, Canada

I. INTRODUCTION

Over the past two decades, the number of immunocompromised patients has continued to rise globally, promoting a dramatic increase in the incidence and variety of fungal infections (1–3). As a result of this mycological crisis, there is a pressing need to improve the accuracy and speed of the diagnosis, to identify the sources of individual cases and outbreaks, and to understand the genetics and distribution of populations of pathogenic fungi. As described in Part 1 of this book, the identification of mycotic species and strains is often problematic. Established methods of classification rely on phenotypic differences in morphology and physiology, and when possible, mating. For many taxa, definitive phenotypic features are difficult to observe or are highly variable. Hence there is considerable scientific and clinical interest in molecular approaches to the identification of species and strains of pathogenic fungi. In recent years there has been substantial progress in the development of innovative methods to analyze organisms at the genomic level. In addition to the practical benefits, the application of molecular methods to pathogenic fungi has reaped much new information about the epidemiology and evolution of human mycoses. Consequently there are compelling reasons to develop molecular methods to identify various taxa of pathogenic fungi.

Most medical fungi lack a sexual cycle, and for many pathogens the concept of a species is poorly defined. Accordingly, there is a crucial need for rapid, precise, and reproducible methods: (1) to identify fungal species accurately in the clinical laboratory, (2) to determine the source of a mycotic infection, (3) to resolve the status of problematic species, (4) to track the transmission of strains involved in nosocomial mycoses, (5) to recognize strains with clinically important phenotypes, such as specific virulence factors or resistance to antifungal drugs, (6) to clarify the origin(s) of diversity and the population genetics of the major pathogens, and (7) to provide or identify genotypes of typical strains for use by basic scientists and by researchers in the pharmaceutical and diagnostics fields.

In this chapter we will review molecular methods currently used for typing species and strains of medically important fungi and discuss the selection of methods to resolve specific questions. The content will emphasize the concept of each method, as well as its advantages and limitations. The literature in this field is rapidly expanding, and we will not attempt to describe every application of each method or the multiple studies applying various methods to the more common medical fungi. References will be cited for broader issues, comprehensive reviews, and detailed protocols.

II. MOLECULAR METHODS

The current revolution in molecular biology has provided techniques to identify numerous taxa of medically important fungi, including pathogenic species, as well as strains of a species. Many new methods exploit the tremendous variation in the DNA of fungi. In this section we will introduce the more common typing techniques and compare their value and drawbacks. The established methods for typing medical fungi entail the comparison of allozymes, electrophoretic karyotypes, hybridization of probes to DNA, PCR-based fingerprints, restriction fragment length polymorphisms, and DNA sequencing. All these approaches are designed to generate molecular markers to compare fungi.

Overall, molecular markers are definitive and more stable than phenotypic observations. A *molecular marker* can be any detectable property that identifies a specific region of the genome. There is no ideal molecular marker for every organism or every purpose. Some markers may be better at discriminating individual strains, separate species, or higher taxonomic groups. For some purposes, it is important to identify markers in specific genes. In other situations, markers in noncoding, usually anonymous portions of the genome are preferable because they are assumed to be neutral or unaffected by selective pressure. Even with the same organism, different markers will be used to address different questions.

A. Isozyme Electrophoresis

The electrophoretic migration of enzymes is among the most cost-effective methods to investigate genetic variation at the molecular level. The four common methods of protein electrophoresis differ both in the nature of the supporting medium or gel and whether they are run horizontally or vertically: starch (including both horizontal and vertical systems), polyacrylamide (vertical), agarose (horizontal), and cellulose acetate. These methods are compared and reviewed in detail by Murphy et al. (4). The migration M of a protein is influenced by many factors, including its net charge Q, its molecular size as measured by its radius r, the strength of the electric field E, and the viscosity of the supporting gel V. The relationship between M and other factors can be described as follows:

$$M = QE/4 \, \pi \, r^2 V$$

Under appropriate conditions, the rate of M increases with the net charge, which is influenced by the pH of the buffering system and the strength of the electric field, and M decreases as the molecular size of the protein and the viscosity of the suspension medium are increased. Since not all proteins are globular, various shapes may affect migration differently.

Most useful *isozymes* are functional enzymes that differ in amino acid sequence. After separation, the variant bands of an enzyme are recognized by adding the appropriate substrate and a detection system. The substrate is often coupled with a dye that is released upon enzymatic activity. The detection of specific enzymatic activity in more than one band or electrophoretic mobility denotes an enzyme with allelic variants, or isozymes.

The accurate application of isozyme data requires that the observed banding patterns on gels are correctly interpreted. There are two commonly held assumptions. The first is that changes in the mobility of an enzyme in an electric field reflect a change in its amino acid and thus encoding DNA sequence. Therefore, if the banding patterns of two individuals differ, these differences are assumed to be genetically based and heritable. The second assumption is that enzyme expression is codominant; that is, every allele at a locus is expressed.

There are two general forms of protein data. Isozymes are functionally similar forms of an enzymatic protein—including all its subunits—that may be produced by different gene loci or by different alleles at the same locus. In contrast, *allozymes* are a subset of isozymes in which polypeptide variants of the enzyme are formed by different allelic alternatives at the same gene locus. Allozyme data are required for correct inferences about population structure. Isozyme data are only useful under appropriate circumstances. The analysis of allelic status is essential for complex isozyme patterns. Obtaining allozyme data from isozyme data requires crosses and analyses of meiotic progeny. In asexual diploid fungi, such as *Candida albicans*, it is impossible to infer allelic status correctly from isozyme patterns that involve either multiple loci and/or enzymes with polymeric structures.

Isozyme analyses have been successfully applied to studies of *C. albicans* (5, 6), to other species of *Candida* (7, 8), and to *Cryptococcus neoformans* (9–11).

B. Electrophoretic Karyotypes (EKs)

Variation in the number and size of fungal chromosomes can be detected by electrophoresis under conditions that provide alternating fields of electric current, referred to as pulsed field gel electrophoresis (PFGE) (12). Several instruments and procedures have been developed for PFGE, and perhaps the most common is the contour-clamped homogeneous electric field (CHEF). In these procedures, intact chromosomes migrate through an agarose gel matrix under the influence of the pulsed fields. Following electrophoresis and optimal separation of chromosomes, the gels can be stained with ethidium bromide and viewed under ultraviolet light to analyze the patterns of chromosomal banding, or *electrophoretic karyotype* (EK). This technique can potentially detect large deletions, insertions, duplications, and translocations among chromosomes (13–15). Identifying genes and chromosomes with these polymorphisms may require additional analyses, such as digestion with endonucleases and analysis of restriction fragments and/ or blotting and probing the chromosomal gels (16, 17). A common method of improving the resolution of the EK is to first digest the chromosomes with rare cutting restriction endonucleases (e.g., the eight-base cutting enzyme, *Sfi*), and then use PFGE or CHEF to separate these large restriction fragments (18, 19).

Depending upon the species, the EK method compares favorably with other typing methods. It may be as good or better than other methods for discriminating among species (20). Among the prominent medically important fungi, EKs have been developed for *Candida albicans* (18, 21–24), other *Candida* species (25, 26), *Cryptococcus neoformans* (27–30), *Aspergillus fumigatus* (31), and others (32–37).

There is evidence that EKs are often unstable in *C. albicans* (38–40) and *C. neoformans* (41–43). They may be too variable and unstable to be useful in species or strain delineation. Diverse EKs have been observed following asexual propagation and subculturing within a single genotype. The loss of dispensable chromosomes also causes variation (44). Because chromosomes must pair during meiosis, it is often assumed that the EKs of sexual species should be less variable than those of asexual ones, but the EKs of sexual species of fungi may indeed vary in size and gene arrangement (45). Another drawback of EKs is the occasional difficulty of scoring homologous chromosomes, which may migrate together and appear as one band. It is also difficult to quantify differences in EKs.

C. DNA–DNA Hybridization

Several increasingly sophisticated methods of molecular typing entail variations of the hybridization of complementary strands of DNA. DNA–DNA hybridization techniques offer a quantitative measure of the similarity between two or among several individuals. This category of molecular typing methods began with DNA reassociation kinetics. Currently, hybridization using oligonucleotide probes and DNA microarrays are becoming powerful tools for assaying genetic differences at defined genomic regions.

1. DNA Reassociation Kinetics

DNA hybridization takes advantage of the double-stranded nature of the DNA molecule in which nucleotides on opposing strands are held together by hydrogen bonds. Two hydrogen bonds are formed between adenine and thymine, and three hydrogen bonds link guanine and cytosine. When double-stranded DNA is heated to around 100°C, the hydrogen bonds between complementary base pairs are broken and the opposing strands separate. During subsequent cooling of the solution, the complementary DNA strands will reanneal. If DNA from two different species are combined, denatured, and then allowed to reassociate, the double-stranded molecules that form between complementary strands from the two species will contain base pair mismatches because of their evolutionary divergence from a common ancestor (46). The conditions of reassociation must be standardized because the amount of base pair mismatches that form in the hybrid molecules is affected by the salt concentration, temperature, viscosity, and DNA fragment size. Under highly stringent conditions of reassociation, which are generally achieved by increasing the temperature and/or decreasing the salt concentration, base pairing between DNA strands from different strains or species will only occur between well-matched sequences. Under conditions of progressively lower stringencies, more mismatches will be tolerated during reassociation.

The extent of mismatching determines the temperature at which these hybrid molecules melt when they are placed in a thermal gradient. The more mismatches, the lower the temperature at which the hybrid strands will separate. The decrease in the melting temperature of a heteroduplex hybrid relative to a homoduplex control provides an index of divergence between the DNAs under consideration. Reassociation analyses have been used to investigate the relationships among several closely related anamorphic species (46–50).

2. Oligonucleotide Hybridization

This technique utilizes known single nucleotide polymorphisms in a species. Oligonucleotides of more than 20 bases are designed with known polymorphic sites placed near the middle of the oligonucleotide, which can be end-labeled with a radioactive tag or fluorescent dye. The labeled oligonucleotide probes are then hybridized by conventional Southern hybridization to either total genomic DNA or specific gene fragments amplified by the polymerase chain reaction (PCR). The presence or absence of a hybridization signal for each probe can be scored as alternative alleles at a specific site. The drawbacks of this technique are that it requires two hybridizing procedures for every locus in diploid individuals, and unknown mutations at any of the 20 or so nucleotides may cause the loss of a hybridizing signal.

3. DNA Chip Technology

The new technology of DNA microarrays on chips represents a miniature but mass version of individual oligonucleotide hybridization. Briefly, DNA chips are glass surfaces to which arrays of specific DNA fragments have been attached at discrete locations. These fragments serve as probes for hybridization. Under conditions suitable for hybridization, the DNA spots on the chip are exposed to a solution containing a complex sample of fluorescent-labeled DNA or RNA. This technique has been established for only a few species; current use has concentrated on *Saccharomyces cerevisiae* (51). A typical array might consist of 20 complementary pairs of oligonucleotides (25-mers) for each gene. In addition, there will be three permutations of each consensus 25-mer, each with a single base change in the central nucleotide position. Thus many base pair substitutions of the gene will be represented on the chip. A DNA sample of each isolate is permitted to hybridize to the array. Comparing the hybridization intensities to the consensus oligonucleotides and the one-base mutants provides internal controls in the search for sequence polymorphisms. The amount of sequence divergence and similarity among strains can be calculated.

Another broad application of DNA chip technology involves the analysis of gene expression. At present this technology is limited to intraspecific comparisons. The high expense currently prohibits its extensive use.

D. PCR-Based Methods

The PCR technology has spawned many procedures for typing strains, some of which have become established methods in systematics and strain typing. PCR methods are easy to set up, rapid, and have the advantage of requiring only minute amounts of starting material or template DNA. Although simple in concept, the PCR entails unrivaled, often overlooked complexity. The source of this complexity is inherent in the PCR itself; the products result from myriad ionic interactions, kinetic constants, and enzymatic activities, which repeatedly affect the reactants in a small volume over an extended time period. Some of the common PCR-based strain typing techniques are described below.

1. Random Amplified Polymorphic DNA (RAPD)

In RAPD analysis, genomic or template DNA is primed at a low annealing temperature (30–38°C) with a single short oligonucleotide (approximately 10 bases) in the PCR. Multiple PCR products of different electrophoretic mobilities are typically generated, and in comparing species or strains, each isolate will yield

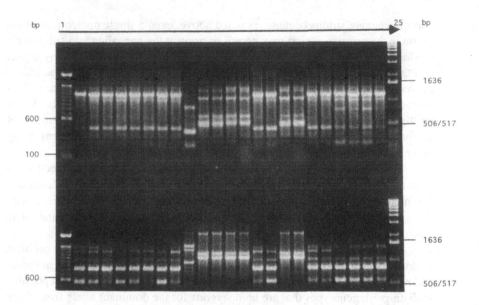

Figure 1 Electrophoretic separation of RAPD fingerprints obtained by amplifying genomic DNA from 23 strains of *Cryptococcus neoformans* with single primers: (A) OPA-03 (5′ AGTCAGCCAC 3′) or (B) OPA-17 (5′ GACCGCTTGT 3′). Lanes 1 and 25 are 100 bp and 1 kbp DNA ladders (GIBCO, BRL), respectively. Lanes 2 through 24 are RAPD profiles of the following strains from the Duke Medical Mycology Research Laboratory: CnA-1, CnA-2, CnA-3, CnA-4, CnA-5, CnA-6, CnA-7, CnA-8, CnB-1, CnD-1, CnD-2, CnD-3, CnD-4, CnA-9, CnA-10, CnD-5, CnD-6, CnA-11, CnA-12, CnA-13, CnA-14, CnA-15, and CnA-16, respectively. (Note: Cn denotes *C. neoformans*, and the letters indicate the serotype, A, B, C, or D.)

several bands of different sizes. (See Fig. 1.) RAPD analysis detects variations in the length between two primer binding sites or sequence length polymorphisms in the fragments between PCR priming regions (52, 53). Nucleotide substitutions in the region of PCR primer binding, particularly at the 3′ ends, can prevent binding of the primer to the DNA template and subsequent PCR amplification, and a band will be missing. Similarities in banding profiles among strains (i.e., the number and mobility, but not the density of the bands) can be calculated and used to infer epidemiological relationships. When multiple primers are screened, RAPD analysis is sensitive enough to detect variation among isolates that cannot be observed by using other methods, although a combination of methods often provides optimal discrimination (54–56).

Although technically fast and simple, there are some disadvantages to RAPD. The major drawback is reproducibility. RAPD analysis can detect minute variation among strains because, as noted above, even a single nucleotide substitution in the priming region may permit or prevent the annealing and subsequent production or absence of a characteristic band. Small differences in any aspect of PCR conditions that affect binding of the primer will have the same effect; consequently, RAPDs are sensitive to the vagaries of the testing procedure. This problem can be minimized if strains under study are treated similarly. When multiple strains are to be compared for distinct RAPD patterns, the same PCR buffer, master mix, and thermal cycler should be used, and the strains being compared should be amplified at the same time.

RAPDs can also be problematic because bands with the same electrophoretic mobility may not share the same sequence. This problem may be common for interspecific studies and is affected by the conditions of electrophoresis. With the usual agarose gels (1–1.5%), it is difficult to distinguish RAPD bands with size differences of less than 20 base pairs.

Another concern with the interpretation of RAPD profiles is the problem of dominant and null alleles. In haploid organisms, both the dominant (presence) and null (absence) alleles can be scored, but in diploid organisms, it is not possible to distinguish genotypes that are homozygous for the dominant allele from those that are heterozygous. Therefore, RAPD data are generally not ideal for inferences of population genetic history (57).

Nevertheless, for comparing the similarities among strains and developing fingerprints for molecular epidemiology, RAPD analyses have been widely applied to a number of medically important fungi (10, 56, 58–68).

2. Amplification of Microsatellite Repeats

This emerging technique exploits the hypervariability of DNA regions composed of 10 or more tandem repeats of di-, tri-, or multiple nucleotides. This hypervariability can be caused by either strain slippage during DNA replication or unequal crossing-over during meiosis. Useful microsatellites have been located by probing a genomic library or searching databases of gene sequences (69, 70). PCR primers flanking these repeat regions can be designed, and PCR products are typically run on polyacrylamide gels to detect differences in a single repeat. This technique can discern levels of variability as high or higher than PCR fingerprinting. Because multiple alleles are typically found at a single locus, the relationships among alleles can be difficult to decipher, and alleles may not be identical by descent.

3. PCR Fingerprinting

This technique is similar to RAPD, except that the primers are longer (>15 bases) and the annealing temperatures are more stringent. Most PCR fingerprinting primers are designed from repetitive DNA sequences, microsatellites (as above), or somewhat longer minisatellites (71, 72). Commonly used primers include

M13, which is derived from the core sequence of phage M13, T3B, which originates from internal sequences of tRNA genes, and TELO1, which is based on fungal telomere repeat sequences. Because of more stringent reaction conditions, PCR fingerprinting is generally more reproducible than RAPDs. Nonetheless, it suffers the same problems of interpretation as RAPDs. However, under standardized conditions, PCR fingerprinting has proven quite reliable for the identification of species and strains (71, 73–75).

4. PCR-RFLP

With increasing knowledge of the genomics of human pathogenic fungi, the number of reported gene sequences is increasing. These gene or intergenic sequences can be used to investigate the variability among strains and the history of populations of a species. One typical application is to design PCR primers to amplify a particular stretch of DNA, and subsequently digest the amplicon with a battery of four-cutter restriction enzymes to screen for variability. Variable restriction sites can then be used to screen a larger sample of isolates. This approach was used to develop multilocus genotypes of *Candida albicans* (43, 76). In a diploid organism, codominant genetic information is obtained, and unlike RAPDs or PCR fingerprinting, both alleles can be scored. This technique is highly reproducible.

5. SSCP and Heteroduplex

Single-strand confirmation polymorphism (SSCP) and heteroduplex are promising techniques that allow efficient detection of nucleotide substitutions in short fragments (<500 bp) of DNA. SSCP analysis typically involves the amplification of a discrete segment of genomic DNA, melting the PCR products, and analysis of the single strands on a nondenaturing polyacrylamide gel (77). The DNA strands are usually visualized by using radiolabeled DNA during PCR or by silver-staining the DNA after gel electrophoresis. Polymorphisms in strand mobility result from the effects of primary sequence differences on the folded structure of the single DNA strands; that is, the primary sequence differences alter the intramolecular interactions that generate a three-dimensional folded structure. The molecules may consequently migrate at different rates through the nondenaturing polyacrylamide gel. Because these conformational variations may in theory reflect a single base difference, they are subtle, and the success of any particular SSCP assay depends highly on the particular DNA fragments being investigated, including their sequence and length, and the optimization of the experimental conditions to maximize the differential migration of the fragments. A variety of methods have been invoked to improve the resolving power of SSCP, including the addition of glycerol to the polyacrylamide gels, reducing the temperature, and increasing the length of the gels or duration of electrophoresis. Nevertheless, differentiation among polymorphic molecules on a polyacrylamide matrix is not entirely predictable, and this method can produce false negative patterns, ambiguous results, and experimental artifacts (77). SSCP works best with fragments ≤300 bp. It has been used to develop markers for several pathogenic fungi (78–83).

In contrast to SSCP, heteroduplex analysis is dependent on conformational differences in double-stranded DNA. In this technique, equal amounts of two PCR products (e.g., from wild-type and mutant DNA samples) are combined in a nondenaturing buffer. The DNA is melted at 95°C and slowly cooled to room temperature. During the cooling process, the complementary single strands from the same sample reanneal to form double-stranded homoduplexes, and the noncomplementary single strands from different samples also reanneal to form heteroduplex DNA. The mismatch in the heteroduplex DNA imparts a different three-dimensional shape or flexibility compared to the homoduplex DNA, and consequently the heteroduplex DNA will have less electrophoretic mobility than the homoduplex DNA. The electrophoresis and detection methods for heteroduplex analysis are similar to those for SSCP. Heteroduplex analysis works well for fragments ranging in length from 200 to 600 bp.

6. Amplified Fragment Length Polymorphism (AFLP)

The relatively recent development of the method of amplified fragment length polymorphisms (AFLPs) has already exerted a significant impact as a genotyping method. The determination of AFLPs is a powerful method for fingerprinting genomic DNA, as well as generating a large number of dominant markers for genotypic analysis (84, 85). AFLP technology combines the strategies of enzymatic digestion and PCR. With AFLP, after digestion with two endonucleases (usually a frequent and a rare cutting enzyme), dsDNA adapters are ligated to the ends of the DNA fragments to create template DNA for the PCR. AFLP adapters consist of a core sequence and an enzyme-specific sequence. AFLP amplification primers consist of the core sequence, the enzyme-specific sequence, and a selective extension of one to three nucleotides, which will amplify a subset of the restriction fragments. AFLP usually involves two PCR steps: (1) in the preamplification step unlabeled primers with a single selective nucleotide are used; (2) after this preliminary PCR, the reaction mixtures are diluted tenfold and used as templates for the second or selective PCR, which uses a longer extension and labeled primers.

AFLP has several powerful advantages over the other methods. Many more fragments can be generated and analyzed. By varying the restriction enzymes and the selective nucleotides, 30 to more than 200 fragments can be produced, depending as well on the complexity of the genome. Different enzymes or extension nucleotides (or both) can be used to create new sets of markers. AFLP can therefore provide an almost limitless set of genetic markers. In addition, the fragments are stable and highly reproducible since they are amplified with two specific primers under stringent conditions. Like RAPD markers, AFLP markers are amplified by using arbitrary sequences, but with greater reproducibility and fidelity. An example of AFLP products is shown in Fig. 2.

In Table 1 comparisons are made of the steps involved with AFLP and two other common approaches—the classic hybridization-based method to identify RFLPs, and PCR-based fingerprinting techniques, which involve amplification of particular DNA sequences using specific or arbitrary primers. Amplification products are separated by electrophoresis and detected by staining the DNA or by using radiolabeled primers and detected by exposure to X-ray film.

E. Restriction Fragment Length Polymorphism (RFLP)

Restriction endonucleases have been used to discriminate among species and strains of fungi as well as other biological taxa. The most common approach is to digest genomic DNA with a four-base or other frequently cutting enzyme and examine the resulting bands on electrophoresis in agarose or polyacrylamide gels (64, 86–91). Depending upon the size of the genome and the number of sites of enzymatic cleavage, it may be possible to compare digests of genomic DNA for different banding patterns, indicating DNA sequence polymorphisms. This method of analysis can only detect differences in DNA molecules that are present in high copy numbers, such as ribosomal or mitochondrial DNA. For genetic elements that are low in number, it is almost impossible to observe restriction site polymorphisms following a simple digestion and electrophoresis on agarose or polyacrylamide gel.

Alternatively, an amplified gene or other PCR product can be digested and analyzed (76), as described in section II.D.4 (page 685). For comparisons between species, differences in the rDNA motif are often discernable by PCR-RFLP (92–94). Because it is frequently difficult to determine accurately the migration of bands (i.e., DNA fragment sizes), comparisons should be made on samples in adjacent lanes of the same agarose gel with size gradients on both sides. Table 2 compares this classic method of DNA fingerprinting with PCR and AFLP fingerprinting.

The most widely used RFLP method is a DNA–DNA hybridization technique that involves digesting genomic DNA with restriction endonuclease(s) followed by electrophoretic separation of the DNA fragments. RFLPs are detected by Southern hybridization with labeled probes targeted to single copy markers or repetitive DNA. Several probes have been well characterized for *C. albicans*, including Ca3, CARE2, 27A, MGL1, and the RPS family (95–100). These probes may be as large as several kb. Other probes are specific for other species of *Candida* (101–106), for *A. fumigatus* (107–109), and for *Cryptococcus neoformans* (110–113).

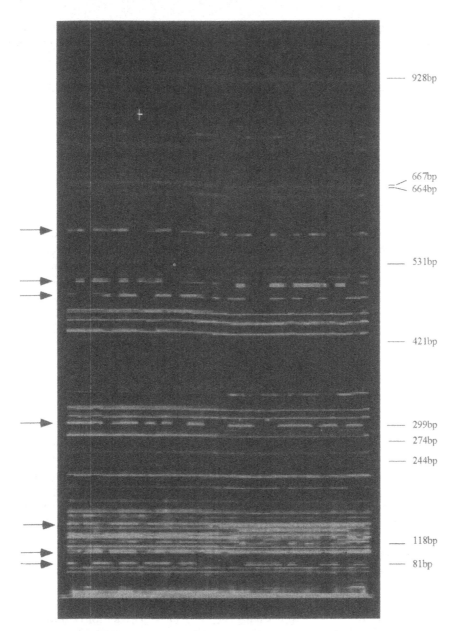

Figure 2 AFLP in *Cryptococcus neoformans* were generated by using the method of Vos et al. (84) with fluorescent-labeled PCR primers. The gel images are analyzed by GeneScan (Perkin-Elmer). Numbers on the right indicate molecular weight standards. Thirty lanes containing AFLP bands from basidiospore progeny are loaded in the parallel lanes. Arrows denote polymorphisms.

Table 1 Comparison of Steps in DNA Strain Typing by Hybridization, RAPD, and AFLP

Conventional hybridization DNA fingerprinting
  1. Extract DNA
  2. Digest DNA with restriction enzyme(s)
  3. Separate DNA fragments by agarose gel electrophoresis
  4. Denature DNA
  5. Blot single-stranded DNA to membrane
  6. Hybridize with labeled oligonucleotide probe
  7. Wash and detect bands (e.g., autoradiography, fluorescence)
RAPD (PCR, AP-PCR, or DAF) fingerprinting
  1. Extract DNA
  2. Randomly PCR amplify with single oligonucleotide primer
  3. Separate products by agarose gel electrophoresis
  4. Photograph and analyze bands
AFLP fingerprinting
  1. Extract DNA
  2. Digest DNA with restriction enzyme(s)
  3. Ligate dsDNA adapters (i.e., core sequence + restriction enzyme-specific sequence) to ends of the DNA fragments
  4. PCR amplify restriction fragments with primers consisting of adapter + enzyme-specific site
  5. Dilute and PCR selectively with primers consisting of adapter + enzyme-specific site + 3' extension of 1-3 nucleotides, thereby amplifying only a subset of the restriction fragments
  6. Separate products by denaturing polyacrylamide gel electrophoresis and analyze bands

---

RFLPs generated from total genomic digests alone or by hybridization with repetitive elements may be difficult to interpret. As the number of bands increases, accuracy may be compromised and affected by the conditions of electrophoresis. These banding patterns are often used to recognize species and to fingerprint strains, but it is difficult to use them to identify the allelic status of individual loci. Conversely, for RFLPs detected by hybridization with probes that target single-copy DNA markers, the interpretation is uncomplicated and the data can be used for multiple purposes. For example, single-copy RFLPs are quite useful for population genetic analyses of diploid fungi (76).

F. DNA Sequencing

The most exacting and laborious method to catalog differences at the DNA level is directly sequencing cloned genes or PCR products. This approach provides the

Table 2 Comparison of Methods Currently Used to Study Genetic Variation in Fungal Populations

Techniques → ↓ Considerations	MLEE	EK (PFGE/CHEF)	DNA–DNA hybridization	DNA probe hybridization	RAPD/AP-PCR PCR fingerprint	AFLP	SSCP	RFLP	DNA sequencing
Typical applications									
Population structure	Yes	No	Usually yes	Yes	Yes or no	Yes	Yes	Yes	Yes
Identify taxa	No	Yes or no	Yes	Yes	Yes	Yes	Yes	Yes	Yes
Phylogenetic analysis	Yes or no	No	No	Yes or no	Yes or no	Yes	Yes	Yes	Yes
Practical factors									
Pure cultures required	Yes	Yes	Yes	Yes or no	Yes	Yes	No	Yes or no	Not with PCR
Sample preparation	Minimal	Medium to high	Medium to high	Medium	Minimal	Minimal	Minimal	Medium	Maximal
Reproducibility	Good	Good	Very good	Good	Good to poor	Very good	Very good	Very good	Best
Cost	Least expensive	Moderate	Expensive	Moderate	Low to moderate	Moderate	Expensive	Low to moderate	Most expensive
Turnaround time	Moderate	Slow	Slow	Slow	Rapid	Moderate	Slow	Moderate	Slowest
Analytical factors									
Sensitivity to detect polymorphism	Low (but often adequate)	Low to high	High	Moderate	High	Very high	Extremely high	Moderate	Highest
Utility of genetic markers (dominant or co-dominant)	Codominant	Chromosome markers codominant	Codominant	Usually codominant	Dominant	Codominant	Codominant	Usually codominant	Codominant

Note: MLEE, multilocus enzyme electrophoresis; RFLP, restriction fragment length polymorphism; EK, electrophoretic karyotype; PFGE, pulsed field gel electrophoresis; CHEF, contour-clamped homogeneous electric field; RAPD, random amplified polymorphic DNA; AP-PCR, arbitrarily primed-PCR; SSCP, single strand conformational polymorphism.

most accurate data for phylogenetic analyses. As with other procedures, several investigations have analyzed medical fungi by direct DNA sequencing (80, 114–123). For phylogenetic analyses, the most common regions to be sequenced and compared are portions of the ribosomal DNA cluster, including the internal transcribed spacers (ITS1 and ITS2), the intergenic spacer (IGS), and the 18S, 5.8S, and 28S rDNA. These regions are highly conserved within species and variable among species.

At the species level, ribosomal RNA genes are under strong evolutionary constraints. Conversely, stable mutations are more common in noncoding sequences, and third-base substitutions are common in protein-encoding genes. Consequently, for comparisons among strains of a species, sequences of protein-encoding genes or nonfunctional DNA are usually more informative than rDNA. For example, nucleotide substitutions were found among strains of *Coccidioides immitis* in genes encoding five proteins: chitin synthase, chitinase, orotidine monophosphate decarboxylase, serine proteinase, and a T-cell reactive protein similar to mammalian dioxygenase (124). Phylogenetic analysis of these variations identified geographically isolated populations of *C. immitis*. In the ensuing years, more studies will employ this type of gene genealogical analysis.

Analyses of this type will also be facilitated by whole genome sequence projects. The genome of *S. cerevisiae* has been fully sequenced, and the genomes of both *Candida albicans* and *Cryptococcus neoformans* are currently being sequenced. These data will permit direct comparisons of specific sequences among isolates within these species. Future evolutionary studies of these and other fungal pathogens will feature comparative genomics across and within taxa.

III. NO UNIVERSAL MOLECULAR METHOD SUITS EVERY PURPOSE

There is not an ideal molecular method for every application, nor is there a best or worst method of molecular typing. Different typing methods are appropriate for different epidemiological or evolutionary studies in different species. Table 2 summarizes the advantages and limitations of the major molecular techniques. In the remainder of this section we will attempt to recommend appropriate molecular methods to address specific questions. However, as indicated below, aside from the technique, the availability of appropriate control samples may be more crucial than the selection of a typing method.

A. Species Identification

Most pathogenic fungi can be identified by conventional laboratory methods, which are based primarily on morphology and physiological tests for molds and yeasts, respectively. However, the interpretation of these tests may be subjective, requiring a skilled mycologist, and some isolates yield variable or atypical results. Phenotypic variation is common among and within fungal species. As the panoply of pathogenic fungi continues to expand, there is a growing need to develop a more efficient, accurate, and rapid approach to identification.

To develop a DNA-based protocol to identify pathogenic fungi, it is feasible to consider signature sequences in the 28S or large nuclear subunit of rDNA. Portions of the 28S rDNA sequences of most human pathogenic yeasts are currently available in Genbank. In addition, many of the molecular methods described in Sec. II are amenable to species identification, and many have been applied to this purpose. Indeed, a variety of DNA-based methods for species identification have been reported, and the results are often reliable, reproducible, and easy to interpret. However, no standard database or protocols have been established for these methods, and it is difficult to compare the results from different laboratories.

B.   Molecular Epidemiology

Besides species identification, molecular methods can help determine the origin of a clinical isolate. Although a few human fungal pathogens are geographically restricted, such as *Coccidioides immitis*, most are highly prevalent. They exist either as members of the normal flora, such as species of *Candida*, or they are ubiquitous in nature, such as *Cryptococcus neoformans* and *A. fumigatus*.

To determine whether the source of a case or outbreak of nosocomial candidiasis can be attributed to a commensal isolate, the local yeast microflora should be sampled. Environmental isolates might be obtained from a variety of clinical settings and health care workers. Control samples could be isolated from vicinal patients, similar body sites, and patients with similar risk factors. After the appropriate isolates and case histories are obtained, a variety of molecular techniques may be employed to determine the similarity of the isolates. For such studies, DNA sequencing of specific genes may have little value. The most useful methods are those that detect numerous polymorphisms throughout the genome and are highly discriminatory, such as AFLP, RAPD, PCR fingerprinting, and probe hybridization methods. Other markers may also be used, especially those that detect polymorphisms that can be unambiguously scored and quantified. Obviously, molecular markers that do not detect intraspecies variation are not helpful, although such invariable species markers have great value for species identification. When the data are collected, appropriate statistical tests should be applied to determine the significance or statistical support for similarity clusters. The subsequent detection of significant clusters can be used to infer the source of the strain responsible for an infection(s). However, because similarity among strains does not prove identity, strain typing can be used with more confidence to exclude certain sources or strains from involvement.

Determining the environmental source of an infection can be more problematic. This is because the extent of genetic variation among environmental populations of pathogenic fungi is usually unknown. As described above, appropriate samples from nature can be collected from the suspected sources of the pathogen and critically evaluated. For comparison, additional control isolates should be obtained from similar environments.

C. Antibiotic Resistance

Methods similar to those for determining the source of an infection (see Section III.B) can be used to investigate the origin of a strain with antibiotic resistance. Controls include isolates from different samples, for which the minimal inhibitory concentration to an antifungal agent must also be determined. In general, if molecular typing determines that several drug-resistant strains are dissimilar, resistance can be assumed to have arisen independently in each strain. Alternatively, a significant clustering of the resistant strains suggests a clonal origin of the resistant genotype and spread of the resistant genotype among patients (125–128).

D. Genetic Diversity and Population Structure

To investigate population structure, it is necessary to have markers that can be interpreted genetically and population samples of sufficient sizes. The selection of markers is influenced by the ploidy of the species and the extent of existing genetic variation in the population. For haploid species, dominant-recessive marker systems can be used, such as AFLP or RAPD. Codominant markers are preferable because multiple alleles can be detected at the same time. For diploid species, only codominant genetic markers provide enough information to infer the mode of reproduction in nature and genetic differences among populations. The analytical methods for understanding the genetic structure of fungal populations are discussed in detail in Chap. 13.

IV. FUTURE CONSIDERATIONS

With the completion of the sequencing of the genome of *S. cerevisiae*, near completion of the *Candida albicans* genome, initiation of the sequencing of the genome of *Cryptococcus neoformans*, and sequences of other microorganisms in public databases, abundant genomic information is now available to design primers to develop systems for the identification of fungal species. The sequence databases can also be used to generate gene-specific products with which to further compare strains within a species. Developing a set of genetic markers from genes with known functions could also facilitate the analysis of interspecific population genetics. Subsequent standardization will promote the exchange of genetic information among laboratories and improve the usefulness of studies of molecular epidemiology.

In this era of fungal molecular biology and genetics, a wealth of genotyping methods is available. Determining the origin and spread of fungal pathogens requires thoughtful selection of control samples and appropriate typing methods. Molecular methods provide new tools, not necessarily the solution, to epidemiological questions.

ACKNOWLEDGMENTS

Support from Public Health Service grants AI 25783, AI 28836, and AI 44975 is greatly appreciated. The authors are members of the Duke University Mycology Research Unit.

REFERENCES

1. V Krcméry Jr, I Krupova, DW Denning. Invasive yeast infections other than *Candida* spp. in acute leukaemia. J Hosp Infec 41:181–194, 1999.
2. F-MC Müller, AH Groll, TJ Walsh. Current approaches to diagnosis and treatment of fungal infections in children infected with human immunodeficiency virus. Eur J Pediat 158:187–199, 1999.
3. DC Coleman, MG Rinaldi, KA Haynes, JH Rex, RC Summerbell, EJ Anaissie, DJ Sullivan. Importance of *Candida* species other than *Candida albicans* as opportunistic pathogens. Med Mycol 36 (suppl. 1):156–165, 1998.
4. RW Murphy, JW Sites Jr, DG Buth, CH Haufler. Proteins: Isozyme electrophoresis. In: DM Hillis, C Moritz, BK Mable, eds. Molecular Systematics. Sunderland, MA: Sinauer Associates, 1996, pp. 51–120.
5. P Boerlin, F Boerlin-Petzold, J Goudet, C Durussel, J-L Pagani, J-P Chave, JL Bille. Typing *Candida albicans* oral isolates from human immunodeficiency virus-infected patients by multilocus enzyme electrophoresis and DNA fingerprinting. J Clin Microbio 34:1235–1248, 1996.
6. J Reynes, C Pujol, C Moreau, M Mallié, F Renaud, F Janbon, J-M Bastide. Simultaneous carriage of *Candida albicans* strains from HIV-infected patients with oral candidiasis: Multilocus enzyme electrophoresis analysis. FEMS Microbio Lett 137: 269–273, 1996.

7. D Lin, L-C Wu, MG Rinaldi, PF Lehmann. Three distinct genotypes within *Candida parapsilosis* from clinical sources. J Clin Microbio 33:1815–1821, 1995.

8. DA Lacher, PF Lehmann. Application of multidimensional scaling in numerical taxonomy: Analysis of isoenzyme types of *Candida* species. Ann Clin Lab Sci 21: 94–103, 1991.

9. ME Brandt, SL Bragg, RW Pinner. Multilocus enzyme typing of *Cryptococcus neoformans*. J Clin Microbio 31:2819–2823, 1993.

10. D Lin, PF Lehmann, BH Hamory, AA Padhye, E Durry, RW Pinner, BA Lasker. Comparison of three typing methods for clinical and environmental isolates of *Aspergillus fumigatus*. J Clin Microbio 33:1596–1601, 1995.

11. E Rinyu, J Varga, L Ferenczy. Phenotypic and genotypic analysis of variability in *Aspergillus fumigatus*. J Clin Microbio 33:2567–2575, 1995.

12. PT Magee. Analysis of the *Candida albicans* genome. In: AJP Brown, MF Tuite, eds. Yeast Gene Analysis. London: Academic, 1998, pp. 395–415.

13. C Thrash-Bingham, JA Gorman. DNA translocations contribute to chromosome length polymorphisms in *Candida albicans*. Curr Genet 22:93–100, 1992.

14. F Fierro, JF Martín. Molecular mechanisms of chromosomal rearrangement in fungi. Crit Rev Microbio 25:1–17, 1999.

15. J Perez-Martin, JA Uria, AD Johnson. Phenotypic switching in *Candida albicans* is controlled by a SIR2 gene. EMBO J 18:2580–2592, 1999.

16. ME Zolan. Chromosome-length polymorphism in fungi. Microbio Rev 59:686–698, 1995.

17. J Pla, C Gil, L Monteoliva, F Navarro-García, M Sánchez, C Nombela. Understanding *Candida albicans* at the molecular level. Yeast 12:1677–1702, 1996.

18. BB Magee, PT Magee. Electrophoretic karyotypes and chromosome numbers in *Candida* species. J Gen Microbio 133:425–430, 1987.

19. WG Merz, C Connelly, P Hieter. Variation of electrophoretic karyotypes among clinical isolates of *Candida albicans*. J Clin Microbio 26:842–845, 1988.

20. LF Di Francesco, F Barchiesi, F Caselli, O Cirioni, G Scalise. Comparison of four methods for DNA typing of clinical isolates of *Candida glabrata*. J Med Microbio 48:955–963, 1999.

21. TJ Lott, P Boiron, E Reiss. An electrophoretic karyotype for *Candida albicans* reveals large chromosomes in multiples. Molec Gen Genet 209:170–174, 1987.

22. T Suzuki, I Kobayashi, I Mizuguchi, I Banno, K Tanaka. Electrophoretic karyotypes in medically important *Candida* species. J Gen Microbio 34:409–416, 1988.

23. BA Lasker, GF Carle, GS Kobayashi, G Medoff. Comparison of the separation of *Candida albicans* chromosome-sized DNA by pulsed field gel electrophoresis techniques. Nucleic Acids Res 17:3783–3793, 1989.

24. S Scherer, PT Magee. Genetics of *Candida albicans*. Microbio Rev 54:226–241, 1990.

25. M Doi, M Homma, A Chindamporn, K Tanaka. Estimation of chromosome number and size by pulsed-field gel electrophoresis (PFGE) in medically important *Candida* species. J Gen Microbio 138:2243–2251, 1992.

26. A Espinel-Ingroff, JA Vazquez, D Boikov, MA Pfaller. Evaluation of DNA-based typing procedures for strain categorization of *Candida* spp. Diag Microbio Infec Dis 33:231–239, 1999.
27. JR Perfect, BB Magee, PT Magee. Separation of chromosomes of *Cryptococcus neoformans* by pulsed field gel electrophoresis. Infec Immun 57:2624–2627, 1989.
28. I Polacheck, GA Lebens. Electrophoretic karyotype of the pathogenic yeast *Cryptococcus neoformans*. J Gen Microbio 135:65–71, 1989.
29. JR Perfect, N Ketabchi, GM Cox, CW Ingram, C Beiser. Karyotyping of *Cryptococcus neoformans* as an epidemiological tool. J Clin Microbio 31:3305–3309, 1993.
30. BL Wickes, TDE Moore, KJ Kwon-Chung. Comparison of the electrophoretic karyotypes and chromosomal location of ten genes in the two varieties of *Cryptococcus neoformans*. Microbiology 140:543–550, 1994.
31. MB Tobin, RB Peery, PL Skatrud. An electrophoretic molecular karyotype of a clinical isolate of *Aspergillus fumigatus* and localization of the MDR-like genes *AfuMDR1* and *AfuMDR2*. Diag Microbio Infec Dis 29:67–71, 1997.
32. CS Kaufmann, WG Merz. Electrophoretic karyotypes of *Torulopsis glabrata*. J Clin Microbio 27:2165–2168, 1989.
33. SC Pan, GT Cole. Electrophoretic karyotypes of clinical isolates of *Coccidioides immitis*. Infec Immun 60:4872–4880, 1992.
34. C Fekete, R Nagy, AJM Debets, L Hornok. Electrophoretic karyotypes and gene mapping in eight species of the *Fusarium* sections Arthrosporiella and Sporotrichiella. Curr Genet 24:500–504, 1993.
35. JD Sobel, JA Vazquez, ME Lynch, C Meriwether, MJ Zervos. Vaginitis due to *Saccharomyces cerevisiae*: Epidemiology, clinical aspects, and therapy. Clin Infec Dis 16:93–99, 1993.
36. T Boekhout, M Kamp, E Guého. Molecular typing of *Malassezia* species with PFGE and RAPD. Med Mycol 36:365–382, 1998.
37. D Senczek, U Siesenop, KH Bohm. Characterization of *Malassezia* species by means of phenotypic characteristics and detection of electrophoretic karyotypes by pulsed-field gel electrophoresis (PFGE). Mycoses 42:409–414, 1999.
38. EP Rustchenko-Bulgac. Variations of *Candida albicans* electrophoretic karyotypes. J Bacteriol 173:6586–6596, 1991.
39. S-I Iwaguchi, M Homma, K Tanaka. Clonal variation of chromosome size derived from the rDNA cluster region in *Candida albicans*. J Gen Microbio 138:1177–1184, 1992.
40. EP Rustchenko-Bulgac, DH Howard. Multiple chromosomal and phenotypic changes in spontaneous mutants of *Candida albicans*. J Gen Microbio 139:1195–1207, 1993.
41. BC Fries, FY Chen, BP Currie, A Casadevall. Karyotype instability in *Cryptococcus neoformans* infection. J Clin Microbio 34:1531–1534, 1996.
42. T Boekhout, A van Belkum. Variability of karyotypes and RAPD types in genetically related strains of *Cryptococcus neoformans*. Curr Genet 32:203–208, 1997.
43. J Xu, RJ Vilgalys, TG Mitchell. Lack of genetic differentiation between two geographically diverse samples of *Candida albicans* isolated from patients infected with human immunodeficiency virus. J Bacteriol 181:1369–1373, 1999.

44. VP Miao, SF Covert, HD VanEtten. A fungal gene for antibiotic resistance on a dispensable ("B") chromosome. Science 254:1773–1776, 1991.

45. DM Geiser, ML Arnold, WE Timberlake. Wild chromosomal variants in *Aspergillus nidulans*. Curr Genet 29:293–300, 1996.

46. CP Kurtzman. DNA–DNA hybridization approaches to species identification in small genome organisms. Meth Enz 224:335–348, 1993.

47. LC Mendonça-Hagler, LR Travassos, KO Lloyd, HJ Phaff. Deoxyribonucleic acid base composition and hybridization studies on the human pathogen *Sporothrix schenckii* and *Ceratocystis*. Infec Immun 9:934–938, 1974.

48. FD Davison, DWR Mackenzie. DNA homology studies in the taxonomy of dermatophytes. Sabouraudia 22:117–123, 1984.

49. E Guého, J Tredick, HJ Phaff. DNA base composition and DNA relatedness among species of *Trichosporon* Behrend. Antonie Van Leeuwenhoek 50:17–32, 1984.

50. M Masuda, W Naka, S Tajima, T Harada, T Nishikawa, L Kaufman, PG Standard. Deoxyribonucleic acid hybridization studies of *Exophiala dermatitidis* and *Exophiala jeanselmei*. Microbio Immun 33:631–639, 1989.

51. EA Winzeler, DR Richards, AR Conway, AL Goldstein, S Kalman, MJ McCullough, et al. Direct allelic variation scanning of the yeast genome. Science 281:1194–1197, 1998.

52. JGK Williams, AR Kubelik, KJ Livak, JA Rafalski, SV Tingey. DNA polymorphisms amplified by arbitrary primers are useful as genetic markers. Nucleic Acids Res 18:6531–6535, 1990.

53. J Welsh, M McClelland. Fingerprinting genomes using PCR with arbitrary primers. Nucleic Acids Res 18:7213–7218, 1990.

54. T Boekhout, A van Belkum, ACAP Leenders, HA Verbrugh, P Mukamurangwa, D Swinne, WA Scheffers. Molecular typing of *Cryptococcus neoformans*: Taxonomic and epidemiological aspects. Internat J Syst Bac 47:432–442, 1997.

55. P Mondon, MP Brenier, F Symoens, ER Rodriguez, E Coursange, F Chaib, et al. Molecular typing of *Aspergillus fumigatus* strains by sequence-specific DNA primer (SSDP) analysis. FEMS Immunol Med Microbio 17:95–102, 1997.

56. ME Brandt, LC Hutwagner, RJ Kuykendall, RW Pinner, and the Cryptococcal Disease Active Surveillance Group. Comparison of multilocus enzyme electrophoresis and random amplified polymorphic DNA analysis for molecular subtyping of *Cryptococcus neoformes*. J Clin Microbio 33:1890–1895, 1995.

57. AG Clark, CMS Lanigan. Prospects for estimating nucleotide divergence with RAPDs. Molec Bio Evol 10:1096–1111, 1993.

58. D Kersulyte, JP Woods, EJ Keath, WE Goldman, DE Berg. Diversity among clinical isolates of *Histoplasma capsulatum* detected by polymerase chain reaction with arbitrary primers. J Bacteriol 174:7075–7079, 1992.

59. PF Lehmann, D Lin, BA Lasker. Genotypic identification and characterization of species and strains within the genus *Candida* by using random amplified polymorphic DNA. J Clin Microbio 30:3249–3254, 1992.

60. A Bostock, MN Khattak, RC Matthews, JP Burnie. Comparison of PCR fingerprinting, by random amplification of polymorphic DNA, with other molecular typing methods for *Candida albicans*. J Gen Microbio 139:2179–2184, 1993.

61. KW Loudon, JP Burnie, AP Coke, RC Matthews. Application of polymerase chain

reaction to fingerprinting *Aspergillus fumigatus* by random amplification of polymorphic DNA. J Clin Microbio 31:1117–1121, 1993.

62. KE Yates-Siilata, DM Sander, EJ Keath. Genetic diversity in clinical isolates of the dimorphic fungus *Blastomyces dermatitidis* detected by a PCR-based random amplified polymorphic DNA assay. J Clin Microbio 33:2171–2175, 1995.

63. ME Brandt, LC Hutwagner, LA Klug, WS Baughman, D Rimland, EA Graviss, et al. Molecular subtype distribution of *Cryptococcus neoformans* in four areas of the United States. J Clin Microbio 34:912–917, 1996.

64. KV Clemons, F Feroze, K Holmberg, DA Stevens. Comparative analysis of genetic variability among *Candida albicans* isolates from different geographic locales by three genotypic methods. J Clin Microbio 35:1332–1336, 1997.

65. D Liu, S Coloe, R Baird, J Pedersen. PCR identification of *Trichophyton mentagrophytes* var. *interdigitale* and *T. mentagrophytes* var. *mentagrophytes* dermatophytes with a random primer. J Med Microbio 46:1043–1046, 1997.

66. JA Kim, K Takizawa, K Fukushima, K Nishimura, M Miyaji. Identification and genetic homogeneity of *Trichophyton tonsurans* isolated from several regions by random amplified polymorphic DNA. Mycopathology 145:1–6, 1999.

67. W Meyer, K Marszewska, M Amirmostofian, RP Igreja, C Hardtke, K Methling, et al. Molecular typing of global isolates of *Cryptococcus neoformans* var. *neoformans* by polymerase chain reaction fingerprinting and randomly amplified polymorphic DNA—A pilot study to standardize techniques on which to base a detailed epidemiological survey. Electrophoresis 20:1790–1799, 1999.

68. MM Lopes, G Freitas, P Boiron. Potential utility of random amplified polymorphic DNA (RAPD) and restriction endonuclease assay (REA) as typing systems for *Madurella mycetomatis*. Curr Microbio 40:1–5, 2000.

69. E Bart-Delabesse, J-F Humbert, E Delabesse, S Bretagne. Microsatellite markers for typing *Aspergillus fumigatus* isolates. J Clin Microbio 36:2413–2418, 1998.

70. D Field, L Eggert, D Metzgar, R Rose, C Wills. Use of polymorphic short and clustered coding-region microsatellites to distinguish strains of *Candida albicans*. FEMS Immunol Med Microbio 15:73–79, 1996.

71. W Meyer, TG Mitchell. PCR fingerprinting in fungi using single primers specific to minisatellites and simple repetitive DNA sequences: Strain variation in *Cryptococcus neoformans*. Electrophoresis 16:1648–1656, 1995.

72. W Meyer, TG Mitchell, EZ Freedman, RJ Vilgalys. Hybridization probes for conventional DNA fingerprinting used as single primers in the polymerase chain reaction to distinguish strains of *Cryptococcus neoformans*. J Clin Microbio 31:2274–2280, 1993.

73. Y Gräser, M el Fari, W Presber, W Sterry, H-J Tietz. Identification of common dermatophytes (*Trichophyton*, *Microsporum*, *Epidermophyton*) using polymerase chain reactions. Brit J Derm 138:576–582, 1998.

74. W Meyer, GN Latouche, H-M Daniel, M Thanos, TG Mitchell, D Yarrow, et al. Identification of pathogenic yeasts of the imperfect genus *Candida* by polymerase chain reaction fingerprinting. Electrophoresis 18:1548–1559, 1997.

75. M Thanos, G Schönian, W Meyer, C Schweynoch, Y Gräser, TG Mitchell, et al. Rapid identification of *Candida* species by DNA fingerprinting with PCR. J Clin Microbio 34:615–621, 1996.

76. J Xu, TG Mitchell, RJ Vilgalys. PCR-restriction fragment length polymorphism (RFLP) analyses reveal both extensive clonality and local genetic differences in *Candida albicans*. Molec Ecol 8:59–73, 1999.

77. M Orita, H Iwahana, H Kanazawa, K Hayashi, T Sekiya. Detection of polymorphisms of human DNA by gel electrophoresis as single-strand conformation polymorphisms. Proc Natl Acad Sci USA 86:2766–2770, 1989.

78. A Burt, DA Carter, GL Koenig, TJ White, JW Taylor. Molecular markers reveal cryptic sex in the human pathogen *Coccidioides immitis*. Proc Natl Acad Sci USA 93:770–773, 1996.

79. DA Carter, A Burt, JW Taylor, GL Koenig, BM Dechairo, TJ White. A set of electrophoretic molecular markers for strain typing and population genetic studies of *Histoplasma capsulatum*. Electrophoresis 18:1047–1053, 1997.

80. A Forche, G Schönian, Y Gräser, R Vilgalys, TG Mitchell. Genetic structure of typical and atypical populations of *Candida albicans* from Africa. Fung Gen Bio 28:107–125, 1999.

81. PM Hauser, P Francioli, J Bille, A Telenti, DS Blanc. Typing of *Pneumocystis carinii* f. sp. *hominis* by single-strand conformation polymorphism of four genomic regions. J Clin Microbio 35:3086–3091, 1997.

82. Y Gräser, M Volovsek, J Arrington, G Schönian, W Presber, TG Mitchell, RJ Vilgalys. Molecular markers reveal that population structure of the human pathogen *Candida albicans* exhibits both clonality and recombination. Proc Natl Acad Sci USA 93:12473–12477, 1996.

83. TJ Walsh, A Francesconi, M Kasai, SJ Chanock. PCR and single-strand conformational polymorphism for recognition of medically important opportunistic fungi. J Clin Microbio 33:3216–3220, 1995.

84. P Vos, R Hogers, M Bleeker, M Reijans, T van de Lee, M Hornes, et al. AFLP: A new technique for DNA fingerprinting. Nucleic Acids Res 23:4407–4414, 1995.

85. P Vos, M Kuiper. AFLP analysis. In: G Caetano-Anollës, PM Gresshoff, eds. DNA Markers: Protocols, Applications, and Overviews. New York: Wiley-Liss, 1997, pp. 115–132.

86. J Zhang, RJ Hollis, MA Pfaller. Variations in DNA subtype and antifungal susceptibility among clinical isolates of *Candida tropicalis*. Diag Microbio Infec Dis 27:63–67, 1997.

87. MA Pfaller, J Rhine-Chalberg, AL Barry, JH Rex, the NIAID Mycoses Study Group, and the Candidemia Study Group. Strain variation and antifungal susceptibility among bloodstream isolates of *Candida* species from 21 different medical institutions. Clin Infec Dis 21:1507–1509, 1995.

88. MR Elias Costa, S Carnovale, MS Relloso. Oropharyngeal candidosis in AIDS patients: An epidemiological study using restriction analysis of *Candida albicans* total genomic DNA. Mycoses 42:41–46, 1999.

89. MJ McCullough, KV Clemons, DA Stevens. Molecular epidemiology of the global and temporal diversity of *Candida albicans*. Clin Infec Dis 29:1220–1225, 1999.

90. R Khatib, MC Thirumoorthi, KM Riederer, L Sturm, LA Oney, J Baran Jr. Clustering of *Candida* infections in the neonatal intensive care unit: Concurrent emergence of multiple strains simulating intermittent outbreaks. Pediat Infec Dis J 17:130–134, 1998.

91. M Birch, MJ Anderson, DW Denning. Molecular typing of *Aspergillus* species. J Hosp Infec 30 (suppl.):339–351, 1995.

92. S Cresti, B Posteraro, M Sanguinetti, P Guglielmetti, GM Rossolini, G Morace, G Fadda. Molecular typing of *Candida* spp. by random amplification of polymorphic DNA and analysis of restriction fragment length polymorphism of ribosomal DNA repeats. New Microbio 22:41–52, 1999.

93. D Dlauchy, J Tornai-Lehoczki, G Peter. Restriction enzyme analysis of PCR amplified rDNA as a taxonomic tool in yeast identification. Syst Appl Microbio 22:445–453, 1999.

94. A Velegraki, ME Kambouris, G Skiniotis, M Savala, A Mitroussia-Ziouva, NJ Legakis. Identification of medically significant fungal genera by polymerase chain reaction followed by restriction enzyme analysis. FEMS Immun Med Microbio 23:303–312, 1999.

95. JE Cutler, PM Glee, HL Horn. *Candida albicans*- and *Candida stellatoidea*-specific DNA fragment. J Clin Microbio 26:1720–1724, 1988.

96. S Scherer, DA Stevens. A *Candida albicans* dispersed, repeated gene family and its epidemiologic applications. Proc Natl Acad Sci USA 85:1452–1456, 1988.

97. S-I Iwaguchi, M Homma, K Tanaka. Variation in the electrophoretic karyotype analysed by the assignment of DNA probes in *Candida albicans*. J Gen Microbio 136:2433–2442, 1990.

98. J Schmid, E Voss, DR Soll. Computer-assisted methods for assessing strain relatedness in *Candida albicans* by fingerprinting with the moderately repetitive sequence Ca3. J Clin Microbio 28:1236–1243, 1990.

99. I Oren, EK Manavathu, SA Lerner. Isolation and characterization of a species-specific DNA probe for *Candida albicans*. Nucleic Acids Res 19:7113–7116, 1991.

100. AR Holmes, YC Lee, RD Cannon, HF Jenkinson, MG Shepherd. Yeast-specific DNA probes and their application for the detection of *Candida albicans*. J Med Microbio 37:346–351, 1992.

101. BL Wickes, JB Hicks, WG Merz, KJ Kwon-Chung. The molecular analysis of synonymy among medically important yeasts within the genus *Candida*. J Gen Microbio 138:901–907, 1992.

102. HGM Niesters, WHF Goessens, JFGM Meis, WGV Quint. Rapid, polymerase chain reaction-based identification assays for *Candida* species. J Clin Microbio 31:904–910, 1993.

103. A Carlotti, A Couble, J Domingo, K Miroy, J Villard. Species-specific identification of *Candida krusei* by hybridization with the CkF1,2 DNA probe. J Clin Microbio 34:1726–1731, 1996.

104. SR Lockhart, S Joly, C Pujol, JD Sobel, MA Pfaller, DR Soll. Development and verification of fingerprinting probes for *Candida glabrata*. Microbio 143:3733–3746, 1997.

105. S Joly, C Pujol, M Rysz, K Vargas, DR Soll. Development and characterization of complex DNA fingerprinting probes for the infectious yeast *Candida dubliniensis*. J Clin Microbio 37:1035–1044, 1999.

106. DJ Sullivan, MC Henman, GP Moran, LC O'Neill, DE Bennett, DB Shanley, DC Coleman. Molecular genetic approaches to identification, epidemiology and taxonomy of non-albicans *Candida* species. J Med Microbio 44:399–408, 1996.

107. HA Fletcher, RC Barton, PE Verweij, EGV Evans. Detection of *Aspergillus fumigatus* PCR products by a microtitre plate based DNA hybridisation assay. J Clin Path 51:617–620, 1998.
108. J-P Debeaupuis, J Sarfati, V Chazalet, J-P Latgé. Genetic diversity among clinical and environmental isolates of *Aspergillus fumigatus*. Infec Immun 65:3080–3085, 1997.
109. H Girardin, J-P Latgé, T Srikantha, BJ Morrow, DR Soll. Development of DNA probes for fingerprinting *Aspergillus fumigatus*. J Clin Microbio 31:1547–1554, 1993.
110. ED Spitzer, SG Spitzer. Use of a dispersed repetitive DNA element to distinguish clinical isolates of *Cryptococcus neoformans*. J Clin Microbio 30:1094–1097, 1992.
111. A Varma, KJ Kwon-Chung. DNA probe for strain typing of *Cryptococcus neoformans*. J Clin Microbio 30:2960–2967, 1992.
112. BP Currie, LF Freundlich, A Casadevall. Restriction fragment length polymorphism analysis of *Cryptococcus neoformans* isolates from environmental (pigeon excreta) and clinical sources in New York City. J Clin Microbio 32:1188–1192, 1994.
113. A Varma, D Swinne, F Staib, JE Bennett, KJ Kwon-Chung. Diversity of DNA fingerprints in *Cryptococcus neoformans*. J Clin Microbio 33:1807–1814, 1995.
114. SP Franzot, IF Salkin, A Casadevall. *Cryptococcus neoformans* var. *grubii*: Separate varietal status for *Cryptococcus neoformans* serotype A isolates. J Clin Microbio 37:838–840, 1999.
115. Y Gräser, M el Fari, RJ Vilgalys, AFA Kuijpers, GS de Hoog, W Presber, H-J Tietz. Phylogeny and taxonomy of the family Arthrodermataceae (dermatophytes) using sequence analysis of the ribosomal ITS region. Med Mycol 37:105–114, 1999.
116. AR Bowen, JL Chen-Wu, M Momany, R Young, PJ Szaniszlo, PW Robbins. Classification of fungal chitin synthases. Proc Natl Acad Sci USA 89:519–523, 1992.
117. K Makimura, Y Tamura, M Kudo, K Uchida, H Saito, H Yamaguchi. Species identification and strain typing of *Malassezia* species stock strains and clinical isolates based on the DNA sequences of nuclear ribosomal internal transcribed spacer 1 regions. J Med Microbio 49:29–35, 2000.
118. GF Sanson, MR Briones. Typing of *Candida glabrata* in clinical isolates by comparative sequence analysis of the cytochrome c oxidase subunit 2 gene distinguishes two clusters of strains associated with geographical sequence polymorphisms. J Clin Microbio 38:227–235, 2000.
119. C Hennequin, E Abachin, F Symoens, V Lavarde, G Reboux, N Nolard, P Berche. Identification of *Fusarium* species involved in human infections by 28S rRNA gene sequencing. J Clin Microbio 37:3586–3589, 1999.
120. T Sugita, A Nishikawa, R Ikeda, T Shinoda. Identification of medically relevant *Trichosporon* species based on sequences of internal transcribed spacer regions and construction of a database for *Trichosporon* identification. J Clin Microbio 37:1985–1993, 1999.
121. P Valente, JP Ramos, O Leoncini. Sequencing as a tool in yeast molecular taxonomy. Can J Microbio 45:949–958, 1999.

122. E Guého, MC Leclerc, GS de Hoog, B Dupont. Molecular taxonomy and epidemiology of *Blastomyces* and *Histoplasma* species. Mycoses 40:69–81, 1997.

123. CP Kurtzman, CJ Robnett. Identification of clinically important ascomycetous yeasts based on nucleotide divergence in the 5′ end of the large-subunit (26S) ribosomal DNA gene. J Clin Microbio 35:1216–1223, 1997.

124. V Koufopanou, A Burt, JW Taylor. Concordance of gene genealogies reveals reproductive isolation in the pathogenic fungus *Coccidioides immitis*. Proc Natl Acad Sci USA 94:5478–5482, 1997.

125. J Xu, AR Ramos, RJ Vilgalys, TG Mitchell. Clonal and spontaneous origins of fluconazole resistance in *Candida albicans*. J Clin Microbio 38:1214–1220, 2000.

126. J-L López-Ribot, RK McAtee, S Perea, WR Kirkpatrick, MG Rinaldi, TF Patterson. Multiple resistant phenotypes of *Candida albicans* coexist during episodes of oropharyngeal candidiasis in human immunodeficiency virus-infected patients. Antimicrob Agents Chemo 43:1621–1630, 1999.

127. D Metzgar, A van Belkum, D Field, RH Haubrich, C Wills. Random amplification of polymorphic DNA and microsatellite genotyping of pre- and posttreatment isolates of *Candida* spp. from human immunodeficiency virus-infected patients on different fluconazole regimens. J Clin Microbio 36:2308–2313, 1998.

128. F Dromer, L Improvisi, B Dupont, M Eliaszewicz, G Pialoux, S Fournier, V Feuillie. Oral transmission of *Candida albicans* between partners in HIV-infected couples could contribute to dissemination of fluconazole-resistant isolates. AIDS 11:1095–1101, 1997.

13

Population Genetic Analyses of Medically Important Fungi

Jianping Xu
McMaster University, Hamilton, Ontario, Canada

Thomas G. Mitchell
Duke University Medical Center, Durham, North Carolina, U.S.A.

I. INTRODUCTION

In the last decade there has been a vast accumulation of information on the population structure and epidemiology of fungi pathogenic to human and other animals. This increase is due both to the rising incidence of mycotic diseases and to the increasing availability of strain-typing techniques. As reviewed in Chap. 12, there are numerous applications of these methods to analyzing the similarity among individual strains (1–10). Equally important are studies of the population structure of medically important fungi. Understanding the population dynamics of a pathogenic fungus will clarify epidemiological trends and assist researchers in the selection of appropriate strains in the quest for virulence factors and target molecules for novel antifungal drugs, vaccines, or diagnostic tests.

In this chapter we will review approaches based on population genetic analyses to understand the patterns and mechanisms of genetic variation. Rather than exhaustively reviewing the population genetic studies of all medically important fungi, in this chapter we will introduce basic concepts, issues, and rationales for population genetic studies of medically important fungi. We will also review the analytical methods, present examples of their application, and indicate their limitations in addressing population genetic questions.

II. WHAT IS A POPULATION?

There are several overlapping definitions of "population": a population can be (1) the organisms that inhabit a particular locality, (2) a group of interbreeding

organisms that represent the level of organization at which speciation begins, or
(3) a group of individual persons, objects, or items from which samples are taken
for statistical measurement. The first definition is based on geographic locations,
although the size and boundaries of individual populations can vary widely and
are often arbitrary. This definition fits most population genetic studies of fungi
and other organisms. The second definition is based on genetics and is the most
restrictive. By this definition, a population refers to groups of individuals that
are genetically isolated but still capable of interbreeding with individuals of other
populations of the same species. For many species, including human pathogenic
fungi, it is usually difficult to establish the precise mating boundaries for groups
of individuals. In species of plants and animals, both geographical and ecological
factors contribute to determining the population breeding boundaries (11). An
obvious problem with this definition is that many species of pathogenic fungi
have not been observed to undergo sexual mating and meiosis.

The third definition is the most versatile. It allows multidimensional analyses of the distribution of genetic variation within a species. This is an operational
definition of a population. For example, a population of *Candida albicans* can
be a collection of individual strains from a particular locality. The locality may
be defined as a geographical entity (e.g., a continent, country, state or province,
city, hospital, or ward) or ecological niche (e.g., isolates from patients with certain medical conditions, specific body sites, or different body sites of the same
host). A population of *C. albicans* isolates could be further circumscribed according to the vital statistics of their hosts (sex, age, race, and ethnic background).

The only limitation on the number of delineating criteria is that the sample
size of the smallest category must be sufficiently large for statistical analysis.
The sample size that is sufficient to detect differences among samples depends
on the patterns of genetic variation of the population. The smaller the actual
difference in allele frequencies, the larger the sample sizes required for their
reliable detection. For populations with small numbers of members, the sample
size N needed to detect a given level of differentiation at a diploid locus among
populations at least 50% of the time (i.e., a power of 0.5) with a statistical significance of 0.05 can be approximated as

$$N = \frac{1}{2}F_{ST} \quad \text{and} \quad F_{ST} = \frac{1}{2}N$$

where F_{ST} represents the proportion of the total genetic variation explained by
the difference between samples. For example, to detect an F_{ST} value of 0.05,
samples of just 10 diploids per locality may be adequate (or 20 strains per locality
for haploids) (12). A more detailed explanation and calculations are presented
in Sec. V. This definition of a population is commonly applied to study the epidemiology and prevention of infectious diseases and to identify risk factors among
humans and potential virulence factors in pathogens.

III. GENETIC MARKERS

A genetic marker can be defined as an allele. For the species under investigation, the types of markers amenable to population genetic-based analysis depend on the ploidy and the mating system. Genetic markers must be unambiguously interpretable as alleles of a specific locus, and all alleles at each locus must be scorable for all isolates. In haploid species, most markers can be easily analyzed by population genetic approaches. Since there is only one set of genetic material (1N) for each strain, all markers can be scored for each strain. However, in diploid species (2N), determining the number of loci and the number of alleles per locus for many fingerprinting-based markers can be tedious and is often prone to error. In a species with a tractable mating system, strains can be crossed and the meiotic progenies can be analyzed to detemine the number of loci and the number of alleles per locus for a marker system. With diploid species lacking in sexual mating and/or meiosis, the interpretation of markers is highly problematic. Generally speaking, codominant, single-copy genetic markers are best suited for allelic assignments in strains of diploid species (13). These markers include allozymes (see Chap. 12), single locus RFLP, and DNA sequence-based single-nucleotide polymorphisms.

Some markers are better at addressing certain questions. Selectively neutral, codominant markers are best suited to investigate recombination and gene flow between populations. Loci that are under selective pressure will have a higher probability of convergent and sometimes diversifying evolution than neutral loci. For example, genes involved in drug resistance and in response to host defenses are under constant selective pressure in clinical settings, and these genes may not be well suited for examining recombination and genetic differentiation in natural populations.

IV. ANALYSIS OF VARIATION WITHIN A POPULATION

A. Issues and Rationales

Many simple measures have been used to describe the genetic variation within a population. Among others, they include: (1) observed heterozygosity (11), (2) gene diversity (11), and (3) observed genotypic diversity (13,14). The observed heterozygosity Ho represents the percentage of observed heterozygotes at each locus, applicable only in diploids. Gene diversity is defined as the expected heterozygosity He as follows:

$$He = 1 - \Sigma p_i^2$$

where p_i is the frequency of the ith allele at a locus. Both the observed heterozygosity and the gene diversity (i.e., both Ho and He) are individual locus-based measures of genetic variation for population samples. The mean observed hetero-zygosity or gene diversity of a sample is estimated as the arithmetic mean of all loci sampled.

A common measure of multilocus population genetic variation is the observed genotypic diversity (G_0): $G_0 = 1/\sum p_i^2$, where p_i is the frequency of a particular multilocus genotype. G_0 ranges from 1 to N, where N is the sample size. This measure was originally proposed as an overall index of within-population genetic diversity (12). Another potentially useful measure of the distribution of genotypes is the probability that random pairs of isolates will have different multilocus genotypes. This probability is calculated as $1 - \sum p_i^2$, where p_i is the frequency of a particular multilocus genotype.

Examining the patterns of genetic variation in natural populations, a major question about within-population genetic variation in fungi (and other pathogenic microbes) is the contribution to this variation, if any, of recombination.

Since all microbes are known to reproduce asexually via mitosis, it is therefore expected that most microbial species would show some evidence of clonality in nature. One focus of fungal population genetic studies has been whether recombination plays any role in generating genetic variation in natural populations.

The amount of recombination that occurs in natural populations of microorganisms effects their evolution. Sexual reproduction and recombination pose an evolutionary paradox. Whereas an organism that reproduces asexually passes on all its genes to each of its progeny, one that reproduces sexually passes on only half of its genes to each progeny. Under uniform conditions, natural selection therefore favors the organism that reproduces asexually because with an equal number of progeny, the asexual individual has double the fitness of the sexually reproducing one.

Sexual reproduction affords microorganisms two possible advantages: panmixia and DNA repair (15, 16). The panmictic argument suggests that without the mixing of genes generated by sexual recombination, adaptive evolution is limited to the accumulation of favorable mutations that occur successively in each independently evolving lineage. Sexual recombination allows favorable mutations that arise in separate lineages to become combined in the same individual, providing an advantage in the adaptation to different environments. The repair argument points out that the two haplotypes associated wth diploidy during sex provide an error-correction mechanism for repairing genetic damage. Genetic damage can occur spontaneously and continuously during replication and possibly transcription. The intact DNA of one haplotype can serve as a template for correcting the damaged DNA in the other haplotype. Moreover, deleterious mutations in one haplotype can be overcome by compensatory dominant mutations in diploids. Whether one or both of these purported advantages of sexual recombiantion accounts for the origin and maintenance of sexual reproduction is subject to considerable debate and investigation (15).

A second reason to assess the role of recombination in natural populations relates to issues of sampling, research, and treatment strategies. If recombination occurs in natural populations, there may be significant ramifications for the evolution and dissemination of genes related to antibiotic resistance, pathogenicity, and host specificity. A recombinant population structure implies that the selected properties reflect distinct genes or nonrecombining genetic elements. (Discreet genes are usually assumed to be nonrecombining units, even though intragenic recombination has been demonstrated in all organisms that were critically examined.) Conversely, a clonal population structure implies that the selected units are clones or clonal lineages. To study medically relevant traits in a clonal population, representatives of every clonal lineage should be sampled. However, in a recombining population, it would be more profitable to focus primarily on individual genes.

To determine whether a microbial population structure is predominantly clonal or recombining, population microbiologists employ genetic tests. The next section will discuss the tests used to dissect the genetic structure within a population. These tests differ somewhat for haploid and diploid species.

Many of the methods used in the analysis of population genetic structure require rigorous statistical measurements. These statistical tests, such as those used to test null hypotheses, are selected to address specific questions and evaluated according to their appropriateness and accuracy. Statistical tests should be examined for their propensity to be erroneous. Statistical tests are prone to two distinct and opposing types of errors, known as type I or type II errors. Type I errors occur if a test is too rigorous for the data, perhaps by not accounting for errors due to random sampling, which results in rejection of the hypothesis, even when it is true. Conversely, with a type II error, the statistical test permits excessive leeway in the data, and the hypothesis will seldom be rejected, even when it is false (11). The balance between type I and type II errors is such that reducing the probability of one type will increase the probability of the other. For null hypotheses, the 5% level of statistical significance is applied, indicating there is a 95% probability that the hypothesis is true a 5% chance that a true hypothesis has been rejected (type I error). The chance of a type II error, failure to reject a false hypothesis, will vary.

B. Analysis of Clonality and Recombination in Haploids

Since a haploid genome contains only one set of chromosomes, each isolate has only one allele for each locus. Consequently, tests of recombination in natural haploid populations involve the examination of the allelic associations between different loci.

For the purposes of population genetics, a genetic locus may be defined as: (1) a single polymorphic nucleotide site, (2) a polymorphic restriction endonuclease recognition site (four to six base pairs), (3) a single nucleotide insertion or deletion, (4) any nucelotide sequence difference(s) within a region of continuous DNA of variable length, possibly constituting a distinct allele, or (5) an enzymatic staining profile. Because of the large variation in the size of the DNA segment or gene being recognized as a locus, analytical methods may be different. All analytical methods rely on two basic assumptions: (1) once a locus is defined, recombination within the locus is assumed to occur very rarely, therefore negligibly or not at all, and (2) each distinct allele is the result of a unique mutational event that occurred only once in the population. Within these assumptions, indistinguishable alleles are considered to be "identical by descent."

In tests for the occurrence of recombination in natural populations, there are two distinct questions. First, is the population panmictic? A panmictic population structure implies that alleles at all loci are randomly associated with each other. If the hypothesis of panmixia is statistically rejected and a clonal population structure is assumed, the second question arises: Can all or part of the genetic variation in natural populations be attributed to recombination? Since all medical fungi are capable of reproducing asexually through mitosis, a clonal component in populations is expected. A widely used approach is to analyze representatives of each different multilocus genotype to distinguish between the null hypothesis of recombination and the alternative hypothesis of clonality.

Often, a small number of genotypes are dominant or overrepresented within a population. It is common in these analyses to include the genotypes of all isolates in the population as well as only the unique genotypes within the population. This smaller group is the "clone-corrected" sample. The rejection of panmixia for the total sample but acceptance of panmixia for the clone-corrected sample is usually regarded as evidence for clonal expansion with evidence of random mating in the genetic structure. There are several caveats related to the truncation of a sample by clone correction before testing. Even though generally assumed, it is usually not confirmed that identical multilocus genotypes are actually clonal in origin. The justification for clonal-correlation of the sample should be based on biological and ecological considerations. The decrease in sample size may also decrease the power of the statistical test for rejection of the null hypothesis of recombination, thus increasing the possibility of type II error (11, 17).

Tests for whether or not the haploid population is panmictic involve comparing observed allelic associations with those derived under the null hypothesis of random mating. The two most widely used population genetic tests for haploid genomes are tests for allelic association (linkage equilibrium) between pairs of loci, and the overall index of association (I_A) involving alleles at all loci. A third test compares the observed genotypic diversity with that expected under the null hypothesis of random mating (14). A fourth test uses phylogenetic analysis of

DNA sequence-based characters (18,19). In this test, incongruences between different gene genealogies for a set of strains are considered to be consistent with hypothesis of recombination.

1. Linkage Disequilibrium

Linkage disequilibrium (gametic phase disequilibrium or gametic disequilibrium) is a measure of the association between alleles at pairs of loci (11). Random association between alleles at different loci in a population is considered to be an indication of recombination between these two loci. The test for random association can be described as follows. If alleles at two loci, A and B, segregate independently, then the expected frequency of the genotype AiBi is simply the product of the frequencies of the two alleles Ai and Bi. A chi-square test or Fisher's exact test can be performed to determine whether or not the observed genotypic counts are significantly different from the expected counts (11,17). If the observed and expected counts are not significantly different, then the population under study is assumed to have a recombining structure. Conversely, if the observed and expected counts are significantly different, the population is assumed to have a clonal structure as inferred by this pair of loci. With balanced gene frequencies, in a completely panmictic population less than 5% of locus pairs are expected to have genotypic counts significantly different from those expected.

A key factor in this test is the independence of the loci. If the paired loci are linked, they are not totally independent. The degree of linkage between loci affects the association of alleles at these loci. However, it has generally been assumed and has been mathematically proven that if the loci under study are selectively neutral, any positive and/or negative allelic association between loci will be broken down rapidly if sexual mating and meiosis are frequent (11). If an organism reproduces clonally (through either mitotic division or homothallic mating), the entire genome is effectively linked since there is no segregation and reassortment of alleles. Both linkage and clonal reproduction can cause deviations from the expected (random) genotypic frequencies for any pairs of loci. The *degree of deviations (D)* or nonrandom association between two loci, each with two alleles, A1 and A2, and B1 and B2, is computed as: $D = pA1B1pA2B2 - pA1B2pA2B1$, where pA1B1 is the observed frequency of genotype A1B1, and so on.

This approach has the drawback that many different tests are required and the results are not easily interpreted. If there are n polymorphic loci in a sample, the number of tests between possible pairs of loci are $n(n - 1)/2$. Even if a population is panmictic, some tests are likely to show significant deviation from random mating by chance. Conversely, even in a strictly clonal population, some tests might still indicate random association, especially when the sample size is small and allelic frequencies are skewed. To avoid some of the problems in the test for linkage disequilibrium, an index was introduced to measure the overall allelic association in a sample, described below.

2. Index of Association (I_A)

To assess overall allelic associations in haploid organisms, an index of association I_A has been used widely in fungal population genetic analysis. This index was first used by Brown et al. (20) to measure population structure in *Hordeum spontaneum*, and it was used later to demonstrate nonrandom association of alleles in *Escherichia coli* (21) and other bacteria (22). I_A is a generalized measure of linkage disequilibrium and has an expected value of zero if there is no association between loci. The calculation of I_A is derived as follows:

> Suppose that M loci have been analyzed in N individuals. Let p_{ij} represent the frequency of the ith alleles at the jth locus. Then the gene diversity h_j $= 1 - \sum p_{ij}^2$, which is the probability that two individuals are different at the jth locus. Let K represent the number of loci having different alleles between two different individuals. The observed variance of K, or Vo, can be calculated based on the distributions of K. Since there are $N(N - 1)/2$ possible pairs of individuals, the mean difference between any two individuals, K', is $\sum h_j$. The expected variance of K is $Ve = \sum h_j(1 - h_j)$. The index of association is $I_A = Vo/Ve - 1$.

There are two ways to test whether or not I_A is significantly different from zero—that is, the null hypothesis of random association of alleles at different loci. The first test assumes that the sampling distribution of the error variance of I_A, which is calculated as

$$Var(Ve) = [\sum h_i - 7\sum h_i^2 + 12\sum h_i^3 - 6\sum h_i^4 + 2(\sum h_i - h_i^2)^2]/N$$

approximates normality, then the upper 95% confidence limit for Var(Ve) is

$$L = \sum h_j - \sum h_j^2 + 2[Var(Ve)]^{1/2}$$

If Vo does not exceed L, then the null hypothesis of random association of alleles at all loci is not rejected. The population under study is therefore concluded to have a population structure not significantly different from panmixia. If Vo is greater than L, the population is assumed to have a significant clonal reproduction component.

    The second statistical test of I_A is to use a randomization approach in which the null distribution of Vo is generated by randomly permuting the alleles among all individuals within each locus and calculating Vo many times. The Vo from the observed sample is then compared to the null distribution from permuted samples to determine whether or not there is a significant difference. If the observed variance Vo is $>(1 - \alpha)$ of the Vo from randomized data sets, where α is the acceptable type I error rate (0.05 or 0.01), then the sample deviates significantly from random mating. Since the sampling distribution of Vo is not known to be normal, randomization tests for significance are preferable, especially when sample sizes are relatively small.

3. Phylogenetic Methods

As shown above, tests for both linkage disequilibrium and asosciation index were against the null nypothesis of random mating. However, testing against a null hypothesis of random mating can create a type II error, the probability of accepting a false hypothesis, especially when the sample size is small and the allele frequencies are skewed. Furthermore, it is often difficult to determine whether or not recombination contributes at all to variation in a population determined to have a predominantly clonal structure.

To overcome these problems associated with testing against the null hypothesis of panmixia, phylogenetic analysis offers tests against the alternative hypothesis of strict clonality. There are two types of phylogenetic tests for clonality and recombination. The first uses population genetic data of allelic information at individual loci for each strain. The premise is that if populations are clonal, the ancestry of isolates will be represented by a phylogenetic tree of good fit. However, for recombining populations there will be little or no phylogenetic consistency because different loci reflect different patterns of descent among individuals. Furthermore, the length of the tree can be compared with the minimum length expected if the population were strictly clonal (2).

The second phylogenetic test uses gene sequences from several genes. By comparing phylogenetic trees built for different genes, clonality can be distinguished from recombination. There is strong evidence for clonality if the trees for different genes are concordant, but if the trees are in conflict, recombination is a likely explanation. The partition homogeneity test (PHT) was developed to assess gene genealogy congruences (23). For congruent gene trees, the sum of the lengths of the most parsimonious tree for each gene should not change significantly if the polymorphic nucleotides in each gene are swapped among genes. In contrast, for incongruent gene trees, the sum of the gene trees for the observed data should be shorter than the sum for gene trees made after polymorphic nucleotides have been swapped among genes. This is true because recombination is assumed to be correlated with linkage relationships in the genome. Intragenic recombination, if it exists at all, would be much rarer than intergenic recombination. The statistical significance of this test is established by making many resampled data sets and comparing the observed sum of length to the distribution of 1000 or more resampled data sets.

4. Examples

Although most human fungal pathogenic species are haploid, few have been critically examined for the patterns of genetic variation in natural populations. Four species of medically important fungi have been examined for the roles of clonality and recombination in natural populations—*Coccidioides immitis, Histoplasma capsulatum, Cryptococcus neoformans,* and *Aspergillus flavus.*

Three populations of *C. immitis* were analyzed with the same set of genetic markers by Burt et al. (2,24). All the populations—one from Tucson, Arizona, one from Bakersfield, California, and the third from San Antonio, Texas—had genetic structures not significantly different from panmixia. The allelic associations between many pairs of loci were in linkage equilibrium (24; J. Taylor, personal communication). The overall association index I_A for the Tucson population was 0.041, not significantly different from zero (2).

Thirty isolates of *H. capsulatum* from Indianapolis, Indiana, were examined for associations of alleles at 11 different loci (25). Every isolate had a unique multilocus genotype. Alleles at many pairs of loci were not significantly associated with each other, and with an I_A close to zero (0.0428), this sample of *H. capsulatum* was concluded to have a recombining population structure (25).

Cryptococcus neoformans is somewhat different from the above two species. The current taxonomy classifies this biological species into two varieties, *C. neoformans* var. *neoformans* and *C. neoformans* var. *gattii.* Each variety has two predominant serotypes, serotypes A, D (or AD) in *C. neoformans* var. *neoformans* and serotypes B and C in *C. neoformans* var. *gattii.* Despite cross-hybridization among all serotypes in the laboratory (26,27), various strain-typing studies have revealed clear differences between the two varieties (1,28–33). Strains within each variety can also be readily sorted according to their serotypes (28,30,32,34). It has been suggested that different serotypes might correspond to different cryptic species reproductively isolated for a significant amount of time (35). When populations of *C. neoformans* were analyzed through multilocus enzyme electrophoresis, abundant evidence supported a predominantly clonal genetic structure (28). Significant clonal components were still observed when only representatives of unique multilocus genotypes were analyzed separately for individual serotypes (36). While the null hypothesis of panmixia is rejected for isolates of *C. neoformans* from natural and clinical sources, it is not known whether sexual recombination contributed to the patterns of genetic variation in this species. As with *C. neoformans,* the genetic structure of *Aspergillus fumigatus* is clonal (37).

The use of phylogenetic methods for detecting recombination in medical fungi was best illustrated with *Aspergillus flavus.* Analyzing the sequences of five genes, Geiser et al. (38) demonstrated recombination in one of the two genetically isolated groups of *A. flavus*. The PHT was highly significant ($p < 0.0001$) for incongruence among five gene genealogies, consistent with recombination (38).

C. Analysis of Clonality and Recombination in Diploid Species

1. Hardy–Weinberg Equilibrium Test (Association of Alleles Within a Locus)

Since each diploid strain has two alleles at each locus, tests for recombination in diploid organisms are somewhat different from haploid species. In diploid species, the association of alleles within a locus is quite often used as a measure of recombination. A chi-square goodness of fit test can be performed to compare the observed and expected genotypic counts at each locus and summed across loci (11,39). This computation is the test for Hardy–Weinberg equilibrium (HWE). To illustrate, for a single locus A with two alleles, A1 and A2, there are three possible genotypes at this locus, A1A1, A1A2, and A2A2. The expected Hardy–Weinberg frequencies of genotypes A1A1, A1A2, and A2A2 are p^2, $2pq$, and q^2, respectively, where p is the frequency of allele A1 and q is the frequency of allele A2, and $A1 + A2 = 1$. If observed counts of genotypes are not significantly different from expected counts, then the population is in HWE and assumed to have a recombining structure. However, Hardy–Weinberg frequencies are valid only when populations comply with several common but crucial assumptions: (1) the population size is large, (2) there is no selective pressure on the markers being analyzed, (3) migration and mutation are negligible, and (4) the population undergoes random mating (11).

It is imprudent to infer population structure from HWE tests alone. First, violation of any of the first three assumptions above may cause significant deviation from the expected frequencies, even if the population is randomly mating. Second, failure to reject the null hypothesis of HWE does not guarantee that the population is indeed randomly mating (type II error). Third, different populations of the same species may have different population structures (24).

2. Composite Genotypic Disequilibrium

The "exact test" for allelic association among loci was developed by Zaykin et al. (40). In this test, the probability of the set of multilocus genotypes in a sample, based on allelic counts, is calculated from the multinomial theory under the hypothesis of no association. Alleles are then permuted and the conditional probability is calculated for the permuted genotypic array. The proportion of arrays no more probable than the original sample provides the significance level of the test. The exact test is not restricted by the number of loci. It also allows the calculation of the probability of multilocus genotypes conditioned on genotypic counts at individual loci. Separating the allelic association test *within* a locus from genotypic association *among* loci can be very useful for diploid species, such as *Candida albicans*, because deleterious recessive mutations may be common and many loci with genotypic counts deviated significantly from Hardy–Weinberg expectations (11).

3. Examples

Pujol et al. (41) used 21 isozyme loci to characterize the genotypes of 55 strains of *C. albicans* isolated from patients infected with the human immunodeficiency virus (HIV). Thirteen of the 21 enzymatic loci were polymorphic among these strains. Six of the 13 polymorphic loci deviated significantly from HWE. The expected counts for some of the multilocus genotypes were also much lower than observed counts. They concluded that *C. albicans* has a clonal population structure (41).

Gräser et al. (42) used 12 nucleotide-based markers to characterize a mixed sample of *C. albicans* isolated from Durham, North Carolina. Among the 52 strains analyzed, 27 unique multilocus genotypes were detected. Similar to the findings of Pujol et al. (41), about half of the markers showed significant deviation from HWE (42). However, when they calculated the associations of alleles at different loci, >70% of the pairwise loci comparisons were not significantly different from random association. They concluded that the population structure of *C. albicans* included both clonal and recombinant components. Similar patterns have been confirmed among populations of *C. albicans* from different groups of hosts, including HIV-infected patients, non-HIV patients, and healthy persons (43).

D. Epidemic Clones: Calculations and Implications

Analyzing the genetic structure of populations to determine the extent of clonality *versus* recombination is only the first step, albeit a critical one, toward understanding population history and the evolution of pathogenicity. The subsequent questions concern the reason(s) for a particular genetic structure for certain populations. Approaches to address this question are beginning to emerge, but definitive answers are lacking. Many factors can contribute to the genetic structure of a population, including the history of the population, the mutation rate, the mutation spectrum (i.e., the distribution of mutations that are lethal, deleterious, neutral, and advantageous), genetic drift, population size, mating system (or lack of), and the selection pressure from the environment. It is often difficult to examine at the same time the relative contributions of these factors to the genetic structure of natural populations.

One approach is to investigate whether and why certain genotypes are more prevalent than expected. This information can be obtained by inspecting the relative frequencies of different multilocus genotypes or by calculating the expected frequencies of individual genotypes based on allelic frequencies and under the hypothesis of random mating (i.e., random associations of alleles within and among loci). When predominant genotypes are identified, the medically important traits of those isolates can be compared with less common genotypes in the population to explore possible associations between the genotypes and relevant phenotypes.

In a recent analysis of different samples of *C. albicans* by codominant, single-copy PCR-RFLP markers, one multilocus genotype predominated in all four samples (43,44). One sample was obtained from HIV-infected patients in Vitória, Brazil, and three from the United States (Durham, North Carolina). The predominant genotype was much more prevalent in all four samples than would be expected under random mating. This genotype might be selectively more advantageous than other genotypes of *C. albicans* in human populations. The pathogenicity, transmission, and antifungal drug susceptibility profiles of isolates with this genotype warrant further investigation.

V. ANALYSIS OF GENETIC VARIATION BETWEEN POPULATIONS

A. Issues and Analytical Methods

Depending on the questions of interest, there are multiple ways to compare genetic variations among populations. The first general approach could be to compare the genetic parameters calculated for each population. These parameters include gene diversity, heterozygosity, genotypic diversity, relative percentage of polymorphic loci, and number of alleles per locus. The extent of recombination and clonality might differ among natural populations of the same species because population histories, mutation rates, environmental conditions, and selection pressure can vary among populations. Any associations between specific environmental factors and genetic elements can be determined through general statistical analysis.

The second approach is the more traditional population genetic analysis. Calculations can determine the extent of population subdivision, genetic differentiation, and gene flow among populations based on differences in the gene and genotype frequency and measures of inbreeding coefficients.

Population subdivision involves an inbreedinglike effect within a population and a lack of interbreeding among strains between populations (11, 39). This effect can be assessed from the decrease in the proportion of heterozygous genotypes. A subdivided population of a diploid organism has three distinct levels of complexity: individual organisms I, subpopulations S, and the total population T. Let

H_I = the observed heterozygosity of an individual averaged over subpopulations

Hs = the expected heterozygosity of an individual in an equivalent random mating subpopulation

Hsa = the average of Hs taken over all subpopulations

H_T = the expected heterozygosity of an individual in an equivalent random mating total population

The effects of population subdivision are measured by a quantity called the fixation index (symbolized Fs_T), which is the reduction in heterozygosity of a subpopulation due to random genetic drift (45).

$$Fs_T = (H_T - Hsa)/H_T$$

Fs_T is always ≥ 0, because the Wahlund effect (11) assures that $H_T > Hsa$. If all subpopulations are in HWE with the same allele frequencies, then $Fs_T = 0$.

In haploid organisms, all the calculations for Fs_T are similar to those for diploid organisms. However, the symbols have different meanings. The expected heterozygosity Hs of an individual in subpopulations of diploids is the same as the gene diversity averaged among subpopulations in haploids, and the H_T of an individual in the total population of diploids is the same as gene diversity in the total population of haploids.

The calculations of gene flow based on Fs_T are different for haploid and diploid organisms because each migrant individual carries two alleles per locus in diploid species but only one allele per locus in haploids. The number of migrants per generation Nm equals $(1 - Fs_T)/4 Fs_T$ for diploids (46) but $(1 - Fs_T)/2 Fs_T$ for haploids. N is the number of individuals in a subpopulation, and m is the migration rate between pairs of populations. The estimation of gene flow based on Fs_T is conditional on many assumptions (46). These assumptions include low mutation rates, stable environments, no selection on the marker(s) under investigation, large population sizes, and genetic equilibrium within each population. If these assumptions are violated, Nm cannot be used as a parameter to estimate the gene flow between populations.

The statistical test of whether or not pairs of populations are significantly different is usually achieved by a direct comparison of gene frequencies (47). This is done by using the chi-square contingency table tests. Alternatively, Fisher's exact test is used when the sample sizes are small or when the lowest expected allele count is smaller than five. These tests are described in general biostatistical and epidemiological texts (e.g., 17).

Population genetic relationships are also frequently described in un-weighted pair-group method using arithmetic means (UPGMA) phenograms (11). These phenograms are based on population genetic distances. There are several measures of genetic distances, all based on population gene frequencies and inbreeding coefficients. The most widely used is Nei's genetic distance D (48). The basis of the measure is a so-called normalized identity, which expresses the probability that a randomly chosen allele from each of two different populations will be identical, relative to the probability that the two randomly chosen alleles from the same population will be identical. ("Identical" in this context means indistinguishable, not "identical by descent.") The calculations can be illustrated as follows:

> Suppose there are two hypothetical populations, A and B. At locus X, the frequency of allele I in population A is p_i and in poulation B is q_i. J_{AA} is defined as the probability that two alleles chosen at random from population A are identical, then,
>
> $$J_{AA} = \Sigma p_i^2$$
>
> This is because with random mating, J_{AA} equals the homozygosity in population A. Likewise, J_{BB} is defined as the probability that the two alleles chosen at random from population B are identical.
>
> $$J_{BB} = \Sigma p_i^2$$
>
> J_{AB} is the probability that two alleles are identical when one allele is chosen from population A and the other is chosen from population B, or
>
> $$J_{AB} = \Sigma p_i q_i$$
>
> Nei defines the normalized identity, I, for this gene as
>
> $$I = J_{AB}/(J_{AA}J_{BB})^{1/2}$$
>
> and the standard genetic distance, D, as
>
> $$D = -\ln(I)$$

D varies from 0 to ∞. This measure of genetic distance is more reliable when data from many genes are available, because with many genes the quantities J_{AA}, J_{BB}, and J_{AB}, are the arithmetic means of the individual values. Using the average of J_{AA}, J_{BB}, and J_{AB}, the same formulas for I and D apply. If two populations are identical, then $J_{AA} = J_{BB} = J_{AB}$, and the normalized identity is 1. Lack of genetic divergence between populations results in a distance $D = 0$. Since the statistic is calculated from samples from populations, the exact value of D between populations can vary from one sample to another. The degree to which different samples produce different D values depends on the amount of genetic divergence, the genes assayed, and the size of the samples.

After pairwise population genetic distances are calculated, population similarities can be described on a UPGMA phenogram (11,17). Genetic distances
are often compared to geographic distances. Assuming roughly constant rate of
migration, one might expect that pairs of populations that are situated farther apart
geographically would show greater genetic distance. This principle is known as
isolation by distance (49).

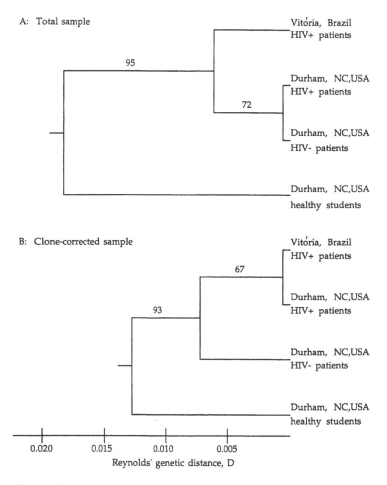

Figure 1 UPGMA phenograms describing genetic similarity among four samples of
Candida albicans (44). The distance measure is based on Reynolds' genetic distance (39).
These two phenograms are based on two different sample types: (A) the total samples;
(B), the clone-corrected samples in which only one representative of each multilocus geno-
type from each population is used for analysis. Bootstrap values from 1000 replicates are
given.

B. Example

At least two studies have estimated the genetic differentiation and gene flow between geographic populations of human pathogenic fungi. In the analysis of three geographic samples of the haploid *Coccidioides immitis* from Bakersfield, Tucson, and San Antonio, the overall F_{ST} between pairs of populations ranged from 0.198 to 0.944 (24). The overall allele frequencies were significantly different between all pairs of populations. Taken together, the results suggest significant genetic differentiation between geographic populations of *C. immitis* (24).

Conversely, the geographically separated populations of the diploid *Candida albicans* from HIV-infected patients in Vitória, Brazil, and Durham, North Carolina, were genetically indistinguishable (44). The overall F_{ST} estimated from 16 loci between these two samples were 0.017 and 0.002 for the total samples and the clone-corrected samples, respectively. The number of migrants estimated from F_{ST} values ranges from 6.89 to ∞ per generation between the two geographic populations. Both samples shared the same most common multilocus genotype for the 16 loci examined (44). Furthermore, these two samples were genetically more similar to each other than either was to the sample from healthy persons in Durham. (See Fig. 1.)

VI. CONCLUSIONS

Although the study of population genetic structure of medically important fungi is still embryonic, clear trends are emerging. Different species and different populations of the same species may have very different genetic structures. Understanding the evolutionary and ecological bases for these differences should contribute to the understanding of fungal pathogenicity and epidemiology. This knowledge can be expected to influence the design of effective strategies for the treatment and prevention of fungal infections.

ACKNOWLEDGMENTS

Support from Public Health Service grants AI 25783, AI 28836, and AI 44975 is greatly appreciated. The authors are members of the Duke University Mycology Research Unit.

REFERENCES

1. ME Brandt, LC Hutwagner, RJ Kuykendall, RW Pinner, and the Cryptococcal Disease Active Surveillance Group. Comparison of multilocus enzyme electrophoresis

and random amplified polymorpyhic DNA analysis for molecular subtyping of *Cryptococcus neoformans*. J Clin Microbio 33:1890–1895, 1995.

2. A Burt, DA Carter, GL Koenig, TJ White, JW Taylor. Molecular markers reveal cryptic sex in the human pathogen *Coccidioides immitis*. Proc Natl Acad Sci USA 93:770–773, 1996.

3. DA Carter, A Burt, JW Taylor, GL Koenig, BM Dechairo, TJ White. A set of electrophoretic molecular markers for strain typing and population genetic studies of *Histoplasma capsulatum*. Electrophoresis 18:1047–1053, 1997.

4. FY Chen, BP Currie, L-C Chen, SG Spitzer, ED Spitzer, A Casadevall. Genetic relatedness of *Cryptococcus neoformans* clinical isolates grouped with the repetitive DNA probe CNRE-1. J Clin Microbio 33:2818–2822, 1995.

5. KV Clemons, F Feroze, K Holmberg, DA Stevens. Comparative analysis of genetic variability among *Candida albicans* isolates from different geographic locales by three genotypic methods. J Clin Microbio 35:1332–1336, 1997.

6. TM Díaz-Guerra, JV Martínez-Suárez, F Labuna, JL Rodríguez-Tudela. Comparison of four molecular typing methods for evaluating genetic diversity among *Candida albicans* isolates from human immunodeficiency virus-positive patients with oral candidiasis. J Clin Microbio 35:856–861, 1997.

7. E Guého, MC Leclerc, GS de Hoog, B Dupont. Molecular taxonomy and epidemiology of *Blastomyces* and *Histoplasma* species. Mycoses 40:69–81, 1997.

8. PR Hunter. A critical review of typing methods for *Candida albicans* and their applications. CRC Crit Rev Microbio 17:417–434, 1991.

9. BB Magee, PT Magee. Electrophoretic karyotypes and chromosome numbers in *Candida* species. J Gen Microbio 133:425–430, 1987.

10. DA Stevens, FC Odds, S Scherer. Application of DNA typing methods to *Candida albicans* epidemiology and correlations with phenotype. Rev Infec Dis 12:258–266, 1990.

11. DL Hartl, AG Clark. Principles of Population Genetics. 2nd ed. Sunderland, MA: Sinauer, 1989.

12. R Chakraborty, O Leimar. Genetic variation within a subdivided population. In: N Ryman, F Utter, eds. Population Genetics and Fisheries Management. Seattle: University of Washington Press, 1987, pp. 90–120.

13. JA Stoddart. A genotypic diversity measure. J Hered 74:489–490, 1983.

14. JA Stoddart, JF Taylor. Genotypic diversity: Estimation and prediction in samples. Genetics 118:705–711, 1988.

15. RE Michod, BR Levin. The Evolution of Sex: An Examination of Current Ideas. Sunderland, MA: Sinauer, 1987.

16. JM Maynard Smith. The Evolution of Sex. Cambridge: Cambridge University Press, 1978.

17. RR Sokal, FJ Rohlf. Biometry. 2nd ed. New York: Freeman, 1988.

18. DE Dykhuizen, L Green. Recombination in *Escherichia coli* and the definition of biological species. J Bacteriol 173:7257–7268, 1991.

19. JW Archie. A randomization test for phylogenetic information in systematic data. Systemat Zool 38:239–252, 1989.

20. AHD Brown, MW Feldman, E Nevo. Multilocus structure of natural populations of *Hordeum spontaneum*. Genetics 96:523–536, 1980.

21. TS Whittam, H Ochman, RK Selander. Multilocus genetic structure in natural populations of *Escherichia coli*. Proc Natl Acad Sci USA 80:1751–1755, 1983.

22. JM Maynard Smith, NH Smith, M O'Rourke, BG Spratt. How clonal are bacteria? Proc Natl Acad Sci USA 90:4384–4388, 1993.

23. JP Huelsenbeck. Combining data in phylogenetic analysis. TREE 11:152–158, 1996.

24. A Burt, BM Dechairo, GL Koenig, DA Carter, TJ White, JW Taylor. Molecular markers reveal differentiation among isolates of *Coccidioides immitis* from California, Arizona and Texas. Molec Ecol 6:781–786, 1997.

25. DA Carter, A Burt, JW Taylor, GL Koenig, JT White. Clinical isolates of *Histoplasma capsulatum* from Indianapolis, Indiana, have a recombining population structure. J Clin Microbio 34:2577–2584, 1996.

26. KJ Kwon-Chung. A new genus, *Filobasidiella*, the perfect state of *Cryptococcus neoformans*. Mycologia 67:1197–1200, 1975.

27. KA Schmeding, S-C Jong, R Hugh. Sexual compatibility between serotypes of *Filobasidiella neoformans* (*Cryptococcus neoformans*). Curr Microbio 5:133–138, 1981.

28. ME Brandt, SL Bragg, RW Pinner. Multilocus enzyme typing of *Cryptococcus neoformans*. J Clin Microbio 31:2819–2823, 1993.

29. ME Brandt, LC Hutwagner, LA Klug, WS Baughman, D Rimland, EA Graviss, RJ Hammill, C Thomas, PG Pappas, AL Reingold, RW Pinner, and the Cryptococcal Disease Active Surveillance Group. Molecular subtype distribution of *Cryptococcus neoformans* in four areas of the United States. J Clin Microbio 34:912–917, 1996.

30. T Boekhout, A van Belkum, ACAP Leenders, HA Verbrugh, P Mukamurangwa, D Swinne, WA Scheffers. Molecular typing of *Cryptococcus neoformans:* Taxonomic and epidemiological aspects. Internat J Syst Bact 47:432–442, 1997.

31. JR Perfect, N Ketabchi, GM Cox, CW Ingram, C Beiser. Karyotyping of *Cryptococcus neoformans* as an epidemiological tool. J Clin Microbio 31:3305–3309, 1993.

32. BL Wickes, TDE Moore, KJ Kwon-Chung. Comparison of the electrophoretic karyotypes and chromosomal location of ten genes in the two varieties of *Cryptococcus neoformans*. Microbio 140:543–550, 1994.

33. W Meyer, TG Mitchell. PCR fingerprinting in fungi using single primers specific to minisatellites and simple repetitive DNA sequences: Strain variation in *Cryptococcus neoformans*. Electrophoresis 16:1648–1656, 1995.

34. SP Franzot, JS Hamdan, BP Currie, A Casadevall. Molecular epidemiology of *Cryptococcus neoformans* in Brazil and the United States: Evidence for both local genetic differences and a global clonal population structure. J Clin Microbiol 35:2243–2251, 1997.

35. SP Franzot, BC Fries, W Cleare, A Casadevall. Genetic relationship between *Cryptococcus neoformans* var. *neoformans* strains of serotypes A and D. J Clin Microbio 36:2200–2204, 1998.

36. JW Taylor, DM Geiser, A Burt, V Koufopanou. The evolutionary biology and population genetics underlying fungal strain typing. Clin Microbio Rev 12:126–146, 1999.

37. J-P Debeaupuis, J Sarfati, V Chazalet, J-P Latgé. Genetic diversity among clinical and environmental isolates of *Aspergillus fumigatus*. Infec Immun 65:3080–3085, 1997.

38. DM Geiser, JI Pitt, JW Taylor. Cryptic speciation and recombination in the alfatoxin-producing fungus *Aspergillus flavus*. Proc Natl Acad Sci USA 95:388–393, 1998.
39. BS Weir. Genetic Data Analysis. 2nd ed. Sunderland, MA: Sinauer, 1996.
40. D Zaykin, L Zhivotovsky, BS Weir. Exact tests for association between alleles at arbitrary numbers of loci. Genetica 96:169–178, 1995.
41. C Pujol, J Reynes, F Renaud, M Raymond, M Tibayrenc, FJ Ayala, F Janbon, M Mallié, J-M Bastide. The yeast *Candida albicans* has a clonal mode of reproduction in a population of infected human immunodeficiency virus-positive patients. Proc Natl Acad Sci USA 90:9456–9459, 1993.
42. Y Gräser, M Volovsek, J Arrington, G Schönian, W Presber, TG Mitchell, R Vilgalys. Molecular markers reveal that population structure of the human pathogen *Candida albicans* exhibits both clonality and recombination. Proc Natl Acad Sci USA 93:12473–12477, 1996.
43. J Xu, TG Mitchell, RJ Vilgalys. PCR-restriction fragment length polymorphism (RFLP) analyses reveal both extensive clonality and local genetic differences in *Candida albicans*. Molec Ecol 8:59–73, 1999.
44. J Xu, RJ Vilgalys, TG Mitchell. Lack of genetic differentiation between two geographically diverse samples of *Candida albicans* isolated from patients infected with human immunodeficiency virus. J Bacteriol 181:1369–1373, 1999.
45. BS Weir, CC Cockerham. Estimating F-statistics for the analysis of population structure. Evolution 38:1358–1370, 1984.
46. CC Cockerham, BS Weir. Estimation of gene flow from F-statistics. Evolution 47:855–863, 1993.
47. RR Hudson, DD Boos, NL Kaplan. A statistical test for detecting geographic subdivision. Molec Bio Evol 9:138–151, 1992.
48. M Nei. Molecular Population Genetics and Evolution. New York: American Elsevier, 1975.
49. S Wright. Isolation by distance. Genetics 28:114–138, 1943.

14

Genetic Instability of
Candida albicans

Elena Rustchenko and Fred Sherman
University of Rochester Medical School, Rochester, New York, U.S.A.

I. INTRODUCTION

Candida albicans is a polymorphic, commensal, and opportunistic pathogenic fungus, often incorrectly referred to as "dimorphic," with a remarkable degree of both genetic and phenotypic variability. Under normal circumstances *Candida* survives on the mucous surfaces of a host in different niches, which include the gastrointestinal tract as well as the genitalia of mammals. The mucus is believed to be a source of a variety of aminosugars, which can be utilized as energy sources (1). As at least eight different mucinlike genes are differently expressed in different body niches (2), it is reasonable to suggest that the content of mucus, as well as the content of body fluids, vary between different regions of the body, between different individuals within the same species depending on the individual's metabolism and diet, and between different mammalian species. In this respect, Singh and Datta (3) demonstrated that, in contrast to *C. albicans*, the nonpathogenic species of the genus *Saccharomyces* cannot utilize the aminosugar *N*-acetylglucosamine, which suggests that the aminosugar metabolic pathway is important in pathogenesis or at least in survival in a mammalian host. Because of the above-mentioned factors, environmental diversity requires phenotypic plasticity in the form of tolerance to different pH, food supply, and coexistence with other natural microflora. One example of the means of *C. albicans* to cope with a diversity of niches is the induced expression of catabolic enzymes on the pathway of *N*-acetylglucosamine (4). Another example is transcriptional induction of *PHR1* and *PHR2* genes, which respond to certain ranges of pH (5–7). Still other potential mechanisms involve genes, which sense the environment, and which are also implicated in morphogenesis. (See, e.g., Refs. 8–25.)

The subject of this chapter will be genetic instability based on mutational events, in contrast to both the transcriptional induction or repression implicated in the above-mentioned forms of regulation and the control of such transitory cellular stages as budding, pseudohyphae, hyphae, chlamydospores, and germ tubes.

The sites of occurrence and mechanisms of genetic instability in the host are still not understood, although they obviously play the major role in the formation of new phenotypes. As emphasized in this review, chromosomal alteration is the main cause of this instability, and we have considered two major aspects of this phenomenon. One is chromosomal instability, which occurs spontaneously (i.e., without an obvious selective condition), resulting in preadaptation of a portion of the population (26). Another aspect is chromosomal instability occurring in response to the changing environment (27). So far, neither of these mechanisms has been addressed in animal models.

The mechanism of *C. albicans* pathogenicity is currently a popular subject of study. As summarized by Odds (28, 29), the traditional understanding of *Candida* virulence has been derived largely from studies in which each of the attributes reported to contribute to pathogenicity, such as adhesion, hyphae formation, and proteinase secretion, was assessed one at a time in vitro or in attenuated variants. Some findings have demonstrated that the *Candida*–host interplay is subtler than previously appreciated. Odds also emphasized that a panel of specialized virulence attributes may play a role at each stage of the infectious process. Cutler (30) also made a special point in his thorough review on putative virulence factors that no single factor accounts for virulence and that some, but not all, virulence factors may be important at specific tissue sites. Similarly, Hube (31), in his review on SAP genes and Sap isoenzymes, emphasized that *C. albicans* pathogenesis is multifactorial. In fact, Odds et al. (32) recently concluded that the pathological factors that ultimately led to the death of animals infected with certain strains of *C. albicans* are unknown, but they were, for example, clearly not absolutely dependent on the capacity of the infecting strain to form hyphae. Although several genes have been suggested to encode virulence factors, there has been a conceptual problem with defining pathogenicity genes. For example, *ura3/ura3* mutants are not pathogenic, and neither are most, if not all, mutants defective in genes affecting growth (33). In this regard, strains with diminished growth are rapidly cleared from the body. Although the pathogenicity of fungi in general is poorly understood, there is a growing body of evidence that genes expressed, for example, by phytopathogens during starvation for either nitrogen or carbon are also expressed during infection in plants (34–36). Some, but not all, positive regulators of metabolic genes of phytopathogens were shown to be directly implicated in pathogenicity (37).

A striking example of importance of secondary carbon utilization for *C. albicans* pathogenicity was recently published by Lorenz and Fink (38). After ingestion in vitro of *C. albicans* cells by cultured mammalian macrophages, two principal enzymes of the glyoxylate cycle that permits utilization of two-carbon compounds, isocitrate lyase and malate synthase, were upgraded in the surviving cells. The mutants lacking isocitrate lyase were markedly less virulent in mice than the parental strain. The authors speculated that glucose-deficient environment of phagolysosome—which, however, is rich in fatty acids and their breakdown products (e.g., acetyl-CoA)—provides alternative nutrients, which can be utilized for the synthesis of glucose through glyoxylate cycle. Some products, such as acetyl-CoA, can be utilized only through glyoxylate cycle. The authors concluded that isocitrate lyase is essential for full virulence, and that glyoxylate pathway is necessary, but not sufficient for the virulence.

Because under regular circumstances the major factor for *Candida* infectivity is undoubtedly the condition of the host, the ability of *C. albicans* to adjust and to survive in a healthy host is as important as the action of so-called pathogenic factors. In fact, *C. albicans* phenotypic diversity, with the rapid ability to alter phenotypes, was recognized a long time ago as an important putative contributing factor for virulence (39; reviewed in Refs. 28,30). The variable phenotypes represent a wide spectrum of the morphological, physiological, and biochemical properties. Some important phenotypes include the resistance to antibiotic (40,41,42), the radiation sensitivity (43), the ability to assimilate different nutrients (38,44), and the virulence in animal model (these and other phenotypes are also reviewed in Refs. 26, 28, 45). We recently showed that adaptation to new environment is achieved through a mechanism of adaptive mutagenesis, which provides a dramatic increase in several order of magnitudes in "fit" mutants (reviewed herein). This study can be viewed as a model of an opportunistic pathogen strategy to adapt to different niches in host.

Since the first report in 1990 of a large repertoire of frequent spontaneous chromosomal alterations within the same population of *Candida* cells, two hypotheses were introduced: (1) chromosomal instability is a means to gain genetic variability in the absence of the more conventional way by meiotic segregation (46); and (2) chromosomal instability observed within the same population of cells under laboratory conditions is a source of genetic and phenotypic variability among isolates in nature (47). During the course of the last decade, we presented evidence supporting these hypotheses (reviewed in Refs. 26, 48). A causal relationship between specific and different chromosomal alteration and different stressfull environments, i.e., presence of secondary nutrient, antibiotic or other toxine, was established (27,44,40; M. Wellington and E. Rustchenko, unpublished data). Most important, we identified chromosome copy number, as provided by the loss or gain of a chromosome, as a novel means for regulating gene expression in microbes. Because this novel regulation controls in *C. albicans* important functions, we believe it may be a general regulatory mechanism in *C. albicans* (44). For the sake of clarity, we will distinguish between natural variability among independent strains and spontaneous instability observed within a particular strain. The genetic variability with emphasis on chromosomal instability will be the main subject of this chapter.

II. MATING

The hypothesis of high-frequency chromosomal rearrangements that lead to genetic and subsequently to phenotypic diversity in *C. albicans* was proposed 10 years ago (46) as a substitute for mating and meiosis. Recent experiments with specially manipulated diploid strains demonstrated that mating could occur, and may therefore function as a means of introducing genetic diversity in *C. albicans*. With the demonstration that mating can be induced, we must now compare the extent to which chromosomal instability and mating is used to generate diversity.

Mating was achieved by two different approaches. Johnson's group (49) analyzed the structures of a mating-type-like (*MTL*) locus, identified *MTLa1*, *MTLα1*, and *MTLα2* alleles similar to the *Saccharomyces cerevisiae MAT* genes, and constructed hemizygous derivatives of the strain with either the *MTLa1* or the *MTLα1* and *MTLα2* alleles disrupted (50). On the other hand, Magee's group (51) eliminated either one of an opposite type of alleles, **a** or α, by passing an original heterozygous *C. albicans* strain through L-sorbose plates. This treatment, based on the study in *C. albicans* of the chromosome copy number regulating utilization of L-sorbose (44), causes the loss of either one of two chromosome 5 homologues, carrying the opposite *MTL* allele in an otherwise diploid cell. The loss of one homologue, however, can be compensated for by the duplication of the remaining homologue, leading to homozygosity of the *MTL* allele, which occurs after growth of cells on a rich medium. The frequency of duplication is approximately 10^{-3} in strain 3153A, but varies with different strains (44). The genetically manipulated strains of Hull et al. (50) did not mate on a plate under the initial conditions tested, but mating occurred when the mixture of cells was passed through a mouse by injection in the tail vein followed by the removal of animals' kidneys, which were homogenized and spread on a plate with selective medium. A total of 12 out of 16 products contained a combination of two alternative functional alleles, **a** and α, and the two alternative disrupted alleles **a** and α present in diploid parents. These results are consistent with mating of the two parental diploid strains. The four other products lacked one *MTL* allele, which could have occurred by the loss of one homologue of chromosome 5 or by the homozygosis of the *MTL* locus after mating. The frequency of mating varied and was not determined, but appeared to be low (A. D. Johnson, unpublished results). On the other hand, Magee and Magee (51) achieved mating by mixing post-L-sorbose cultures containing the opposite *MTL* locus directly on plates. The frequency of the event was not estimated, although it was observed that longer incubation (up to 8 days) and lower temperature increased mating, the latter being consistent with the property of chromosome 5 to increase its instability about one order of magnitude at room temperature (E. Rustchenko, unpublished data). Southern blot analysis showed the presence of both a and α alleles in the products of mating. Earlier we suggested that the unique karyotype of each *C. albicans* strain creates a unique genomic condition (47; also reviewed herein), thus defining the phenotypic differences between strains. The expression of the genes important for mating may significantly differ in the different *Candida* strains, thus diminishing the efficiency of mating. Consistent with this view, many genes have been identified previously in *C. albicans* that are homologous to *S. cerevisiae* genes encoding components of the mating pheromone response and meiosis pathways (e.g., *GPA1*, *STE12*, *DMC1*) (reviewed in Ref. 49). While some of the sexual cycle homologues have been found to be expressed and functional, however, the expression of others has not been detected under laboratory conditions. In this respect, it is important to note that Magee and Magee obtained mating with Sou⁺ cultures from different strains, but not from the same one, with only one tested strain being an exception.

In *S. cerevisiae* mating occurs with high frequencies between a and α strains independently of the ploidy of cells. Hull et al. (50) suggested an explanation of why *Candida* mating appears to occur only rarely, if at all, in nature. The authors suggested that mating requires a and α strains that arise by homozygosis of the *MTL* allele or by chromosome loss. The appropriate pairs of a and α cells may simply arise only rarely in the same host. As suggested above, the "weakness" of genes involved in mating, which requires the combination of different genetic backgrounds, may add to the rarity of the appropriate strains, additionally limiting mating. About 5–7% of *C. albicans* strains are homozygous for the *MTL* allele (B. B. Magee, personal communication), which poses one more obstacle to mating.

The current research focus of Johnson's group is determining the role of the *MTL* locus in the life cycle of *C. albicans*. An unusual aspect of *MTL* is that in addition to three regulatory proteins (MTLa1p, MTLα1p, and MTLα2p), the *MTL* locus contains three additional gene pairs never seen before in fungal mating-type loci: poly (A) polymerases, oxysterol binding proteins, and phosphatidyl inositol kinases (49). This complexity implies some unknown functions of the *MTL* locus. Recent comparison of the X-ray survival curves of the parental strain 3153A heterozygous by *MTL* and sorbose-positive derivatives of this strain, which were represented by mix populations of hemizygous and homozygous by *MTL* cells, showed that contrary to the expectation and by the analogy with *S. cerevisiae*, the heterozygous parental strain did not have higher resistance to ionizing radiation (43). This result indicates that the normal heterozygous strains may not have the properties of heterozygous diploid strains of *S. cerevisiae* dependent on the dimer protein MATa1p-MATα1p, a product of the *MAT* locus that induces the repair genes belonging to the *RAD52* epistatic group (52). The further analysis of the *MTL* locus will elucidate its presumably different functions in *C. albicans*.

One observation of the mating products is relevant to studies of chromosomal instability. Hull et al. (50) established the absence of one allele of *MTL* in some mating products, suggesting the loss of one copy of chromosome 5. Similarly, Magee and Magee reported that not all the mating products contained a full tetraploid DNA content, as established by fluorescence-activated cell sorter, and also that some markers were lost. It remains to be seen whether or not specific chromosomes alternate, as well as whether or not mating depends on these chromosomes' alternation.

Overall, mating does not seem to be a prevailing means for genetic variability. As discussed herein, chromosomal instability was shown to be responsible for the formation of such phenotypes as utilization of various nutrients or primary resistance to drugs. We currently believe that chromosomal instability is the major cause of phenotypic diversity in populations of *C. albicans*, although mating may be responsible to a minor degree.

III. VARIABILITY AND INSTABILITY OF DIFFERENT ASPECTS OF THE *C. ALBICANS* GENOME

The first reports on genomic variability among *Candida* natural isolates included several aspects of the genome. Mitochondrial DNA was analyzed and reviewed by Riggsby (53). Restriction-site polymorphism was uncovered as variations between homologous chromosomes in different *C. albicans* strains with the hybridization probes containing the markers *ADE2*, *URA3*, and possibly *LEU2* (54, 55). Differences were even observed between alleles at the *URA3* locus in certain strains. At the same time, the first separations of electrophoretic karyotypes were providing an insight into a natural chromosomal variability (56–58). Subsequently, other aspects of the genome were shown to be variable, including, for example, the number of rDNA, telomeres, and dispersed repetitive sequences, thus providing a potential to control some still unknown phenotypes. The possible mechanisms of genomic variability including for example, unequal crossing-over or mitotic recombination, have been reviewed by Wickes and Petter (59). We wish to emphasize that during the last decade most of the understanding of the phenotypic diversity came from studies of chromosomal instability within the same population of cells, which will be discussed in a special section in this chapter. (See section on chromosomal instability.) Recently evidence was obtained that *C. albicans* genetic instability, which results in chromosome copy number differences, controls such important functions as utilization of secondary carbon sources and resistance to the drug fluconazole, as well as a drastic increase in the frequency of specific mutation under adverse condition.

A. Chromosomes, Ploidy, and Cell DNA Content

Initially the combined use of several methods, such as the measurement of the total DNA content of a cell, reassociation kinetics, and UV survival curves, provided evidence for the basic diploid nature of *C. albicans*. This approach was important before the introduction of the pulsed field gel electrophoresis (PFGE) technique, a method that allowed the visualization of native chromosomes in the form of ethidium bromide-stained bands on a gel, with each band representing the bulk of a particular chromosome from a population of cells. The improvement of the quality of separation permitted a comparative analysis of chromosomal size and copy number, which was apparently a more discriminative approach than measurement of DNA content. In some situations, as for example the hypothetical ploidy shift, the electrophoretic karyotype may not be helpful in detecting the chromosome copy number, and the determination of the DNA content has to complement the analysis of chromosomal pattern. Another advantage of the use of the chromosomal separation was the opportunity to apply Southern blot analysis to analyze the differences in the distribution of genes over chromosomes.

In the early 1990s a substantial amount of results accumulated showing that the majority of *C. albicans* strains contain a haploid number of eight chromosomes. This genomic organization is not rigid, however. Rearrangements of linkage groups, aneuploidy, and multinuclei conditions were documented. The continuous improvement of the PFGE technique was instrumental in establishing the haploid number of chromosomes and natural chromosomal variability among laboratory strains, as well as spontaneous chromosomal instability of the same strain.

Many earlier versions of pulse electrophoresis, as for example, PFG (pulsed field gradient), FIGE (field inversion gel electrophoresis), Pulsarphor, TAFE (transverse alternating field electrophoresis), or RGE (rotating gel electrophoresis), which have been reviewed in Refs. 60–62, are rarely used. A commercial version of the CHEF (contour clamped homogenous electric field) (63), denoted CHEF-DR and provided by Bio-Rad Laboratories (Hercules, California) is currently popular. The advantages of the CHEF-type system are the straight lanes and the large number of samples that can be tested with one gel. One of the early systems with a nonhomogeneous field, however, OFAGE (orthogonal field gel electrophoresis) (64), produces sharper bands. It is worthy to note that the reliable visual comparison of the amounts of DNA in bands, as allowed by the OFAGE system, proved to be essential for the analysis of changes in chromosome number. (See, e.g., Refs. 40, 44.) The present limits of resolution in OFAGE are approximately 600 bp (65), and probably can be further extended if needed, as it is a matter of the combination of different parameters.

The chromosomes of C. *albicans* were denoted by at least five numbering systems, as reviewed by Pla et al. (66). These systems differed in the use of either roman numerals or arabic numbers, or letters, as well as in how the chromosomes are numbered, either from the smallest to the largest or vice versa. We will use the currently accepted nomenclature, in which the penultimate largest to the smallest chromosome is designated by arabic numbers 1–7, whereas the largest chromosome, containing the ribosomal DNA cluster, is designated R. The following are the equivalencies of one of the earlier and current nomenclatures:

Chromosome	VIII	VII	VI	V	IV	III	II	I
	R	1	2	3	4	5	6	7

1. Natural Isolates and Laboratory Strains

Total DNA Content per Cell and Ploidy. The convincing evidence for the diploid nature of C. *albicans* came from the sequence complexity, which was measured as the reassociation kinetics of the total DNA. These included three independent methods, the susceptibility of single-stranded DNA to S_1 nuclease digestion, DNA separations with hydroxyapatite chromatography, and optical hypochromicity (67). It was concluded that the unrepeated genome size is approximately 18 fg, which was approximately one-half of the cell DNA content (53). The measurements of total DNA per cell using the standard diphenylamine method supported this conclusion. Riggsby (53) comprehensively reviewed the results at that time. These early estimates were corroborated by Rustchenko-Bulgac et al. (46) using the diphenylamine method, and by Suzuki et al. (68) using diaminobenzoic acid. The size of the genome was subsequently estimated to be 16 to 17 Mbp by physical mapping (69). A definitive evaluation of the DNA content will come from the entire genomic DNA sequence, which is essentially completed and only recently available on the Web (70). At the time of this writing, it is estimated that C. *albicans* contains approximately 6,500 genes (S. Scherer, personal communication).

The first indication that C. *albicans* contains a haploid number of eight chromosomes came from Southern blot analysis of then still limited chromosomal separations. One difficulty that had to be overcome was frequent deviation in size of two homologous chromosomes, which appeared as two different bands on the pulsed field electrophoresis gels. Some workers purposely chose strains that lacked deviated homologues for use in representative studies, even though such strains were exceptional. Using three laboratory strains and an elaborate collection of markers, Lasker et al. (71) suggested that the haploid number of

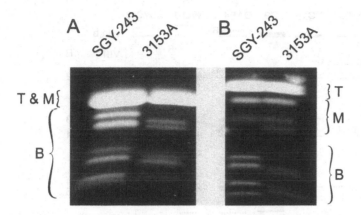

Figure 1 The precise separation of the chromosomes of two *C. albicans* laboratory strains, 3153A and SGY-243. Orthogonal-field alternating gel electrophoresis (OFAGE) conditions were chosen to accentuate the separations of either: (a) only the bottom, B, group; or (b) the bottom, B, and middle, M, groups of chromosomes. These conditions of separation did not resolve the top group, T, of large chromosomes, which can be seen schematically in Fig. 2. When two groups, B and M, were separated on the same gel, there was less resolution in the B group, and as a result, some closely running bands in strains SGY-243 co-migrated. Source: Adapted from Ref. 40.

chromosomes is eight. Two other reports, in which chromosomal separations of the laboratory strains were clearer but the number of markers was limited, corroborated the haploid number of eight chromosomes (46, 72). In one of these reports, in which a strain with five pairs of deviated homologues, as well as its mutants with spontaneously altered chromosome copy number and other alterations were used, the authors demonstrated the variety of ways by which a diploid parental strain could give rise to high frequencies of aneuploidy by either a single or multiple changes in the chromosome copy number (46). This and still other publications, in which more laboratory strains with improved chromosomal separations were used, contributed further evidence for the predominantly diploid state of the laboratory strains with some strains being aneuploid, as exemplified in Fig. 1 and summarized in Fig. 2 (40, 44, 47, 48). As discussed below, the discrepancy between relatively frequent spontaneous aneuploidy within the same strain, and rare cases of the aneuploidy among natural isolates, remains to be clarified.

Natural Chromosomal Polymorphism. Although the early PFGE-separating techniques did not fully resolve *C. albicans* electrophoretic karyotype, the differences in chromosomal patterns among the strains were convincing. Two groups, each using as an example five different laboratory strains, first pointed

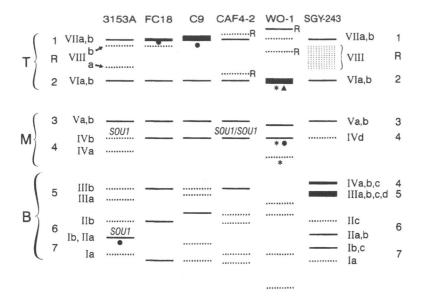

Figure 2 Schematic representation of the chromosomal patterns of some popular *C. albicans* laboratory strains. The following two chromosome numbering systems are presented: 1 to 7 and R; and I to VIII and homologues, a or b. Numeration presented on the right refers solely to strain SGY-243. Dotted, thin, and thick lines correspond, respectively, to one, two, and three or more chromosomes. The array of lines for chromosome R represents a cloud of inseparable weakly stained bands. Three groups of *C. albicans* chromosomes, bottom, B, middle, M, and top, T, can be precisely resolved singly or in various combinations by several different electrophoresis conditions. (See, e.g., Fig. 1.) The symbol ● designates comigration of two to four homologues of two different chromosomes; ▲ designates comigration of the following chromosomes: one homologue of 1, two homologues of 2, and one homologue of the chimeric chromosome; and ✳ designates a chimeric chromosome (see Ref. 47). The chromosomal assignment of *SOU1* gene in strains 3153A and CAF4-2 is indicated. The assignment of chromosomes to individual bands for reference electrokaryotype of 3153A and their approximate sizes, as well as the rationale for identifying the eight pairs of chromosomes, have been previously published (46–48, 65). The chromosomal assignment of FC18 and C9 with the 16 markers was previously published (47). Chromosomes of strain SGY-243 were identified from the similarity of positions and intensity of bands to the reference karyotype of 3153A, as well as by assignment of the seven markers (40). Chromosome copy number of strains SGY-243 and WO-1 was corroborated by densitometry. The assignment of largely rearranged chromosomes of strain WO-1 is not presented in this figure (see 47, 69), except for the R chromosome. The karyotype of strain CA14, which is not shown here, was identical with CAF4-2, except for chromosome R, which was represented by inseparable weak stained bands similar to the chromosome R of strain SGY-243. The relatively precise estimate of the differences between two homologous chromosomes is available for chromosome 7 in strain 1006, 20 kbp (134), in strain NUM114, 100 kbp (H. Chibana and P. T. Magee, unpublished results), and chromosome 7 in strain CAF4-2, 10 kbp (Y.-K. Wang and E. Rustchenko, unpublished results).

out the high natural chromosomal polymorphism of *C. albicans* (57, 58). Their observation was supported by partial separation of four more individual chromosomal patterns of laboratory strains published by Suzuki et al. (73). At the same time, Merz et al. (56) separated chromosomes of clinical isolates from 17 patients and obtained 14 unique patterns. Undoubtedly a poor resolution of the long chromosomes in the early separations was a limiting factor in establishing a unique electrophoretic karyotype for each of the 17 strains. Lasker et al. (71) and Iwaguchi et al. (72) subsequently reported unique patterns of three and 27 more laboratory strains, respectively, combining PFGE separation with the assignment of the chromosomal markers. In a study of 100 clinical isolates by Monod et al. (74), the chromosomal polymorphism was not emphasized, and was regarded as a minor variation. Nevertheless, their work contributed to the understanding of chromosomal variability. Asakura et al. (75) and Doi et al. (76) continued to document natural chromosomal variability by providing statistics on chromosomal separations of approximately 160 clinical isolates and revealing that each strain probably had a unique electrophoretic karyotype. At approximately the same time, two independent groups addressed electrokaryotypic variations of the laboratory strains. Rustchenko-Bulgac (47) analyzed four strains using improved PFGE separation procedures and 16 chromosomal markers, and documented varying sizes of the homologous chromosomes in different strains as well as dramatic chromosomal differences in the popular laboratory strain WO-1. These included multiple translocations, a large truncation of chromosome 5, and a recently analyzed duplication of the longer homologue of chromosome R (D. H. Huber and E. Rustchenko, unpublished data) (Fig. 2). In addition, Rustchenko-Bulgac (47) found it difficult to explain the multiple comigrations of the nonhomologous chromosomes in this strain, which included combined groups of linkages not present in the other strains, as well as common groups of linkages. The *Sfi*I map of WO-1 provided further details on the chromosome structures of this strain (69).

Furthermore, Rustchenko-Bulgac (47) unexpectedly found that in spite of the high karyotypic variability under the condition of lack of meiosis, the chromosomal patterns of different strains retained the same pivotal structure, which was later confirmed by using more strains, as well as by other authors (77). For example, as presented schematically in Fig. 2, the chromosomes in most of the strains seemed to fluctuate around certain modal sizes. The range of sizes and the three major groups of sizes—short, medium, and long chromosomes—remained approximately the same. This paradox still requires an explanation.

Thrash-Bingham et al. (77) used 22 markers to analyze six laboratory strains and reported that three strains contained translocations, which were suggested to be a natural source of variation in *C. albicans*. Perhaps more translocations would be uncovered with a larger number of probes. The same authors hypothesized that translocations might arise by recombination between any of several families of the repeated dispersed sequences; for example Rel-1, Rel-2, 27A, or Ca3 (77, 78). Iwaguchi et al. (79) suggested that the RPS1 sequence, highly homologous to Ca3/27A, may be involved in chromosomal re-arrangements and may in part explain chromosomal polymorphism. Finally, Chu et al. (69) proposed that translocations have a tendency to occur at or near *Sfi*I sites within RPS1, and that such a mechanism may be a general means of generating translocations in *C. albicans*. It has to be noted, however, that the later hypothesis was based on the analysis of three translocations from a single strain WO-1, which has many unique features not observed in the other strains (47, 48), as for example, two different sizes of the rDNA unit (65). In another strain 1006, no translocation associated with *Sfi*I sites was observed. It would be of interest to determine if translocations occur at *Sfi*I sites in other strains. The most recent hypothesis attributed the chromosomal variability to the recombination between retrotransposon-like sequence kappa (80), which is found in both CARE-2 and Rel-2 repeats. The suggestion that repeats like Ca3/27A, RPS1 or kappa may be the source of chromosomal variability does not agree with the analysis of mutants with various spontaneous chromosomal alterations. When DNA digests of these mutants were hybridized with a Ca3 probe, their signal patterns were no different from the parental strain (46, 81; see also section III.C).

One more rare condition of the electrokaryotype can be illustrated with the example of strain SGY-243, which has a banding pattern that has more complexity than in the other strains in that it contains six additional chromosomes in the positions of chromosomes 4, 5, 6, and 7. The chromosome R in this strain is represented by an inseparable cloud of weakly staining bands instead of two homologues (Fig. 2). This type of diffused band has been described previously for chromosome 6 in strain 300 (48), and indicates a cumulative karyotype of a highly heterogeneous population of corresponding molecules. Apparently there is some poorly understood high instability in the genome that is limited to a single chromosome. Although the instability is detectable on a gel and this approach reveals various sizes, it somehow fails to determine possible changes in the copy number of the corresponding chromosome, which can vary in different cells. Another kind of single chromosome instability leading to what can also be called a cumulative karyotype, occurs when the band on a gel is represented by a different number of the homologues of the same size in a mixed population of cells. In this situation, the corresponding band has an intermediate brightness between, for example, single and double copies (48).

The multiple aneuploidy exemplified by the strain SGY-243 could be either natural or induced by exposure to UV light (82) formed under laboratory cultivation, or acquired by improper storage. (See, e.g., Refs. 46, 48.) In this regard, three chromosomal alterations—the loss of two nonhomologous chromosomes and the change in the length of chromosome R, which occurred in strain WO-1 during regular maintenance in the laboratory—is significant (83). One possible explanation could be a subcloning of the culture after removal from the $-70°C$ freezer. This method, when applied to *Candida*, often leads to the selection of a mutant with altered chromosome R (47; E. Rustchenko, unpublished results). For example, we currently have two *C. albicans* CAI4 strains received from different laboratories. These strains differ by the appearance of their chromosomes R. Another example is strain 3153A, in which alteration of several chromosomes occurred. Strains 3153A and 300 were originally derived from the same strain, but after maintenance and preservation in different laboratories exhibited phenotypic differences. Identical patterns of restriction fragments revealed with the Ca3 probe confirmed their common identity, however (48). These two strains were compared for their electrophoretic karyotypes and three differences were found (48), a result that is consistent with a number of differences in assimilating profiles of these strains (26). Because of the unstable condition of one of the chromosomes (see above), 300 was assumed to be an unstable mutant of 3153A. The mutagenic genetic manipulations, such as transformation, could also affect the chromosomal pattern by producing chromosomal instability, as may be the situation for CAI4 and SGY-243. The implication of supposedly genetic manipulations was independently noted by several groups that used different strains (84; F. Navarro-Garcia, L. Monteoliva, unpublished data). Larriba and colleagues reported alterations of size of chromosome R due to gene disruption procedure, as well as due to the exposure to 5-fluoro-orotic acid (5-FOA) (85), which is a required step when two copies of gene are disrupted using the URA3-blaster. We recently found that short exposure of 24 hrs to 5-FOA induced nonspecific chromosomal alterations. The prolonged exposure, however, resulted, on one hand, in the death of the majority of the cells, and on the other hand, in the formation of 5-FOA-resistant mutants having specific alterations of either chromosome 5 or chromosome 4 (M. Wellington and E. Rustchenko, unpublished data). Although transformation procedure itself is also considered mutagenic by many researchers, a reliable experimental proof has not been presented. In the work of Ramsey et al. (86) using Southern blot analysis, UV light was shown to induce instability of chromosome R in strain 3153A. The poor separation did not allow an evaluation of the other chromosomes, however.

So far only a few cases of aneuploidy of $2n + x$ type have been observed in laboratory strains, but none in clinical isolates. Nevertheless, it is premature to conclude that this alteration is less frequent in a natural environment, and is preferably occurring, for example, in the laboratory condition. As we recently showed, the cultivation of the monosomic cells in a rich liquid medium selects for the genetically balanced diploid of the 2n type (44). One explanation for the lack of aneuploidy in clinical isolates thus can be simply the selection of a balanced diploid state after maintenance in the laboratory. It is still possible, however, that laboratory strains, which may have been converted to aneuploidy, such as WO-1 or SGY-243, have acquired an effective gene balance. Another explanation could be a lack of resolution in the PFGE in early separations, which provided most of the documentation on the electrokaryotypes of clinical isolates. This conclusion needs further clarification.

Overall, the comparison of a large number of *C. albicans* strains (72, 75, 76) established that there is a high degree of variability of every chromosome. Frequent differences include deviated homologues of the same chromosome, which can be due to either a deletion or a translocation. The deletion can be identified by Southern blot analysis as lacking one or more chromosomal markers on the shorter homologue (40, 87), whereas the translocation would give two signals with two nonhomologous chromosomes (44, 46, 47, 77). As highly homologous repetitive sequences Ca3/27A and RPS-HOK-RB2—also called major repetitive sequence (MRS), representing probably the same large segment—were shown to vary in size and copy number per chromosome within the same strain and between different strains (H. Chibana, unpublished data), these repetitive sequences have to be considered as one of the sources of chromosomal polymorphism as well. The combination of several different events also has to be considered, although the current methods reveal only the final result of rearrangements, and not the sequence of events. Obviously, electrokaryotyping and Southern blot analyses are crude methods that probably detect only a portion of structural changes in chromosomes. For example, an inversion, a small deletion or insertion, as well as sequence changes may remain undetected. It is important to note that because the native chromosomes are prepared from a cell mass, the cumulative chromosomal pattern is produced on the PFGE gel. Only the electrokaryotype representing the major portion(s) of the cells is visualized by staining, however, and underrepresented chromosomal patterns are not observed.

It can be considered at this time that reports on two identical electrokaryotypes of independently derived strains have not been published. In light of this knowledge, it is important to correct the notion that some strains have ''typical'' and others have ''atypical'' electrophoretic karyotypes. A correct view would be that each strain is represented by an individual karyotype. Another popular misconception is to regard different strains as phenotypically uniform. In fact, strains differ by multiple phenotypes, including, for example, the appearance of streak culture or colonial appearance and assimilation profile (26, 45, 47, 48, 88, 89), to mention a few, in accordance with their unique genomic condition.

The complex relationship between stress and chromosomal instability, possibly leading to the diversification of strains, is just beginning to be understood. (See Sec. III, ''The Role of Chromosomal Instability in Adaptation to the Changing Environment.'')

In summary, there are many results indicating that chromosomal polymorphism arises from different genomic rearrangements. The recent accumulation of data suggests an important role of environmental stresses in chromosomal variability. The mechanisms producing this variability are still obscure, however, as is the mechanism(s) of strain diversification.

2. Spontaneous Mutants

Instability in Historical Perspective. The study of the *C. albicans* instability was initiated in 1935, when Negroni (90) reported a "rough" variant, as opposed to a "smooth" or "normal" one, with micro- and macromorphological differences, as compared to the parental population under the same condition. A series of similar reports followed, describing the individual morphology of either the colonies or streaks of the mutants derived from the same strain (reviewed in Ref. 45), as well as sectors developed within giant colonies (91). Soon after it became obvious that a large number of possible morphological forms can arise in the same population (45, 92). Early workers, however, attempted to produce simplified classifications, and either did not analyze large collections of mutants or did not report the full extent of the variations, as they tended to consider only some types but not all forms. This approach systematically led to the assumption that a certain defined number of colonial morphologies were produced from a given strain. The extreme situation of this kind was probably that reported by Slutsky et al. (39), who studied colonial morphology of mutants in lightly UV-irradiated population of cells of strain 3153A, and who assigned colonial forms to just seven types. On the other hand, Rustchenko-Bulgac (47) showed that the same strain 3153A spontaneously produced a very large number of the colonial morphologies, and the frequency of recovery of the particular form probably depended on the rate of growth of the corresponding mutant. The comparative study of a number of *Candida* species, including a total of 100 strains, suggested that colony instability is a common property among isolates of many *Candida* species (93). Early workers also systematically described different combinations of multiple changes in other features related to these morphological mutants, including traits used for taxonomic assignment, such as germination or carbon utilization, and traits believed to be correlated with putative pathogenic factors, such as adhesion and colonization, as well as virulence for laboratory animals (reviewed in Ref. 26). Despite a substantial amount of data on phenotypic variability in morphological mutants, no underlying mechanism was proposed (reviewed in Refs. 28, 45, 91). From today's prospective, the main result of the early studies was a comprehensive documentation of multiple altered phenotypes associated with a change in macromorphology.

It was also previously observed that much higher frequencies of various colonial forms could be recovered from the oral cavity of patients than from healthy carriers (94). The author discussed whether the altered morphologies occurred in the oral cavity or were induced by the transfer. Subsequently, using the Ca3 probe, the question of whether different morphologies belong to the same strain and occur as a result of the instability or represent different strains in the same patient, was addressed (95–97). At that time, however, the high instability of *C. albicans* related to environmental stresses was not taken into account, and care was not taken for protecting cells from low temperature, aging during storage, or working with the whole population of cells versus traditional subcloning.

For the purpose of this review, it is important to note that early workers often reported the increased frequencies of morphological mutants under stressful conditions, such as excessively alkaline medium, immune serum, lithium chloride (94), aging (45, 98), cold shock, heat shock, UV light, or benzazoles (99). More recently two additional significant stressful factors were suggested. Jones et al. (100) obtained data on the increased instability of colony morphology in strains recovered from patients with the invasive infection versus strains isolated from superficial infection, and Odds (101) pointed out the gravity of general changes in the environment, which *Candida* experiences after transfer from the human microniche to a laboratory plate. We would like to note in this connection that among the multiple differences between these two milieu (i.e., in vivo versus in vitro), the excessive amount of primary sources of carbon and nitrogen under laboratory conditions is an obvious contrast that would be expected to produce changes in the expression of numerous genes. (See, e.g., Refs. 36, 102.)

The scope of the colonial instability and its possible relation with virulence was the subject of some speculations, whose main idea was that multiple changeable phenotypes serve for adaptation, as summarized by Odds (101).

Currently the prominent phenotypic instability in *C. albicans* is explained by the high frequency of chromosomal instability. The series of papers by Rustchenko et al. (26) and Rustchenko-Bulgac et al. (46, 48) established a link between colonial appearance, multiple associated phenotypes, and alterations in electrophoretic karyotype. Unstable and highly unstable morphological mutants were shown to have correspondent levels of chromosomal instability (47). Karyotypic alterations consisted of a large repertoire of single and multiple changes, as documented with the following *C. albicans* strains: 46 mutants from 3153A, 307 and 310; one mutant and its several phenotypic "revertants" from NUM961; and 15 mutants from ATCC 32077 (reviewed in Ref. 26; M. J. McEachern, unpublished data). It is important to note that we never encountered a situation in which the electrokaryotype of a morphological mutant was not altered. Some mutants altered solely in chromosome R are discussed in "Chromosome R Instability" below. The simplest assumption is thus that the morphological changes and the multiplicity of the related phenotypes are caused by chromosomal alterations due to changes in dose or level of expression of a large number of genes. Presently, differently appearing colonies can serve as a convenient means to identify chromosomal instability.

Because important physiological functions, such as the utilization of different secondary nutrients, was associated with chromosomal instability manifested by various chromosomal alterations, it was concluded that *C. albicans* developed an unusual but effective means for controlling gene expression (26). This is in direct support of the earlier hypothesis that chromosomal alterations in *Candida* act as a means to achieve genetic variability (46).

Although recent findings on the increased frequency of chromosomal non-disjunction during adaptation directly support early ideas on the scope of morphological instability, the role of instability in infection requires a more careful analysis. In this connection the uncertainty of the contribution of instability through the handling of clinical samples needs to be clarified.

Chromosomal Instability. Both Suzuki et al. (68) and Rustchenko-Bulgac et al. (46) initiated the analysis of spontaneous chromosomal instability. Suzuki et al. (68) found a different looking colony, which was called a semirough variant, among approximately 3×10^5 colonies of a parental strain. When cells from the variant were subcloned, "revertants" to the original colonial appearance were derived with a high frequency of 5×10^{-3}. The parental strain, semirough variant, and revertants were compared for the cell DNA content, the production of acid proteinase, and for their electrophoretic karyotypes. Of approximately 10 resolved bands in their chromosomal patterns, the sizes of five did not change in any of the strains, whereas the remainder varied. The suggested interpretation of the mechanism of chromosomal instability was based on the observation of multiple spindles in the nucleus of the semirough mutant and changes in the DNA content in the mutant and some of the revertants. It was assumed that chromosome reorganization in the derivatives of the original strain was coupled with the ploidy shift. It was also suggested that chromosomal instability plays a role in natural chromosomal polymorphism. Similarly, Rustchenko-Bulgac et al. (46), who studied spontaneous chromosomal instability with numerous morphological mutants or variants, found one mutant with an approximate twofold increase of DNA content. This finding could be explained, however, by the combined effect of the presence of 2–5% dikaryotic cells in the population with multiple chromosomal duplication. Although the authors also speculated that multiple chromosomal duplication could be the result of polyploidization followed by partial restoration of the original ploidy, the hypothesis that ploidy shift was responsible for the chromosomal rearrangements was neither rigorously proved nor dismissed.

The rich YPD medium used by Suzuki et al. (68) is for the most part unable to distinguish morphological mutants. For example, some mutants give rise to smooth colonies on YPD medium, but rough ones on LBC medium. Similarly, the laboratory strains have identical-appearing colonies on YPD medium, but have individual morphologies on LBC plates (47). It is likely that the semirough variant was a highly unstable mutant dissociating into various morphological forms, similar to the many mutants analyzed by Rustchenko-Bulgac (47). These mutants are known to segregate large numbers of the new colonial forms, as presented in Fig. 3, each of which possesses unique spontaneous changes in their electrokaryotype.

Rustchenko et al. (46, 47) obtained a collection of 384 spontaneous colony form mutants or so-called morphological mutants from five separate cultures, each derived in turn from a separate clone of the original strain. This procedure ensured that mutants arising from separate clones would be the result of independent mutational events, and not arising from siblings. The mutants were detected on a glucose-containing synthetic medium, LBC (103), due to their unusual looking colonies, which included a wide variety of shapes and sizes (46–48), and sometimes, after repeated subcloning, segregated colonies of opaque, yellow, brown, or dark red colors (E. Rustchenko, unpublished results). It is of interest to note that some mutants with small colony size analyzed earlier by Bianchi (104) had slow rates of growth, but were not, for example, "petite" mutants having mitochondrial defects. The coloration was due to unknown factors, and was not due to auxotrophy (E. Rustchenko, unpublished results 1989). The spectrum of morphologies produced by each of the five original clones included some common forms, as well as unique forms. Overall each clone seemed to display an individual spectrum of morphologies, which might indicate that clones in *C. albicans* populations have unique genomic conditions. In order to corroborate that the mutants derived from the same strain and to exclude contamination, 14 of them, representing all five original clones and having the most rare morphologies, were analyzed with the molecular probe Ca3 (105). These results revealed the authenticity of the original strain 3153A and fourteen mutants (46). The mutants appeared at approximate frequencies of 3.1×10^{-4} to 1.4×10^{-2}. Mutation rates were not determined, however, and a quantitative analysis needs to be undertaken. As it became gradually evident, multiple factors, especially temperature, aging, or nutrient depletion, could influence the frequency of mutation (46, 48). For example, the highest mutation frequency of 1.4×10^{-2} was obtained after prolonged incubation at 4°C. The data on the role of aging or low temperature are consistent with the reports by early authors on the increased frequencies of morphological mutants in aged cultures (45, 98) or after cold shock (104). More examples of the environmental factors inducing *Candida* instability are described in section III.A.2a. A subset of 223 mutants was analyzed for stability of their colonial morphologies by repeated subcloning, and it was found that one-third of the mutants were unstable or highly unstable, segregating different colonial forms in an individual manner (46). Some of the instability patterns were reported in detail (47).

A systematic study of a total of 46 colony morphology mutants revealed that each electrokaryotype was individually altered, containing a broad range of single or multiple chromosomal changes, which resembled the natural karyotypic

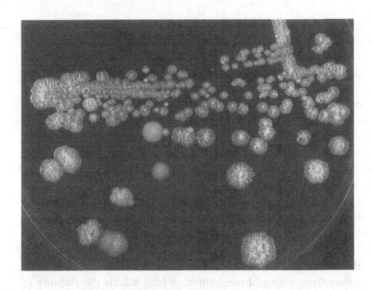

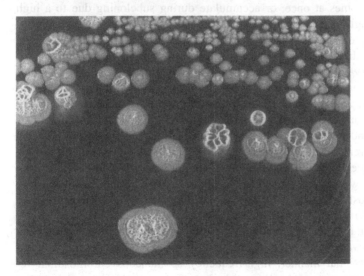

Figure 3 The appearance of colonies of two unstable morphological mutants, m504 (top) and m505 (bottom), derived from independent original subclones of *C. albicans* strain 3153A. The LBC plates were incubated at 23°C for 3 weeks.

variability among laboratory strains and clinical isolates described above in Sec. "Chromosomal Polymorphism" (46–48). Spontaneous changes included aneuploidy, deletions, translocations, and variation in the length of the rDNA cluster and sometimes complex rearrangements of the whole chromosomal pattern. We wish to note that aneuploidy was mostly of a type 2n + x, with only two instances of 2n − x type reported in early studies (47). Subsequently it became clear that growing spontaneous mutants with a reduced chromosome copy number in a rich liquid medium leads to selection of the compensatory duplication of the remaining homologue (44). The corresponding cells have higher rates of growth, thus overgrowing other cells in the culture. It is reasonable to assume that the spontaneous loss and gain of a chromosome is based on nondisjunction, in which both events occur with the same probability. The future use of controlled growth conditions may further demonstrate the role of selection.

Multiple changes, which included either chromosomal duplications or a combination of different kinds of alterations, occurred in one-half of the mutants at frequencies higher than expected for independent single events. Multiple alterations can result either from a special mechanism, which affects the stability of several chromosomes at once, or accumulate during subcloning due to a high frequency of single sequential alterations. Ploidy shift, followed by a partial restoration of the diploid state by the reduction or duplication of some chromosomal copies, can also be considered as an explanation of multiple aneuploidy.

The observation of unstable and highly unstable mutants adds to the complexity of the phenomenon of spontaneous chromosomal alterations. Of two such unstable mutants analyzed in more detail, both gave rise on subsequent replating to mutants that had diverse colony morphologies and unique chromosomal patterns (47). In addition, these two unstable mutants differed from each other by producing different arrays of morphological mutants. Some—but not all—of the mutants subsequently stabilized after subcloning, a property reminiscent of mobile elements. It was demonstrated that in some cases a single spontaneous mutant could continuously produce an overwhelming range of altered karyotypes, resulting in a wide variety of different phenotypes.

Because at the time of initiation of the studies of the electrophoretic karyotypes, *Candida* was not known to mate and to possibly produce genetic variability in a conventional manner, high-frequency spontaneous chromosomal rearrangements were suggested to be a substitute for generating new phenotypes in populations (46), and thus to be a source of natural variability (47). As mating was recently demonstrated with *Candida* diploid strains, which were either genetically manipulated or "forced" to mate, we evaluated the possible role of this process in the *Candida* life cycle in the section "Mating." We came to the conclusion that although the mating can occur at a low frequency, thus contributing to genetic variability, the chromosomal rearrangements remain the prominent means for introducing the diversity. Consistent with this view, a large variety of phenotypes was found to be associated with spontaneous chromosomal alterations. Some features, such as cellular or colonial morphology, were always changed, although they did not appear to be advantageous for the organism. Some other traits seemed to represent either destroyed or diminished functions; for example, the inability to form germ tubes or chlamydospores, or the diminution of adherence or virulence that appeared in some but not all mutants. The phenotypic changes were usually multiple, and arose in different combinations, reflecting large genomic changes due to chromosomal rearrangements.

A particular class of phenotypic changes was especially significant. A study of utilization of 21 carbon and three nitrogen sources at three different temperatures in more than 100 spontaneous mutants showed multiple changes in their assimilation profiles compared with the parental strain (26). The differences included both the gain and loss of multiple assimilation functions, as well as temperature dependence, with an almost endless array of new combinations. Each of the spontaneous mutants had a differently altered chromosomal pattern and a different assimilation pattern, a finding that established for the first time a relationship between chromosomal alterations and vital functions.

The natural diversity among *C. albicans* strains in utilization of different carbon sources is well documented. (See, e.g., Ref. 89.) The repertoire of chromosomal differences, and the related differences in assimilation profiles in naturally occurring strains and spontaneous mutants is similar, consistent with the proposed view that natural chromosomal variability results from chromosomal instability. Nevertheless, diversification under laboratory conditions is considerably less pronounced than that observed among different naturally occurring strains. For example, probing with the repetitive sequence Ca3, as well as other repetitive sequence probes, clearly distinguishes among independent strains, but fails to reveal differences between the parental strain and spontaneous or selected mutants, in spite of karyotypic changes. (See section C on regular repeated sequences.) Scherer and Stevens (106), however, reported that probing with 27A, which is identical with Ca3, showed one instance of polymorphism during laboratory subculture. Also, Franz et al. (107) observed slight differences in series of genetically related clinical fluconazole-resistant isolates tested with epidemiological probe CARE-2.

In summary, genetic instability is a means to achieve a wide range of phenotypes within a population. The available data provided an insight on "regular" diversity within populations. Perhaps a certain permanent level of instability within a population can be viewed as a resource of preadapted variants, which will be ready to proliferate in different environments. Although extensive results on spontaneous instability have accumulated, and some means of instability were elucidated, its mechanisms are still unknown.

Chromosome R Instability.   Sadhu et al. (105) were the first to report that the chromosomes carrying rDNA fluctuate in size at a very high rate from one colony to another in the same strain. Iwaguchi et al. (108) subsequently reported that the alteration of chromosome R occurred in approximately 10% of the sub-clones in the same population and is due to different lengths of rDNA clusters. Rustchenko et al. (65) reported a physiological condition that controls this type of instability. In a slowly growing population, chromosome R of all the subclones were identical. In a rapidly growing population the distribution among subclones was shifted to an increase in the number of rDNA units. The authors suggested that there normally is a distribution in the number of rDNA units per cell in population due to unequal crossing over (109) or gene conversion within the rDNA cluster during mitotic growth (H. Zou, S. Gangloff, and R. Rothstein, unpublished results), as seen with *S. cerevisiae*. Perhaps a certain optimum number of rDNA units exists for every growth condition. Enrichment of a small number of cells with the appropriate number of rDNA units is expected after a shift to a different growth condition. In other words, the representative karyotypes of a cell population can be controlled by growth conditions.

The comparison of spontaneously altered karyotypes in morphological mutants showed that chromosome R, which carries rDNA, was at least twice as unstable as any other chromosome (47). More detailed analysis demonstrated that in approximately 92% of cases the change in the length of chromosome R resulted from a change in the length of its rDNA cluster (60). The other rearrangements can be exemplified by a mutant with three chromosomes that each carries an rDNA cluster. One of the chromosomes was about 700 kb shorter in the region outside the cluster and failed to hybridize with four genes available as probes for chromosome R. This rearrangement can be interpreted either as a large deletion of the additional third copy of chromosome R or a transposition or translocation of the entire rDNA cluster to another unknown duplicated chromosome.

There is an apparent difference between rDNA instability in regular sub-clones of normal strains, which preserve the parental colonial morphology and the instability of rDNA in morphological mutants. For example, approximately one-third of the morphological mutants analyzed by Rustchenko-Bulgac (47) had alterations only in chromosome R, even though there were distinct changes in colonial morphology (46, 47). Also, the rDNA cluster lengths in morphological mutants had a wide distribution under growth conditions, which produced no instability among the regular subclones (Fig. 4). Although it is unknown if the differences in rDNA cluster length are directly responsible for the phenotypic changes in morphological mutants, it is reasonable to assume that rearrangements occurring in their rDNA cluster differ in that they lead to the altered expression of rDNA or other genes. In fact, morphological mutants can arise as a result of the insertional-deletional events affecting all or a certain portion of units similar to many cases of rDNA rearrangements in other organisms described below. (See

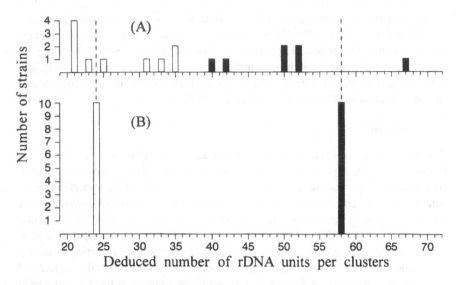

Figure 4 The comparative distribution of the deduced number of genomic rDNA units in different morphological mutants and subclones from population of *C. albicans* 3153A that were grown under the same conditions: (A) Morphological mutants m3, m7, m16, m17, m20, m500, and m500-3; (B) Random subclones C.a.1 to C.a.10. The number of subclones with the designated number of rDNA units in shorter homologue of R (open bars) and longer homologue of R (filled bars) are indicated. The number of rDNA units was deduced from the size of a single unit and the lengths of *Hin*dIII fragments that encompassed tandem rDNA units. The strains were grown on LBC plates at 22°C for 4 weeks. Source: Adapted from Ref. 65.

Sec. III.B.2, "rDNA Unit Size and Number of Units in Genome.") One of these examples is the regulation of drug resistance in *Candida* (41, 42). In contrast, rearrangements in rDNA cluster occurring in the subclones of a population may be caused by looping out of a portion of the cluster or unequal crossing-over, which changes the number of units in the cluster but not their structure. Such a type of innocuous change is expected to affect overall protein synthesis, but presumably not expression of specific genes.

B. Regular Repetitive Sequences

Similar to other lower eucaryotes (110, 111), about 13% of the *C. albicans* genome is identified as repetitive sequences. Regular repetitive sequences, which are present in all cells as rDNA, mitochondrial DNA, or telomeres, are a source of natural variability in many microbes, including *C. albicans*. In many lower eukaryotes, this variability is associated with some important regulatory functions in addition to the normal cellular functions of these sequences or organelles. Examples include age-associated accumulation of rDNA circles, which are released from the cluster of chromosomal rRNA genes in *S. cerevisiae* (112), and senescence in fungi of the genus *Neurospora*, which is determined by the integration of plasmids in mitochondria (reviewed in Ref. 113).

1. Telomeres

C. albicans telomeres were analyzed by McEachern and Hicks (114), who cloned them as a middle repetitive sequence designated Ca7. The telomeres consisted of unusually large and not so common repeats of 23 bp, which superficially resembled the mitochondrial telomeres of members of the genus *Tetrahymena* more than other known chromosomal termini. This result places *C. albicans* in a group of budding yeasts such as *C. tropicalis* and *Kluyveromyces lactis* that also have large telomeric repeats (114, 115). The telomeric repeats had a remarkable uniformity among different strains, as well as clones from the same strain. In marked contrast, the adjacent subtelomeric DNA sequences were the subject of interstrain variability in the form of frequent deletions and changes. Although overall the telomeric repeats were stable, a physiological condition controlling their instability in a population was identified. Their number increased when cells were grown at a higher temperature. The functional implication of the temperature-controlled number of telomeric repeats is not known.

2. rDNA Unit Size and Number of Units in Genome

rDNA units exist as a cluster of multiple repeated copies in the genomes of the majority of eukaryotes. The variation in rDNA cluster length and composition were reported in earlier studies of *S. cerevisiae* and many other organisms. The mechanisms involved were reported to be unequal crossing-over (109), gene conversion during mitotic growth (H. Zou, S. Gangloff, and R. Rothstein, unpublished data), short deletions, and insertions at regular intervals within the nontranscribed spacers of rDNA units (116), as well as the insertion of a single large segment of DNA (117) or Ty element (118). It is noteworthy that rDNA can be a recipient of mobile elements in different species. Xiong and Eickbush (119) reported the precise insertion of functional retrotransposons that inactivated a fraction of 28S RNA genes in *Drosophila melanogaster* and *Bombyx mori*. The insertions of mobil group I intron Pp LSU3 in all rDNA units followed the cross between intron-bearing and intron-lacking strains of the slime mold *Physarum polycephalum* (120). Taken together these results suggest that the rDNA cluster of many different organisms is a site of dynamic rearrangements that can control the expression of the genes.

Genomic rDNA from *C. albicans* has been studied less than that from *S.*

cerevisiae. The results obtained so far, however, are indicative of the similarly dynamic nature of rDNA cluster in *C. albicans* and of its high potential for producing new phenotypes. A total of 120 clinical isolates of *C. albicans* were studied for the presence of the group I intron in the large subunit 25S rDNA precursor molecule, which was subsequently denoted CaLSU (41). About 40% of the strains harbored CaLSU in each copy of their 25S rDNA coding sequences. Intron-bearing *C. albicans* strains all exhibited a high degree of susceptibility to antifungal drugs, 5-fluorocytosine, or 5-fluorouracil. The other groups demonstrated the heterogenous condition of rRNA genes, with and without CaLSU intron, in some of *C. albicans* strains (42, 121, 122), as well as sensitivity of the intron-bearing strains to other antibiotics, pentamidine (42), and flucytosine (122).

The variation in the number of *C. albicans* chromosomal rDNA units was first analyzed with *Hind*III digest of genomic DNA in unrelated studies on the natural chromosomal variability (123). Currently the number and organization of rDNA units can be conveniently and accurately assessed by digesting either agarose blocks or agarose beads with the intact chromosomes with restriction endonucleases that do not cut within the rDNA unit, as for example *Hind*III and *Xho*I, followed by examining the length of the fragments with PFGE (48, 108). On the other hand, a restriction enzyme such as *Not*I, which cleaves at a single site within an rDNA unit, allows the determination of the size of the single unit (108). Subsequently the data from two groups demonstrated the large variability in sizes of the clusters from different laboratory strains (65, 108). The estimated number of rDNA units can vary between 9 and 176 per cluster (65, 108), thus determining the cluster size and ultimately the size of chromosome R.

The size of a single unit is also variable. Rustchenko et al. (65) determined two different units, 12.2 kbp and 11.6 kbp, in three different strains. It is interesting that not only laboratory strains differ among themselves by the length of their rDNA units, but also that a single strain can contain two different sizes of rDNA units, as exemplified by WO-1, which contains 11.5 and 12.5 kbp units in approximately equal proportion. Iwaguchi et al. (108) also reported two different sizes among three subclones of the same *C. albicans* TCM297 laboratory strain. The difference in the ranges of size reported by two groups, however, 14.3 to 15.2 kbp (108) and 11.5 to 12.5 kbp (65), was relatively large and could not be easily attributed solely to the differences between strains, but may have been due to differences in the methods of measurement. There are preliminary results on the sources of differences in unit sizes, which probably do not include all possible sources of the variability and instability. Mercure et al. (41) found a 379-bp group I intron, denoted CaLSU, in a highly conserved region, 25S nuclear rRNA-encoding gene, in which most large ribosomal subunit introns of other organisms have also been mapped. The presence or absence of the intron is reflected in a well-characterized *Eco*RI digest banding pattern, as a 4.2 kbp band or 3.7 kbp band dimorphism, respectively. It is interesting to note that the CaLSU intron does not closely resemble other introns in these generally conserved regions, which makes it unusual. The question of whether or not CaLSU might be a mobile sequence has not been addressed. On the other hand, the study of the 25S rRNA in two *C. albicans* strains revealed the difference of 3361 bp versus 3363 bp in length, as well as seven nucleotides substitutions (124). These differences cannot account for the differences in rDNA units mentioned above.

3. Extrachromosomal rRNA Genes

The presence of rRNA genes on extrachromosomal molecules has been reported in organisms throughout the phylogenetic tree, including circular or linear plasmids in lower eukaryotes (reviewed in Ref. 125). Recently extrachromosomal copies of rRNA genes were found in two types of plasmids of *C. albicans*, which had circular and linear forms, and which coexisted in laboratory strains (125). The number of extrachromosomal rRNA genes varied with the growth cycle. The copies of the linear plasmid per cell accumulated in abundance in actively growing cells, and strongly declined during the stationary phase of growth. It was suggested that the total copy number of rDNA units in *C. albicans* cells is controlled in part by variation in the copy number of the linear extrachromosomal plasmid. It was also suggested that rapid change in the total number of rDNA copies is important for adaptation, and ultimately, better survival in the mammalian host.

The preliminary data indicate that the copy number of extrachromosomal rRNA genes may be regulated and not determined simply by selection. The variation of rRNA genes in *C. albicans* is reminiscent of the well-known cases of physiologically related amplification of the cellular copy number of rDNA in other organisms (reviewed in Ref. 125).

4. Mitochondrial DNA

The mitochondrial genome of *C. albicans* is circular with a contour length of approximately 41 kbp, thus representing a middle-size range for fungi (126–128; see also 70). The restriction site polymorphism of mitochondrial DNA was analyzed in 300 strains of *C. albicans*, and the variability found was less than expected by comparison with similar analysis of *S. cerevisiae* (53). It was suggested that the inverted repeat of 5 kbp in *C. albicans* mtDNA was acting as a stabilizing element in a manner similar to the stabilizing element in chloroplasts (126, 129, 130).

C. Dispersed Repetitive Sequences

The work of Hicks and co-workers (95–97), and Scherer and coworkers (106), who cloned *C. albicans* middle-repetitive identical sequences, Ca3 of 12 kbp and

27A of 15 kbp, initiated the use of molecular epidemiological probes to distinguish between species of the genus *Candida*, as well as strains of the species *C. albicans*. Both probes were also successfully used to address an important clinical problem of the identity of the strains from multiple sites of infection in the body. In the decade that followed, several more repetitive sequences were reported, and some of them were also used as molecular probes.

Three sequences, so-called RPSs whose sizes varied between 2.1 and 2.9 kbp (79, 131, 132), HOK of 5.3 kbp, and RB2 of undetermined size (133), were homologous to a portion of Ca3/27A (79, 133, 134). In addition, the following sequences have been reported: the so-called MspI fragment of 2.9 kbp (135); CARE-1 and CARE-2 of 0.47 kbp and 1.6 kbp, respectively, with no homology between them (136, 137); a Ca24 family of a size comparable to that of Ca3 and sharing some sequences in common but made up mostly of non-Ca3 sequences (105); EOB1 and EOB2 of 1.0 kbp and 0.38 kbp, respectively (138); and Rel-1 and Rel-2 of 0.22 kbp and 2.8 kbp, respectively, which are different but share two short subrepeat sequences (78). Recently it was found that CARE-2 and Rel-2 share a retrotransposonlike sequence-designated kappa, which is believed to be partially responsible for their repetitious nature (80). The MspI fragment is thought to be present as tandem copies rather than dispersed. Additional data on interstrain variability related to MspI are needed to clarify this suggestion, however. It appears that Ca24, CARE-1, Rel-1, MspI fragment, and Ca3/27A/ RPS/HOK/RB2 represent independent families of the repeats. Some of the above-mentioned sequences hybridized only to *C. albicans* DNA, whereas others produced various degrees of weak signals with different species of the genus *Candida* (105, 132, 135, 138). This result simply confirms relatedness of species within the genus. The copy number of repetitive sequences ranged from few to approximately 100 per cell, and the number of the repeats differed from strain to strain. For example, there are 80 copies of RPS1 in the *C. albicans* FC18 strain, but much less in NUM812 (79). The repetitive DNA of Rel-2 is present on most and perhaps all chromosomes in every strain studied, suggesting that this sequence is important for chromosome functioning. The unique portion of Rel-2, however, maps to only a few nonhomologous chromosomes, which are variable between strains (77). The repeats are usually dispersed over the majority of the chromosomes, but they also could be tandemly amplified on the same chromosome (78, 105, 132, 136; H. Chibana and P. T. Magee, unpublished results). When repetitive sequences are labeled and hybridized to the blots of restriction digests of the genomic DNA from independent strains, they produce specific patterns of multiple signals for each strain, as illustrated in Fig. 5a for sequence CARE-2. It was found that morphological mutants with various types of spontaneously altered electrokaryotypes preserved the same pattern of signals as their parental strain (46, 48, 81), thus allowing distinction from independent strains with different unique electrokaryotypes. Similarly, identical hybridization patterns were reported for morphological mutants and a parental strain by Lasker

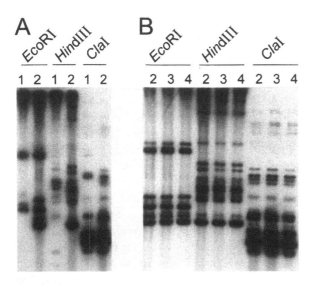

Figure 5 The comparison of signal patterns by molecular epidemiological probes. (A) The two unrelated *C. albicans* laboratory strains (1) 3153A and (2) SGY-243, as analyzed with the probe CARE-2. (B) The fluconazole resistant mutants fzE5 (3) and fzD5 (4) derived from strain SGY-243, as analyzed with the probe 27A. See section IV.A.2 for the chromosomal alterations in the mutants. See Ref. 40 for the origin of the strains. Total DNA was digested with restriction enzymes *Eco*RI, *Hin*dIII or *Cla*I, as indicated at the top of the figure, and was separated by conventional electrophoreses. The Southern blot analysis was performed by hybridization with probes CARE-2 and 27A labeled with Digoxigenin (Genius Kit, Boehringer-Mannheim, Indianapolis, IN).

et al. (137), although electrokaryotypes were not presented. Also, as presented in Fig. 5b, a parental strain and its fluconazole resistant mutants with altered chromosomes (see IV.A.2) showed identical patterns with molecular probe. In one instance, DNA polymorphism was observed among mutants selected on 5-fluorocytosine solid minimal medium with subsequent overnight incubation in a rich medium, as probed with 27A (106). A total of two signals out of 12 was changed. Another group published several series of the sequential isolates from patients undergoing fluconazole treatment, in which some members of the series exhibited one or two changes in CARE-2 pattern (107, 139). Despite a basic similarity of patterns of the signals between the original strain and the mutant or the original isolate and its sequential derivatives, such cases may require the use of several epidemiological probes; for example, 27A and CARE-2, combined with several enzymatic digests of the total DNA to rigorously exclude possible contamination. Recently, PCR techniques have been used for species and strain identification (140). These techniques allow rapid screening of a larger number of isolates. Because repetitive-sequence probes have been more carefully studied and controlled, however, they remain a practical tool for strain delineation.

The function, if any, of the above-mentioned repetitive sequences in the *Candida* genome remains obscure. (See Sec. III.A.1b, and, e.g., Ref. 132.)

During the past 10 years several repetitive sequences with homology to the known transposable elements in *S. cerevisiae* were identified in the *Candida* genome. Tca1 consists of two direct repeats of the alpha element separated by approximately 5.5 kb of DNA (141) containing three open reading frames that are no longer functional (142). Tca1, however, is transcriptionally active, producing an abundant mRNA at room temperature but not at 37°C, which suggests that this element might be involved in the mechanism of sensing environmental changes. It was argued that the above-mentioned property of Tca1 is significant for virulence, and also that such elements as Tca1 may have a role in the evolution of genomic diversity in *C. albicans* (141). In relation with the latter suggestion it is interesting to note that the frequency of chromosomal rearrangements in morphological mutants substantially increases at low temperatures. (See Sec. III. A.2b.) Tca1 was also used for the identification of *C. albicans* (143). Another retrotransposonlike element, pCal (or Tca2), is considered to be unusual, as it produces a high level of its preintegrative double-stranded DNA, which can be identified as a band on the conventional electrophoresis gel (142). It is important to note that this unusual property was found in a strain subjected to the action of two mutagens, which might have created a rare genomic condition. The same laboratory characterized one more retrotransposonlike sequence, kappa (see above) (80). A recent screen of a *C. albicans* genomic sequence provided by the Stanford University database helped to identify more than 350 insertions homologous to retrotransposons (144). At present the function of *Candida* retrotransposonlike elements remains unknown. It is also not clear if all of them are truly transposable elements that change positions in the genome. The specific regulation of the tRNA genes, by the analogy with other organisms, was suggested by the identification of a 399-bp beta element, 9-bp upstream of a seryl-tRNA$_{CAG}$ gene, in several fresh clinical isolates, as well as immediately adjacent to tRNA^{SER}, tRNA^{ASP}, tRNA^{ALA}, and tRNA^{ILE} genes in one laboratory strain, but not in others (145).

IV. THE ROLE OF CHROMOSOMAL INSTABILITY IN ADAPTATION TO THE CHANGING ENVIRONMENT

Under regular circumstances, *Candida* infectivity is undoubtedly defined by the condition of the host, and is restricted to the mucous surfaces of the gastrointestinal tract and genitalia by a healthy immune system. Nevertheless, within this comparatively restricted habitat, a wide variety of conditions occur, requiring microbial phenotypic plasticity in the form of tolerance to different pH, nutrients, and coexisting natural microflora and mucous conditions. Although the pathogenicity of the fungi is poorly understood, the ability of the opportunistic species to survive on mucosal surfaces in a healthy host can be compared with the survival of the pathogens, whose pathogenetic factors are well defined. Furthermore, in the course of migration from mucosal surfaces to deeper tissues of the immunocompromised host, *Candida* encounters unfamiliar environments, which require rapid adaptation. In the course of the past few years, as the methodology and the separation of chromosomes continued to improve, a role of one process, nondisjunction, emerged as an important means in adaptation to the new environments. It was found that the copy number of different specific chromosomes is altered in different environments, perhaps as a response to the different specific stresses. Because the chromosome copy number was involved in such important functions as utilization of nutrients or resistance to antibiotics (44, 40, 87), this novel means for regulating gene expression in microbes may be general in *C. albicans*. On the other hand, the role of other types of chromosomal rearrangements remains to be elucidated, including, for example, deletions and translocations, largely documented in altered electrokaryotypes of spontaneous morphological mutants (46, 47). Finally, the detailed investigation of the fate of *Candida* cells on a selective plate changed our understanding of this microbe's reaction to the new environment (27). As will be described below, the appearance of the advantageous mutants having a specific chromosomal nondisjunction highly increased after the contact with either an alternative carbon source or a drug. These types of experiments may serve as a model of *Candida* behavior in the course of progressive dissemination in the body or under the conditions of the antibiotic treatment.

A. Chromosome Copy Number: A New Principle of Gene Regulation

The understanding of the causal relationship between chromosome alteration and gene expression came from the study of a specific chromosomal nondisjunction, which occurred in mutants selected on L-sorbose medium (42). In the cases analyzed so far, including utilization of the alternative carbon sources L-sorbose and D-arabinose, as well as resistance to fluconazole, many mutants, each derived as an independent mutational event, had the gain or loss of chromosomes specific to the given change in the environment (40, 44, 78). We speculate that the well-known phenomenon of nondisjunction (146), which is frequent and well tolerated in lower fungi, is responsible for the decrease or increase of the copy number of a specific chromosome.

The aneuploidy condition created by the change in the chromosome copy number obviously leads to gene imbalance and diminution of many functions. We suggest that additional compensatory mutations partially or fully correct the functional damages under the continuing selective pressure. In case the pressure is removed, the opposite second nondisjunction may occur as a relatively frequent event, thus recovering euploidy, and the mutant regains a competitive rate of growth. The evolutionary possibility of the monosomy was discussed by Janbon et al. (27). It was concluded that new strains could evolve by selection in the changing environment.

Because the diminution of the copy number of a particular chromosome was observed, the concept of the negative regulation was introduced (44). This hypothesis was recently substantiated by cloning a number of the negative-regulators genes of the utilization of L-sorbose that are distributed over different chromosomes (147; Y.-K. Wang et al., unpublished results). In addition, an independent group reported two negative regulators of fluconazole resistant (148, 149). It was suggested that *C. albicans* contains a resource of potentially beneficial genes, as well as their positive and negative regulators, which are distributed over chromosomes in such a way that their expression can be activated by changes in the chromosome number (40). This mechanism is based on the ratio between the structural and regulatory gene number, and is controlled by a simple but effective process; that is, the change in the chromosome copy number that is tolerable in this and other lower fungi. It remains to be seen if this novel means of regulation by changes in the chromosome number operates in organisms other than *C. albicans*.

1. Utilization of the Alternative Carbon Sources L-sorbose and D-arabinose

Some but not all strains of *C. albicans* cannot assimilate L-sorbose as a carbon source. Such normal strains give rise to L-sorbose utilizers at high frequencies, however. The mutants, which acquired the ability to utilize L-sorbose, were selected in such a way that each mutant was derived as an independent mutational event (44, 87). The *SOU1* gene (named for sorbose utilization) responsible for the utilization of L-sorbose was subsequently cloned from the *C. albicans* genomic library (44). It was shown that the copy number of chromosome 5 controls the expression of *SOU1*, so that strains disomic and monosomic for this chromosome were, respectively, nonutilizers and utilizers of L-sorbose (Fig. 6). This conclusion was based on the examination of 69 electrophoretic karyotypes of sorbose-positive and -negative derivatives of two laboratory strains, 3153A and CAF4-2, including four complete series; that is

Sou⁻ $\rightarrow$ Sou⁺ $\rightarrow$ Sou⁻ $\rightarrow$ Sou⁺
Disomic   Monosomic   Disomic   Monosomic

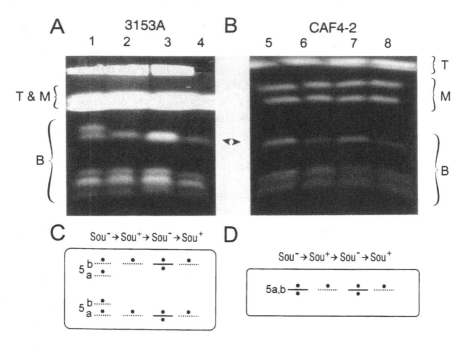

Figure 6 Chromosomal patterns from two *C. albicans* laboratory strains, 3153A and CAF4-2, showing sequential sorbose-utilizing and sorbose-non-utilizing derivatives. (a) and (b): OFAGE separation of chromosomes of typical-representative sequential-series of derivatives from strains 3153A and CAF4-2, respectively. (1) 3153A (Sou⁻); (2) Sor55 (Sou⁺); (3) Sor55-1 (Sou⁻); (4) Sor55-1-1 (Sou⁺); (5) CAF4-2 (Sou⁻); (6) Sor19 (Sou⁺); (7) Sor19-1 (Sou⁻); (8) Sor19-1-1 (Sou⁺). For the origin of mutants, see Ref. 44. Arrows indicate alternating chromosome 5. For explanations, see Fig. 1. (c) and (d): Schematic representation of chromosomes 5 from the Sou⁻ and Sou⁺ strains shown in (a) and (b), respectively. The gene *CSU51*, a hypothetical negative regulator carried on chromosome 5, is designated by the symbol ●. Source: Adapted from Ref. 44.

as well as nine incomplete series of the type Sou⁻ (parental) → Sou⁺ → Sou⁻. The analysis of these series also proved that this relationship could be continuously perpetuated with each of the mutants derived from the opposite type. In our original report on chromosome 5 alteration leading to the Sou⁻ phenotype, we did not appreciate the monosomy of this chromosome, because the cultures were routinely grown in liquid medium. This condition allowed for the selection of reconstituted chromosome 5 disomics. Of the examined independent Sou⁺ mutants, 47 retained a single copy of either one of two homologues of chromosome 5, as summarized in Fig. 7. One additional mutant, however, retained both copies,

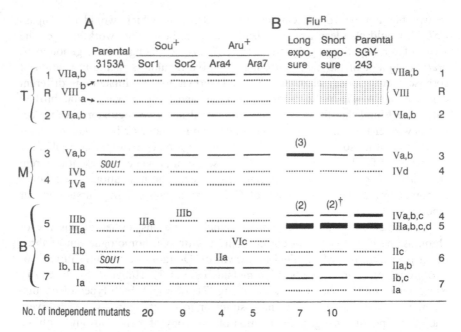

Figure 7 Schematics of representative electrophoretic karyotypes of independently derived sorbose- and arabinose-utilizing, as well as fluconazole-resistant mutants and two *C. albicans* parental strains. (A) Chromosomal patterns of strain 3153A and its sorbose (Sou⁺)- and arabinose (Aru⁺)-utilizing mutants. The chromosomal assignment of *SOU1* gene, which preserved the same positions in the Sou⁺ mutants (44), is indicated only in strain 3153A. (B) Chromosomal patterns of strain SGY-243 and its fluconazole-resistant mutants (FluR). See Fig. 2 for explanations. The chromosomal numbers in Sor1, Sor2, and Ara4 indicate a remaining homologue; in Ara7 an additional homologue. The numbers in parentheses above chromosomes of the fluconazole-resistant mutants indicate the copy numbers. The numbers at the bottom of the figure denote the number of independently derived mutants that were analyzed. These numbers do not include mutants from the series of sequential derivatives presented in Fig. 6. †One mutant, fzE5, obtained after a short exposure, contained only one shorter homologue of chromosome 4 instead of two.

but instead had an approximately 300-kbp deletion on one of the chromosome 5 homologues. Another mutant of strain CAF4-2 acquired alteration of chromosomes R and presumably 3 but had no visible change in any other chromosome (E. Rustchenko, unpublished data). While there may be other genetic changes that produce a sorbose-positive phenotype, the monosomy of chromosome 5 is a major means. Because *SOU1* does not reside on alternating chromosome 5, it was suggested that chromosome 5 contains a hypothetical negative regulatory factor, *CSU51* (control of sorbose utilization), which inhibits expression of *SOU1* in two but not one copy. Obviously, the deletion, which would encompass *CSU51*, would also lead to the utilization of L-sorbose. This work initiated the study of the negative regulation in *C. albicans*. By using a total genomic library a number of the genes encoding negative regulators of the utilization of sorbose were cloned, characterized, and shown to be situated on different chromosomes (147; Y.-K. Wang, G. Janbon, and E. Rustchenko, unpublished data). We found that these genes are additional negative regulators, whose action is weaker than the action of major negative regulator *CSU51*, which is controlled by chromosome 5. This result revealed the complexity of the control of the utilization of food supplies. Apparently these different genes represent different cellular levels of repression of structural genes for the utilization of secondary carbon sources. The identification of *CSU51* is currently being pursued in our laboratory.

Another case of specific chromosomal alterations, either monosomy of chromosome 6 or trisomy of chromosome 2 with one homologue out of three being greatly truncated, resulted in growth on D-arabinose medium in each of two groups of D-arabinose-positive mutants (87; D. H. Huber, J. Smith and E. Rustchenko, unpublished observation) (Fig. 7). Although both types of nondisjunction—either leading to the monosomy or to the trisomy—were observed, one of the specific changes, the addition of one copy of a chromosome with the large deletion of approximately 1 Mbp, cannot be viewed simply as a trivial increase in the copy number of all the genes carried on this chromosome. Although less studied, the case of D-arabinose utilization by means of monosomy of a specific chromosome is a direct parallel of the causal relationship uncovered in mutagenesis on L-sorbose. The understanding of this mechanism, which is alternative to the monosomy and, respectively, negative regulation, will be the subject of a future study.

2. Resistance to Fluconazole

Fluconazole is the main antifungal agent used to control candidosis. The high frequency of occurrence of fluconazole-resistant mutants has hampered its use, however. The possibility of electrokaryotypic changes associated with fluconazole resistance in clinical *C. albicans* isolates was investigated in several laboratories (reviewed in Ref. 40), and no differences in chromosomal patterns were reported. The lack of revealing altered electrophoretic karyotypes apparently was due to the limitations in the resolution of the chromosomal separation. As subsequently found by Perepnikhatka et al. (40), exposure of a *C. albicans* laboratory strain SGY-243 to fluconazole on a petri plate resulted in the nondisjunction of two specific chromosomes in 17 drug-resistant mutants, each obtained as an independent mutational event. The changes were related to the duration of the drug exposure. The loss of one homologue of chromosome 4 occurred after incu-

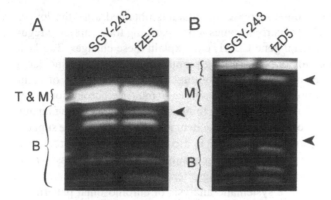

Figure 8 Precise separation of the chromosomes of the representative fluconazole-resistant mutants fzE5 and fzD5, obtained after 7 and 35 days of exposure to fluconazole on the plates, respectively. Electrophoretic karyotypes obtained by two running conditions are presented in (A) and (B). (See Fig. 1 for explanations.) Arrows indicate chromosome 4, in which two out of three copies are lost, and chromosome 3, in which one copy was acquired. No other detectable chromosomal changes are seen in these separations. Source: Adapted from Ref. 40.

bation on fluconazole medium for 7 days (Figs. 7, 8a). A second change, the gain of one copy of chromosome 3, which carried the genes *CDR1* and *CDR2* affiliated with fluconazole resistance, was observed after exposure for 35 or 40 days (Figs. 7, 8b). The fluconazole-resistance phenotype, electrokaryotypes, and transcript levels (see below) of mutants were stable after growth for 112 generations in the absence of fluconazole. For the first time the resistance to fluconazole was reported to be commonly caused by chromosomal changes. Throughout the work of many groups, fluconazole resistance in clinical isolates of *C. albicans* has been associated with a combination of several distinct mechanisms (139, 150, 151). These include mutations at the active site of the drug target 14α-sterol demethylase, encoded by the *ERG11* gene; overexpression of *ERG11* (152–154); alterations in the ergosterol biosynthetic pathway (152, 155); and overexpression of the genes involved in energy-dependent drug efflux (139, 151, 156, 157). The overexpression of two relevant genes, *CDR1* and *CDR2*, which encode proteins homologous to ATP-binding cassette (ABC) drug pumps, as well as *MDR1* (*BEN^R*), which encodes a protein similar to pumps of the major facilitator superfamily (MFS), has been associated with fluconazole resistance in clinical isolates (156, 158, 159). When the transcription levels of the candidate fluconazole-resistance genes, *ERG11*, *CDR1*, *CDR2*, and *MDR1*, were measured in the mutants with the specific chromosomal nondisjunctions, it was found that they either remained the same or were diminished, except that expression of *CDR1*, carried on the duplicated chromosome, increased in mutants obtained after the longer exposure. The copy number of chromosomes 4 or 3 carrying transcriptional regulators, as well as the structural gene CDR1, can explain these changes. The lack of substantial overexpression of putative drug pumps or the drug target indicated that some other mechanism(s) might be operating. Significantly, none of eight mutants of strain 3153A possessing various single or multiple chromosomal rearrangements were altered in their sensitivity to fluconazole. Because the mutants analyzed in our study were obtained as independent mutational events, we suggest that a change in chromosomal copy number is a common means for producing resistance. This mechanism might be based on a similar response to stress, which lead to the formation of the mutants utilizing alternative nutrients.

Despite the importance, a systematic analysis of chromosomal patterns in clinical isolates has not been undertaken. Such a study requires reliable series of genetically related strains, which have developed resistance, and which have been adequately collected, handled, and preserved in order to avoid the induction of irrelevant chromosomal instability. In addition, the chromosomal separations have to be of a high quality.

It is important to note that there is a remarkable similarity between the results obtained under laboratory conditions by Perepnikhatka et al. (40) and the results obtained by several other groups from patients undergoing treatment with fluconazole. A number of independent studies revealed that a series of *C. albicans* strains isolated sequentially in time from patients showed a progressive increase of resistance, and that overexpression of efflux pumps or point mutations of the target gene, *ERG11*, was not initially manifested, or in fact did not appear at all during the study (107, 157, 160, 161). In one of these studies of five different patients (160), sequentially obtained fluconazole-resistant mutants from one patient did not exhibit overexpression of *MDR1*, *CDR1*, *CDR2*, nor did the mutants have point mutations in the *ERG11* gene, even though strains were isolated and tested for a period extending to nearly 10 months. Strains from the other patients showed an alternation in overexpression of the *MDR1* and *CDR* genes. In addition, a recent study showed that cross-resistant mutants, including resistance to fluconazole, were obtained in strains with deleted genes for drug-pumps after exposure to the itraconazole (162). Another recent study with clinical isolates demonstrated that high constitutive expression of the *MDR1* gene in fluconazole-resistant strains was due to a transregulatory factor, but not, for example, a mutation in the promoter region of *MDRI* (163). Perhaps the latter result might be interpreted in the context of chromosomal alterations. We similarly found a substantial overexpression of *MDR1* gene in highly fluconazole-resistant mutant FR2 having five chromosomal alterations, including a chromosome carrying *MDR1* (40). Another explanation might be that the change of *MDR1* expression was due to a change of copy number of a regulatory gene on an altered chromosome. A combined effect of several chromosomal alterations, as well as gene mutations, also cannot be excluded. Thus, the laboratory study conducted by us and the clinical studies undertaken by the other groups both revealed that at early stages of infection, fluconazole resistance arose by mechanisms other than overexpression of known efflux pumps or mutations of a target gene. Because chromosomal nondisjunction produced high frequencies of relatively low-level fluconazole-resistant mutants, it is reasonable to suggest that a similar alteration is causing primary resistance under clinical treatments. The higher resistance levels of the latter clinical isolates can be viewed as a secondary event, which occurred as a result of the accumulation of gene mutations under prolonged selective pressure. The identification of the resistance genes controlled by chromosomal nondisjunction will be the subject of future research. The importance of the discovery of chromosome copy number as a general response controlling drug resistance lies in revealing new genes for the resistance, which in combination with mutations in the target gene, genes for the efflux pumps, or alone, can allow a single strain to adapt to antifungal treatment and establish a persistent infection. Together with the fact that the same type of regulation was observed for the utilization of different nutrients, these data further establish the hypothesis that change in the copy number of a chromosome is a general means of regulating physiologically important genes in *Candida*.

3. Adaptive Mutagenesis as a Means to Cope with New Environment

The ability of the microbial cells to adapt to the changing environment has been the subject of numerous studies. A number of characteristic features of microbial adaptability to selective conditions have been investigated as a result of renewed interest during the last decade, primarily due to a prominent paper by Cairns et al. (164). It was originally suggested that mutations arise in nondividing microorganisms subjected to selective pressures, and this phenomenon was denoted as "directed mutation" or "adaptive mutation." Currently, however, adaptive mutations denote mutations that are formed at higher frequencies in response to the selective conditions, even though they may not necessarily be "directed" (165). Under the current usage, mutants resistant to and induced by UV light would be considered to arise by adaptive mutations, as would nutritionally selected mutants induced by starvation or stress. The first attempt to document the adaptive mutagenesis in *Candida* involved experiments on the resistance to heavy metals (166). A detailed analysis, as well as crucial evidence for adaptive mutagenesis, came from studies of the comparison of the daily rates of formation of Sou$^+$ mutants (27). As presented in Fig. 9, there was a striking increase of four orders of magnitude in the rates of mutant formation immediately after contact with the selective medium to days 4 and 8, clearly suggesting that the vast majority of the mutational events appeared gradually after contact with the selective

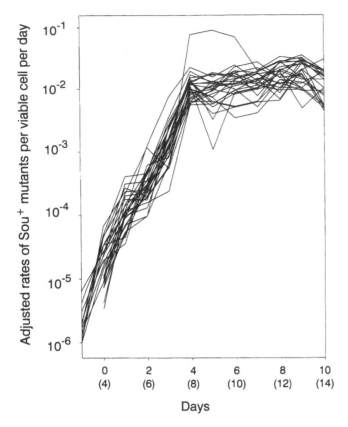

Figure 9 The adjusted rates of sorbose-utilizing mutants of each of the 27 clones of strain 3153A per viable cell per day. The adjusted values were calculated from the daily number of colonies, divided by the number of viable cells 4 days prior to the appearance of colonies. Both the deduced time of formation of the mutation (top row) and the time of the appearance of the corresponding colonies (bottom row, in parentheses) are presented. Source: Adapted from Ref. 27.

condition. Taken together with the time-dependent manner of the appearance of the mutants, the increased rates of formation of at least the late mutants were indicative of adaptive mutagenesis. The comparison of the electrokaryotypes of the mutants, which appeared on the experimental plate on different days within a 2-week period, showed that, as expected, all of them had a common alteration, a monosomy of chromosome 5. Some random additional changes in a few of the mutants were not considered relevant. Most important, the control experiments designed to test in several different ways if there was any turnover, residual growth, or formation of microcolonies on sorbose-containing plates, all gave negative results.

Even if one assumes some reasonable turnover during the death of the Sou⁻ cells, in which the rate of death exceeds the rate of growth and that this turnover is required for producing Sou⁺ mutants, the high rate of formation of approximately 10^{-2} Sou⁻ mutants per viable cell per day clearly established that these mutants arose by adaptive mutagenesis. With a normal mutation rate of 10^{-5} per cell per generation, one would have to assume that the dying cells are dividing every minute in order to accumulate the number of Sou⁺ mutants observed during the sharp increase of the rates. At the present time, the study of the _C. albicans_ cells' behavior on L-sorbose plates is the most striking example of the adaptive mutagenesis. Other, mostly bacterial models, showed approximately a tenfold difference between the frequencies of classic and adaptive mutagenesis. Earlier, Borst and Greaves (158) pointed out that the reaction of a microbe to major changes in the nutrient supply is expected to be of two kinds, a preadaptation of a fraction of the population through DNA rearrangements, or adaptation after an environmental change. Because starvation is not immediately lethal, it is more efficient for the organism to use alternative genes when needed. The study of the adaptive mutagenesis to the condition of a new source of carbon can be considered as a model of _Candida_ reaction to the outside stress and change. Such controlled laboratory studies may represent situations in the host during a progressive disseminate infection or treatment with the antibiotic.

As was previously realized, "_C. albicans_ should be considered as a fungus possessing a multiplicity of properties, each with a low propensity for enhancing fungal infectivity but none necessarily dominant, and all, even in combination, unlikely to overcome fully intact host defenses" (28). The genes, which promote survival in the host, and particularly genetic diversity and plasticity, are as important as virulence factors of other pathogens under these circumstances. Any of the potentially unstable features of the genome listed in this chapter could change the expression of the relevant genes.

V. PHENOTYPIC FEATURES AS A SUBJECT OF VARIABILITY AND INSTABILITY

A. "Switching" Colonial Morphologies

The term _switch_ was introduced in 1985 by Soll and co-workers (37) to describe reversible changes of colonial morphologies in populations of the same strain. Subsequently Soll (159) defined all reversible high-frequency mutations occurring in _C. albicans_ and related species as "switching," independent of any mechanism. As originally used, switching refers to two or a few alternative genetic states that are reversible. The most extensively studied examples include mating-type switching in *S. cerevisiae* (160), inversion control of gene expression in *Salmonella* (161), and P$^+$ and P$^-$ antigenic variation in *Neisseria gonorrhoeae* (162). Clearly, denoting any high rates of mutation switching can be inappropriate. For example, Das et al. (163) uncovered *cyc1* mutations in *S. cerevisiae* that reverted with the frequency of 10^{-4} and had a reversion rate of over 10^{-5} per generation.

Soll and co-workers (159) have investigated and applied the term switching primarily to two types of instability, colony morphology mutants studied mainly with strain 3135A, as described below, and to the white–opaque transition studied with strain WO-1 and its derivatives, as discussed in the next section.

Spontaneously arising colonial forms, as well as those induced by mild UV irradiation, were postulated to switch with certain defined frequencies between seven morphologies (37). The normal parental colony type is "smooth," whereas variants arising from the parental strain 3153A were denoted as "star," "ring," "irregular wrinkle," "stipple," "hat," and "fuzzy." This first report already contained data inconsistent with the proposed nomenclature, as two independent photographs of the "ring" morphology were not exactly the same. In later papers some visibly different morphologies were given the same name introduced by Slutsky et al. (37); compare, for example, the photograph of "fuzzy" reported by Soll (159). "Revertant smooth" denoted the "smooth" revertants from each of the variants. The normal strain was reported to give rise to variants at a frequency of approximately 10^{-4} spontaneously or approximately 10^{-2} after UV irradiation. "Revertant smooth," however, spontaneously gave rise to variants at a frequency of approximately 10^{-2}, indicating the lack of true reversion. The frequencies of a variant producing another type of variant varied from approximately 10^{-2} to approximately 10^{-4}, although one type could not be interconverted (37). In addition, "revertant smooth" types segregated morphologies, which the authors could not fit into the above-mentioned seven types.

The spontaneous switching of colony morphology studied with 3153A by Soll and co-workers is obviously the same phenomenon systematically studied by Rustchenko and co-workers, who, however, demonstrated an association with chromosomal alterations. Whelan (164) also suggested in his review the basic similarity between the results of the Soll group and the early author Brown-Thomsen. Our contention is that Soll and his co-workers have not used rigorous criteria to define specific phenotypes, and their results do not fulfill the concept of "reversibility." Furthermore, our studies with the same strain 3153A established that morphological mutants, including highly unstable representatives, do not return to the truly original form, nor do they alternate between a limited number of forms. The number of distinguishable types of colony morphologies certainly exceeds seven. In addition, the unstable morphological mutants studied by Rustchenko and co-workers gave rise to a large number of different forms specific to each mutant. Our conclusion, based not only on our own extensive work and numerous reports published since 1935 (see Ref. 28), but also on the work of Pomés et al. (165) and on published work by Soll and his co-workers, is that colonial forms do not readily revert, and therefore do not switch. Furthermore, the variation of colony forms can be simply explained by the gene imbalance due to chromosomal alterations.

Pérez-Martín et al. (166) recently reported that double disrupted *sir2-Δ/ sir2-Δ* strains of *C. albicans* produced mutants with altered chromosomal patterns and colony morphologies at frequencies as high as one in 10. In the yeast *S. cerevisiae*, the *SIR2* gene maintains the inactive chromatin domains required for transcriptional repression at the silent mating-type loci and telomeres. Such repression is associated with specialized chromatin structures whose integrity depends upon a complex combination of cis-acting sites, several shared transacting factors, such as Sir1p, Sir2p, Sir3p, and Sir4p, as well as histones. Because *sir2-Δ/sir2-Δ* could affect the expression of any of a number of genes, the increased frequencies of mutants with altered karyotypes can be explained by numerous models (166). More important, J. Pérez-Martín (unpublished result) has been unable to consistently reproduce the effect of *SIR2* disruptions on colony morphologies in *C. albicans*. This variation of instability resembles the findings of Rustchenko-Bulgac (47), who observed the periodic stabilization of initially unstable chromosomal alterations in morphological mutants. Chromosomal instability adequately explains the results reported by Pérez-Martín et al. (175) and J. Pérez-Martín (unpublished result).

More recently, Lachke et al. (167) reported that the related species, *Candida glabrata*, undergoes a reversible, high-frequency switching between three colony types that were distinguished on $CuSO_4$ indicator plates. The three colony types, white (Wh), light brown (LB), and dark brown (DB), were associated with different levels of *MT-II* (metallothionen) and *HLP* (hemolysinlike protein) mRNAs. The statistical nature of the transitions, the electrophoretic karyotypes, and the mechanisms responsible for producing the three types of colony forms were not presented. Because of an earlier report on the ability of a natural haploid species *C. glabrata* to undergo chromosomal rearrangements (168), it would be of particular interest to examine the electrophoretic karyotypes of the above-mentioned mutants.

B. White-Opaque Transition

As introduced above, the extensively studied white-opaque "switching" in the strain WO-1 and its derivatives consists of transitional changes between the white phase, in which colonies appear as smooth and white on solid media, and the opaque phase, in which colonies appear flattened and gray, (178–180). Opaque cells differ from white cells in a number of properties, including the following: cell shape (in which the opaque cells appear elongated and rod shaped, superficially resembling pseudomycelium); budding pattern; response to different temperatures; growth rate; frequency of transition; sensitivity to UV irradiation, sulfometuron methyl, and 5-fluoroorotic acid; accumulation of bismuth sulfide on bismuth medium; interaction with host cells; and virulence (179, 181). Interconversion of the white and opaque phases occurs spontaneously and can be induced by temperature shifts. For example, temperature shift from 24°C to 34°C results in mass conversion of the opaque form to the white form (179) within two cell divisions (J. Hicks, personal communication). Conversely, at 24°C a population of white cells produces white colonies with opaque sectors at the periphery (179). These opaque sectors become visible after 5 days of incubation at 24°C on YPD medium and subsequently develop into large clusters of opaque cells (Huber and Rustchenko, unpublished results). The frequency of mutual interconversion at 24°C was estimated by Anderson et al. (178) to be between 10^{-2} and 10^{-3}, and the rates were estimated by J. Hicks (personal communication) to be less than 5 $\times$ 10^{-3} per cell division for opaque-to-white conversion and 5×10^{-2} per cell division for white-to-opaque conversion. On the other hand, Rikkerink et al. (179) reported that opaque cells converted to white cells at 24°C with a frequency of 10^{-4} to 10^{-5} and that from 0.1% to 5% of white colonies arise from population of opaque cells formed at room temperature. Superimposed on the interconversion between opaque and white morphologies is the formation of various colony morphology mutants (180), presumably due to chromosomal alteration, as described in the previous section. There are no systematic reports that the interconversion between white and opaque phenotypes involves all of the other multiple phenotypes, including estimated frequencies of the white-opaque transition. Furthermore, genes that are specifically expressed in the white phase, *WH11*, or in the opaque phase, *OP4*, *SAP1*, and *CDR3*, have been reported, and portions of the regulatory regions within the promoter regions have been characterized (182–187). Srikantha et al. (187) presented evidence that common regulatory processes determine white-opaque switching and budding-hyphal dimorphism of *C. albicans*. Wh11p, an abundant cytoplasmic protein, is induced both by the opaque-white transition, and the hypha-budding morphogenesis in WO-1, and both of these repressional events require identical promoter segments (187). Furthermore, opaque-phase cells of WO-1 and hyphal cells both contain certain specific antigens not found in white-phase cells (169).

In addition, Sonneborn et al. (188) suggested that the Efg1p transcription factor, which is required for hyphal induction (17, 23), is also an essential element for white-opaque switching in WO-1. *EFG1* is transcribed as a 3.2-kb mRNA in white-phase cells and as a less abundant 2.2-kb mRNA in opaque-phase cells, due to different transcription start sites (189). After the commitment event, the *efg1* null mutant forms daughter cells that have the smooth (pimpleless) surface of white-phase cells, but maintain the elongate morphology of opaque-phase cells, indicating that Efg1p is not essential for the switch event per se, but is essential for a subset of phenotypic characteristics necessary for the full expression of the phenotype of white-phase cells. Srikantha et al. (189) concluded that *EFG1* is not the site of the switch event, but that it is controlled downstream of the switch event.

Klar et al. (190) and Srikantha et al. (191) have suggested that white-to-opaque transitions in strain WO-1 is mediated through histone deacetylases, which regulates chromatin structure by selective histone deacetylation, leading to changes in chromatin folding and interactions between DNA and DNA-binding proteins. Klar et al. (190) demonstrated that the frequency of white-to-opaque transitions in WO-1 was greatly enhanced by exposing the strain for 48 hr on agar containing 10 µg/l of the histone deacetylase inhibitor trichostatin-A (TSA). In addition, Srikantha et al. (191) examined the expression of five histone deacetylase genes in white and opaque phases of the white-opaque transition, and they examined the frequency of switching and the expression of white-phase and opaque-phase specific genes in mutants deleted in one or an other of the histone deacetylase genes. The results of their study suggested that the two deacetylase genes, *HDA1* and *RPD3*, play distinct roles in the suppression of switching, and that the two distinct and selective roles in the regulation of phase-specific genes. Furthermore, they suggested that the white-opaque transition is due to epigenetic changes in chromatin structure, presumably involving the inhibition of gene expression by chromatin modification through the deacetylation of histones at a key gene, which has not yet been identified.

Although the molecular mechanisms that determine white-opaque switching is still unclear, because opaque-phase cells are elongated and resemble pseudohyphae, and because common components are required for both processes, it appears that white-opaque transition and hyphal-budding morphogenesis are related phenomena that are conceptually distinct from the other types of "general instability" discussed in this review.

ACKNOWLEDGMENTS

We thank Dr. M. Wellington for obtaining signal patterns with molecular probes 27A and CARE-2. The studies from our laboratories cited in this review were supported by grants AI29433 and GM12702 from the National Institutes of Health.

REFERENCES

1. A Datta. What makes *Candida albicans* pathogenic? Curr Sci 62:400–404, 1992.
2. E Seregni, C Botti, S Massaron, C Lombardo, A Capobianco, A Bogni, E Bombar-

dieri. Structure, function and gene expression of epithelial mucins. Tumori 83:625–632, 1997.

3. B Singh, A Datta. Regulation of *N*-acetylglucosamine uptake in yeast. Biochim Biophys Acta 557:248–258, 1979.

4. K Natarajan, A Datta. Molecular cloning and analysis of the NAG1 cDNA coding for glucosamine-6-phosphate deaminase from *Candida albicans*. J Bio Chem 268: 9206–9214, 1993.

5. F De Bernardis, FA Mühlschlegel, A Cassone, WA Fonzi. The pH of the host niche controls gene expression and virulence of *Candida albicans*. Infec Immun 66:3317–3325, 1998.

6. FA Mühlschlegel, WA Fonzi. *PHR2* of *Candida albicans* encodes a functional homolog of the pH-regulated gene *PHR1* with an inverted pattern of pH-dependent expression. Molec Cell Bio 17:5960–5967, 1997.

7. SM Saporito-Irwin, CE Birse, PS Sypherd, WA Fonzi. *PHR1*, a pH-regulated gene of *Candida albicans*, is required for morphogenesis. Molec Cell Bio 15:601–613, 1995.

8. LA Alex, C Korch, CP Selitrennikoff, MI Simon. COS1, a two-component histidine kinase that is involved in hyphal development in the opportunistic pathogen *Candida albicans*. Proc Natl Acad Sci USA 95:7069–7073, 1998.

9. BR Braun, AD Johnson. Control of filament formation in *Candida albicans* by the transcriptional repressor *TUP1*. Science 277:105–109, 1997.

10. JA Calera, X-J Zhao, R Calderone. Defective hyphal development and avirulence caused by a deletion of the SSK1 response regulator gene in *Candida albicans*. Infec Immun 68:518–525, 2000.

11. C Csank, K Schröppel, E Leberer, D Harcus, O Mohamed, S Meloche, D Thomas, M Whiteway. Roles of the *Candida albicans*: Mitogen-activated protein kinase homolog, Cek1p, in hyphal development and systemic candidiasis. Infec Immun 66:2713–2721, 1998.

12. CA Gale, CM Bendel, M McClellan, M Hauser, JM Becker, J Berman, MK Hostetter. Linkage of adhesion, filamentous growth, and virulence in *Candida albicans* to a single gene, *INT1*. Science 279:1355–1358, 1998.

13. JR Köhler, GR Fink. *Candida albicans* strains heterozygous and homozygous for mutations in mitogen-activated protein kinase signaling components have defects in hyphal development. Proc Natl Acad Sci USA 93:13223–13228, 1996.

14. E Leberer, D Harcus, ID Broadbent, KL Clark, D Dignard, K Ziegelbauer, A Schmidt, NAR Gow, AJP Brown, DY Thomas. Signal transduction through homologs of the Ste20p and Ste7p protein kinases can trigger hyphal formation in the pathogenic fungus *Candida albicans*. Proc Natl Acad Sci USA 93:13217–13222, 1996.

15. E Leberer, K Ziegelbauer, A Schmidt, D Harcus, D Dignard, J Ash, L Johnson, DY Thomas. Virulence and hyphal formation of *Candida albicans* require the Ste20p-like protein kinase CaCla4p. Curr Bio 7:539–546, 1997.

16. H Liu, JR Köhler, GR Fink. Suppression of hyphal formation in *Candida albicans* by mutation of a *STE12* homolog. Science 266:1723–1725, 1994.

17. HJ Lo, JR Köhler, B DiDomenico, D Loebenberg, A Cacciapuoti, GR Fink. Non-filamentous *C. albicans* mutants are avirulent. Cell 90:939–949, 1997.

18. JDJ Loeb, M Sepulveda-Becerra, I Hazan, H Liu. A G1 cyclin is necessary for maintenance of filamentous growth in *Candida albicans*. Molec Cell Bio 19:4019–4027, 1999.

19. RA Monge, F Navarro-García, G Molero, R Diez-Orejas, M Gustin, J Pla, M Sánchez, C Nombela. Role of the mitogen-activated protein kinase Hog1p in morphogenesis and virulence of *Candida albicans*. J Bacteriol 181:3058–3068, 1999.

20. T Nakazawa, H Horiuchi, A Ohta, M Takagi. Isolation and characterization of *EPD1*, an essential gene for pseudohyphal growth of a dimorphic yeast *Candida maltosa*. J Bacteriol 180:2079–2086, 1998.

21. LL Sharkey, MD McNemar, SM Saporito-Irwin, PS Sypherd, WA Fonzi. HWP1 functions in the morphological development of *Candida albicans* downstream of *EFG1*, *TUP1*, and *RBF1*. J Bacteriol 181:5273–5279, 1999.

22. A Sonneborn, DP Bockmühl, JF Ernst. Chlamydospore formation in *Candida albicans* requires the Efg1p morphogenetic regulator. Infec Immun 67:5514–5517, 1999.

23. VR Stoldt, A Sonneborn, CE Leuker, JF Ernst. Efg1p, an essential regulator of morphogenesis of the human pathogen *Candida albicans*, is a member of a conserved class of bHLH proteins regulating morphogenetic processes in fungi. EMBO J 16:1982–1991, 1997.

24. P Singh, S Ghosh, A Datta. A novel MAP-kinase kinase from *Candida albicans*. Gene 190:99–104, 1997.

25. PJ Riggle, KA Andrutis, X Chen, SR Tzipori, CA Kumamoto. Invasive lesions containing filamentous forms produced by a *Candida albicans* mutant that is defective in filamentous growth in culture. Infec Immun 67:3649–3652, 1999.

26. EP Rustchenko, DH Howard, F Sherman. Variation in assimilating functions occurs in spontaneous *Candida albicans* mutants having chromosomal alterations. Microbiology 143:1765–1778, 1997.

27. G Janbon, F Sherman, E Rustchenko. Appearance and properties of L-sorbose utilizing mutants of *Candida albicans* obtained on a selective plate. Genetics 153:653–664, 1999.

28. FC Odds. Candida and Candidosis. 2nd ed. Philadelphia: Saunders, 1988.

29. FC Odds. Candida species and virulence. ASM News 60:313–318, 1994.

30. JE Cutler. Putative virulence factors of *Candida albicans*. Ann Rev Microbio 45:187–218, 1991.

31. B Hube. *Candida albicans* secreted aspartyl proteinases. Curr Top Med Mycol 7:55–69, 1996.

32. FC Odds, L Van Nuffel, NAR Gow. Survival in experimental *Candida albicans* infections depends on inoculum growth conditions as well as animal host. Microbio 146:1881–1889, 2000.

33. J Lay, LK Henry, J Clifford, Y Koltin, CE Bulawa, JM Becker. Altered expression of selectable marker *URA3* in gene-disrupted *Candida albicans* strains complicates interpretation of virulence studies. Infec Immun 66:5301–5306, 1998.

34. M Coleman, B Henricot, J Arnau, RP Oliver. Starvation-induced genes of the tomato pathogen *Cladosporium fulvum* are also induced during growth in planta. MPMI 10:1106–1109, 1997.

35. CM Pieterse, AM Derksen, J Folders, F Govers. Expression of the *Phytophthora*

infestans ipiB and *ipiO* genes *in planta* and *in vitro.* Molec Gen Genet 244:269–277, 1994.

36. NJ Talbot, DJ Ebbole, JE Hamer. Identification and characterization of *MPG1*, a gene involved in pathogenicity from the rice blast fungus *Magnaporthe grisea.* Plant Cell 5:1575–1590, 1993.

37. G Lau, JE Hamer. Regulatory genes controlling *MPG1* expression and pathogenicity in the rice blast fungus *Magnaporthe grisea.* Plant Cell 8:771–781, 1996.

38. MC Lorenz, GR Fink. The glyoxylate cycle is required for fungal virulence. Nature 412:83–86, 2001.

39. B Slutsky, J Buffo, DR Soll. High frequency switching of colony morphology in *Candida albicans.* Science 230:666–669, 1985.

40. V Perepnikhatka, FJ Fisher, M Niimi, RA Baker, RD Cannon, Y-K Wang, F Sherman, E Rustchenko. Specific chromosomal alterations in fluconazole-resistant mutants of *Candida albicans.* J Bacteriol 181:4041–4049, 1999.

41. S Mercure, S Montplaisir, G Lemay. Correlation between the presence of a self-splicing intron in the 25S rDNA of *C. albicans* and strains susceptibility to 5-fluorocytosine. Nucleic Acids Res 21:6020–6027, 1993.

42. KE Miletti, MJ Leibowitz. Pentamidine inhibition of group I intron splicing in *Candida albicans* correlates with growth inhibition. Antimicrob Agents Chemother 44:958–966, 2000.

43. G Janbon, F Sherman, E Rustchenko. UV and X-ray sensitivity of *Candida albicans* regular strains and mutants having chromosomal alterations. Revista Iberoamericana de Micologia 18:12–16, 2001.

44. G Janbon, F Sherman, E Rustchenko. Monosomy of a specific chromosome determines sorbose utilization in *Candida albicans.* Proc Natl Acad Sci 95:5150–5155, 1998.

45. J Brown-Thomsen. Variability in *Candida albicans.* Hereditas 60:355–398, 1968.

46. EP Rustchenko-Bulgac, F Sherman, JB Hicks. Chromosomal rearrangements associated with morphological mutants provide a means for genetic variation of *Candida albicans.* J Bacteriol 172:1276–1283, 1990.

47. EP Rustchenko-Bulgac. Variation of *Candida albicans* electrophoretic karyotypes. J Bacteriol 173:6586–6596, 1991.

48. EP Rustchenko-Bulgac, DH Howard. Multiple chromosomal and phenotypic changes in spontaneous mutants of *Candida albicans.* J Gen Microbio 139:1195–1207, 1993.

49. CM Hull, AD Johnson. Identification of a mating type-like locus in the asexual pathogenic yeast *Candida albicans.* Science 285:1271–1275, 1999.

50. CM Hull, RM Raisner, AD Johnson. Evidence for mating of the "asexual" yeast *Candida albicans* in a mammalian host. Science 289:307–310, 2000.

51. BB Magee, PT Magee. Induction of mating in *Candida albicans* by construction of MTLa and MTLα strains. Science 289:310–313, 2000.

52. M Heude, F Fabre. a/α-control of DNA repair in the yeast *Saccharomyces cerevisiae*: Genetic and physiological aspects. Genetics 133:489–498, 1993.

53. WS Riggsby. Physical characterization of the *Candida albicans* genome. In: DR Kirsch, R Kelly, MB Kurtz, eds. The Genetics of *Candida.* Boca Raton, FL: CRC Press, 1990, pp. 125–146.

54. MB Kurtz, DR Kirsch, R Kelly. The molecular genetics of *Candida albicans*. Microbio Sci 5:58–63, 1988.

55. MB Kurtz, R Kelly, DR Kirsch. Molecular genetics of *Candida albicans*. In: DR Kirsch, R Kelly, MB Kurtz, eds. The Genetics of *Candida*. Boca Raton, FL: CRC Press, 1990, pp. 21–73.

56. WG Merz, C Connelly, P Hieter. Variation of electrophoretic karyotypes among clinical isolates of *Candida albicans*. J Clin Microbio 26:842–845, 1988.

57. RG Snell, IF Hermans, RJ Wilkins, BE Corner. Chromosomal variations in *Candida albicans*. Nucleic Acids Res 15:3625, 1987.

58. BB Magee, PT Magee. Electrophoretic karyotypes and chromosome numbers in *Candida* species. J Gen Microbio 133:1–6, 1987.

59. BL Wickes, R Petter. Genomic variation in *C. albicans*. Curr Top Med Mycol 7: 71–86, 1996.

60. R Anand. Pulsed field gel electrophoresis: A technique for fractionating large DNA molecules. Trends Genet 2:278–283, 1986.

61. I Bancroft, CP Wolk. Pulsed homogeneous orthogonal field gel electrophoresis (PHOGE). Nucleic Acids Res 16:7405–7418, 1988.

62. S Ferris, L Sparrow, A Stevens. Megabase DNA electrophoresis: Recent advances. Austr J Biotech 3:33–35, 1989.

63. G Chu, D Vollrath, R Davis. Separation of large DNA molecules by contour-clamped homogeneous electric fields. Science 234:1582–1585, 1986.

64. GF Carle, MV Olson. Separation of chromosomal DNA molecules from yeast by orthogonal-field-alternation gel electrophoresis. Nucleic Acid Res 12:5647–5664, 1984.

65. EP Rustchenko, TM Curran, F Sherman. Variations in the number of ribosomal DNA units in morphological mutants and normal strains of *Candida albicans* and in normal strains of *Saccharomyces cerevisiae*. J Bacteriol 175:7189–7199, 1993.

66. J Pla, C Gil, L Monteoliva, F Navarro-Garcia, M Sanchez, C Nombela. Understanding *Candida albicans* at the molecular level. Yeast 12:1667–1702, 1996.

67. AF Olayia, SJ Sogin. Ploidy determination of *Candida albicans*. J Bacteriol 140: 1043–1049, 1979.

68. T Suzuki, I Kobayashi, T Kanbe, K Tanaka. High frequency variation of colony morphology and chromosome reorganization in the pathogenic yeast *Candida albicans*. J Gen Microbio 135:425–434, 1989.

69. W-S Chu, BB Magee, PT Magee. Construction of an *Sfi*I macrorestriction map of the *Candida albicans* genome. J Bacteriol 175:6637–6651, 1993.

70. S Scherer. *Candida albicans* information. 2000; http://alces.med.umn.edu/Candida.html.

71. BA Lasker, GF Carle, GS Kobayashi, G Medoff. Comparison of the separation of *Candida albicans* chromosome-sized DNA by pulsed-field electrophoresis technique. Nucleic Acid Res 17:3783–3793, 1989.

72. Sh-I Iwaguchi, M Homma, K Tanaka. Variation in the electrophoretic karyotype analysed by the assignment of DNA probes in *Candida albicans*. J Gen Microbio 136:2433–2442, 1990.

73. T Suzuki, I Kobayashi, I Mizuguchi, I Banno, K Tanaka. Electrophoretic karyo-

types in medically important *Candida* species. J Gen Appl Microbio 34:409–416, 1988.

74. M Monod, S Porchet, F Baudraz-Rosselet, E Frenk. The identification of pathogenic yeast strains by electrophoretic analysis of their chromosomes. J Med Microbio 32:123–129, 1990.

75. K Asakura, S-I Iwaguchi, M Homma, T Sukai, K Higashide, K Tanaka. Electrophoretic karyotypes of clinically isolated yeasts of *Candida albicans* and *C. glabrata*. J Gen Microbio 137:2531–2538, 1991.

76. M Doi, I Mizuguchi, M Homma, K Tanaka. Electrophoretic karyotypes of isolates of *Candida albicans* from hospitalized patients. J Med Vet Mycol 32:133–140, 1994.

77. C Thrash-Bingham, JA Gorman. DNA translocations contribute to chromosome length polymorphisms in *Candida albicans*. Curr Genet 22:93–100, 1992.

78. C Thrash-Bingham, JA Gorman. Identification, characterization and sequence of *Candida albicans* repetitive DNAs Rel-1 and Rel-2. Curr Genet 23:455–462, 1993.

79. S-I Iwaguchi, M Homma, H Chibana, K Tanaka. Isolation and characterization of a repeated sequence (RPS1) of *Candida albicans*. J Gen Microbio 138:1893–1900, 1992.

80. TJD Goodwin, RTM Poulter. The CARE-2 and Rel-2 repetitive elements of *Candida albicans* contain LTR fragments of a new retrotransposon. Gene 218:85–93, 1998.

81. N Dabrowa, EP Rustchenko-Bulgac, JB Hicks, DH Howard. Karyotype variation among morphological mutants of *Candida albicans*. In: Abstr. FEMS Symposium on Candida and Candidamycosis, April 24–28, 1989, Alanya, Turkey, 1990.

82. R Kelly, SM Miller, MB Kurtz, DR Kirsh. Directed mutagenesis in *Candida albicans*: One step gene disruption to isolate *ura3* mutants. Molec Cell Bio 7:199–208, 1987.

83. BB Magee, PT Magee. WO-2, a stable aneuploid derivative of *Candida albicans* strain WO-1, can switch from white to opaque and form hyphae. Microbiology 143:289–295, 1997.

84. WA Fonzi, MY Irwin. Isogenic strain construction and gene mapping in *Candida albicans*. Genetics 134:717–728, 1993.

85. E Andaluz, T Ciudad, MS Garcia de la Marta, V Salguero, G Larriba. An evaluation of the role of LIG4 in genomic instability and adaptive mutagenesis in *Candida albicans*. FEMS Yeast Research (in press).

86. H Ramsey, B Morrow, DR Soll. An increase in switching frequency correlates with an increase in recombination of the ribosomal chromosomes of *Candida albicans* strain 3153A. Microbiology 40:1525–1531, 1994.

87. EP Rustchenko, DH Howard, F Sherman. Chromosomal alterations of *Candida albicans* are associated with the gain and loss of assimilating functions. J Bacteriol 176:3231–3241, 1994.

88. PR Hunter. A critical review of typing methods for *Candida albicans* and their applications. Microbiology 17:417–434, 1991.

89. API Anaylab Products. API 20C Clinical Yeast System. API Laboratory Products Ltd., St. Laurent, Quebec 445 1M5, 1990.

90. P Negroni. Variacion hacia el tipo R de *Mycotorula albicans*. Rev Soc Argent Bio 11:449–453, 1935.
91. CG Saltarelli. Morphological and physiological variations between sectors isolated from giant colonies of *Candida albicans* and *C. stellatoidea*. Mycopath Mycol Appl 34:209–220, 1968.
92. JE Mackinnon. Dissociation in *Candida albicans*. J Infec Dis 66:59–77, 1940.
93. FC Odds, LA Merson-Davies. Colony variations in *Candida* species. Mycoses 32: 275–282, 1989.
94. ME di Menna. Natural occurrence of rough variant of a yeast, *Candida albicans*. Nature 169:550–551, 1952.
95. DR Soll, CJ Langtimm, J McDowell, J Hicks, R Galask. High frequency switching in *Candida* strains isolated from vaginitis patients. J Clin Microbio 25:1611–1622, 1987.
96. DR Soll, M Staebell, J Langtimm, M Pfaller, JB Hicks, TVG Rao. Multiple *Candida* strains in the course of a single systemic infection. J Clin Microbio 26:1448–1459, 1988.
97. DR Soll, R Galask, S Isley, TVG Rao, D Stone, JB Hicks, K Mac, C Hanna. Switching of *Candida albicans* during successive episodes of recurrent vaginitis. J Clin Microbio 27:681–690, 1989.
98. RA Vogel, RS Sponcler. The study and significance of colony dissociation in *Candida albicans*. Sabouraudia 7:273–278, 1970.
99. DE Bianchi. A small colony variant of *Candida albicans*. Amer J Bot 48:499–503, 1961.
100. S Jones, G White, PR Hunter. Increased phenotypic switching in strains of *Candida albicans* associated with invasive infections. J Clin Microbio 32:2869–2870, 1994.
101. FC Odds. Switch of phenotype as an escape mechanism of the intruder. Mycoses 40:(suppl. 2) 9–12, 1997.
102. GA Marzluf. Genetic regulation of nitrogen metabolism in the fungi. Microbio Mol Bio Rev 61:17–32, 1997.
103. KL Lee, HR Buckley, CC Campbell. An amino acid liquid synthetic medium for the development of mycelial and yeast forms of *Candida albicans*. Sabouraudia 13:148–153, 1975.
104. DE Bianchi. Small colony variant in *Candida albicans*. J Bacteriol 82:101–105, 1961.
105. C Sadhu, MJ McEachern, EP Rustchenko-Bulgac, J Schmid, DR Soll, JB Hicks. Telomeric and dispersed repeat sequences in *Candida* yeasts and their use in strain identification. J Bacteriol 173:842–850, 1991.
106. S Scherer, DA Stevens. A *Candida albicans* dispersed, repeated gene family and its epidemiologic applications. Proc Natl Acad Sci 85:1452–1456, 1988.
107. R Franz, M Ruhnke, J Morschhauser. Molecular aspects of fluconazole resistance development in *Candida albicans*. Mycoses 42:453–458, 1999.
108. S-I Iwaguchi, M Homma, K Tanaka. Clonal variation of chromosome size derived from the rDNA cluster region of *Candida albicans*. J Gen Microbio 138:1177–1184, 1992.
109. JW Szostak, R Wu. Unequal crossing over in the ribosomal DNA of *Saccharomyces cerevisiae*. Nature 284:426–430, 1980.

110. WS Riggsby, LJ Torres-Bauza, JW Wills, TM Towns. DNA content, kinetic complexity, and the ploidy question in *Candida albicans*. Molec Cell Bio 2:853–862, 1982.

111. JW Wills, BA Lasker, K Sirotkin, WS Riggsby. Repetitive DNA of *Candida albicans*: Nuclear and mitochondrial components. J Bacteriol 157:918–924, 1984.

112. DA Sinclair, L Guarente. Extrachromosomal rDNA circles—A cause of aging in yeast. Cell 91:1033–1042, 1997.

113. J Hermanns, A Asseburg, HD Osiewacz. Evidence for giant linear plasmids in the ascomycete *Podospora anserina*. Curr Genet 27:379–386, 1995.

114. MJ McEachern, JB Hicks. Unusually large telomeric repeats in the yeast *Candida albicans*. Molec Cell Bio 13:551–560, 1993.

115. MJ McEachern, EH Blackburn. A conserved sequence motif within the exceptionally diverse telomeric sequences of budding yeasts. Proc Natl Acad Sci USA 91: 3453–3457, 1994.

116. R Jemtland, E Maehlum, OS Gabrielsen, TB Oyen. Regular distribution of length heterogeneities within non-transcribed spacer regions of cloned and genomic rDNA of *Saccharomyces cerevisiae*. Nucleic Acids Res 14:5145–5158, 1986.

117. E Rustchenko, F Sherman. Physical constitution of ribosomal genes in common strains of *Saccharomyces cerevisiae*. Yeast 10:1157–1171, 1994.

118. A Vincent, TD Petes. Isolation and characterization of a Ty element inserted into the ribosomal DNA of the yeast *Saccharomyces cerevisiae*. Nucleic Acids Res 4: 2939–2949, 1986.

119. Y Xiong, TH Eickbush. Similarity of reverse transcriptase-like sequences of viruses, transposable elements, and mitochondrial introns. Molec Bio Evol 5:675–690, 1988.

120. DE Muscarella, VM Vogt. A mobile group I intron from *Physarum polycephalum* can insert itself and induce point mutations in the nuclear ribosomal DNA of *Saccharomyces cerevisiae*. Molec Cell Bio 13:1023–1033, 1993.

121. KV Clemons, F Feroze, K Holmberg, DA Stevens. Comparative analysis of genetic variability among *Candida albicans* isolates from different geographic locales by three genotypic methods. J Clin Microbiol 35:1332–1336, 1997.

122. MJ McCullough, KV Clemons, DA Stevens. Molecular and phenotypic characterization of genotypic *Candida albicans* subgroups and comparison with *Candida dubliniensis* and *Candida stellatoidea*. J. Clin Microbiol 37:417–421, 1999.

123. B Wicks, J Staudinger, BB Magee, KJ Kwon-Chung, PP Magee, S Scherer. Physical and genetic mapping of *Candida albicans*: Several genes previously assigned to chromosome 1 map to chromosome R, the rDNA-containing linkage group. Infec Immun 59:2480–2484, 1991.

124. S Mercure, N Rougeau, S Montplaisir, G Lemay. The nucleotide sequence of the 25S rRNA–encoding gene from *Candida albicans*. Nucleic Acids Res 21:1490, 1993.

125. DH Huber, E Rustchenko. Large circular and linear rDNA plasmids in *Candida albicans*. Yeast 18: 261–272, 2001.

126. JA Shaw, WB Troutman, BA Lasker, MM Mason, WS Riggsby. Characterization of the inverted duplication in the mitochondrial DNA of *Candida albicans*. J Bacteriol 171:6353–6356, 1989.

127. Ch-Sh Su, SA Meyer. Characterization of mitochondrial DNA in various _Candida_ species: Isolation, restriction endonuclease analysis, size, and base composition. Internat J Systemat Bacteriol 41:6–14, 1991.

128. RTM Poulter. Genetics of _Candida_ species. In: AH Rose, AE Wheals, JS Harrison, eds. The Yeasts. vol. 6, 2nd ed. London: Academic, 1995, pp. 285–308.

129. JW Wills, WB Troutman, WS Riggsby. Circular mitochondrial genome of _Candida albicans_ contains a large inverted duplication. J Bacteriol 164:7–13, 1985.

130. Y Miyakawa, T Mabuchi. Characterization of a species-specific DNA fragment originating from the _Candida albicans_ mitochondrial genome. J Med Vet Mycol 32:71–75, 1994.

131. H Chibana, S Iwaguchi, M Homma, A Chindamporn, Y Nakagawa, K Tanaka. Diversity of tandemly repetitive sequences due to short periodic repetitions in the chromosomes of _Candida albicans_. J Bacteriol 176:3851–3858, 1994.

132. A Chindamporn, Y Nakagawa, M Homma, H Chibana, M Doi, K Tanaka. Analysis of the chromosomal localization of the repetitive sequences (RPSs) in _Candida albicans_. Microbio 141:469–476, 1995.

133. A Chindamporn, Y Nakagawa, I Mizuguchi, H Chibana, M Doi, K Tanaka. Repetitive sequences (RPSs) in the chromosomes of _Candida albicans_ are sandwiched between two novel stretches, HOK and RB2, common to each chromosome. Microbiology 144:849–857, 1998.

134. H Chibana, BB Magee, S Grindle, Y Ran, S Scherer, PT Magee. A physical map of chromosome 7 of _Candida albicans_. Genetics 149:1739–1752, 1998.

135. JE Cutler, PM Glee, HL Horn. _Candida albicans_ and _Candida stellatoidea_-specific DNA fragment. J Clin Microbio 26:1720–1724, 1988.

136. BA Lasker, LS Page, TJ Lott, GS Kobayashi, G Medoff. Characterization of CARE-1; _Candida albicans_ repetitive element-1. Gene 102:45–50, 1991.

137. BA Lasker, LS Page, TJ Lott, GS Kobayashi. Isolation, characterization, and sequencing of _Candida albicans_ repetitive element 2. Gene 116:51–57, 1992.

138. AR Holmes, YC Lee, RD Cannon, HF Jenkinson, MG Shepherd. Yeast-specific DNA probes and their application for the detection of _Candida albicans_. J Med Microbio 37:346–351, 1992.

139. R Franz, SL Kelly, DC Lamb, DE Kelly, M Ruhnke, J Morschhäuser. Multiple molecular mechanisms contribute to a stepwise development of fluconazole resistance in clinical _Candida albicans_ strains. Antimicrob Agents Chemo 42:3065–3072, 1998.

140. A van Belkum. DNA fingerprinting of medically important microorganisms by use of PCR. Clin Microbio Rev 7:174–184, 1994.

141. J-Y Chen, WA Fonzi. A temperature-regulated, retrotransposon-like element from _Candida albicans_. J Bacteriol 174:5624–5632, 1992.

142. GD Matthews, TJ Goodwin, MI Butler, TA Berryman, RT Poulter. pCal, a highly unusual Ty1/copia retrotransposon from the pathogenic yeast _Candida albicans_. J Bacteriol 179:7118–7128, 1997.

143. J Chen, Z Fu. A retrotransposon-like element Tca1 was used for taxonomic determination of _Candida albicans_. Acta Microbiol Sinica 36:161–167, 1996.

144. TJD Goodwin, RTM Poulter. Multiple LTR-retrotransposon families in the asexual yeast _Candida albicans_. Genome Res 10:174–191, 2000.

145. VM Perreau, MA Santos, MF Tuite. Beta, a novel repetitive DNA element associated with tRNA genes in the pathogenic yeast *Candida albicans.* Molec Microbio 25:229–236, 1997.

146. JRS Fincham, PR Day. Fungal Genetics. Oxford and Edinburgh: Blackwell Scientific Publications, 1971.

147. Y-K Wang, B Das, DH Huber, M Wellington, A Kabir, F Sherman, E Rustchenko. The protein 14-3-3 is required of L-sorbose utilization and hyphal and cell wall formation in *Candida albicans.* 2002 (manuscript submitted).

148. D Talibi, M Raymond. Isolation of a putative *Candida albicans* transcriptional regulator involved in pleiotropic drug resistance by functional complementation of a *pdr1 pdr3* mutation in *Saccharomyces cerevisiae.* J Bacteriol 181:231–240, 1999.

149. AM Alarco, M Raymond. The bZip transcription factor Cap1p is involved in multidrug resistance and oxidative stress response in *Candida albicans.* J Bacteriol 181: 700–708, 1999.

150. AM Alarco, I Balan, D Talibi, N Mainville, M Raymond. AP1-mediated multidrug resistance in *Saccharomyces cerevisiae* requires *FLR1* encoding a transporter of the major facilitator superfamily. J Bio Chem 272:19304–19313, 1997.

151. TC White, KA Marr, RA Bowden. Clinical, cellular, and molecular factors that contribute to antifungal drug resistance. Clin Microbio Rev 11:382–402, 1998.

152. D Sanglard, F Ischer, L Koymans, J Bille. Amino acid substitutions in the cytochrome P-450 lanosterol 14α-demethylase (CYP51A1) from azole-resistant *Candida albicans* clinical isolates contribute to resistance to azole antifungal agents. Antimicrob Agents Chemo 42:241–253, 1998.

153. H Vanden Bossche, P Marichal, J Gorrens, D Bellens, H Moereels, PAJ Janssen. Mutation in cytochrome P450-dependent 14α-demethylase results in decreased affinity for azole antifungals. Biochem Soc Trans 18:56–59, 1990.

154. TC White. The presence of an R467K amino acid substitution and loss of allelic variation correlate with an azole-resistant lanosterol 14α-demethylase in *Candida albicans.* Antimicrob Agents Chemo 41:1488–1494, 1997.

155. SL Kelly, DC Lamb, DE Kelly, NJ Manning, J Loeffler, H Hebart, U Schumacher, H Einsele. Resistance to fluconazole and cross-resistance to amphotericin B in *Candida albicans* from AIDS patients caused by defective sterol Δ (5,6) desaturation. FEBS Lett 400:80–82, 1997.

156. GD Albertson, M Niimi, RD Cannon, H Jenkinson. Multiple efflux mechanisms are involved in *Candida albicans* fluconazole resistance. Antimicrob Agents Chemo 40:2835–2841, 1996.

157. D Sanglard, K Kuchler, F Ischer, JL Pagani, M Monod, J Bille. Mechanisms of resistance to azole antifungal agents in *Candida albicans* isolates from AIDS patients involve specific multidrug transporters. Antimicrob Agents Chemo 39:2378–2386, 1995.

158. M Niimi, RD Cannon, M Arisawa. Multidrug resistance genes in *Candida albicans.* Jpn J Med Mycol 38:297–302, 1997.

159. D Sanglard, F Ischer, M Monod, J Bille. Cloning of *Candida albicans* genes conferring resistance to azole antifungal agents: Characterization of *CDR2*, a new multidrug ABC transporter gene. Microbiology 143:405–416, 1997.

160. JL Lopez-Ribot, RK McAtee, LN Lee, WR Kirkpatrick, TC White, D Sanglard,

TF Patterson. Distinct patterns of gene expression associated with development of fluconazole resistance in serial *Candida albicans* isolates from human immuno-deficiency virus-infected patients with oropharyngeal candidiasis. Antimicrob Agents Chemo 42:2932–2937, 1998.

161. TC White. Increased mRNA levels of *ERG16*, *CDR*, and *MDR1* correlate with the increases in azole resistance in *Candida albicans* isolates from a patient infected with human immunodeficiency virus. Antimicrob Agents Chemo 41:1482–1487, 1997.

162. M Niimi, CY Shin, FJ Fischer, K Niimi, RD Cannon. Azole cross-resistant *Candida albicans* variants isolated from drug-pump deleted strains. ASM Conference on Candida and Candidiasis 52, 1999.

163. S Virsching, S Michel, G Kohler, J Morschhauser. Activation of the multiple drug resistance gene *MDR1* in fluconazole-resistant, clinical *Candida albicans* strains is caused by mutation in a *trans*-regulatory factor. J Bacteriol 182:400–404, 2000.

164. J Cairns, J Overbaugh, S Miller. The origin of mutants. Nature 335:142–145, 1988.

165. PL Foster. Adaptive mutation: Has the unicorn landed? Genetics 148:1453–1459, 1998.

166. MJ Malavasic, RL Cihlar. Growth response of several *Candida albicans* strains to inhibitory concentrations of heavy metals. J Med Vet Mycology 30:421–432, 1992.

167. P Borst, DR Greaves. Programmed gene rearrangements altering gene expression. Science 235:658–667, 1987.

168. DR Soll. Dimorphism and high frequency switching in *Candida albicans*. In: DR Kirsch, R Kelly, MB Kurtz, eds. The Genetics of *Candida*. Boca Raton, FL: CRC Press, 1990, pp. 147–176.

169. I Herskowitz, J Rhine, JN Strathern. Mating-type determination and mating-type interconversion in *Saccharomyces cerevisiae*. In: EW Jones, JR Pringle, JR Broach, eds. The Molecular and Cellular Biology of the Yeast *Saccharomyces*. vol. 2. Gene Expression. Cold Spring Harbor, NY: Cold Spring Harbor Laboratory Press, 1992, pp. 583–656.

170. M Simon, M Silverman. Recombinational regulation of gene expression in bacteria. In: J Beckwith, J Davies, JA Gallant, eds. Gene Expression in Procaryotes. Cold Spring Harbor, NY: Cold Spring Harbor Laboratory Press, 1983, pp. 211–227.

171. E Segal, P Hagblom, HS Seifert, M So. Antigenic variation of gonococcal pilus involves assembly of separated silent gene segments. Proc Natl Acad Sci USA 83: 2177–2181, 1986.

172. G Das, S Consaul, F Sherman. A highly revertible *cyc1* mutant of yeast contains a small tandem duplication. Genetics 120:57–62, 1988.

173. WL Whelan. The genetics of medically important fungi. CRC Crit Rev Microbio 14:99–170, 1987.

174. R Pomés, C Gil, C Nombela. Genetic analysis of *Candida albicans* morphological mutants. J Gen Microbio 131:2107–2113, 1985.

175. J Pérez-Martín, JA Uria, AD Johnson. Phenotypic switching in *Candida albicans* is controlled by a *SIR2* gene. EMBO J 18:2580–2592, 1999.

176. SA Lachke, T Srikantha, LK Tsai, K Daniels, DR Soll. Phenotypic switching in *Candida glabrata* involves phase-specific regulation of the metallothionein gene *MT-II* and the newly discovered hemolysin gene HLP. Infec Immun 68:884–895, 2000.

177. P Marichal, H Vanden Bossche, FC Odds, G Nobels, DW Warnock, V Timmerman, C Van Broeckhoven, S Fay, P Mose-Larsen. Molecular biological characterization of an azole-resistant *Candida glabrata* isolate. Antimicrob Agents Chemo 41: 2229–2237, 1997.

178. J Anderson, R Mihalik, D Soll. Ultrastructure and antigenicity of the unique cell wall pimple of the *Candida* opaque phenotype. Bacteriol 172:224–235, 1990.

179. E Rikkerink, BB Magee, P Magee. Opaque-white phenotype transition: A programmed morphological transition in *Candida albicans*. J Bacteriol 170:895–899, 1988.

180. B Slutzky, M Staebell, J Anderson, L Risen, M Pfaller, DR Soll. "White-opaque transition": A second high-frequency switching system in *Candida albicans*. J Bacteriol 169:189–197, 1987.

181. DR Soll, B Morrow, T Srikantha. High frequency phenotypic switching in *Candida albicans*. Trends Genet 9:61–65, 1993.

182. I Balan, A-M Alarco, M Raymond. The *Candida albicans CDR3* gene codes for an opaque-phase ABC transporter. J Bacteriol 179:7210–7218, 1997.

183. SR Lockhart, M Nguyen, T Srikantha, DR Soll. A MADS box protein consensus binding site is necessary and sufficient for activation of the opaque-phase-specific gene OP4 of *Candida albicans*. J Bacteriol 180:6607–6616, 1998.

184. B Morrow, T Srikantha, DR Soll. Transcription of the gene for a pepsinogen, *PEP1*, is regulated by white-opaque switching in *Candida albicans*. Molec Cell Bio 12: 2997–3005, 1992.

185. T Srikantha, A Chandrasekhar, DR Soll. Functional analysis of the promoter of the phase-specific WH11 gene of *Candida albicans*. Molec Cell Bio 15:1797–1805, 1995.

186. T Srikantha, B Morrow, K Schröppel, DR Soll. The frequency of integrative transformation at phase-specific genes of *Candida albicans* correlates with their transcriptional state. Molec Gen Genet 246:342–352, 1995.

187. T Srikantha, LK Tsai, DR Soll. The *WH11* gene of *Candida albicans* is regulated in two distinct developmental programs through the same transcription activation sequences. J Bacteriol 179:3837–3844, 1997.

188. A Sonneborn, B Tebarth, JF Ernst. Control of white-opaque phenotypic switching in *Candida albicans* by the Efg1p morphogenetic regulator. Infec Immun 67:4655–4660, 1999.

189. T Srikantha, LK Tsai, K Daniels, DR Soll. *EFG1* null mutants of *Candida albicans* switch but cannot express the complete phenotype of white-phase budding cells. J Bacteriol 182:1580–1591, 2000.

190. AJ Klar, T Srikanta, DR Soll. A histone deacetylation inhibitor and mutant promote colony-type switch of the human pathogen *Candida albicans*. Genetics 158:919–924, 2001.

191. T Srikantha, LK Tsai, K Daniels, AJS Klar, DR Soll. The histone deacetylase genes *HDA1* and *RPD3* play district roles in regulation of high-frequency phenotypic switching in *Candida albicans*. J Bacteriol 183:4614–4625, 2001.

Index

Printed in the United States
by Baker & Taylor Publisher Services